Handbook of Bioplastics and Biocomposites Engineering Applications

Scrivener Publishing
3 Winter Street, Suite 3
Salem, MA 01970

Handbook of Bioplastics and Biocomposites Engineering Applications

Edited by
Srikanth Pilla
Wisconsin Institute for Discovery
University of Wisconsin-Madison, USA

Scrivener
WILEY

Co-published by John Wiley & Sons, Inc. Hoboken, New Jersey, and Scrivener Publishing LLC, Salem, Massachusetts.
Published simultaneously in Canada.

Cover design by Russell Richardson
Front cover photos supplied by Joseph G. Lawrence

Library of Congress Cataloging-in-Publication Data:

ISBN 978 0-470-62607-8

Printed in the United States of America

10 9 8 7 6 5 4 3 2 1

Contents

Foreword

The sky rocketing price of petroleum along with its dwindling nature and coupled with climate change concern and continued population growth have drawn the urgency for the plastic industries in adapting towards sustainability. The use of bio- or renewable carbon, as opposed to petro-carbon, for manufacturing bioplastics and biobased materials, is moving forward for a reduced carbon footprint. The goal is to use biobased materials containing the maximum possible amount of renewable biomass-based derivatives to secure a sustainable future. Bioplastics, biofibres, biocomposites and related biomaterials will serve as substitutes for materials and products traditionally made from petroleum resources. The research and development on these biobased materials are an emerging area of research that focuses on a low-carbon economy, through revolutionary use of agricultural products and many other bio-renewable resources for new industrial uses, ranging from car parts to consumer products, and packaging materials to green building products.

The incorporation of bio-resources, e.g. plant derived biofibres and bioplastics into composite materials are gaining prime importance in designing and engineering green composites. Biocomposites derived from natural fibers and traditional polymers like polypropylene, polyethylene, epoxy and polyesters have been developed for industrial uses and are still under development for diversified applications.

Thus, lots of research activities have been started by academic institutions and research centers along with their industrial partners, for the development of innovative bioplastics and biocomposites to cater for an increasing range of applications. With these efforts, it is envisioned that the billion-dollar market of plastics and composites will be equaled by bioplastics and biocomposites in a decade's time. In this view, this Handbook is a timely reference work for the scientific community.

The *Handbook of Bioplastics and Biocomposites Engineering Applications* is an application-oriented book in the field of bioplastics and biocomposites. This Handbook is a perfect resource for research professionals, technology investors, and industrial engineers. It showcases the engineering practices of biomaterials in the fields of packaging, civil, biomedical, automotive, etc. In addition, the Handbook presents two studies on processing of these sustainable materials. One chapter especially is focused on modeling the theories in processing which is helpful for cost-effective research and development in process design of equipment. I hope that this Handbook will inspire many current and future generations of academic and industrial researchers to develop more novel renewable materials that will play their part in making a sustainable society and helping to conserve life on this planet.

Amar K. Mohanty
Premier's Research Chair in Biomaterials and Transportation
Professor, Department of Plant Agriculture and School of Engineering
Director, Bioproducts Discovery & Development Centre
University of Guelph, Ontario, Canada

Preface

Plastics have been one of the most highly valued materials mainly because of their extraordinary versatility and low cost. Their usage span a wide range of applications – packaging, structural (building materials), transportation (automobiles, watercraft, aircraft parts), electrical components, biomedical (gloves, gowns, masks, coverings etc.) and consumer products such as toys, utensils, cameras and watches. However, the widespread use of plastics has become a significant concern due to their negative impact on the environment; specifically, the sources from which plastics are derived and their biodegradability. Almost all synthetic plastics are made from petroleum and its allied components. These natural resources take millions of years to form and are finite in quantity. In addition, plastics derived from fossil resources are largely non-biodegradable. The increased use in plastics over the years has resulted in an increase in plastic waste, which often is dumped as municipal solid waste. Thus, there is an immediate need to develop non-petroleum-based and sustainable feed stocks, and this has predominantly shifted the attention of many researchers, academic and industrial, towards biobased and biodegradable plastics.

Biobased plastics or *bioplastics* are sustainable, largely biodegradable and biocompatible. They reduce our dependency on depleting fossil fuels and are CO_2 neutral. But in spite of providing timely and essential need for environmental sustainability, bioplastics are yet to gain a strong position in the plastics world. This is because of less than superior properties of bioplastics compared to their synthetic counterparts. Hence, scientists and engineers around the world have been exploring ways to improve the properties of bioplastics by blending/compounding them with other polymers and fibers. Blending of bioplastics with natural fibers, termed *green composites* provide a sustainable alternative with 100% biodegradability. However, research is also being carried out to blend bioplastics with synthetic fillers and/or blend synthetic plastics with natural fibers. Generally termed as *Biocomposites*, such materials not only provide outstanding properties to meet the target application, but also reduce the carbon foot print on the environment.

This Handbook is believed to be the first application oriented book in the field of bioplastics and biocomposites. The Handbook presents various studies related to different engineering fields such as packaging, civil, biomedical, and automotive. In addition, it contains a section on processing aspects of bioplastics and biocomposites. Though life cycle analyses of bioplastics and biocomposites are not included, they will be included in future editions.

I hope that the Handbook will serve mainly as direction and important reference for engineers, scientists, academicians, students and other researchers who are working in the fields of bioplastics, biomaterials, biochemistry, materials science, engineering etc. Particularly, I believe that the Handbook will play a key role as a professional reference and in teaching.

During of the course of my journey in developing renewable materials for sustainable society I have interacted with many academicians, scientists, engineers, and students and I would like to extend my heartiest thanks to all; particularly, I would like to thank Profs. Sarah Gong, Tom Turng, Chul Park, Sarah Billington, and Curtis Frank who have guided and directed me during my graduate and postgraduate studies. Also, I am highly grateful to all contributing authors for their extraordinary efforts in writing the chapters presented in this Handbook. I would like to express my sincere gratitude to Prof. Amar Mohanty of University of Guelph for kindly agreeing to write the Foreword. On behalf of the contributing authors, I also thank all the publishers and authors who granted copyright permissions to use their illustrations and figures in this Handbook. I would also like to acknowledge the help and support provided by Martin Scrivener of Scrivener Publishing in the timely publication of this Handbook. Finally, I would like to thank my parents, Mohan and Padma, and my wife Ashwini for their continuous encouragement and support, as well as my daughter, Varsha, for her love in this exciting journey.

Srikanth Pilla
Wisconsin Institute for Discovery
University of Wisconsin-Madison

List of Contributors

Hazizan Md Akil is Associate Professor of Polymer Composites at the School of Materials & Mineral Resources Engineering, Universiti Sains Malaysia Engineering Campus (USM), Malaysia. He received his Bachelors in Polymer Engineering from North London, UK at 1996 and completed his PhD from University of Liverpool, United Kingdom in Polymer Composites Engineering at 2002.

Séverine A.E. Boyer received her PhD (2003) in polymer physics and chemistry from the Blaise Pascal University and the French Institute of Petroleum (France). She worked as a postdoctoral fellow at the Tokyo Metropolitan University (Japan), and as an associate researcher at the Ecole des Mines de Paris, France. Currently, she is an associate researcher at the Ecole des Mines de Douai, France. Her research interests include the thermo-diffuso-mechanics and the patterns formation of polymers under extreme conditions.

Venkata S. Chevali received his PhD in Materials Science (2009) from The University of Alabama in Birmingham. He has been in the position of Postdoctoral Research Associate in the Mechanical Engineering and Applied Mechanics Department at North Dakota State University since May, 2009. His primary research interests include the processing, material characterization, and mechanical analysis of synthetic and bio-based forms of polymer matrix composites (PMCs).

Alan Fernyhough graduated from Liverpool University with a PhD in Polymer Chemistry. He has more than 25 years industrial experience with BP, The Kobe Steel Group and Scion in developing new polymer technologies for plastics and composites. Since 2002 his work has focused on biobased options and in particular on developing high performance bioplastics, thermoset bioresins, biopolymer foams, and on wood and other biofibre technologies for high performance biocomposites. He is a Team Leader (Biopolymer & Green Chemical Technologies) at Scion, Rotorua, New Zealand.

Michael A. Fuqua received his B.E. degree in Mechanical Engineering from the University of Delaware (2006) and MS degree in Mechanical Engineering from North Dakota State University (2008). He is currently a PhD candidate in the Mechanical Engineering Department at North Dakota State University. Mike's research focus is in polymer matrix composite (PMC) manufacturing and materials development.

Jean-Philippe Garancher graduated from Ecole Nationale Supérieure des Mines de Saint-Etienne, France with a MSc degree in mechanical engineering. He is currently enrolled as a PhD student at the University of Auckland, New Zealand investigating mechanical properties of biopolymer foams. He previously worked at Scion, New Zealand investigating mechanical properties of biobased composites.

Shaoqin "Sarah" Gong is an Associate Professor in the Department of Biomedical Engineering and Wisconsin Institute for Discovery at the University of Wisconsin-Madison. She received her PhD degree from the University of Michigan-Ann Arbor. Her current main research interests include nanobiomaterials, polymer nanocomposites, microcellular biobased plastics, and biosensors. She has authored more than 130 technical papers and is the recipient of National Science Foundation CAEER Award.

K M Gupta received his PhD degree from Allahabad University, India where he is a Professor in the Department of Applied Mechanics. His research interests are in the fields of Green Composite Materials, Solid Mechanics, and Stress Analysis of Plates. He has authored more than 24 books on engineering subjects and has more than 75 research papers to his credit in international and national journals and conferences.

Ali Harlin is Professor for Bio-based materials and is leading the Industrial Biomaterials program in VTT, the Technical Research Centre of Finland, which is targeting industrial application of materials produced using renewable raw materials. He is also tutor at the Finish Academy, Centre of excellence – White Biochemistry and Green Chemistry in the field of biomass-based monomers and polymers were he aims to integrate these new value chains into existing bio-refineries.

Alireza Javadi is a postdoctoral research associate in the Department of Biomedical Engineering and Wisconsin Institute for Discovery at the University of Wisconsin-Madison. He received his BS degree from Tehran Polytechnic University, Iran and earned his MS with honors from Chalmers Institute of Technology (Gothenburg, Sweden). He received his PhD from the University of Wisconsin-Milwaukee. His current research is mainly focused on polymer nanocomposites, microcellular biobased plastics, and surface modification of various organic and inorganic nanoparticles.

B. S. Kaith is Professor & Head, Department of Chemistry, National Institute of Technology, Jalandhar in India. He gained his PhD from the University Chandigarh in 1990. He has more than 80 research papers in international journals and 160 research papers in the proceedings of international and national conferences.

Susheel Kalia is Assistant Professor in Department of Chemistry, Bahra University, Shimla Hills, India. He received his PhD from PTU Jalandhar. He has 35 articles in international journals & books and 50 chapters in proceedings of international and national conferences. He is editing two books on polymer composites and biopolymers. His current research interests are in the fields of polymer composites & nanocomposites, hydrogels and cryogenics.

Prashantha Kalappa received his PhD degree in Industrial Chemistry from Kuvempu University, India in 2002. Currently he is Assistant Professor in the Polymers and Composites Technology & Mechanical Engineering Department of Ecole des Mines de Douai (France). Before joining Ecole des Mines, he was a post doctoral fellow at Chonbuk National University (Korea). He has published over 25 research papers in peer-reviewed journals. His research mainly focuses on polymer nanocomposites, blends, conducting composites and functional polymeric materials.

Hisatoshi Kobayashi is group leader of Biofunctional Materials at Biomaterials Centre, National Institute for Materials Science, Tsukuba, Japan. He is also affiliated with universities in Tokyo, Allahabad, India, and the Kanazawa Institute of Technology. He has published about 150 publications, books and patents in the field of biomaterials science and technology. He has also edited/authored three books on the advanced state-of-the-art of biomaterials. His recent research interest is focus on designing and development of the biodegradable biomaterials scaffolds for ophthalmologic devices and nanocomposites for medical devices.

Andy Kovats is a Sales and Application Engineer with Brabender Technologie, Inc, Toronto, Canada, manufacturers of dry ingredient feeding equipment. He holds a BSc (Chem Eng) from Queen's University, Kingston and an MBA from York University in Toronto, He is a Registered Professional Engineer in the Province of Ontario. He is a member of SPE, has 25 years of experience in dry material handling, and has given many professional courses and authored numerous papers and publications in this field.

Patricia Krawczak is Professor of Plastics and Composites Engineering at Ecole des Mines de Douai, France. She obtained a PhD in polymer science in 1993 and a qualification as Research Director in physics in 1999, both from University of Lille (France), gaining extensive scientific and technical expertise in the field of composites and plastics engineering through numerous collaborations with industrial companies. She has been the Director of the Polymers and Composites Technology & Mechanical Engineering Department of the Ecole des Mines de Douai since 2000. Her current research interests cover processing technologies, physics and mechanics of polymer and composite materials, including materials from renewable resources and nanocomposites

M.N.Satheesh Kumar completed his Master of Science (Polymer Science) from University of Mysore, India. He joined the Corporate Research and Innovation Centre of Raman Boards Limited, Mysore, after his master degree. He obtained his PhD (Polymer Science) in the year 2007. After doctoral work, he worked at the University of Guelph, Canada and National University of Malaysia as a postdoctoral research fellow. His area of research is in fibre reinforced polymer composites for structural and non-structural applications.

Marie-France Lacrampe is Professor of Plastics Processing at Ecole des Mines de Douai, France. She obtained a PhD in fluid mechanics from University of

Valenciennes in 1993 and a qualification as Research Director in physics from University of Lille in 2006. She has been the Head of the Polymer Group within the Polymers and Composites Technology & Mechanical Engineering Department of the Ecole des Mines de Douai (France) since 2006. Her research interests cover processing technologies (in particular injection, extrusion and rotational molding), rheology, polymer blends, including materials from renewable resources and nanocomposites.

Rafael Marangoni received his Chemistry MS degree in 2005 and Chemistry PhD in 2009 at the Federal University of Paraná., Brazil. Currently he is a researcher at the same University where he works at the Cepesq (Research Center of Applied Chemistry) laboratory. His main areas of research are in nanocomposites and layered materials.

Martin Markotsis received his PhD from the University of New South Wales, Australia, and then undertook postdoctoral research into bioplastics at the University of Queensland and the Canadian NRC's Industrial Materials Institute. He is currently employed as a Polymer Scientist/Chemist at Scion (Rotorua, New Zealand).

Adlan Akram Mohamad Mazuki is a Chemistry Metrologist at National Metrology Laboratory (NML-SIRIM) of Malaysia. He received his Bachelor in Materials Technology from Universiti Teknologi Mara Malaysia at 2008 and Master in Science from School of Materials & Mineral Resources Engineering, Universiti Sains Malaysia (USM), Malaysia at 2011. His research focuses on fabrication of kenaf fibers, and reinforced composites using pultrusion method for engineering applications.

Reza Masoodi received his PhD in mechanical engineering from University of Wisconsin-Milwaukee (UWM) in 2010. His PhD dissertation was on "Modeling Imbibition of Liquids into Rigid and Swelling Porous Media". He has done some research on flow of swelling-induced liquids such as bio-resins, organic liquids, and water-based liquids into natural fibermats. Dr. Masoodi has done some theoretical study on single-phase flow modeling in swelling porous media, capillary pressure, and permeability in swelling porous materials. Currently, he is studying flow of bio-resin in green composites made through Liquid Composite Moulding in Laboratory for Flow and Transport Studies in Porous Media at UWM.

Ajay Kumar Mishra is Associate Editor of *Advanced Materials Letters* and currently working as Senior Lecturer at Nanomaterials Research Centre, Department of Chemical Technology, University of Johannesburg, South Africa. He pursued his MPhil and PhD in bio-inorganic Chemistry at Department of Chemistry, University of Delhi, India. In 2006, he moved to the University of Free State, South Africa for Postdoctoral studies in Materials Science. He was also awarded the prestigious AVI Award in 2009 for his outstanding contribution. His expert areas include synthesis of multifunctional nano-materials, nano-composites, biopolymer and/or petrochemical based biodegradable polymers and polymers based materials/composites.

Shivani Mishra received her BSc and MSc degree in chemistry from University of Madras, India. She went to the Jamia Millia Islamia, New Delhi, India where she obtained her PhD degree in chemistry in 2003. She pursued several postdoctoral research at University of Free state, CSIR and University of Johannesburg. Currently, she is working as Senior Lecturer at the Department of Chemical Technology, University of Johannesburg, South Africa. Her research interests include inorganic and materials chemistry especially in the area of smart materials and composites and nanocomposites for various applications.

Kate Parker graduated from the University of Auckland, New Zealand with a PhD in Chemistry. She has a background in environmental and waste treatment technologies, in addition to polymer processing. She is Science Leader for biopolymer foam developments at Scion.

Krishna M. Pillai is Associate Professor at University of Wisconsin-Milwaukee. He is also the director of Laboratory for Flow and Transport Studies in Porous Media at UWM. His research interests lie in several areas of porous media transport including flow and transport in fibrous media, wicking in rigid and swelling porous media, and evaporation modeling using network and continuum models. He has published extensively in reputed journals and presented his work in numerous international conferences and workshops. He was awarded the prestigious CAREER grant in 2004 by the National Science Foundation of USA to model and simulate flow processes during mold filling in liquid molding processes used for manufacturing polymer composites.

Luiz Pereira Ramos is an Associate Professor at the Chemistry Department of the Federal University of Paraná, Brazil. He received his BSc in Chemistry (1978-82) at the Catholic University of Paraná, MSc in Biochemistry at the Federal University of Paraná (1983-87) and PhD from the Ottawa-Carleton Institute of Biology, University of Ottawa, Canada (1988-92). At the University he teaches Chromatography, Spectrometry, Biocatalysis and Biomass Chemistry. His research lines involve wood and carbohydrate chemistry, second generation biofuels (biodiesel and bioethanol) and enzyme technologies.

Daniela Rusu is Associate Professor of Polymers and Biomaterials at Ecole des Mines de Douai, France. She obtained a PhD in materials science and engineering from Ecole des Mines de Paris/Mines in 1997. She worked as Research Scientist at the Ecole des Mines de Paris and the University of Mainz, Germany and as Associate Professor of Polymers and Biomaterials at the University of Medicine and Pharmacy of Iasi, Romania. Her scientific expertise includes understanding of the mechanisms and processing-structure-properties relationships in multiphase polymer systems (polymer blends and composites), and tailoring them for different industrial or biomedical applications. Her current research focuses on bioplastics, biocomposites and polymer biomaterials.

Kestur Gundappa Satyanarayana received BSc and MSc Degrees (1965 and 1968) from Mysore and Bangalore Universities respectively and PhD (1972) from

Banaras Hindu University, India. He is a Consulting Chief Scientist at the Acharya R&D Center, Research Consultant in the College of Engineering, and Honorary Professor at the Poornaprajna Institute for Scientific Research, Bangalore, India. Prior to this, he worked as Visiting Professor and Researcher from 2003 to 2009 in the Department of Chemistry at the University of Parana, Brazil. His areas of interest include utilization of agro-industrial wastes and development of new materials including composites.

Samir Shah gained a Post Graduate Diploma in Plastics Processing and Testing from S.P.University, India. He is currently a Scientist at Scion, New Zealand. His research interests include biopolymer foams, developing bioplastic composites, and extrusion and injection moulding processes.

Dr. Siddaramaiah completed his Master of Science (Chemistry) (1986) and PhD (1993) from University of Mysore, India. He is now a Professor in the Department of Polymer Science and Technology, Sri Jayachamarajendra College of Engineering, Mysore. He worked with Chonbuk National University, South Korea, and Institute of Macromolecules, University of Federal, Rio de Janeiro, Brazil, as a postdoctoral research fellow. He has published more than 200 research articles in reputed journals. His area of research is in interpenetrating polymer networks of polyurethanes; polymer composites–modification, tribological, molecular transport; conducting polymers; biopolymers/biodegradable polymers- for drug delivery.

Jérémie Soulestin obtained his PhD in polymer science in 2004 from University of Lille, France in the field of processing and plastic behavior of polymer nanocomposites. After a 1 year postdoctoral position in the Université Catholique de Louvain-La-Neuve, Belgium in the field of polymer composites based on renewable resources, he joined the Polymers and Composites Technology & Mechanical Engineering Department of the Ecole des Mines de Douai as assistant professor in 2006. His current research topics focus on processing and mechanical characterization of polymer nanocomposites & composites and polymer based on renewable resources.

Dohiko Terada studied polymer engineering at Kyoto Institute of Technology in Japan and received his doctorate in 2005 for work in the field of structural analysis and processing of polymers. During a postdoctoral stay in National Cardiovascular Center and Osaka Institute of Technology, he worked in the biomaterial and the regenerative medicine and tissue engineering. In 2009 he moved to National Institute for Materials Science and started to work in nanotechnology and nanoprocessing of biomaterials. His research interests include engineering and nanotechnology of synthesis and natural polymers for biomaterial use and green material use.

Ashutosh Tiwari, is the Editor-in-Chief, *Bio-Medical Materials and Devices* and *Advanced Materials Letters*. He is Secretary-General of the International Association of Advanced Materials. A materials chemist, he graduated from the University of Allahabad, India before moving to the National Physical Laboratory, India, and University of Wisconsin-Milwaukee, USA. Currently, he is Invited Professor at School of Chemistry and Chemical Engineering, University of Jinan, Adjunct Professor at Department of Materials Science and Engineering, Jiangsu University, China and Foreign Researcher at Biomaterials centre, National Institute for Materials Science, Japan. He has published 150-plus publications and patents, as well as edited/authored ten books, in the field of materials science and technology. His recent research interest is focused on designing and development of the smart materials for biomedical and engineering applications.

Lih-Sheng (Tom) Turng is a Professor at the Department of Mechanical Engineering at the University of Wisconsin-Madison. He is the Director of the Tissue Engineering Scaffolds Theme at the Wisconsin Institute for Discovery and the Co-Director of the Polymer Engineering Center. He has published 200 technical papers on microcellular injection molding, nanocomposites, bio-based polymers, and tissue engineering scaffolds. He is an elected Fellow of the Society of Plastics Engineers (SPE) and the American Society of Mechanical Engineers (ASME).

Chad Ulven received his BS degree in Mechanical Engineering from North Dakota State University (2001) and MS and PhD in Materials Engineering from the University of Alabama in Birmingham (2003 & 2005). He has been involved in the research of polymer matrix composites (PMCs) for various commercial and defense applications for the past 11 years. He has co-authored 26 journal articles, 9 U.S. Department of Defense technical reports, 4 book chapters, and over 60 conference papers related to PMCs. He has spent the past 5 years studying biobased PMCs and recycling of PMCs.

Stephanie Weal graduated from Waikato University, New Zealand in Environmental Technology (BSc(tech) and Materials Science (MSc). As a Scientist at Scion, New Zealand, she has 10 years experience with biopolymers and composites with a focus on utilisation of waste streams and biodegradation. More recently she has been with the Biopolymer Network Ltd researching biopolymer foams.

Marco Aurelio Woehl graduated in Chemical Engineering in 1995. He received his Chemistry MSc degree in 2009 at the Federal University of Paraná. Currently, he is a PhD student at the same University where he works at the BioPol (Biotechnology and Polysaccharide - Based Materials) and Cepesq (Research Center of Applied Chemistry) laboratories. His main areas of interest involve the interactions and technological applications of polysaccharides, polysaccharide-based nanocomposites and bacterial cellulose.

Fernando Wypych graduated in Chemistry (1980-84) at the Federal University of Paraná; MSc in Inorganic Analytical Chemistry at the Catholic University of Rio de Janeiro (1985-87) and PhD at the same university, in a joint project with the Technical University of Berlin, Germany (1988-92). In 1992 he was appointed Professor at the Chemistry Department of the Federal University of Paraná. His research interests involve layered materials, chemical modification of layered materials surfaces, immobilization of metallocomplexes, catalysts for esterification/transesterification reactions and polymer nanocomposites reinforced with cellulose nanofibers and layered materials.

Zahira Yaakob completed her BSc from University of Toledo, Ohio, USA in Chemical Engineering in 1988. She obtained her MSc (Chemical Engineering, 1991) and PhD (Chemical Engineering) from Department of Chemical Engineering, University of Science and Technology Manchester, UK (UMIST) in 1995. After that she joined the Department of Chemical and Process Engineering at The National University of Malaysia as lecturer. Her primary research interest is Reaction Engineering and Catalysis. She has more than 50 research publications, 4 Malaysian patents and 1 Korean patent.

1

Engineering Applications of Bioplastics and Biocomposites – An Overview

Srikanth Pilla

Wisconsin Institute for Discovery, University of Wisconsin-Madison, Madison, WI, USA

Abstract
Human society has benefited tremendously from the use of plastics due to their extraordinary versatility and manufacturability. However, this prosperity comes at the price of depleting fossil fuels and adverse effects on the environment. To minimize these undesirable consequences, scientists have been finding new sources (plastics) that are renewable, sustainable and biodegradable. This has led to the development of biobased plastics. Thus, bioplastics help reduce dependency on petroleum-based polymers, reduce the accumulation of persistent plastic waste, and better control the emission of CO_2 in the environment. On the other hand, biocomposites can substitute for petroleum based composites and provide equivalent strength to weight ratios. Biocomposites made from bioplastics and natural fibers such as hemp, wood, kenaf, coir, sisal, grasses etc are termed as green composites. They are 100% biobased and provide end-of-life options such as biodegradability and/or compostability. On the flip side, biocomposites, made from either synthetic plastics impregnated with natural fibers or bioplastics reinforced with synthetic fibers will reduce the carbon footprint on the environment. In either case, biocomposites offer sustainable alternatives for glass fiber reinforced composites. This chapter provides a general overview of bioplastics and biocomposites and their engineering applications. The engineering applications discussed are packaging, civil, biomedical, automotive etc. Also, the chapter discusses some introductory concepts about processing of bioplastics and biocomposites. Detailed discussions about all these studies are given in subsequent chapters of this handbook.

Keywords: Bioplastics, biocomposites, engineering applications, processing of bioplastics

1.1 Introduction

Plastics have been one of the most highly valued materials mainly because of their extraordinary versatility and low cost [1]. Most of the plastics are made from polyolefins such as poly(propylene) (PP), poly(carbonate) (PC), poly(vinyl chloride) (PVC), poly(ethylene) (PE), poly(styrene) (PS) etc. All these synthetic polymers are derived from petroleum and its allied components. These natural resources take

Srikanth Pilla (ed.) Handbook of Bioplastics and Biocomposites Engineering Applications, (1–16)

millions of years to form and are finite in quantity. In addition, plastics derived from fossil resources are largely non-biodegradable. Thus, the depletion of petroleum resources and increasing environmental awareness and regulations have triggered for the development of next generation materials that are environment-friendly and/or available resourcefully to meet the ever-increasing demand for plastics.

1.1.1 Bioplastics

Biobased plastics or simply *bioplastics* made from renewable resources can be naturally recycled by biological processes, thus conserving limited natural resources (fossil fuels) and reducing greenhouse gas emission (CO_2 neutral) [2–3]. Henceforth, bioplastics are sustainable, largely biodegradable and biocompatible [4–6]. Today, bioplastics have become a necessity in many industrial applications such as food packaging, agriculture, composting bags, and hygiene. Apart from these, it is foreseeable that with improved material performance, bioplastics will be used in biomedical, structural, electrical and other consumer products. So far the world's consumption of bioplastics has increased from 15,000 tons (in 1996) to 225,000 tons (in 2008) [4,7]. With increasing demand for the world's plastic consumption, it is predicted that the demand for biodegradable plastics will grow by 30% each year [6]. Hence to meet the ever increasing demand for biobased and biodegradable polymers, lot of research is being dedicated towards exploring new *green polymeric materials*. Some of the most commonly known bioplastics in todays' world are polylactic acid (PLA), polyhydroxybutyrate (PHB), soy based plastics, cellulose polyesters, starch based bioplastics, vegetable oil derived bioplastics, poly (trimethylene terephthalate), biopolyethylene etc.

Though bioplastics greatly interests many scientists and engineers throughout the world, they possess inferior properties compared to their synthetic counterparts. Hence, their application is limited in areas that currently are dominated by fossil fuel-based plastics. To improve the properties of bioplastics, polymer blends and composites are commonly investigated. For polymer composites or *biocomposites*, various types of fillers have been studied, including inorganic fillers (e.g., calcium carbonate, nanoclay), natural fibers (both wood and plant fibers), and other types of fillers such as carbon nanotubes (CNTs) [8–10]. In general, adding fillers to polymers will improve properties such as stiffness, strength, gas barrier properties, melt strength, thermal stability, etc.

1.1.2 Biocomposites

Biocomposites are of great importance to the material world because they provide unique properties that do not exist naturally. Also, their properties can be tailored based on selective design composition and processing. This leverages the use of biocomposites in different sectors such as aerospace, automotive, building and construction, marine, consumer products, electronic components etc. The design of composites using fiber reinforced polymers (FRP) is an age-old study

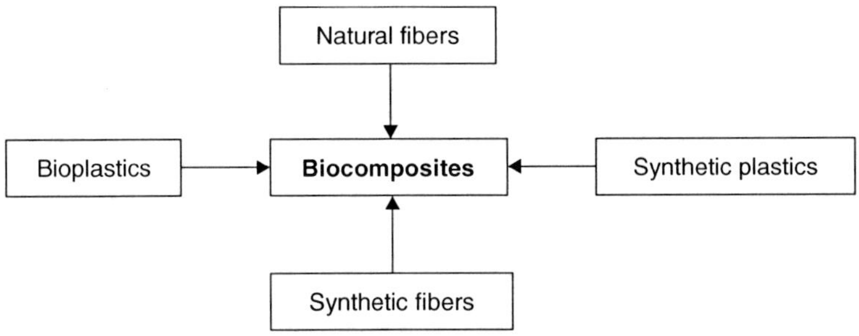

Figure 1.1 Different routes to make biocomposites.

dating back to 1908 where glass fibers were impregnated in synthetic plastics [11]. However in 1941, Henry Ford introduced biocomposites made from hemp, sisal and cellulose based plastics. Since then, lot of research is dedicated towards biocomposites and much advancement has taken place in widening its usage in various sectors, as mentioned above. Recently, scientists and engineers around the world are also focusing on reducing the carbon footprint of all the existing products by either blending bioplastics and synthetic plastics and/or reinforcing them with natural/synthetic fibers. Henceforth, the term biocomposites refers to composites made from both bioplastics and synthetic plastics impregnated with natural fibers or synthetic fibers or both (see Figure 1.1). Though synthetic fibers offer superior reinforcement capability compared to natural fillers, the latter are gaining renewed interest owing to the following advantages: renewable nature, low cost, low density, low energy consumption, high specific strength and stiffness, CO_2 sequestration, biodegradability, and less wear on machinery [12–13]. Thus, biocomposites made from bioplastic and natural fibers are also termed as '*green composites*' and are more environment friendly compared to one made from synthetic plastics and/or fillers.

1.2 Engineering Applications of Bioplastics and Biocomposites

For about half a century, vast amount of research is being carried in the field of bioplastics and biocomposites, which illustrates their significance. However, research and development is just part of a product life cycle. The real engineering starts when the science that is developed, is being applied to a specific application. Thus, the engineering process either introduces a new material (or product) into the market or supplements an existing one. Some of the applications that bioplastics and/or biocomposites are applied include packaging, civil, construction and building, biomedical, automotive etc. This handbook is focused on applications of bioplastics and biocomposites. The handbook is divided into 19 chapters. Chapters 2 and 3 focus on various parameters related to processing of bioplastics and biocomposites. Chapters 4-8 discuss packaging, 9–11 civil engineering, 12–13 biomedical, 14–15 automotive and 16–19 general engineering applications of bioplastics and biocomposites. These applications and related studies are elaborately discussed below.

1.2.1 Processing of Bioplastics and Biocomposites

Processing is a critical step in engineering of bioplastics and/or biocomposites. Especially, for industrial applications, mass production is a requirement which mandates that any processing step that is newly developed to be robust. Thus, this handbook has two chapters (chapters 2 and 3) exclusively focused on the processing aspects of bioplastics and bicomposites. Chapter 2 discusses how to handle various forms of dry ingredients in bioplastics manufacturing. The chapter will review the current technologies to handle the dry ingredients in plastics processing and finally presents challenges associated with biocomposites feedstock handling.

Chapter 3 presents modeling of the processing of natural fiber composites made using liquid composite molding. The investigative results from this chapter will help the industries to expand the application horizon of bioplastics and biocomposites from engineering and processing points of view.

In general, plastics processing begins by either mixing or compounding followed by shaping and finishing. Some of the equipment widely used for mixing or compound are:

- Blenders
- Extruders (single-screw and/or twin-screw)
- Pulverizers
- Mills (open/two-roll)
- Mixers

Of these, extruders are notably used for mass production in industrial set-ups. Though there isn't any significant difference in terms of the processing methods used for conventional plastics vs bioplastics and/or biocomposites, care should be taken while designing the process-conditions for bioplastics and biocomposite since they have narrow processing windows. As such, any small deviation from the process-design conditions might thermally degrade bioplastics and biocomposites thereby deteriorating their properties.

Bioplastics/Biocomposites 'shaping' yields a spectrum of finished products. It is typically done at appropriate temperatures for thermoplastics and at both temperatures and curing agent(s) for thermosets. The shaping methods are categorized based on the inherent properties of bioplastics and the state at which they are more suitable for the transformation into final products. These methods are classified as:

1. *Shaping in molten state:* This is also known as melt-processing and constitutes injection molding, compression molding, melt spinning, blow molding, calendaring and/or extrusion. Based on the application needs, suitable technique is employed.

2. *Shaping in rubbery state:* This is done using techniques such as thermoforming and calendaring.
3. *Shaping in wet state:* This is conducted for polymer solutions using wet-spinning, fiber-spinning, spreading and dipping.

Though the above three classifications provide a broad spectrum of processing methods for bioplastics and biocomposites, not all of them are industrially relevant for mass production. For large scale production, robust techniques need to be used. As such, the methods described above are further classified into the following three categories:

a. *Molding:* This is defined as a shaping process wherein either pressure or both pressure and temperature are applied simultaneously in a closed space viz. mold. This includes methods such as injection-molding, compression-molding, blow-molding and transfer-molding. A wide variety of products for different applications (e.g. automotive, consumer, electronics etc) are currently manufactured using any of these processing methods.
b. *Forming:* This includes techniques such as extrusion, calendaring, thermoforming, casting, slush-molding and rotomolding. Most of the packaging products are made using these techniques. In this handbook, chapters 4–8 describe studies related to bioplastics based packaging materials processed using any of these methods[1].
c. *Foaming:* Foaming is a process wherein small pores or cells are being created with the aid of a foaming or blowing agent. It typically reduces density, provides cushioning and insulation properties while structurally integrating mechanical properties, in case of foamed injection-molding. Foaming is widely classified as three types: conventional, microcellular and nanocellular foaming.

Conventional foaming is an age old technique wherein foaming was primarily done using chemical foaming agents. In conventional foams, the cell sizes were typically ≥100μm and cell densities in the order of 10^3 to 10^6 cells/cm^3 [14]. The microcellular foamed plastics possess cell densities on the order of 10^9 cells/cm^3 and cell diameters of 10μm or less [15]. Nanocellular foams possess cell sizes ≤10mnm and are a recent invention in foams [16–17]. Currently, nanofoams exist only in batch processes and investigations are going on to develop a continuous nano foam process that could be used for large scale production in industries.

In this handbook, chapters 6, 9 and 14 exclusively talk about foams made from bioplastics and biocomposites while several other chapters discuss the act of foaming as applied to respective application that the investigation has been carried out to.

---

[1] Some of the studies use techniques that are not listed here which may or may not be scalable. Further investigations are needed for such studies.

1.2.2 Packaging Applications of Bioplastics and Biocomposites

Among the total plastics usage, 'packaging' occupies the top position with 41%, of which about 20% is used in food industry. Since most of the packaging materials are made up of non-renewable and non-degradable synthetic plastics, packaging waste also occupies the top position in landfills [18]. Also, there have been many health related issues for using synthetic polymers for packaging, especially in food division. This has mandated the use of biobased and biodegradable or compostable materials in the packaging sector.

Design of materials for packaging applications is a multi-step process. It mandates a meticulous engineering of the material to obtain target properties. Some of the properties that good packaging materials possess are permeability (gas and vapor), sealing, resistance to chemicals, UV and light, transparency, mechanical properties, machinability etc. Additionally, cost considerations and availability should be taken into account while proposing a new material into market. Finally, the material should follow a 'cradle to grave' cycle. Synthetic plastics that currently dominate the packaging sector possess most of the specifications listed above except for sustainability which bioplastics offer. Thus interest and research in using bioplastics in the packaging sector has increased both in academia and industries. The use of bioplastics not only provides a sustainable alternative for packaging but also biodegradability and/or compostability. These properties will leverage the reduction of landfill waste in addition to providing high-valued gases as compost products.

Some of the notable companies that have been developing bioplastics are Dow chemicals (EcoPLA), DuPont (Sorona and Hytrel), BASF (Ecoflex and Ecovio) etc. Especially BASF which is a world leader in polymers and chemicals has developed several biobased and biodegradable plastic lines based on starch, PLA, PBAT etc. [19]. Ecoflex or PBAT is a fully biodegradable plastic material and Ecovio is a blend of Ecoflex and PLA. These two have found several applications in packaging especially in shopping bags, compost bags etc. Ecoflex is resistant to water and grease making it an ideal choice for disposable wrapping. Also, both Ecoflex and Ecovio have found significant applications in agricultural sector e.g. in making mulch films etc. Additionally, many other companies such as Nature Works, Environmental Polymers, Novamont, Mirel, Tianan, Innovia etc are manufacturing bioplastics from renewable sources.

In spite of the unique features of bioplastics, it is not imperative to say that they will dominate the packaging sector in the current domain. This is due to not-so-superior properties of bioplastics compared to synthetic ones. However, the inferiorities could be eliminated by modifying the formulation design to suit the target application. For instance, PLA is a brittle polymer and hence could not be aptly used for thermoforming. However when blended with processing aids and impact modifiers such as starch, Ecoflex etc, we could impart toughness for PLA, making it suitable for such applications [20]. Thus unique materials designs are needed that will impart the best possible properties for bioplastics and a lot of research has already been in place to address these kinds of issues.

Some of the investigations are presented in this handbook in chapters 4–8. Chapter 4 discusses recent advancements in biodegradable polymer nanocomposites with a focus on developing cost-effective bio-based packaging materials. Some of the biopolymers and fillers reviewed in this chapter are PLA, PHAs and starch and nanoclays, starch nanocrystals, cellulose and chitin nanofibers, respectively. The nanofillers are added to improve the mechanical and the barrier properties of the biopolymers.

Chapter 5 presents research results from various investigations related to food packaging applications of biopolymers: their origin, structure, development, processing and characterization.

Chapter 6 reviews PLA based foams used in packaging applications. It discusses extrusion foaming, expanded particle (bead) foaming, and other foam processes for making polylactic acid foams and describes some key performance features of such foams.

Chapter 7 summarizes key concepts of PVs graft copolymerization onto guar gum by highlighting its properties and applications in the packaging science and technology and chapter 8 focuses on starch based bioplastics and biocomposites and their application to packaging.

With the advancement of the science and technology, it is envisioned that the future of commodity plastics sector, especially packaging, is going to be fully sustainable. This will help to reduce the waste in the environment and our dependence on fossil resources. Thus a balanced approach towards ecosystem will be maintained from 'cradle to grave'.

1.2.3 Civil Engineering Applications of Bioplastics and Biocomposites

Civil engineering, especially building and construction materials utilize about 23% of the world's total plastic usage. Also, many of these materials are energy intensive to produce. Besides packaging, the construction and demolition debris constitute a large percentage of landfill waste. These practices make the construction materials to occupy a large carbon footprint in the ecosystem. Thus it is important to look for opportunities that practice sustainable approaches for providing an ecological balance. The use of bioplastics and biocomposites would provide such opportunities.

Though bioplastics provide a sustainable alternative for building and construction materials, the nature of application necessitates the use of biocomposites owing to their superior strength properties. Thus, biocomposites in addition to being environment-friendly, offer many advantages such as light-weight, low material costs, high specific properties [21–22]. For instance, as shown in Table 1.1, natural fibers possess higher specific properties compared to E-glass.

Hence, lots of work is dedicated in designing novel biocomposites as per the layout shown in Figure 1.1. Some of the building and construction applications where biocomposites are potentially applied include formwork, scaffolding, decking, railing, fencing, framing, walls and wallboard, window frames, doors, flooring, decorative paneling, cubicle walls and ceiling panels. Additionally foamed biocomposites are investigated for housing insulation applications [18]. Other

Table 1.1 Comparison of properties of natural fibers with E-glass (reproduced from [23]).

Fiber Type	Density (g/cm^3)	Elastic modulus (GPa)	Specific modulus
E-glass	2.55	73	29
Hemp	1.48	70	47
Flax	1.4	60–80	43–57

Figure 1.2 Temporary housing made from wood-plastic composites set-up for displaced Haitians (courtesy: http://www.innovida.com/).

than these, major area where biocomposites find critical application is in building a temporary housing. Generally, temporary housings, made from wood plastic composites, are set-up whenever a major catastrophe occurs such as earthquake, hurricane etc [shown in Figure 1.2]. During such times, temporary rehabilitation centers are built to provide shelter for people who lost their homes. Once the situation recovers to normalcy, the temporary housings are dismantled and the waste is dumped in landfills. Thus, to eliminate this type of landfill waste, such application necessitates the use of biocomposites (more specifically *green composites*) that can potentially be composted after their service life.

In spite of the aforementioned advantages that biocomposites offer, there exist few critical issues in their design i.e. hydrophilicity of natural fibers and weak interfacial bonding. Hydrophilicity of natural fibers will result in uptake of water/moisture during the service-life of the composite thereby making it structurally weak. A weak interface creates voids which will also add to the failure of the structure. Thus it is important to do an interfacial engineering of the fibers that will not only eliminate the hydrophilicty of the fiber but also make it bond with the hydrophobic polymer, perfectly. As such, interfacial engineering provides a strong interface i.e. perfect bonding between the fibers and polymers thereby leveraging for efficient stress transfer between the polymer and the fiber and enhancing the strength properties of composites, significantly. These 'engineered' biocomposites possess high stiffness and strength-to-weight ratios making them comparable to traditional composites at much lower cost.

This handbook discusses three studies on 'engineered' biocomposites. Chapter 9 focuses on vegetable-oil derived rigid polymeric foam composites. More specifically, the chapter discusses the routes to make these biofoams, potential applications and their environmental impacts.

Chapter 10 presents kenaf based biocomposites fabricated by pultrusion process that will potentially replace either steel-based or synthetic components in construction applications. Two types of kenaf fibers, with and without NaOH treatment, were used. Compared to untreated, NaOH treated kenaf fiber reinforced biocomposites showed higher flexural properties and lower water absorption with good interfacial adhesion between fiber and matrix. This type of engineering is very important for construction applications as discussed earlier.

Chapter 11 discusses various applications of starch, both as a bioplastic and as a biocomposite, in construction and civil engineering sectors. Some of the applications discussed include starch as a binder, plaster, thickening agent, acoustic material, composite, additive for reducing mortar stickiness, additive for rheological transformations in concrete and mortar etc.

The present research efforts in biocomposites for building and construction applications are limited for non-structural engineered parts as listed above. However, work is underway to extend this to structural parts such as beams and columns. Also, long-term performance and durability under indoor and outdoor environments and all weather conditions should be evaluated. Successful completion of these projects would ensure a strong position for bioplastics and biocomposites in building and construction applications.

1.2.4 Biomedical Applications of Bioplastics and Biocomposites

With the advent of innovations in the field of medicine, new materials are being explored for usage into the broad spectrum of biomedical applications such as implants, tissue engineering, drug delivery etc. In this context, bioplastics and biocomposites play vital role since they are biobased, biodegradable and biocompatible (the most critical aspect for biomedical applications). In fact the biomaterials that are used in human body must be compatible with the tissue and other related organs that they are found or fitted into. Also, the prime reason for the biomaterials to biodegrade inside the body is to eliminate any further surgical or medical intervention for the removal of the part that was made from the biomaterial. Though bioplastics either independently or as a blend offer feasible solutions for biomedical industry, biocomposites, especially the ones where bioplastics are impregnated with hydroxyapatite (HAP), are finding suitable application in implant making and/or tissue engineering. However, the emergence of new generation of hybrid nanostructured materials has not only opened a new route to make biomaterials but also widened the range of applications in biomedical industry. These materials that constitute bioplastics embedded with nano particles (both inorganic and organic) form a closed loop between the fields of materials science, life science and nanotechnology, bringing outstanding innovations to biomedical engineering.

Similar to bioplastics, bionanocomposites exhibit biocompatibility and biodegradability, the key requirements for biomaterials. One of the prominent applications of bionanocomposites in biomedical field is in the regeneration of damaged tissues and in implants [24]. In addition to being biocompatible and biodegradabe, bionanocomposites should provide mechanical stability to avoid collapse of the implant and possess open-pore structure (macroporosity) for efficient transportation of nutrients and metabolic wastes [25–26]. Some of the common biopolymers used for this application are PLA, chitin and cellulose. Figure 1.3 shows the SEM micrograph of PLA based bionanocomposite processed via microcellular processing technology for tissue engineering scaffolds.

For drug delivery and other related applications, the biomaterials, in addition to being biocompatible, need to be at reduced dimensions i.e. nanoscale [24]. Thus, bionanocomposites will aptly fit into this division of biomedical applications. Over the past few years, several researchers around the world have been investigating to bring new formulations for the design of novel drug delivery systems. Other critical divisions of biomedical engineering where bioplastics and/or bionanocomposites are applied include cancer therapy and diagnosis, gene vectors, biosensors and dental applications such as dental implants.

This handbook presents reviews on the application of two biopolymers viz. cellulose, chitin and chitosan, to the biomedical field. Chapter 12 presents the application of cellulose and derived composites to tissue and neural engineering, pharamaceutical engineering and implants. It was inferred that the target applications is governed by specific form of cellulose i.e. microcrystalline, powder, sponges or nano-structure.

Chapter 13 highlights chitin and chitosan nanofiber structures, nanofibrous membranes and their biocompatible nanocomposites. The chapter reviews potential biomedical applications of chitin and chitosan, especially in the areas of drug

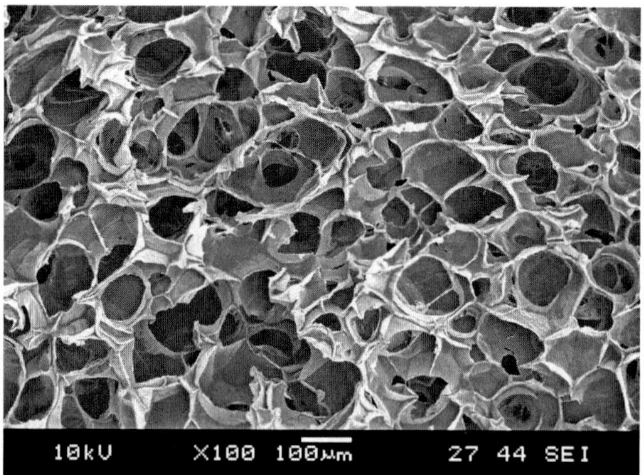

Figure 1.3 PLA based microcellular porous structure used as a tissue engineering scaffold.

release, dental, bone tissue engineering, catalyst and enzyme carriers, wound healing, skin regeneration, biosensors, medical implants, and liver functioning.

Due to the versatility in bioplastics and nanoparticles and the synergy that exists between them, the field materials and life sciences are envisaged to see more explorations in biomaterials. Especially, the multifunctionality of bionanocomposites will open up new research arenas with plentiful of opportunities for great innovations. These will definitely help to revolutionize the field of biomedical engineering.

1.2.5 Automotive Applications of Bioplastics and Biocomposites

The history of transportation system shows how advances in science and technology have played a crucial role in its growth [27–28]. Especially, automobiles have been an integral part of human-beings. They are used not only for transporting people, individual and groups, but also goods. Large-scale manufacture of these automobiles, operation of the transportation (road) network system, and construction of the required infrastructure has dominant impacts on the economic growth of a country. Thus the growths of automotive and economic sectors are inter-related.

Due to the complexity involved in the making of automobiles, there exists a considerable need for fabricated materials in this sector. According to a study conducted by department of transportation (DOT) in 2007, there are more than 250 million passenger vehicles in use in USA and about 5 million new cars are produced every year [29]. Assuming that an average of 1500 kg of material is required for each automobile, the total consumption of materials is 10 million tons per year. It means that this quantity has to be produced, used and ultimately recycled or disposed at the end of useful life. Thus efforts are on for developing newer or alternative (sustainable) materials to achieve: (a) fuel efficiency and cost savings; (b) reduced emissions and (c) future ability to recycle or biodegrade. In a competitive environment among domestic and international auto manufacturers, there is significant emphasis on attaining these three objectives so as to pioneer the auto industry market and lead its economy. Thus from materials point of view, these three objectives are very critical for the growth of both the transportation and economic sectors of USA.

As discussed earlier, increasing societal and environmental concerns due to the use of non-biodegradable and non-recyclable materials coupled with US government legislation which states that the deposition fraction of a vehicle in landfills should be ~5% by year 2015 [30] have prompted automobile researchers in US to enhance the content of recyclable or biodegradable materials in automobiles. Though recycling is a much desirable option as it reuses the raw materials and reduces the amount of plastic waste in landfills, there are certain limitations associated with it such as: (1) some plastics, such as those with a complex formulation or thermosets, are not easily recyclable; (2) multiple recycling cycles may result in deterioration of material properties; and (3) recycling generally suffers from unfavorable economics. This necessitates the need for exploring newer materials such as bioplastics and biocomposites that are environment-friendly (or reduce carbon footprint) and either biologically degrade or compost.

The use of biocomposites makes the automobiles not only environment-friendly but also 'lighter' due to the use of low-density natural fibers for high-density glass fibers [31]. In fact, the rising cost of fuel has triggered the use of lightweight biocomposites for all transport vehicles. This will result both in reduction of fuel consumption and CO_2 emissions. According to a study, about 25% reduction in the weight of the vehicle is equivalent to a savings of 250 million barrels of crude oil and reduction in CO_2 emissions to the tune of 220 billion pounds per annum [27].

In spite of the above-mentioned advantages that biocomposites offer to the automobiles, there exist few critical issues in their design i.e. hydrophilicity of natural fibers and weak interfacial bonding (discussed in detail in section 1.2.3). Thus, lots of investigations are being carried for using bioplastics and biocomposites in automobiles.

In this handbook, chapters 14 and 15 discuss the application of bioplastics and biocomposites in automotive sector. Chapter 14 presents an overview of the synthesis, processing, properties, and applications of biobased and biodegradable PHBV and its blends and composites. Though the chapter presents the application of the PHBV based composites is various sectors, it lays special emphasis on automotive sector. The chapter also introduces a novel polymer processing method viz. microcellular injection molding which is an environmentally-friendly polymer processing technology that is capable of mass-producing components with minimally compromised material properties while consuming less energy and materials, as compared to components produced by the conventional injection-molding process [33].

Chapter 15 discusses a variety of thermoplastic and thermosetting bioplastics and biocomposites with a focus on the automobile industry. It presents the drawbacks and necessary improvements of these renewable materials while discussing the potentiality of their expected future evolution.

In the current scenario, the application of bioplastics and biocomposites for automotive sector is limited only to low to mid-level load-bearing parts (non-structural). However, with further advancements in R&D, it is foreseen that the structural (high load-bearing) automobile components are also being designed and fabricated using bioplastics and biocomposites. This would restructure the automotive sector both in terms of sustainability and economics.

1.2.6 General Engineering Applications of Bioplastics and Biocomposites

As the research horizon of bioplastics and biocomposites is expanding, studies are also being conducted wherein the objective is to have fundamental understanding of the science without any boundary of applications. Chapters 16–19 of this handbook present some of the studies carried out under this category. They focus on expanding the horizon of bioplastics i.e. to develop various synthesis routes, processing methods and characterization techniques so as to obtain optimized properties. This will allow for those bioplastics and biocomposites to be applied to any generic engineering application, depending on specific properties requirements.

Chapter 16 reviews the synthesis and properties of cellulose nanofibers and their applications in bioplastics. Specifically, the chapter describes various approaches for the preparation and extraction of cellulose nanofibers from plant resources. Then, it presents various thermo-physical properties of cellulose nanofibers while illustrating the applications of bionanocomposites made from cellulose nanofibers and bioplastics taking into consideration the interfacial affects between the fibers and polymer.

Chapter 17 discusses the fundamental aspects of bionanocomposites based on starch and nanosized biofibers. Besides providing discussions on various processing methods used, the chapter presents some results on structure, properties, and applications of these bionanocomposites. Finally the chapter underlines possible business opportunities for these bionanocomposites in different sectors of industries such as construction, food packaging, transportation (e.g. automotives) etc.

Chapter 18 briefly discusses various biogenic precursors for the making of variety of polyphenol, polyester and polyurethane resins. The chapter starts by listing different reactive polymers and the precursors (monomers) needed to synthesize them. Then it describes various platforms such as sugar, lipid, biogenic olefin, acid, lignin etc to make chemicals, promoters for reactions and additives. Special emphasis is made on lignin which is a single source for producing renewable aromatic compounds through pyrolysis, cracking and oxidation.

Chapter 19 presents the investigations carried out to produce high performance bioplastics and biocomposites by impregnating long biofibers and engineered pulp. Especially, the chapter focuses on thermoplastic materials and processes thereof i.e. manufacturing methods as applied to natural and wood fiber reinforced thermoplastics. The chapter starts with introductory explanations about biofibers, bioplastics and biocomposites. Then it details different processing techniques such as sheet-molding and injection-molding used to fabricate the said biocomposites with engineering examples from Toyota, Ford, and Scion Research.

The emergence of bioplastics and biocomposites has contributed significantly in building an economically and ecologically advanced sustainable society. Especially, the stringent measures that nations across the world are taking to promote biobased content in their products and reduce carbon footprint on the environment, would soon help to realize a sustainable world that does not depend to a large extent on petroleum. In this front, the current investigations carried at both fundamental and application levels are critical. Understanding the fundamental theories would help to tailor the properties of the bioplastics and biocomposites for specific applications.

1.3 Conclusions

The research and development of biobased and biodegradable plastics has been catalyzed by the scarcity of oil, increases in the cost of petroleum-based commodities, and growing environmental concerns with the dumping of non-biodegradable plastics in landfills. *Green composites*, made from bioplastics and natural fibers such as hemp, kenaf, wood, agricultural residue etc are 100% biodegradable and provide a sustainable alternative for synthetic, glass fiber reinforced, composites. On the other hand, biocomposites, made from synthetic/bioplastics impregnated with natural/synthetic fibers, respectively, will reduce the carbon footprint on the environment. These approaches are mandated for maintaining the sustainability of the ecosystem. The current system of design methods has limited the application of bioplastics and biocomposites only to certain sections of industries. However, with the advent of new 'engineering' methods, it is believed that the application base of bioplastics and biocomposites will be widened. Also, with innovations in existing processing technologies and with development of novel processing methods, the production costs will decrease without affecting the rates of production. Thus, the use of bioplastics and biocomposites will not only provide a renewable approach but also an economical alternative for petroleum based plastics and composites, thereby contributing for an ecologically balanced sustainable society.

References

1. E.S. Stevens, *Green Plastics: An Introduction to the New Science of Biodegradable Plastics*, Princeton University Press, 2002.
2. S.S. Ray, and M. Bousmina, M., *Progress in Materials Science*, Vol. 50, p. 962, 2005.
3. V.A. Fomin, and V.V Guzeev, *Progress in Rubber and Plastics Technology*, Vol. 17, p. 186, 2001.
4. R.A. Gross, R. A., and B. Kalra, *Science*, Vol. 297, p. 803, 2002.
5. C. Bastioli, *Starch/Staerke*, Vol. 53, p. 351, 2001.
6. R. Leaversuch, *Plastics Technology*, Vol. 48, p. 66, 2002.
7. "Bioplastics consumption to reach 2 mln tons by 2018", http://www.plastemart.com/Plastic-Technical-Article.asp?LiteratureID=1454, July 14, 2010.
8. R. Andrews, and M.C. Wisenberger, *Current Opinion in Solid State and Materials Science*, Vol. 8, p. 31, 2004.
9. T. Li, L-S. Turng, S. Gong, and K. Erlacher, *Polymer Engineering and Science*, Vol. 46, p. 1419, 2006.
10. A.K. Mohanty, M. Misra, and G. Hinrichsen, *Macromolecular Materials and Engineering*, Vol. 276–277, p. 1, 2000.
11. A.K. Mohanty, M. Misra, Drzal, L.T., S.E. Selke, B.R. Harte, and G. Hinrichsen, "Natural Fibers, Biopolymers, and Biocomposites: An Introduction," in A.K Mohanty, M. Misra, and L.T. Drzal, eds., *Natural Fibers, Biopolymers, and Biocomposites*, Taylor and Francis, Florida, pp. 1–36, 2005.
12. A.K. Mohanty, M. Misra, and L.T. Drazel, *Journal of Polymers and the Environment*, Vol. 10, p. 19, 2002.
13. M.A.S.A. Samir, F. Alloin, and A. Dufresne, *Biomacromolecules*, Vol. 6, p. 612, 2005.
14. J. Throne, *Science and Technology of Polymer Processing*, N. P. Suh and N. Sung, eds., MIT Press, Cambridge, pp. 77, 1979.
15. S. Gong, M. Yuan, A. Chandra, H. Kharbas, A. Osorio, and L.S. Turng, *International Polymer Processing*, Vol. 20, p. 202, 2005.
16. B. Krause, H.J.P. Sijbesma, P. Munuklu, N.F.A. van der Vegt, and M. Wessling, *Macromolecules*, Vol. 34, p. 8792, 2001.
17. B. Krause, K. Diekamann, N.F.A. van der Vegt, and M. Wessling, *Macromolecules*, Vol. 35, p. 1738, 2002.
18. S. Pilla, *Processing and Characterization of Novel Biobased and Biodegradable Materials*, PhD Dissertation, The University of Wisconsin, Milwaukee, 2009.
19. V.A. Fomin, *Progress in Rubber and Plastics Technology*, Vol. 17, p. 186, 2001.
20. Biopolymers present new market opportunities for additives in packaging, Plastics Additives and Compounding, pp. 22–25, May/June 2008.

21. A.K. Mohanty, M. Misra, and G. Hinrichsen, *Macromolecular Materials and Engineering*, Vol. 276/277, p. 1–24, 2000.
22. W.D. Brouwer, *Sampe Journal*, Vol. 36, p. 18, 2000.
23. R. Burgueno, M.J. Quagliala, G.M. Mehta, A.K. Mohanty, M. Misra, and L.T. Drzal, *Journal of Polymers and the Environment*, Vol. 13, p. 139, 2005.
24. M. Darder, P. Aranda, and E. Ruiz-Hitzky, *Advanced Materials*, Vol. 19, p. 1309, 2007.
25. V. Thomas, D. R. Dean, and Y. K. Vohra, *Current Nanoscience*, Vol. 2, p. 155, 2006.
26. M.S. Widmer, and A. G. Mikos, "Fabrication of biodegradable polymer scaffolds," in C.W. Patrick, Jr., A. G. Mikos, and L. V. McIntire, eds., *Frontiers in Tissue Engineering*, Elsevier, Oxford, pp. 107, 1998.
27. Tomorrow's Plastic Cars, ATSE Focus #113, July–Aug 2000.
28. K. Hess, "The growth of Automotive Transportation", http://www.klhess.com/car_essy.html, June 9, 1996.
29. http://www.bts.gov/publications/national_transportation_statistics/html/table_01_11.html.
30. N. N., Directive 2000/53/EC of the European Parliament and the Council of end-of-life vehicles, Office Journal of the European Communities, ABl. EG Nr. L 269 S. 34L 269/34, 21 October 2000.
31. P. Lammers, K. Kromer, Competitive Natural Fiber Used in Composite Materials for Automotive Parts, ASAE Paper No. 026167, Chicago, Illinois, 2002.
32. Current, Michigan State University Newsletter, Vol 93, Summer 1999.
33. S. Gong, L.S. Turng, C. Park, and L. Liao, "Microcellular Polymer Nanocomposites for Packaging and other Applications," in: A. Mohanty, M. Misra, H.S. Nalwa, eds., *Packaging Nanotechnology*, American Scientific Publishers, pp.144, 2008.

PART 1
PROCESSING OF BIOPLASTICS AND BIOCOMPOSITES

2

The Handling of Various Forms of Dry Ingredients in Bioplastics Manufacturing and Processing Applications

Andy Kovats

Brabender Technologie, Canada

Abstract

The handling and feeding of various forms of dry ingredients (pellets, powders, flakes, etc) in plastic compounding and similar manufacturing processes is nothing new and has been evolving for decades. The recent rise in interest in BioComposites and BioPlastics production, however, has challenged producers and equipment suppliers alike as they strive to handle raw ingredients and feedstock in fibrous and other forms, up until now not common in extrusion processes. This paper will review the conventional technology as it exists today from characterizing dry bulk solids to reviewing typical equipment and process operation guidelines. Finally the unique challenges of handling BioComposite feedstock will be presented along with suggestions for process optimization.

Keywords: Feeders, powders, fibers, agitation, flexible hopper, loss-in-weight feeders, fiber feeder, extrusion, dry ingredients, volumetric feeders, screw feeders

2.1 Introduction

The recent rise in interest in the production of BioPlastics has of necessity resulted in a requirement to review the basic principles of storage, handling and feeding of dry ingredients. Although the focus of this chapter shall be feeders that transfer dry (and liquid) ingredients into a starve fed, co-rotating twin screw extruder, the basic principles can equally be applied to other processes involving dry ingredients such as continuous mixing or batching.

For Starve fed extruders, feeders control the total extrusion rate and the preparation of each ingredient to the total, all within very close tolerances (Figure 2.1).

Since the form of ingredients, the flowability of the ingredients and the feed rate of the ingredients vary, feeders for such ingredients vary in design and size. For example, the flow properties of dry ingredients vary from extremely good flowing plastic pellets to poor flowing titanium dioxide powder. Feeder selection becomes an important design consideration depending on the ingredient of interest. Since

Srikanth Pilla (ed.) Handbook of Bioplastics and Biocomposites Engineering Applications, (19–42)

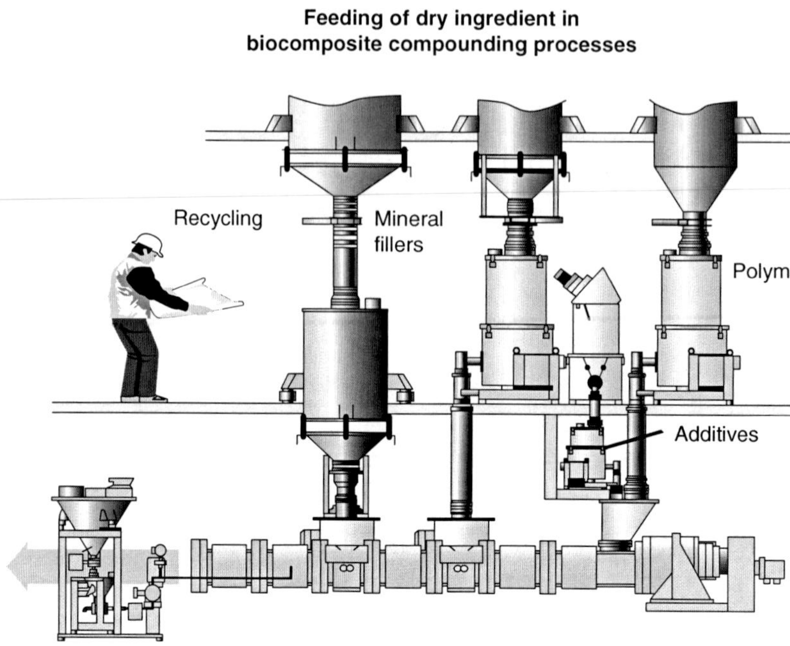

Figure 2.1 Feeders used in extrusion.

the rate and formulation totally depends on the feeders, they become vital partners in the extrusion process.

2.2 Ingredient Properties Affecting Feedrates and Dry Ingredients Handling

When choosing a right feeder for an ingredient, it is important to have as much information as possible about the ingredient. The following ingredient characteristics affect "flowability". Flowability of an ingredient affects feeder performance ranging from "good flowing" (plastic pellets for example) to "poor flowing" such as titanium dioxide (a sticky, bridging powder). Some poor flowing powders have flooding characteristics as well. Flooding is a phenomenon where air is entrapped in the powder particles and the powder behaves like a liquid.

2.2.1 Name

Name define trade name, chemical name and manufacturer if possible.

2.2.2 Bulk Density

Bulk density: measure loose and packed bulk density (weight per unit volume). Note that particle specific gravity (as defined on MSDS sheets, for example) is useless here. Bulk, not particle density is required for feeder sizing.

2.2.3 Compressibility

Compressibility is defined as:

$$\text{Compressibility} = 1 - \left(\frac{\text{loose bulk density}}{\text{packed bulk density}} \right) \times 100\%$$

Materials that have a large difference between loose and packed bulk density (and therefore higher compressibility) are usually poorer flowing materials requiring care in hoppering and agitation. Typical compressibility values are between 3 and 40%.

2.2.4 Particle Form

Particle form choose from:

Powder	Prill (or bead)	Crumb
Pellet	Chunk	Dust
Flake	Fiber	Irregular shape
Granule	Crystalline	

2.2.5 Particle Size

Particle size is important to know particle size from several standpoints.

a. Fine powders, to define flowability and floodability;
b. For fibers (e.g. fiberglass) flowability varies widely with particle size;
c. For pellets, granules, and irregulars, to define physical feeder characteristics such as clearance between screw and tube when selecting feeder size.

2.2.6 Angle of Repose

Angle of repose is defined for a stockpiled ingredient as that angle between a horizontal line and the sloping like from the top of the pile to the base. The lower the angle of repose, the better flowing of the material.

2.2.7 Angle of Slide

Angle of slide is defined as that angle to the horizontal of an inclined flat surface on which an amount of material will slide downward due to its own weight. Care must be taken not to confuse material sliding upon itself with material actually moving on the plate.

2.2.8 Packing and Compaction

It is important to distinguish between the two:

2.2.8.1 *Packing, By Pressure*

If you squeeze some in your hand, you can make a snowball. These materials will bridge or rathole, and require hopper agitation.

2.2.8.2 *Compacting, By Vibration*

If squeezed, will shoot out between your fingers. Has a high compressibility, but may flow out of 75° sloped hopper.

For example, titanium dioxide tends to pack and compact. Many plastic resin powders, on the other hand, have high compaction but do not pack under their own weight.

2.2.9 Moisture Content

Moisture content is the weight of water that can be evaporated by drying compared with the total weight of the ingredient. For certain ingredients this is very critical to define; for example, starch with 7% moisture content flows very well while the same material with 30% moisture content is sluggish.

It is clear that while not all of the above information may be available for a particular ingredient, the more that is available, the less the risk in the feeder installation.

2.3 Storage Hoppers and Ingredient Activation

Application of correct activation techniques is vitally important to ensure a constant supply of homogeneous, pre-conditioned dry ingredient into the feed mechanism. Numerous designs which have all stood the test of time are available from various manufactures, each of which has its own unique advantages and disadvantages. Some of the most common designs are discussed below:

2.3.1 Vibration

Vibration as a means to activate the ingredient into the feed is very effective. However, with the advent of Loss-In-Weight control for feeders, vibration is less favorable due to its effect on the scale that weighs the feeder.

2.3.2 Internal Stirring Agitation

Please refer to the screw hopper internal stirring agitator shown in Figures 2.2–2.4. In this design a "screw hopper" located directly above the screw trough houses a horizontal shaft agitator, which rotates within the hopper promoting ingredient flow and conditioning it to a constant density. The usual agitator design is a 4-blade type with a pointed triangle at the end of each blade.

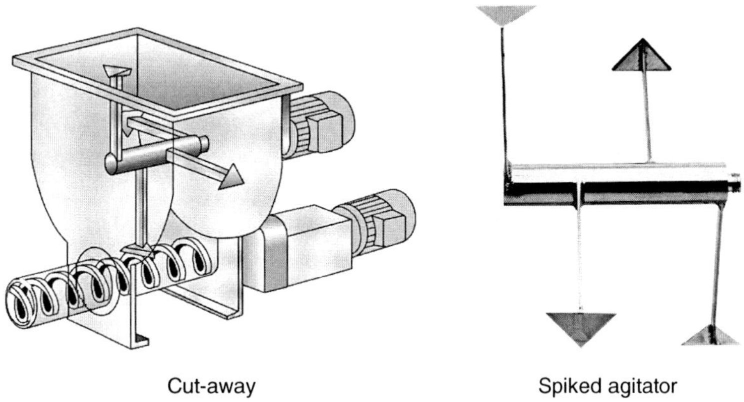

Figure 2.2 Internal stirring agitator screw feeder.

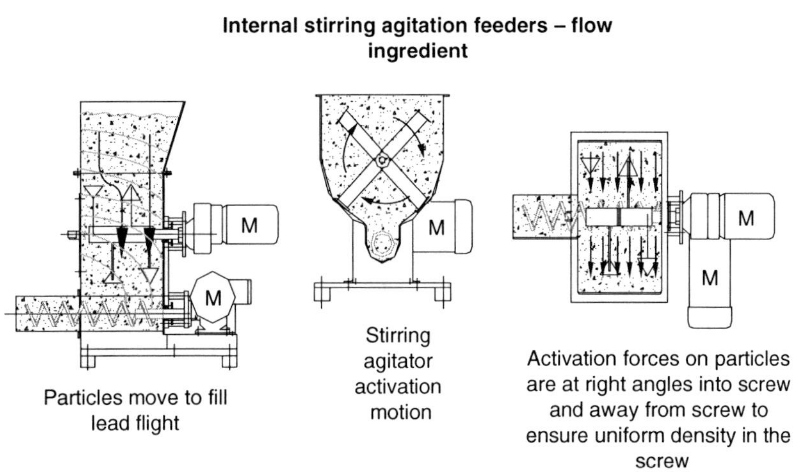

Figure 2.3 Internal stirring agitation feeders.

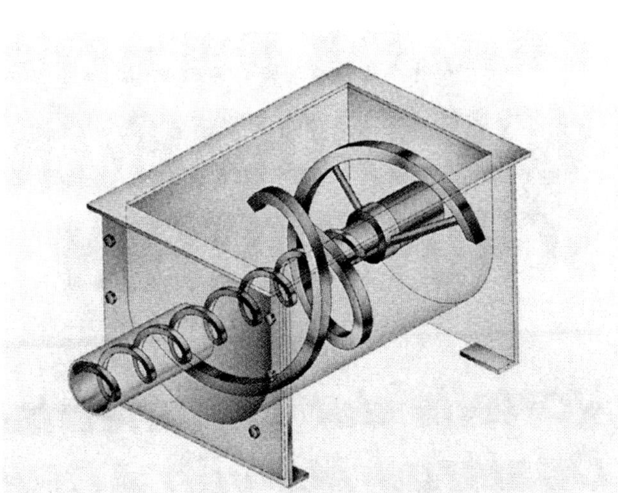

Figure 2.4 Concentric screw agitated feeder.

This can be increased to 8 blades for poor flowing ingredients that have a tendency to arch over the top of the screw. Alternatively, the triangles are removed for ingredients such as soft rubber chunks that would normally be impaled by the triangle points. If ingredient "bridging" can occur in the extension hopper above the screw hopper, a vertical shaft agitator in a circular extension hopper is used.

Advantage: Direct interaction with ingredients, can be used in Loss-In-Weight Feeders.

Disadvantage: Screw hopper inlet dimension should be large enough to prevent bridging above, doesn't activate entire ingredient mass, dangerous for hands, difficult to clean.

2.3.3 Concentric Screw Agitation

The concentric agitated feeder uses an internal agitator blade (or blades) that surround the screw. This agitator helps the flow within the screw hopper by increasing the cross-sectional area with the agitation and by positively moving the ingredient into the screw.

2.3.4 External Agitation (Flexible Hopper)

Please refer to Figures 2.5, 2.6 and 2.7. The feeder utilizes a flexible hopper. The hopper is massaged from the outside by two massaging paddles that undulate against the sides of the hopper, breaking the material bridge and massaging it into the screw.

Advantage: Large inlet into the flexible hopper, entire ingredient mass is activated, activation is external so disassembly of feeder and cleaning is easy.

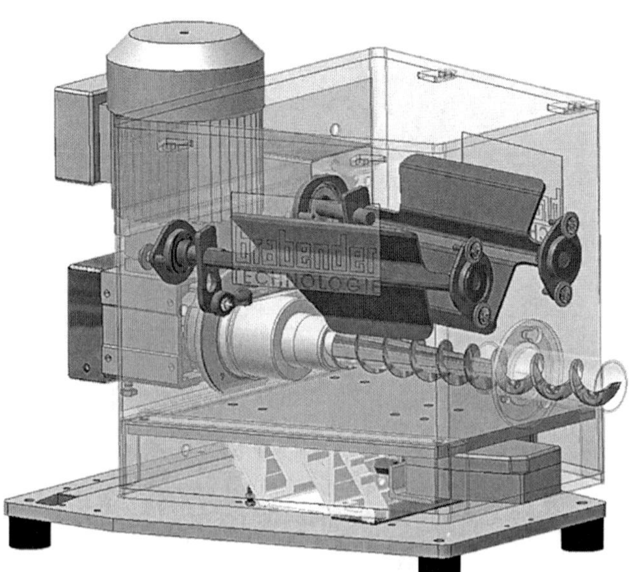

Figure 2.5 Internal components of flexible hopper feeder.

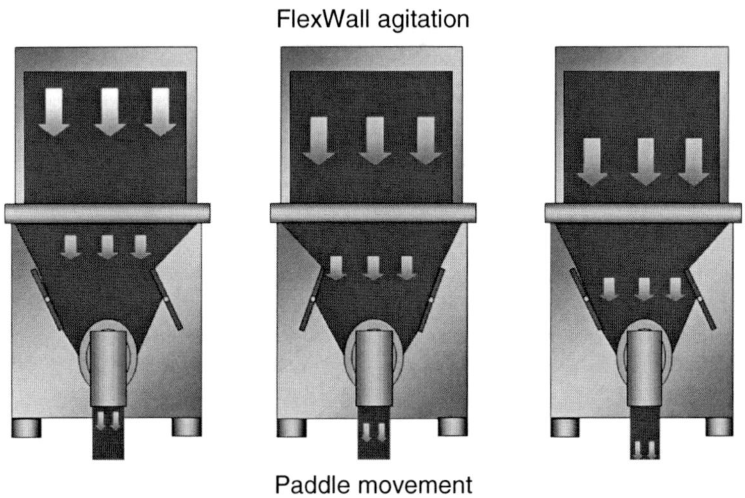

Figure 2.6 Flexible hopper wall with external paddle feeder.

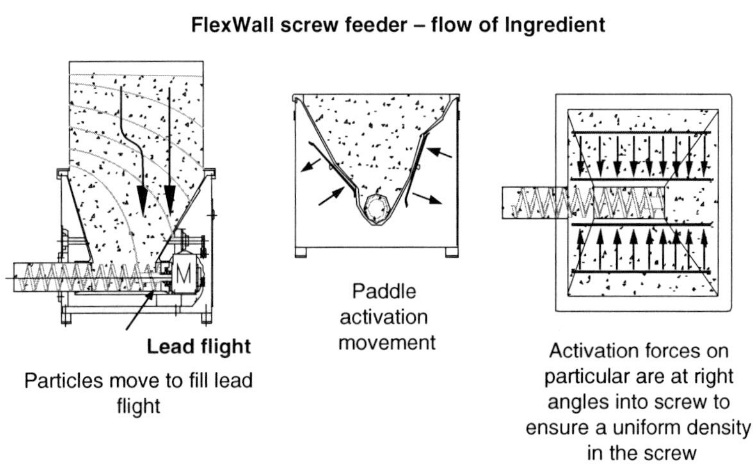

Figure 2.7 Flow of ingredient in a flexible hopper feeder.

A feeder storage hopper is an extension hopper, integral with the feed mechanism, which can hold from 5–30 minutes ingredient capacity in automatic refill situations to a few hours capacity for manual refill. This is in contrast with silos and large day storage hoppers, the designs of which are not covered here.

The extension hopper must be carefully designed to ensure that it is not the weak link in the feed system. Perfect selection of the feed device and agitation technique is useless if the ingredient bridges in the hopper above the agitator. Rather, the storage hopper, agitator and feeder must work in unison to smoothly deliver ingredient from day storage to the process.

The hopper designs may be, cylindrical, rectangular, conical, or slope-sided (Figures 2.8 and 2.9).

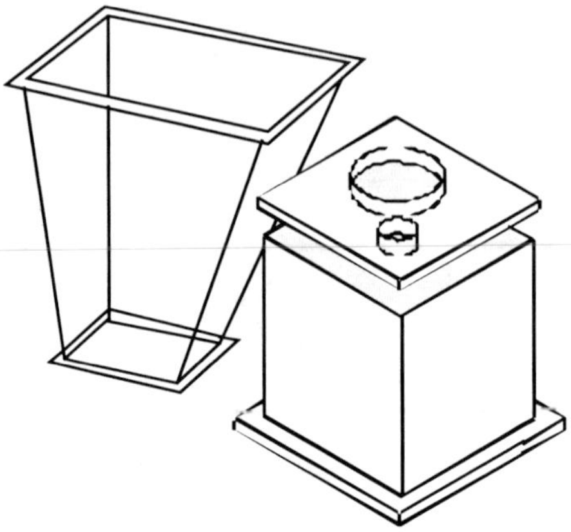

Figure 2.8 Storage hopper.

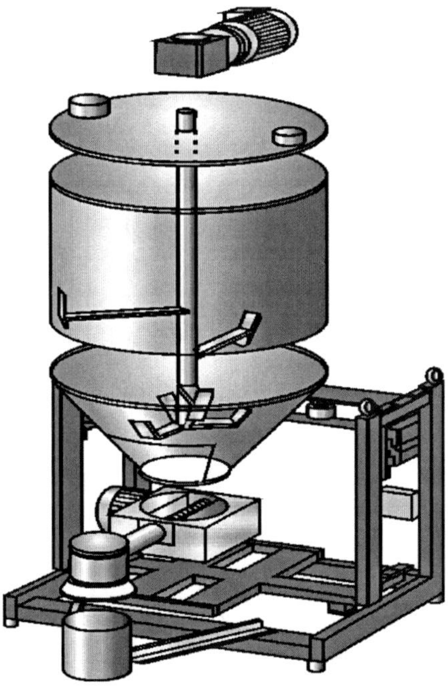

Figure 2.9 Storage hopper with vertical stirring agitator.

2.4 Volumetric Feeders

A volumetric feeder is a device, which, at a given motor speed, dispenses a certain volume of ingredient over a period of time. Capacities of these units are expressed in terms of units such as cubic feet per hour, cubic feet per revolution, gallons per minute, etc. Example of volumetric feeders is as follows:

- Single screw (Spiral and blade)
- Twin screw (Concave, spiral and blade)

- Vibrating trays and tubes
- Rotary valves
- Disk feeders
- Metering pumps (for liquid feeder)

By definition, volumetric feeders control only the volume of the bulk ingredient discharged. The vast majority of process requirements; however, are expressed in units of weight over time (for example, continuously feed at 100lb/hr). The calculation of the federate is not simply a matter of multiplying the volumetric rate by the bulk density as will be learned further.

The bulk density of an ingredient is its weight per unit volume. It is clear, then, that volumetric feeders can achieve good weight flow accuracy only when the bulk density is constant – a rare condition in the process environment due to the ingredient flow properties presented earlier. The key, then, to accurate and reliable volumetric feeder performance is to minimize ingredient bulk density variations leading to consistent and repeatable filling of the feed mechanism.

2.4.1 Single Screw Feeders – Sizing and Feed Rate Calculation

Screws are very commonly used as feed devices due to their linearity (feed rate/screw speed), flexibility (screw diameter and pitch can be changed), totally enclosed design (no dust), simple control (motor with speed control).

2.4.1.1 Screw Sizing

For example, to calculate the volumetric feed rate of a 2-inch diameter screw with a 2-inch pitch at 200rpm, the following calculation is used:

$$\text{Feedrate} = (\pi\ r^2)(p)s$$

r- radius of the screw
p- pitch of the screw
s- screw speed (rpm)

Solving the example:

$$\text{Feedrate} = \frac{\pi(1)^2(2)\times 200\times(1\text{cu.ft}\times 60\text{min})}{12\times 12\times 12/\text{in}\ \ 1\text{hr}}$$

$$= 43.6\ \text{cuft/hr}$$

2.4.1.2 Screw Fill Efficiency

Consider also two different ingredients of identical bulk densities (say, 50lb/cu.ft.), one ingredient having the flow characteristics of plastic pellets and the other of titanium dioxide. If both ingredients were fed with the same feeder at the same speed, one might achieve an actual measured output of 2,250lbs/hr of pellets and only 1,250lbs/hr of TiO_2. Compared to the theoretical expected value of

2,500lb/hr. (50cu.ft./hr×50lbs/cuft) We would say that the screw filling efficiency of 0.9 in the first case and 0.5 in the second.

In theory this filling efficiency has to be calculated for a given screw type and size and material via actual testing in the field or test laboratory. In reality most feeder manufactures can make an intelligent selection based on their feeding experience, although testing should never be ruled out to be absolutely certain the proper screw is being supplied.

2.4.1.3 *Feed Rate Calculation*

The initial screw selection must always be made based on maximum feed rate, ingredient filling efficiency and minimum bulk density (loose). Ideally, the screw speed at these parameters should be in the area of 75-80% of maximum. If the ingredient does not fall from a screw as a constant stream, it pulses. The negative effects of pulsing in the process can be reduced by increasing the screw speed. For example:

Maximum feed rate = 1,000 lbs./hr
Minimum feed rate = 200 lbs./hr
Bulk density, Loose = 25 PCF
Bulk density, Packed = 32 PCF
Screw filling efficiency = 0.7 (estimated)

Assuming we want to run at the maximum feed rate at 75% motor speed, the screw theoretical rating at 100% motor speed must be:

$$\frac{1{,}000\text{ lbs./hr}}{25\text{PCF}} \times \frac{1}{0.7} \times \frac{1}{0.75} = 76\text{ cu.ft./hr}$$

The minimum feed rate requirement now has to be considered. There are several types of variable speed drives possible with the most common being DC motors with SCR control or AC motors with variable frequency control. Although some drives are capable of turndown to 30:1 or more it makes no sense to operate a screw at 2–3 % motor speed if this can be avoided. At this low speed, a screw sized for a higher feed rate would make a revolution every 10–15 seconds which results in pulsing possible intolerable to the process. A better solution, if the process allows it, is to substitute a smaller screw (in diameter and/or pitch) to allow operation at faster speeds. Look at utilizing extra smaller screws to improve process accuracy any time the turndown exceeds 10:1.

2.4.1.4 *Feeder Selection*

A single screw can be of open spiral or blade construction and operates by capturing ingredient at the inlet to the screw channel and pushing it towards the discharge through an enclosed tube (Figure 2.10).

Screw pitch is generally between one half to one screw diameter. If the pitch is too small, the area between the shaft and flights is so small that the shear forces acting on the ingredient cause it to start rotating with the screw rather than moving towards the discharge end. The same phenomenon occurs if the pitch is too large, this time caused by the angle of the flight with respect to the shaft being too shallow.

2.4.1.5 *Spiral Screw*

See Figures 2.11 and 2.12 below.

- Most common first choice selection for most ingredients
- Use over-sized tube when feeding pellets to prevent screws jamming in tube
- Large diameter spiral screws are good for ingredients that stick to screws. The smaller contact surface area of a spiral screw vs. a blade permits less material adherence to the screw
- It has a low resistance to flow. As a result, if an ingredient is aerated, it will flow without content of the screw rotations
- Requires a center rod for heavy metal powders as the coil may compress
- Consider relief grinding the screw when powder builds up on a hard layer on the inside of the screw tube. This reduces friction by reducing the screw surface area in contact with the tube

Figure 2.10 Single screw feeder with internal stirring agitator.

Figures 2.11 & 2.12 Spiral screws with different pitches.

2.4.1.6 Blade Screw

See Figure 2.13 below

- For most powders which do not stick to metal
- Has high shear and increased resistance to floodable flow

2.4.2 Twin Screw Feeders

A common type of twin screw is the twin concave (Figure 2.14). This device consists of two co-rotating solid screws placed side by side which form advancing pockets which are filled with ingredient and progresses around the outside of the screws. Twin screw feeder is often best for feeding powder at feed rate below 20 lbs/hr. Operating the feeder at high screw speed increases flow pulse frequency, yet the screw fill efficiency is uniform due to low screw volume per rotation.

2.4.2.1 Twin Concave Screws

- Feeds powders only due to small clearance between screw and tube.
- Produces highest flow resistance.

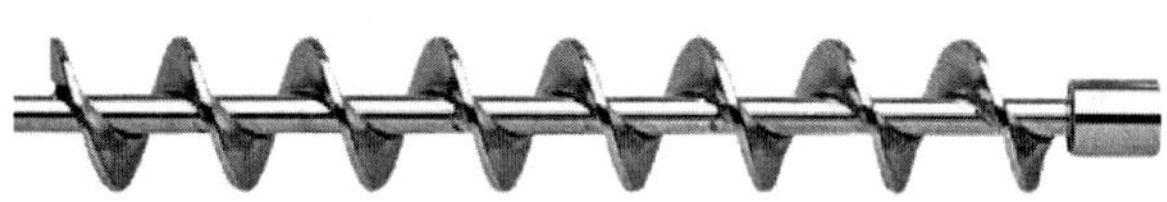

Figure 2.13 Blade screw.

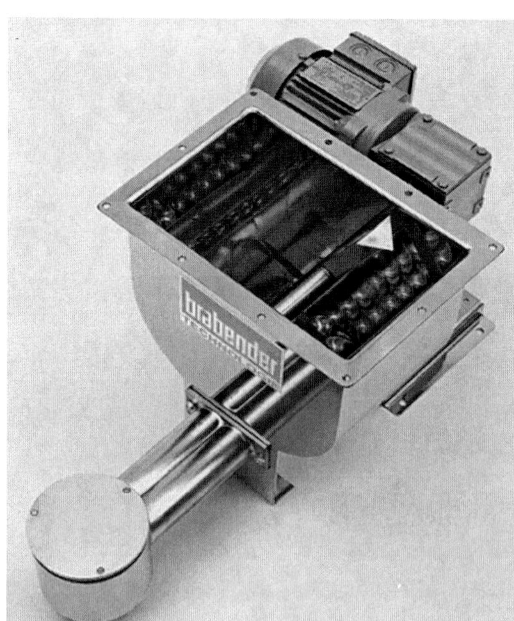

Figure 2.14 Twin screw feeder.

- Large inlet into screw reduces possibility of bridging above the screws.
- Since screws are self-wiping, the flight volume remains constant even with sticky powders, increasing accuracy.

The other types of twin screws are fully interchangeable with the twin concave and generally applied where a feeder must feed more than one ingredient, at lease one of which requires a twin concave screw (Figure 2.15).

2.5 Vibrating Tray Feeders

Vibrating tray feeders are very popular for feeding plastic pellets, particularly at low feedrates (from 3 to 20 lbs/hr) and other large particle ingredients assuming such ingredients can be conditioned to flow onto the tray (Figures 2.16 and 2.17).

Tray feeders for Loss-In-Weight feeder application in the plastic industry have characteristics as follows:

- Flat tray, circular tube and V tray for low feed rates
- Sizes 1 inch to 12 inches width (dia.)

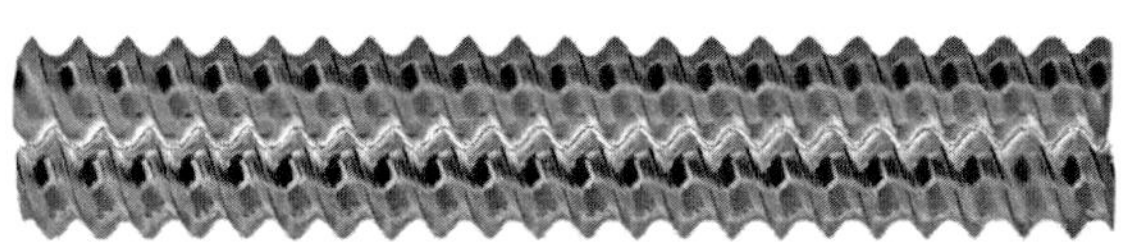

Figure 2.15 Twin concave screw.

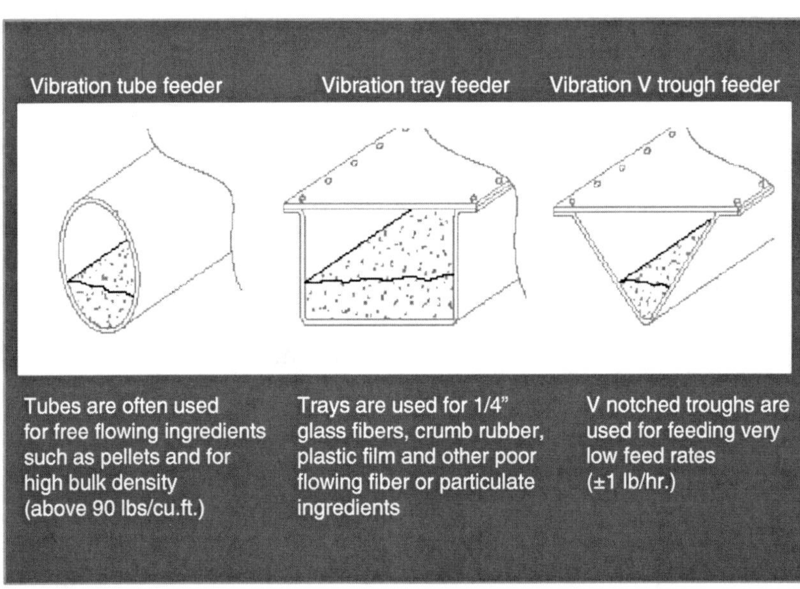

Figure 2.16 Type of vibrating tray.

Figure 2.17 Vibrating tray.

- Feed rate adjusted by changing amplitude
- Frequency remains constant close to resonant frequency
- Amplitude sensed with control feed back to resonant frequency drive
- Good for pellets, granules and flakes and some powders (mostly above 40lb/cu.ft.)
- Flow level on tray controlled from position of supply hopper outlet
- Reliable flow out of supply hopper critical
- Frequency is close to 60 Hertz (3600 pulses per minute) – smooth flow
- Low shear but ingredient stratifies on tray

2.6 Belt Feeders

A belt feeder is a good feeder when feeding free flowing ingredients at the feed rate higher than 100lbs./hr. Belt widths ranges from 6 inches to 32 inches. The belt feeder can be used to feed or meter dry ingredients, and has low shear (Figures 2.18 and 2.19).

If continuous feeding application is required, the ingredient is fed to the feeder by gravity overhead supply bin in a flood fed condition ("choke application"). The ingredient is introduced to the belt through the inlet chute. As the belt moves, the ingredient is sheared by an adjustable gate, which sets the ingredient bed depth for optimum feeder control. As the ingredient then passes over the highly sensitive weigh section a belt load signal is generated. The belt load is integrated by the control system with the belt speed signal from the tachometer to yield a feed rate by weight.

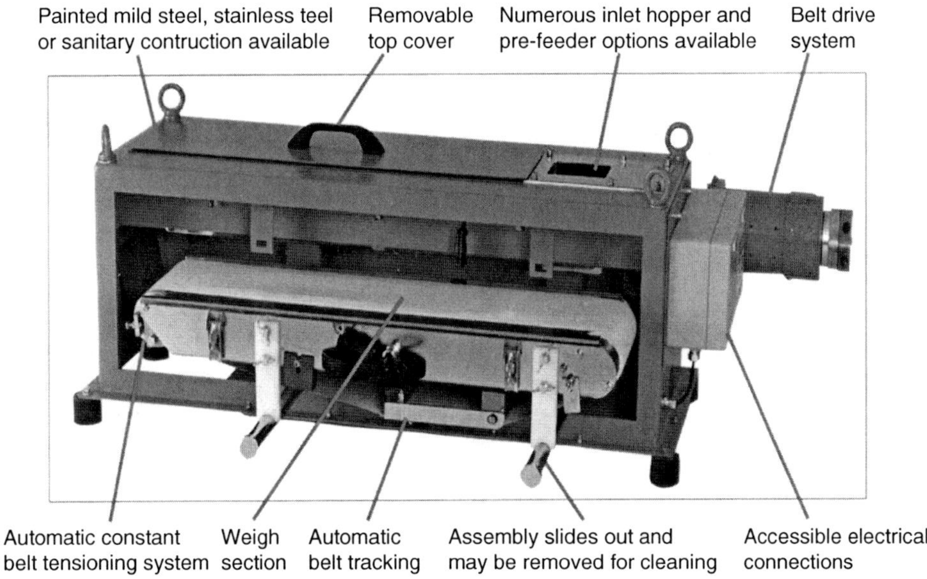

Figure 2.18 Belt feeder.

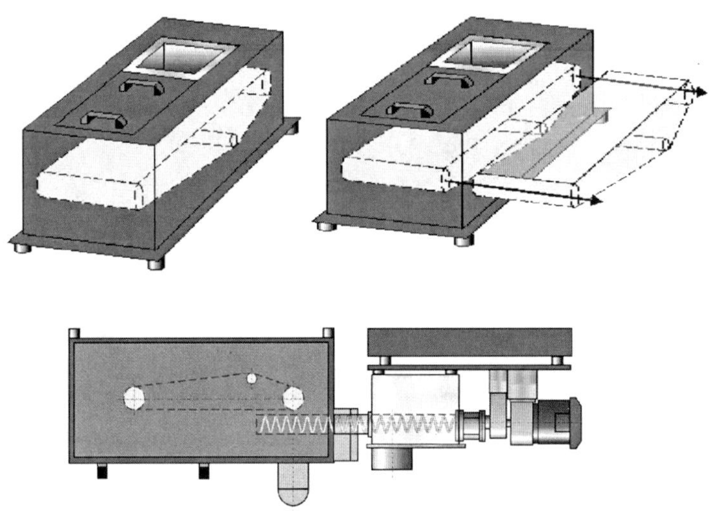

Figure 2.19 Weigh-belt feeder.

In case of poor flowing ingredients that do not lend themselves to direct feeding from supply bins, a volumetric pre-feeder (screw feeder for example) with a variable speed drive can be used directly upstream of the weigh belt feeder.

- Flow level on belt controlled by shear gate or pre-feeder
- Scrapers needed on belt
- Weigh belt feeders require more housekeeping than screw feeders since belt transport and weighing is exposed to dust
- Low shear, good for low melt point ingredients

2.7 Loss-In-Weight Feeders

A LIW feeder has five basic components including the feed device, weigh hopper (integral to feed device), scale, controller and refill. Refer to the Figures 2.20, 2.21 and 2.22.

2.7.1 Scale

A scale is a device that can measure and display the weight of a load while containing and supporting it. The precision of the scale is a function of the accuracy of weight measurement required and the stabilization time available between load change and movement reading. The scale incorporates one, two, or three load cells, which can be of different types. Strain gauge (analog with digital amplifier), LVDT (analog with digital amplifier), vibrating wire (single or twin) digital and other).

2.7.2 Feed Device

This is mechanically identical to a volumetric feeder as described previously, but when used in this context it is mounted on the scale and used as a "take away" device to dispense material at a controlled, precise rate into the process. The feed mechanism must be equipped with variable speed drive. The proper feed mechanism is usually determined from experience (plant and supplier), accuracy requirements and testing.

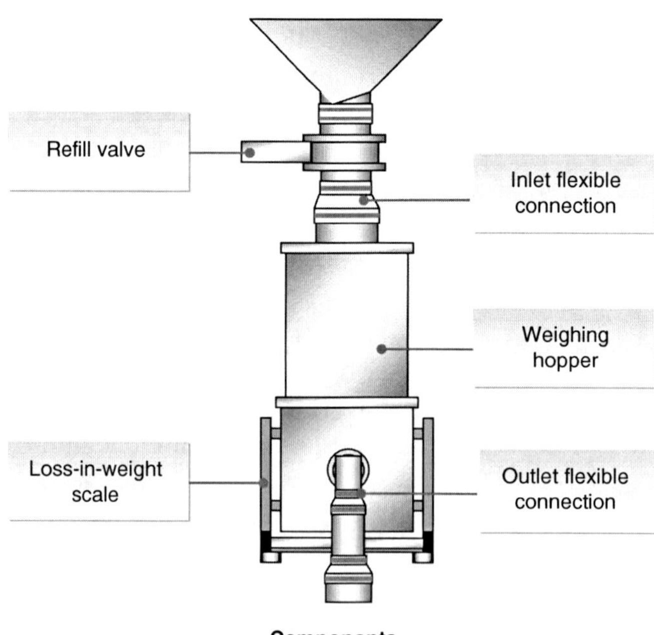

Figure 2.20 Components of LIW feeder.

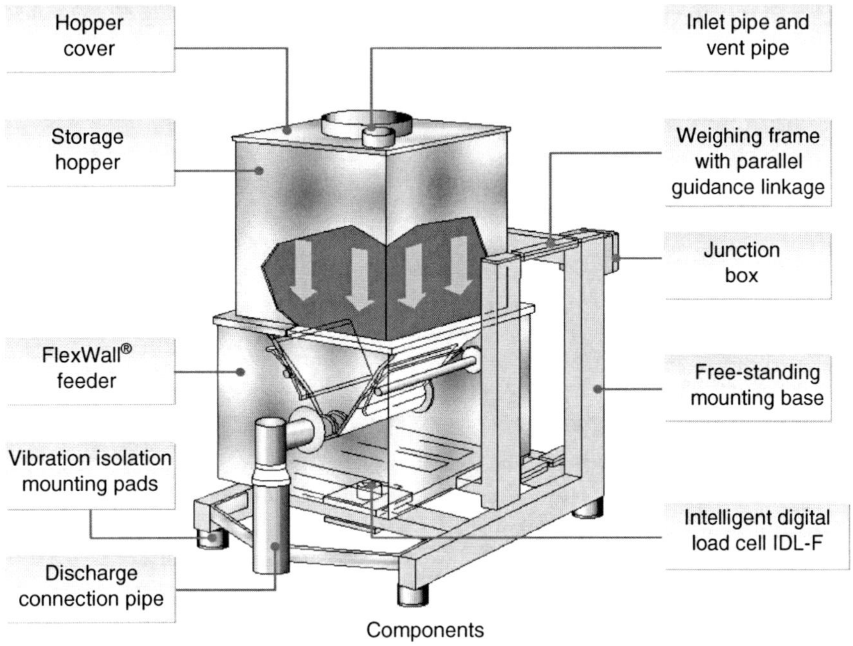

Figure 2.21 LIW feeder components.

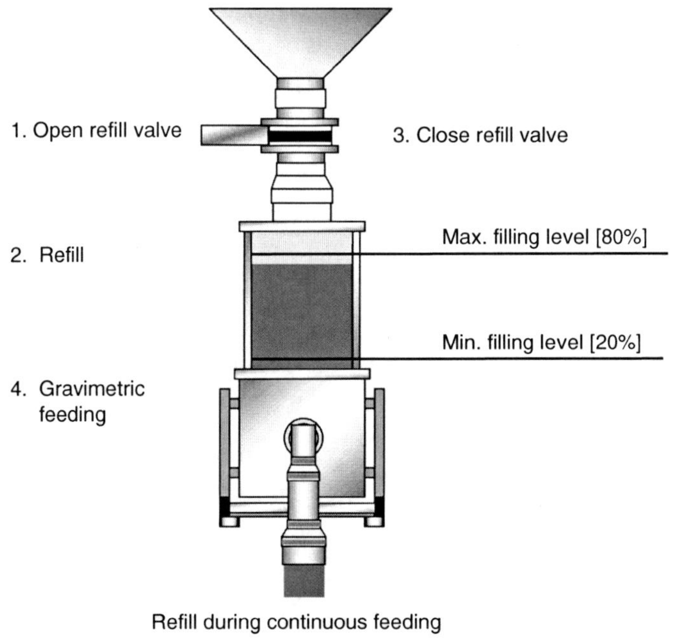

Figure 2.22 LIW feeder.

2.7.3 Weigh Hopper

The weigh hopper stores a pre-determined amount of material (typically 4–5 minutes of feed time at maximum feed rate) directly above the feed mechanism for introduction to the process. It is integral with the feeder. Care must be taken to supply the hopper with the proper geometry and agitation (if required) to ensure the material does not bridge or rathole and flows in a uniform manner into the feed mechanism as outlined earlier.

2.7.4 Feeder Controller

This utilized a PID based algorithm to accept a setpoint input (whether local or remote), compares the actual feed rate to the setpoint, generates a control signal output to the feed mechanism to maintain or change the motor speed, and stores and makes accessible the total weight of material fed over the previous time period. Modern day processes generally require the feeder controller(s) to tie in with the overall plant control system. Some ways this is done are:

- Local control direct from controller keypad, perhaps with hard wired interlocks;
- Remote control, usually utilizing a PLC interface;
- Central control from graphics screen including data manipulation and software SPC packages.

2.7.5 Refill Device

This valve (or conveyer) refills the LIW feeder when low level (weight) is required in the weigh hopper. The controller automatically initiates and stops refill.

2.7.6 Principle of Operation–Continuous Feeding from a Loss-In Weight Feeder

A continuous LIW feeder feeds ingredient into the extruder continuously at the required rate (Figure 2.23). For example, using LIW feeder as a feed device feeding a screw feeder, the screw speed is determined by the change in weight as sensed by the scale and compared to the change in weight that should have occurred at the desired feed rate. The change in weight is measured over a fixed time, normally less than 1 second. The controller compares these two weights and generates a signal for the screw feeder variable speed drive to either increase its speed, decreases its speed or remain at the same speed.

As shown by the feeding graph in Figure 2.23, the LIW feeder cycles from a full hopper to a preset low level in the hopper to initiate a refill. During refill, the weight calculation of feed rate is not available; as a result, the screw speed is maintained at speed(s) determined during the controlled feeding cycle. Since feed rate calculation does not occur during refill, it is let to refill quickly (10 seconds for example). Refill turn is normally reduced if the weigh hopper is as small as possible, preferably representing 2 to 4 minutes of maximum feed rate.

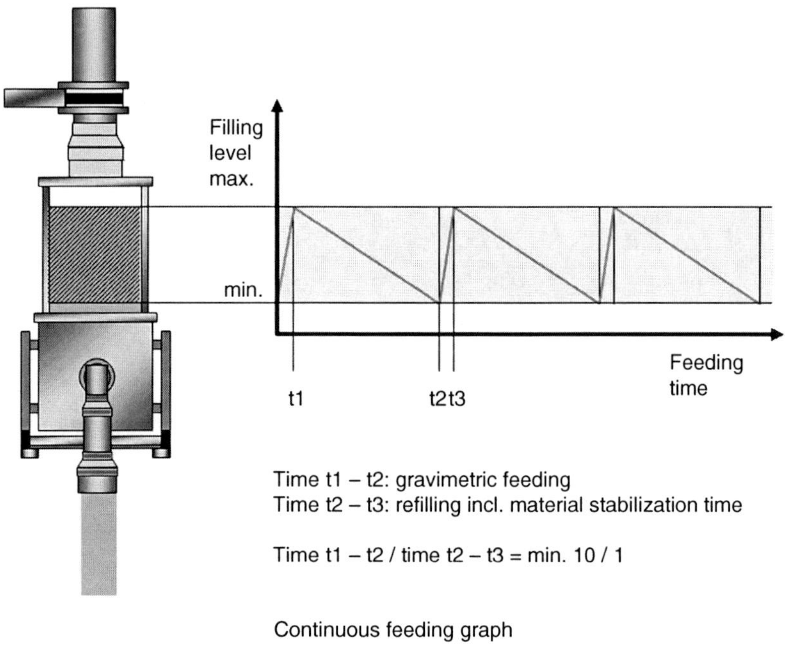

Figure 2.23 LIW feeder cycle.

2.7.7 Loss-In-Weight Feeding Helpful Comments

2.7.7.1 *Refilling a Loss-In-Weight Feeder*

- Extrusion applications are continuous
- For LIW feeders to feed continuously they require a refill when the feeder hopper empties
- During refill, the feed rate is not gravimetrically controlled
- For best control, it is best to refill quickly, 10–20 seconds
- Hopper volumes are sized, typically for 15 refills per hour with 60% of the total volume refilled
- Plastic pellets can have 30 or more refills per hour
- Hoppers are smaller with more frequent refills
- As a rule of thumb, size refill rate at 20×max. feed rate
- Best operation occurs if refill time is consistent from refill to refill
- Refill flow velocity entering feeder should be low, avoid discharging from high above feeder
- The refill connection is sealed and flexible

2.7.7.2 *Venting a Loss-In-Weigh Feeder*

- Loss-In-Weigh Feeder must be vented (Figure 2.24).
- Gas pressure inside the feeder should be the same as the atmospheric pressure outside the feeder.
- Clean air (or gas) must be allowed to exhaust the feeder during refill.

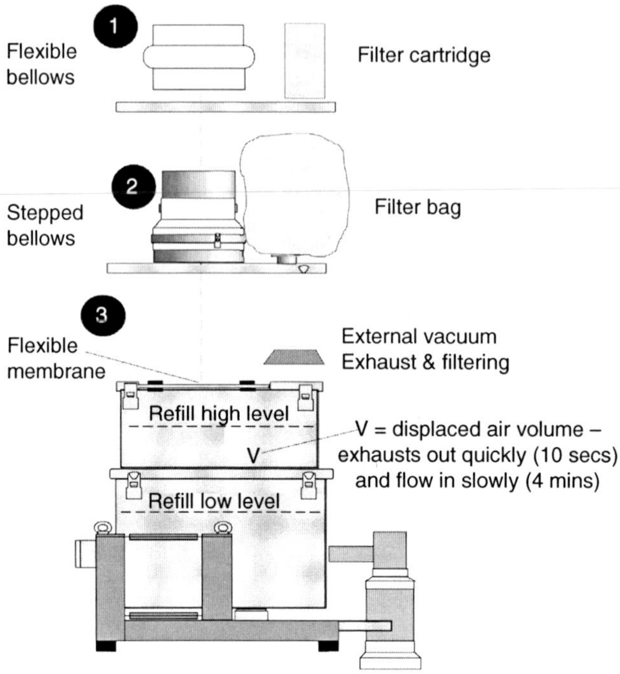

Figure 2.24 Venting a loss-in-weight feeder.

- Exhaust rate is calculated knowing refill volume and refill time.
- The vent (exhaust) connection must be either a dust stock or a non-dusty ingredients) directly mounted on the hopper or a non-touching vacuum exhaust connection.

2.7.7.3 *In Plant Vibration Effects on Feeder Performance*

The scale senses vertical components of undesired in-plant vibration. Load cells are filtered; however, filtering cannot always eliminate the erroneous readings caused by vibration.

2.7.7.4 *Temperature Effects in Feeder Performance*

Changes in surrounding temperature can cause the load cell heavy to light. Load cells require temperature compensation capabilities.

2.7.7.5 *Scale Stabilization Time*

During a refill, new ingredient rushes into the weigh hopper. This causes a disturbance in the scale for a few seconds. Gravimetric feeding control cannot resume until the scale readings have stabilized. Some scales, particularly those with load cells with small movement due to load change, stabilize quicker than other types.

2.7.7.6 *Flexible Connections*

The feed device and integral weight hopper are all mounted on the scale and hence are weighed. External components cannot come in contact with the feed device without affecting the weighing integrity. In order to provide a sealed (dustless) feed, the inlet (refill), vent and process connection must be sealed. These seals are specially designed to have no influence on weighing. These should be checked regularly to ensure they are functioning properly. Also, the wiring from the feed device motor to the non-weighed junction box also needs to be carefully connected to ensure it doesn't affect weighing.

2.8 Special Feeders for BioPlastics Ingredients

2.8.1 Bio Ingredients-Typical Physical Characteristics

Bio ingredients are normally fiberous, with varying dimensions to the fiber strands, and all clumped together. A hay stack is a good example. The bulk density is typically low. This means that the feed device is feeding high volumes of fibers to achieve a relatively low mass feed rate. A fiber weighing 3 lbs/cuft is typical. 100 lbs has a volume of 33.3 cuft (approximately a cube 3 ft $\times$ 3 ft $\times$ 3 ft).

2.8.2 The Physical Characteristics Aggravate Controlled Rate Feeding

- Fibers are often entangled. Conventional feeders rely on granules or powders releasing from the feed mechanism as a flow of discrete particles. Fibers do not do this. Fibers release in clumps, which for most processes is very undesirable since the clumps don't separate well in the process itself. A product comprising bio fibers has its best quality when the fibers are all separated from one another in an evenly disbursed random direction configuration.
- The bridging dimension is quite large, particularly if a converging head of fibers is above. Flow will occur through hoppers if the cross-sectional area of the hopper remains constant. It generally will not flow through converging hopper sections.
- Bulk density variation is random. It varies in two ways, a) with head and b) with clumping variations within its own mass.
- Head—fibers are normally compressible. As a result, bulk density at the bottom of a hopper can be 10% higher than it is at the top. Feed rate varies as head changes.
- Clumping—fiber bundles within close proximity to one another can have widely ranging bulk densities due to the tendency of fiber to form "clumps". Clumps are small volumes of fiber bundles which are closely bound to one another, normally of a higher bulk density than surrounding fibers and normally held together in such a way that the entire clump moves (flows) instead of individual fibers.

2.8.3 Fibers Need to be Tested in Feeders to Determine How They Can Be Fed

2.8.3.1 Start with a Traditional Feeding Device, Example a Screw Feeder

A screw feeder is a well known feeding device. For poor flowing ingredients, screw feeders are designed with flow enhancement devices in the screw trough (converging hopper section above the screw). These devices include stirring agitators that promote flow into the screw or massaging paddles outside the screw trough (screw trough manufactured from a flexible elastomer such as polyurethane). The flexing of the screw trough breaks bridges and promotes flow into the screw.

Some fiberous ingredients will flow uniformly into the screw flights and feed reliably.

Often the fibers flow randomly in the screw flights and the feed from the screw is random from no flow to a complete flow. This is not acceptable.

If fibers need to enter the downstream process as a controlled feed of separated fibers, then a special fiber feeding device is required.

To overcome the erratic feed characteristic, a metering device that secures the fiber mass as it flows by gravity (and mechanical assistance) into the metering pinned rolls has been developed (Figure 2.25).

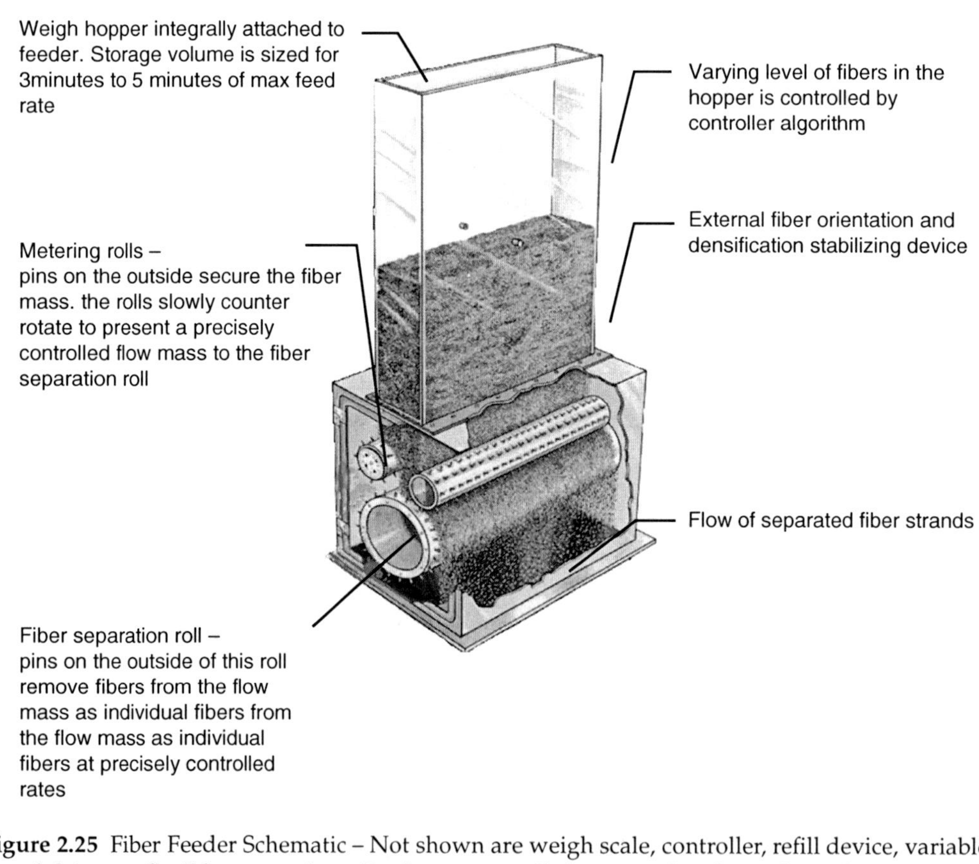

Figure 2.25 Fiber Feeder Schematic – Not shown are weigh scale, controller, refill device, variable speed drives or flexible connections. Feeders are supplied mounted and wired on a base.

The pinned rolls' rotational speed is precisely controlled. The controlled flow mass of fibers is presented to the fiber separation pinned roll. Fibers are released individually at a precisely controlled rate.

To overcome the bridging characteristic, a specially designed hopper with constant cross-sectional area is used. Gravity flow is induced by a mechanical enhancement system.

To overcome the varying bulk density due to "clumping", the fibers are "conditioned" in the hopper section directly above the pinned metering rolls to produce a stable bulk density. Head effect variations are compensated by a feed rate weight feed back and control system (loss-in-weight control). See further Feed Rate by Weight Control.

Refilling the hopper requires that the refill device produces a flow of fibers in a flow cross-section no larger than the flow cross-section of the feeder hopper and at a rate approximately 20 times the maximum feed rate required by the process.

2.8.4 Feeder Control and Checking the Feed Rate

In either the screw feeder or the Fiber Feeder, feed rate is varied by varying the speed of the metering element (screw or metering rolls).

If a constant feed rate is required, the speed can be set to achieve a reasonably consistent feed rate.

It is often desired to set the feed rate in weight units (lbs/hr). To achieve this, the feeder is mounted on a scale. As the feeder feeds out fibers, the weight of the feeder (the fibers in the hopper) starts to reduce. This weight reduction (weight loss) is measured over short selected time periods, the actual feed rate is calculated and the screw/metering roll speed is set at the correct speed to achieve set point. This feedback control is performed in a loss-in-weight feeder controller.

As the hopper level (weight) reduces, to a preset low level, an automated refill is initiated and the feeder hopper is refilled. Typical feed/refill time cycles are 4 minutes feed/20 seconds refill. During refill, the speed of the screw/metering rolls is maintained at a pre-determined speed.

Feed rate checking can be performed on line, in production. Assuming the feeder scale is calibrated (checked when the feeder is off using calibrated weights), the feeder scale is used and the feed rate is determined independent of the feeder controller by a feed test algorithm and a laptop that analyzes the feed rate, screw/metering rolls speed, weight in the hopper, and deviation from set point. This method uses a special program for PC's called *SmartService*.

2.8.5 Ingredient Storage and Keeping the Feeder Full

This is a serious consideration for low bulk density (below 5 lbs/cuft) fibers and high feed rate – say 1,000 lbs/hr. In an 8 hour shift, 1,600 cuft of fibers will have been processed. That is a volume approximately 12 ft × 12 ft × 12 ft.

This volume has to be placed in storage nearby the feeder. This may be difficult in itself and special machinery may be necessary to achieve this.

Then, fibers need to be transferred from storage to the feeder and reliably refill the feeder during the refill cycle.

An inclined, cleated belt, no wider than the feeder hopper, is one method of conveying the fibers into the feeder hopper. This assumes the fibers can be delivered onto the belt in a reliable method.

2.9 Conclusions

As composites, natural fibers begin to gain a place in the production of biocomposites, the material handling of the fibers, the local (to production) storage of the fibers, the transfer from storage and into feeding device, and a feeding device able to separate the fibers into a stream of separated fibers will be necessary.

This chapter has presented some challenges that need overcoming and some solutions.

The steps include:

- Analyzing the physical characteristics;
- Determine the volumes to be used;
- Testing for a reliable feed device;
- Design a material handling system to maintain the feed device full.

3

Modeling the Processing of Natural Fiber Composites Made Using Liquid Composite Molding

Reza Masoodi and Krishna M. Pillai

Laboratory for Flow and Transport Studies in Porous Media
Department of Mechanical Engineering
University of Wisconsin, Milwaukee, WI, USA

Abstract

Natural fibers are being used increasingly to substitute artificial glass and carbon fibers in polymer composites. Liquid Composite Molding (LCM) processes are an important set of "liquid molding" technologies to manufacture net-shaped composites parts that involve filling a dry, fiber-packed mold with a thermosetting resin. However, not much is known about the flow of resins, bio-resins and test liquids through a preform made from natural fibers. The swelling of natural fibers due to liquid absorption adds a new dimension to the conventional mold-filling simulation in LCM conducted to optimize mold-design. Unlike the glass or carbon fiber mats, the swelling of natural fibers causes the permeability and porosity of the LCM fiber mats to reduce with time during the mold-filling process. This chapter presents some recent developments in the science of LCM flow-modeling with natural fibers used as reinforcements. Some studies on measuring the permeability of natural-fiber preforms using the conventional LCM flow model and employing the organic and inorganic test liquids are presented first. Later some recent attempts to include the swelling and absorption into the LCM flow physics are also discussed where some analytical solutions for simple 1-D flows under constant pressure and constant flow-rate conditions are discussed. Some recent approaches to numerically simulate the LCM mold-filling type processes in swelling, natural-fiber based materials are also presented.

Keywords: Natural fibers, resin transfer molding, RTM, LCM, polymer composites, natural fiber composites, permeability, swelling, flow modeling, porous media

3.1 Introduction to Liquid Composite Molding (LCM) Processes

Fiber reinforced polymer composites are made of polymer resins as the matrix and fibers as the reinforcement. Due to their light weight, high strength, excellent corrosion resistance, and design flexibility, polymer composites have now been widely used in fields as diverse as automobile manufacturing, aerospace,

Srikanth Pilla (ed.) Handbook of Bioplastics and Biocomposites Engineering Applications, (43–74)

civil constructions, shipbuilding, and military. The demands for advanced composite materials in the last few decades have been greatly pushing forward the development of polymer composites.

There are several methods for manufacturing fiber reinforced polymer composites such as hand lay-up, spray-up, filament winding, pultrusion, liquid composite molding (LCM), compression molding, automatic fiber placement, and autoclave processing [1, 2]. Each of the above-mentioned manufacturing techniques has some or all of the following four steps[1]: laying-up, wetting, consolidation, and solidification. In the lay-up step, fiber reinforcement materials are formed to make the final shape of the part; the forming process can be done either manually (e.g. hand lay-up, spray-up) or automatically (e.g. filament winding, LCM, automatic fiber placement). In the wetting step, the liquid resin is applied to impregnate the fibrous preform. In the consolidation step, some external compression is used on the impregnated fibrous structure to remove the entrapped air and reach the designed fiber-volume ratio of the final composite products. In the solidification step, both thermoset and thermoplastic resins are solidified through curing[2] and cooling.

Among the various manufacturing processes used for making composites parts, liquid composite molding (LCM) has been recognized as a cost-effective and promising process for making net-shaped parts [3]. The LCM processes, which include technologies such as resin transfer molding (RTM), vacuum-assisted resin transfer molding (VARTM), Seeman Composite Resin Infusion Molding Process (SCRIMP), and Structural Reaction Injection Molding (SRIM) entail the following generic steps (see Figure 3.1). First, a preform is created from reinforcing fibers, typically in the form of random, woven, or stitched fiber mats made from carbon, glass, or other materials. Next, the preform is inserted in a mold that matches the dimensions of the desired part and the mold is closed (in the case of the rigid mold processes such as RTM) or covered with a flexible sheet (in the case of the soft mold processes such as VARTM or SCRIMP). Then, a low viscosity thermoset resin such as epoxy, polyester, phenolic, or vinyl ester resin is mixed with a hardener and injected under pressure (in the case of the former) or imbibed under vacuum (in the case of the latter) into a closed mold containing the perform. The resulting part is cured at room temperature or under a strictly controlled mold-temperature cycle till the end of the curing reaction. Finally, the cured hardened part is extracted and is ready for use after some machining.

LCM processes have several advantages over other composite manufacturing techniques [4]. The injection pressure in LCM is much lower as compared to the compression and injection molding processes, which means that the tool costs and

---

[1] It is not applicable to the compression molding process when it deals with sheet molding compounds (SMCs) or bulk molding compounds (BMCs). The considered compression molding pertains to the infusing of liquid resin into a dry fiber prefom through compression of the mold cavity.

[2] Curing is a term in polymer chemistry and process engineering that refers to hardening of a polymer resin by cross-linking of polymer chains. Curing may be induced by chemical additives, ultraviolet radiation, electron beam or heat.

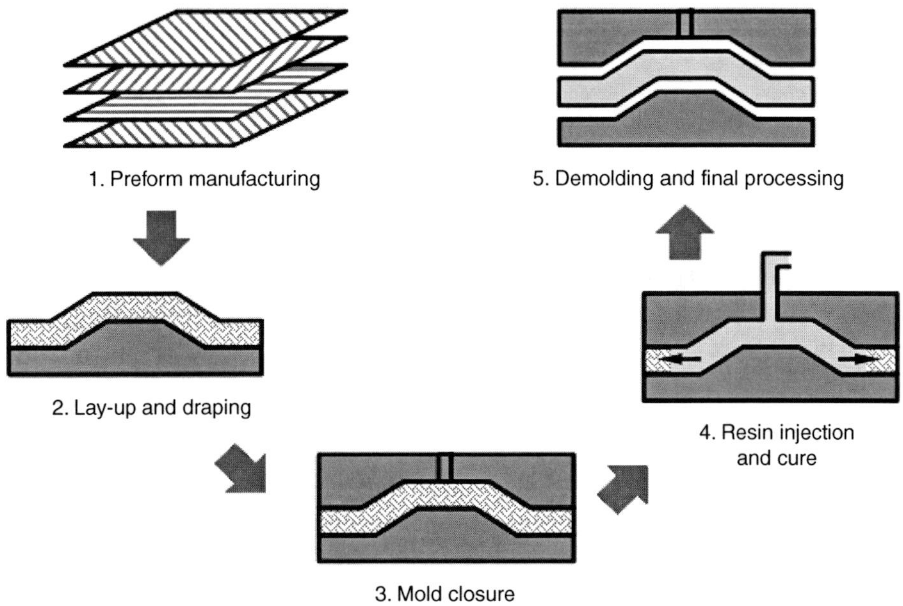

Figure 3.1 Process steps in resin transfer molding (RTM).

operating expenses are lower. The production rate of composite parts through LCM processes such as RTM can be moderately high, and hence quite suitable for the high-volume automotive sector. LCM allows production of composites parts with high fiber volume fractions[3] and a good control over fiber orientation (and hence over directional properties as well). LCM processes lead to the production of net-shape parts, so material wastage and machining cost are reduced. Because the closed molding processes offer low volatile emissions during processing, LCM processes are environment-friendly.

The quality of the LCM product and the efficiency of the process depend strongly on the mold filling stage of LCM. The mold filling in the hard-mold LCM processes such as RTM and SRIM is affected by several parameters including the location of resin inlet-gates and air vents, the permeability of fiber mats, the resin infusion pressure, the applied clamp force, and the temperature of the resin mixture. The traditional trial-and-error methods to optimize the mold and process design can be too time-consuming and expensive. As a consequence, the numerical mold-filling simulations are used as one of the most effective ways to optimize the LCM technology. Successful computer simulations are able to improve the mold design in virtual space without the need for the expensive and time-consuming trial-and-error approach to mold design.

[3] In general, the higher the fiber volume fraction of a composite is, the higher is its performance.

3.2 Introduction to the Use of Bio-fibers and Bio-resins in Polymer Composites

Growing environmental awareness worldwide has aroused an interest in the use of environmentally benign materials in engineering. Since the 1990s, natural fiber composites have emerged as an alternative to the glass-reinforced or carbon-reinforced polymer composites. Natural fiber polymer composites such as curauá fiber-polypropylene (Figure 3.2), hemp fiber-epoxy, flax fiber-polypropylene, and china reed-polypropylene, are particularly attractive for use in the automotive industries because of their lower cost and lower density (leading to lighter weights), and acceptable specific strength and modulus [5-8]. Other advantages of natural plant fibers over traditional glass fibers are biodegradability, CO_2 sequestration, economic viability, reduced tool wear in machining operations, enhanced energy recovery, and reduced dermal and respiratory irritation. Such advantages have been verified by several life cycle assessment studies conducted with these fibers [6].

Bio-composites are combination of bio-fibers such as Kenaf, Hemp, Flax, Jute [see Figure 3.3], Henequen, Pineapple leaf fiber and Sisal with resin matrices that can come from both non-renewable and renewable resources. Recently, natural fiber reinforced polypropylene (PP) composites have widely used in automotive industries to make body and some cosmetic parts of automobiles [9]. However, polymer composites made of natural fiber – PP or natural fiber – polyester are not very eco-friendly since the polymer matrix is still petro-based and non-biodegradable [9]. Using natural fibers with those plastics that are based on renewable resources can improve the eco-friendliness of such composites.

Some commercially available biopolymers are sourced from renewable plant resources such as corn, lactic acid, soy-bean oil, linseed oil, pine and vegetable oils. As seen from Table 3.1, most of these resins are of the thermoset type and hence are usable in LCM processes for making natural-fiber based polymer composites. (Note that the thermoset resins in the initial monomeric form are small molecular weight liquids that are Newtonian in nature and are of low viscosity. Hence, unlike the highly viscous thermoplastics melts, the thermoset resins can

Figure 3.2 Curauá is favored by automobile part manufacturers due to its superior mechanical properties. (left) Curauá plant, (right) Curauá fibers [5].

Figure 3.3 Jute is a long, soft, and shiny vegetable fiber.

Table 3.1 Details about a few commercially available bio-based resins used for manufacturing the plant-based, 'green' polymer composites.

Resin Name	Commercial Name	Manufacturer	Type of Resin	Resin Base
AESO	Ebecryl 860	UCB Chemicals	Thermoset	Soy-bean oil
Polylactic acid	PLA	Cargill Dow LLC	Thermoplastic	Lactic acid
	Ingeo	NatureWorks LLC	Thermoplastic	Corn
Epoxidized soybean oil	Vikoflex® 7170	Atofina Chemicals Inc.	Thermoset	Soybean Oil
	Sorona® EP	DuPont Engineering Polymers	Thermoplastic	Corn
	Super Sap 100 Epoxy	Entropy Resins	Thermoset	Pine and vegetable oils

Table 3.1 (cont.) Details about a few commercially available bio-based resins used for manufacturing the plant-based, renewal polymer composites.

Resin Name	Commercial Name	Manufacturer	Type of Resin	Resin Base
Polyester bio-resin	Envirez 5000	Ashland Chemical Co.	Thermoset	Soy
Epoxidized soybean oil	Vikoflex 7170	Atofina Chemicals Inc.	Thermoset	Soybean oil
Epoxidized linseed oil	Vikoflex 7190	Atofina Chemicals Inc.	Thermoset	Linseed oil

easily flow through microscopic gaps of a fibrous porous medium created in an LCM mold packed with reinforcing fibers.)

A literature survey reveals that in a majority of natural composites studies involving thermoset resins, the composites were produced using the hand lay-up or press molding techniques. Very little research has been done on the use of LCM processes, such as RTM and VARTM, in making natural fiber polymer composites. However, the trend of using LCM to make natural fiber polymer composites is beginning to grow [9-11] due to the short cycle time and automation-friendliness of the process that lends itself particularly well for use in the high-volume automotive sector.

3.3 Physics for Modeling Mold-filling in LCM Processes

Since the advent of several LCM technologies in the last few decades, significant research has been done to study and model these processes in order to minimize defects such as incomplete mold filling due to the improper placement of gates and vents, and the creation of voids due to the residual air bubbles formed during the mold filling process. Because of the significant interest in LCMs by automotive, aerospace and other industries, hundreds of papers, numerous book chapters, and several books have been written about the physics of fiber wetting and resin flow during injection/imbibition of resin into an LCM mold [12–25].

In any LCM process, complete filling of the mold with adequate wetting of fibers is the mold designer's primary objective. Incomplete filling in the mold leads to production of defective parts with dry spots. It is also important for the mold designer to minimize fill-time and fluid pressure buildup during the filling process (especially in RTM) to make the technology cost-effective in a manufacturing environment[4]. Many factors affect the filling of a mold including the permeability of the fiber mats, presence of gaps in the mold, position of inlet and outlet gates,

[4] Lower mold pressure implies lower mold-wall thicknesses, lower clamping and sealing requirements, and consequently lower mold cost.

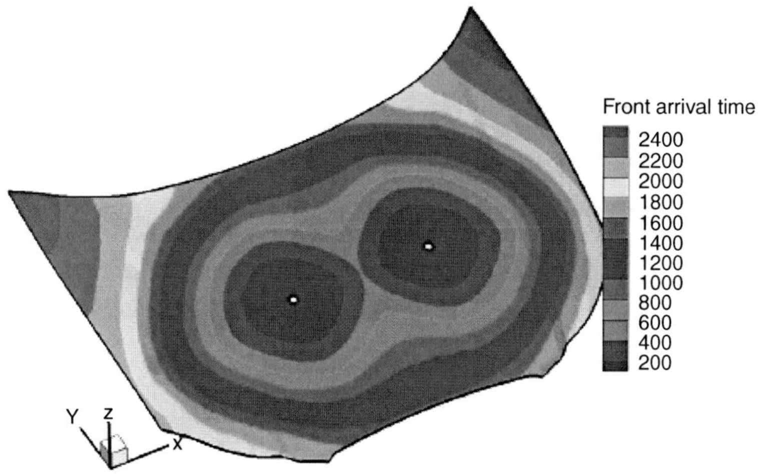

Figure 3.4 A typical RTM mold filling simulation showing progression of flow fronts in a car hood. (The simulation was conducted using PORE-FLOW [26].)

and rates of resin injection from different inlet ports [12]. Often, it is not possible for the mold designer to visualize and design an adequate system for resin infusion by intuition alone; as a result, mold-filling simulations are used to optimize mold performance (See Figure 3.4) [14-25]. Numerical simulations allow designers to optimize mold design in virtual space quickly and economically.

3.3.1 Modeling Single-phase Fluid Flow in Porous Media

The thermosetting resin is a low viscosity, Newtonian liquid that is assumed to fill up all pore space behind a moving front in an LCM mold. Hence, the single-phase flow (i.e., only one fluid flowing through the porous medium) is a most common assumption employed to model resin flow in LCM. The single-phase flow of a Newtonian liquid in an isotropic and rigid porous medium is governed by the following forms of Darcy's law and the continuity equation:

$$\text{Darcy's Law:} \quad \langle \vec{V} \rangle = -\frac{K}{\mu} \vec{\nabla} \langle P \rangle^f \tag{3.1}$$

$$\text{Continuity Equation:} \quad \vec{\nabla} \cdot \langle \vec{V} \rangle = 0 \tag{3.2}$$

Here $\langle \vec{V} \rangle$ and $\langle P \rangle^f$ are volume-averaged liquid velocity and pore-averaged modified pressure, respectively, while K is the permeability of the porous medium. The averaged variables $\langle \vec{V} \rangle$ and $\langle P \rangle^f$ are obtained after integrating the point-wise liquid velocity and pressure in an averaging volume several times bigger than the particles of a porous medium [27, 28].

Using the terminology of the well-known volume averaging method used for deriving the volume-averaged flow and transport equations in porous media [29], the volume average (also called the phase average) and the pore average (also called the intrinsic phase average) for any flow quantity q_f in a porous medium are defined, respectively, as

$$\langle q_f \rangle = \frac{1}{Vol} \int_{Vol_f} q_f dV \tag{3.3}$$

$$\langle q_f \rangle^f = \frac{1}{Vol_f} \int_{Vol_f} q_f dV \tag{3.4}$$

where q_f is integrated over an averaging volume (*Vol*) called the representative elementary volume or REV[5]. Vol_f is the volume of pores within a REV.

3.3.2 Modeling LCM Mold Filling in Synthetic Fiber Mats

Traditionally, the fiber preforms are viewed as the porous media with unimodal pore-size distributions. Assuming that the pores in the fiber preform behind the flow front are fully saturated with resin, the liquid resin flow impregnating the dry fiber preform during the mold-filling stage of LCM can hence be modeled using the Darcy's law, Eq (3.1), where the permeability is a tensor for the usually anisotropic fiber preforms. (When the fiber preform is an isotropic porous medium such as the medium created by randomly laid out fibers, the permeability becomes a scalar instead of a tensor.) The resin is assumed to be incompressible, hence the continuity equation, Eq (3.2), can be employed in our flow model.

Inserting Eq. (3.1) into Eq. (3.2) leads to an elliptic-type partial different equation (Laplace equation) that has only one unknown variable, the resin pressure. The resulting Laplace equation governs the pressure field in the region wetted by the resin. Introducing the proper boundary conditions, the pressure as well as the flow velocity can be obtained by solving the governing equations.

It is clear from Eq. (3.1) that the permeability of fibrous preform is a key parameter based on the preform microstructure, which relates the resin pressure distribution to the average resin velocity. The permeability of the fiber preform plays an important role in flow analysis through numerical simulations—for a successful numerical simulation, one needs to characterize the permeability of the fiber preform accurately so that the filling pattern, injection pressure, resin velocity, as well as mold fill-time can be predicted correctly.

As mentioned above, the permeability of porous medium is a property that has a significant impact on the accuracy of any LCM flow simulation. Though several theoretical models [30–34] exist to estimate K based on idealized fiber arrangement in an RTM mold, they are not very useful because the fiber arrangement in a real mold depends on how the preform is packed by a worker and marked by randomness and clustering; such models merely provide an order-of-magnitude

[5] REV is typically much bigger than the solid constituents (particles or fiber cross-sections) in a porous medium [29].

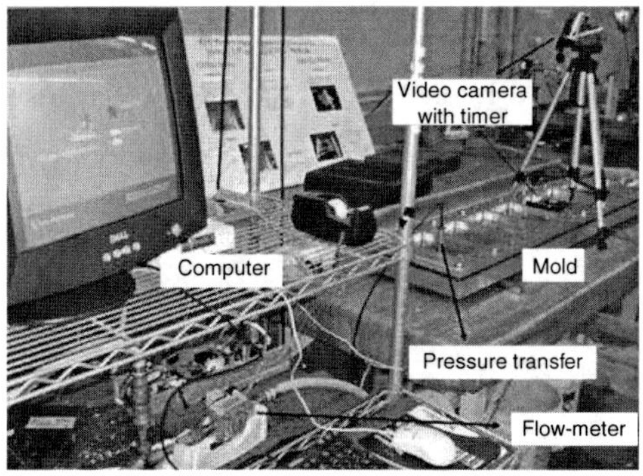

Figure 3.5 A 1-D and radial flow set-up available at UWM for measuring permeability as well as for studying LCM mold-filling flows.

estimate of this important quantity. As a result, several experimental methods exist to estimate the real permeability of fiber preforms [4, 35–37]. Primarily, these can be characterized either as the 1-D flow method (where a resin-like test liquid is injected uniformly from one side of the rectangular domain) [30, 38], or the radial flow method (where the test liquid, injected from a small hole in the packed mat, radiates out) [39, 40]. Either of these methods typically use a horizontal mold with a thin, flat cavity packed with fibers; the mold is connected to a pumping unit that injects the test liquid either at a constant injection-pressure or at a constant flow rate (see Figure 3.5). The attached data-acquisition unit measures flow rate and pressures at the inlet or other locations. Later, Darcy's law, Eq (3.1), reduced to a much simpler form for such simplified flow geometry, is employed for estimating the permeability.

Once the details of permeability and porosity of fiber mats are known the flow simulation is straightforward. In recent years, significant progress has been made in modeling flows in dual-scale porous media created by the packing of woven or stitched fiber mats in LCM molds. The theoretical model for resin flow in duel-scale porous media, developed by Pillai [41] and Pillai and Munagavalsa [42], has also been applied to model resin flow in LCM molds [43]. Details of simulation physics is given in subsequent sections.

3.3.3 Modeling LCM Mold Filling in Natural Fiber Mats

A fundamental difference between the LCM done with artificial fibers and the LCM done with natural fibers is that in the latter case, the fibers absorb liquid resin and swell. As a result, the porosity of the wetted fiber-preform, and hence its permeability, reduce with time. (Note that the permeability of a porous medium is directly related to its porosity.) Liquid absorption and swelling by the fibers, as we shall see in the subsequent sections, leads to a fundamental change in the governing equations as well—the equation for mass balance or the continuity equation is fundamentally altered. As a result, the conventional mold-filling physics is not adequate to model the flow of resin in natural-fiber mats.

3.3.3.1 *Swelling of Natural Fiber Mats in Organic Resins*

By definition, a solid swells when three conditions are met [44]: 1) Its dimensions are increased as a result of absorption of another phase. 2) It remains homogeneous at the microscopic level. 3) Its cohesion is decreased but not destroyed. Extensive work has been done in the forestry community to study the swelling of wood when exposed to various organic liquids [44]. Matanis *et al.* [45] through extensive experimentation showed that the natural fibers also swell by a significant amount when exposed to various organic liquids with different functional groups such as amines, alcohol and benzene rings. Since the thermosetting resins used in LCM along with the bio-resins described in Table 3.1 are organic liquids with similar molecular weights as well as similar functional groups, they are also expected to cause significant swelling in the natural fibers during the manufacture of composites. The presence of cellulose molecules is the main reason for swelling and absorption in natural fibers. Table 3.2, lists the cellulose percentage of various natural fibers. As the large weight-percentages of natural fibers are cellulose, so the swelling is expected when bio-resins, which are organic liquids, come in contact with natural fibers.

Parameters that affect swelling of natural fibers as cellulose-based materials in organic, swelling-inducing liquids are: 1) Hydrogen bonding capability of the liquid, 2) molecular size (both weight and volume) of the liquid, 3) cohesive energy density of the liquid, 4) surface coating and treatment of the fibers, 5) density of the fibers, 6) ambient temperature, 7) crystallinity structure of the fibers, 8) basicity of the liquid, 9) percentage of cellulose in fibers and 10) Steric effects [44, 45, 49, 50].

Table 3.2 The cellulose percentages of some natural fibers [46–48].

Fiber	Cellulose Percentage (wt%)
Sisal	66–77.2
Banana	61.5
Bowstring Hemp	69.7
Caroa	60
Cebumaguey	75.8
Henequen	77.6
Phormium	63
Pineapple	71.6
Piteria	75.6
Tulaistle	73.48
Ramie	91
Cango Jute	75.3

Table 3.2 (cont.) The cellulose percentages of some natural fibers [46–48].

Fiber	Cellulose Percentage (wt%)
Hemp	77.07
Jute	63.24
Kenaf	65.7
Sunhemp	80.4
Cotton	90
Bamboo	50
Flax	82
Sugar cane	50
Coir	43

3.3.3.2 Some Recent Studies on Changes in Permeability of Natural-Fiber Mats Due to Liquid Absorption and Swelling

Recently, Rodriguez *et al.* [51], conducted several tests to estimate the steady-state and transient permeabilities of jute fiber mats. It was discovered that the jute fiber-mat has a higher permeability under steady-state (saturated flow) conditions compared with the transient (unsaturated flow) conditions. Rodriguez *et al.* [51] reasoned that the permeability decreases under unsaturated flow conditions because the liquid 'disappears' in jute fiber mats and slows the flow as a result. It was conjectured that there can be two reasons for liquid absorption: 1) Individual jute fibers absorb a tremendous amount of liquid. 2) Bundles of jute fibers, from which jute mats are woven, absorb liquid as well due to the dual-scale nature of the resultant porous medium. Rodriguez *et al.* [51] used the governing equation for flow in rigid porous media, Eqs (3.1) and (3.2), to analyze the flow.

Rodriguez *et al.* [52], also conducted some tests to estimate the permeability-porosity relationship for glass, sisal, and jute fiber mats. (Figure 3.6 shows the structure of different fiber mats, similar to the ones used in their work. Figure 3.7 describes a permeability measuring setup.) They used glycerin as a test liquid and added some water to it to decrease its viscosity to about 1.2 Pa.s, which is close to the viscosity of commercial LCM resins. It was discovered that the permeability of natural fiber mats is higher than permeability of glass fiber mats. Rodriguez *et al.* [52] used the governing equation for flow in *rigid* porous media, Eqs (3.1) and (3.2), to derive the following equations for saturated permeability in 1-D flow:

$$K_{sat} = \frac{Q}{A}\mu\frac{\Delta L}{\Delta P} \tag{3.5}$$

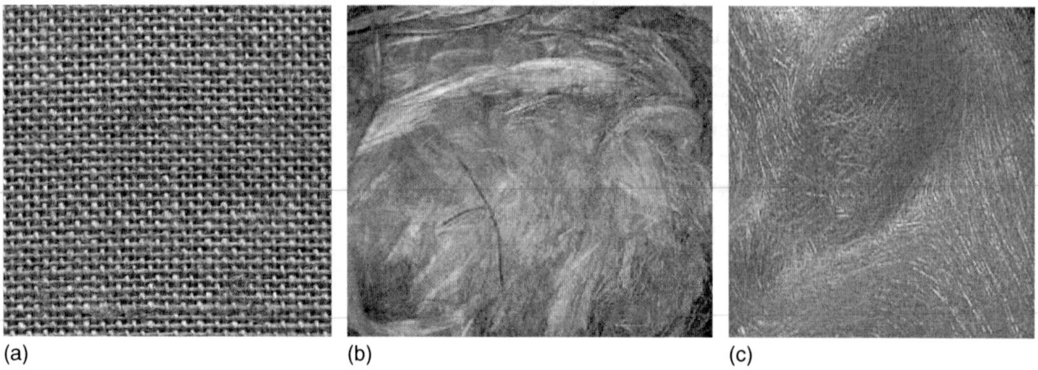

Figure 3.6 Photograph of the three different fiber mats: a) jute; b) sisal; c) glass.

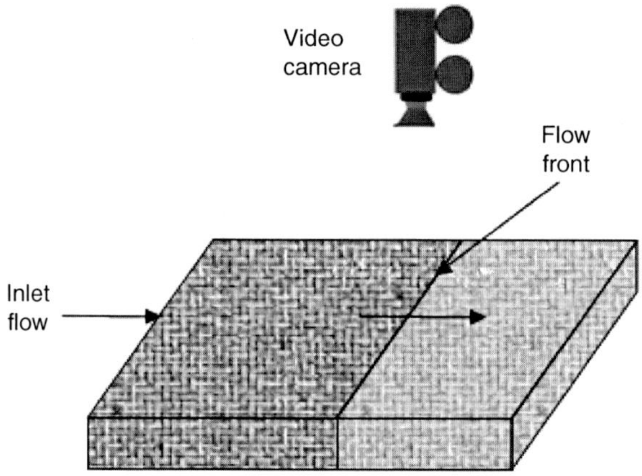

Figure 3.7 A schematic of the permeability measurement setup.

Here K_{sat} is the saturated permeability in m^2, Q is the volumetric flow rate in m^3/s, $\Delta P/\Delta L$ is the pressure drop per unit length in Pa/m, μ is the flow viscosity in $Pa.s$, A is the cross-sectional area of the medium in m^2. The following modified Carman-Kozeny equation was suggested to relate the permeability to the fiber-preform porosity

$$K = \frac{\varepsilon^{n+1}}{C(1-\varepsilon)^n} \tag{3.6}$$

where C and n are the empirical parameters found by curve fitting. (Table 3.3 lists the Carman-Kozeny parameters pertaining to Eq (3.6) for various fibrous porous media.) Figure 3.8 shows a plot of the predicted permeability as a function of the porosity obtained using Eq (3.6).

Some similar permeability characterization for wood-fiber mats was conducted recently by Umer *et al.*, [53–55]. Two different test-liquids, the water-diluted glucose syrup and the mineral oil, were used in their experiments. Four different

Table 3.3 Carman-Kozeny parameters estimated by Rodriguez *et al* [52]

Fiber Mats	$C\left[\times 10^{8}\,m^{-2}\right]$	n
Sisal	4.8	1.48
Jute	5.3	1.48
Glass	7.4	0.9

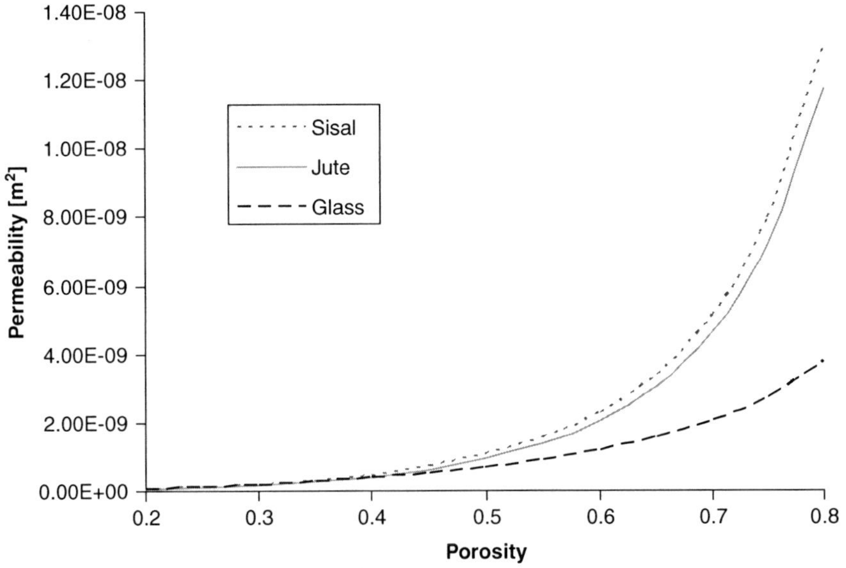

Figure 3.8 The relation between permeability and porosity as predicted by Eq (3.6).

wood fiber manufacturing techniques were employed to create the fiber mats: Papier Dynamic Former mats (PDF), Handsheets (HS), Dry Mat Former (DMF), and Medium Density Fiber (MDF). They also used a glass-fiber mat, the chopped fiber mat, (CFM) as a reference.

The permeability of wood fibers was found to be smaller when measured with the water-diluted syrup than when measured with the oil. The reason for this difference was explained in terms of a qualitative hypothesis: the permeability decreases for the syrup because it induces swelling in fibers due to the absorption of water, which in turn reduces porosity and hence the size of flow paths between fibers. Figure 3.9 shows comparisons between measured permeabilities for the wood fibers using the two test liquids.

The predicted permeability data for wood fiber mats and CFM are compared in Figure 3.10, in which Eq (3.6) along with the coefficients suggested by Umer *et al.* [54] were used. It shows that the type of test liquid used does not influence

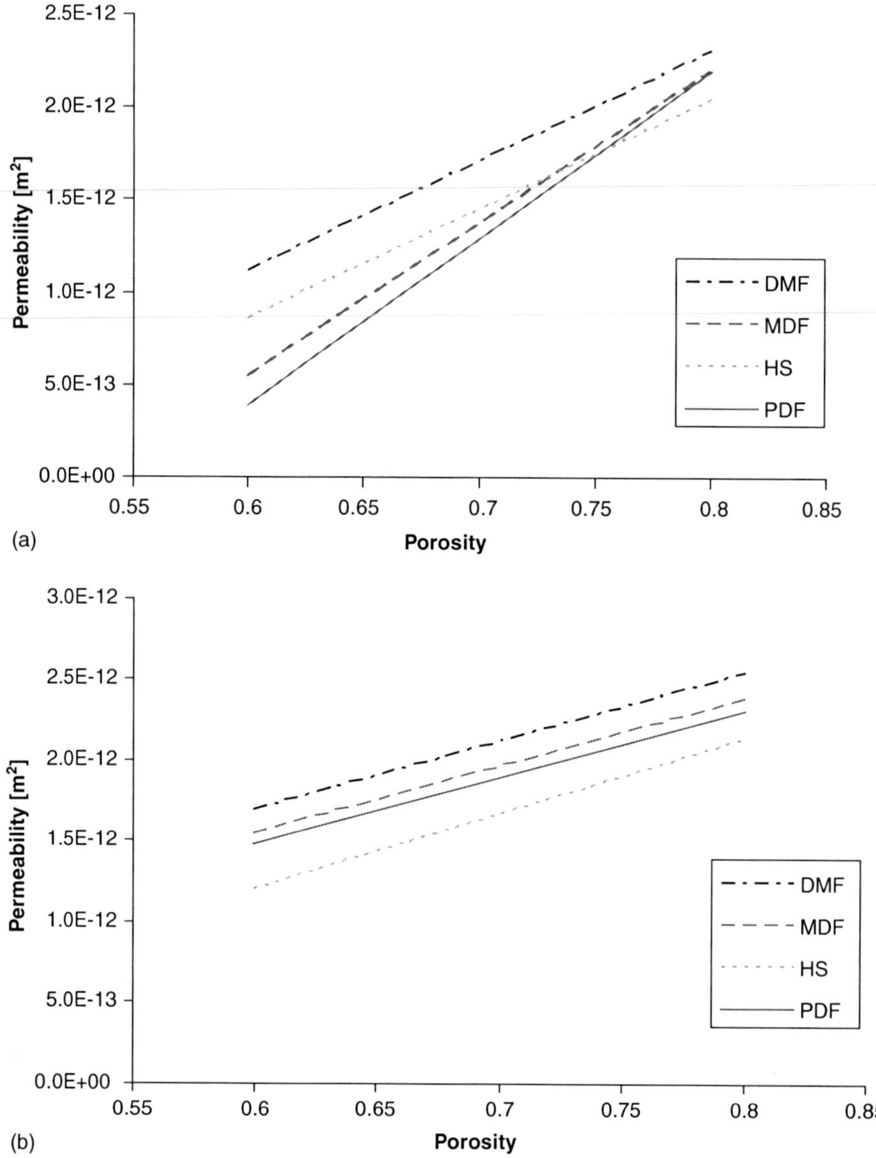

Figure 3.9 A permeability versus porosity plot for different wood fiber mats using a) glucose syrup and b) mineral oil (based on the data published by Umer *et al.* [54]).

the permeability of glass-fiber mat. It also depicts that the permeability of wood fiber mats is approximately two orders of magnitude lower than that of the CFM. One of the reason proffered for such a difference is that the wood fibers are made from several short fibers while glass fibers made from a continuous bundle of fibers—as a result, the compressed wood fiber leads to extremely torturous flow paths while CFM has a more efficient fiber packing and provide better, less tortuous flow paths for test liquids or resins.

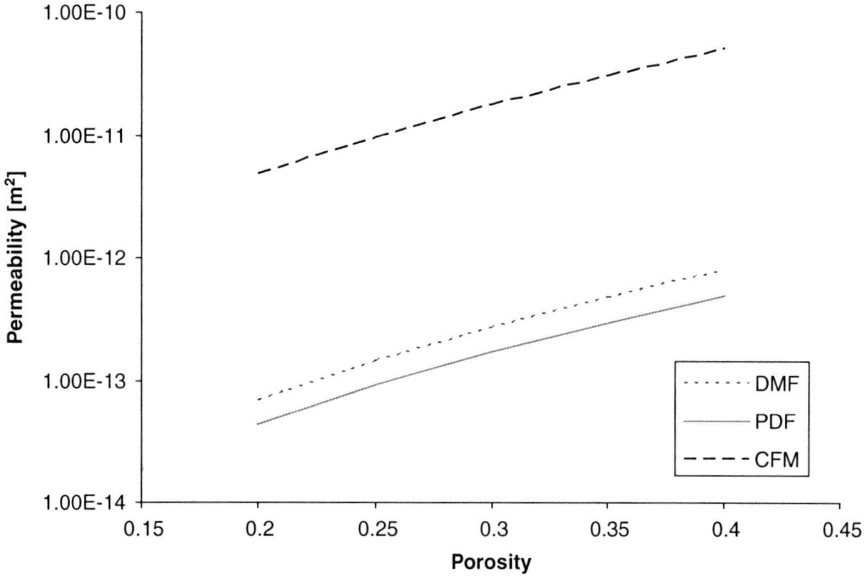

Figure 3.10 Permeability versus porosity plot for different fibers as predicted by Eq (3.6), and using the coefficients suggested by Umer *et al.* [54].

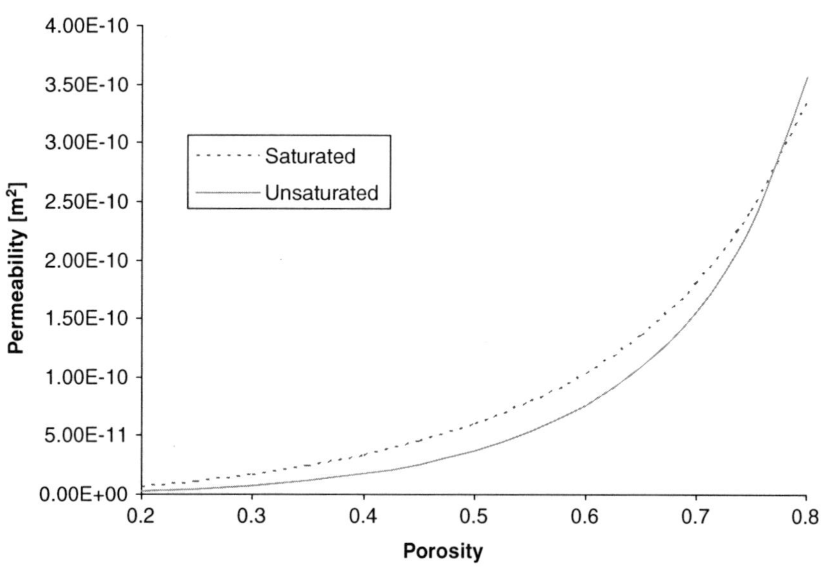

Figure 3.11 A plot of saturated and unsaturated permeabilities against the porosity for jute fibers predicted by Eq (3.6), using the coefficients suggested by Francucci *et al.* [56].

In an another study, a slight difference between the saturated and unsaturated permeabilities for natural fiber mats has been observed [56]—Figure 3.11 shows such a difference for jute fibers.

3.3.3.3 *Mold Filling Modeling in Natural-fiber Mats After Including the Swelling of Fibers Due to Liquid Absorption*

The absorption of liquid and subsequent swelling is expected during the flow of liquids (test liquids as well as resins and bio-resins) in natural fiber mats. The conventional LCM flow model (Eqs [3.1] and [3.2]) is derived for rigid, non-swelling porous media [28] and does not take these crucial phenomena into account. Moreover, the permeability is often used a fitting parameter in such equations during the transient 1-D and radial flow experiments to fit the pressure versus time plots. However, in the theoretical literature on single-phase flows, the permeability is deemed a geometrical property of porous medium and hence is a strong function of particle size and pore volume fraction only. Since the porous medium geometry in a hard-mold process such as RTM is fixed, any permeability that is made to change during the course of resin flow in an RTM mold without linking it to fiber swelling is clearly incomplete and is merely an ad-hoc effort to satisfy the deviant flow variables such as pressure measured during the standard 1-D and radial flow experiments.

The first attempts to include the swelling effect in the single-phase fluid flow LCM model using the Darcy's law was done by Masoodi and Pillai [57, 58] where they modified the continuity equation to include the effects of swelling and liquid absorption through additional terms. The modified version of continuity equation is given as

$$\vec{\nabla}\cdot\langle\vec{V}\rangle = -S - \frac{\partial \varepsilon}{\partial t} \tag{3.7}$$

where S is the sink term, which is the rate of liquid absorption by solid phase in the porous medium, and ε is the porosity. It was postulated [57] that the sink term is directly proportional to the rate of change of the porosity, and hence the above equation simplifies to

$$\vec{\nabla}\cdot\langle\vec{V}\rangle = (b-1)\frac{\partial \varepsilon}{\partial t} \tag{3.8}$$

Later, it was discovered that b, the constant of proportionality or the absorption coefficient, has to be very close to unity to satisfy the experimental observations for a capillary-suction driven flow in a swelling porous media [57, 58]. Hence $b=1$ is a good assumption for LCM flow model as well (it implies that the volumetric rate of liquid absorption in natural fibers is equal to the volumetric rate of solid-phase expansion). As a result, the continuity equation, Eq (3.8), simplifies to a form that is identical to the traditional form of the continuity equation, i.e., Eq (3.2).

Masoodi *et al* [59] used the thus modified continuity equation, with the assumption that permeability in the wetted preforms changes as a function of time only, to model the flow of test liquids in jute fiber mats in an LCM process. There the flow is considered to be one-dimensional, and hence the governing equations (Darcy's law and the modified continuity equation) simplify to

$$\langle u\rangle = -\frac{K}{\mu}\frac{d\langle P\rangle^f}{dx} \tag{3.9}$$

$$\frac{d\langle u\rangle}{dx} = 0 \tag{3.10}$$

If we substitute Eq (3.9) in Eq (3.10), and assume the permeability to be a function of time only, then we obtain

$$\frac{d^2\langle p\rangle^f}{dx^2} = 0 \tag{3.11}$$

which yields a linear spatial distribution for pressure. If the process is a constant injection-rate process, then the boundary conditions are

$$\langle p\rangle^f (x = 0) = p_{in} \tag{3.12a}$$

$$\langle p\rangle^f (x = x_f) = 0 \tag{3.12b}$$

Note that the capillary suction pressure at the flow front was neglected. After integrating Eq (3.11) two times and applying the boundary conditions Eqs (3.12a) and (3.12b), the final expression for pressure reduces to

$$\langle p\rangle^f = p_{in}\left(1 - \frac{x}{x_f}\right) \tag{3.13}$$

The liquid-front velocity and the Darcy (filtration) velocity are related through the equation

$$\frac{dx_f}{dt} = \left.\frac{\langle u\rangle}{\varepsilon_0}\right|_{x=x_f} \tag{3.14}$$

where ε_0 is the surface porosity at the liquid front, which incidentally is the initial porosity of fiber mats before swelling. Substitution of Eq (3.9) in Eq (3.14) along with the usage of Eq (3.13) for pressure and doing some further algebraic manipulations yields the final relation for liquid front as

$$x_f = \sqrt{\frac{2p_{in}}{\varepsilon_0\mu}\int_0^t K(t')dt'} \tag{3.15}$$

where x_f is the liquid-front location, p_{in} is the injection pressure (which is a constant here), and K is the time-dependent global permeability of the wetted fiber mats behind the moving resin-front in an LCM mold. It is important to keep in mind the limitations of the analytical solution given by Eq. (3.15); this can be ascertained by studying the following main assumptions used in the derivation of the equation:

1. The absorption rate of liquid into fibers per unit volume is equal to the rate of change of fiber volume within the same volume.
2. The porosity and permeability are considered uniformly decreasing in the whole of the liquid-wetted preform; thus they are assumed functions of time only.

3.3.4 Constant Inlet-Pressure Injection Solution

Masoodi *et al.* [59] used a 1-D flow test setup and a RTM machine (Figure 3.12) to conduct the constant inlet-pressure injection into a natural-fiber (jute) preform to test the validity of the derived analytical solution. The RTM machine used to

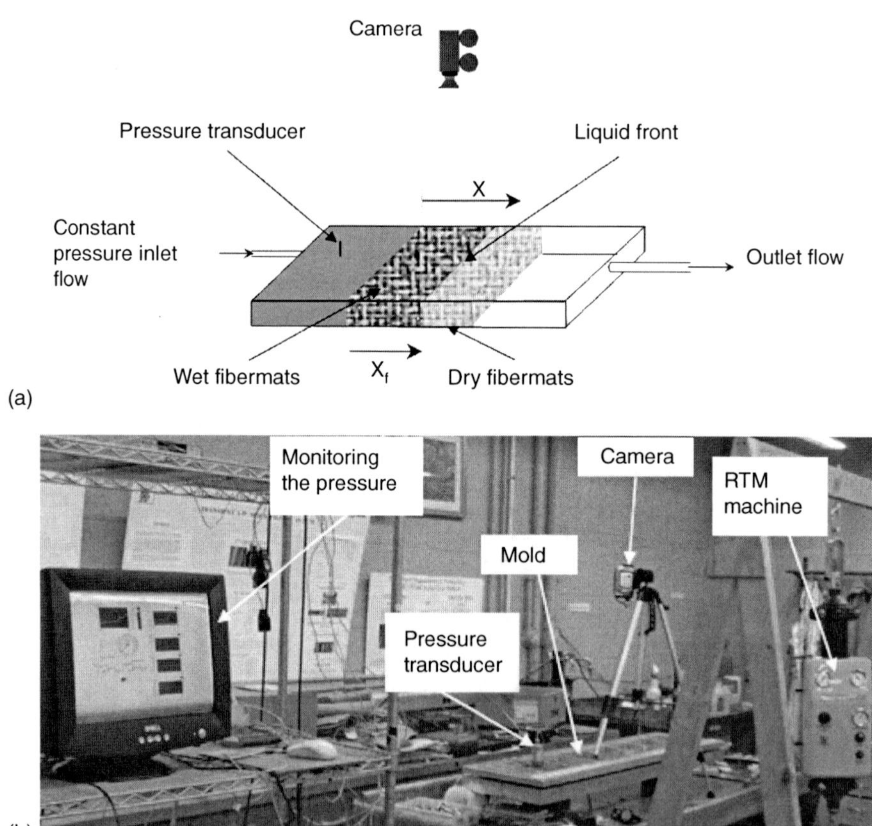

Figure 3.12 The experimental setup used for permeability measurement and flow studies during 1-D flow in a flat RTM mold. a) A schematic of the 1-D flow test setup. b) A photo of the test setup.

provide a constant inlet-pressure injection was a Hypajet model MK1 that was connected to a shop air source of 6 bars. Two different test liquids, one diluted corn syrup and the other the motor oil, were used in the experiments. Both test liquids were drawn into the machine, and then injected into the mold at a constant pressure. Fiber mats used in the experiment were sections of jute gunny bags. As shown schematically in Figure 3.12a, several layers of jute fiber mats were stacked in the middle of the flow channel to create a fiber preform. The liquid under a constant pressure entered the front area of the mold and moved through the fiber mats. A transducer was used to record the pressure of the liquid entering the fiber preform. A camera, placed on top of the transparent top-plate of the mold, was used to track the liquid-front location, so the liquid-front location was recorded as a function of time by reviewing the movies.

The jute fibers do not swell in motor oil while they swell in the diluted corn syrup due to the presence of water in the latter. The reason for using diluted corn-syrup was to match the viscosity of the syrup with that of the motor oil (the two viscosities are close to that of a thermosetting resin used in RTM) while at the same time having the fiber-swelling property that mimics the swelling property of organic liquids such as bio-resins. The viscosity of motor oil at the test temperature was found to be 245 mPa.s—in order to reach this viscosity, some water was added to the corn syrup such that 20% of the final solution was water.

Porosity is the ratio of void volume to the whole volume of the compressed fiber-preform. The importance of porosity is that it indicates the percent of the mold volume that should be filled by resin. Masoodi *et al* [59] measured the porosity of the preform when the mold was stacked with the fiber mats. One layer of jute fiber mat with known dimensions was inserted into a burette filled with a known volume of motor oil (a non-polar liquid that does not induce swelling and liquid absorption in fibers). The difference between the volume before and after inserting the material into the oil was measured. Since the number of layers in the mold is known, the total volume of the jute fibers can thus be computed and the overall porosity can hence be estimated. Eight layers of jute fiber mats were used in the mold and the preform porosity was estimated to be 0.5.

Since jute fibers do not swell in motor oil, so the permeability of jute layers was expected to remain constant during the 1-D flow tests with the oil. The inlet pressure and volume flow-rate of the passing liquid under steady-state conditions were first measured, and the permeability was subsequently estimated by using the following relation for steady flow:

$$K_0 = \frac{Q\mu L_f}{AP_{in}} \tag{3.16a}$$

The reason for using the subscript zero for permeability in this measurement is that K_0 is also equal to the initial ($t=0$) permeability for the case of the diluted corn-syrup, a swelling-inducing liquid. In the case of the diluted corn syrup, the permeability was estimated right after the liquid-front had reached the end of the fiber-preform length L_f at time t_{end}. If this permeability is represented as K_{end}, then one can estimate it as

$$K_{end} = \frac{Q\mu L_f}{AP_{in}} \tag{3.16b}$$

In this estimation, steady-state flow conditions were assumed at the end of the filling process[6], and the effect of liquid absorption on the overall flow-rate Q during the transient mold-filling process was neglected.

As a first approximation, Masoodi *et al.* [59] assumed the permeability to be a linear function of time, i.e., $K(t)=C_1+C_2t$. The values of the measured K_0, K_{end}, and t_{end} were used to find the constants in this permeability function and the final relation for such time-dependent permeability was found to be

$$K(t) = K_0 + \frac{K_{end} - K_0}{t_{end}} t \tag{3.17}$$

The measured values for K_0, K_{end}, and t_{end} were 4.816 e-10 m^2, 2.51e-10 m^2, and 44 s, respectively.

The liquid-front tracking was done by reviewing the recorded movies. A scale alongside the fiber mats was used in the flow mold (Figure 3.13), and by comparing the liquid-front location with the scale, it was possible to find the x coordinate (along the flow direction) of the liquid front as a function of time. (A stop watch was used to keep time, which was filmed along with the liquid front in the movie.)

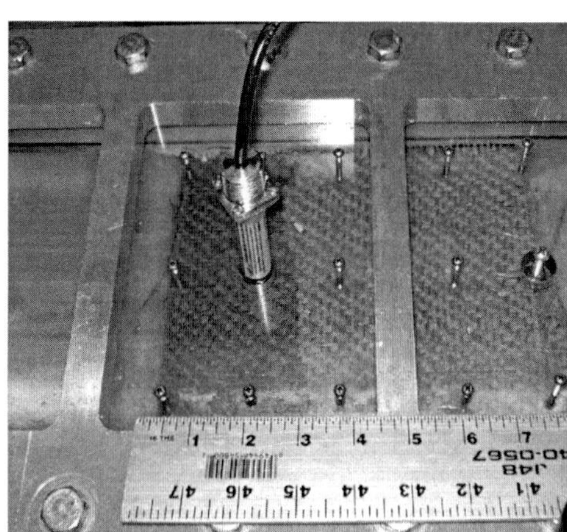

Figure 3.13 Visual estimation of the liquid-front location in jute fiber mats during the 1-D flow experiments in the test setup [59].

---

[6] Such an assumption can be justified since the slow flow of viscous liquids in porous media is often treated as a quasi-steady-state flow.

The theoretical prediction of the liquid-front position for the case of using motor oil as a test liquid (which does not induce swelling in jute fibers) is shown in Figure 3.14. As can be observed, a good validation of the theoretical model through experiments is achieved. However, there are some variations in the match as the time passes, but the predictions are still very good. The plausible reason for some variations and differences could be the uneven nature of flow front (Figure 3.14) due to 1) inherent inhomogeneity in the jute mats, and 2) slight differences in the structure of the fiber mats used in various layers.

Later, Masoodi *et al.* [59] investigated the effect of rendering the permeability variable as a result of the fiber swelling phenomena on the theoretical solution. Figure 3.15 shows the difference between the predictions of the

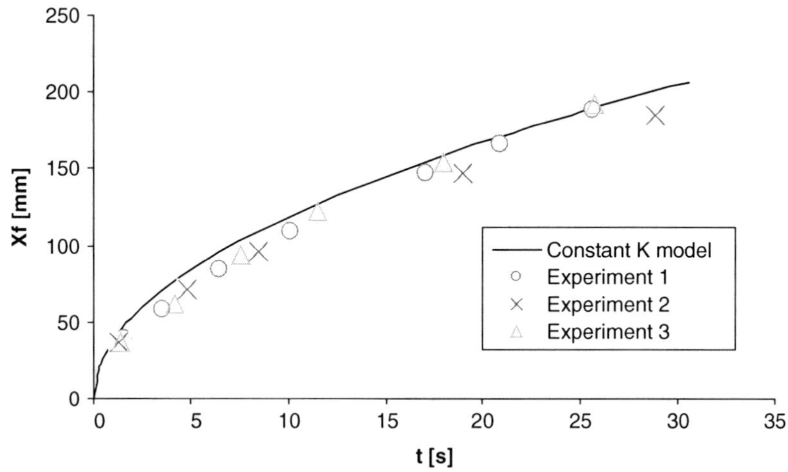

Figure 3.14 A comparison of the theoretical prediction of liquid-front location as a function of time with experimental observations for the case of using motor oil as test liquid in the absence of fiber swelling [59].

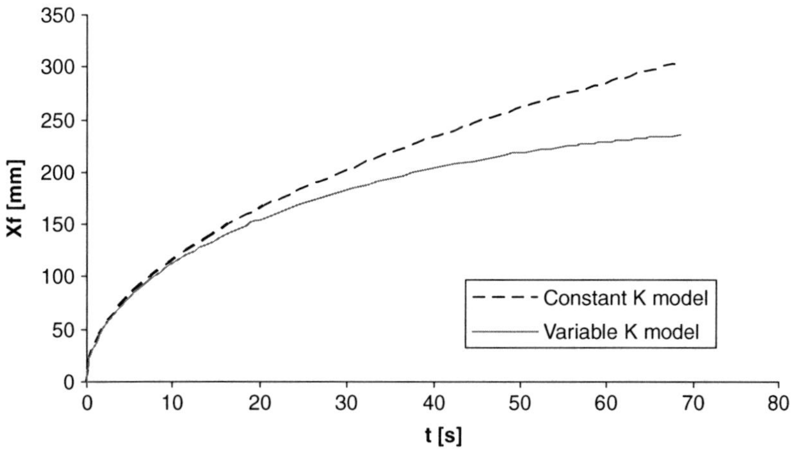

Figure 3.15 A comparison of theoretical predictions for the cases of non-swelling and swelling fibers using the constant and variable K models, respectively [59].

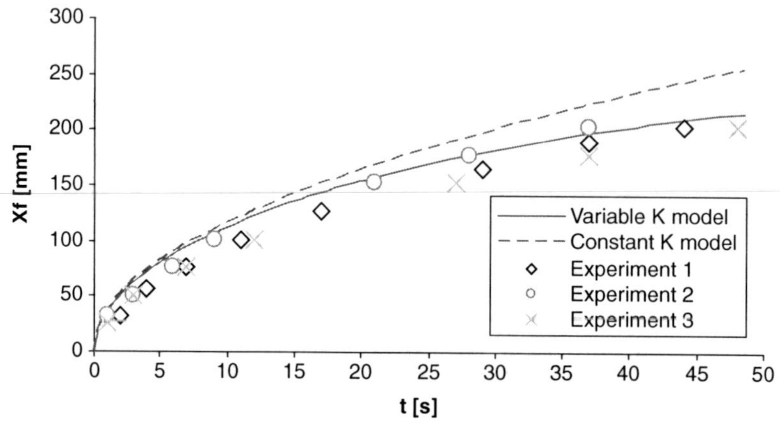

Figure 3.16 A comparison of the theoretical prediction of liquid-front location as a function of time with the experimental observations for the fiber-swelling inducing test-liquid diluted corn-syrup [59].

constant permeability model and the variable permeability model, Eq (3.14). In the beginning, both models behave in an identical manner; but as the time passes, the variable-permeability model predictions are lower than that of the fixed permeability model. As the front progresses, fibers swell and the permeability deceases, and one can expect a deceleration of front—this is what is shown by the variable permeability model.

The variable permeability model is compared with the experimental results in Figure 3.16. It shows that the predictions are in general quite accurate—although there is some inaccuracy initially, but this difference decreases with time. It was also observed that the predictions of the fixed K model are a little higher than the experimental data, while the predictions of the variable K model are closer to the experimental data, and hence are more accurate.

The variable K model, which has been proven to be quite accurate, assumes a simple, linear variation of permeability with time. Because of its simplicity, one may not expect to have very accurate results; however, these results are very good as they are quite close to the experimental observations. Note that the two constants in the linear model, Eq (3.17), were easy to estimate through the use of just two experimentally-obtained permeability values.

3.3.5 Constant Flow-rate Injection Solution

Another approach to study flow in natural fiber mats is to study the fiber swelling first and then relate it to changes in the permeability and porosity. Languri *et al* [60] used such an approach in studying the 1-D flow under constant flow-rate. To derive the governing equation, they used Eq (3.14) and consider this fact that the injection pressure is not constant, so the integration leads to

$$x_f = \sqrt{\frac{2}{\varepsilon_0 \mu} \int_0^t P_{in}(t') K(t') dt'} \tag{3.18}$$

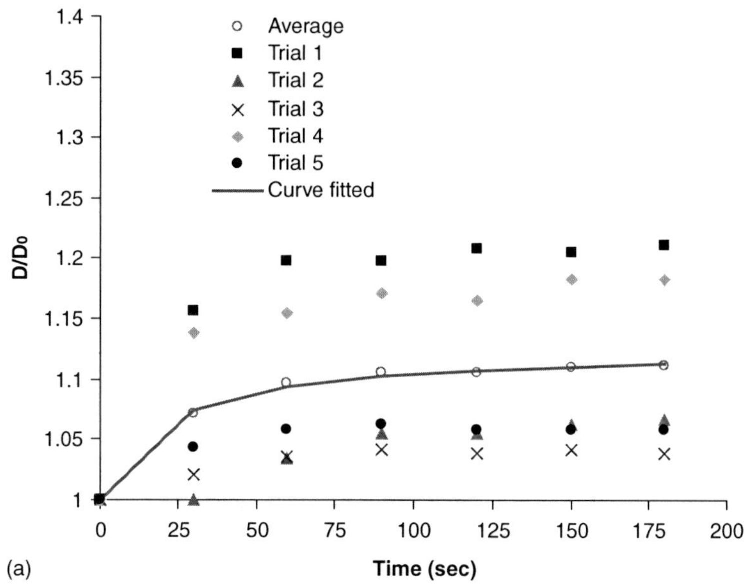

(a)

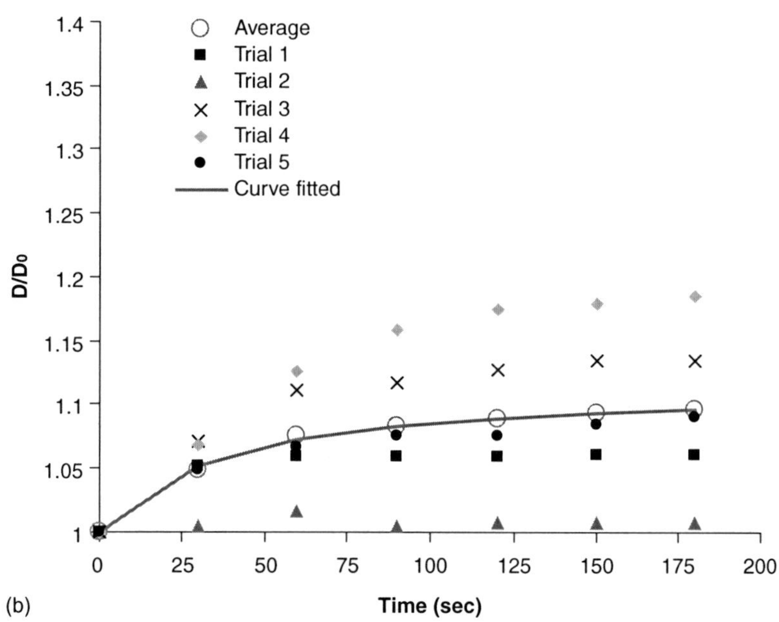

(b)

Figure 3.17 Swelling fiber-diameter D as function of time for a) broken wood fibers and b) unbroken wood fibers. (The fibers are wetted by the water-diluted corn syrup.) D_o represents the initial diameter at time t=0 [60].

The natural fiber mat made of 100% Kenaf was used along with two different liquids, the motor oil and the diluted corn-syrup with 40% water. To experimentally study the diameter change in the wetted natural fibers due to swelling, five random fibers (either broken or unbroken) were picked from a Kenaf wood fiber mat. A microscope was used at *10*× magnification to capture the growth of fiber diameter every thirty seconds for the total duration of three minutes. (Figure 3.17 shows the experimental data on fiber diameters as they were swelling with time. Table 3.4 describes the details on the fitted curves used in Figure 3.17.) It is clear that the fiber diameter increases about 10% after first three minutes. The figure also shows that the broken fibers (after removing the surface covering) swell slightly more than the unbroken fibers[7]. Another typical fiber-growth pattern due to swelling is shown in Figure (3.18), which was measured for jute fibers in the diluted corn syrup [61].

Initial porosity of kenaf fibers, ε_0, was found to be 0.85. Since the porosity of a swelling porous medium is a function of time, Masoodi and Pillai [57] derived the following relation for porosity by using the time-dependent fiber-diameter:

$$\varepsilon(t) = 1-(1-\varepsilon_0)\left(\frac{D_f(t)}{D_{f0}}\right)^2 \tag{3.19}$$

Table 3.4 Swelling fiber diameter as function of time: the following parameters correspond to a fitted curve of the form $D_f(t) = a.\exp(b/c+t)$ for our experimental data.

Parameter	Broken Kenaf Fiber	Unbroken Kenaf Fiber	Average Values
a	1.126	1.117	1.1215
b	–2.313	–3.935	–3.124
c	19.516	35.432	27.474

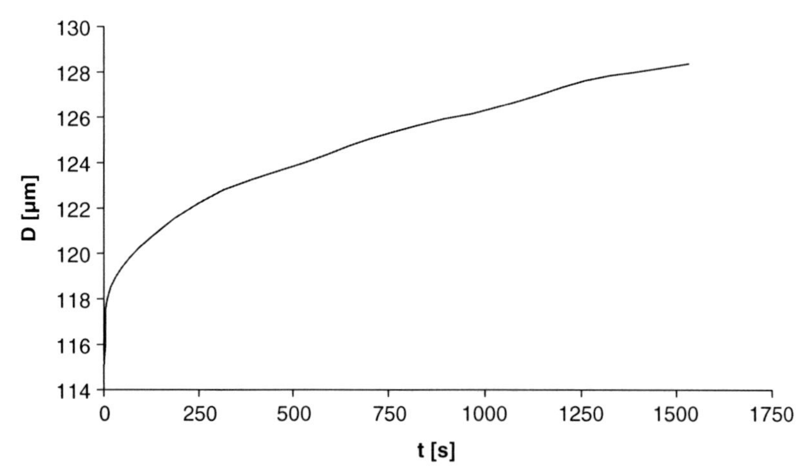

Figure 3.18 The measured jute-fiber diameter D increasing with time when the fiber comes in contact with the diluted corn syrup [61].

[7] It is generally acknowledged by the forestry research community that broken fibers lead to broken cell walls inside the fibers, and as a result, the cellulose present inside the cells finds it easier to expand as compared to the situation of intact cell walls where the lignin reinforced cell walls (which swell much less compared to the cell matter inside) can essentially 'choke' the cell inside from swelling.

Since the porosity ε is not constant in a swelling porous medium, an expression for the permeability that varies with porosity, i.e. $K = K(\varepsilon)$, is needed as well for solving the Darcy law in the LCM flow model. Substituting $\varepsilon(t)$ from Eq. (3.19) and the fiber diameter $D_f\ (t)$ from Table (3.4) in the Kozeny-Carman model for permeability [27], the permeability as function of time [57] can be worked out to be

$$K = K_0 \left(\frac{\varepsilon_f}{\varepsilon_{f0}} \right)^3 \frac{1-\varepsilon_{f0}}{1-\varepsilon_f} \tag{3.20}$$

The steady-state 1-D flow experiment with the motor oil (which causes no swelling in fibers) was used to determine K_0, the initial permeability. The flow-front positions were tracked by reviewing the mold-filling video along with a stop watch. Figure 3.19 shows the comparison of flow-front locations at different

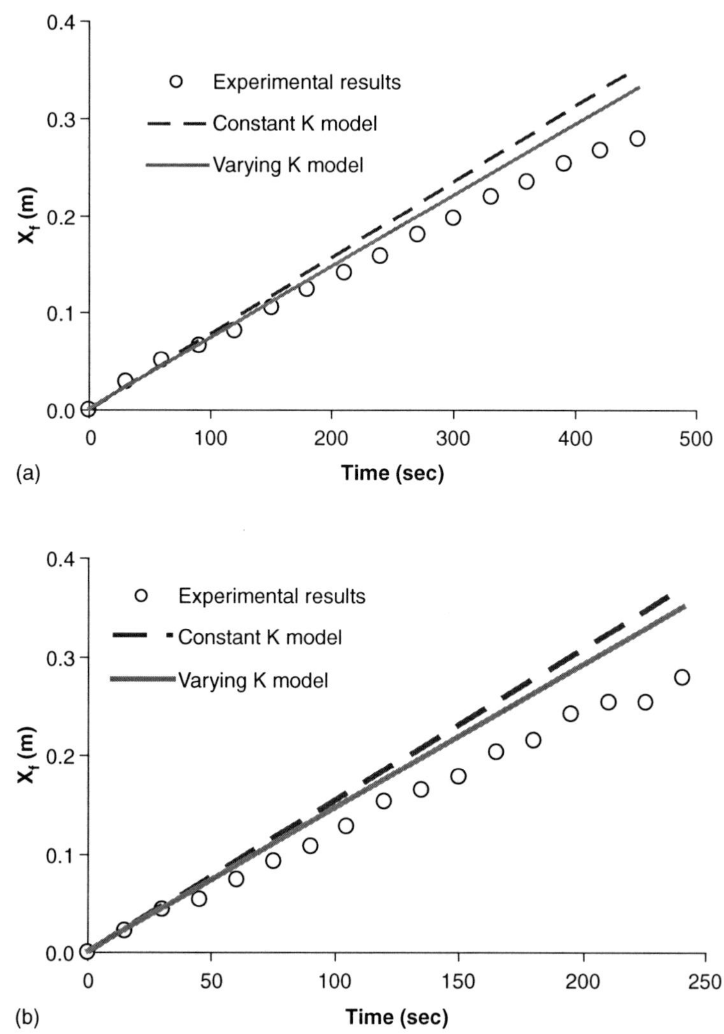

Figure 3.19 Flow-front position versus time plot during the 1-D constant injection-rate LCM experiment [60]. a) $Q = 1\ mL/s$; b) $Q = 2\ mL/s$.

times obtained from the theoretical model and the experimental results: it is clear that the varying permeability model predicts the flow-front more accurately than the constant *K* model. This clearly is another affirmation of the theory proposed in [57].

3.4 Numerical Simulation

3.4.1 Mold Filling Simulation in Non-swelling Fiber Mats

The numerical modeling of LCM mold-filling flow in non-swelling fiber mats such as glass and carbon fibers has been extensively studied during the last two decades [62–66]. Several computer programs are available for the LCM mold-filling simulations that include research codes such as LIMS by University of Delaware, RTMFLOT by Ecole Polytechnique de Montréal, CRIMSON by NIST as well as commercial codes such as PAM-RTM by ESI Group and Plastics Advisor by Moldflow (Autodesk). PORE-FLOW©, a research code developed at University of Wisconsin—Milwaukee, can also model the resin flow through duel-scale porous media such as the stitched and woven fiber mats [42, 43, 67–71]. In the algorithm employed by these mold-filling simulations, the transient fluid-flow in porous media involving a moving-boundary (i.e., a flow front) is divided into multiple time steps. After assuming a quasi-steady condition during each time step, the Laplace equation for pressure is first solved for the modified pressure using the hybrid FE/CV algorithm in the wet region saturated by the moving liquid-front. Then the pressure field is computed at FE nodes using the Galerkin weighted residual method; later, this pressure, in conjunction with the proper boundary conditions, is used to estimate the velocity field through Darcy's law at the surfaces of CVs described around FE nodes; later the velocity field is used to find the new location of the liquid front at each time-step [72].

3.4.2 Recent Developments in LCM Mold Filling Simulation in the Swelling Natural-fiber Mats

Currently, there is no commercial software that can model LCM mold-filling flow in swelling fiber mats; however, PORE-FLOW© [26] has the capability of including the effects of swelling in porous media by using the variable, wetting-time-dependent permeability and porosity. Figure 3.20 describes the example of using PORE-FLOW© for simulating liquid flow during wicking in a strip of paper composed of 10% CMC, a highly swelling and liquid-absorbing porous medium [73]. To the best knowledge of the authors, the PORE-FLOW© is the only available numerical simulation that has the capability of predicting liquid flow through swelling bio-fibers packed in an LCM mold. Work is currently going on to validate the code for modeling mold-filling flows in LCM molds packed with natural fibers [61].

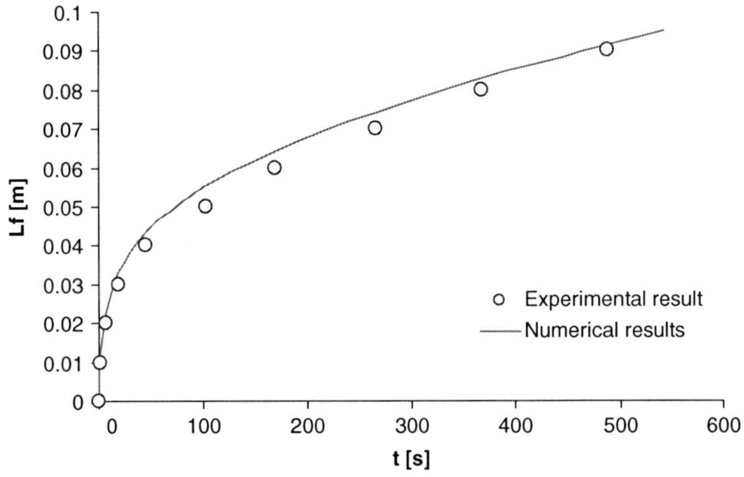

Figure 3.20 Achievement of a good match between the numerical prediction by the code PORE-FLOW® [26] and the experiments for wicking in a strip of 10% CMC paper [73].

3.5 Summary and Conclusions

LCM processes can be used to produce near net-shaped parts from natural fibers and bio-resins on a commercial scale. Optimization of LCM molds through LCM mold-filling simulations is an accepted practice in the composites industry. We discussed the science behind modeling the flow of the thermosetting bio-resins during mold-filling in LCM where natural fibers are used as reinforcements. The swelling of natural fiber mats in the presence of organic liquids similar to the thermosetting bio-resins is briefly discussed. The two approaches for modeling the LCM mold-filling process through natural-fiber preforms are presented: 1) Neglecting the fiber-swelling phenomenon and applying the conventional flow-model as applicable to the glass and carbon fiber mats. 2) Inclusion of the fiber-swelling phenomenon caused by liquid absorption and a suitable modification of the governing equations. Both approaches are discussed with the help of the published literature. Accuracy of the recently developed analytical models, based on the new flow physics for the 1-D flow in simple LCM molds, is discussed in detail; it is quite clear that the practice of using time-dependent permeability as well as the assumption of equating the volumetric rate of liquid absorption into natural fibers with the volumetric rate of fiber swelling yields more accurate solutions. After describing the currently available mold-filling simulations based on conventional physics, the current trends in modeling LCM flows in natural-fiber based preforms is presented.

References

1. S.K. Mazumdar, *Composites Manufacturing: Materials, Product, and Process Engineering*, New York, CRC Press LLC, 2001.
2. T.G. Gutowski, ed. *Advanced Composites Manufacturing*, New York, John Wiley & Sons, 1997.

3. Hua Tan and Krishna M. Pillai, "Processing Composites for Blast Protection" In N. Uddin ed., *Blast Protection of Civil Infrastructures and Vehicles Using Composites*, Cambridge, Woodhead Publishing Limited, 2010.
4. R.S. Parnas, *Liquid Composite Molding*, Cincinnati, Hanser Gardner Publications, 2000.
5. P.A.D. Santos, J.C. Giriolli, J. Amarasekera, and G. Moraes, Natural fibers plastics composites for automotive applications, 8th Annual SPE Automotive Composites Conference and Exhibition, Troy, Michigan, Sept. 16–18, 2008.
6. S.V. Joshi, L.T. Drzal, A.K. Mohanty, and S. Arora, "Are Natural Fiber Composites Environmentally Superior to Glass Fiber Reinforced Composites?" *Composites, Part A: Applied Science and Manufacturing*, Vol. 35, pp. 371–376, 2004.
7. A. K. Mohanty, M. Misra, and L. T. Drzal, "Sustainable Bio-Composites from Renewable Resources: Opportunities and Challenges in the Green Materials World," *Journal of Polymers and the Environment*, Vol. 10 (1/2), pp. 19–26, 2002.
8. M.A.S.A. Samir, F. Allioin, and A. Dufresne, "Review of Recent Research into Cellulosic Whiskers, Their Properties and Their Application in Nanocomposite Field," *Biomacromolecules*, Vol. 6, pp. 612–626, 2005.
9. Lawrence T. Drzal, A. K. Mohanty, and M. Misra, Bio-composite Materials as Alternatives to Petroleum-based Composites for Automotive Applications, http://www.speautomotive.com/SPEA_CD/SPEA2001/pdf/e/E1.pdf, 2001.
10. Richardson, M.O.W. and Zhang, Z.Y. "Experimental Investigation and Flow Visualization of the Resin Transfer Mould Filling Process for Non-Woven Hemp Reinforced Phenolic Composites," *Composites: Part A*, Vol. 31, pp. 1303–1310, 2000.
11. Richard Wool and X. Susan Sun, *Bio-Based Polymers and Composites*, Academic Press, Amsterdam, 2005.
12. K.M. Pillai, "Unsaturated Flow in Liquid Composite Molding Processes: A Review and Some Thoughts," *Journal of Composites Materials*, Vol. 38(23), pp. 2097–2118, 2004.
13. K.M. Pillai, T. L. Luce, M. V. Bruschke, R. S. Parnas, and S. G. Advani, "Modeling the Heterogeneities Present in Preforms During Mold Filling in RTM," *Advanced Materials: Expanding the Horizons*, Vol. 25, 25th International SAMPE Technical Conference, Boston, MA, USA, 1993.
14. C.A. Fracchia, J. Castro, and C.L. Tucker, III, A finite element/control volume simulation of resin transfer molding, Proceedings of the American Society for Composites Fourth Technical Conference, pp. 157–166, Lancaster, Pa, 1989.
15. J. Molnar, L. Trevino, and L. J. Lee, "Liquid Flow in Molds with Prelocated Fiber Mats," *Polymer Composites*, Vol. 10, pp. 414–423, 1989.
16. M. V. Bruschke and S. G. Advani, "RTM Filling Simulation of Complex Three Dimensional Shell-like Structures," *SAMPE Q.*, Vol. 23(1), pp.2–11, October 1991.
17. W. B. Young, K. Rupel, K. Han, L. J. Lee, and M. J. Liou, "Analysis of Resin Injection Molding in Molds with Preplaced Fiber Mats, II: Numerical Simulation and Experiments Of Mold Filling," *Polymer Composites*, Vol. 12, pp. 30–38, 1991.
18. W. B. Young, K. Han, L. H. Fong, L. J. Lee, and M. J. Liou, "Flow Simulation in Molds with Preplaced Fiber Mats," *Polymer Composites*, Vol. 12, pp. 391–403, 1991.
19. F. R. Phelan Jr, "Simulation of the Injection Process in Resin Transfer Molding," *Polymer Composites*, Vol. 18(4), pp. 460–476, 1997.
20. Wu et al., "Simulation of Reactive Liquid Composite Molding Using an Eularian-Lagrangian Approach," *International Polymer Processing*, Vol. 4, p.398, 1998.
21. R.V. Mohan, N.D. Ngo, and K.K. Tamma, "On a Pure Finite-Element-Based Methodology for Resin Transfer Mold Filling Simulations," *Polymer Engineering and Science*, Vol. 39(1), pp. 26–43, 1999.
22. F.D. Dungan and A.M. Sastry, "Saturated and Unsaturated Polymer: Microphenomena and Modeling," *Journal of Composite Materials*, Vol. 36(13), pp. 1581–1603, 2002.
23. Shojaei et al., "Numerical Simulation of Three-Dimensional Mold Filling Process in Resin Transfer Molding Using Quasi-Steady State and Partial Saturation Formulations," *Composites Science and Technology*, Vol. 62(6), pp. 861–879, 2002.

24. S. Soukane and F. Trochu, "Application of the Level Set Method To The Simulation Of Resin Transfer Molding," *Composites Science and Technology*, Vol. 66(7&8), pp. 1067–1080, 2006.
25. S. Jiang et al., "PCG Solver And Its Computational Complexity for Implicit Control-Volume Finite-Element Method of RTM Mold Filling Simulation," *Composites Science and Technology*, Vol. 67(15&16), pp. 3316–3322, 2007.
26. POREFLOW, http://www4.uwm.edu/porous/, 2011.
27. Jacob Bear, *Dynamics of Fluids in Porous Media*, Elsevier Science, New York, 1972.
28. C.I. Tucker and R.B. Dessenberger, "Governing Equation for Flow and Heat Transfer in Stationary Fiber beds" In SG Advani ed., *Flow and Rheology in Polymer Composites Manufacturing*, Ch. 8, Elsevier Science, Amsterdam, 1994.
29. S. Whitaker, *The Method of Volume Averaging*, Springer, Dordrecht, 1998.
30. J.R. Weitzenbock, R.A. Shenoi, and P.A. Wilson, "Measurement of Principal Permeability with the Channel Flow Experiment," *Polymer Composites*, Vol. 20(2), pp. 321–335, 1998.
31. B.R. Gebart, "Permeability of Unidirectional Reinforcements for RTM." *Journal of Composite Materials*, Vol. 26(8), pp. 1100–1133, 1992.
32. Bruschke, M. V. and Advani, S. G., "Flow of Generalized Newtonian Fluids Across a Periodic Array of Cylinders," *Journal of Rheology*, Vol. 37(3), May/June 1993.
33. Westhuizen, J. V. and Plessis, J. P. D., "Quantification of Unidirectional Fiber Bed Permeability," *Journal of Composite Materials*, Vol. 28(7), pp. 619–637, 1993.
34. Skartis, L., Kardos, J. L., and Khomami, B., "Resin Flow through Fiber Beds during Composite Manufacturing Processes. Part 1: Review of Newtonian Flow through Fiber Beds," *Polymer Engineering and Science*, Vol. 32(4), pp. 221–230, 1992.
35. H. Tan, T. Roy, and K.M. Pillai, "Variations in Unsaturated Flow with Flow Direction in Resin Transfer Molding: an Experimental Investigation," *Composites Part A: Applied Science and Engineering*, Vol. 38(8), pp. 1872–1892, 2007.
36. S.G. Advani, ed., *Flow and Rheology in Polymer Composites Manufacturing*, Elsevier, Amsterdam,1994.
37. C. Lekakou, M.A.K. Johari, D. Norman, and M.G. Bader, "Measurement Techniques and Effects on in-Plane Permeability of Woven Clothes in Resin Transfer Moulding," *Composites: Part A*, Vol. 27A, pp. 401–408, 1996.
38. B.R. Gebart and P. Lidstrom, "Measurements of in-plane permeability of Anisotropic Fiber Reinforcements," *Polymer Composites*, Vol. 17(1), pp. 43–51, 1996.
39. K.L. Adams and L. Rebenfeld, "In-plane Flow of Fluids in Fabrics: Structure/Flow Characterization," *Textile Research Journal*, Vol. 57(11), pp. 647–654, 1987.
40. H. Tan and K.M. Pillai, A new method to estimate permeability of isotropic fiber mats through radial flow, Proceedings of SAMPE2007 Conference, Baltimore, MD, 2007.
41. Krishna M. Pillai, "Governing Equations for Unsaturated Flow in Woven Fiber Mats: Part 1 Isothermal Flows," *Composites Part A: Applied Science and Manufacturing*, Vol. 33, pp. 1007–1019, 2002.
42. K.M. Pillai and M.S. Munagavalsa, "Governing Equations for Unsaturated Flow Through Woven Fiber Mats, Part 2: Nonisothermal Reactive Flows," *Composites Part A: Applied Science and Manufacturing*, Vol. 35, pp. 403–415, 2004.
43. H. Tan, Simulation of Flow in Dual-Scale Porous Media, PhD Thesis, Dept. of Mechanical Engineering, University of Wisconsin-Milwaukee, 2010.
44. Mantanis, G.I., Young, R.A., and Rowell, R.M., "Swelling of Wood, Part II. Swelling in Organic Liquids," *Holzforschung*, Vol. 48, 6, pp. 480–490, 1994.
45. Mantanis, G.I., Young, R.A., and Rowell, R.M., "Swelling of Compressed Cellulose Fiber Webs in Organic Liquids," *Cellulose*, Vol. 2, pp. 1–22, 1995.
46. Chand, N and Rohatgi P.K. *Natural Fibers and Their Composites*, Periodical Experts Book Agency, Delhi, India, 1994.
47. Bailey, L.H., *Cyclopedia of American Agriculture V2: Crops*, Macmillan Company, London, 1911.
48. Rowell, R, O'Dell, J, Basak, R.K, and Sarkar, M, Applications of Jute in Resin Transfer Molding, http://www.fpl.fs.fed.us/documnts/pdf1997/rowel97h.pdf

49. Mantanis, G.I., Young, R.A., and Rowell, R.M., "Swelling of Wood, Part I. Swelling in Water," *Wood Science Technology*, Vol. 28, pp. 119–134, 1994.
50. Boluk, Y, "Acid–base Interactions and Swelling of Cellulose Fibers in Organic Liquids," *Cellulose*, Vol. 12, 6, pp. 577–593, 2005.
51. E.S. Rodriguez, F. Giacomelli, and A. Vazquez, Study of saturated and unsaturated permeability in natural fiber fabrics, In The 9th International Conference on Flow Processes in Composite Materials, Montreal, QC, Canada, 2008.
52. E.S. Rodriguez, F. Giacomelli, and A. Vazquez, "Permeability-Porosity Relationship in RTM for Different Fiberglass and Natural Reinforcement," *J. of Composite Materials*, Vol. 38, 3, pp. 259–268, 2004.
53. R. Umer, S. Bickerton, and A. Fernyhough, "Characterizing Wood Fiber Mats as Reinforcements for Liquid Composite Moulding Processes," *Composites: Part A*, Vol. 38, pp. 434–448, 2007.
54. R. Umer, S. Bickerton, and A. Fernyhough. "Chapter 23: Wood Fiber Mats as Reinforcements for Thermosets" In Stoyko Fakirov and Debes Bhattacharya eds., *Engineering Biopolymers: Homopolymers, Blends and Composites*, Hanser, Munich, 2007.
55. R. Umer, S. Bickerton, and A. Fernyhough, "Modelling the Application of Wood Fiber Reinforcements within Liquid Composite Moulding Processes," *Composites: Part A*, Vol. 39, pp. 624–639, 2008.
56. Francucci, G., Rodriguez, E.S., and Vazquez, A., "Study of Saturated and Unsaturated Permeability in Natural Fiber Fabrics," *Composites: Part A*, Vol. 41, pp. 16–21, 2010.
57. Masoodi, R. and Pillai, K.M., "Darcy's Law Based Model for Wicking in Paper-like Swelling Porous Media," *AIChE Journal*, Vol. 59, 9, pp. 2257–2267, 2010.
58. Masoodi, R., Modeling Imbibition of Liquids into Rigid and Swelling Porous Media, PhD Thesis, Dept. of Mechanical Engineering, University of Wisconsin-Milwaukee, 2010.
59. Masoodi, R., Pillai, K.M. and Verhagen, M.A., Flow Modeling in Natural-Fiber Preforms used in Liquid Composite Molding, Proc. 1st joint American-Canadian International Conference on Composites, Delaware, USA, September 15–17, 2009.
60. Languri, E.M., Moore, R.D., Masoodi, R., Pillai, K.M., and Sabo, R., An Approach to Model Resin Flow through Swelling Porous Media made of Natural Fibers, Proc. 10th International Conference on Flow Processes in Composite Materials (FPCM10), Monte Verità, Ascona, Switzerland, July 11–15, 2010.
61. Masoodi, R. and Pillai, K.M., Flow modeling in jute fiber mats during liquid composite molding, manuscript is under preparation, 2010.
62. K.M. Pillai, F. R. Phelan Jr. and C. L. Tucker III, "Numerical Simulation of Injection/Compression Liquid Composite Molding. Part 2: Preform Compression," *Composites Part A: Applied Science and Manufacturing*, Vol. 32(2), pp. 207–220, 2001.
63. K.M. Pillai, A. Benard, K.I. Jacob and S.G. Advani, "Numerical Simulation of Crystallization in High Density Polyethylene Fibers," *Polymer Engineering and Science*, Vol. 40(11), pp. 2356–2373, 2000.
64. K.M. Pillai, F.R. Phelan Jr. and C.L. Tucker III, "Numerical Simulation of Injection/Compression Liquid Composite Molding," *Part 1: Mesh Generation, Composites Part A: Applied Science and Manufacturing*, Vol. 31(1), pp. 87–94, 2000.
65. K.M. Pillai and S.G. Advani, "Numerical Simulation of Unsaturated Flow in Woven or Stitched Fiber Mats in Resin Transfer Molding," *Polymer Composites*, Vol. 19(1), pp. 71–80, 1998.
66. K.M. Pillai and S.G. Advani, "Numerical and Analytical Study to Estimate the Effect of Two Length Scales upon the Permeability of a Fibrous Porous Medium," *Transport in Porous Media*, Vol. 21, pp. 1–17, 1995.
67. Hua Tan, Pillai, K.M., "Fast LCM Simulation of Unsaturated Flow in Dual-Scale Fiber Mats Using the Imbibition Characteristics of a Fabric-Based Unit Cell," *Polymer Composites*, Vol. 31, 10, pp. 1790–1807, 2010
68. Hua Tan and Pillai, K.M., "Numerical Simulation of Reactive Flow in Liquid Composite Molding Using Flux-Corrected Transport (FCT) Based Finite Element/Control Volume (FE/CV) Method," *International Journal of Heat and Mass Transfer*, Vol. 53, 2256–2271, 2010.

69. Hua Tan, K.M. Pillai, "Finite Element Implementation of Stress-Jump and Stress-Continuity Conditions at Porous-Medium, Clear-Fluid Interface," *Computers & Fluids*, Vol. 38 (6), 1118–1131, 2009.
70. Hua Tan, K.M. Pillai, "A Method to Estimate the Accuracy of Radial Flow-Based Permeability Measuring Devices," *Journal of Composite Materials*, Vol. 43 (2009).
71. Hua Tan, K.M. Pillai, "Effect of Fiber-Mat Anisotropy on 1D Mold Filling in LCM: a Numerical Investigation," *Polymer Composites*, Vol. 29 (8), 869–882, 2008.
72. Hua Tan, Krishna M. Pillai, "Processing Composites for Blast Protection" In: *Blast Protection of Civil Infrastructures and Vehicles Using Composites*, edited by N Uddin, Woodhead Publishing Limited, Cambridge, 2010.
73. Masoodi, R., Tan. H., and Pillai, K.M., "Numerical Simulation of Liquid Absorption in Paper-Like Swelling Porous Media," submitted to *AIChE Journal*, 2010.

PART 2
PACKAGING APPLICATIONS

4

Bioplastics Based Nanocomposites for Packaging Applications

J. Soulestin[1,2*], K. Prashantha[1,2], M.F. Lacrampe[1,2] and P. Krawczak[1,2]

[1]*Univ. Lille Nord de France, Lille, France*
[2]*Ecole des Mines de Douai, Department of Polymers and Composites Technology & Mechanical Engineering, Douai, France*

Abstract
Development of packaging materials based on bio-nanocomposites for food and other food contact surfaces is expected to grow in the next decade with the current focus on exploring alternatives to petroleum and emphasis on reducing environmental impact. In this context, this chapter reviews recent advancements related to biodegradable polymer nanocomposites. The chapter discusses various techniques that have been used for developing cost-effective bio-based packaging materials with optimum material properties. The biodegradable polymers addressed in this chapter include polylactide (PLA), poly (hydroxyalkanoate)s (PHA) such as poly(β-hydroxybutyrate) (PHB), poly(3-hydroxybutyrate-co-3-hydroxyvalerate) (PHBV), and natural renewable polymers such as starch, and proteins. Special emphasis is given to the advantages of using clays as nanofiller, in order to improve the mechanical and the barrier properties of these biopolymeric matrices. New natural nanofillers such as cellulose and chitin nanofibers or starch nanocrystals are also addressed.

Keywords: Nanocomposites, bio-based polymer, clay, cellulose whiskers, starch nanocrystals, PLA, plasticized starch, PHA, proteins, mechanical properties, barrier properties

4.1 Introduction

Food products are primarily packaged to protect them from environment and to provide ingredient and nutritional information to the consumers. Traceability, convenience, and tamper identification are secondary functions of increasing importance. Materials that have been traditionally used in food packaging include glass, metal, paper and paperboard, and plastics. However, food packaging has become a central focus of waste reduction efforts because proper waste management is important to protect human health and environment [1].

Nowadays, the largest parts of materials used in packaging industries are produced from fossil fuels and are practically un-degradable. For this reason,

Srikanth Pilla (ed.) Handbook of Bioplastics and Biocomposites Engineering Applications, (77–120)

packaging materials for foodstuff, like any other short-term storage packaging material, represent a serious global environmental problem [2]. A big effort to extend the shelf life and enhance food quality while reducing packaging waste has encouraged the exploration of new bio-based packaging materials, such as edible and biodegradable films from renewable resources [3]. Nevertheless, like conventional packaging, bio-based packaging materials must fulfill a number of important criteria, including containment and protection of food, maintaining its sensory quality and safety, and communicating information to consumers [4].

The use of bio-based materials, due in some cases to their biodegradable nature, could at least to some extent solve the waste problem. However, poor mechanical and water vapor barrier properties as compared to plastics produced from petrochemicals limit their industrial use. Therefore, research has been geared to develop techniques to improve above-mentioned properties so as to obtain suitable bio-based packaging materials. Some of the techniques developed so for include chemical modification of biopolymers, addition of plasticizer to overcome brittleness, incorporation of other biodegradable polymers with improved properties into biopolymers to produce material with intermediate properties, and addition of compatibilizers to increase miscibility of incompatible polymers to decrease interfacial energy and stabilize polymer blends [5, 6].

Recently, a great attention has emerged around the polymer nanocomposites, which are proven to be a promising option in order to improve barrier and mechanical properties of polymers. They are thus of high interest for bio-based polymers. The polymer nanocomposites consist of a polymer matrix reinforced with fillers having at least one dimension in the nanometer range and possess very unusual properties, very different from their microscale counterparts. They often show improved mechanical and oxidation stability, decreased solvent uptake, self-extinguishing behavior and, eventually, tunable biodegradability due to high aspect ratio and high surface area of nanofillers [7–9].

Nanofillers can be three-dimensional spherical and polyhedral particles such as colloidal silica, two-dimensional nanofibers such as nanotubes, or one-dimensional disc like clay platelets. The most common class of material used as nanofillers are layered inorganic solids such as clay minerals, graphite and metal phosphates. Clay minerals such as montmorillonite (MMT), hectorite, saponite, and laponite have been proved to be very effective due to their unique structure and properties [7-9]. Thanks to the natural origin of clay, bio-based polymers can be reinforced with these clay minerals in order to enhance their mechanical and barrier properties while maintaining their biodegradability. In a same way, cellulose nanofibers or whiskers [10] are of high interest for bio-based polymers giving the opportunity to make use of renewable resources. The biopolymer-based nanocomposites with improved properties could potentially replace conventional packaging materials such as plastics obtained from oil (or petro-chemicals).

This chapter presents recent developments in bio-based polymers nanocomposites for packaging applications, analysis of their potentiality and discussion of the problems encountered with these emerging materials. The chapter starts with a brief introduction to definitions and categories of biopolymers, followed by detailed description of bio-based polymers of interest for packaging applications. Various techniques used for developing cost-effective bio-based, generally biodegradable, packaging materials with optimum mechanical and barrier properties are discussed. Whereas polymer blends have resulted in commercialization of bio-based compounds, bio-nanocomposites are the most promising way to further improve their properties.

4.2 Definitions and Classification

This chapter focuses on biodegradable polymers obtained from renewable resources. Indeed, the biodegradability may bring a solution to the waste management issue caused by the oil-based non biodegradable materials. Even if biodegradable polymers obtained from oil are also available their environmental impact is an issue because of CO_2 emissions coming from fossils resources.

According to ASTM D996-04 'biodegradable' is defined as: "capable of undergoing decomposition into carbon dioxide, methane, water, inorganic compounds, or biomass in which the predominant mechanism is enzymatic action of microorganisms, that can be measured by standard tests, in a specified period of time, reflecting available disposal [11]." Biodegradable plastics are polymeric materials in which at least one step in the degradation process is through metabolism in the presence of naturally occurring organisms. Under appropriate conditions of moisture, temperature and oxygen availability, biodegradation leads to fragmentation or disintegration of the plastics with no toxic or environmentally harmful residue [12].

Biodegradable polymers obtained from renewable resources presented as suitable matrices for bio-nanocomposites in the following can be classified according to their source:

- Polymers derived from renewable resources such as polysaccharides (starch and cellulose), proteins (wheat gluten and other proteins), and polylactic acid (PLA)
- Polymers produced by living organisms such as bacteria. For ex. polyhydroxyalkanoates (PHAs)

In the same way, the scope of this chapter will be limited to nanofillers of natural origin reducing the environmental impact of the materials and simplifying the waste management. For instance, carbon nanotube nanocomposites which have proven their high potential in terms of mechanical and electrical properties will not be considered.

4.3 Biopolymers Based Packaging Materials

4.3.1 Poly Lactic Acid (PLA)

Lactic acid can be obtained by the fermentation of carbohydrate material, usually glucose derived by hydrolysis from starch. The fermentation route can provide either enantiomer of lactic acid in high purity and dominates over chemical routes. The structure of lactic acid contains one asymmetric carbon, and can therefore exist as two stereoisomers. L-Lactic acid is present naturally in numerous organisms, whilst the mirror image D-lactic acid is very rare in nature.

Two methods are currently used to obtain polylactic acid, via polycondensation of lactic acid or via lactide (a dimer of lactic acid) ring opening. The direct synthesis of PLA by polycondensation features the typical drawbacks of step growth polymerization, i.e. achievement of low molecular weight PLA, unsuitable for most thermoplastic applications. High molecular weight polymers are obtained in relatively low yields, and these are very sensitive to the presence of impurities such as ethanol or acetic acid arising from the fermentation process. Nevertheless, high molecular weight PLA (300,000 g/mol) can be attained by employing highly pure lactic acid and removing the water formed during the polycondensation. Another solution has been provided by the use of chain extenders to couple oligomers to provide high molecular weight products.

The production of PLA by ring opening polymerization of lactic acid was started by Carothers in 1932 and was further developed by Dupont and Ethicon [13]. Since lactic acid can be produced by fermentation of carbohydrate by lactobacillus, PLA is considered a renewable material. Compared to the other biodegradable polyesters, PLA is a preferred product because of its availability and low cost. Cargill-Dow offers a series of PLA grades (NatureWorks®) manufactured using renewable agricultural resources such as corn or sugar beets. The company has the production capacity of 180,000 T/yr. Different companies such as Mitsui Chemicals (Japan) and Shimadzu (Japan) also manufacture PLA with smaller production capacity.

PLA has good mechanical and thermal properties similar to poly(ethylene-terephtalate) (PET) or polystyrene (PS) depending on the considered properties. However, properties of PLA are highly related to the ratio between two meso-forms (D and L). L-PLA has higher crystallinity, which can lead to higher melting temperatures and brittleness. Furthermore, PLA can be plasticized using polyethylene glycol (PEG), triethyl citrate (TC), and partial fatty acid esters [14-15]. PLA has moderate barrier properties (water vapor permeability and oxygen permeability) as compared to those of polystyrene (PS). However, high density, high polarity, poor heat resistance, and brittleness limit its use. PLA is currently used in packaging as films, thermoformed and blow molded containers, food service ware, and short shelf-life bottles [16] competing with PS and/or PET.

4.3.2 Starch Based Materials

Among the different bio-based polymers used in the industry, starch based polymers are specific because of their easy achievement directly by plasticizing of a renewable resource, the native starch, contrary to most of the other bio-based polymers which need expensive synthesis steps. Moreover, the use of starch as a polymeric material is a good opportunity for starch industry to extend the growth of non-food applications sector. Starch is generally extracted from corn, wheat, potato, cassava, tapioca and rice. It is a polysaccharide constituted by two different macromolecules, amylose and amylopectin, based on glucose units. The amylose is a linear homopolymer with a molecular weight ranging from 100,000 to 500,000 g/mol. The amylopectin is a highly branched polymer with very high molecular weight higher than millions. The amylose/amylopectin ratio depends on the starch botanical origin ranging from rich-amylose starch to rich-amylopectin starch (waxy corn). Native starch after extraction presents a granular structure composed by the alternating of amorphous and crystalline zone and cannot be used directly as a polymeric material. A preliminary step is needed to obtain the destructurization either by gelatinization or by applying thermal combined with mechanical energy using a continuous processing method (i.e. generally extrusion). In both cases a plasticizer is needed so as to break up the hydrogen-bonds existing between the macromolecules leading to a lowering of the melting and glass temperatures below the decomposition temperature. Among the different kinds of plasticizers, the most currently used are water and glycerol. The obtained material is homogeneous and mainly amorphous, even if some crystalline zones may remain. This polymeric material is usually named thermoplastic starch (TPS) or plasticized starch. TPS is a thermoplastic polymer which can be used like oil-based thermoplastics for different industrial applications, particularly in the packaging industry.

From the packaging industry point of view, starch based plastics represent a great potential because of their biodegradability, their combustibility but also the natural abundance and the renewability of starch. Moreover, due to its relatively low cost, it represents an attractive alternative to polymers based on petrochemicals (especially polyolefins). Since the beginning of its industrial exploitation in the plastics industry, starch plastics have became one of the most important polymers in the bio-based polymer market. In Europe, the production capacity of starch plastics increased from 30,000 T in 2003 to 130,000 T in 2007, representing an average annual growth of nearly 50% [17]. The most common applications of starch based polymers are for packaging industry with applications such as soluble films, films for bags and sacks, and loose fills. The most important starch materials producers are Novamont (60,000 T/yr, Italy) Rodenburg (40,000 T/yr, Netherlands) Biotec (20,000 T/yr, Germany), Limagrain (10,000 T/yr, France) and Cereplast (10,000 T/yr, USA) [17].

4.3.3 Poly Hydroxyalkanoates (PHA)

PHAs are the polymers of hydroxyalkanoates which are accumulated as a source of carbon or energy in various microorganisms under the condition of limiting nutritional elements. More than 300 different microorganisms are known to synthesize and accumulate PHAs [18]. The best known biopolymer types are the polyhydroxyalkanoates, mainly polyhydroxy-butyrate (PHB) and polyhydroxybutyratevalerate (PHBV). PHBV is a copolymer of hydroxyl-butyrate (HB) and hydroxyl-valerate (HV).

Poly-3-hydroxybutyrate (PHB) is a biopolyester accumulated as a reserve of carbon and energy by a number of bacteria. It is located in the cytoplasm in the form of granules of approximately 0.5 μm size. Under suitable conditions, up to 90%

polymer can be accumulated with respect to bacteria dry mass. Isolation of the PHB requires breaking the cell walls by means of mechanical shear or enzymatic digestion followed by extraction of the polymer. This can be performed by means of washing in a centrifuge. PHB was produced on the kilogram scale in the 1960s, but its stereo-chemical regularity led to progressive crystallization with aging, thus making it brittle. This has been overcome by incorporation of co-monomers by grafting or by the use of suitable formulations.

In the 1970s, PHBV (polyhydroxybutyrate-co-3-hydroxyvalerate) was successfully produced by using specific additives in the growth medium. Such an approach, whilst it improves the properties of PHB, is not cost effective, because the copolymer costs are higher, and its toxicity to the bacterium leads to lower production yields and also its presence affects PHB crystallization kinetics, which results in longer processing cycle times. Nevertheless, PHB could be toughened by the process of annealing by conditioning in an oven, a process that widens its application possibilities. By comparison to PHB, which melts at 180 °C, the melting point of PHBV can be lowered to 137 °C by the introduction of 25% hydroxyvalerate. This greatly improves thermoplastic processability. In addition, mechanical stability is improved by an order of magnitude.

PHAs are produced by Metabolix under the trademark Biopol™. Manufacturing of blow-molded bottles using Biopol™ for shampoo packaging was started by Wella AG (Germany) [19]. Other companies producing bacterial PHBV include PHB Industrial SA (Brazil) and Tianan (China). Recently, Procter and Gamble has begun to develop a large range of polyhydroxybutyrate co-hydroxyalkanoates with the trademark Nodax™. Packaging materials made from PHA possess excellent film forming and coating properties. PHAs have properties close to that of polypropylene (PP) [20]. The properties of the film can be adjusted by changing the ratio of HB and HV. A high content of polyhydroxybutyrate (PHB) gives a strong and stiff material whereas polyhydroxyvalerate (PHV) improves flexibility and toughness. Properties of PHBV properties can be improved by using plasticizers [21]. The polyalkanoates are more hydrophobic than polysaccharide-based materials resulting in their better moisture barrier properties. PHAs are biodegradable in soil and have excellent processability. Higher cost of production, brittleness, and poor gas barrier properties limit the use of PHAs [22]. Several processes for producing PHA from cheap carbon sources have been developed which have been reviewed by Lee *et al.* [18].

4.3.4 Proteins

One of the most interesting protein is wheat gluten (WG) which is a low cost byproduct of the wheat starch industry and represents thus a cost effective opportunity for packaging industry [23]. It has very interesting viscoelastic properties and low water solubility. Gluten is a mixture of two main proteins, gliadins and glutenins. Apart from wheat gluten, soy, pea and whey proteins among all the protein sources have attracted attention for bio-based packaging materials because of their excellent film forming properties. As it is the case for starch or gluten, useable films are only obtained after the addition of plasticizers leading to a significant decrease of the mechanical properties and high sensitivity to water which may be overcome using nanofillers.

4.4 Structure of Bio-nanocomposites

4.4.1 Bio-nanocomposites for Packaging Applications

The mechanical and thermal properties of nanocomposites are the key factors in designing the material for packaging application. Adequate mechanical strength throughout the service life of a packaging application is necessary to ensure the integrity of a film. Thermal properties are important not only because of polymer processing technologies but also because of food preparation conditions (sterilization steps), storage conditions (for freeze packed food), and cooking conditions (in the case of microwave packed food). As previously mentioned, the main purpose of packaging is not only to protect the product from its surroundings, but also to maintain the quality of the product for its shelf-life, while addressing communication, legal and commercial demands [24].

The barrier properties of packaging materials are as important to the application as are the thermo-mechanical properties. A key characteristic of glass and metal packaging materials is their high barrier properties to gases and vapors. While polymers can provide an attractive balance of properties such as flexibility, toughness, lightweight, formability and printability, they do allow the transport of gases and vapor to some extent. The selection of a barrier polymer for a particular application typically involves tradeoffs between permeation, mechanical and aesthetic properties as well as economic and recycling considerations [25, 26]. Quality and shelf-life are reduced when the packaged product, through interactions with the outside environment, gains or looses moisture or aroma, takes up oxygen (leading to oxidative rancidity) or becomes contaminated with microorganisms.

Reinforcement using nanofillers to enhance the polymer performance has been a subject of interest in recent times. Nanofillers currently being used are layered silicates [27], cellulose nanowhiskers [28, 29], ultra fine layered titanate [30], and carbon nanotubes [31]. Among these, the natural nanofillers, layered silicates such as clay, cellulose nanofibers or starch nanocrystals have attracted great attention by the packaging industry because of their environmental friendliness, natural abundance, and their potential for improving in-use properties of packaging materials.

Nanocomposites exhibit remarkable enhancement in properties with very low nanofiller content ($\leq$ 5wt%). These improvements can include high tensile modulus, increased strength and heat resistance, and superior barrier properties. Nanocomposites also offer other benefits such low density, transparency, better surface properties, and recyclability [26]. Potential applications of these nanocomposites include automobiles (gasoline tanks, bumpers, interior and exterior panels etc.), construction (building sections, structural panels), aerospace (flame retardant panels, high performance components), electronics (printed circuit boards, electric components), and pigments. In order to take advantage of their substantially enhanced properties, polymer nanocomposites have also been studied for food packaging applications including injection blow-molded bottles for beverage or beer, coatings for paperboard juice cartons, and cast and blown films [32]. Use of these nanocomposites for food packaging with oxygen scavenging, reduced flavor scalping, increased heat resistance, and better gas barrier properties has resulted in shelf-life of 3 to 5 years for packaged food [33].

In contrast with polymer-based nanocomposites, biopolymer-based nanocomposites (bio-nanocomposites) have received little attention. However, several research groups have reported preparation and characterization of various kinds of bio-nanocomposites showing potential for a wide range of applications. In the present chapter, the review is restricted to natural nanofillers. In the subsequent sections, a detailed discussion on the properties of renewable resources based polymer nanocomposites will be presented.

4.4.2 Structure of Nanocomposites Based on Natural Nanofillers

4.4.2.1 *Layered Silicate Filled Nanocomposites*

The most commonly used silicates in the preparation of polymer nanocomposites are clays such as montmorillonite (MMT), hectorite and saponite, and their various modifications. These layered silicates belong to the general family of 2:1 layered silicates or phyllosilicates [26]. Their crystal lattice structure consists of two-dimensional, 1 nm thick layers which are made up of an octahedron sheet of aluminum sandwiched in between two tetrahedral sheets of silicon (Figure 4.1). The lateral dimensions of these layers vary from 30 nm to several microns or larger, depending on the particular layered silicate. Stacking of the layers leads

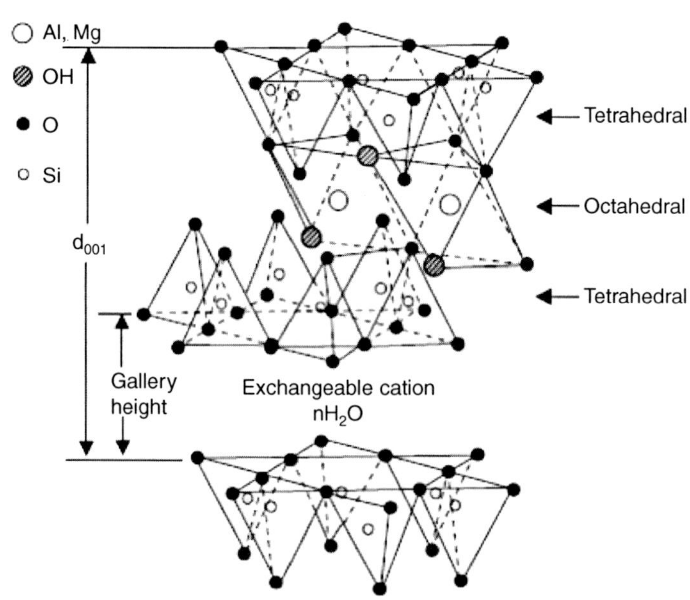

Figure 4.1 Structure of layered silicates.

to a regular van der Waals gap between the layers called the interlayer or gallery. Isomorphic substitution within the layers (Al^{3+} replaced by Mg^{2+} or Fe^{2+} in octahedron sheet, or Si^{4+} replaced by Al^{3+} in tetrahedron sheets) results in net negative charge that are counterbalanced by alkali and alkaline earth cations such as Na^+ residing in the galleries.

In natural layered silicates, the interlayer cations are usually hydrated Na^+ or K^+, showing hydrophilic surface properties. In this natural state, layered silicates are only miscible with hydrophilic polymers. To render layered silicates miscible with hydrophobic polymers, one must convert the hydrophilic silicates surface to an organophilic one. This is done by ion-exchange reactions with various organic cations (e.g. alkylammonium cations, cationic surfactant etc.) leading to organo-modified layered silicates (organoclay). The organic cations lower the surface energy of the silicate surface and result in a larger interlayer spacing. Additionally, the organic cations may contain various functional groups that react with the polymer to improve interaction between the silicates and the polymer matrix.

Layered silicates have a very high aspect ratio (e.g. 10–1000). A low weight percent of layered silicates that are properly dispersed throughout the polymer matrix thus create much higher surface area for polymer/filler interaction as compared to conventional composites. Depending on the surface properties, level of dispersion and the strength of interfacial interactions between the polymer matrix and layered silicate (modified or not), three different types of polymer/layered silicate composite microstructure are achievable (Figure 4.2).

(i) Phase separated microcomposites: conceptually the unmodified silicate layers are stacked together and the polymer molecules cannot penetrate into the galleries. The silicates are a kind of fillers that stay as agglomerates.
(ii) Intercalated nanocomposites: the insertion of a polymer matrix into the layered silicate structure occurs in a crystallographically regular fashion, regardless of the clay to polymer ratio. Intercalation occurs when a small amount of polymer penetrates into the galleries, resulting in finite expansion of the silicate layers. This leads to a well-ordered multilayered structure with a repeat distance of a few nanometers.
(iii) Exfoliated nanocomposites: the individual clay layers are separated in a continuous polymer matrix by an average distance that depends on clay loading. Usually, the clay content of an exfoliated nanocomposite is much lower than that of an intercalated nanocomposite.

The complete dispersion of clay platelets in a polymer optimizes the number of available reinforcing elements for carrying an applied load and deflecting cracks. The coupling between the tremendous surface area of the clay and the polymer matrix facilitates stress transfer to the reinforcement phase, allowing for such mechanical improvements. In addition, the impermeable clay layers mandate a tortuous pathway for a permeant to transverse the nanocomposites. The enhanced barrier properties, chemical resistance, reduced solvent uptake,

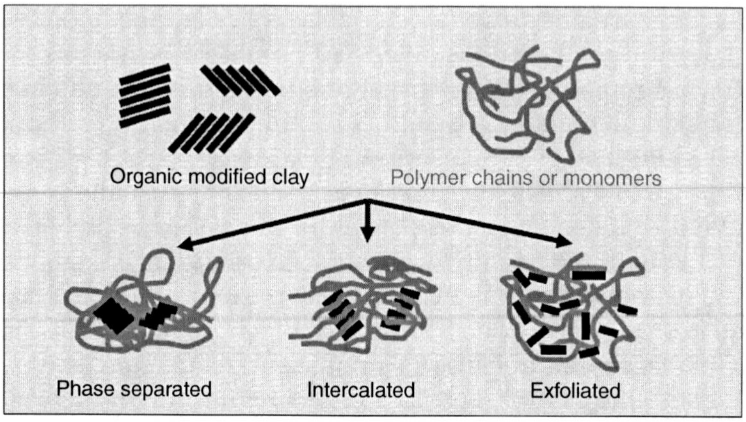

Figure 4.2 Micro-/nano-sctructures of polymer:layered silicate composites.

and flame retardancy of polymer/clay nanocomposites result from the hindered diffusion pathway through the nanocomposites.

Different approaches may be used to prepare polymer/clay nanocomposites: in-situ polymerization, solution and melt intercalation. This last preparation method is the most interesting as it makes use of regular polymer processing equipment and has the greatest industrial potential.

4.4.2.2 Cellulose Nanoparticles Filled Nanocomposites

Cellulose is one of the most naturally abundant polymer and derived from annually renewable resources. The most exploited natural resource containing cellulose is wood but other plants also contain a large amount of cellulose, including hemp, flax, jute, ramie and cotton. Some other non-plant sources of cellulose exist; for instance, cellulose produced by bacteria (bacterial cellulose BC) and cellulose produced by tunicates (Tunicin). Compared to layered silicates, cellulose nanofibers or whiskers have advantages such as their renewability, low cost, low density, high specific strength and modulus, very high aspect ratio (100–1000), easy processability (non abrasive filler). However, some drawbacks have to be considered such as low thermal stability and low production yield. The low thermal stability is an issue when melt blending for nanocomposites elaboration needs high processing temperature.

Considering the microstructure of wood or plants, cellulose is found in the cell walls and is a polysaccharide based on a saccharide unit named glucose. Cellulose has a monoclinic primitive crystalline cell constituted by two cellulose macromolecular chains. These primitive cells of the cellulose connect to each other in order to form supramolecular structures being the elementary fibril having a cross section of 3.5×3.5 nm. It contains around 40 cellulose chains. These elementary fibrils are gathered in a larger entity called microfibril, whiskers or nanoparticle. Its diameter goes up to 30 nm depending on the considered plant or wood.

Thanks to the crystalline structures of cellulose, nanofibers have a high strength in the direction of the chain axis. Young's modulus of the nanofibers has been

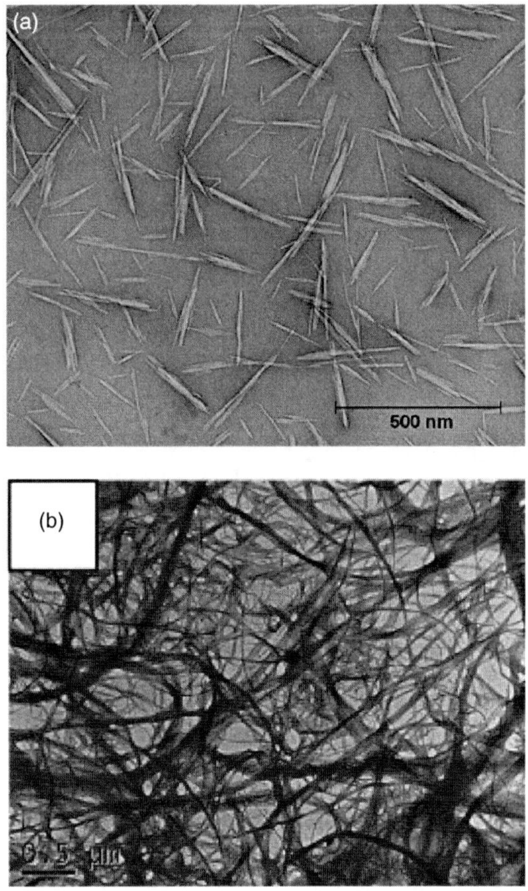

Figure 4.3 TEM images of a) cellulose nanowhiskers obtained from sisal by chemical treatment [37] and b) a dispersion of parenchymal cell cellulose after mechanical treatment [38].

evaluated within the range 130–170 GPa [34–36]. The extraction of the nanofibers from natural fibers may be obtained using different type of extraction methods based on chemical (acid hydrolysis), mechanical (homogenization, microfluidization) and combination of mechanical and chemical or enzymatic treatments. Depending on the chosen method, the properties and the aspect ratio of the nanofibers may change (Figure 4.3a,b). However, for every method the yield is limited and the production time consuming, and it still has to be optimized to be used in an industrial production.

Apart from cellulose nanofibers, similar polysaccharides based nanofibers can be extracted from other sources (outer skeleton of insects, crabs, shrimps or mantle of tunicates) such as chitin or tunicin.

4.4.2.3 *Starch Nanocrystals Filled Nanocomposites*

As already discussed, native starch has a granular structure. These starch granules are composed by alternating crystalline and amorphous zones. This structure is

Figure 4.4 TEM images of individual waxy maize starch nanocrystals obtained after 6 weeks of acid hydrolysis (scale bars: 50 nm) [40].

highly complex and its description is still incomplete. Using a similar approach as the one used for cellulose it has been possible to extract nanocrystals from native starch. Dufresne *et al.* first reported a method for producing "microcrystalline starch" which is claimed to be agglomerated particles of a few tens of nanometres in diameter [39]. These nanocrystals were obtained using an acid hydrolysis treatment. Putaux *et al.* [40] first reported the morphology of starch nanocrystals resulting from the disruption of the waxy maize starch granules by acid hydrolysis. TEM observations (Figure 4.4) showed (a) a longitudinal view of lamellar fragments consisting of a stack of elongated elements with a width of 5–7 nm and (b) a planar view of an individualized platelet after hydrolysis.

Shapes and lateral dimensions were derived from the observation of individual platelets in planar view: a marked 60–65° acute angles for parallelepipedal blocks with a length of 20–40 nm and a width of 15–30 nm. However, the shape and the size of starch nanocrystals is related to the starch origin. Contrary to cellulose nanofibers, starch nanocrystals structure is not 100% crystalline but rather 45% crystalline depending on the botanic origin. As in the case for cellulose nanofibers, the main drawback of this natural nanofiller is its low thermal stability, which limits its use for elaboration of nanocomposites using melt blending techniques. Moreover, starch nanocrystals are still at the early research step and are far from industrialization.

4.5 Properties of Bio-nanocomposites

Success of the nanocomposite concept in the area of synthetic polymer has stimulated new research on nanocomposites based on biodegradable bio-based polymers matrices. So far, the most studied bio-nanocomposites are based on polylactide (PLA), thermoplastic or plasticized starch, polyhydroxy-alkanoates (PHA) and proteins (wheat gluten and others proteins). In this section, the properties will be discussed individually.

As already mentioned previously, one way to improve the performances of bioplastics is to develop nanostructured materials by adding nanofillers. This kind of nanocomposite is known to improve greatly the in-use properties of the polymeric materials and particularly the barrier and mechanical properties by adding a very low content of nanofiller allowing preserving the optical properties (i.e. transparency) of the product. Nanocomposites represent thus a great opportunity to extend the possibilities of packaging applications by limiting the main problems encountered with bio-based polymers, for instance the lower mechanical properties, or high hydrophilicity. Moreover, it is interesting to notice that a wide part of the nanofillers generally used for polymer nanocomposites elaboration are based on natural components such as clay platelets and cellulose nanofibers which are the most commonly used. The resulting nanocomposite is then fully based on natural resources being of great interest in terms of environmental footprint optimizing the exploitation of renewable resources to make a packaging which will be biodegradable, limiting its impact on the environment.

4.5.1 PLA Based Bio-nanocomposites

Use of nanoclay as a reinforcement agent has the potential to expand the application of PLA. Large numbers of studies [41–55] have reported preparation and characterization of PLA-clay bio-nanocomposites. For example, Ogata *et al.* [41] prepared PLA-organoclay blends by dissolving the polymer in hot chloroform in the presence of dimethyl disteary ammonium modified montmorillonite (MMT). The results showed a strong tendency of tactoids formation by solvent-cast method. Bandyopadhyay *et al.* [42] reported the successful preparation of PLA/organoclay nanocomposites by melt extrusion with much improved thermal and mechanical properties. Ray *et al.* [45–47] used melt extrusion for the preparation of PLA/organoclay bio-nanocomposites with improved properties. XRD patterns and TEM observations established that the silicate layers were intercalated and randomly distributed in the bio-nanocomposite matrix. The intercalated bio-nanocomposites exhibited significant improvement in properties in both solid and melt states as compared to those of PLA matrices without clay. Table 4.1 summarizes the tensile properties, and when available, other mechanical and barrier properties of PLA based nanocomposites

4.5.1.1 *Mechanical Properties*

Maiti *et al.* [43] prepared a series of PLA/layered silicate bio-nanocomposites with three different types of natural and modified layered silicates (saponite, MMT, and synthetic mica). Layered silicates were modified with alkylphosphonium salts having different chain lengths. The study showed that miscibility of an organic modifier (phosphonium salt) and PLA is enhanced as the chain length of modifier is increased. These authors also studied the effects of dispersion, intercalation, and aspect ratio of the clay on the material properties.

Alternatively, bio-nanocomposites of blends of PLA and polycaprolactone (PCL) were obtained by melt mixing with a modified kaolinite [53]. Blending of

Table 4.1 Tensile and other properties of PLA/Clay nanocomposites.

Authors	Filler type	Plasticizer	Preparation Method	Tensile properties (Relative increase/decrease compared to neat polymer matrix)			Other properties
				Modulus, MPa	Strength, MPa	Elongation at Break, %	
J H Chang *et al* [62, 69]	C_{16}-MMT (4 wt.%)	–	Solvent casting	285 (37%)	28 (47%)	1146 (35%)	O_2 barrier (cc/m^2/day) at 23°C: 449 (–46%)
Ratto *et al* [70]	Modified MMT, Cloisite* 25A (5 wt.%)	Citroflex A–2 (10 wt.%)	Melt mixing	NR	30 (36%)	8 (100%)	O_2 barrier (cc/m^2/day) at 23°C: 425 (–50%) WVP[g]:125 (–50%)
Hassok *et al* [71]	Modified MMT (4.76 wt.%)	–	Melt mixing	4448 (19%)	40 (–12%)	1.74 (–13%)	–
	Modified MMT (4.76 wt.%)	PCL[a] (4.7 wt.%)	Melt mixing	4237 (13%)	40 (–12%)	2.0 (0%)	
Li *et al* [72]	Modified MMT, Cloisite* 20A (5wt.%)	CSR[b] (10wt.%)	Melt mixing	2068 (14%)	56 (–8%)	3.9 (–65%)	Izod impact strength (kJ/m^2): 2.24(2%)
	Modified MMT,Cloisite* 20A (5wt.%)		Melt mixing	1525 (–15%)	56 (–8%)	11 (66%)	Izod impact strength (kJ/m^2): 4.2 (81%)

	Modified MMT, Cloisite* 30B(5wt.%)		Melt mixing	2183 (21%)	56.6 (–7%)	4.46 (–32%)	Izod impact strength (kJ/m²): 2.10(–4%)
	Modified MMT, Cloisite* 30B(5wt.%)	CSR[b](10wt.%)	Melt mixing	1916 (7%)	49.6 (–18%)	7 (0%)	Izod impact strength (kJ/m²): 3.37 (53%)
Shibota *et al* [75]	Natural MMT (3wt.%)	PL710[c](10wt.%)	Solvent casting	3500 (34%)	68 (5%)	25 (30%)	–
Jiang *et al* [73]	Natural MMT (2.5wt.%)	–	Melt mixing	60 (9%)	67 (8%)	7 (300%)	–
Solarski *et al* [74]	O-Bentonite (4wt.%)	–	Fibre spinning	6500 (0%)	350 (–17%)	-NR-	–
Jiang *et al* [75]	Natural MMT(5wt.%)	PBAT[d](5wt.%)	Melt mixing	3950 (15%)	53 (–17%)	18 (350%)	–
E. Nieddu *et al* [76]	Modified MMT, Cloisite* 30B (5 wt.%)	–	Melt mixing	1981(45%)	35 (–35%)	2.1 (–130%)	
Ozkok and Kemaloglu [77]	Natural MMT (3wt.%)	–	Melt mixing	1884 (34%)	32 (–4%)	2.76 (–43%)	Water uptake (%) :6.1 (100%)
	Natural MMT (3wt.%)	PEG[e](20wt.%)	Melt mixing	1124 (–20%)	24.95 (–26%)	63.2 (1150%)	Water uptake (%) :15 (350%)
Wu *et al* [78]	Natural MMT (8wt.%)	–	Solvent casting	178.8 (–88%)	5.04 (–300%)	35.07 (1000%)	–

Table 4.1 (cont.) Tensile and other properties of PLA/Clay nanocomposites.

Authors	Filler type	Plasticizer	Preparation Method	Tensile properties (Relative increase/decrease compared to neat polymer matrix)			Other properties
				Modulus, MPa	Strength, MPa	Elongation at break, %	
Rhim *et al* [79]	Modified MMT, Cloisite* 20A (2 phr)	–	Solvent castin	NR	53 (5%)	3.2 (25%)	WVP[g]:1.5 (–5%)
Balakrishna *et al* [62]	Natural clays (2 phr Clay)	–	Melt mixing	3500 (20%)	55 (–8%)	NR	Flexural modulus (MPa): 3500 (20%) Izod impact strength (J/m): 38 (–4%)
	Natural clays (2 phr)	LLDPE[f]	Melt mixing	2800 (–6%)	43 (–25%)	NR	Flexural modulus (MPa): 3000 (10%) Izod impact strength (J/m):40 (0%)

*Cloisite, Trademark from Southern Clay Product
[a]Poly(caprolactone)
[b]Core shell rubber particles
[c]Diglycerine tetraacetate
[d]Poly(butylene adipate-co-terephthalate)
[e]Linear low density poly (ethylene)
[f]Poly(ethylene glycol)
[g]Water vapour permeability
NR: Not reported
phr: Part per hundred resin

PCL was aimed at decreasing the brittleness of PLA. Bio-nanocomposites with 4wt% modified kaolinite showed better processability, thermal stability, and improvement in mechanical properties as compared to the polymer and blends without clay.

Lee *et al.* [54] reported the MMT content dependence of tensile modulus of Poly(L-lactide acid) (PLLA) nanocomposites scaffolds. The authors suggested that the layered silicates of MMT could act as a mechanical reinforcement of polymer chains. The crystallinity and the glass transition temperature of PLLA nanocomposites were lower than neat PLLA, but the modulus was significantly increased due to the addition of the clay. The glass transition temperature depends primarily on chain flexibility, molecular weight, branching/crosslinking, intermolecular attraction and steric effects, etc [55–56]. Due to their low crystallinity and glass transition temperature, the nanocomposites system seems to be disturbed by the charged MMT layers, and subsequently the PLLA backbone chains additionally gain segmental mobility.

Dynamic mechanical analysis (DMA) of melt blended polylactide PLA/layered silicate nanocomposites plasticized with 20 wt% of 1000 g/mol poly(ethylene glycol) (PEG) has been reported by Pluta *et al* [57]. Three kinds of commercial organo-modified montmorillonites such as Cloisite™ (20A, 25A, 30B from Southern Clay Product, the most important modified and unmodified clay provider for polymer nanocomposites applications) were used as fillers at a concentration level varying from 1–10 wt%. The dynamic mechanical properties were reported to be sensitive to the sample composition. Generally, the storage modulus increased with the filler content. Glassy PEG, well dispersed within unfilled PLA matrix, also showed reinforcing effect, since the storage modulus of this sample was higher than for unplasticized reference at temperatures below the glass transition of PEG. Moreover, loss modulus of all plasticized samples revealed an additional maximum ascribed to the glass transition of PEG–rich dispersed phase, indicating partial miscibility of organic components. This mechanical loss also occurred within plasticized nanocomposites and exhibited an increasing tendency with the filler content. This increase was somewhat correlated with the intercalation magnitude, giving an evidence on the ability of the PEG molecules to penetrate the silicate gallery. Same authors, prepared PLA-based systems composed of an organoclay (Cloisite 30B™) and/or a compatibilizer (maleic anhydride) by melt blending [58]. The X-ray investigations showed the presence of exfoliated nanostructure in 3 wt% MMT nanocomposite. The reported results indicated that compatibilization noticeably enhanced the degree of exfoliation of the organoclay due to combined interactions of the organoclay surfactant with polylactide chains and maleic anhydrite groups of the compatibilizer. In the 10 wt% MMT nanocomposite, mixed – intercalated and exfoliated nanostructures were detected due to high concentration of the filler. Rheological properties suggested a sort of silicate network formation.

Shibota *et al.* [59] prepared using melt intercalation method nanocomposites made of poly(lactide) (PLA) and plasticized-PLA with diglycerine tetraacetate and ethylene glycol oligomer containing montmorillonite organo-modified by the protonated ammonium cations of octadecylamine (ODA-MMT) and poly(ethylene glycol) stearylamine (PGS-MMT) [42]. The PLA and plasticized-PLA composites containing ODA-MMT showed a higher tensile strength and modulus than the corresponding composites with PGS-MMT. The plasticized-PLA (10 wt%) composite containing ODA-MMT showed considerably higher elongation at break than the pristine plasticized PLA, and had a comparable tensile modulus to pure PLA.

Fukushima *et al.* [60], reported the effect of addition of different types of clays on the thermal and mechanical properties of PLA. The results indicate that nanocomposites filled Cloisite 30B™ (Southern Clay Product) showed better thermo-mechanical properties than that of Nanofil 804™ (Süd-Chemie) filled nanocomposites because of the high levels of dispersion of the former clay into the PLA compared to the latter as observed in TEM. Nanocomposites based on Cloisite 30B™ show indeed a high level of intercalation and exfoliation of the silicate layers, as small stacks of swollen clay layers and single dispersed layers can be observed in the TEM micrograph. The materials based on Nanofil 804™ show certain level of intercalation as well as the occurrence of micro-aggregates of the silicate layers.

Recently, Balakrishna *et al.* [61] developed novel nanocomposites of PLA/organo-modified montmorillonite (MMT) toughened using linear low density polyethylene (LLDPE) by melt mixing technique. XRD and TEM studies revealed an intercalated structure in LLDPE toughened PLA nanocomposite. The mechanical properties such as the Young's and flexural modulus improved with increasing loadings of MMT and the impact strength of PLA and PLA/MMT nanocomposites increased with addition of LLDPE as an impact modifier. However, the tensile and flexural strengths decreased with addition of MMT and LLDPE. Thermal analysis through differential scanning calorimetry (DSC) revealed that the crystallization temperature (Tc) of PLA in both PLA/MMT and LLDPE toughened PLA/MMT nanocomposites and decreased with increasing content of MMT, which is an indication of nucleating effect of MMT. Thermogravimetric analysis (TGA) revealed that the incorporation of MMT and LLDPE had improved the thermal stability of PLA in both PLA/MMT and LLDPE toughened PLA/MMT nanocomposites, respectively. DMA analysis showed that the storage modulus (E') improved with increasing content of MMT below and above Tg due to the reinforcing effect of MMT in both PLA/MMT and LLDPE toughened PLA/MMT nanocomposites.

4.5.1.2 Barrier Properties

Effect of different kinds of organically modified layered silicates on the oxygen gas permeability of PLA nanocomposites (prepared by melt intercalation) has been studied by Chang [62] *et al.* An increase of the tortuous paths in nanocomposites is observed in the presence of the clay. The permeability value of the nanocomposites decreased to half of the PLA one, regardless of the nature of organically modified layered silicates. This was attributed to the increase in the lengths of the tortuous paths in nanocomposites in the presence of high clay content [63]. Whereas it was clear that polymer nanocomposites show enhanced barrier properties, the dependence on factors such as the relative orientation and dispersion (intercalated, exfoliated or some intermediate) is not still well understood. Later, Bharadwaj [64] addressed the modeling of barrier properties in polymer/layered silicates nanocomposites, and also gave explanations regarding the state of delamination of the sheets in the polymer matrix. Exfoliation appears to be the critical factor in determining the maximum performance of polymer nanocomposites for barrier applications. A significant decrease in the permeability of O_2 gas through the nearly exfoliated PLA-nanocomposite relative to the neat polymer (PLA) was clearly observed, which is based completely upon the tortuosity arguments described by Nielsen [65].

Gusev and Lustic [66] deduced a rational design of nanocomposites for barrier applications. They considered that the presence of high aspect-ratio atomic-thickness nanoplatelets can lead to some molecular level transformations in the polymer matrix. It would be very interesting to understand the effect of the changes in the local gas permeability coefficients on the overall barrier properties of the nanocomposites. The favorable interactions between PLA and silicate layers gave disordered intercalated system of PLA/saponite [67]. As a result of the formation of phosphonium oxide by the reaction between the hydroxyl edge group of PLA and alkylphosphonium cation, it was concluded that the barrier property of PLA/saponite shot up unexpectedly higher than that of the other systems [67].

Influence of different types of montmorillonite nanofillers (Cloisite 30B™ and Nanofil 2™, Southern Clay Product), with two kinds of organic modifiers (poly(methyl methacrylate) and ethylene/vinyl alcohol copolymer) and two types of compatibilizers (polycaprolactone and poly(ethylene glycol)) on transmission rates of water vapor, oxygen, and carbon dioxide through polylactide films has been reported by Marien and Richert [68]. Cloisite 30B™ decreased the film permeability much more than Nanofil 2™. All the modifiers and compatibilizers reduced the carbon dioxide transmission rate, while only the modifiers reduced the transmission rates of water vapor and oxygen. The sample containing 75, 5, and 20 wt% of polylactide, Cloisite 30B™, and poly(methyl methacrylate), respectively, has been shown to be the film with the best barrier properties among 27 studied materials. The permeabilities of this film to water vapor, oxygen and carbon dioxide decreased by 60, 55 and 90%, respectively, as compared to those of the neat polylactide film. Unfortunately, the authors did not explain the physical and chemical processes associated with the transmission rates of various substances through the polylactide nanocomposite films.

4.5.2 Starch Based Nanocomposites

Thermoplastic starch (TPS) is unsuitable for a broader range of applications due to various drawbacks, the most important ones being its brittleness and hydrophilicity. This section will focus on the elaboration of starch-based nanocomposites using different natural nanofillers and on the properties of interest for packaging applications meaning mainly barrier, mechanical, surface and optical properties.

One of the advantages of starch-based materials is that they have a very low processing temperature or may be obtained using a gelatinization process. Thus, it extends the possibly used nanofillers to the ones of lower thermal stability such as cellulose nanofibers or chitin fibers, and starch nanocrystals in addition of classically used clay nanofillers (montmorillonite and sepiolite).

4.5.5.1 *Elaboration Processes*

The most widely studied nanofiller for the compounding starch-based nanocomposites is montmorillonite. Its advantage is to be thermally stable allowing the nanocomposites to be elaborated using extrusion process even if solution casting method is also reported in the literature.

The solution casting method has been used to elaborate some starch-based nanocomposites. The method is based on the gelatinization of starch by water. Classically, starch powder, clay and eventually the plasticizer are dispersed in water. The aqueous mix is then boiled during the chosen time. Thanks to the natural hydrophilicity of clay, this method was successfully implemented by different authors to disperse unmodified montmorillonite [80–83]. This method leads essentially to intercalation with few exfoliations. However, the type of starch may influence the dispersion of unmodified montmorillonite using solution casting as evidenced by Mondragon *et al.* [84]. The dispersion is better in the case of waxy maize starch. The authors proposed that the highly branched structure of amylopectin is more favorable for the diffusion of macromolecules within the interlayer galleries compared to the highly linear, and thus entangled, structure of amylose. Besides, Pandey *et al.* [80] showed the influence of the sequence of addition of the components (starch, clay and plasticizer) during the gelatinization process. It appears that the diffusion of the plasticizer within the clay gallery is easier than the diffusion of starch macromolecules.

The elaboration of starch/clay nanocomposites is possible by a solution casting method using either unmodified or modified clays. However, rich amylopectin and modified montmorillonite with a polar surfactant seem to favor exfoliation. Moreover, the competition between the intercalation of the plasticizer and the starch macromolecules may lead to poor dispersion. Consequently, it is better to intercalate first starch macromolecules and add the plasticizer in a second step to obtain better dispersion of the clay platelets [80, 83, 85]. Even if solution casting is a method allowing elaborating starch-based nanocomposites films, melt blending is a more efficient method from an industrial point of view.

Besides standard extrusion compounding, which is widely used in the packaging industry for the production of materials and films, the melt blending of starch-based nanocomposites can be carried out using an internal mixer at the laboratory scale to study and optimize the elaboration process. Park *et al.* [86–87] prepared TPS/clay nanocomposites using a roll mixer. Dried TPS and clays were mixed in the mixer at 110 °C, 50 rpm for 20 minutes. Chivrac *et al.* [88] optimized this method to study more deeply the elaboration step of starch/clay nanocomposites. The native starch was dried to remove water and then dry mixed with the plasticizer (glycerol) using a turbomixer at high rotation speed (1700 rpm) to obtain a homogeneous dispersion. The mixture was then heated (170 °C during 40 min) to eliminate water and favor diffusion of glycerol into the starch granule. The clay was dispersed in water to obtain a swollen clay. It was blended with the dry mix in the internal mixer (70 °C for 20 min, 150 rpm). This method is efficient to simulate what may be obtained by extrusion compounding even if the shearing applied is lower (which can be compensated using longer residence times).

Some authors also studied the possibility to compound starch/ clay nanocomposites using extrusion which is more appropriate for continuous production of polymer based materials. Both single screw and twin screw extruders have been used for the elaboration. Single screw extruder is more commonly used for industrial production when twin screw is generally preferred for the elaboration of polymer nanocomposites as it allows improving the dispersion of the nanofillers because of a higher shear.

Huang *et al.* in 2006 [89] proposed a two steps processing to elaborate starch-based nanocomposites. First the plasticizers (urea and formamide) are premixed with corn-starch using a high speed mixer and are transferred to the single screw extruder (screw ratio L/D 25:1) to obtain plasticized starch pellets. In a second step the plasticized starch pellets are mixed with the citric acid activated montmorillonite prior to extrusion with the same single screw extruder leading to homogeneously plasticized starch/clay nanocomposites. The montmorillonite layers are well dispersed in the starch matrix thanks to the good affinity between the citric acid activated montmorillonite and starch. Wang *et al.* [90, 91] proposed a similar two steps approach. However, in their studies, the MMT is activated using the plasticizer, namely glycerol, using either a high speed mixer (HSM) or manual mixing. It appears that the pre-processing step is crucial to improve the dispersion. Moreover, citric acid added during the first step allows increasing the plasticization of TPS and dispersion of MMT in nanocomposites. Dai *et al.* [92] chose the same approach to elaborate starch/clay nanocomposites using N-(2-Hydroxyethyl)formamide (HF) which act as both plasticizer for TPS and swelling agent for MMT. From the structural analysis (SEM), it appears that starch granules were completely disrupted and a continuous phase was obtained. Partially exfoliated TPS/MMT nanocomposites were formed as shown by the atomic force microscopy (AFM) analysis. Ma *et al.* [93] proposed an original dual melt extrusion processing method. First the sorbitol is blended with montmorillonite using a high speed mixer; the obtained mixture is extruded using a single screw extruder and then pulverized using a disk-mill. In a second step, this MMT-sorbitol powder is mixed with starch and plasticizer (sorbitol and formamide) in a high speed mixer and manually fed into the single screw extruder to obtain the starch nanocomposites. This dual-melt extrusion is a novel and effective processing method to prepare starch-based nanocomposites without any organic modification of MMT.

Finally, other authors [94–97] also considered the possibility to use twin screw extrusion, which is generally known to promote a better dispersion of nanofillers compared to single screw extrusion because of a higher shearing and the possibility of using specific screw design optimized for nanofillers dispersion [98].

4.5.2.2 *Effect of the Surfactant and Plasticizer on the Structure*

Considering the clay filled polymer nanocomposites, it is well admitted that the role of the surfactant is essential to promote the better dispersion of clay platelets.

The surfactant allows improving the dispersion by enlarging the gallery height between the clay platelets and by providing better interaction with the polymer matrix. However, the choice of the appropriate surfactant has to be done carefully otherwise it will lead to a microcomposite or a nanocomposite with few exfoliations. In the case of starch based materials, when a plasticizer has to be used to get a plasticized starch, the situation is much more complex because the plasticizer may also act as a surfactant. Thus, the role of the surfactant in the presence of a plasticizer is an issue of high interest studied by different authors.

As a pioneers in the elaboration of starch/clay nanocomposites studied Park *et al.* [86–87] the influence of the surfactant on the dispersion of the montmorillonite platelets. Different commercial organo-modified montmorillonites (Cloisite™, Southern Clay Product) were used. The different Cloisite™ were tested with surfactants of different polarity. Cloisite 30B™ has the highest polarity and Cloisite 10A/6A™ the lower polarity. Natural montmorillonite (Cloisite Na+™), which is a highly hydrophilic unmodified montmorillonite, was also tested. TEM combined with XRD analysis demonstrated that the nanocomposite filled with unmodified montmorillonite (Cloisite Na+™) exhibits a multilayer nanostructure whereas the nanocomposites containing modified montmorillonite (Cloisite™ 30B, 10A, 6A), because of a lack of compatibility between TPS and organoclays, present neither intercalation nor exfoliation and large particle agglomerates. The possibility to obtain exfoliation or at least intercalation is correlated to the compatibility and interaction between polymer, silicate layer, and the ammonium cations located within the gallery. Because of the polar interactions existing between the hydroxyl group of the TPS chain and the silicate layer of Cloisite Na+™, TPS chains have a driving force to intercalate into the interlayer spacing of Cloisite Na+™. In the case of the Cloisite™ 6A, 10A, because of the hydrophobicity of the surfactant, unfavorable interactions with TPS chains exist and limit the possiblity of intercalation. For Cloisite 30B™, because of the presence of hydroxyethyl groups on the surfactant, more favorable interactions may be developed with TPS chains. However, it also enhances the interaction between the surfactant and the silicate surface. As a result, replacement of the surface contacts by TPS chains will be less favorable, impeding the extensive intercalation explaining that no exfoliation is observed. Majdzadeh-Ardakani *et al.* [85] observed a similar trend comparing nanocomposites containing Cloisite 30B™ and Cloisite Na+™. Chiou *et al.* also demonstrated the negative effect on the dispersion of the hydrophobic surfactant using rheological measurements. It appears clearly that starch gel based on highly hydrophilic unmodified montmorillonite (Cloisite Na+™) exhibits a gel-like behavior when those based on modified montmorillonite (Cloisite™ 30B, 10A, 15A) are characterized by a more liquid-like behavior. The gel-like behavior is known to be related to the existence of well-dispersed (exfoliation) nanoclay platelets and attests that unmodified montmorillonite is more efficient to elaborate starch/clay nanocomposites.

Due to the lack of compatibility between starch and regular surfactant used for polymer/clay nanocomposites some authors proposed to use new surfactants specially dedicated to TPS. Actually, Huang *et al.* [89] and Majdzadeh-Ardakani *et al.* [85] chose to use citric acid to modify montmorillonite. Citric acid is able to swell montmorillonite leading to an increase of the interlayer spacing needed to ease the starch macromolecules to intercalate. Majdzadeh-Ardakani *et al.* also evidenced that the interlayer distance of citric acid modified montmorillonite is higher compared to Cloisite 30B ™. Moreover, the authors supposed that citric acid molecules act as a bridge between clay surface and starch molecules through hydrogen bonding leading to higher efficiency of intercalation and thus to a large interlayer spacing of silicate layers in corresponding nanocomposites. Considering the polysaccharides structure of starch, some authors proposed to use other new surfactants, based on similar chemical structure as starch, such as chitosan based surfactant [83] and cationic starch surfactant [99]. Chung *et al.* proposed to take advantage of the compatibility of chitosan with starch and its ability to be ion-exchanged in the clay. Chitosan activated montmorillonite was used to elaborate starch-based nanocomposites using the solution casting method. As evidenced by XRD, TEM and SEM, the obtained nanocomposites using chitosan activated montmorillonite present a well dispersed microstructure with partial exfoliation and small agglomerates attesting the efficiency of the method. Chivrac *et al.* [99] proposed a similar approach using cationic starch, which is a starch by-product. Compared to unmodified montmorillonite which present mainly intercalated structure, the nanocomposites based on cationic starch modified montmorillonite present an homogeneous structure with few large aggregates (Figure 4.5a). Nevertheless, the authors observed an heterogeneous dispersion with rich MMT domains composed of tactoïds having less than 5 layers, and regions without clay (Figure 4.5b). This heterogeneity was attributed to the high glycerol content of the plasticized starch, which induces a phase separation between low and high glycerol content domains. The XRD analyses show that glycerol is easily and preferentially intercalated into the MMT platelets, because glycerol has a strong affinity with them.

It clearly appears that the plasticizer used to obtain a plasticized starch may interfere during the elaboration and decrease the quality of the dispersion. Indeed, the plasticizers are generally highly hydrophilic and thus have favorable interactions with layered silicates. Moreover, thanks to the small size of the plasticizer molecules these interactions are preferred compared to interactions between starch macromolecules and silicates surfaces, which are the driving force for the intercalation process.

Pandey *et al.* [80] studied the plasticizer effect on the structure of starch based nanocomposites. Glycerol was used as plasticizer, and the influence of preparation methods and sequence of addition of components (starch, glycerol, montmorillonite) on the nanocomposite structure was evaluated. Based on this study, it appears that due to polar interactions with clay, starch and glycerol are attracted within the clay galleries. However, glycerol is preferred over starch in this competition due to its smaller molecular size and due to mutual attraction between starch and glycerol. Thus, it delays the migration inside the interlayer spacing leading to limited dispersion of the clay platelets. Considering, this competition between starch and glycerol, the best nanocomposite is obtained when starch and clay are mixed in a first step followed by plasticization. In this case, the starch chains intercalate without interference thanks to polar interaction with clay. During the plasticizing step,

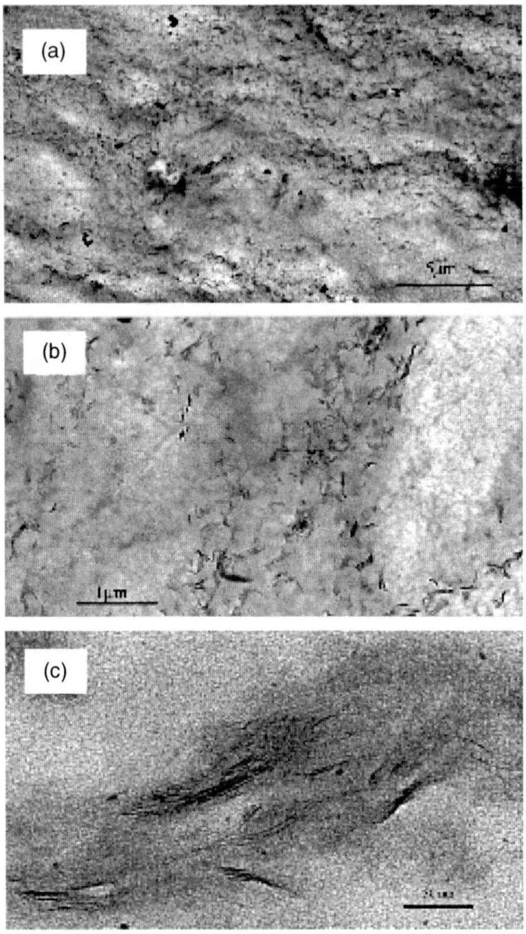

Figure 4.5 TEM Pictures of wheat starch nanocomposites based on cationic starch modified montmorillonite at a) low magnification, b) medium magnification and c) high magnification [99].

the plasticizer molecules are attracted towards the gallery space due to interaction with clay and to electrostatic hydrogen bonding formation with starch chains. Different authors confirm this effect of the glycerol as plasticizer [90, 91] and try to use glycerol to promote the dispersion of layered silicates by swelling the montmorillonite (increased d-spacing and destructed tactoïds are favorable to form intercalated or exfoliated nanocomposites) before melt blending with starch. However, the possible competition between starch and glycerol for the intercalation between MMT layers can deteriorate the plasticization of starch.

Apart from these observations some authors focused on the influence of the glycerol content on the structure of resulting starch-based nanocomposites [85, 94, 100]. All these authors note that at higher glycerol content only intercalation is observed while at lower glycerol content (around 5%) the exfoliation of montmorillonite is obtained. If glycerol is the most commonly used plasticizer for TPS, some authors tried to use other plasticizers to optimize the plasticizing step in presence of montmorillonite. Thus, Ren *et al.* proposed carbamide and ethanolamine as combined plasticizers [101]. Wang *et al.* demonstrated that citric acid in combination with glycerol (used to activate MMT) results in efficient plasticizing [91], whereas Ma *et al.* used the same method combining sorbitol (used to activate MMT) and formamide [93]. Dai *et al.* successfully utilized N-(2-Hydroxyethyl) formamide (HF) which act both as plasticizer for TPS and swelling agent for MMT [92]. These approaches use successfully the activation of the montmorillonite with the plasticizer allowing increasing the gallery height making the intercalation easier. Chivrac *et al.* studied the influence of the plasticizer (glycerol, Polysorb™, sorbitol) in the case of unmodified and cationic starch modified montmorillonite [88]. For unmodified montmorillonite the intercalation of the plasticizer is observed leading to a microcomposite. In the case of modified montmorillonite, exfoliation is obtained with glycerol when Polysorb and sorbitol leads to an intercalated/exfoliated structure.

If most of the studies focused on layered silicates, mainly on montmorillonite, some other silicates have been tested. Chivrac *et al.* [102] prepared well dispersed nanocomposites filled with sepiolite, which is a needle-like clay. These nanocomposites were prepared using an internal mixer. The use of cationic starch at high content promotes the dispersion of the sepiolite leading to a nanoscaled dispersion whereas a low cationic starch content leads to limited dispersion with large aggregates. It appears that sepiolite has a great potential for compounding well dispersed nanocomposites more easily compared to clay based nanocomposites.

Apart from the mineral clay platelets such as montmorillonite and needle-like clay such as sepiolite, other polysaccharides nanofillers have been studied. The most common are cellulose nanofibers [103] and starch nanocrystals [104]. The extraction yield of this kind of nanofillers using chemical or enzymatic treatment is rather low, limiting their use for industrial applications. However, because of the high interest of this kind of natural nanofillers, the industrial optimization of the extraction method will probably allow achieving in the future more appropriate yield opening the way to finding some application possibilities. In both cases, because of the limited thermal stability of these nanofillers, the nanocomposites films were prepared using the solution casting method. Even, if this method is not directly applicable in the packaging industry, it has been the only one currently reported.

4.5.2.3 Mechanical properties

Park *et al.* studied the tensile properties of nanocomposites containing unmodified (Cloisite Na+™) and modified montmorillonite (Cloisite™ 30B, 10A, 6A). In their first study [86] reported a promising improvement of the tensile properties of the nanocomposites compared to neat plasticized starch. The tensile strength is increased significantly for the unmodified montmorillonite (Cloisite Na^{+}™) and for the most polar modified montmorillonite (Cloisite 30B™), respectively by +27% and +7% for a clay content of 5 wt. %. For the nanocomposites containing the most hydrophobic modified montmorillonite a decrease of the tensile strength is observed. The most interesting is that the elongation at break is also improved in the case of nanocomposites based on unmodified montmorillonite (Cloisite Na+TM), which is quite surprising as usually the addition of a filler (even a nano-filler) results in a decrease of the ductility of the material. In the case of nanocomposites containing modified montmorillonite a decrease of the elongation at break is observed. Park *et al.* confirmed [87] these results and showed the influence of the montmorillonite content (2.5, 5 and 10 wt. %) in the case of unmodified montmorillonite (Cloisite Na+TM) and hydrophilic modified montmorillonite (Cloisite 30BTM). The tensile strength increases as the montmorillonite content increases in both cases, and the ductility is maintained in the case of modified montmorillonite nanocomposites (Cloisite 30BTM) and slightly increased for unmodified montmorillonite nanocomposites even at high montmorillonite content (10 wt. %). Since this pioneer work, a lot of studies have reported similar tendencies. Table 4.2 gathers the reported tensile properties (Young modulus, tensile strength and elongation at break) of some starch/clay nanocomposites. Generally, in the case of a well-dispersed montmorillonite nanocomposite, the Young modulus and tensile strength increase. Concerning the elongation at break, which represents the ductility of the material, the decrease is generally limited. Interestingly, for some nanocomposites (usually obtained via melt blending) [86-87, 89, 90, 105] different authors report an improvement of the ductility. These nanocomposites combine a higher rigidity with a better ductility compared to plasticized starch, which is of high interest for industrial applications. Chivrac *et al.* reported recently some interesting results using sepiolite, a new needle-like type of clay [102]. The results obtained using sepiolite compared to nanocomposites filled with unmodified montmorillonite (Na MMT) show a higher Young modulus and tensile strength. This behavior is related to the good affinity between the nanofiller and the polysaccharide chains, and on the new crystalline structure induced by the sepiolite dispersion, which increases the overall material crystallinity. As already observed for montmorillonite nanocomposites, the ductility of the nanocomposites containing sepiolite is preserved compared to plasticized starch.

Apart from clay/starch nanocomposites, which are widely studied, recent numerous studies focused on cellulose/starch nanocomposites. Even if those nanocomposites are far from a useable industrial production process in the packaging industry because of the processing method (solution casting), it is interesting to explore the potential of these materials. It clearly appears, compared (Table 4.2) with starch/clay nanocomposites, that the improvement of the tensile properties of the cellulose reinforced nanocomposites is significantly higher when cellulose nanofillers are used. This effect is attributed to the formation of a percolated network thanks to the very high aspect ratio of the cellulose nanofibers. However, this aspect ratio depends on the extraction method. Among these methods, bacterial cellulose gives surprising results with an increase of 2200% for the Young modulus and 850% for the tensile strength [106]. Concerning, the elongation at break the opposite trend is observed with a large decrease of the ductility compared to starch/clay nanocomposites. One can notice the exception of nanocomposites containing Pea hull nanowhiskers for which a large increase of the elongation at break is reported [107]. The reported tensile properties of starch nanocrystals filled nanocomposites are intermediate between the properties of cellulose based

Table 4.2 Tensile properties of starch-based nanocomposites.

Authors	Preparation method	Plasticizer	Nanofiller type	Young modulus (MPa)	Tensile strength (MPa)	Elongation at break (%)
				(Relative increase/decrease compared to neat polymer matrix)		
Clay/starch nanocomposites						
Park *et al.* [86]	Internal mixer	Glycerol, Water	Na MMT Cloisite* Na+ (5%)	–	3.32 (+27%)	57.2 (+22%)
			Modified MMT Cloisite* 30B	–	2.80 (+7%)	44.5 (–5%)
Park *et al.* [87]	Internal mixer	Glycerol, Water	Na MMT Cloisite* Na+ (5%)	–	3.2 (+23%)	52 (+11%)
			Modified MMT Cloisite* 30B	–	3.00 (+15%)	45.7 (–3%)
Cyras *et al.*[81]	Solution casting	Glycerol, water	Na MMT Cloisite* Na+ (5%)	195.6 (+550%)	5.2 (++57%)	46,8 (–25%)
Wang *et al.* [91]	Extrusion	Glycerol, Citric acid	Na MMT (5%)	–	~8 (+60%)	~6 (–45%)
Dai *et al.* [92]	Extrusion	HF	HF-MMT (5%)	–	3.75 (+50%)	40 (–27%)
Chung *et al.* [83]	Solution casting	Glycerol	Na MMT (5%)	1390 (+65%)	–	–
Huang *et al.* [89]	Extrusion	Urea, Formamide	Citric Acid MMT		21.1 (+370%)	134.5 (+23%)
Dean *et al.* [96]	Extrusion	Water	Na MMT	–	58 (+115%)	7 (–14%)
Xuechen *et al.* [90]	Extrusion	Glycerol	Glycerol MMT (5%)		7.8 (+73%)	80 (+18%)
Majdzadeh-Ardakani *et al.* [85]	Solution casting	Glycerol (10%)	Citric acid MMT (5%)	187.5	28.1	57.3

Table 4.2 (cont.) Tensile properties of starch-based nanocomposites.

Authors	Preparation method	Plasticizer	Nanofiller type	Young modulus (MPa)	Tensile strength (MPa)	Elongation at break (%)
				(Relative increase/decrease compared to neat polymer matrix)		
Tang *et al.* [94]	Extrusion	Glycerol	Na MMT (6%)		18.6 (+31%)	4.44 (–16%)
		Formamide			26.64 (+87%)	3.25 (–38%)
Chivrac *et al.* [99]	Internal mixer	glycerol	Na MMT (6%)	39 (+40%)	1.8 (–20%)	21 (–32%)
			Cationic Starch MMT (6%)	46.5 (+66%)	2.6 (+16%)	33 (+6%)
Ma *et al.* [93]	Extrusion	Formamide, Sorbitol	Sorbitol MMT (6%)	38 (+90%)	7 (+75%)	12 (–15%)
Maksimov *et al.* [82]	Solution casting	Glycerol	Na MMT (6%)	450 (+140%)	13 (+75%)	22 (–65%)
Mondragon *et al.* [84]	Solution casting	glycerol	Natural MMT (5%)	275 (+15%)	10 (+100%)	28 (–20%)
			Natural MMT (15%)	400 (+75%)	20.5 (+300%)	15 (–55%)
Chivrac *et al.* [102]	Internal mixer	Glycerol	cationic starch Sepiolite (6%)	28.1 (+66%)	34.6 (+6%)	3.19 (+42%)
Cellulose/starch nanocomposites						
Chang *et al.* [110]	Solution casting	Glycerol	Cellulose nanoparticles (5%)		10.5 (+250%)	3 (–73%)
Chen *et al.* [107]	Solution casting	Glycerol	Pea Hull nanowhiskers (10%)		8 (+100%)	60 (+100%)
Cao *et al.* [135]	Solution casting	Glycerol	cellulose nanocrystals (10%)	180.4 (+460%)	7.6 (+95%)	35.9 (–47%)
Alemdar *et al.* [137]	Solution casting	Glycerol	Wheat straw nanofibers (10%)	271 (+145%)	7.71 (+73%)	

Woehl *et al.* [106]	Solution casting	Glycerol	bacterial cellulose nanofibers (2.5%)	575 (+2200%)	8.5 (+850%)	
Wan *et al.* [138]	Solution casting	Glycerol	Bacterial cellulose (7.8%)	328.3 (+110%)	26.7 (+104%)	
Cao *et al.* [136]	Solution casting	Glycerol	Hemp cellulose nanocrystals (10%)	112.1 (+240%)	6.1 (+56%)	50.4 (–26%)
Starch nanocrystals/starch nanocomposites						
Angellier *et al.* [139]	Solution casting	Glycerol	Waxy maize starch nanocrystals (5%)	80 (+630%)	3.6 (+260%)	97 (–67%)
Ma *et al.* [111]	Solution casting	Glycerol	Citric acid modified starch nanoparticles (4%)	120 (+140%)	7.5 (+85%)	35 (–15%)
Viguié *et al.*[108]	Solution casting	Sorbitol	Waxy maize starch nanocrystals	36.6 (+113%)	0.99 (+160%)	57 (–10%)
	After ageing			120.7 (3%)	4.47 (+0%)	28 (–20%)
Chitin and tunicin nanocrystals/starch nanocomposites						
Chang *et al.* [110]	Solution casting	Glycerol	Chitin nanoparticles (5%)		8 (+200%)	2.75 (–65%)
Neus Angles *et al.* [28]	Solution casting	Glycerol	Tunicin Whiskers (16.7%)			

*Cloisite, Trademark from Southern Clay Product

and clay filled nanocomposites in terms of rigidity (Young modulus) and tensile strength. As observed for cellulose the elongation at break decreases in spite of the good compatibility with the plasticized starch matrix. It is also worth noticing that Viguié *et al.* [108] showed that the improvement of the mechanical properties remains rather limited after aging.

4.5.2.4 Barrier Properties

One of the main drawbacks of plasticized starch is its sensitivity to water because of its high hydrophilicity. It represents a main issue for packaging applications because it can expose the content of the packaging to humidity, which is generally undesired. Nanocomposites represent a great potential to overcome this problem. Table 4.3 summarizes *the barrier properties of starch-based nanocomposites.*

Park *et al.* [86, 87] focused their attention on water permeability. It appears that the relative water vapor transmission rate (WVTR) of the TPS nanocomposites was reduced by nearly a half compared to the neat TPS at only 5 wt% of montmorillonite. The observed decrease in WVTR is of great importance in evaluating TPS based compounds for use in food packaging, protective coatings, and other applications where efficient polymeric barriers are needed. For these applications, significant reduction in WVTR can result in either increased barrier efficiency, or reduced thickness of the barrier layer for the same efficiency. This significant decrease of WVTR in the nanocomposites is attributed to the presence of large aspect ratio platelets in TPS matrix. The same effect is generally reported for clay/nanocomposites based on others polymers [10]. While diffusing through the film the water molecules follow a tortuous path through the polymer matrix surrounding the silicate particles. Thereby, it increases the effective diffusion path length, thus decreasing the WVTR. Park *et al.* [87] compared nanocomposites based on unmodified montmorillonite (Cloisite Na^{+TM}) and modified montmorillonite (Cloisite 30B^{TM}). The barrier property to water vapor is better in the case of unmodified clay, regardless of the clay contents, because of the better dispersion.

Most of the other author considered the water vapor permeability (WVP) instead of the WVTR. The relation between WVP and WVTR may be found elsewhere [109]. Compared to WVTR, the water vapor permeability takes into account the thickness of the film and partial pressure difference across the film. These studies (Table 4.3) confirm the results obtained by Park *et al.* showing a correlation between the dispersion degree of clay silicate layered and the decrease of the water vapor permeability. Some authors also considered the water uptake and evidenced a reduced sensitivity to water diffusion (Table 4.3).

A similar phenomenon is observed in the case of nanocomposites reinforced with cellulose nanofibers with a decrease of the WVP compared to neat TPS. This effect is attributed to the increase of the tortuosity induced by the presence of the nanofibers [110]. However, one can notice that the decrease is less pronounced compared to clay nanocomposites probably because of the shape of the nanofibers, which is less appropriate to increase the path length. Table 4.3 also gathers the results obtained for starch nanocrystals for which a decrease of the WVP has

Table 4.3 Barrier properties of starch-based nanocomposites.

Authors	Preparation method	Plasticizer	Nanofiller type	Water vapor permeability (WVP)	Water uptake
				(Relative increase/decrease compared to neat polymer matrix)	
Clay/starch nanocomposites					
Park *et al.* [75]	Internal mixer	Glycerol, Water	Na MMT Cloisite* Na+ (5%)	– (–40%)	
			Modified MMT Cloisite* 30B	– (–15%)	
Park *et al.* [76]	Internal mixer	Glycerol, Water	Na MMT Cloisite* Na+ (5%)	– (–40%)	
			Modified MMT Cloisite* 30B	– (–30%)	
Wang *et al.* [80]	Extrusion	Glycerol, Citric acid	Na MMT Cloisite Na+ (5%)	2.2 (g*mm/kPa*h*m^2)	
Dai *et al.* [81]	Extrusion	HF	HF-MMT (5%)		– (+80% 25 days)
Tang *et al.* [83]	Extrusion	Glycerol	Na MMT (6%)	~0.5 (g*mm/kPa*h*m^2)	
Mondragon *et al.* [73]	Solution casting	glycerol	Natural MMT (5%)		– (+40% equilibrium)
Cellulose/starch nanocomposites					
Chang *et al.* [99]	Solution casting	Glycerol	Cellulose nanoparticles	3.5 (x10^{-10} g/m*s*Pa)	
Chen *et al.* [96]	Solution casting	Glycerol	Pea Hull nanowhiskers (10%)		– (+62% 7 days)

Table 4.3 (cont.) Barrier properties of starch-based nanocomposites.

Authors	Preparation method	Plasticizer	Nanofiller type	Water vapor permeability (WVP)	Water uptake
				(Relative increase/decrease compared to neat polymer matrix)	
Cao *et al.* [122]	Solution casting	Glycerol	cellulose nanocrystals (10%)		– (+65% 70h)
Wan *et al.* [125]	Solution casting	Glycerol	Bacterial cellulose (7.8%)		+13% (8 Days) (+14% Starch)
Cao *et al.* [123]	Solution casting	Glycerol	Hemp cellulose nanocrystals (10%)		+60% (72h) +68% (Starch)
Starch nanocrystals/starch nanocomposites					
Ma *et al.* [100]	Solution casting	Glycerol	Citric acid modified starch nanoparticles (4%)	2.75 ($x10^{-10}$ g/m*s*Pa)	
Garcia *et al.* [140]	Solution casting	Glycerol	Waxy starch nanocrystals	2.7 ($x10^{-10}$ g/m*s*Pa)	
Chitin and tunicin nanocrystals/starch nanocomposites					
Chang *et al.* [99]	Solution casting	Glycerol	Chitin nanoparticles (5%)	3.3 ($x10^{-10}$ g/m*s*Pa)	

* Cloisite, Trademark from Southern Clay Product

also been reported [111, 112] and is attributed to the high aspect ratio of the starch nanocrystals leading to higher barrier to water molecules.

4.5.2.5 *Optical Properties*

Few studies focused on optical properties of starch-based bio-nanocomposites although these properties are of high interest for some packaging applications. The advantage of the nanofillers versus conventional fillers is that the nanoscale size, which is lower than the wavelength of the visible light, allows avoiding the light diffusion induced by the fillers which results in the opacification of the materials. In the case of well dispersed nanocomposites, the material may have a perfect clarity.

Chen *et al.* reported some measurements of the transmittance of TPS, TPS/cellulose composites and TPS/cellulose nanocomposites [107]. It is worth noticing that, whereas the transmittance of the composites films is significantly reduced because of the micrometer scale of the fillers, the transparency of the nanocomposites films is close to the one of TPS films. Therefore, the interest of the use of starch-based nanocomposites for films packaging is confirmed, as it is possible to obtain a transparent material with higher mechanical properties and lower sensitivity to water vapor.

4.5.3 PHA Based Bio-Nanocomposites

Nanocomposites consisting of a polyhydroxyalkanoate (PHA) matrix reinforced with layered silicates are promising packaging materials due to their low cost with high aspect ratio and exceptional barrier properties [113]. Nanocomposites based on PHAs are relatively new. However, PHAs polymers have to be modified significantly to exhibit better matrix properties. It is hoped that nanocomposites will enable PHAs to compete more effectively with petroleum-based plastics. Several PHAs were incarcerated in layered silicates to improve the neat polymer properties. Table 4.4 summarizes the tensile properties of PHA/clay nanocomposites. Maiti *et al.* [114] reported the preparation of polyhydroxybutyrate PHB-organoclay bio-nanocomposites using melt extrusion. XRD results showed formation of well-ordered intercalated bio-nanocomposite structure. However, bio-nanocomposites based on organically modified MMT showed thermal degradation because PHB is very unstable and degrades at temperatures near its melting point. Same research group [115] further prepared bio-nanocomposites based on PHB and clay and reported a significant improvement in thermal and mechanical properties of bio-nanocomposites as compared to the neat polymer. The rate of biodegradation of PHB was also enhanced significantly in the bio-nanocomposites. Moreover, when PHB was used as the host matrix for octadecylammonium modified montmorillonite (MMT) and fluoromica, it was found that the rate of degradation of PHB during nanocomposite preparation was higher in MMT than in the fluoromicas. It is still unclear how the fluoromicas help to protect PHB. The presence of clay particles might have decreased the degradation rates in these nanocomposites. In addition, the occurrence of Al

Lewis acid sites in MMT was thought to be the reason for the higher degradation rate of this composite as the Al Lewis acid sites catalyze the ester linkage hydrolysis [113].

Sanchez-Garcia *et al* [116] studied the structure and barrier properties of PHB and PHB/clay nanocomposites. The addition of highly intergallery swollen organomodified montmorillonite clays to the PHB led to a highly dispersed morphology of the filler, but this simultaneously increased to a significant extent the melt instability of the biopolymer. Particularly at 4% clay loading, enhanced barrier properties to oxygen, D-limonene, and water were observed. D-limonene and specially water molecules were, however, found to sorb in both hydrophobic and hydrophilic sites of the filler, respectively, hence diminishing the positive barrier effect of an enlarged tortuosity factor in the permeability.

Influence of clays addition on the thermal stability of PHB nanocomposites has been recently reported by Erceg *et al.* [117]. The authors demonstrate that the addition of organo-modified montmorillonite (OMMT) in amounts higher than 1 wt% shifts the establishment of constant mass plateau to longer degradation times compared with neat PHB, i.e., improves its thermal stability. The most pronounced effect is observed for the addition of 7 wt% of OMMT when the establishment of a constant mass plateau is shifted for 25–35 min toward longer degradation times compared to neat PHB.

Very recently, Botana *et al.* [118] prepared a melt mixed polymer nanocomposite of PHB, and two commercial montmorillonites, pristine Cloisite Na^{+TM} (Na-MMT) and organo-modifed Cloisite 30B^{TM} (OMMT). The authors reported that intercalated/partially exfoliated structure as observed by TEM and XRD was more pronounced for PHB/OMMT than for PHB/Na-MMT, indicating the better compatibility of OMMT with the PHB matrix. An increase in crystallization temperature and a decrease in spherulites size were observed for PHB/OMMT. The intercalated/exfoliated structure also increased the modulus of the nanocomposites.

PHB is the most popular polymer among PHAs because it possesses mechanical properties similar to synthetic thermoplastics, such as poly(propylene). However, its drawbacks of brittle behavior and lack of melt stability have seriously limited its application. These disadvantages have been conquered to a certain extent when PHB was substituted by poly(3-hydroxybutyrate-co-3-hydroxyvalerate) (PHBV), which has been recognized as a potentially environment-friendly substitute for traditional plastics. Even though, the use of PHBV presents some problems, such as a high cost, a slow crystallization rate, a high degree of crystallinity, and difficulty in processing. The addition of nanoparticles to PHBV may contribute to overcome some of these issues [119-122]. Overall mechanical properties in the literature indicate an improvement in modulus and strength at the expense of ductility upon nanoclay inclusions as indicated by Table 4.4.

Choi *et al.* [121] reported the preparation of PHBV/montmorillonite nanocomposites, through a melt intercalation method using Cloisite 30B^{TM} as the organoclay. An intercalated structure was determined by XRD and TEM analyses. The temperature and rate of crystallization of PHBV increased as a result of the effective nucleating effect of the organoclay. Moreover, the nanocomposites showed significant increases in tensile strength and thermal stability. However,

Table 4.4 Tensile properties of PHA/Clay nanocomposites.

Authors	Matrix	Nanofiller type	Preparation Method	Tensile properties (Relative increase/decrease compared to neat polymer matrix)		
				Modulus, MPa	**Strength, MPa**	**Elongation at break, %**
Choi *et al* [121]	PHBV	O-MMT, Cloisite* 30B (3 wt.%)	Melt mixing	795 (65%)	33 (13%)	5.6 (–34%)
Bordes *et al* [123]	PHB	O-MMT, Cloisite* 30B (5wt. %)	Melt mixing	2201 (14%)	28(16%)	1.47(–6%)
	PHBV	O-MMT, Cloisite* 30B (4.4 wt.%)	Melt mixing	1971(17%)	29(30%)	1.97(23%)
Botarna *et al* [118]	PHB	MMT, Cloisite* Na (5 wt.%)	Melt mixing	3200 (16%)	25(–28%)	NR
	PHB	O-MMT, Cloisite* 30B (5 wt.%)	Melt mixing	3440 (25%)	25(–23%)	NR
Zhang *et al* [124]	PHBV	O-MMT, Cloisite* 20A (15 wt.%)	Solvent casting	1400 (40 %)	30 (15%)	12 (–40%)
	PHBV	O-MMT, Cloisite* 25A (15 wt.%)	Solvent casting	1200 (20%)	27 (12%)	10 (–50%)

*Cloisite, Trademark from Southern Clay Product

Table 4.5 Tensile and barrier properties of wheat gluten and proteins/Clay nanocomposites.

Authors	**Preparation method**	**Matrix**	**Nanofiller type**	**Young modulus (MPa)**	**Tensile strength (MPa)**	**Elongation at break (%)**
				(Relative increase/decrease compared to neat polymer matrix)		
Zhang *et al.* [112]	Internal mixer	Wheat gluten	Na MMT (3%)	106.2 (+15%)	13.9 (+32%)	63 (–50%)
Tunc *et al.* [115]	Solution casting	Wheat gluten	Na MMT (5%)	10.68 (+185%)	4.70 (+150%)	16 (–73%)
Angellier-Coussy *et al.* [113]	Internal mixer	Wheat gluten	Na MMT (5%)	23.6 (+260%)	2.5 (+57%)	41.7 (–51%)
Zheng *et al.* [121]	Internal mixer	Soy protein	Starch nanocrytals (2%)	325 (+195%)	10.39 (+51%)	13 (–92%)
Sothornvit *et al.*[119]	Solution casting	Whey protein	Cloisite* 30B (5%)	162.6 (–5%)	3.29 (–3%)	51.7 (+2%)
Chang *et al.* [118]	Solution casting	Pea protein	Na MMT (4%)	275 (+60%)	9 (+100%)	10 (–98%)
Kumar *et al.* [120]	Extrusion	soy protein	Na-MMT (5%)		6.28 (+180%)	64.6 (+450%)
Chen *et al.* [117]	Solution casting	Soy protein	Na MMT (4%)	275 (+60%)	10.5 (+25%)	300 (–45%)
Authors	Preparation method	Matrix	Type of nanofiller	Water vapor permeability		Water uptake (%)
Tunc *et al.* [115]	Solution casting	Wheat gluten	Na MMT (5%)	6 ($\times 10^{-12}$ mol/m*s*Pa)		113
Guilherme *et al.* [116]	Solution casting	Wheat gluten	Na MMT (5%)	18 ($\times 10^{-12}$ mol/m*s*Pa)		
Angellier-Coussy *et al.* [113]	Internal mixer	Wheat gluten	Na MMT (5%)			100

Table 4.5 (cont.) Tensile and barrier properties of wheat gluten and proteins/Clay nanocomposites.

Authors	Preparation method	Matrix	Nanofiller type	Young modulus (MPa)	Tensile strength (MPa)	Elongation at break (%)
				(Relative increase/decrease compared to neat polymer matrix)		
Zheng *et al.* [121]	Internal mixer	Soy protein	Starch nanocrytals (2%)			32 (Matrix 29%)
Sothornvit *et al.*[119]	Solution casting	Whey protein	Cloisite* 30B (5%)	5.56 (×10–10 g/m*s*Pa) (–15%)		
Kumar *et al.* [120]	Extrusion	soy protein	Na-MMT (5%)	2.96 (×10–10 g/m*s*Pa) (–33%)		

* Cloisite, Trademark from Southern Clay Product

Wang *et al.* [122] found that the biodegradability of PHBV/organo-modified montmorillonite nanocomposites in soil suspension decreased with an increase in the amount of clay.

4.5.4 Proteins Based Nanocomposites

Proteins based materials have focused attention for a long time in packaging industry. However, few studies on proteins based nanocomposites are reported. Table 4.5 summarizes the tensile properties, and when avalaible the barrier properties, of various protein based nanocomposites

Recently, some authors reported the possibilities to elaborate wheat gluten nanocomposites films either using an internal mixer [125-127] or a solution casting method [128-129]. All these studies are based on unmodified montmorillonite except for Zhang *et al.* who used a modified montmorillonite (Cloisite 30B™) [125]. In this last case, TEM and XRD show some exfoliation and homogeneous dispersion. Tunc *et al.* [128] also reported well-dispersed nanocomposites in the case of unmodified montmorillonite.

Tensile strength and Young modulus of the nanocomposites are improved (Table 4.5) even if the elongation at break is reduced. Angellier-Coussy *et al.* reported that a thermal treatment of the films at high temperature (120°C) improves the tensile properties of the wheat gluten films which are over increased by the addition of montmorillonite [126]. As shown in Table 4.5, the water sensitivity of gluten films is globally reduced by the addition of the montmorillonite platelets as it is observed in the case of starch-based nanocomposites (see 5.2. section). Tunc *et al.* reported a slight decrease of the barrier properties to O_2, CO_2 and aromas [128].

Some studies on other types of proteins such as soy and whey proteins are also reported in the literature. The incorporation of montmorillonite has been successfully carried out using solution casting [130-132] or extrusion [133] using modified [132] or unmodified montmorillonite [130-132]. Zheng *et al.* also reported the elaboration using an internal mixer of soy protein nanocomposites filled with starch nanocrystals [134]. These studies focused on tensile properties of the nanocomposites films and their sensitivity to water. Generally, the films have a higher Young modulus and tensile strength whereas a decrease of the ductility is observed (Table 4.5). The water sensitivity is also generally slightly reduced.

4.6 Conclusion

A high concern is currently focused in the environmental impact of the materials for which the packaging industry is directly concerned. For this reason, bio-based polymers represent a good opportunity to reduce the environmental footprint of packaging materials as they are generally biodegradable or compostable. However, these materials are not able to fulfill all the requirements of the packaging industry yet, particularly because of their poor thermo-mechanical and barrier properties. Therefore, many techniques have been developed to improve the in-use properties of these biodegradable materials so as to match the constraining requirements of engineering applications: modification of biopolymers, blending with other materials, and reinforcement of nanofillers to form bio-nanocomposite. Among these, bio-nanocomposite have been identified as the one of the promising option to improve the material properties without sacrificing biodegradability.

The most promising developments in that field make use of PLA, starch, PHA and proteins as polymer matrices. Regarding the reinforcing nanofillers, clay, and particularly montmorillonite, is of high interest because of its natural origin. Though, some emerging other natural nanofillers such as cellulose, chitin or tunicin nanofibers or starch nanocrystals represent also great new opportunities to elaborate bio-nanocomposite being 100% natural. Numerous research studies have demonstrated that nanocomposites technology may improve the in-use properties of the bio-based polymers. Particularly, it has been proven that materials combining higher rigidity and better toughness are possibly obtained upon addition of nanofillers. Barrier properties are also potentially improved with an O_2 barrier permeability lowered by two times and less water sensitivity.

Even if the bio-nanocomposite existing today will not be able to replace at a short term all the oil-based polymer materials used in the packaging industry because the range of in-use properties covered by these new materials is still limited, recent advances in the area of polymer materials using nanocomposites technologies promise great potential in terms of opportunities to extend the range of usability of bio-based polymers.

Wide spread applications of new bio-based polymer nanocomposites in the packaging industry however still need to reduce production and materials costs to make them cost-effective against synthetic polymers. Besides, one should keep in mind that using nanofillers represents a cause of concern for human health both at the elaboration step in the materials production workshops and during the life cycle of the products. Especially, in the case of food packaging industry a high concern exists because of the potential transfer of the nanofillers from the packaging to the food product. Studies on these challenging issues are also going on.

References

1. K. Marsh, and B. Bugusu, *Journal of Food Science*, Vol. 72, p. 39, 2007.
2. M.J. Kirwan, and J.W. Strawbridge, *Food Packaging Technology*, Vol. 24, p. 174, 2003.
3. R.N. Tharanathan, *Trends in Food Science & Technology*, Vol. 14, p. 71, 2003.
4. G.L. Robertson (ed.), *Food packaging. Principles and practice*. New York, NY: Marcel Dekker, 1993.
5. S. Guilbert, B. Cuq, and N. Gontard, *Food Additives and Contaminants*, Vol. 14, p. 741, 1997.
6. K. Petersen, P.V. Nielsen, G. Bertelsen, M. Lawther,M.B. Olsen and N.H. Nilssonk, *Trends in Food Science & Technology*, Vol. 10, p. 52, 1999.
7. M. Alexandre, and P. Dubois, *Material Science and Engineering*, Vol. 28, p. 1, 2000.
8. S. Sinha Ray, and M. Bousmina, *Progress in Material Science*, Vol. 50, p. 962, 2005.
9. S. Sinha Ray, and M. Okamoto, *Progress in Polymer Science*, Vol. 28, p. 1539, 2003.
10. S.J. Eichhorn, A. Dufresne, M. Aranguren, N.E. Marcovich, J.R. Capadona, S.J. Rowan, C. Weder, W. Thielemans, M. Roman, S. Renneckar, W. Gindl, S. Veigel, J. Keckes, H. Yano, K. Abe, M. Nogi, A.N. Nakagaito, A. Mangalam, J. Simonsen, A.S. Benight, A. Bismarck, L.A. Berglund, and T. Peijs, *Journal of Materials Science*, Vol. 45, p. 1, 2010.

11. Standard terminology of packaging and distribution environments, ASTM Standard ASTM D996-04, ASTM International, West Conshohocken, PA, 2005.
12. R. Chandra, and R. Rustgi, *Progress in Polymer Science*, Vol. 23, p. 1273, 1998.
13. K. Peterson and P.V. Nielsen, G. Bertelsen, M. Lawther, M.B. Olsen, N.H. Nilsson, and G. Mortensen, *Trends in Food Science and Technology*, Vol. 10, p. 52, 1999.
14. S. Jacobsen, and H.G. Fritz, *Polymer Engineering and Science*, Vol. 39, p. 1303, 1999.
15. T. Ke, and X. Sun, *Transactions of the ASAE*, Vol. 44, p. 945, 2001.
16. R.A. Gross, and B. Kalra, *Green Chemistry*, Vol. 297, p. 803, 2002.
17. L. Shen, J. Haufe, and M.K. Patel, Product overview and market projection of emerging bio-based plastics, Group Science, Technology and Society (STS) Copernicus Institute for Sustainable Development and Innovation Utrecht University, Utrecht, The Netherlands, 2009.
18. S.Y. Lee, J. Choi, and H.H. Wong, *International Journal of Biological Macromolecules*, Vol. 25, p. 316, 1999.
19. A.K. Mohanty, M. Misra, and G. Hinrichsen, *Macromolecular Materials Engineering*, Vol. 276/277, p. 1, 2000.
20. H. Brandl and P. Puchner, *Biodegradation*, Vol. 2, p. 237, 1992.
21. M.A. Kotnis, G.S. O'Brien and J.L. Willett, *Journal of Environment and Polymer Degradation*, Vol. 3, p. 97, 1995.
22. K. Peterson and P.V. Nielsen, G. Bertelsen, M. Lawther, M.B. Olsen, N.H. Nilsson, and G. Mortensen, *Trends in Food Science and Technology*, Vol. 10, p. 52, 1999.
23. H. Angellier-Coussy, S. Torres-Giner, M. Morel, N. Gontard, and E. Gastaldi, *Journal of Applied Polymer Science*, Vol. 107, p. 487, 2008.
24. T.H. McHugh, and J.M. Krotcha, *Journal of Agricultural Food Chemistry*, Vol. 48, p. 3913, 2000.
25. M. Zhu, C.L. Chu, S.L. Wang, and R.W. Lencki, *Journal of Food Science*, Vol.66, p. 30, 2001.
26. E.P. Giannelis, *Advanced Materials*, vol. 8, p. 29, 1996.
27. T.J. Pinnavaia, and G.W. Beall, Polymer-Clay Nanocomposites. John Wiley & Sons, Inc., NewYork, 2001.
28. M. Neus Angles, and A. Dufresne, *Macromolecules*, Vol. 33, p. 8344, 2000.
29. M. Neus Angles, and A. Dufresne, *Macromolecules*, Vol. 34, p. 2921, 2001.
30. R. Hiroi, S.S. Ray, M. Okamoto, and T. Shiroi, *Macromolecular Rapid Communications*, Vol.25, p. 1359, 2004.
31. S. Kumar, *Polymeric Materials Science and Engineering*, Vol. 90, p. 59, 2004.
32. S.S. Ray, S.Y. Quek, A. Easteal, and X.D. Chen, *International Journal of Food Engineering*, Vol. 2, p. 1, 2006.
33. M.G. Sajilata, K. Savitha, R.S. Singhal, and V.R. Kanetkar, *Comprehensive Reviews in Food Science and Safety*. Vol. 6, p. 17, 2007.
34. I. Sakurada, Y. Nukushina, and T. Ito, *Journal of Polymer Science*, Vol. 57, p. 651, 1962.
35. A. Sturcova, Geoffrey R. Davies, and S.J. Eichhorn, *Biomacromolecules*, Vol. 6, p. 1055, 2005.
36. K. Tashiro, and M. Kobayashi, *Polymer*, Vol. 32, p. 1516, 1991.
37. A.J. de Menezesa, G. Siqueiraa, A.A.S. Curvelob, and A. Dufresne, *Polymer*, Vol. 50, p. 4552, 2009.
38. G. Agoda-Tandjawa, S. Durand, S. Berot, C. Blassel, C. Gaillard, C. Garnier, J.L. Doublier, *Carbohydrate Polymers*, Vol. 80, p. 677.
39. A. Dufresne, J.Y. Cavaille, and W. Helbert, *Macromolecules*, Vol. 29, p. 7624, 1996.
40. J.L. Putaux, S. Molina-Boisseau, and T. Momaur, *Biomacromolecules*, Vol. 4, p. 1198, 2003.
41. N. Ogata, G. Jimenez, H. Kawai, and T. Ogihara, *Journal of Polymer Science Part B: Polymer Physics*, Vol. 35, p. 389, 1997.
42. S. Bandyopadhyay, R. Chen, and E.P. Giannelis, *Polymer Materials: Science and Engineering*, Vol. 81, p.159, 1999.
43. P. Maiti, K. Yamada, M. Okamoto, K. Ueda, and K. Okamoto, *Chemistry of Materials*, Vol. 14, p. 4654, 2002.
44. M. Pluta, A. Galeski, M. Alexandre, M.A. Paul, and P. Dubois, *Journal of Applied Polymer Science*, Vol. 86, p.1497, 2002.
45. S.S. Ray, K. Okamoto, K. Yamada, and M. Okamoto, *Nano Letters*, Vol. 2, p. 423, 2002.

46. S.S. Ray, K. Yamada, M. Okamoto, and K. Ueda, *Nano Letters*, Vol. 2, p. 1093, 2002.
47. S.S. Ray, P. Maiti, M. Okamoto, K. Yamada, and K. Ueda, *Macromolecules*, Vol. 35, p. 3104, 2002.
48. S.S. Ray, K. Yamada, M. Okamoto, and K. Ueda, *Polymer*, Vol. 44, p. 857, 2003.
49. S.S. Ray, K. Yamada, M. Okamoto, A. Ogami, and K. Ueda, *Chemistry of Materials*, Vol. 15, p. 1456, 2003.
50. V. Krikorian, and D..Pochan, *Chemistry of Materials*, Vol. 15, p. 4317, 2003.
51. J.H. Lee, T.G. Park, H.S. Park, D.S. Lee, Y.K. Lee, and S.C. Yoon, *Biomaterials*, Vol.24, p. 2773, 2003.
52. M.A. Paul, M. Alexandre, P. Degee, C. Calberg, R. Jerome, and P. Dubois, *Macromolecular Rapid Communications*, Vol. 24, p. 561, 2003.
53. L. Cabedo, J.L. Feijoo, M.P. Villanueva, J.M. Lagaron, and E. Gimenez, *Macromolecular Symposia*, Vol. 233, p. 191, 2006.
54. J.H. Lee, T.G. Park, H.S. Park, D.S. Lee, Y.K. Lee, S.C. Yoon, and J.D. Nam, *Biomaterials*, Vol. 24, p. 2773, 2003.
55. J.M.G. Cowie, Polymers: Chemistry and Physics of Modern Materials, New York, Chapman and Hall, 1991.
56. R.H. Burton, and M.J. Folkes, Mechanicals Properties of Reinforced Thermoplastics, ed. D.W. Clegg, and A.A. Collyer, London: Elsevier, 1986.
57. M. Pluta, M.A. Paul, M. Alexandre, and P. Dubois, *Journal of Polymer Science: Part B: Polymer Physics*, Vol. 44, p. 299, 2006.
58. M. Pluta , J.K. Jeszka, and G. Boiteux, *European Polymer Journal*, Vol. 43, p. 2819, 2007.
59. M. Shibata, Y. Someya, M. Orihara, and M. Miyoshi, *Journal of Applied Polymer Science*, Vol. 99, p. 2594, 2006.
60. K. Fukushima, D. Tabuani, and G. Camino, *Materials Science and Engineering C*, Vol. 29, p. 1433, 2009.
61. H. Balakrishnan, A. Hassan, M.U. Wahit, A.A. Yussuf and S.B.A. Razak, *Materials & Design*, Vol. 31, p. 3298, 2010.
62. J.-H. Chang, Y. Uk-An,. and G.S. Sur, *Journal of Polymer Science Part B: Polymer Physics*, Vol. 41, p. 94, 2003.
63. S.S. Ray, and M. Okamoto, *Macromolecular Rapid Communication*, Vol. 24, p. 815, 2003.
64. R.K. Bharadwaj, A.R. Mehrabi, C. Hamilton, C. Trujillo, M.F. Murga, and A. Chavira, *Polymer*, Vol. 43, p. 3699, 2002.
65. L. Nielsen, *Journal of Macromolecular Science: Chemistry: A*, Vol.1, p. 929, 1967.
66. A.A. Gusev, and H.R. Lusti, *Advanced Materials*, Vol. 13, p. 1641, 2001.
67. S.S. Ray, K. Yamada, M. Okamoto, Y. Fujimoto, A. Ogami, and K. Ueda, *Polymer*, Vol. 44, p.6633, 2003.
68. M. Żenkiewicza, and J. Richert, *Polymer Testing*, Vol. 27, p.835, 2008.
69. J-H. Chang, Y. Uk-An, D. Choa and E.P. Giannelis, *Polymer*, Vol. 4, p. 3715, 2003.
70. C. Thellen, C. Orroth, D. Froio, and D. Ziegler, J. Lucciarini, R. Farrellb, N. D'Souza and JA. Ratto, *Polymer*, Vol. 46,p. 11716, 2005.
71. A. Hasook, S. Tanoue, Y. Iemoto, and T. Unryu, *Polymer Engineering and Science*, Vol. 46, p.1001, 2006.
72. T. Li, L.S. Turng, S. Gong, and K. Erlacher, *Polymer Engineering and Science*, Vol. 46, p. 1419, 2006.
73. L. Jiang, J. Zhang, P. Michael,and P. Wolcott, *Polymer*, Vol. 48, p. 7632, 2007.
74. S. Solarski, M. Ferreira, E. Devaux, G. Fontaine, P. Bachelet, S. Bourbigot, R. Delobel, P. Coszach, M. Murariu, A.D.S. Ferreira, M. Alexandre, P. Degee, and P. Dubois, *Journal of Applied Polymer Science*, Vol. 109, p. 841, 2008
75. L. Jiang, B. Liu, and J. Zhang, *Industrial and Engineering Chemistry Research*, Vol. 48, p. 7594, 2009.
76. E. Nieddu, L. Mazzucco, P. Gentile, T. Benko, V. Balbo, R. Mandrile, and G. Ciardelli. *Reactive & Functional Polymers*, Vol. 69, p.371, 2009.
77. G. Ozkoc, and S. Kemaloglu, *Journal of Applied Polymer Science*, Vol. 114, p. 2481, 2009
78. X. Wu, J. Yuan, Y. Yu, and Y. Wang, *Journal of Wuhan University of Technolotgy-Material Science Edition*, Vol. 24, p. 562, 2009.
79. J-W. Rhim, S.i. Hong and C.S. Ha, LWT - *Food Science and Technology*, Vol. 42, p.612, 2009.

80. J. Pandey, and R. Singh, *Starch/Staerke*, Vol. 57, p. 8, 2005.
81. V. Cyras, L. Manfredi, M. Ton-That, and A. Vázquez, *Carbohydrate Polymers*, Vol. 73, p. 55, 2008.
82. R. Maksimov, A. Lagzdins, N. Lilichenko, and E. Plume, *Polymer Engineering and Science*, Vol. 49, p. 2421, 2009.
83. Y.L. Chung, S. Ansari, L. Estevez, S. Hayrapetyan, E.P. Giannelis, and H.M. Lai, *Carbohydrate Polymers*, Vol. 79, p. 391, 2010.
84. M. Mondragón, J. Mancilla, and F. Rodríguez-González, *Polymer Engineering and Science*, Vol. 48, p. 1261, 2008.
85. K. Majdzadeh-Ardakani, A. Navarchian, and F. Sadeghi, *Carbohydrate Polymers*, Vol. 79, p. 547, 2010.
86. H. Park, X.L. Li, C. Jin, C. Park, W. Cho, and C. Ha, *Macromolecular Materials and Engineering*, Vol. 287, p. 553, 2002.
87. H. Park, W. Lee, C. Park, W. Cho, and C. Ha, *Journal of Materials Science*, Vol. 38, p. 909, 2003.
88. F. Chivrac, E. Pollet, P. Dole, and L. Avérous, *Carbohydrate Polymers*, Vol. 79, p. 941, 2010.
89. M. Huang, J. Yu, and X. Ma, *Carbohydrate Polymers*, Vol. 63, p. 393, 2006.
90. W. Xuechen, X. Zhang, H. Liu, and N. Wang, *Starch/Staerke*, Vol. 61, p. 489, 2009.
91. N. Wang, X. Zhang, N. Han, and S. Bai, *Carbohydrate Polymers*, Vol. 76, p. 68, 2009.
92. H. Dai, P. Chang, F. Geng, J. Yu, and X. Ma, *Journal of Polymers and the Environment*, Vol. 17, p. 225, 2009.
93. X. Ma, J. Yu, and N. Wang, *Macromolecular Materials and Engineering*, Vol. 292, p. 723, 2007.
94. X. Tang, S. Alavi, and T. Herald, *Carbohydrate Polymers*, Vol. 74, p. 552, 2008.
95. S. Kalambur, and S. Rizvi, *Polymer International*, Vol. 53, p. 1413, 2004.
96. K. Dean, L. Yu, and D. Wu, *Composites Science and Technology*, Vol. 67, p. 413, 2007.
97. D. Chaudhary, *Journal of Polymer Science, Part B: Polymer Physics*, Vol. 46, p. 979, 2008.
98. H.R. Dennis, D.L. Hunter, D. Chang, S. Kim, and J.L. White, *Polymer*, Vol. 42, p. 9513, 2001.
99. F. Chivrac, E. Pollet, M. Schmutz, and L. Avérous, *Biomacromolecules*, Vol. 9, p. 896, 2008.
100. B. Chiou, D. Wood, E. Yee, S. Imam, G. Glenn, and W. Orts, *Polymer Engineering and Science*, Vol. 47, p. 1898, 2007.
101. P. Ren, T. Shen, F. Wang, X. Wang, and Z. Zhang, *Journal of Polymers and the Environment*, Vol. 17, p. 203, 2009.
102. F. Chivrac, E. Pollet, M. Schmutz, and L. Avérous, *Carbohydrate Polymers*, Vol. 80, p. 145, 2010.
103. S.J. Eichhorn, A. Dufresne, M. Aranguren, N.E. Marcovich, J.R. Capadona, S.J. Rowan, C. Weder, W. Thielemans, M. Roman, S. Renneckar, W. Gindl, S. Veigel, J. Keckes, H. Yano, K. Abe, M. Nogi, A.N. Nakagaito, A. Mangalam, J. Simonsen, A.S. Benight, A. Bismarck, L.A. Berglund, and T. Peijs, *Journal of Materials Science*, Vol. 45, p. 1, 2010.
104. D. Le Corre, J. Bras, and A. Dufresne, *Biomacromolecules*, Vol. 11, p. 1139, 2010.
105. F. Chivrac, O. Gueguen, E. Pollet, S. Ahzi, A. Makradi, and L. Averous, *Acta Biomaterialia*, Vol. 4, p. 1707, 2008.
106. M.A. Woehl, C.D. Canestraro, and A. Mikowski, *Carbohydrate*, Vol. 80, p. 866, 2010.
107. Y. Chen, C. Liu, P. Chang, D. Anderson, and M. Huneault, *Polymer Engineering and Science*, Vol. 49, p. 369, 2009.
108. J. Viguie, S. Molina-Boisseau, and A. Dufresne, *Macromolecular Bioscience*, Vol. 7, p. 1206, 2007.
109. X. Tang, S. Alavi, and T. Herald, *Cereal Chemistry*, Vol. 85, p. 433, 2008.
110. P. Chang, R. Jian, J. Yu, and X. Ma, *Food Chemistry*, Vol. 120, p. 736, 2010.
111. X. Ma, R. Jian, P. Chang, and J. Yu, *Biomacromolecules*, Vol. 9, p. 3314, 2008.
112. N. Garcia, L. Ribba, A. Dufresne, M. Aranguren, and S. Goyanes, *Macromolecular Materials and Engineering*, Vol. 294, p. 169, 2009.
113. J.K. Pandey, A.P. Kumar, M. Misra, A.K. Mohanty , L.T. Drzal and R.P. Singh , *Journal of Nanoscience and Nanotechnology*, Vol. 5, p.497, 2005.
114. P. Maiti, C.A. Batt, and E.P. Giannelis, *Polymer Materials: Science and Engineering*, vol.88, p. 58, 2003.
115. P. Maiti, C.A. Batt, and E.P Giannelis, *Biomacromolecules*. Vol.8, p. 3393, 2007.
116. 4.M.D. Sanchez-Garcia, E. Gimenez, and J.M. Lagaron, *Journal of Applied Polymer Science*, Vol.108, p. 2787, 2009.

117. M. Erceg, T. Kovačić, and P. Sanja, *Polymer Composites*, Vol. 31, p. 272, 2010.
118. A. Botana, M. Mollo, and P. Eisenberg, *Applied Clay Science*, Vol. 47, p. 263, 2010.
119. C.Y. Tang, D.Z. Chen, C.P. Tsui, P.S. Uskokovic, P.H. F. Yu, and M.C. P. Leung, *Journal of Applied Polymer Science*, Vol. 102, p. 5388, 2006.
120. D. Dubief, E. Samain, and A. Dufresne, *Macromolecules*, Vol. 32, p. 5765, 1999.
121. W.M. Choi, T.W. Kim, O.O. Park, Y.K. Chang, and J.W. Lee, *Journal of Applied Polymer Science*, Vol. 90, p.525 , 2003.
122. S.F. Wang, C.J. Song, G.X. Chen, T.Y. Guo, J. Liu, B.H. Zhang, and S. Takeuchi, *Polymer Degradation and Stability*, Vol. 87, p. 69, 2005.
123. P. Bordes, E. Pollet, and L. Averous, *Macromol. Chem. Phys.*, Vol. 209, p. 1473, 2008.
124. X. Zhang, G. Lina, A.H. Red, M.K. Hassan, I. Noda and J.E. Mark, *European Polymer Journal*, Vol. 43, Issue 8, August 2007, Pages 3128–3135.
125. X. Zhang, M. Do, K. Dean, P. Hoobin, and I. Burgar, *Biomacromolecules*, Vol. 8, p. 345, 2007.
126. H. Angellier-Coussy, S. Torres-Giner, M. Morel, N. Gontard, and E. Gastaldi, *Journal of Applied Polymer Science*, Vol. 107, p. 487, 2008.
127. M. Mauricio-Iglesias, S. Peyron, V. Guillard, and N. Gontard, *Journal of Applied Polymer Science*, Vol. 116, p. 2526, 2010.
128. S. Tunc, H. Angellier, Y. Cahyana, P. Chalier, N. Gontard, and E. Gastaldi, *Journal of Membrane Science*, Vol. 289, p. 159, 2007.
129. M. Guilherme, L. Mattoso, N. Gontard, S. Guilbert, and E. Gastaldi, *Composites Part A: Applied Science and Manufacturing*, Vol. 41, p. 375, 2010.
130. P. Chen, and L. Zhang, *Biomacromolecules*, Vol. 7, 2006.
131. P. Chang, Y. Yang, J. Huang, W.L. Xia, L. Feng, and J. Wu, *Journal Of Applied Polymer Science*, Vol. 113, p. 1247, 2009.
132. R. Sothornvit, J. Rhim, and S. Hong, *Journal of Food Engineering*, Vol. 91, p. 468, 2009.
133. P. Kumar, K.P. Sandeep, S. Alavi, V.D. Truong, and R.E. Gorga, *Journal of Food Engineering*, Vol. 100, p. 480, 2010.
134. H. Zheng, F. Ai, P. Chang, J. Huang, and A. Dufresne, *Polymer Composites*, Vol. 30, p. 474, 2009.
135. X. Cao, Y. Chen, P. Chang, A. Muir, and G. Falk, *Express Polym. Lett.*, Vol. 2, p. 502, 2008
136. X. Cao, Y. Chen, P. Chang, M. Stumborg, and M.A Huneault, *J. Appl. Polym. Sci.*, Vol. 109, p. 3804, 2008.
137. A. Alemdar, and M. Sain, *Compos. Sci. Technol.*, Vol. 68, p. 557, 2008.
138. Y. Wan, H. Luo, F. He, H. Liang, Y. Huang, and X. Li, *Compos. Sci. Technol.*,Vol. 69, p. 1212, 2009.
139. H. Angellier, S. Molina-Boisseau, P. Dole, and A. Dufresne, *Biomacromolecules*, Vol. 7. p. 531, 2006.
140. N.L. Garcia de Rodriguez, W. Thielemans, and A. Dufresne, *Cellulose*, Vol. 13, p. 261, 2006.

5

Biobased Materials in Food Packaging Applications

M.N. Satheesh Kumar[1], Z. Yaakob[1] and Siddaramaiah[2]

[1]Department of Chemical and Process Engineering, Faculty of Engineering and Built Environment, National University of Malaysia (UKM), Bangi, Malaysia
[2]Department of Polymer Science & Technology, Sri Jayachamarajendra College of Engineering, Mysore, India.

Abstract

The anticipated diminution of fossil fuel reserves dictates the need for the utilisation of biobased resources for various applications. Until 1970 various industries have utilised biobased resources for the manufacture of a wide range of products from dyes to synthetic fibers. The replacement of petroleum derived materials by biorenewable resources and its enhanced uses have created a demand in the market. Due to increased demand and the anticipated diminution of petroleum derived materials, biorenewable resources are expected to be a major contributor for the production of industrial products. Currently, various attempts have been in progress to develop the technology to reduce the costs and to improve the performance of biobased products. In the meantime, the environmental concerns are intensifying the interest in agricultural and forestry resources as alternative feedstock. The steady and sustained growth of these industries depend on the development of new markets. A potential new market for these materials is food packaging, a highly competitive area with great demands for performance and cost.

Keywords: Biopolymers, biobased materials, PLA, PHB, food packaging, MAP

5.1 Introduction

The introduction of polymer-based structures as packaging materials for food stuffs has been increasing over the last few decades. The main commercial appeal of these materials lies on their ability to offer a broad variety of tailor-made properties, low cost and easily be processed and conformed into a myriad shapes and sizes. The different food products need different packaging requirements. As a result, a large number of packaging technologies such as multilayer structures, modified/equilibrium modified atmosphere packaging, active packaging, etc., have been developed [1, 2].

Srikanth Pilla (ed.) Handbook of Bioplastics and Biocomposites Engineering Applications, (121–160)

Biopolymers are the polymers produced by living organisms. Cellulose, starch, proteins, peptides, DNA and RNA are few examples of biopolymers in which the monomeric units respectively are sugars, amino acids, and nucleotides. Biobased materials or biomaterials fall under the broad category of bioproducts or biobased products which includes materials, chemicals and energy derived from renewable biological resources. Since its invention in the 1930s, plastic packaging has initiated two challenges: its dependence on petroleum and the problem of waste disposal. The package is the culmination of a series of innovations that have been successfully bundled into a consumer product. Designing and manufacturing a packaging material is a multistep process and involves careful and numerous considerations to successfully engineer the final package with all the required properties. The properties to be considered in relation to food distribution may include gas/water vapour permeability, mechanical properties, sealing capability, thermoforming properties, resistance (towards water, grease, acid, UV light, etc.), machinability (on the packaging line), transparency, anti fogging capacity, printability, availability and cost. The process of disposal of the package at the end of its useful life must also be taken into consideration [3–5].

The food packages are divided into three categories: (i) primary, (ii) secondary and (iii) tertiary packaging. The packaging material when it is in direct contact with the food is called as primary packaging. The functions of primary packaging are protection and safety of foods [6]. Secondary packaging is used for physical protection of the product; for example, a box containing a number of primary packages. The inclusion of a secondary package provides easy handling during storage/distribution and safety against mechanical damage. Tertiary packaging incorporates the secondary packages in a final transportation package system. The purpose is to protect the product from the mechanical damage, weather conditions, etc.

Today, the synthetic plastics such as polypropylene (PP), polyethylene (PE), aromatic polyesters, etc., are being used for food packaging applications because of the advantages such as low cost, processing flexibility, durability and structures that resulted in wide ranges of strengths and shapes [7, 8]. These synthetic polymers are petroleum-derived (non-degradable) polymers. The purpose of food packaging is to preserve the quality and safety of the food till it reaches the consumer (after production) [9]. The polymers used for food packaging should have the combination of better moisture/gas barrier, mechanical and thermal properties [10]. The protection of packaged food against water and oxygen is one of the most important requirements and they can be blocked by the use of coatings on the packaging materials. A conventional barrier coating on packaging materials typically consists of expensive and synthetic polymers such as ethylene vinyl alcohol (EVOH), polyvinylidene chloride (PVDC) and polyesters [11].

Though, various synthetic coatings can be done on biobased materials, the associated disadvantage is its recycling. The recycling becomes difficult as the coated product contains multilayers. The recycling is feasible only with single component plastics. In addition to this, the growing reliance on these coated packaging films has raised a number of environmental concerns [12, 13]. According to Pira

International Private Limited, the biodegradable packaging would grow at a compound annual growth rate (CAGR) of 22% with the introduction of low cost polyhydroxyalkanoates (PHA) in 2011 [14]. In order to reach this goal, the poor barrier properties of uncoated biodegradable materials need to be improved. Further, in a report it is revealed that, 41% plastics production is being used for the packaging industry and 47% of this is used for food packaging [15]. Based on these facts, the use of biobased materials for food packaging applications appear to be an excellent alternative for reducing current environmental problems and dependency on petroleum based raw materials. The purpose of this chapter is to discuss the reported research results of various authors related to the materials, processes and properties of food packaging.

5.2 Biobased Packaging Materials

The packaging made using the materials derived from biorenewable resources is known as biopackaging. These materials can be utilised in food packaging applications. Figure 5.1 shows the three main categories of biobased polymers based on their origin and production. The three main categories are; (i) polymers directly extracted or isolated from the biomass. Proteins like casein/gluten and polysaccharides such as cellulose/starch are the few examples. These polymers are semicrystalline in nature and hydrophilic which may provide excellent gas barrier properties, (ii) polymers obtained after the chemical synthesis using renewable biobased monomers. Polylactic acid (PLA) obtained after the polymerisation of lactic acid monomer is the one example. The monomers themselves may be produced

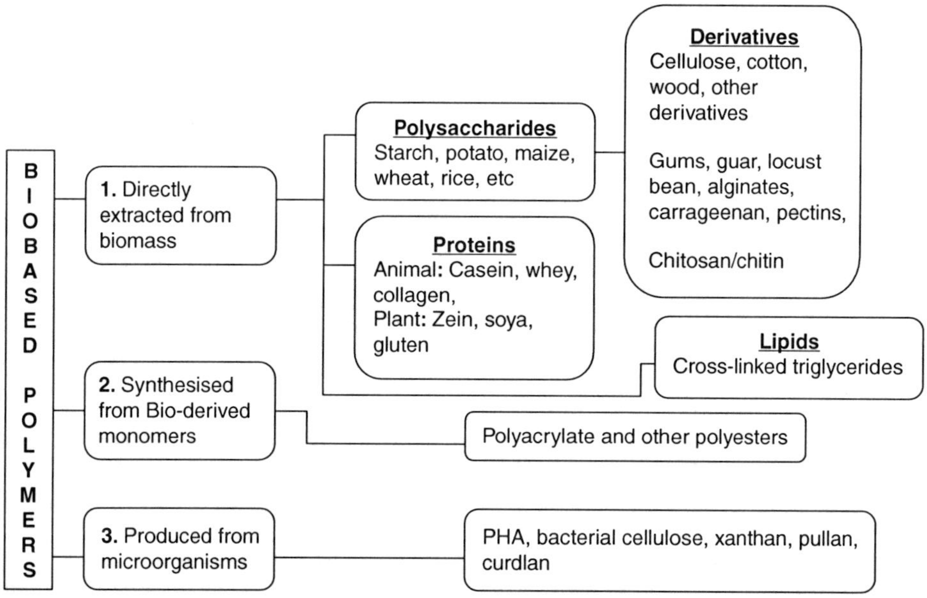

Figure 5.1 Classification of biobased materials based on their origin and production.

through fermentation of carbohydrate feedstock and (iii) polymers produced by micro-organisms or genetically modified bacteria. PHA, bacterial cellulose and polyhydroxybutyrates (PBA) are the few examples of such biobased polymers [16–18]. The increased market potential for food packaging materials made the industrialists to put new efforts or renew their previous efforts to develop commercially viable and sustainable packaging materials using starch, cellulose, and microbially produced biopolymers. A few examples of these companies are listed in the Table 5.1 which reflects the diversified applications of bioplastics in food packaging area. The comparative evaluation of a few biobased polymers with conventional petroleum based polymers which are used in food packaging applications are given in Table 5.2 [19].

Table 5.1 List of few companies engaged in the production of biobased materials for packaging application.

Company	Country	Product Details
National Starch Company	United Kingdom	Packing applications for shipping and distribution
FKur Kunststoffe GmbH	Germany	Product name: Biograde, Disposable catering items
NODAX	USA	Product name: Nodax, includes aerobic and anaerobic degradability, barrier properties, printability and mechanical properties
Metabolix	USA	Product name: Mirel, high-performance bioplastic alternative in consumer goods, compost bags, business equipment, packaging, agriculture/horticulture, and marine/aquatic applications
Nature Works LLC	USA	Product name: Ingeo, a biopolymer makes it well suited for a broad range of packaging applications including high-value films, rigid thermoformed food and beverage containers, coated papers and boards and other packaging applications.
Starch Tech Inc.	USA	Packaging material is a starch-based loose fill packing peanut, biodegradable in water or a compost setting.
Ever Corn Inc.	Japan	Modified starch for food packing applications, production of biodegradable multi film or food wrapping film
VTT Chemical Technology	Finland	Special materials from renewable sources including for packaging applications

Table 5.2 Comparative evaluation of the properties of few biobased polymers with conventional synthetic polymers reprinted with permission from [19.], K. Petersen *et al., Trends in Food Science and Technology*, 10, 52 (1999) © 2010, Elsevier.

	Polymer	Moisture Permeability	Oxygen Permeability	Mechanical Properties
Biobased	Cellulose acetate	Moderate	High	Moderate
	Starch/polyvinyl alcohol	High	Low	Good
	Proteins	High-medium	Low	Good
	Cellulose/cellophane	High-medium	High	Good
	Polyhydroxyalkanoates (PHA) Polyhydroxybutyrate/valerate (PHBA)	Low	Low	Good
	Polylactate	Moderate	High-moderate	Good
Synthetic	Low density polyethylene	Low	High	Moderate-good
	Polystyrene	High	High	Poor-moderate

5.2.1 Polymers Produced from Biomass

Biomass is carbon based and is composed of a mixture of organic molecules containing hydrogen usually including atoms of oxygen, often nitrogen and also small quantities of other atoms including alkali, alkaline earth and heavy metals. The polymers obtained from different biomass are discussed herein below.

Starch and its derivatives: Starch is composed of carbon, hydrogen and oxygen in the ratio of 6:10:5 $(C_6H_{10}O_5)_n$ placing it in the class of carbohydrate organic compounds. It can be considered to be a condensation polymer of glucose and yields glucose when subjected to hydrolysis by acids and certain enzymes. Owing to its economical advantage, starch based materials have received considerable attention of industrialists and scientists to select it as a material for packaging application. Starch is a known brittle polymer and stored in granules as a reserve in most plants. It is composed of repeating 1, 4-α-d glucopyranosyl units, amylose and amylopectin. The relative amounts of amylose and amylopectin depend upon the plant source. Corn starch granules typically contain approximately 70% amylopectin and 30% amylose. The ratio of these two components is very important to tailor the required properties [20]. Starch can be converted into a thermoplastic material with the application of thermal and mechanical energy in an extruder. The elimination of poor film forming ability and inadequate mechanical properties of starch needs plasticization or blending with other polymers or materials. Various attempts have been made to develop the modified and un-modified starch for the packaging applications [21–26].

Cellulose and derivatives: Among all the available packaging materials, cellulose based packaging materials account for high proportion and it is most familiar in the form of paper or cardboard. Because of the regular structure and array of hydroxyl groups, cellulose tends to form strong hydrogen bonded crystalline microfibrils and fibers. The crystallanity and structural organisation of cellulose vary according to its origin and processing [27]. The associated inherent properties of cellulose such as hydrophilic nature, insolubility and crystalline structure are the few disadvantages of cellulose in view of its processing. The cellophane film produced after the xanthation of cellulose has been widely used in packaging applications. The cellophane attracts moisture due to the presence of hydrophilic groups but has good mechanical properties. It is however not a thermoplastic material because of the fact that, the theoretical melt temperature is above the degradation temperature and therefore cannot be heat-sealed. Cellophane is often coated with nitrocellulose or PVDC to improve the barrier properties and in such form it is used for packaging of baked goods, processed meat, cheese and candies. Few examples of commercially available cellulose derivatives are carboxymethyl cellulose (CMC), methyl cellulose, ethyl cellulose, hydroxyl ethyl/propyl cellulose and cellulose acetate.

Chitosan/Chitin: Chitin is a naturally occurring macromolecule present in the exo-skeleton of invertebrates and represents the second most abundant polysaccharide resource after cellulose [28]. The chemical composition of chitin consists of repeating units of 1, 4-linked-2-deoxy-2-acetoamido-α-D-glucose and chitosan refers to a family of partially acetylated 2-deoxy-2-amino-α-glucan polymers derived from chitin. Chitosan has several uses as flocculants, thickeners, clarifier, etc. The film forming ability of chitosan coupled with its very high gas barrier property has lead to the development of edible coating technology [27]. The material development with the combination of chitosan and other polymers would be beneficial to tailor the specific properties based on the end use. The specific properties may be tailored due to the presence of cationic charges in chitosan which offers electron interactions with numerous compounds during processing. Chitosan has received considerable attention to use it as a material for food packaging due to its two main properties such as antimicrobial property and the ability to absorb heavy metal ions. The antimicrobial property of chitosan is expected to enhance the microbial shelf life and safety of the food product. The ability of chitosan to absorb heavy metal ions is expected to diminish the oxidation processes in the food catalyzed by free metals. Significant amount of research has been done on chitosan as a material for food packaging [29–34]. Recently, Ojagh *et al* [35] have developed a novel biodegradable film made from chitosan and cinnamon essential oil (CEO). It is reported that, the properties of CEO added chitosan film can be improved through the crosslinking of CEO component in chitosan.

Proteins: Proteins are built up of various combinations of 20 different amino acids. The different combination of amino acids within the protein provides a unique structure and function. Proteins can vary greatly in chain length giving molecular weights from a few hundreds to several millions. A protein is considered to be a random copolymer of amino acids and the side chains are highly suitable for chemical modification which will be helpful to the material engineer when tailoring the required properties for packaging materials. Proteins can be divided into two types, namely proteins from plant and animal origin. The polar characteristics of protein films determine the barrier properties. They have high permeability to polar substances, such as water and low permeability to non-polar substances such as oxygen, many aroma compounds and oils. The good gas barrier/selectivity properties (oxygen and carbon dioxide) of protein films could help to preserve fresh and minimally processed vegetables by achieving a modified atmosphere effect. The excellent gas barrier properties of protein based films made them to select it as a material for food packaging applications. Like starch based films, protein based films are also sensitive to moisture due to the presence of hydrophilic groups and as a result the mechanical and gas barrier properties are influenced by the relative humidity. The moisture sensitivity of protein films can be reduced by blending it with other biobased or synthetic materials (in a small quantity) and by lamination technique. Morillon *et al.*, [36], McHugh [37] and Callegarin *et al.*, [38] have thoroughly reviewed the aspects concerning to this strategy. The presence of a wide variety of chemical moieties in protein-based materials is expected to help tailoring the properties towards specific applications. Casein and gluten are the two protein based materials studied extensively as edible films for food packaging applications.

Casein is a milk-derived protein. The two milk proteins such as casein and whey have been used in the production of edible films. Owing to its excellent mechanical and barrier properties, casein finds the application in food packaging area. In addition to this, casein bears excellent emulsifying ability, high nutritional value and solubility in water. Gluten is the main storage protein in wheat and corn. The mechanical treatment of gluten leads to the formation of a disulphide bridge which is responsible for the creation of strong, viscoelastic and voluminous dough. Hence, the processing of gluten may become difficult and requires proper reducing agent to break the disulphide bridge. Gluten exhibits high gloss and show good resistance to water under certain conditions but do not dissolve in water. The commercially available soy proteins are termed as soy flour, soy concentrate and soy isolate depending on the amount of protein content present in them. Like gluten, the disulphide bridges are also present in the soy proteins. The properties of soy proteins are almost similar to gluten. The protein based materials such as wheat- gluten (WG) films exhibit humidity dependent gas permeabilities (O_2 and CO_2) and water vapour permeability (WVP); allowing optimal gas composition for food preservation [39]. Collagen is another fibrous structural protein present in animal tissue. It has traditionally been used for preparing edible sausage casing. It is the basic raw material for the production of gelatin which is a common food additive with potential film and foam forming ability.

Zein comprises a group of alcohol soluble proteins found in corn endosperm. Zein is mostly used in formulations of speciality food and pharmaceutical coatings. The films are brittle and needs plasticizers to make them flexible. Lee *et al* [40] have characterized the protein coated PP films as a novel composite structure for active food packaging application. The authors have studied the effect of different proteins (soy, whey and corn zein) and plasticisers (propylene glycol, glycerol (GLY), polyethylene glycol (PEG), sorbitol and sucrose) on the gloss behaviour

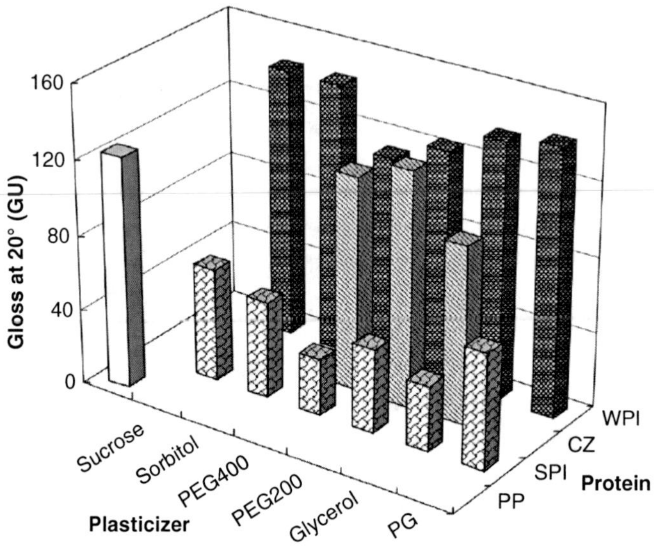

Figure 5.2 Effect of different plasticisers and protein type on the gloss behaviour of protein coated polypropylene films. Reprinted with permission from [40], Lee *et al.*, *Journal of Food Engineering*, 86, 484 (2008) ©2010, Elsevier.

of protein coated PP films. Figure 5.2 shows the effect of different proteins and plasticisers on the gloss behaviour of novel films produced through simple casting technique. The coatings made out of whey protein, corn-zein protein and sucrose plasticised whey protein have showed the highest gloss surface.

5.2.2 Polymers from Bio-derived Monomers

Polylactic acid (PLA): Today, PLA based packaging materials are most popular to replace petroleum based plastics worldwide. Unlike petroleum-based plastics, PLA is made from fermented plant starch (100% renewable resource), carbon neutral and is compostable. By comparison, plastic is not biodegradable, takes over a 1,000 years to break down and only 1–3% of plastic is recycled. Compared to synthetic plastics, PLA uses around 65% less energy for its production and contains no toxins.

PLA is a bio-polyester obtained after the polymerisation of lactic acid (monomer) falls into this category. Lactic acid is produced principally by microbial fermentation of carbohydrate feed stock [41] and is then polymerised to produce PLA. The carbohydrate agricultural feed stock may be wheat, maize or waste products from food or agriculture industry. Grade *et al* [42] have showed that, a cost effective PLA can be produced from the green juice (a waste product from the production of animal feeds). Depending on the isomers of lactic acid feedstock or its intermediates, PLA can be semi-crystalline or totally amorphous. As a result, the number of potential structures for PLA is substantial. L-lactic acid is the natural and most common form. D-lactic acid can also be produced by micro-organisms or through racemisation. Adding this D-lactic acid co-monomer to the polymer backbone behaves similarly to a co-monomer in other polyester polymer and influences the kinetics of crystallisation (critical to fabrication processes and applications) [43]. PLA has been widely accepted material for packaging applications. The ratio between the two mesoforms (L or D) of lactic acid decides the properties of PLA. High crystallanity and melting point can be obtained with 100% L-PLA. The mixture of D- and L- PLA can be used instead of L-PLA to obtain an amorphous PLA polymer with a glass transition temperature (T_g) of 60°C [44]. Depending on the ratio of D- to L- lactic acid in the polymer, the processing temperature of PLA could be 60 – 125°C. The T_g of PLA can be reduced by plasticisation with its monomer or alternatively oligomeric lactic acid in presence of external plasticisers. PLA can be formed into blown films, coatings and injection-moulded products. Significant amount of research has been done on the food packaging applications of PLA. Recently, Guinault *et al* studied the influence of crystallinity on gas barrier and mechanical properties of PLA food packaging films [45]. In food packaging applications, a major research gap is the development of packaging materials that can provide the release of active compounds at rates suitable for a wide range of food packaging applications. The aforementioned gap has been addressed recently by Mascheroni *et al* [46].

Other Bio-Derived Monomers: Bio-polyesters can also be derived from fermentation of plant sugars. Commercial materials derived from 1, 3-propanediol offers an alternative to nylon and polyethylene terephthalate (PET) for fiber and fabric manufacture [42]. Other developments include the possible production of biodegradable polymers currently derived from petroleum sources. An example of these developments is the work of Bionolle from renewable feedstock [47]. Presently, biobased monomers may not be commercially attractive, however biobased monomers derived by biotechnological pathways may become as an alternative to petrochemical polymer routes.

5.2.3 Polymers Produced from Micro-organisms

Polymers produced by microorganisms or genetically modified bacteria would fall into this category. To date, this group of biobased polymers consists of polyhydroxyalkanoates (PHAs) and bacterial cellulose [48].

Polyhydroxyalkanoates: PHAs are linear polyesters produced in nature by bacterial fermentation of sugar or lipids. They are produced by the bacteria to store carbon and energy. More than 150 different monomers can be combined within this family to give materials with extremely different properties. These plastics are biodegradable and are used in the production of bioplastics. PHAs can be either thermoplastic or elastomeric materials with melting points ranging from 40 to 180 °C.

Polyhydroxybutyrate (PHB) is the most common material found in the category of PHAs. As the properties of PHAs are dependent on their monomer composition, in addition to PHB, a large variety of PHAs can be synthesized using microbial fermentation technique. Based on the selection of carbon source and micro-organisms, the monomer composition of PHAs can be altered. The poly-3-hydroxybutyrate (P3HB) is the most common type of PHAs but many other polymers of this class are produced by a variety of micro-organisms; these include polyhydroxyvalerate (PHV) and its polyhydroxy butyrate valerate (PHBV) copolymer (Figure 5.3). Water insolubility and resistant to hydrolytic degradation differentiates PHB from most other currently available biodegradable plastics which are either water soluble or moisture sensitive. The melting point and T_g of PHB is around 175 and 15 °C respectively. Like LDPE, PHB exhibits low water permeability. PHB resembles isotactic PP (iPP) in relation to melting temperature and mechanical behaviour (tensile strength of 40 MPa). The biocompatibility of PHA has attracted scientists to explore this material for medical applications. The interesting property of PHAs with respect to food packaging applications is their low WVP which is close to that of LDPE (a known polymer widely used in food packaging applications). The major draw back of PHB in commercial use is its unfavourable ageing process. The annealing process is expected to overcome the aforementioned draw back of PHB by changing its lamellar morphology [49]. It has also been reported in the literature that, the incorporation of 3HV or 4HB co-monomers produces remarkable changes in the mechanical properties of PHB. Unlike PHB or its copolymers, medium chain length PHAs behave as elastomers with crystals acting as physical crosslinks and therefore can be regarded as a class of its own with respect to mechanical properties. In a study [50], it was concluded that PHB has a different resistance to dynamic compression in relation to PP which reflects its deformation value around 50% lower than that of PP (characterizing as a more rigid and less flexible material). Under normal freezing and refrigeration conditions, the performance of PHB tends to be inferior to that of PP whereas at higher temperatures, the performance of PHB was better than PP. These results showed the future possibility of packaging made from biobased materials such as PHB. The major limiting factor of PHB is its relatively expensive production costs when compared with plastics produced from petrochemicals. Because, it offers the benefits of biodegradability, PHB based materials have good potential for replacing PP in bottles, bags, and film applications. As the large-scale production of PHA depends on its production cost, Chen *et al* [51] have made an attempt to develop the low cost PHA production technology using continuous and non-sterile processes.

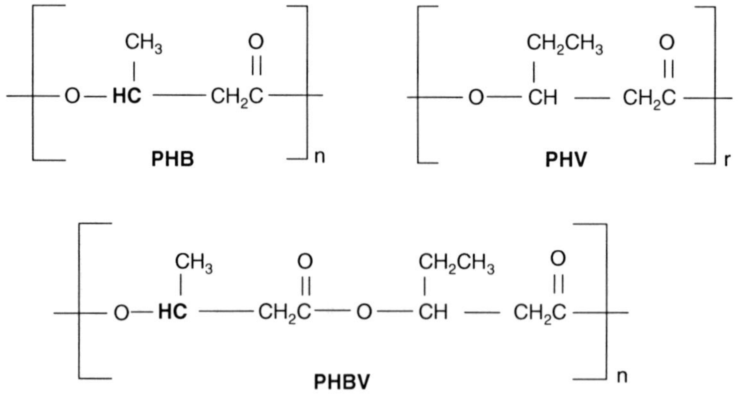

Figure 5.3 Structure of polyhydroxybutyrate (PHB), polyhydroxyvalerate (PHV) and polyhydroxybutyrate valerate (PHBV) copolymer.

Bacterial Cellulose: Bacterial cellulose (BC) is an unexploited material but represents a polymeric material with major potential. Bacterial strains of *Acetobacter xylinum and A. pasteurianus* are able to produce an almost pure form of cellulose (homo-beta-1,4-glucan). Its chemical and physical structure is identical to the cellulose formed in plants [52]. However, plant cellulose has to undergo a harsh chemical treatment to remove lignin, hemicellulose and pectins. This treatment severely impairs the material characteristics of plant cellulose (degree of polymerisation decreases almost ten-fold and affects crystallainity).

The degree of polymerisation of BC is 15 times higher than cellulose of wood pulp. Bacterial cellulose is highly crystalline. The BC has 70% crystalline and 30% amorphous regions. BC features smooth texture and high water-holding capacity. These properties function positively in food systems; BC functions as a heat-stable suspending agent as well as a filler to reinforce the body of fragile food hydrogels, improves the quality of pasty foods by reducing their stickiness and could be applied to meat products as a fat substitute and to jam as a non-caloric bulking agent. These results showed that, BC would be more effective to improve the quality of processed foods [53]. The production cost of BC is very high due to the low efficiency of bacterial processes. Approximately, 10% of the glucose used in the process is incorporated in the cellulose. The material has the potential application in making artificial skin, as a food grade non-digestible fiber, as an acoustic membrane and as a separation membrane [54]. Nguyen *et al* [55] have developed a nisin (polycyclic antibacterial peptide used as a food preservative) containing BC film to pack the processed meats. George *et al* [56] have studied the physico-mechanical properties of chemically treated BC membranes. Compared to un-treated BC membranes, the reported tensile strength and percentage elongation was higher in case of 0.1 M sodium carbonate and potassium carbonate treated BC membranes. Based on the high mechanical properties and comparatively low oxygen transmission rates, the authors have concluded that, the chemically treated BC membranes may find use as a biopackaging material in controlled atmosphere packaging applications.

5.3 Properties of Packaging Materials

Packaging is a means of providing the correct environmental conditions for food during the length of its time stored and/or distributed to the consumer. A good package has to perform the following functions:

(i) It must keep the product clean and provide a barrier against dirt and other contaminants.
(ii) It should prevent losses. Its design should provide protection and convenience in handling during transport, distribution and marketing. In particular, the size, shape and weight of the packages must be considered.
(iii) It must provide protection to the food against physical/chemical damage (example water and water vapour, oxidation and light), insects and rodents. It must provide identification and instruction so that the food is used correctly and have sales appeal.

The selection of biobased packaging materials depends on the product characteristics (applicable food product to be packed). The effect of chemical, microbial, physical, enzymatic and biological changes can cause deteriorative reactions inside the food and can damage the food. In addition to this, the food also needs protection from insects, pets and rodents. The enzymatic changes in foods are determined by temperature, water activity and alteration of substrate (ex. oxygen availability in oxygen-dependent reactions catalyzed by enzymes) [57].

Table 5.3 Details of existing different packaging materials.

Textiles	Poor gas and moisture barrier properties and have a poorer appearance than plastics. Used to transport a wide variety of bulk foods including grain, flour, sugar and salt. Example, Woven jute sacks
Cotton	Inexpensive and is satisfactory as a wrapper for flour, grains, legumes, coffee beans and powdered or granulated sugar. It can be re-used as many times as the material withstands washing and is easily marked to indicate the contents of the bag.
Kenaf	It is chiefly used for making ropes and string but can be spun into a yarn which is fine enough to make a coarse canvas.
Sisal	Sisal is resistant to salt water and therefore makes an ideal natural material from which to make rope. The nets in which hard fruits are transported are often hand-made from vegetable fibre.
Wood	Has been used for a wide range of solid and liquid foods including fruits, vegetables, tea and beer.
Traditional packaging materials	
Leaves	Banana or plantain leaves are the most common and widespread leaves used for wrapping foods, Cornhusk is used to wrap corn paste or block brown sugar, and cooked foods of all sorts are wrapped by leaves. 'Pan' leaves are used for wrapping spices (India), they are an excellent solution for products that are quickly consumed, as they are cheap and readily available
Vegetable fibres	The lightweight is an advantage in handling and transport.
Bamboo and rattan	These are widely used materials for basket making.
Coconut palm	Palyra palm leaves are used to weave boxes in which items such as cooked foods are transported.
Treated skins	Water and wine are frequently stored and transported in leather containers (camel, pig and kid goat hides). Manioc flour and solidified sugar are also packed in leather cases and pouches.
Flexible films	Cellulose, polypropylene, polyethylene, polystyrene, coated films, laminated films, co-extruded films, etc.

The different packaging materials that have been used in past and at present for the packaging applications are listed in Table 5.3. As the food packaging requires specific atmospheric condition to sustain their freshness and overall quality, it is very essential to understand the various properties of the materials used for packaging. The various properties of packaging materials are discussed below in detail. The packaging material properties that are required to pack different foods are given in Table 5.4.

5.3.1 Gas Barrier Properties

To ensure a constant gas composition inside the package, the packaging material needs to possess gas barrier properties. Barrier properties in polymers are necessarily associated to their inherent ability to permit the exchange (high or low extent) of low molecular weight substances through mass transport process like permeation. Foods are being packed in protective atmosphere with a specific mixture of gases to ensure the optimum quality and safety of the food during storage. In most of the packaging applications, the gas mixture inside the package consists of carbon dioxide, oxygen, nitrogen and their combinations. Extensive research studies have been conducted to provide the information on the barrier properties of the biobased materials. However, the comparison between different biobased materials is complicated and sometimes may not be possible due to the use of different types of equipment and dissimilar conditions for the measurements. As the oxygen permeability of biobased materials are quite similar to the wide range of conventional petroleum based materials, it is easy to choose the appropriate biobased material (as discussed in the previous section) for food packaging applications. The oxygen diffusion barrier properties of transparent oxide coatings on polymeric substrates have been reviewed by Chattam [58]. The conventional approach to produce high barrier films for food packaging in protective atmosphere is to use multi layers of different films.

The high barrier properties of EVOH and excellent mechanical properties of LDPE have been used to construct a laminate of EVOH/LDPE for food packaging applications [59]. Rubino *et al* [60] evaluated the effects of gaseous chlorine dioxide (ClO_2) treatment on properties and performance of different polymeric packaging materials, including PE, biaxially oriented PP, polystyrene (PS), poly vinyl chloride (PVC), PET, PLA, nylon and a multilayer structure of ethylene vinyl acetate (EVA)/EVOH. The authors have noticed a reduction in tensile properties of ClO_2 treated PE samples compared to untreated ones. A reduction in moisture, oxygen and/or carbon dioxide barrier properties were observed in the treated PE, PET, and multilayer (EVA/EVOH/EVA) samples. Hult *et al* [61] reported on the possibility of enhancing the barrier properties of paper and paper board using microfibrillar cellulose and shellac. The influence of crystallinity on gas barrier and mechanical properties of PLA food packaging films have been investigated by Guinault *et al* [45]. The authors have related the crystallinity and morphology to the gas barrier properties of the films. The modified atmospheric packaging (MAP) with specific mixture of gases has been widely used to ensure the optimum food quality. Raei *et al* [62] have evaluated the effect

Table 5.4 Properties of packaging materials to pack different foods.[a] Reprinted with permission from [19], K. Petersen *et al.*, Trends in Food Science and Technology, 10, 52 (1999) © 2010, Elsevier.

Food	Typical Temp.	Quality Factors Determining Shelf-life						Required Gas Composition	Optimal Required Barrier Properties				
		Microbiology	Colour	Oxidation	Structure	Flavour	Others		O_2	CO_2	H_2O	High	Other
Animal derived products													
Red meats	0–5°C 6–14d	X	X	–	–	–	–	High oxygen (70~80%), high carbon dioxide (30–20%)	High	High	High	–	–
	0–5°C-1–4d	X	X	–	–	–	–	No control: atmospheric	Low	–	High	–	–
Other meats	0–5°C 1 d–6w	X	(x)	–	–	–	–	Low oxygen, high carbon dioxide	High	High	High	–	–
Cured meat products	5°C 4w	X	X	X	–	–	–	No oxygen	High	High	High	High	–
Cured fish products	5°C 4w	X	X	X	–	–	–	high carbon dioxide	High	High	High	High	–
Fish, high fat	0–5° C 1–7d	X	–	X	–	X	Enzymatic	carbon dioxide (40%) Oxygen (30%) and nitrogen (30%)	High	High	Low	–	Odours
Fish, low fat	0–5° C 1–7d	X	–	–	–	X	Enzymatic	Carbon dioxide (40–60%) and nitrogen (60–40%)	High	High	Low	–	Odours
Eggs	2–12°C 25 d–4w	X	–	–	–	–	Water loss	–	–	–	–	–	Aroma
Fluid milk	2–5°C 2 d–8 md	X	–	X	–	X	Photooxidation	–	High	–	High	High	Aroma
Fermented milk	2–5°C 16–18 d	X	–	X	X	–	Photooxidation	–	High	High	High	High	Aroma
Fresh cheese	<5°C1–8 w	X	–	–	–	X	Photooxidation	–	High	High	High	High	–

Semi soft and hard cheese	< 5°C 1w–18w	X	–	X	–	–	Photooxidation	–	High	Low	High	Low or high	–
Mould ripened cheese	<5°C 1–8 w	X	–	–	–	X	Photooxidation	–	High	High	High	High	–
Butter, oil, dairy spreads, etc	<5°C 6 w or more	X	–	X	–	X	Chemical, enzymatic	–	High	–	High	High	Odours grease
Pastas (fresh]	2–5°C 4 w	X	–	(x)	–	–	–	Low oxygen	High	High	High	High	–
Fruits and vegetables													
Root crops	0–25°C for 1w	X	–	X	X	–	–	No oxygen, no carbon dioxide	High	High	High	High	Odours
Most vegetables	0–25°C for 1w	X	–	X	X	–	–	Oxygen (1–5%), no carbon dioxide	High	High	High	High	Odours
Most fruits and some vegetables	0–18°C for 1w	X	–	X	X	–	–	Oxygen (1–5%), carbon dioxide (0–5%)	High	High	High	High	Odours
Dry products													
Flour/ grains	2°C–room temp. 3 yr.	X	–	X	–	–	Enzymatic	–	–	–	High	–	–
Powders, high fat	Room temp. 1 yr.	–	–	X	–	X	Staling	Low oxygen	High	–	High	High	–
Powders, low fat	Room temp. > 1 yr.	–	–	–	–	X	Stale flavour	Low oxygen	High	–	High	High	–
Breakfast cereals	Room temp. 1 yr.	–	–	X	–	X	Humidity	Atmospheric / low oxygen	High	Low	High	–	–

Table 5.4 (cont.) Properties of packaging materials to pack different foods.[a] Reprinted with permission from [19], K. Petersen *et al.*, Trends in Food Science and Technology, 10, 52 (1999) © 2010, Elsevier.

Food	Typical Temp.	Quality Factors Determining Shelf-life						Required Gas Composition	Optimal Required Barrier Properties				
		Microbiology	Colour	Oxidation	Structure	Flavour	Others		O_2	CO_2	H_2O	Light	Other
Pastas (dried)	Room temp. 1-6 md	-	X	X	-	-	Enzymatic	-	-	-	High	-	-
Spices and herbs	Room temp, from 6 md	-	X	X	-	X	-	-	High	-	High	High	-
Snack foods	Room temp. 1 yr	-	X	X	-	X	Enzymatic loss of cripness, Hydrolytic rancidity	Low oxygen	High	-	High	High	-
Caffees/teas	5°C-room temp 1 yr	-	-	X	-	X	Staling	Low oxygen	High	Low	High	High	-
Breads	5°C-room temp, 1d-12w	X	-	-	X	-	Staling	Low oxygen, high carbon dioxide	High	High	High	-	-
Cakes	5°C-room temp,, 1d-4 md	X	X	X	X	-	-	Atmospheric/ low oxygen, high carbon dioxide	High	High	High	High	-
Crackers and cookies	Room temp. 4 md	-	X	X	x	X	Loss of crispness	-	-	-	High	-	-
Chocolates	2°C-room temp. 6-12 md	-	X	X	-	X	-	-	-	-	-	High	Odours, greese
Beverages													
Water	Room temp. 1 yr	X	-	-	-	X	Migration, absorption	-	-	-	High	-	Aroma

Juice	5°C-room temp. 5d–1 yr	X	X	X	–	X (Scalping)	Photo oxidation	Atmospheric, low oxygen	High	–	High	High	–
Cafbonated drinks (beer & soft drinks)	5°C-room temp. 6 md–1 yr	X	–	X	–	X	Absorption, migration	High carbon dioxide, low oxygen	High	High	High	High	–
Frozen foods													
High fat products	−18° C 1 Yr	–	(x)	X	(x)	–	-	Low oxygen	High	–	High	High	–
Low fat products	−18°C 2yr	–	(x)	(x)	(x)	–	Enzymatic	Low oxygen	High	–	High	High	–
Others													
High fat products (dressing, sauce, etc)	5°C-room temp, 1Yr	–	(x)	x	–	–	–	Low oxygen	High	–	High	High	Aroma, grease
Ready meals	2–5°C 3–21 d	x	–	X	–	(x)	–	Low oxygen, hjgh carbon dioxide	High	High	High	–	–
[a]d=Day(s), w=week(s), md=month(s), yr=year(s)													

of different packaging materials at modified atmospheric conditions and storage temperature on physico-chemical properties of roasted pistachio nut (non oil export item of iron). The authors have reported that, the metallised film, five layers, gases of N_2/CO_2 and vacuum conditions can keep the quality of pistachio in better condition with improved shelf-life. The test temperature of 40°C was a favourable condition for pistachio quality as compared to 20°C. In a review, Khwaldia and co-workers [63] have discussed the possibility of using renewable biopolymers as a barrier coating on paper based packaging materials which can replace the conventional synthetic materials. Hirvikorpi *et al* [64] compared different coating techniques such as atomic layer deposition (ALD), electron beam evaporation, magnetron sputtering and a sol/gel method to fabricate barrier layers on packaging materials. The different coating techniques were used to deposit thin aluminum oxide coatings on to two different fiber based packaging materials of with synthetic and biodegradable polymers. The research results revealed that, biodegradable PLA coated paper board having 25 nm thick layer of aluminum oxide (deposited by ALD) showed promising barrier characteristics against water vapor and oxygen.

It was reported in a literature that, high pressure thermal (HPT) processing has the potential to deliver quality benefits to a range of processed foods. The main requirements of packaging materials for thermally processed foods are the preservation of oxygen and water barrier characteristics during processing and duration of the shelf life of the food. Bull and co-workers investigated eleven commercially available packaging materials for thermally processed foods and noticed that, barrier properties of vapor deposited oxide and nylon containing films were found to be good for thermally processed foods [65].

5.3.2 Moisture Barrier Properties

The biobased materials are generally hydrophilic in nature. As the food packaging application demands the moisture resistance, it is a challenging task for one to maintain the same with biobased materials. The moisture barrier properties of packaging materials are evaluated by water vapor transmission rate (WVTR) measurements. Based on the comparative study by Rindlav-Westing *et al* [66] and Butler *et al* [30] on the WVTR of biopolymer based materials with conventional petroleum based material, it is clear that the biobased materials can occupy the place of present conventional food packaging materials. Surface drying of certain fresh or frozen foods and moisture regain of dry foods can be reduced using films that have good barriers to moisture migration. The hydrophilic nature of biopolymer based materials can be overcome using few existing techniques such as; (i) external coating of hydrophobic/water resistant materials (waxes, polyester, fatty acid ester derivatives, etc.,) on biobased packaging materials [67] and (ii) cross-linking with inorganic fillers [68], blending with moisture resistant materials [69] and reinforcement with natural fibers (hemp, sisal, jute, coir, flax, etc.,) [70]. Gas and water vapor barrier properties of edible films obtained from protein and cellulosic materials have been investigated by Park and Chinnan [71]. The authors have noticed high water vapor permeabilities of edible films as compared to plastic

Table 5.5 Effect of wheat gluten coating on the surface treated and untreated paper on the gas and water vapor permeability values of paper samples. Reprinted with permission from [79], Guillaume *et al.*, *Food Research Intl.*, 43, 1395 (2010), © 2010 Elsevier.

Materials	Gas Permeability (10^{-18} mol m^{-1} $s^{-1}Pa^{-1}$		Water Vapour Permeability (10^{-11} $m^{-1}s^{-1}$ Pa^{-1}
	O_2	CO_2	
TP	Out of range		5.91
UTP	Out of range		3.70
WG-TP	49689	59000	3.27
WG-UTP	8328	16927	1.27

films. The potential uses of edible films (e.g. wrapping various products, individual protection of dried fruits, meat and fish, control of internal moisture transfer in pizzas, pies, etc.) based on biobased materials have been discussed by Guilbert *et al* [72]. Gontard *et al* [73] and Schultz *et al* [74] investigated the moisture permeability of films composed of demethoxylated pectins, gluten and various lipids (waxes, fatty acids, etc.). The authors have noticed that, two successive layers followed by a coating of dispersion (solvent based) would be better to achieve the moisture barrier property in protein based films. Kester and Fennema [75] have developed a composite edible film consisting of lipid and cellulose ethers. The authors have studied the moisture vapor transmission behaviour of the films. The edible film effectively retarded the transport of moisture at water activities (a_w) up to at least 0.97 and maintained a good barrier properties even when the a_w on the low-humidity side of the film was relatively high. The apparent activation energy for WVTR through the edible film was 14.2 ± 2.5 kcal/mole. Guillaume [76] discussed the effect of wheat gluten (WG) coating on the untreated paper (UTP) substrate and surface treated paper (TP) substrate for food packaging applications. The gas and water vapor permeabilities of TP, UTP, WG-TP and WG-UTP are presented in Table 5.5. From the table, it can be concluded that the WG coating on UTP plays a dominant role to reduce the water vapor permeability of the films. Water vapor permeability, thermal and wetting properties of WPI based edible films have been discussed by Kokoszka *et al* [77]. The authors have noticed a good film barrier properties for the lowest WPI (7 %) and glycerol (40 %) contents. It was reported in the literature that the coating of WPI, PVB and zein on paper board improves the water barrier properties of paper board by decreasing the MNS by 77–78 % [78].

5.3.3 Mechanical and Thermal Properties

The package of unsuitable quality can adversely affect the quality and hygiene of packed food. In particular, mechanical properties of the packaging materials are the parameters to ensure food protection against mechanical damage and duration of the guarantee term. The mechanical and thermal properties are very important from the view point of processing and end use of the products derived from the materials. The modulus of biobased materials ranges from 2500 to 3000 MPa and lower for stiff polymers like thermoplastic starches (50 MPa) [79]. Butler *et al* [30] have studied the mechanical and barrier properties of edible chitosan films. The effects of plasticizer concentration (0.25 and 0.50 mL/g chitosan) and storage time on the films have been studied. A reduction in tensile strength (15–30 MPa) and an increased percentage elongation (25–45%) were noticed as a function of storage time. Gennadios and co-workers [80] have been investigated the mechanical and barrier properties of egg-albumen films. Egg albumen films were casted from aqueous egg albumen solutions containing different amount or of GLY, PEG, sorbitol (S) and plasticizers. It was found by the authors that, the PEG plasticised films possess relatively high tensile strength as compared to GLY plasticised films. The plasticizing effect of GLY and the crosslinking effect of formaldehyde and calcium chloride on the mechanical and thermo-mechanical properties of whey protein based films were studied by Galietta *et al* [81]. Increased plasticizer content increased the percentage solubility in water and reduced the mechanical resistance, apparent modulus, and glass transition temperature (T_g) of whey protein films. The incorporation of formaldehyde as a crosslinking agent enhanced the mechanical properties, T_g and insolubility behaviour. The effect of processing condition and different amounts of plasticiser (GLY) content on SPI films have been studied by Guerrero *et al* [82]. Films containing 40% GLY and obtained through compression moulding technique has showed higher tensile strength and elongation at break than the one obtained through solvent casting technique (Figure 5.4). Analysis of the thermal properties revealed that, denaturation temperatures of the two main globulin fractions present in SPI were influenced by the

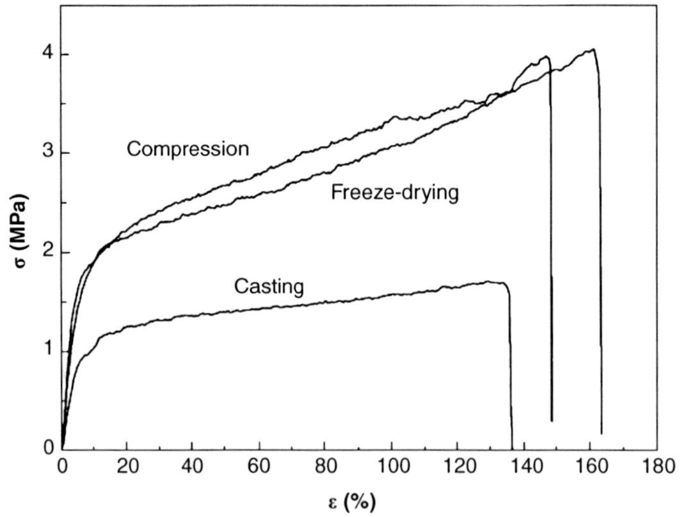

Figure 5.4 Effect of different processing condition and plasticiser (GLY) content on the tensile strength and elongation at break of soy protein isolate (SPI) films. Reprinted with permission from [82], Guerrero *et al., J. Food Eng.*, 100, 145 (2010) © 2010, Elsevier.

moisture content of the sample but not by the processing method employed. The authors also revealed the effect of plasticiser content on the thermal degradation of SPI films. Thermogravimetric analysis (TGA) indicated that, the SPI films exhibit substantial thermal degradation at temperatures above 180°C. Further, Guerrero and Caba [83] studied the effect of pH on the mechanical properties of SPI films. Both tensile strength and elongation at break found to be higher at basic pHs. Mechanical properties remained invariable after having been stored under specific conditions for two months. The characterisation of various polymer materials such as PP (0.03 mm), PE (0.1 and 0.03 mm), PHB, two-layered PP (0.064 mm) and two-layered PP with PVDC (0.012/0.021) used in food packaging materials were investigated by Kljusuri [84] using differential scanning calorimeter (DSC).

5.3.4 Biodegradability

The biodegradability is often used in synonym with compostability. According to the biodegradable products institute (BPI), a biodegradable plastic is the one in which degradation results from the action of naturally occurring micro-organisms such as bacteria, fungi or algae. This takes place in two-steps: (i) degradation/de-fragmentation initiated by heat, moisture or microbial enzymes and (ii) is biodegradation. During the process of biodegradation, the large molecules of the substance are transformed into smaller compounds by enzymes and acids that are naturally produced by micro-organisms. Once the molecules are reduced to a suitable size, the substances can be absorbed through the organism cell walls where they are metabolized for energy.

The increasing garbage mountain is recognized as an ecological threat. Space for landfill is limited and additional incineration capacities require high capital investments and pose additional environmental problems. It was reported in a literature that, household waste of 700,000 tones (out of 14 million tones) was made from non-biodegradable plastics [polyolefines, PS and polyvinylchloride (PVC)] especially used for packaging [85]. The strategies adopted to reduce 5% house hold waste plastics are through recycling, chemical valorization, incineration and use of degradable polymers. As the term biodegradable does not imply a fast process, it is important to couple the term biodegradable with the specification of the particular environment where the biodegradation is expected to happen and of the time scale of the process. The composting may be considered to be a recycling process only if reintegration of the recycled material is being allowed in the market. Because of too many uncontrollable factors in outdoor landfills, the bioreactor appears to be more effective to study the biodegradation process [86].

5.4 Packaging Products from Biobased Materials

The market prices of large-scale (commodity) biobased industrial products would depend on two primary factors: (i) the cost of biobased raw material from which a product is made and (ii) the cost of processing technology to convert the raw material into a desired biobased product. The repeating chemical units of biobased

Table 5.6 Different process adopted to manufacture different packaging products for food packaging applications.

Products	**Process**
Film	Extrusion, co-extrusion, casting
Sheets, trays, cups	Thermoforming
Pots, drinking beakers, cups, plates, bottles, trays	Injection (blow) moulding
Diapers, hygiene products, medical plastics, clothing (nonwovens)	Needle punching, wet laid, etc
Laminated paper or films	Extrusion coating

materials are similar to those of conventional plastics. For example, (i) the peptide functionality of protein is similar to synthetic polyamides, (ii) ester functionality of polylactic acid is similar to polyesters and (iii) acetal functionality of polysaccharides is similar to polyacetals, etc. The same processing technique adapted to process the synthetic plastics can be adapted to process biobased materials to obtain different shape and types of food packaging. The challenging task is to meet the requirements of the food packaging with biobased materials. The packaging products that can be produced or manufactured from different plastic processing techniques are summarised in Table 5.6.

5.4.1 Blown Films

The blown film process is the most diverse conversion system used for PE. The blown film process can produce materials as thick as 0.5 mm (20 mils). Monolayer, laminated and multilayer co-extrusion technologies lay the groundwork for many possibilities to approach an application's need. Compared to industrial packaging applications, food and retail packaging applications needs high optical clarity. Some food packaging needs strength and others need tearablility. Petroleum derived biodegradable polyesters are the first product categories developed to obtain the blown films. The developed films have been successfully used as garbage bags and for related applications. PLA based bio-renewable polymers have been developed as a substitute to petroleum derived materials. Blown films of PLA have shown an excellent transparency and cellophane like mechanical properties. For food packaging applications, the films must have excellent water vapor and gas barrier properties. It may not be possible to achieve these required properties with the use of a single biobased polymer. The utilisation of co-extrusion can produce laminates which can meet the required properties. For example, the materials based on thermoplastic starch film can be film blown in a co-extrusion set up with polymers like PLA and PHB/V as coating materials. The resultant products have been successfully used for the packaging of cheese [87]. In this way, starch based materials could become as cheap alternatives to presently available expensive EVOH and polyamides. George *et al* [88] have developed and characterised the rice bran filled biodegradable LDPE films using twin-screw extruder for packaging applications. Aerobic biodegradation tests using municipal sewage sludge and biodegradation studies using specific microorganism (streptomyces species) revealed that the developed films are biodegradable. Biotec research team [89] has developed a film named 'BIOFLEX' from thermoplastic starch granules (BIOPLAST) using film extrusion process. The research study revealed that, the BIOFLEX is permeable to water vapor and has good barrier properties to oxygen. BIOFLEX can be used in the same way as conventional foils, for instance for garbage sacks, shopping bags, packing up, nets for food packaging, diapers, agricultural or technical uses. It was reported in the literature that the mechanical properties of blown films can be improved by coating a glass-like ultra thin layer of SiO_x or by producing nanocomposites. The incorporation of nanoparticles during the processing of films is expected to produce films having high gas and water vapor barrier properties [90].

5.4.2 Foamed Products

The foam products in food packaging industry has been popular due to its stable form, light in weight, stackable nature, ecologically sound material and its suitability for packaging machines. Single-use packaging articles made from expanded polystyrene (EPS) are currently used to serve and pack a variety of food and non-food products. Recently, there have been efforts to develop and commercialize materials from renewable resources such as starch to replace EPS. Starch has been processed in various ways to make products with some properties similar to petroleum-based plastics. Starch may also be made into foam with insulating properties and densities that are similar to PS foam. Starch based foams are brittle, sensitive to water, and require expensive coating steps when expose to cold or hot liquid. The information in the literature revealed the possibility of making alternative food containers to EPS from mixtures of starch fibers and water using processes such as vacuum filtration [91] or thermo-pressing [92]. Shogren *et al* [93] showed that, starch foam tensile strength and density increased while foam flexibility decreased with increasing starch concentration, molecular weight and amylose content. The authors reported that tuber starch such as potato can produce trays with lower densities and higher flexibilities compared to starch obtained from cereals such as corn. Others reported that foams made from chemically modified starches exhibited lower baking time, lower weight and higher deformability than unmodified starches, while foams made from genetically modified (waxy) starches added with polyvinyl alcohol (PVA) showed higher elongations at break. Salgado and co-workers [94] studied on the biodegradable food packaging trays made from cassava starch, sunflower proteins and cellulose fibers. The authors revealed that, the increment of fiber concentration from 10 to 20 % improves the mechanical properties and slightly reduces the post-pressing moisture content. From the study of Shogren *et al* [95], it appears that, the baked foams made from potato amylopectin, PVA, aspen fiber and monostearyl citrate could have adequate flexibility and water resistance to function as clamshell-type hot sandwich containers. Cinelli *et al* [96] investigated the possibility of making starch based foamed plates using corn fiber (as a reinforcement and a by-product from bioethanol industries). The authors revealed that, the starch based trays with a high corn fiber ratio and PVA improves the water resistance. A study of Glenn *et al* [97] described an in situ method of baking and laminating the starch based foams in a single step. The procedure involved a dough formulation consisting of starch fiber and magnesium stearate followed by heating in a mould at 160°C between two sheets of laminate. The authors have noticed that, the laminated foams generally had a higher density, tensile strength, percentage elongation at break and flexural strength compared to non-laminated sample. In a comparative study, Zabaniotou and Kassidi [98] investigated the life cycle assessment (LCA) of egg packages made from PS and recycled paper. The egg cups made out of recycled paper seems to have less environmental impact than the PS. Pimpa *et al* [99] prepared the starch based foam using sago starch and PVA or polyvinyl pyrrolidone (PVP). The prepared foams have been exposed to electron beam irradiation to achieve crosslinking. The research results revealed that, the good foams with high linear expansion and closed cell structure can be produced from the blends having different ratio (25:15 and 30:10) of sago starch using electron beam irradiation of 15 or 10 kGy for the crosslinking of the blends. Preechawong *et al* [100] studied the preparation and characterisation of starch/PLA hybrid foams. Based on the study, the authors have concluded that, the hybrid foams have better water resistance than pure starch foams. Glenn and Orts [101] developed a compression/explosion method for making moulded starch based foam with physical and mechanical properties similar to the foam used in commercial food packaging. The scanning electron microscopic (SEM) image of wheat starch (prepared using compression/explosion method) and commercially available EPS foam is shown in Figure 5.5. The SEM images revealed that, the cells of starch foams were generally larger than those of EPS foam. Starch

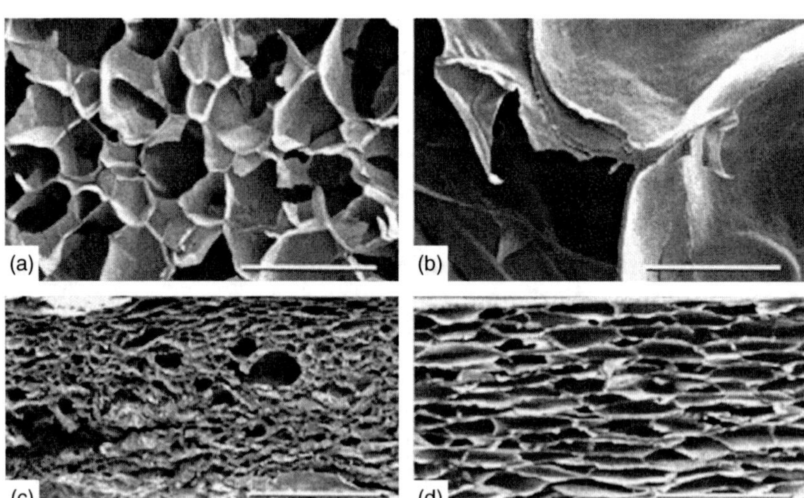

Figure 5.5 Scanning electron micrographs of wheat starch foam made by the compression/explosion process (a–c) and extruded polystyrene foam (d). Scale bars; a, 1 mm, b, 0.2 mm, c, 0.5 mm, d, 1 mm. Reprinted with permission from [101], Glenn and Orts., *Industrial Crops and Products*, 14, 201 (2001) © 2010, Elsevier.

acetate-based foams have been prepared by Guan and Hanna [102] using an extrusion process. The effect of different degree of substitution (DS) of starch acetate, cellulose content, process (barrel) temperature and screw speed revealed that, the melting temperatures changed significantly when higher DS starch acetate was used. Cellulose content, barrel temperature and screw speed showed significant effects on thermal, physical and mechanical properties of extruded foams. The incorporation of calcium carbonate into the starch foam formulation/process did not improve the mechanical properties of the foams [92]. Mali *et al* [103] used the extrusion process to prepare biodegradable foams from starch, bagassefiber, and PVA. It was noticed by the authors that, the addition of PVA in high proportions (40%) increased the expansion index and led to a significant reduction in water adsorption of starch foams.

5.4.3 Thermoformed Containers

The containers obtained through the process of thermoforming are known as thermoformed containers. Thermoforming is a manufacturing process where a plastic sheet is heated to a pliable forming temperature in a mould followed by trimming to create a specific shape product. In food packaging applications, these thermoformed products are utilized to pack cakes, muffins, cookies, ready-to-eat salads, meat, cheese, chickens, sandwich wedges, etc. A single polymer is often unable to provide a suitable barrier and hence, most food packaging materials are multilayer constructions. In multilayer constructions, polymers with different barrier properties are combined where at least one layer acts as an oxygen barrier and other layer acts as water barrier and polymers provides sealing properties. A commonly used converting method to obtain multilayer polymer structures is thermoforming where the thermoplastic sheet/film is heated to a temperature above its softening temperature and then formed/stretched [104]. In thermoforming process, thinning of the material may be a problem when a thin oxygen barrier is included in the composite sheet to be thermoformed [105, 106]. The effect of stretching during the thermoforming process on the oxygen transmission rate (OTR) of the cream cheese packing has been studied by Petersen *et al* [107]. Lim *et al* [108] evaluated the thermoforming technology for PLA and compared the thermal properties (important for thermoforming process) of PLA with other known conventional polymers (Table 5.7) such as PS and polyethyleneterephthalate (PET). Compared to other conventional polymers, PLA has low thermal conductivity and T_g.

5.4.4 Adhesives

Adhesive in packaging area are used in three ways; to form (i) the structure of the packaging material by combination of substrates, (ii) the geometric shape of the package and (iii) to apply labels, for printing and other miscellaneous operations such as extrusion coating, lacquering and the manufacture of adhesive tapes. Canellas *et al* [109] have discussed on the identification of the non-volatile compounds as potential migrants from adhesives used in food packaging. Choi *et al* [110] developed a biodegradable hot melt adhesive based on polycaprolactone (PCL)

Table 5.7 Comparative properties of polylactic acid (PLA), polystyrene (PS) and polyethyleneterepthalate (PET). Reprinted with permission from [113], Lim *et al.*, *Progress in Polymer Science*, 33, 820 (2008) © 2010, Elsevier.

Properties	PLA	PS	PET
Thermal conductivity ($\times 10^{-4}$ cal.cm^{-1} s^{-1} ^{0}C^{-1})	2.9	4.3	5.7
Heat capacity (cal.g^{-1} ^{0}C^{-1})	0.39	0.54	0.44
Glass transition temperature (^{0}C)	55	105	75
Thermal expansion coefficient ($\times 10^{-6}$ ^{0}C^{-1})	70	70	70

and SPI for food packaging system. The elongation of the PCL/SPI hot melt adhesive was varied with the type of plasticizer used. Lap shear strength of PCL/SPI hot melt adhesive was about 1.9 MPa. The suitability of biodegradable copolymers based on poly (L-lactide) (PLLA) and PCL for a packaging hot-melt adhesive was evaluated by Viljanmaa *et al* [111]. Most of the measured properties were comparable to a conventional non-biodegradable EVA based hot melt adhesive which was used as a typical commercial reference. Dahmane [112] discussed on the development and advantages of new range of lower application temperature hot melt adhesives. In canned food packaging, the rapid determination of the durability and quality of a coating to protect corrosion is a practical problem. The problem has to be considered by both can making and the food industry. Barilli and co-workers [113] evaluated the adhesion of three types of lacquers (stoving and UV curing) applied on different tinplate substrate. In detail, the methods used to evaluate the performance are dry adhesion, wet adhesion and electrochemical impedance spectroscopy (EIS) measurements. The EIS method was useful to identify the influence of both types of lacquers and tinplate supports.

5.4.5 Coated Paper

Biobased materials are expected to stay as an important packaging materials. Paper and paper board materials have excellent mechanical properties but the gas permeabilities are too high for many food applications. The water sensitivity of paper and paper board is a major challenge to use them to pack moist foods. In order to make water resistant paper, it has been coated with a thin layer of synthetic plastic which is capable to provide the required gas permeability. It is also possible to explore the biobased coating in place of synthetic one to produce 100% biobased packaging material. In literature, several animal and vegetable proteins such as caseinates [114], WPI [115–118], corn zein [114] and SPI [119, 120] have already been considered as coatings for paper and paper board with the goal of improving surface and/or mass transfer properties of the cellulosic substrate. It has been reported that, the WPI coating on paper board improves the barrier properties against water vapor, oil resistance, printability of water-based ink without changing the mechanical and optical properties [116, 117]. The paper board

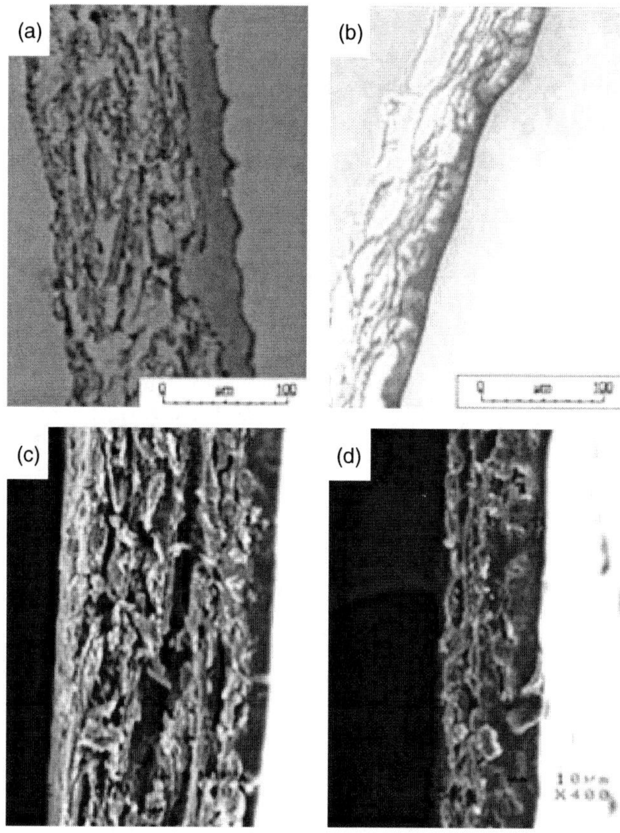

Figure 5.6 Cross sectional views of wheat gluten- coated papers obtained by optical microscopy using amido black as protein dying: (a) WG–TP; (b) WG–UTP and by scanning electron microscopy: (c) WG–TP, (d) WG–UTP. The WG- coated layer is located at the right of all micrographs. Reprinted with permission from [76], Guillaume *et al., Food Research Intl.*, 43, 1395 (2010) ©2010, Elsevier.

is expected to become grease and water resistant with soy protein coat [121]. The effect of WG coating on the treated paper (TP) substrate and untreated paper (UTP) substrate on the structural, surface, water vapor and gas barrier properties was investigated by Guillaume *et al* [76]. The optical microscopic images of WG coated paper (both on TP and UTP) are shown in Figure 5.6. From the images, the surface of both TP and UTP appears to be homogeneous. The authors have noticed a maximum penetration of WG into UTP compared to TP. Due to the maximum penetration, a composite like structure was observed in case of UTP whereas, a bi-layer structure was observed for WG–TP.

Triantafyllou *et al* [122] studied on the migration of organic pollutants from recycled paper board packaging materials to solid food matrices. The proportion of substances migrated to food was found to depend on the nature of the paper samples, fat content of the food, chemical nature and volatility of the migrant. The highest level of migration of organic pollutants was observed for the substrate with the highest fat content. Furthermore, it is shown that, contact time and temperature have a significant effect on migration of model contaminants into foods. The effect of hydroxypropyl methylcellulose (HPMC), GLY and beeswax (BW) coatings on paper have been investigated by Sothornvit [123]. It was reported that, HPMC coating on paper improves the physical properties as compared to uncoated paper.

5.5 Food Applications

In recent years, though there are a lot of interest and research activities on biobased materials for the application of food biopackaging, the unavailability of biobased packaging materials in the market is evident and it looks that more scientific studies on these materials are very much required. Food manufacturers and packaging producers are currently testing biobased packaging materials for foods. Because of the confidential nature of the work, it may be difficult to get more information on these findings.

As the various types of food needs to be packed, stored and transported, the biobased packaging materials should meet the food packaging requirements as similar to conventional food packaging materials. The packaging requirements of different food products in terms of deteriorative reactions which limits their shelf lifes are discussed below;

Snacks: The loss of crispiness and fat rancidity development are the most common type of deterioration in case of snack foods. The packaging material should have low oxygen and water vapor permeability property, good mechanical strength and resistance to light [124]. In order to retard the development of rancidity and protect against moisture vapor, the air has to be removed from the snack product package and flushed with nitrogen [125]. Most of the snack foods are being packed using form fill sealing method [126]. Multi layer structures are known materials used to pack extruded, fried and puffed snack food. Food items such as chips and nuts have been packed using spiral wound paper board cans lined with aluminum foil or a barrier of polymer. Metal cans are used for fried nuts and the container is usually gas flushed with nitrogen. Siripatrawan and Jantawat [127] investigated a novel method of shelf life prediction for a package (for snack packaging) sensitive to moisture. Artificial neural network (ANN) based on multilayer perceptrons (MLP) with back propagation algorithm was developed to predict the shelf life of packaged rice snack stored at different temperatures and relative humidity (RH) comparable to tropical storage conditions. Palou *et al* [128] have studied the effect of different temperatures (25, 35 and 45 °C) on the moisture sorption isotherms of some cookies and corn snacks. It was found that, the isotherms of each product were different ($p<0.05$) and significantly affected by temperature. Various biobased materials such as zein, hydroxypropyl cellulose (HPC), WPI, etc., have already been studied for the application of roasted peanut packaging [129]. The snacks traditionally packed in paper cans or cartons (less resistance to oxygen and water permeability) can be replaced with PHB, PLA or modified starch coated paper board which has high resistance to oxygen and water vapor permeability.

The use of pigments in the formulation of biobased packaging materials can minimise the penetration of light which in turn provides safety to food.

Dairy Products: The dairy products such as milk products, processed cheese, cream and milk requires a packaging material having low oxygen permeability to avoid oxidation and also resistance towards the growth of undesirable microorganisms. In addition, light initiates the oxidation of fats in dairy products and leads to discoloration, off-flavour formation and nutrient loss even at temperatures found in refrigerated display cabinets. Once the oxidative reaction is initiated by light, it may continue even if the products are protected from light. The absorption of odours from the surroundings, water evaporation and maximisation of the shelf life are the few important points to be considered to protect the dairy products. Simon and Hansen [130] evaluated the suitability of different packaging material to pack the dairy products. Milk from three different sources were standardized to 2% fat and pasteurized at different temperatures for 25 sec and packaged into six different packaging boards. Standard plate count (SPC) was used to test the microbial quality of milk. Light-induced degradation reactions in milk create a serious problem for the dairy industry because of the development of off-flavours, the decrease in nutritional quality and the rate by which these phenomena develop. Mestdagh *et al* [131] made an attempt to study the effect of different packaging material on the light oxidation of milk. The milk was packed in 3 types of PET bottles; one transparent bottle provided with active oxygen binding inner layer, one bottle with perfect light barrier, and one transparent bottle provided with a UV absorbing additive. The results of the studies showed that, an adequate light barrier was apparently sufficient to avoid the light induced oxidation of milk during extended storage. Oxygen or UV barriers did not provide a significant protection. Zygoura and co-workers [132] revealed that, the packaging materials such as a mono/multi layer high density polyethylene (HDPE) and PET found to provide sufficient protection against microbial growth as well as degree of lipid oxidation, lipolysis and proteolysis in pasteurized milk stored under fluorescent light and refrigeration for a period of 7 days. The biobased packaging appears to be interesting to pack the cheese due to the fact that cheese releases carbon dioxide during storage. Due to this fact of respiration, packaging materials must have a relatively high carbon dioxide transmittance to avoid inflation of the packages. Plackett *et al* [133] have developed modified and unmodified PLA and PLA/PCL copolymer film for cheese packaging applications. The research investigation revealed that, the copolymer films are relatively efficient to control the fungi development on the packaged cheese. Cerqueira *et al* [134] investigated on functional polysaccharides as an edible coating for cheese. The research investigation revealed that, the uncoated cheese had an extensive mould growth at the surface when compared with the coated cheese. The results are encouraging to use these biobased coatings for cheese as an alternative to synthetic coatings.

Beverages: A beverage is a liquid which is specifically prepared for human consumption. The microbial growth, oxidation of flavour components, loss of carbonation (in case of carbonated beverages) and migration are the few limiting factors which limits the shelf life of beverages. Considering the aforementioned limiting factors, the packaging material must possess low gas transmission, light permeability and resistance towards scalping (migration from food product to package). As most of the beverages contain water, the packaging material must have high water vapor barrier properties to prevent penetration of the beverage through the package. The acidic beverages need a packaging material resistant to acids [19]. Hot/cold filling and aseptic packaging with or without nitrogen injection are the known methods of the packaging beverages [135, 136]. Active packaging is a packaging that performs a role other than as an inert barrier to the outside environment. An example of active packaging is wine skins that collapse with removal of the wine to maintain a minimal headspace in the package. Dawson [137] discussed in detail on the application of active food packaging for beverages. The existing conventional packaging materials for beverage are HDPE, paper, glass, metal, PET (for carbonated drinks, beer), etc. Huggard and Fesersen [138] suggested that, PLA and PHB bottles or cups can be used for packaging orange juice. In a comparative study, PLA cups have exhibited low water vapor permeability and high resistance to scalping as compared to PE. The coating of PHB on PLA or paper board can provide low oxygen transmission rate and water resistance to PLA or paper substrate and such biobased packaging system can be used as packaging materials for beverages.

Ready to Eat Foods: is a self-contained individual ration in a light packaging. Fresh, healthy, safe/easy to eat and of good quality are the few important highlights which contributed to the growth of this sector. Moreover, consumers prefer frozen ready-meals because of their long shelf-life and these products offers better manufacturing/distribution flexibility, food safety and extended storage time. The main challenge for packaging the ready to eat foods is maintaining their shelf life. The extent of oxidative changes and growth of microorganisms dictates the shelf life of ready meals in chill storage [125]. A characteristic off flavour termed as 'warmed over flavour' (WOF) can arise in ready to eat meals due to the rapid oxidative changes [139]. Modified atmospheric packaging (MAP) with nitrogen to replace oxygen and carbon dioxide to inhibit microorganisms growth are well known methods. MAP is a common technical definition that describes the practice of modifying the composition of the internal atmosphere of a package (commonly food packages, but this technique is also used for drugs) in order to improve the shelf life. The study of Murcia and co-workers [140] confirmed that, the vegetables stored using MAP technique has showed good antioxidant capacity and no significant losses in antioxidant activity or scavenging capacity. In a review, Belcher [141] discussed in detail about the processing and packaging of ready to eat foods in particular ready to eat meat. The PHB can be selected as a biobased material to pack ready to eat food due to its low oxygen and water vapor permeability. Though the coating of PHB on the paper board appears as a potential packaging material for ready to eat meals, the coating of PHB seems to be difficult on paper board compared to PE. The edible films and coatings appear to have potential applications to pack ready to eat foods. For example, an edible film/coating composed of alginate or pectin between the base and sauce component of pizza could reduce the water migration between the sauce and base. The edible coatings containing antioxidant components can minimize the WOF in cooked meats.

Fruits and Vegetables: Packing fresh fruits and vegetables is one of the more important steps in the long and complicated journey from grower to consumer. In case of fruits and vegetables, the increased respiration rate even after harvesting can make them more susceptible to the development of disease organisms. In addition to this, the changes in the concentrations of water, carbon dioxide and ethylene can be expected inside the storage packs. Lee *et al* [142] revealed that the changes in gas composition may have a positive influence on the color and flavour of the product but they may also induce negative effects on the texture, shelf life, nutritional quality and color. The review article of Sandhya [143] was discussed on the MAP specifically for fresh fruits and vegetables. Lin and Zhao [144] in their review articles discussed the possibility of edible coatings for fresh and minimally processed fruits or vegetables. The review article highlighted the recent development of coating technology and analytical techniques for measuring some important coating functionality. The physico-chemical properties and application of pullulan (polysaccharide polymer) edible films and coatings in fruit preservation have been reported by Diab *et al* [145]. The authors have noticed a big change in internal fruit atmosphere composition which was beneficial for extending the shelf-life with pullulan based coating on strawberries. Garcia *et al* [146] showed that chitosan coatings can improve the efficiency of osmotic dehydration process in ripening stages of papaya. In general, a good packaging material for fruits and vegetables must consider the respiration processes of the products so that an optimum atmospheric balance (CO_2/O_2 ratio) inside the packaging is maintained. In addition to this, the packaging materials should provide protection against light and mechanical damage, retain desirable odours and prevent odour pickup.

Fresh Meat: Meat is defined as the flesh of animals used as food. The term 'fresh meat' includes the meat from just processed animals as well as packed meat. The diverse nutrient composition of meat makes the growth and propagation of meat spoilage micro-organisms and common food borne pathogens. A high oxygen level over the meat product surface is required to preserve the red color of the fresh meat which is attributed to oxymyoglobin. The color, low water permeability and microbiology are the critical factors have to be considered while selecting the material for packaging the fresh meat. Oxygen permeable packs, vacuum packs or MAP are known packaging systems for fresh meats. Vacuum packaging is expected to minimise the color and flavour defects due to oxidation. The prevention of microbial growth can be expected with MAP and useful to pack red meats. Active packaging consists of oxygen absorbing sachets found to be useful to pack cured meat products [147]. Skandamis and Nychas [148] discussed in detail about the MAP and active packaging systems for the preservation of fresh meats. The potential of bioactive packaging technologies for the preservation of meat and meat products was discussed by Coma [149]. In a review, Zhou and Liu [150] have discussed in detail about the various preservation technologies for fresh meat. The discussed technologies are MAP, vacuum packaging (VP), active packaging (AP), antimicrobial packaging and bioactive edible coatings. The research investigation of Seydim *et al* [151] revealed that, the VP was better compared to MAP (high N_2 and air) technology for ground ostrich meat packaging. The investigations of

Cannarsi *et al* [152] suggested that, the biodegradable starch, polyester and their blend films could be used to replace PVC films to pack fresh processed meat. Sekar and co-worker [153] revealed that, the buffalo meat packed under modified atmosphere and vacuum can keep the meat safely up to 14 days of storage at 4 ± 1 °C. The opportunities for bio-based packaging technologies to improve the quality and safety of fresh and further processed muscle foods was reviewed by Cutter [9].

5.6 Nanotechnology

Nanotechnology is a method of controlling the matter at near atomic scales to produce unique or enhanced materials, products and devices. The commercialization of nanocomposite materials was started by Toyota in the late 1980s. In the 1990s, research on use of nanocomposites for food packaging started using montmorillonite (MMT) clay as the nano component in a number of polymers like PVC, PE, nylon and starch [154]. According to Principia Markets, the nanocomposites market would reach 1 billion pounds in a near future.

The incorporation of nanofillers into synthetic polymers enhanced the properties of polymers and opened up a new avenue/market for nanomaterials. The grand success of nanomaterials development using synthetic polymer motivated for a new research on nanocomposites development using biodegradable polymers as matrix. Nanotechnology has the capability to transform the nature of food packaging materials in future. Certain nano scale innovations could bring amazing improvements to food packaging sector in the forms of detection of pathogens, improved mechanical/barrier properties and smart/active packaging technologies. The incorporation of nanomaterials can produce a smart packaging material which means that it can respond to environmental conditions or repair itself or alert consumer to contamination/pathogens. Nanomaterials technology offers several extraordinary benefits to improve food packages like advancements in fundamental characteristics of food packaging materials such as antimicrobial properties, barrier properties, strength and stability to heat/chilled environment [155, 156].

For packaging applications, relatively poor mechanical and barrier properties of biopolymers (compared to synthetic ones) limit their industrial use. As discussed in the earlier sections, the main challenge with biopolymers is to provide moisture barrier properties due to the presence of hydrophilic groups in biopolymers. However, it has been suggested that inherent shortcomings of biopolymer based packaging materials may be overcome by nanocomposite technology [157]. Montmorillonite (MMT) and kaolinite clays are the most promising nanoscale size fillers. The research and development with respect to graphite nanoplates is in progress. The various nanocomposites available in nature may be inspired scientists to develop nanotechnology using biobased materials. The few nanocomposites available in nature are; (i) seashell has superior mechanical strength and toughness. It consists of carbonate mineral, aragonite and only 1% of organic biopolymer by volume and (ii) bone in an orderly layered array [25].

The literature revealed that, the nanocomposites were produced from different biobased materials such as starch [158–160], protein [161, 162], cellulose [163, 164] and PLA [165, 166] for food packaging applications. Sorrentino [167] discussed in detail on potential perspectives of nanobiocomposites for food packaging applications. The main kinds of nanoparticles which have been studied for use in food packaging systems are overviewed by Azeredo [168]. In a comparative study, Cava *et al* [169] revealed that, the nanobiocomposites of PCL and PLA can enhance the oxygen barrier but are not sufficient to out perform as high-oxygen barrier grades of PET. Morphology and barrier properties of nanobiocomposites of PHB and layered silicates were investigated by Garcia *et al* [170]. High water barrier nanobiocomposites of methyl cellulose and chitosan for film and coating applications was investigated by Lagaron and Fendler [171]. Based on the morphological study, the authors indicated that, a good dispersion with intercalation of the fillers can be obtained in both cellulose and chitosan matrices. The water barrier properties of the nanobiocomposites were found to be enhanced to a significant extent particularly for the high aspect ratio filler as compared to pure matrix materials.

Nanobiocomposites of carrageenan, zein and mica were investigated by Garcia *et al* [172] for food packaging and coating applications. Further Garcia *et al* [173] investigated the synergistic effect of active technologies (antimicrobials) and nanotechnologies to obtain a bioplastic formulation with balanced properties and functionalities for packaging applications. Investigations of Vartiainen *et al* [174] revealed that, the hybrid nanocomposite films made from pectin and MMT exhibits improved barrier properties against oxygen and water vapor. The authors further revealed that, the incorporation of fluidised nanoclay compared to ultrasonically treated nanoclay would be more effective in dispersing the gel like

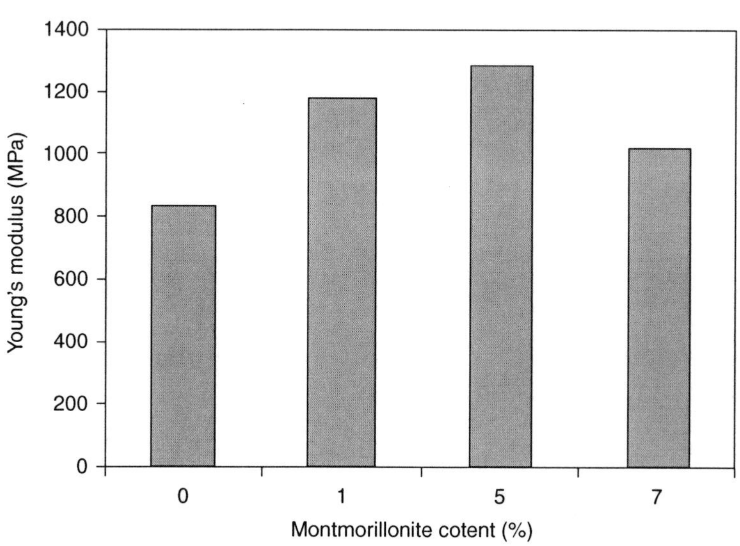

Figure 5.7 Young's modulus of starch–montmorillonite (clay) nanocomposites containing different amounts of clay. The samples were conditioned at 43% RH before measurement. Reprinted with permission from [175], Chung *et al.*, *Food Carbohydrate Polymers*, 79, 391 (2010) © 2010, Elsevier.

pectin matrix. Achieving a high level of clay dispersion in the polymeric matrix using suitable plasticizers and controlling the interfacial strength between starch and clay is a main challenge in preparing high performance starch–clay nanocomposites. Chung *et al* [175] have prepared well-dispersed starch-clay nanocomposites by adding dilute clay (MMT) dispersion to a solution of starch followed by co-precipitation in ethanol. The effect of incorporation of different amounts of MMT on the modulus of starch-clay nanocomposites (Figure 5.7) revealed that, the incorporation of MMT up to 5% can significantly enhance the modulus up to 65 % and a reduction in modulus was reported above 5 % and at 7 % MMT incorporation.

5.7 Conclusions

The utilisation of biobased materials received a considerable attention in world wide to create a market for environmentally friendly packaging materials. The biodegradable packaging materials has greatest potential in countries where landfill is the main waste management tool. Though, it appears that, the packaging products such as trays, cups, films, bottles, etc., can be manufactured using the same technology as conventional materials, the properties of biobased packaging materials needs to compete with highly developed and sophisticated materials used today.

Food packaging is a challenging task because food materials are complex and diverse in nature. Apart from starch/cellulsoe based packaging material for food packaging applications, the biologically derived polymers have also been well explored and commercially introduced for food packaging applications. The properties of biobased materials discussed in comparison with conventional synthetic polymers reveals a great potential of biobased materials in food packaging applications. The inherent property of biobased materials is compostability/biodegradability which is an added advantage and may help in commercialization of these materials. However, a breakthrough research is required to provide moisture resistance to biobased packaging materials. Further, the high cost of biobased materials due to low production volumes compared to synthetic ones needs to be addressed. It is also very essential to produce a biobased material with improved properties than the ones available today to completely eliminate the dependency on petroleum based raw materials. The final and full success of biobased materials for food packaging applications should comply with the safety and quality requirements of the food product and meet legal standards (FDA). The inclusion of small nanoparticles into the biobased materials produces nanobiocomposites which appears to be helpful to improve the properties of packaging materials.

Acknowledgements

The authors are thankful to National University of Malaysia (UKM) for the financial support under the grant UKM-OUP-TK-16-72/2009 to carry out this project.

References

1. R. Ahvenaines, *Novel Food Packaging Techniques*, Boca Raton, CRC Press, 2003.
2. A.L. Bredy, and K.S. Marsh, *The Wiley Encyclopedia of Packaging Technology* (2nd ed.), Wiley, New York, 1997.
3. S. Guilbert, Potential of the protein based biomaterials for the food Industry, Proceed. of the Food Biopack Conf., Copenhagen, pp. 13–18, 2000.
4. J.H. Han, *Innovations in Food Packaging*, Elsevier Academic Press, San Diego, California, 2005.
5. V. Siracusa, P. Rocculi, S. Romani, and M. Dalla Rosa, *Trends in Food Sci. & Tech.*, Vol. 19, p. 634, 2008.
6. W.E. Brown, *Plastics in Food Packaging Properties, Design and Fabrication*, Marcel Dekker, New York, 1992.
7. R.C. Hermand, *Biopolymers-Making Materials Natures' Way*, John Wiley, New York, 1993.
8. C.S.K. Reddy, R.Ghai, and V.C. Rashmi, *Biores. Technol.*, Vol. 87, p. 137, 2003.
9. C.N. Cutter, Meat Science, Vol. 74, p. 131, 2006.
10. C.J. Weber, *Biobased Packaging Materials for the Food Industry: Status and Perspectives*, p. 15, KVL, London, 2000.
11. S.I. Hong, and J.M. Krochta, *J. Food Eng.*, Vol. 74, p. 739, 2004.
12. V.K. Haugaard, A.M. Udsen, G. Mortensen, L.H. Egh, K. Petersen, and F. Monahan, *Starch/Strake*, Vol. 53, p. 189, 2000.
13. R.N. Tharanathan, *Trends Food Sci. Technol.*, Vol. 14, p. 17, 2003.
14. Packaging Digest, Biodegradable packaging to grow at CAGR of 22 percent, http://www.packagingdigest.com/article/CA6490177.html,2007.
15. V.A. Fomin, and V.V. Guzeev, "Biodegradable Polymers, their present state and future prospects," *Progress in Rubber and Plastics Tech.*, p. 186, 2001.
16. R. Chandra, and R. Rustgi, "Biodegradable Polymers," *Progress in Polym. Sci.*, Vol. 23, p. 1273, 1998.
17. R. Montgomery, *Biores. Tech.*, Vol. 91, p. 1, 2004.
18. J.M. Krochta, De Mulder-Johnston, G. Fuller, T.A. McKeon, and D.D. Bills, In "Agricultural materials and renewable resources," *Am. Chem. Soc.*, Washington, DC, p. 120, 1996.
19. K. Petersen, P.V. Nielsen, M.B.G. Lawther, M.B. Olsen, N.H. Nilsson, and G .Mortensen, *Trends in Food Sci. & Tech.*, Vol; 10, p. 52, 1999.
20. J.F. Lambert and G. Poncelet, *Topics in Catalysis*, Vol. 4, p. 43, 1997.
21. P. Ren, T. Shen, F. Wang, X. Wang, and Z. Zhang, *J. Polymers and the Environment*, Vol. 17, p. 203, 2009.
22. L.P.B.M. Janssen, and L. Moscicki, *Thermoplastic starch: A Green Material for Various Industries*, Wiley-VCH, Verlag GmbH & Co., KGaA, Weinheim, 2009.
23. M.A.L. Russo, Cathryn O'Sullivan, B. Rounsefell, P.J. Halley, R. Truss, and W.P. Clarke, *Biores. Tech.*, Vol 100, p. 1705, 2009.
24. I. Arvanitoyannis, C.G. Biliaderis, H. Ogawa, and N.Kawasaki, *Carbohydrate Polymers*, Vol. 36, p. 89, 1998.
25. M.Avella, J.J. De Vlieger, M.E.Errico, S.Fischer, P. Vacca, and M.G. Volpe, *Food Chemistry*, Vol. 93, p. 467, 2005.
26. E. Psomiadou, I.Arvanitoyannis, C.G. Biliaderis, H. Ogawa, and N.Kawasaki, *Carbohydrate Polymers*, Vol 33, p. 227, 1997.
27. M. Itävaara, M. Siika-aho, and L.Viikari, *J., Polymers and the Environment*, Vol. 7, p. 1566, 1999.
28. D. Knoor, *Food Technology*, Vol. 45, p. 114, 1991.
29. J.M. Krochta, and C.L.C. De Mulder-Johnston, "Biodegradable Polymers from Agricultural Products," *ACS Symp. Series*, Vol. 647, 1996.
30. B.L. Butler, P.J. Vergano, R.F. Testin, J.M. Bunn, and J.L. Wiles, *J. Food Sci.*, Vol. 61, p. 953, 1996.
31. P. Appendini, and J.H. Hotchkiss, *Innovative Food Sci. & Emerging Technologies*, Vol. 3, p. 113, 2002.
32. Stefania Quintavalla, and Loredana Vicini, *Meat Science*, p. 373, 2002.

33. Blaise Ouattara, Ronald E. Simard, Gabriel Piette, André Bégin, and R.A. Holley, *Intl. J. Food Microbiology*, Vol. 62, p. 139, 2000.
34. J. Duan, Y. Jiang, G. Cherian, and Y. Zhao, *Food Chem.*, Vol. 122, p. 1035, 2010.
35. S.M. Ojagh, Masoud Rezaei, Seyed Hadi Razavi and S.M.H. Hosseini, *Food Chemistry*, Vol. 122, p. 161, 2010.
36. V. Morillon, F. Debeaufort, G. Blond, M. Capelle, and A. Voilley, *Critical Reviews in Food Science and Nutrition*, Vol. 42, p. 67, 2002.
37. T.H. McHugh, *Nahrung* Vol. 44. p. 148, 2000.
38. F. Callegarin, J.A. Quezada Gallo, F. Debeaufort, and A. Voilley, *J. of the Am. Oil Chemists' Society*, Vol. 74, p. 1183, 1999.
39. N. Gontard, R. Thibault, B. Cuq, and S. Guilbert, *Journal of Agricultural and Food Chemistry*, Vol. 44, p. 1064, 1996.
40. Joo-Won Lee, Seok-Min Son, and Seok-In Hong, *J. Food Eng.*, Vol. 86, p. 484, 2008.
41. R.P. Wool and X.A. Sun, *Bio-based Polymers and Composite*, Elsevier, Amsterdan, The Netherlands, 2005.
42. A. Garde, A.S. Schmidt, G. Jonsson, M. Andersen, A.B. Thomsen, B.K Ahring, and P. Kiel, Agricultural crops and residuals as a basis for polylacetate production in Denmark. Proceed. of the Food Biopack Conf., Copenhagen, p. 45, 2000.
43. A. Mohanty, M. Misra, and L.T. Drazal, *Natural Fibre, Bio-Polymers and Bio-composites*, CRC Press, Taylor & Francis Group, Boca Raton, FL, USA, 2005.
44. R.G. Sinclair, "The case for polylactic acid as a commodity packaging plastic," *Polymeric Materials: Sci. and Eng.*, p. 133, 1996.
45. A.Guinault, C. Sollogoub, S. Domenek, A. Grandmontagne, and V. Ducruet, *Intl. J. Material Forming*, Vol. 3, p. 603, 2010.
46. E. Mascheroni, V. Guillard, F.Nalin, L.Mora, and L.Piergiovanni, *J. Food Eng.*, Vol. 98, p. 294, 2010.
47. Showa Denko, High Polymer, Company information. Showa Denko (Europe) GmbH, Uhlandstr, Düsseldorf, Germany, http://www.showa-denko.com/index. php/id=65 (accessed date 26.08.2010).
48. C.N. Cutter, *Critical Reviews in Food Science and Nutrition*, Vol. 42, p. 151, 2002.
49. G.T.M. De Koning, and P.J. Lemstra, *Polymer*, Vol. 34, p. 4089, 1993.
50. D.Z. Bucci, L.B.B. Tavares, and I. Sell, *Polymer Testing*, Vol. 24, p. 564, 2005.
51. G.Q. Chen, *Plastics from Bacteria, Microbiology Monographs*, p. 121, 2010.
52. R.M. Brown, *J. Macromol. Sci.*, Pure Appl. Chem., Vol. A33, p. 1345, 1996.
53. A. Okiyama, M. Motoki and S. Yamanaka, *Food Hydrocolloids*, Vol. 6, p. 503, 1993.
54. E.J. Vandamme, S. De Baets, A. Vanbaelen, K. Joris, and P. De Wulf, *Polym. Degradation and Stability*, Vol. 59, p. 93, 1998.
55. V.T. Nguyen, M.J. Gidley, and G.A. Dykes, *Food Microbiology*, Vol. 25, p. 471, 2008.
56. Johnsy George, K.V. Ramana, S.N. Sabapathy, and A.S. Bawa, *World J. Microbiology and Biotechnology*, Vol. 21, p. 1323, 2005.
57. O.R. Fennema, and S.R. Tannenbaum, *Food Chemistry*, Marcel Dekker, New York, USA. 1996.
58. Hood Chatham, *Surface and Coatings Technology*, Vol. 78, p. 1, 1996.
59. K.S. Johansson, Improved barrier properties of renewable and biodegradable polymers by means of plasma deposition of glass-like SiOx coatings, Proceed. of the Food Biopack Conf., Copenhagen, Denmark, p. 110, 2000.
60. M. Rubino, S. Netramai, R. Auras, and B.A. Angus, *J. Appl. Polym. Sci.*, Vol. 115, p. 1742.
61. Eva-Lena Hult, M. Iotti, and M.Lenes, *Cellulose*, Vol. 17, p. 575, 2010.
62. M. Raei, A. Mortazavi, and H.Pourazarang, *Food Analytical Methods*, Vol. 3, p. 1291, 2010.
63. K. Khwaldia, E. Arab-Tehrany, and S. Desobry, *Comprehensive Reviews in Food Science and Food Safety*, Vol. 9, p. 82.
64. T. Hirvikorpi, M. Vähä-Nissi, A. Harlin, and M. Karppinen, *Thin Solid Films*, Vol. 518, p. 5463, 2010.
65. M.K. Bull, K. Zerdin, E. Howe, D. Goicoechea, P. Paramanandhan et al., *Innovative Food Sci. & Emerging Technologies*, Vol. 5, p. 135, 2004.

66. A. Rindav-Westling, M. Stading, A.M. Her-Mansson, and P. Gatenholm, *Carbohydrate Polym.*, Vol. 36, p. 217, 1998.
67. N. Gontard, and S. Guilbert, "Biopackaging technology and properties of edible and or biodegradable material of agricultural origin," in: *Food Packaging and Preservation*, Mathlouti M, Blackie, eds., London, p. 159, 1994.
68. J.U. Otaigbe, *Plastics Engineering*, p. 37, 1998.
69. P.J. Stenhouse, J.M. Mayer, M.J. Hepfinger, E.A. Costa, P.A. Dell, and D.L. Kaplan, "Starch based blown films," in: *Biodegradable Polymers and Packaging*, C. Ching Technomic Pub. Comp., Lancaster, p. 151, 1993.
70. M.H.B. Snijder, and J.E.G. van Dam, "Process for continuously manufacturing composites of commodity plastics and jute/allied bast fibres and compounded materials obtained thereof," *Packaging India*, p. 23, 1999.
71. H.J. Park, and M.S. Chinnan, *J. Food Eng.*, Vol. 25, p. 497, 1995.
72. S.Guilbert, N.Gontard, and B.Cuq, *Packaging Tech. and Sci.*, Vol. 8, p. 339.
73. N. Gontard, S. Marchesseau, J.L. Cuq, and S. Guilbert, *Intl. J. Food Sci. Techol.*, Vol. 30, p. 49, 1995.
74. T.H. Schultz, J.C. Meter, H.S. Owens, and W.D. Maday, J. Phys. *Colloid Chem.*, p. 1320, 1999.
75. J.J. Kester, and O. Fennema, *J. Food Sci.*, Vol. 54, p. 1383.
76. Carole Guillaume, Jeremy Pinte, N. Gontard, and E. Gastaldi, *Food Research Intl.*, Vol. 43, p. 1395, 2010.
77. Sabina Kokoszka, Frederic Debeaufort, Andrzej Lenart, and Andree Voilley, *Intl. Dairy J.*, Vol 20, p. 53, 2010.
78. Jaejoon Han, Stphane Salmieri, Canh Le Tien, and Monique Lacroix, *J. Agric. Food Chem.*, Vol. 58, p. 3125, 2010.
79. M. Lguchi, S. Yamanaka, and A. Budhioni, *J. Materials Sci.*, Vol. 35, p. 1, 2000.
80. L.A. Gennadios, C.L. Weller, M.A. Hanna, and G.W. Froning, *J. Food Sci.*, Vol. 61, p. 585. 1996.
81. G. Galietta, L. Di Gioia, S. Guilbert, and B. Cuq, *J. Dairy Sci.*, Vol. 81, p. 3123, 1998.
82. P. Guerrero, A. Retegi, N. Gabilondo, and K. de la Caba, *J. Food Eng.*, Vol. 100, p. 145, 2010.
83. P. Guerrero, and K.de la Caba, *J. Food Eng.*, Vol. 100, p. 261, 2010.
84. J.G. Kljusuri, *Sadhana*, Vol. 28, p. 991, 2003.
85. H.H.Roper, and H. Koch, *Starch - Starke*, Vol. 42, p. 123.
86. L.R. Krupp, and W. Jewell, *Environ. Sci. Technol.*, Vol. 26, p. 193, 1992.
87. R. Tuil van, G. Schennink, B.H. de, Heemst, J. Van, and R. Jaeger, Converting biobased polymers into food packaging. Proceed. of the Food Biopack Conf., Copenhagen, Denmark, p. 28, 2000.
88. J. George, R. Kumar, C. Jayaprahash, A. Ramakrishna, S.N. Sabapathy, and A.S. Bawa, *J. Appl. Polym. Sci.*, Vol. 102, p. 4514, 2006.
89. J. Lorcks, *Polym. Degradation and Stability*, Vol. 59, p. 245, 1998.
90. S. Fischer, J. Vlieger de, T. Kock, J. Gilberts, H. Fischerand, L. Batenburg, Green composites—the materials of the future—a combination of natural polymers and inorganic particles. Proceed of the Food Biopack Conf, Copenhagen, Denmark, p. 109, 2000.
91. K.N. Matsui, F.D.S. Larotonda, S.S. Paes, D.B. Luiz, A.T.N. Pires, and J.B. Laurindo, *Carbohydrate Polymers*, Vol. 55, p. 237, 2004.
92. G.M. Glenn, W.J. Orts, and G.A.R. Nobes, *Industrial Crops and Products*, Vol. 14, p. 201, 2001.
93. R.L. Shogren, J.W. Lawton, W.M.Doane, and K.F.Tiefenbacher, *Polymer*, Vol. 39, p. 6649, 1998.
94. Pablo R. Salgado, Vivian C. Schmidt, Sara E. Molina Ortiz, Adriana N. Mauri, and J. Laurindo, *J. Food Eng.*, Vol. 85, p. 435, 2008.
95. R.L.Shogren', J.W.Lawton, and K.F.Tiefenbacher, *Industrial Crops and Products*, Vol. 16, p. 69, 2002.
96. P. Cinelli, E. Chiellini, J.W. Lawton, and S.H.Imam, *Polym. Degradation and Stability*, Vol. 91, p. 1147, 2006.
97. G.M. Glenn, W.J. Orts, G.A.R. Nobes, and G.M. Gray, *Industrial Crops and Products*, Vol. 14, p. 125, 2001.
98. A. Zabaniotou, and E. Kassidi, *J. Cleaner Production*, Vol. 11, p. 549 , 2003.
99. B.Pimpa, K.Muhammad, Zulkafli Ghazal, K.Hashim, M.Ali Hassan, and D.M. Hashim, *Carbohydrate Polymers*, Vol. 68, p. 751, 2007.

100. D. Preechawong, M. Peesan, P. Supaphol, and R. Rujiravanit, *Carbohydrate Polymers*, Vol. 59, p. 329, 2005.
101. G.M. Glenn,' and W.J. Orts, *Industrial Crops and Products*, Vol. 13, p. 135, 2001.
102. J. Guan, and M.A. Hanna, *Bioresource Technology*, Vol. 97, p. 1716, 2006.
103. S.Mali, Flávia Debiagi, Maria V.E. Grossmann, and Fábio Yamashita, *Industrial Crops and Products*, Vol. 32, p. 353, 2010.
104. A. Aroujalian, M.O. Ngadi, and J.P. Emond, *Polym. Eng. and Sci.*, Vol. 37, p. 178, 1997.
105. M. Bellet, M.H. Vantal, and B. Monasse, "Thermal effects in the numerical simulation of the thermoforming of multilayered polymer sheets," *International Polymer Processing XIII*, p. 299, 1998.
106. J.K. Lee, T.L.Virkler, and C.E. Scott, *Polym. Eng. and Sci.*, Vol. 41, p. 240, 2001.
107. M.K. Pettersen , T. Eie, and A. Nilsson, *Intl. Dairy J.*, Vol. 15, p. 355, 2005.
108. L.T. Lim, R.Auras, and M.Rubino, *Progress in Polym. Sci.*, Vol. 33, p. 820, 2008.
109. E. Canellas, C.Nerín,' R. Moore, and P. Silcock, *Analytica Chimica Acta*, Vol. 666, p. 62, 2010.
110. W.Y. Choi, C.M. Lee, and H.J. Park, *LWT Food Sci. and Tech.*, Vol. 39, p. 591, 2006.
111. M. Viljanma, A. Södergård, and P. Törmälä, *Intl. J. Adhesion and Adhesives*, Vol. 22, p. 219, 2002.
112. H. Dahmane, *Int. J. Adhesion and Adhesives*, Vol. 16, p. 43, 1996.
113. F. Barilli, R. Fragni, S. Gelati, and A. Montanari, *Progress in Organic Coatings*, Vol. 46, p. 91, 2003.
114. E. Gastaldi, P. Chalier, A. Guillemin, and N. Gontard, *Colloids and Surface*, Vol. 301, p. 301, 2007.
115. M. Gallstedt, A. Brottman, and M. Hedenqvist, *Packaging Tech. and Sci.*, Vol. 18, p. 161, 2005.
116. J.H. Han, and J.M. Krochta, Wetting properties and water vapour permeability of whey protein coated paper, Transactions of the ASAE, p. 1375, 1999.
117. J.H. Han, and J.M. Krochta, *J. Food Sci.*, Vol. 66, p. 294, 2001.
118. S.Y. Lin, and J.M. Krochta, *J. Food Sci.*, Vol. 68, p. 229, 2003.
119. P. Chalier, A. Ben Arfa, L. Preziosi-Belloy, and N. Gontard, *J. Appl. Polym. Sci.*, Vol. 106, p. 611, 2007.
120. A. Ben Arfa, Y. Chrakabandhu, L. Preziosi-Belloy, P. Chalier and N. Gontard, *Food Research Intl.*, Vol. 40, p. 22, 2007.
121. J.W. Rhim, J.H. Lee, and S.I. Hong, *LWT – Food Sci. and Tech.*, Vol. 39, p. 806, 2006.
122. V.I. Triantafyllou, K.Akrida-Demertzi, and P.G. Demertzis, *Food Chemistry*, Vol. 101, p. 1759, 2007.
123. R. Sothornvit, *Food Research Intl.*, Vol. 42, p. 307, 2009.
124. D.G. Quast, and M. Karel, *Journal of Food Science*, Vol. 37, p. 584, 1972.
125. T.P. Labuza, *Shelf-life Dating of Foods*, Food & Nutrition Press: Westport, Connecticut. Matz, S.A. (1993).
126. S.A. Matz, *Snack Food Technology*, Van Nostra Reinhold, New York, 1993.
127. U. Siripatrawan, and P. Jantawat, *Expert Systems with Applications*, Vol. 34, p. 1562, 2008.
128. E. Palou, A. López-Malo, and A. Argaiz, *J. Food Eng.*, Vol. 31, p. 85, 1997.
129. J.I. Maté, E.N. Frankel, and J.M. Krochta, *J. Agricultural and Food Chem.*, p. 1736, 1996.
130. M. Simon, and A.P. Hansen, *J. Dairy Sci.*, Vol. 84, p. 767, 2001.
131. F. Mestdagh, B. De Meulenaer, J. De Clippeleer, F. Devlieghere, and A. Huyghebaert, *J. Dairy Science*, Vol. 88, p. 499,2005.
132. P. Zygoura, T. Moyssiadi, A. Badeka, E. Kondyli, I. Savvaidis, and M.G. Kontominas, *Food Chemistry*, Vol. 87, p. 1, 2004.
133. D.V. Plackett, V.K. Holm, P. Johansen, S. Ndoni, P.V. Nielsen, T.Sipilainen-Malm, A. Södergard, and S. Verstichel, *Packaging Tech. and Sci.*, Vol. 19, p. 1, 2006.
134. Miguel A. Cerqueira, lvaro M. Lima, Bartolomeu W.S. Souza, J.A. Teixeira, and Renato A. Moreira, *J. Agric. Food Chem.*, p. 1456, 2009.
135. C.E. Sizer, P.L. Waugh, S. Edstam, and P. Ackermann, Maintaining flavour and nutrient quality of aseptic orange juice, *Food Tech.*, p. 154, 1988.
136. M.R.McLellan, L.R. Lindand, and R.W. Kime, A shelf life evaluation of an oriented polyethylene terephthalate package for use with hot filled apple juice. *J. Food Sci.*, p. 365, 1987.
137. P.L. Dawson, *Active Packaging for Beverages, Beverage Quality and Safety*, edited by T. Foster, and P.C. Vasavada, CRC Press.
138. V.K. Haugaard, and R.M. Festersen, Biobased packaging materials for foods. In: Conf. Proceed. The Food Biopack Conf. Denmark, p. 119, 2000.

139. H. Stapelfeldt, H. Bjørn, and L.H. Skibsted, Effect of packaging and storage conditions on development of warmed over flavour in sliced, cooked meat, *Zeitschrift für Lebensmittel Untersuchung und Forschung*, p. 131, 1993.
140. M.A. Murcia, A.M. Jiménez-Monreal, L. García-Diz, M. Carmona, L. Maggi and M. Martínez-Tomé, *Food and Chemical Toxicology*, Vol 47, p. 2103, 2009.
141. J.N. Belcher, *Meat Science*, Vol. 74, p. 143, 2006.
142. L. Lee, J.Arul, R. Lencki, and F. Castaigne, *Packaging Tech. and Sci.*, Vol. 9, p. 315, 1995.
143. Sandhya, *LWT - Food Sci. and Tech.*, Vol. 43, p. 381, 2010.
144. D. Lin, and Y. Zhao, *Comprehensive Reviews in Food Sci. and Food Safety*, Vol. 6, p. 60. 2007.
145. T. Diab, C.G Biliaderis, D. Gerasopoulos, J. Evangelos, *J. the Sci. of Food and Agriculture*, Vol. 81, p. 988, 2001.
146. M. García, R. Díaz, Y. Martínez and A. Casariego, *Food Research Intl.*, Vol. 43, p. 1656, 2010.
147. H.J. Andersen, and M.A. Rasmussen, *International Journal of Food Science & Technology*, Vol. 27, p. 1, 1992.
148. P.N. Skandamis, and G.J.E. Nychas, *Intl. J. Food Microbiology*, Vol. 79, p. 35, 2002.
149. V. Coma, *Meat Science*, Vol. 78, p. 90, 2008.
150. G.H. Zhou, X.L. Xu, and Y. Liu, *Meat Science*, Vol. 86, p. 119, 2010.
151. A.C. Seydim, J.C. Acton, M.A. Hall and P.L. Dawson, *Meat Science*, Vol. 73, p. 503, 2006.
152. M. Cannarsi, A. Baiano, R. Marino, M. Sinigaglia, and M.A. Del Nobile, *Meat Science*, Vol. 70, p. 259, 2005.
153. A. Sekar, K. Dushyanthan, K.T. Radhakrishnan, and R. Narendra Babu, *Meat Science*, Vol. 72, p. 211, 2006.
154. M. Kawasumi, *J. Polym. Sci., Part A: Polym. Chem.*, Vol. 42, p. 819, 2004.
155. S. Sinha Ray, and M. Okamoto, *Prog. Polym. Sci.*, Vol. 28, p. 1539, 2003.
156. S. Sinha Ray, A. Easteal, S.Y. Quek, and X.D. Chen, *Int. J. Food Eng.*, Vol. 2, p. 1, 2006.
157. Amit Arora ad G.W. Padua, *J. Food Sci.*, Vol. 75, p. 43, 2010.
158. P. Fratzl, H.S. Gupta, E.P. Paschalis, P. Roschger, *J. Mater. Chem.*, Vol. 14, p. 2115, 2004.
159. B.Q. Chen, and J.R.G. Evans, *Carbohydr. Polym.*, Vol. 61, p. 455, 2005.
160. H.M. Park, X. Li, C.Z. Jin, C.Y. Park, W.J. Cho, and C.S. Ha, *Macromol. Mater. Eng.*, Vol. 287, p. 553, 2002.
161. J.J. Zhou, S.Y. Wang, and S. Gunasekaran, *J. Food Sci.*, Vol. 74, p. 50, 2009.
162. L. Shi, J. Zhou, and S. Gunasekaran, *Mater. Lett.*, Vol. 62, p. 4383, 2008.
163. H.M.C. Azeredo, L.H.C. Mattoso, D. Wood, T.G. Williams, R.J. Avena-Bustillos, and T.H. McHugh, Nanocomposite edible films from mango puree reinforced with cellulose nanofibers. *J. Food Sci.*, p. 31, 2009.
164. M.R. De Moura, R.J. Avena-Bustillos, T.H. McHugh, J.M. Krochta, and L.H.C. Mattoso, *J. Food Sci*, Vol. 74, p. 31, 2009.
165. C. Thellen, C.Orroth, D. Froio, D. Ziegler, J. Lucciarini, R.Farrell, N.A. D'Souza, and J.A. Ratto, *Polymer*, Vol. 46, p. 11716, 2005.
166. J.W. Rhim, S. Hong, and C. Ha, "Tensile, Water Vapour Barrier and Antimicrobial Properties of PLA/Nanoclay Composite Films," *LWT—Food Sci. Technol.*, Vol. 42, p. 612, 2009.
167. A.Sorrentino, G.Gorrasi, and V.Vittoria, *Trends in Food Sci. & Tech.*, Vol. 18, p. 84, 2007.
168. M.C. Henriette de Azeredo, *Food Research Intl.*, Vol. 42, p. 1240, 2009.
169. D. Cava, E. Giménez, R. Gavara and J.M. Lagaron, *J. Plastic Film and Sheeting*, Vol. 22 , p. 265, 2006.
170. M.D.Sanchez-Garcia, E.Gimenez, and J.M.Lagaron, *J. Appl. Polym. Sci.*, Vol. 108, p. 2787, 2007.
171. J.M. Lagaron, and A. Fendler, *J. Plastic Film and Sheeting*, Vol. 25, p. 47, 2009.
172. M.D. Sanchez-Garcia, L. Hilliou, and J.M. Lagaron, *J. Agric. Food Chem.*, Vol. 58, p. 6884, 2010.
173. M.D. Sanchez-Garcia, M.J. OcioE. Gimenez, and J.M. Lagaron, *J. Plastic Film and Sheeting*, Vol. 24, p. 239, 2008.
174. J. Vartiainen, T. Tammelin, J. Pere, U.Tapper, and A. Harlin, *Carbohydrate Polymers*, Vol. 82, p. 989, 2010.
175. Yi-Lin Chung, S. Ansari, L. Estevez, S. Hayrapetyan, E.P. Giannelis, and Hsi-Mei Lai, *Carbohydrate Polymers*, Vol. 79, p. 391, 2010.

6

Polylactic Acid (PLA) Foams for Packaging Applications

Kate Parker, Jean-Philippe Garancher, Samir Shah, Stephanie Weal and Alan Fernyhough

Biopolymer Network Limited / Scion, Rotorua, New Zealand

Abstract

Polylactic acid (PLA) foams are emerging as one of the most promising new material options for sustainable packaging applications. The inherent renewable carbon content and optional end of life compostability for PLA foams, together with an increasing range of new polymer grades and additives for enhancements in the performance and/or processability of PLA materials are making PLA foams increasingly attractive. In particular, they are being evaluated as alternatives to conventional polymer foams used in packaging such as polystyrenes and, also, increasingly, as alternatives to polyolefin and other foams used in packaging. This article reviews extrusion foaming, expanded particle (bead) foaming, and other foam processes for making polylactic acid foams and describes some key performance features of such foams.

Keywords: Polylactic acid, foam, packaging, extrusion, expanded bead, particle

6.1 Introduction

Polymer foams are widely used in packaging, and indeed many other applications, due to their unique combination of properties such as mechanical integrity, protective cushioning, insulation and low density. They can be a very cost effective material option for packaging applications where light weight and protective functionality are often critical.

As with other industrial polymers, the polymers used in foams and foam packaging are primarily derived from petroleum. The current global consumption of industrial polymers in plastics applications is more than 200 million tons [1] - emphasizing how dependent the polymer industry is on oil. The most common polymers used in packaging are polyethylene terephthalate (PET), polyvinylchloride (PVC), polyethylene (PE), polypropylene (PP), and polystyrene (PS) [2]. They are available at low cost and have a good balance of performance features. Of these, polystyrene, and to a lesser extent, polyethylene and polypropylene, are the most commonly used in polymer foam packaging.

Srikanth Pilla (ed.) Handbook of Bioplastics and Biocomposites Engineering Applications, (161–176)

The high growth in foamed polymer materials has led to an increasing interest in developing materials which are more environmentally friendly [3]. Lower carbon footprints and favorable life cycle assessments, while still retaining cost-effectiveness and functional performance, are desired for packaging materials. Many factors influence the impact of a material or product on the environment. Weight is a key aspect, hence low density foams are often a very desirable packaging material option as less polymer is required for its production. Start-of-life factors (the resources/inputs used) and end-of-life (disposal/reuse options) are also important. In addition to environmental concerns, the effects of fluctuating crude oil and natural gas pricing on polymer markets are considerable.

Consequently, it is becoming increasingly important to use alternatives to non-renewable oil derived polymers. Biopolymers, that is polymers derived from renewable biobased resources, are increasing in popularity. A number of biopolymers offer compostability as an added attribute and this can be an attractive feature for some biopolymer materials or applications since it offers an alternative route of disposal to land filling. A packaging market report in 2010 indicated that many suppliers, converters, and users (such as retail chains) are making huge investments in biopolymers, and this will impact the packaging industry widely with a prediction that global biopolymer consumption will increase by more than 40% per year during the next five years [4].

Of the available renewable biopolymers today, polylactic acid (PLA) is increasingly preferred since it has a favorable environmental profile and has many processing and performance features which allow it to compete with some petroleum polymers [5–7]. It is also one of the most available and more cost competitive bioplastics with the price dropping lower than polystyrene on several occasions in the last five years [8]. PLA has been approved for use for food contact and is generally recognized as safe (GRAS) [5]. Furthermore, it is compostable according to international standards such as the European standard EN 13432. As a result, the use of PLA for packaging applications is becoming increasing popular, most obviously in the food packaging arena where the discarded packaging does not require separation or decontamination of food waste prior to composting. However, PLA foam is suitable for a wide range of packaging applications due to its excellent insulation properties and good mechanical properties and, also, due to other performance enhancements such as improved heat resistance or flame retardancy which are becoming increasingly available through developments in polymer manufacture, modification or compounding for example.

This Chapter will report on recent technical developments relating to PLA foams for packaging applications, primarily focusing on extrusion foaming and particle (bead) foaming of PLA.

6.2 Polylactic Acid (PLA) Foam Overview

6.2.1 Extruded Foam

Extrusion foams are a major source of packaging foams. Considerable effort has been applied to the development of extrusion foams based on PLA. Although PLA has many attributes, and is in some ways similar to existing petroleum derived polymers, it also has some shortcomings such as brittleness, thermal resistance, and melt strength and stability. The latter is particularly important in extrusion processing and extrusion foaming.

One avenue which has been pursued to overcome some of PLA's processing sensitivities in extrusion foaming is that of using a low shear screw technology such as that developed by Plastics Engineering Associates with their Turbo-Screws® technology [9]. The technology has been trialed in industry facilities and has shown promising results. It has been successful in foaming amorphous and semi-crystalline PLA resins. The Turbo-Screws® technology has reportedly been licensed in the US and Europe. Turbo-Screw® technology has rectangular holes through the screw flights at the root, which move melt from the root to the barrel wall for faster cooling and higher output. The technology appears to works best within a tandem foam extrusion system. The ability to obtain and maintain melt temperature is important. Turbo-Screws® use a "technologically superior screw technology." Turbo-Screws® technology is listed as a preferred equipment partner on NatureWorks®, LLC's website. Sheet densities of ~ 50 g/L are claimed to be readily achieved, although for meat tray type materials higher densities appear to have been more typical. Closed cell counts of 90% and greater are achievable, as are higher percentages of open cell structures. A thickness range down to 0.5 mm and up to 5 mm have been claimed. These cover much of the range for thermoformed food packaging. The material is apparently easily thermoformed and trimmed into containers. Many of the traditional blowing agents used in direct gas foaming have been used, although most of the experience to date has been with hydrocarbons. Higher heat resistant grades are under development and some have been trialed with second and third generation Turbo-Screws® technology screws more specifically designed for PLA foam extrusion. The applications targeted are for hot food contact and uses higher crystallinity grade resins under development at Natureworks®.

In other work on extrusion foaming, using more conventional screw designs, a stable and high melt viscosity is desirable. A common approach to achieving this, via increasing the molecular weight or chain morphology of the polylactic acid, is the use of chain-extenders, the principles of which are well known [10, 11]. Chain-extenders react with a polymer (PLA in this case) to form a higher molecular weight polymer by means of reactions most often occurring during extrusion. Chain-extenders typically have two or more reactive functionalities to link existing polymer chains together. Examples of reactive functionalities are hydroxyl, amine, epoxy, carboxylic acid/anhydride, ionomer and isocyanate. In practice acrylic, styrenic and/or polyethylene polymers, copolymers, terpolymers, oligomers or core-shells, usually with epoxy and/or acid/anhydride or ionomeric groups appear to be preferred. This is due to a combination of the wide range of available options, their suitable reactivity profiles with polyesters (which polylactic acid is) and ease of handling (health & safety issues), with a number being food grade certified.

A number of research publications and patents exist for such options and related routes to chain extend and/or controllably crosslink PLA for films, foams and other applications [12–14]. Indeed an increasing number of commercial suppliers are producing such additives, including those from Clariant, BASF, Johnson, Arkema, Dow (Rohm & Hass), DuPont and Teknor. Many have evolved from

PET (polyethylene terephthalate) technologies and related developments. Few are wholly biobased options, although biobased options are under development at some companies, including Scion in New Zealand. Scion has also undertaken various laboratory extrusion foaming trials with PLA and variously modified PLA compounds to produce a range of extrusion foamed PLA compounds. Other approaches to alter the rheology and/or performance of PLA for extrusion foaming include the use of nanofillers such as nanoclays [15, 16] and the use of peroxides [17] which, if appropriately selected and processed, can also introduce branching and/or light crosslinking.

Reignier *et al.*, have reported on extrusion foaming of PLA using CO_2 as a blowing agent [18]. They found good quality extruded PLA foams with densities of 20–25 g/L with 7–9 wt % CO_2 concentration but the processing window for good quality foam was reported as quite narrow. Other approaches to producing PLA foams for packaging applications include the use of chemical blowing agents [19], though densities appear to be somewhat higher than with physical blowing agents, and the processing of PLA blends in extrusion foaming. In particular, PLA–starch blends have been studied in extrusion foaming by a number of research groups [20, 21]. Good quality foams were obtained with CO_2 blowing. PLA-PHBV (polyhydroxybutyrate valerate) foams have also been studied [22] and it is likely that an increasing number of other PLA blends, with other biopolymers and/or with petroleum based polymers, will be developed for extrusion foams and packaging applications. Various patents or patent applications describe enhanced extrusion foaming, for example, using functional ethylene copolymers as modifiers for PLA in extrusion foaming [23, 24].

Commercial examples of PLA foam packaging made via extrusion include Coopbox (Italy) who developed biodegradable trays based on PLA foam. "Naturalbox" is used for packaging fresh foodstuffs [25].

Sealed Air Corporation has developed a line of environmentally sensitive foam trays for use in the packaging of fresh beef, pork, poultry, and fish: Cryovac® NatureTRAY™ Foam Trays. The PLA foam trays are certified by the Biodegradable Products Institute (BPI) [26]. The NatureTRAY™ is an Ingeo™ PLA based product line.

The Dyne-A-Pak Inc. Ingeo™ PLA foam tray is another PLA tray on the market and represents the product of a multi-year research and development effort [27]. This product has received a QSR Magazine-FPI Foodservice award for manufacturing innovation.

Figure 6.1 shows images of (a) commercial foam using Turbo-Screws® technology and (b) PLA foam meat trays.

6.2.2 Particle (Bead) Foam

Molded particle foam products are frequently used in packaging of goods such as electronics, whiteware, furniture, and biomedical supplies as they allow complex shapes to be made [28]. Regardless of the polymer being foamed, particle foams are generally produced using a similar method. The process starts with an expandable polymer bead, that is, a bead containing a blowing agent such as a volatile

Figure 6.1 Extruded PLA Foam (top) and PLA Foam meat trays (bottom). Photos courtesy of Plastic Engineering Associates Licensing, Inc., USA.

liquid or a compressed gas. Heat, often in the form of steam, is applied to the bead causing it to expand, or foam, to low density. These beads are then conditioned and then placed into a mold where further heat is applied allowing the bead to fuse to form a solid article. The process is most well known for producing EPS (expanded polystyrene particle foam). The first patent for the production of EPS was published about 60 years ago. Today, in the case of expanded polystyrene particle foam (EPS), typically a polystyrene granulate, which contains a foaming agent such as pentane, is pre-foamed at temperatures typically above 90°C. The foamable granulates are usually made by either a suspension polymerization process of styrene monomer with blowing agent incorporation, or by an extrusion process of polystyrene polymer with blowing agent incorporation. The pre-foaming of such granulates causes the foaming agent to evaporate and inflate the thermoplastic polymer to 20 to 50 times its original size, forming expanded PS foam particles or beads. Molding processes then convert these into fused blocks, panels or shaped elements in a superheated steam process, typically at temperatures between 110°C and 120°C.

This process has been modified by several groups investigating the production of PLA particle foam intermediates and molded products. The following is a review of some of the technologies that have been described.

JSP Corporation (Japan) has filed several patents describing PLA foam moldings [29, 30]. The claims for the patent are titled "Expanded polylactic acid resin beads and foamed molding obtained therefrom" [29], focus on the characteristics in heat flux of the pre-foamed bead, as measured by scanning differential calorimetry, and not on the foaming process. However, the examples given in the patent describe the foaming and molding process used by JSP. To summarize, semi-crystalline PLA beads containing small amounts of talc (to provide nucleation sites to enhance foaming) were impregnated with gaseous CO_2 at pressures of 30 or 40 bar for up to 3.5 hours. The amount of CO_2 impregnated in the bead was up to 14 wt.%. Steam was then applied to the beads for five seconds (the expansion temperature was reported to be 64 to 70°C) giving an apparent density of the expanded beads of 50 to 60 g/L. The expanded beads were then subjected to a second impregnation step prior to being placed into the mold and fused using steam (molding temperature around 120°C). JSP Corporation do not as yet appear to be producing commercial products using their technology.

The Biopolymer Network Ltd. (BPN), New Zealand, has developed a PLA particle foam technology which uses commercial PLA polymer without the addition of a nucleating agent [31]. The BPN E-PLA (expanded polylactic acid) process uses moderate temperature and pressure conditions with liquid CO_2 as blowing agent. This has advantages in processing compared to gas impregnation (longer times in the overall process) and supercritical CO_2 impregnation (much higher equipment cost). The technology is also noteworthy for using conventional (commercially available) PLA grades as feedstock, and can use various modified PLAs. This approach potentially allows for easy "drop-in" for existing EPS manufacturing plants, whereas other E-PLA technologies described may require specialized PLA grades, integrated coating processes or other special integrated PLA polymer production and foam production approaches for the manufacture of expanded PLA particle foam products. The claims in the initial BPN patent state that PLA polymer beads were impregnated with CO_2 at 60 bar for up to four hours at which time the CO_2 content of the beads was between 18 and 35 wt.%. The examples describe pre-foaming of the impregnated beads at temperatures of 50 to 70°C before transferring these into a mold for fusing. The final density of the molded articles was reported as between 30 and 60 g/L. More recently BPN has described successful advances such as flame retardant E-PLA versions and also commercial-scale production runs of foamed PLA beads and E-PLA mouldings with various partners. The trials "demonstrated the potential of expanded PLA beads as a realistic alternative to EPS." Industrial scale trials were performed at several existing EPS molding manufacturers located in New Zealand, Europe and USA [32]. Figure 6.2 shows images of (a) loose beads and (b) molded particle foams from the BPN process.

Synbra Technology bv, a Dutch company traditionally specializing in Expandable Polystyrene (EPS), has developed an expanded PLA particle foam material called BioFoam®. A patent [33], which focuses mainly on coatings of the PLA beads (to improve fusion of the PLA beads during molding) describes the method of producing expandable PLA, as well as the technology used for the production of the expanded and fused article. Semi-crystalline PLA beads containing low

Figure 6.2 Loose PLA Foam beads (top, left) and PLA Foam molded particle foam products. Photos courtesy of Biopolymer Network Ltd., New Zealand, and partners.

concentrations of a nucleating agent can be either impregnated with a blowing agent and expanded prior to a coating step, or coated first before impregnation and pre-foaming. The coating can be a number of materials but must be biodegradable and present in an amount between 0.5 and 15 wt% based on the weight of the PLA bead. Gaseous CO_2 is the blowing agent used for all the examples given and is impregnated at 20 bar using a range of times from 20 minutes to 16 hours. The resulting CO_2 concentration in the beads was between 5 and 8 wt.%. Pre-foaming, when used, was carried out using hot air for one minute at 90 or 110°C to a density of 60 or 45 g/L, respectively. In some cases the beads were re-impregnated with CO_2 to bring the CO_2 level back to about 3.5 wt.%. Finally, the PLA beads were molded using steam. Beads which had been through a pre-foaming step resulted in a lower density molded foam than beads which had not. Only beads which had been coated gave a molded article with "good/sufficient" fusion, comparative examples without coating disintegrated. Synbra planned to launch their first generation BioFoam® products early 2010 [34] with integrated production of the PLA polymer. However, at the time of writing BioFoam® does not yet appear to be in use commercially. They expect one of the main markets to be specialty packaging [35].

A patent application by Sekisui Plastics Company Limited (Japan) [36] describes a slightly different process to produce particle foam. Their technology does not use existing EPS-type equipment; instead the pre-foamed beads are prepared using an extruder. The semi-crystalline PLA is feed into the extruder along with the blowing agent, the examples in the patent show this to usually be a mixture of butane isomers. The PLA strand is cut at the die face and allowed to expand to densities of 20 to 600 g/L according to the claims, although the lowest density given in the examples was 140 g/L. These foam beads need to be re-impregnated using an inert gas (usually CO_2) using pressures of two to 16 bar prior to molding. In some cases an additional pre-foaming step (using hot air) followed by an additional re-impregnation step is undertaken prior to fusing, which lead to a lower density pre-foamed bead, the lowest density reported was 48 g/L. Pre-foamed beads were placed in a mold which was immersed in hot water to achieve fusing. The lowest density reported for the fused articles was 48 g/L. Sekisui have reportedly just begun to market their PLA bead foam, Bioceller™, with plans to build a new six hundred ton annual capacity plant depending on the market situation [37]. They envisage Bioceller™ could potentially be used in any application EPS is currently used.

Kaneka Corporation (Japan) has developed a PLA foam product which they call Kanepearl PLA. They have several Japanese patents of which only the abstracts can be found in English [38–40]. PLA was impregnated with a low molecular weight hydrocarbon. It appears that the beads may then be pre-foamed prior to impregnation with CO_2 at pressures of 1 to 20 bar for up to 10 minutes. Beads were then placed in a mold for fusing. Due to limited information available in English, it is difficult to establish the commercial availability of Kanepearl PLA. However, images of Kanepearl PLA in packaging applications, for example wine boxes, are shown on their website [41].

6.2.3 "Sheet" Foam

MicroGREEN Polymers, Inc., a US company, has developed a technology similar to some of the particle foam approaches for producing expanded polymeric articles but uses, primarily, semi-crystalline thermoplastic sheets or films. The technology, called Ad-air®, takes films/sheets of varying material and impregnates them with a plasticizing gas (such as CO_2) before heating them. Upon heating, the polymer sheet expands. In general, this process results in a decrease in density, the final density being typically about 20% relative to the original solid. It is stated that the resulting material can be used in many applications, including general packaging and food and beverage packaging. In the later case a thermoforming step is often required [42].

The Ad-air® technology, described by several patents [43–46], was not specifically developed for PLA, although PLA is listed by name as a semi-crystalline thermoplastic which could be processed by this technology. An example of foaming sheets of PLA is given [45]. PLA sheets of 0.6 and 1.32 mm thickness were impregnated with CO_2 at 30 bar for 4 and 13 hours, respectively. The samples were expanded in an infrared oven until the surface temperature reached 100°C. The 0.6 mm sheets doubled in thickness to 1.2 mm while the 1.32 mm sample more than quadrupled in thickness to 6 mm.

6.3 Foam Properties

Thermal insulation, mechanical properties and heat deflection temperature are discussed in the following section. Most of the data that could be found for insulation and mechanical properties was for PLA particle foams and not extrusion foams.

6.3.1 Thermal Insulation

Thermal conductivity of the material is very important for packaging of some products such as those requiring cool chain packaging, including foodstuffs, such as fresh fish, or for pharmaceutical and medical supplies. Thermal conductivity (also known as the "k" value) specifies the rate of heat transfer in any homogeneous material. The lower the "k" value is, the less heat the material will transfer.

Both BPN and Synbra have reported thermal conductivities for their particle foams. Both sets of results show that PLA foams are excellent thermal insulators, comparable to EPS which itself is considered an excellent insulator. A thermal conductivity value of 0.03 W/(m·k) was obtained for all samples tested (EPS and PLA foam) by both BPN and Synbra. The densities of moldings were all 25 to 35 g/L. Synbra used the European standards EN-12667 and EN-12939 to measure thermal conductivity whereas BPN used an adaptation of the British Standard Method BS 4745: 1986.

6.3.2 Mechanical Properties

Stiffness and strength properties which are above certain minimum values are necessary so that foam packaging retains its shape while it is handled and put under load. Shock absorption, or impact resistance, are also important in packaging applications due the protective role of the product [47].

The mechanical properties (i.e. stiffness and strength) of foam material are strongly related to its density, and the importance of this characteristic is often dominant. However, for particle foams, other attributes do have an effect on the final properties of the foam, such as the cellular structure of the foam [48] or the degree of fusing between the particles [49]. Since PLA particle foams are a relatively new class of material, experimental and commercial data on mechanical properties is relatively rare in the current literature. BPN reported the mechanical properties (in compression, shear and cross-breaking) of their foams over a density range from 30 to 160 g/L [50]. Synbra have reported mechanical properties at densities around 35–40 g/L [51]. Some interesting data can also be found in patents published by Sekisui Plastics [52] and JSP Corporation [53], both using samples with a density of approximately 60 g/L.

Compression is a mode of deformation extremely common for packaging foam products when they are subject to impact or are load bearing. Synbra measured compression properties of their BioFoam® at 40 g/L and found a compressive modulus of 4.0 MPa and a stress at 10% deformation of 200 kPa (following European standard EN-823). Synbra also tested compressive properties of EPS which gave similar results to their BioFoam®. Examples described in Sekisui Plastics patents gave compressive strength of about 300 kPa at a density of 59 g/L using the Japanese Industrial Standard (JIS) testing method A9511 (1995). BPN tested the compression properties of their PLA foam following the ASTM standard D1621-00. They found that the relationship between foam density and modulus, strength and stress at 10% deformation was linear for the range of densities tested and the following relationships were derived:

$$\text{Compressive modulus (MPa)} = 0.30 \cdot \text{density (g/L)} - 5.7$$

$$\text{Compressive strength (kPa)} = 6 \cdot \text{density (g/L)} - 95$$

$$\text{Stress at 10\% strain (kPa)} = 8 \cdot \text{density (g/L)} - 120$$

Using these relationships, BPN would calculate a compressive modulus of 6.3 MPa and stress at 10% deformation of 200 kPa at a density of 40g/L and a compressive strength of about 260 kPa at 59 g/L showing the foams produced by all three companies (BPN, Sekisui, Synbra) do have relatively similar compressive properties. BPN also compared compression properties between PLA foam and EPS at a similar density and found these materials to show similar properties. These results indicate that PLA foams are an excellent alternative to EPS for packaging applications where compression properties are important.

Another mode of deformation common with molded packaging foams is bending. Despite the relative simplicity to perform this kind of method, bending is a rather complex test [52]. Several deformation modes are locally involved, with one face of the tested specimens being under tension and the other under compression. The contact points between specimen and testing machine can generate high compression loads, causing the specimen to be locally crushed. In the case of a foam specimen, these effects can impact on the quality of data for the bending test. This bending test offers valuable information on the fracture mode of the material, as expanded particle-foams can fail either between the particles (and the cracks propagate along the particles boundaries) or within the particles themselves [53, 54]. This reveals the strength of the inter-particle fusion. Synbra followed the European Standard EN 12089 to determine the bending strength of their product. They obtained a value of 300 kPa at 35 g/L which is comparable to the value they measured on EPS samples (300 kPa at 30 g/L). Sekisui Plastics reported a flexural strength of 530 kPa at 59 g/L in their patents following JIS testing method K7221 (1999). JSP Corporation reported flexural strengths in the range of 550 to 590kPa for samples with densities of about 60g/L, also tested in accordance with JIS K7221 (1984). The BPN used the Australian Standard AS 2498.4 (1993) to measure the cross-breaking strength of their material. Values measured across the 40 to 160 g/L range can be described by the following relationship:

$$\text{Cross-breaking strength (kPa)} = 10.1 \cdot \text{density (g/L)} + 165$$

This relationship gives a strength of 520kPa at a density of 35g/L and 760kPa at a density of 59g/L. Again, all companies appear to have produced PLA particle foams with similar mechanical properties. BPN also measured the properties of EPS samples which again showed similar results to their PLA foam with a strength of 327kPa for EPS and 418kPa for their PLA foam, both tested at a density of 25g/L.

Synbra [34] and the BPN have also provided some data about the shear properties of their foams, even though this test is probably less relevant for packaging applications than compression. BPN measured shear modulus and strength using the ASTM standard C273-00 of their PLA foam over the same range of densities as used for their other mechanical testing. Results from BPN studies indicated a linear relationship between shear properties and foam density, and this could be described using:

$$\text{Shear modulus (MPa)} = 0.068 \cdot \text{density (g/L)} + 1.3$$

$$\text{Shear strength (kPa)} = 4 \cdot \text{density (g/L)} + 5$$

Using these relationships, BPN would calculate a shear modulus of 3.7 MPa and shear strength of 145 kPa at a density of 35 g/L. Synbra measured these properties for their BioFoam® at 35 g/L and obtained similar values (2.7 MPa and 140 kPa for modulus and strength, respectively). Both BPN and Synbra also tested EPS and again showed comparable results between EPS and PLA foams.

Due to the protective role of foams used in packaging application, cushioning performances are particularly important. Little data has been reported on this subject for PLA particle foams. Drop tests were conducted by Synbra, using an in-house method, to assess the ability of their products to absorb shocks. They found that their BioFoam® is an excellent potential shock-absorbing material, with better performance under high static stresses than EPS [51]. Sekisui Plastics reported impact resistance values in the range of 1.9 to 2.6 using the ASTM standard D-3763. Unfortunately these values are unitless as absorption energy was taken as an index of impact resistance, and no values for EPS controls were reported.

The limited information of mechanical properties of extruded foams was from a study by Reignier *et al.* [18], described earlier, who reported on the compression modulus of their extruded foam. They reported a compression modulus of 2.1 MPa for foams with a density of 20.8 g/L. This almost doubled, to 4.1 MPa, with just a small increased in density, to 24.2 g/L.

6.3.3 Heat Deflection Temperature

One of the major shortcomings of PLA, be it foamed or not, is its lack of resistance to deformation at higher temperatures. Heat deflection temperature (HDT) is often used to determine short-term heat resistance of materials and is the temperature at which a test sample deflects under a constant load [55]. Semi-crystalline (unfoamed) PLA has a HDT of 55–60°C, while highly crystalline PLA has a HDT of 100–148°C [56].

As with other polymers there are several routes to improve thermal resistance of the PLA which depend on its structure [57–59]. For example, increasing the HDT of semi-crystalline polymers can be achieved by increasing the crystallinity of the polymer. Methods to do this include blending in a nucleating agent which initiates and promotes crystallization of the polymer. Optimizing the processing conditions, especially the temperatures and times, and adding nucleating fillers at optimal levels can increase the HDT nearer to the melting temperature of the polymer. PLA is also able to be blended and reactively modified with other polymers including other biopolymers such as PHBV (polyhydroxybutyrate valerate) which can improve PLA's heat resistance [60]. Various commercial reactive or blendable additives, or fillers, for improved heat resistance of PLAs are now emerging.

Synbra propose to improve the thermal properties of their PLA BioFoam® by using a stereo-complex PLA blend. They polymerize PLA for their foams using technology developed by Sulzer Chemtech and Purac Biochem. Through Purac, they will have access to the pure D-lactic acid to produce polymeric D-lactide (PDLA) which, for example, can be blended 50:50 with PLLA (polymeric L-lactide) to form a stereo complex PLA (scPLA). The addition of 1–10% sc-PLA increases the HDT of PLA to above 100°C. 80–100% sc-PLA (40–50% PDLA in PLA) will increase the HDT to above 150°C. Synbra's BioFoam® particle foam is heat stable to 80°C [34] and microwavable [61]. The relative cost of using such specialized PLA polymers is, as yet, unclear.

Sekisui uses a relatively highly crystalline PLA to produce their particle foam. They have reported "excellent" heat resistance [36]. In addition, Sekisui have sourced PLA from Unitika Ltd. In one development, Unitika have modified the molecular structure of the PLA to produce a material with a HDT of 120°C [62, 63]. This material, TERRAMAC®, is mainly made of Ingeo™ PLA supplied by NatureWorks® [64]. TERRAMAC® may also contain nano-additives, plant fibers, mineral fillers or be a blend of PLA with PMMA [60]. Sekisui Plastics Co. Ltd. has used TERRAMAC® to develop a heat-resistant PLA foam, Bioceller™. Bioceller™ has dimensional stability up to 150°C [65] for articles with a density of 60 g/L and a thickness of 10 to 30 mm [66].

BPN has also developed proprietary approaches to improving the heat resistance (and flame retardancy) of PLA foams and uses, again, more readily available commercial grades of PLA rather than special stereoisomers.

In the area of extrusion foaming, the PLA trays made by both Coopbox and Cryovac have a HDT of 49°C, which is similar to the HDT of the PLA resin [66]. Cereplast Inc.'s PLA foam has a HDT of 54°C [68]. To improve HDT, Cereplast Inc. blended PLA with a thermoplastic starch in a developmental foam grade. The material has a heat resistance above 90°C [69].

6.4 Conclusions

PLA foam is an ideal material for many packaging applications and is becoming increasingly preferred as a renewable biopolymer alternative to the traditional oil derived polymer foams currently widely used for packaging. As described in this chapter, technologies are available to produce both extrusion and particle foams from PLA. These foams have mechanical and thermal properties comparable to the existing (non-renewable) polymer foams. Limitations with PLA such as its lack of performance at higher temperatures are being addressed and overcome through various approaches and these are areas where much research is currently being undertaken. Current commercial packaging applications have focused on cool temperature options, such as the PLA extruded meat trays. Early studies have shown that PLA is an economically feasible material to use as a packaging polymer [68].

References

1. "The Compelling Facts about Plastics 2009". Available from: http://www.plasticseurope.org/plastics-industry/market-data.aspx.
2. "An Introduction to the History of Plastics: Plastic Packaging Products," Packaging Today Available from: <http://www.packagingtoday.com/introplasticproducts.htm>.
3. Goldberry, C. "Market Snapshot: Packaging," Plastics Today, May 5th, 2010 Available from: <http://www.plasticstoday.com/imm/articles/market-snapshot-packaging-0510>.
4. "Biopolymers in Packaging: 2010 to 2014, Allied Development™ Global Markets, Environmental Impact, and Technologies." Available from: <http://www.allied-dev.com/Studies2/Biopolymers09Brochure.pdf>.
5. Garlotta, D., "A Literature Review of Poly(Lactic Acid)," *Journal of Polymers and the Environment*, 2001, Vol. 9(2): p. 63–84.
6. Verespej, M., "Sustainability Issue Again Driving Growth for PLA Maker." *Natureworks, Plastics News*, June 17 2010; Available from: <http://plasticsnews.com/headlines2.html?id=18905>.
7. Queiroz, A.U.B. and F.P. Collares-Queiroz, "Innovation and Industrial Trends in Bioplastics," *Polym. Rev.*, 2009, Vol. 49(2): p. 65–78.
8. Clydesdale, P. in 3rd Bioplastics Market Conference, 2009.
9. Plastic Engineering Associates Licensing Inc., Available from: http://www.turboscrews.com.
10. Villalobos, M., et al., "Oligomeric Chain Extenders for Economic Reprocessing and Recycling of Condensation Plastics," *J. Energy*, 2006, Vol. 31(15): p. 3227–3234.
11. Inata, H. and S. Matsumura, *Journal of Applied Polymer Science*, 1987, Vol. 33(8): p. 3069–3079.
12. Di, Y., *et al.*, "Reactively Modified Poly(Lactic Acid): Properties and Foam Processing," *Macromolecular Materials and Engineering*, 2005, Vol. 290(11): p. 1083–1090.
13. Pilla, S., et al., "Microcellular Extrusion-Foaming of Polylactide with Chain-Extender," *Pol. Eng. Sci.*, 2009, Vol. 49(8): p. 1653–1660.
14. Mihai, M., M.A. Huneault, and B.D. Favis, "Rheology and Extrusion Foaming of Chain-Branched Poly(Lactic Acid)," *Polym. Eng. Sci.*, 2010, Vol. 50(3): p. 629–642.
15. Ema, Y., M. Ikeya, and M. Okamoto, "Foam Processing and Cellular Structure of Polylactide-Based Nanocomposites," *Polymer*, 2006, Vol. 47: p. 5350–5359.
16. Di, Y., et al., "Poly(lactic acid)/Organoclay Nanocomposites: Thermal, Rheological Properties and Foam Processing." *J. Polym. Sci. Part B: Polym. Phys.*, 2005, Vol. 43: p. 689–698.
17. Kawada, T., A. Iwamoto, and K. Takase, Foamed Sheet Of Polylactic Acid Resin, Foam Molding of Polylactic Acid Resin and Method of Preparing Foam Molding, 2010 JSP Corporation: US Patent application 20100086758.
18. Reignier, J., R. Gendron, and M.F. Champagne, *Cell. Polym.*, 2007, Vol. 26: p. 83.
19. Matuan, L.M., O. Faruk, and C.A. Diaz, "Cell Morphology of Extrusion Foamed Poly(Lactic Acid) Using Endothermic Chemical Foaming Agent," *Biores. Tech*, 2009, Vol. 100(23): p. 5947–5954.
20. Preechawong, D., *et al.*, *Carbohydrate Polymers*, 2005, Vol. 59(3): p. 329–337.
21. Mihai, M., *et al.*, *Macromol. Biosci.*, 2007. Vol. 7: p. 907–920.
22. Richards, E., *et al.*, *J. Pol. Environ.*, 2008, Vol. 16(4): p. 258–266.
23. Ajioka, M., *et al.*, Degradable foam and use of same, 1995: US Patent 5,447,962.
24. Oka, Y., T. Yoshioka, and M. Ohyama, Polylactic Acid Foam, 2010: US Patent application number 20100120932.
25. Coopbox. Available from: http://www.coopbox.it/portal/page?_pageid=436,1544342&_dad=portal&_schema=PORTAL.
26. Cryovac® NatureTRAY™ Foam Tray. Available from: http://www.sealedair.com/products/food/trays/nature_tray.html.
27. Dyneapak, Available from: http://www.dyneapak.com.
28. Mills, N., "Bead Foam Microstructure and Processing," in *Polymer Foams Handbook*, 2007, Butterworth-Heinemann: Oxford, p. 69–83.
29. Shinohara, M., T. Tokiwa, and H. Sasaki, Expanded polylactic acid resin beads and foamed molding obtained therefrom, in European patent office, 2002, JSP Corporation: Japan.

30. Haraguchi, K. and H. Ohta, Expandable polylactic acid resin particles, 2005: European Patent application EP 1 683 828 A2.
31. Witt, M.R.J. and S. Shah, Methods of manufacture of polylactic acid foams, 2007: WO 2008/093284 A1.
32. Garancher, J.P., Parker, K., Shah, S., Weal, S., Fernyhough, A. "Expanded Polylactide (E-PLA): A Realistic Alternative to Expanded Polystrene (EPS)". Available SP Conference Proceedings, Biofoams 2010, Seattle, Washington USA. September 28-October 1, 2010. Pub: Society of Plastics Engineers (SPE). Newtown, CT 06470 USA. www.4spe.org.
33. Noordegraaf, J., *et al.*, Particulate expandable polylactic acid, a method for producing the same, a foamed moulded product based on particulate expandable polylactic acid, as well as a method for producing the same, 2007: WO 2008/130226 A2.
34. Noordegraaf, J., P. Matthijssen, and J de Jong, BioFoam®: PLA Particle Foam Expanding in Europe, in 2nd International Biofoams Conference, 2009, Niagara Falls, Canada.
35. Anonymous, "Expanded PLA as a Particle Foam," Bioplastics, 2009, Vol. 4(01): p. 22–23.
36. Hirai, T., K. Nishijama, and T. Ochiai, Polylactic acid resin foam particle for in-mold foam forming, process for producing the same, and process for producing polylactic acid resin foam molding, 2008, Sekisui Plastics Co Ltd., Japan: European Patent Office EP 2 135 724 A1.
37. Thielen, M., "Heat-resistant PLA Bead Foam," *Bioplastics*, 2010, Vol. 5(1): p. 29.
38. Keisuke, O., Method for producing polylactic acid-based foam-molded article 2007, Kaneka Corporation: Japanese Patent application JP20070051293 20070301.
39. Keisuke, O., Method for producing polylactic acid-based foam-molded article Kaneka Corporation Japanese Patent application JP20070051294 20070301 2007.
40. Keisuke, O., F. Tomonori, and Y. Yoshitito, Method for producing polylactic acid foamed molding, 2007, Kaneka Corporation: Japanese Patent application JP20070080700 20070327.
41. Kaneka, Available from: http://www.kaneka.co.jp.
42. MicroGREEN Polymer Inc. Ad-air® Technology: A mechnical expansion resulting in source reduction, 2010; Available from: www.microgreeninc.com.
43. Branch, G., Deep Drawn Microcellularly Foamed Polymeric Containers Made via Solid-State Gas Impregnation Thermoforming, 2010, MicroGREEN Polymers, Inc: United States.
44. Branch, G. and T. Wardle, Method of Producing Thermoformed Articles from Gas Impregnated Polymer, 2005, MicroGREEN Polymers, Inc: United States.
45. Nadella, K., Foamed Cellular Panels and Related Methods, 2009, MicroGREEN Polymers, Inc: United States.
46. Powers, S., Apparatus and Method for Interleaving Polymeric Roll for Gas Impregnation and Solid-State Foam Processing, 2009, MicroGREEN Polymer, Inc: United States.
47. Mills, N., "Product Packaging Case Study," in *Polymer Foams Handbook*, 2007, Butterworth-Heinemann: Oxford, p. 281–306.
48. Arora, K.A., A.J. Lesser, and T.J. McCarthy, "Compressive Behavior of Microcellular Polystyrene Foams Processed in Supercritical Carbon Dioxide," *Polymer Engineering & Science*, 1998, Vol. 38(12): p. 2055–2062.
49. Järvelä, P., *et al.*, "A Method to Measure the Fusion Strength Between Expanded Polystyrene (EPS) Beads," *Journal of Materials Science*, 1986, Vol. 21(9): p. 3139–3142.
50. Parker, K., *et al.*, Expanded Polylactic Acid - an Eco-Friendly Alternative to Polystyrene Foam, in 2nd International Biofoams Conference, 2009: Niagara Falls, Canada.
51. Noordegraaf, J., *et al.*, BioFoam®: PLA Particle Foam Expanding in Europe, in 2nd International Biofoams Conference, 2009, Niagara Falls, Canada.
52. Moosa, A.S.I. and N.J. Mills, "Analysis of Bend Tests on Polystyrene Bead Foams," *Polymer Testing*, 1998, Vol. 17(5): p. 357–378.
53. Stupak, P.R., W.O. Frye, and J.A. Donovan, "The Effect of Bead Fusion on the Energy Absorption of Polystyrene Foam, Part I: Fracture Toughness," *Journal of Cellular Plastics*, 1991, Vol. 27(5): p. 484–505.
54. Mills, N.J. and P. Kang, "The Effect of Water Immersion on the Mechanical Properties of Polystyrene Bead Foam Used in Soft Shell Cycle Helmets," *Journal of Cellular Plastics*, 1994, Vol. 30(3): p. 196–222.

55. ASTM D648 - 07 Standard Test Method for Deflection Temperature of Plastics Under Flexural Load in the Edgewise Position, 2007.
56. Schut, J.H., "PLA Biopolymers: New Copolymers, Expandable Beads, Engineering Alloys & More, Plastics Technology," *Plastics Technology*, 2008.
57. Li, H. and M.A. Huneault, "Effect of Nucleation and Plasticization on the Crystallization of Poly(Lactic Acid)," *Polymer*, 2007, Vol. 48(23): p. 6855–6866.
58. Lim, L.T., R. Auras, and M. Rubino, "Processing Technologies for Poly(Lactic Acid)," *Progress in Polymer Science*, 2008, 33(8): p. 820–852.
59. Michael, T.T., "Towards an Understanding of the Heat Distortion Temperature of Thermoplastics," *Polymer Engineering & Science*, 1979: p. 1104–1109
60. "Blends of PLA being developed to improve viability of PLA." Article from plastemart.com (2010): http://www.plastemart.com/Plastic-Technicle-Article.asp?LiteratureID=1359.
61. "Expanded PLA as a Particle Foam." Article in Bioplastics, 2009, Vol. 4(01): p. 22–23. Ed: Michael Thielen. Polymedia Pub Gmbh., Monchengladbach, Germany.
62. "Polylactic Acid Foam (Environmental Issues)," Article in: *High Performance Plastics, International Newsletters*, 2005.
63. "Unitika Technology Improves Polylactic Acid Foam Properties (Plastics)." Article in: *New Materials Japan*, March 1, 2005.
64. Unitika Ltd. 2010; Available from: http://www.unitika.co.jp/terramac/e/.
65. Thielen, M., "Heat-Resistant PLA Bead Foam," *Bioplastics*, 2010, 05(01): p. 29.
66. Sekisui Plastics Co Ltd., "Success in Developing the Foamed Plastics by Beads Method-Employed Plant-Derived Resin, with a Dimensional Stability Under 150°C," Press release, 2007.
67. Schut, J.H., "Foamed PLA Shows Promise in Biodegradable Meat Trays," *Plastics Technology*, 2007.
68. Cereplast Inc., Cereplast Compostables®, Compostable 5001, Extrusion Foam Sheet Grade, Property Guide. 2010; Available from: http://www.cereplast.com/cmspage.php?pgid=90.
69. Auras, R., B. Harte, and S. Selke, "An Overview of Polylactides as Packaging Materials." *Macromolecular Bioscience*, 2004. Vol. 4: p. 835–864.

7

Polyvinyl Modified Guar-gum Bioplastics for Packaging Applications

Ashutosh Tiwari*, Dohiko Terada and Hisatoshi Kobayash*

Biomaterials Center, National Institute for Materials Science, Tsukuba, Ibaragi, Japan

Abstract

Poly(vinyls) or PVs modified with guar gum constitute a fascinating class of bioplastics derived from natural and synthetic biodegradable polymers. Bioplastics are part of an emerging and promising interdisciplinary scientific arena that bridges biology, chemistry, materials science, and engineering. Recent applications of bioplastics impact numerous diverse areas, particularly in the packaging industries and research. The present chapter summarizes key concepts of current PVs graft copolymerization onto guar gum, characterizations, properties and emerging applications in the packaging science and technology.

Keywords: Polyvinyl, guar gum, graft polymerization, packaging

7.1 Introduction

Guar gum is the most important seed gum among all the industrial gums and is derived from the endosperm of the two annual leguminous plants *Cyanaposis teragonalobus* and *Cyanaposis psoraloides* [1]. Structure and properties of many seed gums from different leguminous and convolvulous plants have been explored owing to tremendous consumption and importance of the guar gum. Many others seed gums reported to have similar structure to guar gum and were subjected to detailed phytochemical investigations, viz. seed gum of *Cassia nodosa, Cassia marilandica, Cassia absus, Cassia fistula, Cassia tora, Cassia occidentalis, Cassia multijuga, Cassia siamea, Cassia marginata, Cassia corymbosa, Cassia laevigata, Cassia angustifolia, Cassia ovata, Cassia javanica, Cassia grandis, Cassia sophera, Cassia renigera, Cassia surattensis, Cassia spectabilis, Cassia alata, Cassis glauca, Cassia javahikai, Cassia fisluosa, Cassia podocarpa, Cassia pudibanda, Cassia reticulata, Crotalaria mucronata, Crotalaria juncea, Crotalaria medicaginea, Ipomoea muricata, Ipomoea campanulata, Ipomoea turpethum, Ipomoea tridenteta, Ipomoea purga, Ipomoea palmata, Ipomoea quamoclit, Ipomoea dasysperma, Ipomoea murucoides, Ipomoea reptans, Teramnus*

* Corresponding authors: Ashutosh Tiwari (E-mail: ashunpl@gmail.com) and Hisatoshi Kobayashi (E-mail: kobayashi.hisatoshi@nims.go.jp)

Srikanth Pilla (ed.) Handbook of Bioplastics and Biocomposites Engineering Applications, (177–188)

labialis, Tamarind, Parkinsonia aculeata, Abutilon indicum, Astragalus lehmannianus Bunge, Borassus flabellifer, Stychnos, Mimosa seabrella, Caesalpinia Phoenix datylifera. They have been reported to have similar galactomannan structures as the guar gum and are inexpensive, eco-friendly, non-toxic and considered as generally recognized as safe, i.e., GRAS [2–5].

In other hand, poly(vinyls), PVs including poly(vinyl chloride), poly(vinyl acetate), poly(acrylates), poly(ethylene), poly(propylene), poly(vinyl alcohol), and poly(styrene) are called general purpose plastics [6]. The features of the plastic are determined by the chemical composition and type of molecular structure, viz. molecular formation, i.e., crystalline or amorphous structure. PVs have normally unique amorphous structure with polar atoms in the molecular structure [7, 8]. Typically, PVs with polar atoms and the amorphous molecular structure are inseparably related. Although plastics seem very similar in the daily use context, PVs have completely different features in terms of performance and functions as compared with olefin plastics, which have only carbon and hydrogen atoms in their molecular structures. Thus, graft copolymerization of PVs with guar gum could have potential for the development of bioplastics as a promising eco-friendly packaging alternative. In the present chapter, we have reviewed general structure, graft copolymerization and properties of PVs-graft-guar gum. The chemical structure, reactions, reaction mechanism, characterization and surface behavior of PVs copolymers are extensively descried.

7.2 Structure and Physical Properties of Guar Gum

Guar gum is an edible and cold water swelling carbohydrate polymer, which is useful as a thickening agent for water and as a reagent for adsorption and hydrogen bonding with materials and cellulosic surfaces [9]. Guar gum is galactomannan and consists of a straight chain of mannose units joined by β-D-(1→4) linkages, having α-D-galactopyranosyl units attached to this linear chain by (1→6) linkages [10]. In the pure polysaccharide "Guran" from Guar, the ratio of D-galactose to D-mannose units is 1: 2 (Fig. 7.1). The molecular weight of guran is 220,000 and guar gum is a natural alternating co-polymer.

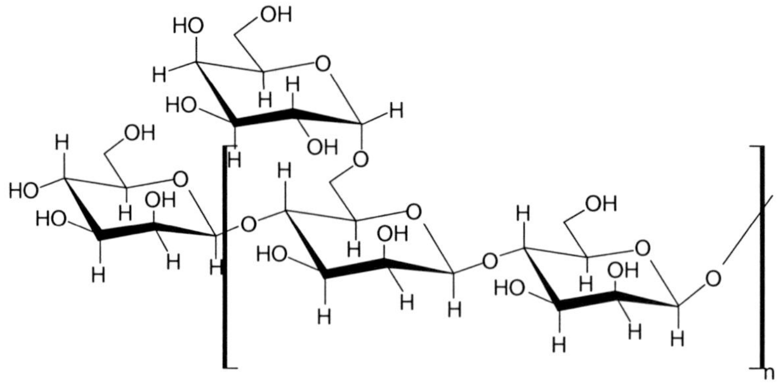

Figure 7.1 Chemical structure of guar gum repeating units.

For the commercial guar gum (i.e., in powder form), the rate of thickening and the final viscosity of the gum solutions reflect the process history of the product, including the particle size of the powder. Heating a guar gum solution reduces the time needed to reach its full viscosity potential. Water is the only common solvent for guar gum, although it will tolerate limited concentrations of water miscible solvents such as alcohols. Dimethylformamide and dimethylsulfoxide are solvents, as they are for most of the other polymers. Commercial guar gum solution is typically turbid. The turbidity is almost caused by the presence of insoluble portions of the endosperm.

Guar gum is one of the most highly efficient water thickening agents. Solutions of guar gum resist shear degradation when compared to other water-soluble polymers, but they will degrade with time under high shear [11]. Guar gum swells in hot and cold water to form a colloidal solution at very low concentrations. The solutions are somewhat cloudy due to the small amount of insoluble fiber and cellulosic materials present in flour. Viscosity of guar gum solutions can be measured with a rotational shear viscometer such as the Brookfield Syncro-lectic or the Haake Rotoviscometer. Less than 1000 ppm (mg/L) viscosity can also be measured with pipette type instrument such as Dudley, Engler and Ostwald viscometers. Viscosity is dependent on time, temperature, concentration, pH, ionic strength and type of agitation. Solutions of guar gum are stable for a shorter time [12]. It is fermented and enzymatically hydrolyzed by micro-organisms. Some preservatives like formaldehyde, chlorinated phenol, benzoic and phenyl mercuric acetate etc. may be used. Which preservative will be used, depends upon the application of guar gum solution.

The viscosity of a fully hydrated 1% (w/v) guar gum solution varies almost directly with changes in temperature over the range of 20–80 °C. The viscosity of a guar gum solution increases by short-term heating but decreases when heated for longer times due to thermal degradation. Higher viscosity is obtained in the temperature range 25–40 °C. In dilute solutions, the viscosity of guar gum increases linearly with concentrations up to about 0.5% (w/v), thereafter, guar gum solutions behave as non-newtonian solutions mainly as a result of complex surface attractions at higher concentration.

Guar gum is stable over a wide pH range. The non-ionic nature of the molecule is responsible for the almost constant viscosity of solutions in the pH range 1–10.5. The pH of a typical 1% (w/v) solution of guar gum is observed in a range from 5.5 to 6.1 and it tends to become more acidic upon standing. The modified guar gums that develop high viscosity in concentrated caustic solutions have recently become available. Because guar gum is non-ionic, its compatibility with salts is exhibited over a wide range of electrolyte concentration. High concentrations of multivalent salts affect hydration and produce gels.

Hydrated guar gum converts into gel due to cross-linking. Borate ion acts as cross-linking agent [13, 14]. The formation and strength of these gels are dependent on pH, temperature and concentration of the reactants. Borate ion inhibits the hydration of guar gum, if it is present at the time of adding powdered gum in water. Guar gum can be insolubilized or gelled by transition metal ions to form commercially useful gels.

A highly orientated tough, pliable film that is sensitive to water can be formed from guar gum. Guar gum with its long, straight chain configuration reduces and minimizes the frictional pressure losses of water when in turbulent flow [15]. The powdered commercial grade of guar gum is stable in the dry form. In common with other water-soluble polymers, the one condition that adversely affects the gum is allowing it to get wet. Many techniques are available to form smooth uniform solutions from these cold water-swelling powders. The greater number of branches in guar gum is responsible for both easier hydration and its different hydrogen bonding activity as compared to that of locust bean gum.

Guar gum derivatives are widely used in printing and dyeing applications as thickeners to control the mobility of dyestuffs [16]. The guar gum controls the rheological characteristics of the dye formulations and permits complex multiple dyes patterns that are sharp, bright and controlled as to penetration. It is also used in textile printing paste extenders. Various patents report its uses as thickeners in powdered plaster, clay, drilling muds and dyes. Its uses in slurried blasting compositions are given in various patents. Prakash *et al.* prepared caustic proof products of guar gum by modifying it with sodium peroxide [17]. These were tested and found suitable for printing cellulosic fibers with colors using alkaline print paste.

Adsorption behavior of guar gum derivatives for removing dyes from the textile effluents has also been studied by Kapoor *et al.* Guar gum and guar derivatives are used in the fracturing of oil and gas wells because of their ability to thicken water efficiently at low concentrations. [18]. They carry the sand or propping agent needed to keep the fracture from closing, when the pressure is released. They are reliable polymers that will hydrate in field water under many conditions. In comparison to other natural polymers like starch, guar gum hydrates more easily and readily. Guar gum can give very high viscosity even at 1% (w/v) concentration in water. Guar gum solution is an eco-friendly material due to its quick biodegradation at room temperature, while modified guar gum is quite stable [19].

7.3 Modification of Guar Gum

Modification of guar gum has been done using two ways:

(1) Derivatization of functional groups
(2) Graft copolymerization with PVs

7.3.1 Derivatization of Functional Groups

Derivatization of guar gum includes the methylation, carboxymethylation, ethylation, phosphorylation, nitration, sulfonation etc. Guar gum was derivatized through *O*-methylation with dimehtyl sulphate in the presence of NaOH in aqueous medium and resulting polymer was yellowish glassy mass with high shelf life. Guar gum was modified by carboxymethylation with chloroacetic acid in the presence of NaOH in alcoholic medium and the resulting polymer has high viscosity, good dye reactivity and is useful for textile printing [20]. Carboxymethylated guar is also used as non-lumbing agent. Aminoethyl guar gum was prepared by reaction of ethylenediamine with substantially dry polygalactomannan [21]. The gum ether is useful in the settling the particles in concentrated ore and increasing the dry strength of paper.

Nitration of guar gum with HNO_3–H_2SO_4 mixture gave the nitrate esters of guar gum for use in a thickening agent for nitroalkane explosive compositions [22]. Sulphonation of guar gum has been done in various systems and the effect of various parameters on the substitution degree has been observed [23]. The maximum substitution degree was found with DMF–SO_3 combination at the temperature 90 °C.

Etherifying guar gum with allyl chloride gave galactomannan allyl ether gels with potential stability, useful for oil well acidizing borehole plugging and for the preparation of explosive slurries [24]. Propyle guar gum is used in conditioning shampoo as the active ingredients. Periodate oxidized guar gum was found to be useful in paper production. Esterification of guar gum with H_3PO_4 and oxidation with alkaline H_2O_2 gave the water-soluble phosphate guar product for use in decreasing porosity and increasing the solvent resistance of paper. Guar gum phosphate as sizing agent essentially decreased the porosity of paper and made the surface of paper almost completely resistant against PhMe. Acrolein adducts and bisulphite adducts have been reported with guar gum. Bisulphite adducts are useful in increasing the wet burst strength of papers [25].

7.3.2 PVS Modified Guar Gum

Modification of guar gum by grafting of vinyl monomers, results in the retention of desirable properties and incorporation of favorable properties. The grafted guar gum is used in the preparation of flocculants for the industrial effluents treatment and as biodegradable drag reducing agents [26]. Cross-linked guar-graft-PVs based anionic micro gels for a pH sensitive drug delivery, has been recently reported [27]. It is used in water transport and drug release. Lokhande prepared water supersorbent guar modified polymers by graft co-polymerization of acrylonitrile onto guar gum through γ-radiation [28]. Conventionally, acrylonitrile has been grafted onto guar gum using various redox systems [29]. Grafting of methylmetharylate, onto guar gum has also been reported using a variety of redox initiators. Concurrent homopolymer formation is the main constraint in graft co-polymerization, which leads to a low grafting yield.

The percentage and efficiency of grafting were calculated according to Kojima *et al.* using Eq. (7.1) and (7.2)

$$\%\ \text{Grafting}\ (\%\text{G}) = \frac{W_1 - W_0}{W_0} \times 100 \tag{7.1}$$

$$\%\ \text{Efficiency}\ (\%\text{E}) = \frac{W_1 - W_0}{W_2} \times 100 \tag{7.2}$$

Where, W_1, W_0 and W_2 denoted the weight of the grafted guar gum, the weight of original guar gum and weight of the vinyl monomer used respectively.

Guar gum was grafted with various vinyl monomer(s) using primary free radicals of $K_2S_2O_8$/ascorbic acid redox initiator. Typically, Mehrotra and Mushran studied the kinetics of the redox system containing ascorbic acid and peroxydisulfate and a mechanism involving $SO_4^{-\cdot}$, $OH^\cdot$, and ascorbate radical intermediates has been proposed [30]. This redox system has been exploited for polymerization of vinyl monomers by several workers and has been shown to initiate vinyl copolymerization with guar gum [31]. The reaction between persulfate and ascorbic acid involves a chain mechanism due to the formation of sulfate ion radicals, which are well known ion chain carriers. The mechanism may be written as shown in Scheme 7.1. The SO_4^{-}, $OH^\cdot$ and $AH^\cdot$ (ascorbate radical) are the primary radicals, generated in the sequence of the redox reactions, and are expressed as $R^\cdot$ in the Scheme 7.2. They initiate the vinyl polymerization as the vinyl polymerization is reported to be faster than the H abstraction from the guar gum backbone. The macroradical $GGO^\cdot$ may be generated by abstraction of H by the growing vinyl polymer radical, which may add onto the vinyl monomer (M), generating new radical $GGOM^\cdot$, and this chain will grow till it combines with other such chains to give a graft copolymer (Scheme 7.2).

One more grafting technique, e.g., grafting of butylacrylate, acrylic acid and acrylonitrile onto the starch, of acrylamide on to low density poly(ethylene) films

$S_2O_8^{2-} \longrightarrow 2SO_4^-$		Initiation
$2SO_4^- + H_2O \longrightarrow SO_4 + OH + H^+$		Propagation
$OH + AH^- \longrightarrow AH + OH^-$		
$AH + S_2O_8^{2-} \longrightarrow A + SO^{2-} + SO_4 + H+$		
$AH + OH \longrightarrow A + OH^- + H+$	or $A + H_2O$	Termination
$AH + SO_4^- \longrightarrow A + SO^{2-} + H+$		

Where AA stands for ascorbic acid is equivalent to AH⁻, A stands for dehydroascorbic acid

Scheme 7.1 Generations of primary free radicals of $K_2S_2O_8$/ascorbic acid redox initiator.

$R + M \longrightarrow RM$	Initiation
$RM + nM \longrightarrow RM_{n+1}$	
$GGOH + RM_n \longrightarrow GGO + RM_nH$	Homopolymer
$GGOH + R \longrightarrow GGO + RH$	
$GGO + M \longrightarrow GGOM$	Propagation
$GGOM + nM \longrightarrow GGOM_n$	
$2GGOM_n \longrightarrow$ Graft copolymer	Termination

Where R is primary radicals, i.e., generated from Scheme 7.1, GGOH stands for guar gum and M for vinyl monomer

Scheme 7.2 Graft copolymerization initiated by the primary radicals of redox initiator.

and of butyl methacrylate on to the wool fibers has been studied under microwave irradiation using redox initiators [32]. Yields were high and the grafting could be achieved in very short time. As microwaves are reported to generate free radicals we were prompted to study if grafting is possible without using any initiator or catalyst under microwave irradiation and surprisingly we could successfully achieve grafting of PVs onto the GG under microwave irradiation without the use of any radical initiators or catalyst.

Using microwave irradiation (MW) grafting of PVs such as poly(acrylates) on to the guar gum was done without using any radical initiator or catalyst in a very short reaction time [33]. The extent of grafting could be adjusted by controlling the reaction conditions and maximum percentage grafting about 190% was obtained under optimum conditions in 2 min. A plausible mechanism for the grafting under MW is proposed. GG molecule is quite large molecule with pendent –OH groups. –OH groups attached to large GG molecule may behave as if were anchored to an immobile raft and its localized rotations therefore will be observed in the microwave region and resulting dielectric heating of the GG molecule may result in an enhancement of reaction rates specifically at these groups. The dielectric heating will involve rapid energy transfer from these groups to neighboring molecules (vinyl monomers and polar solvent such as water) as it is not possible to store the energy in a specific part of the molecule [34, 35]. In the GG molecules there a large numbers of –OH groups so this energy amount is expected to be enormous and this may be responsible for –OH bond breaking. Further, MW are also reported to have special effects of lowering of Gibbs energy of activation of the reactions and in view of the above two affects a plausible free radicals mechanism for the grafting under the microwave irradiation has been proposed. –OH groups of the GG being polar will absorb MW energy and will cleave generating monomer free radical and macro radical. The mechanism under microwave irradiation can be illustrated as shown in Scheme 7.3.

Moreover, PVs were also grafted successfully onto the guar gum with using any radical initiator/or catalyst using microwave irradiation in good yield and in a very short reaction time and the grafting yields comparable to redox, i.e., potassium

$$\mathrm{GOH + M \xrightarrow{MW} GO^{\bullet} + M^{\bullet}}$$
$$\mathrm{GO^{\bullet} + M \rightarrow GOM^{\bullet}}$$
$$\mathrm{GOM^{\bullet} + M \rightarrow GOMM^{\bullet}}$$
$$\mathrm{GOMM^{\bullet}_{\mathit{n}-1} + M \rightarrow GOM^{\bullet}_{\mathit{n}}}$$
$$\mathrm{GOM^{\bullet}_{\mathit{n}} + GOM^{\bullet}_{\mathit{n}} \rightarrow Grafted\ polymer}$$
$$\mathrm{M^{\bullet} + M \rightarrow MM^{\bullet}}$$
$$\mathrm{M^{\bullet}_{\mathit{n}-1} + M \rightarrow M^{\bullet}_{\mathit{n}}}$$
$$\mathrm{M^{\bullet}_{\mathit{n}} + GOH \rightarrow GO^{\bullet} + M^{\bullet}_{\mathit{n}}H\ (Homopolymer)}$$

Scheme 7.3 Graft copolymerization of poly(vinyls) onto guar gum under microwave irradiation.

persulphate/ascorbic acid initiated conventional heating graft copolymerization in very short reaction time. The grafting efficiency up to 20% was further increased when initiators and catalyst were used with microwave. The effect of microwave power and exposure time on the grafting yields, both in presence and in absence of redox initiator and catalyst were studied. Maximum grafting efficiency achieved under microwave was about 70% in within one min in comparison to 50% in 90 min by the conventional method.

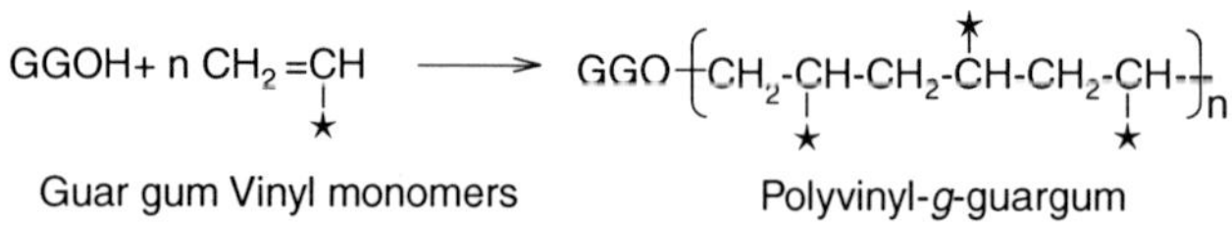

7.4 Characterization

Guar gum-graft-PVs (GG-*g*-PVs) can be characterized using FT-IR, NMR, XRD, TGA and some cases elemental analysis [36]. A) FTIR spectra: Infrared spectrum of pure Guar gum is always has a broad strong band at about 3400 cm^{-1}, a band at 2900 cm^{-1} indicating –OH group, C-H linkages respectively, while FT-IR spectra of GG-*g*-PVs are showing additional peaks of PVs, e.g., absorption peak at 1650 cm^{-1} for >C=O stretching, N-H stretching peaks at 3350 cm^{-1} and 3200 cm^{-1} and C-N stretching at 1450 cm^{-1}. B) NMR spectra: ^{1}H NMR of the pure guar gum is showed a peak at δ 4.67(s), anomeric protons, and at δ 3.5–3.9 (m) due to other sugar protons while the grafted gum is appeared additional peaks like at δ 2.0–2.1, δ 1.5–1.6, etc. due to protons at grafted chains of PVs(s) on the guar gum backbone that indicates the presence of PVs chain attached to the guar gum. C) XRD spectra: XRD of guar gum is compared with the grafted gum (Figure 7.2). XRD spectra of the grafted gum showed additional sharp peaks corresponding to the crystallinity of grafted PVs on guar gum backbone. The other broader peak indicates the decreased crystallinity of pure guar gum after grafting or in other wards the amorphous property of guar gum enhanced after grafting with PVs. D) TGA Spectra: The TGA of the grafted guar gum showed that decomposition onsets at 270 °C and only 55–65% weight loss up to 800 °C. It indicated that PVs grafted guar gum; bioplastics have ability to comfortably resist temperature about 250 °C. E) SEM: The surface topography of PVs grafted and ungrafted guar gum bioplastic is shown in Figure 7.3. The obvious textural properties like spongy morphology favor the packaging applications and elemental analysis: The elemental profile of PVs grafted and native guar gum shows that the grafting is scattered or not, i.e., degree of association, etc. For example, microwave assisted grafting, in absence of radical initiators and catalyst, results into the copolymers with lower %N in comparison to the polymer obtained with conventional thermal grafting indicating that the grafted chains in the former are either of smaller size or more widely scattered on to the GG backbone in case of nitrogen containing PVs.

Other polysaccharides, thermoset acrylate(s) based bioplastics are used commonly as packaging restorative materials, because of a relatively high cure efficiency by free-radical polymerization and excellent artistic qualities [37].

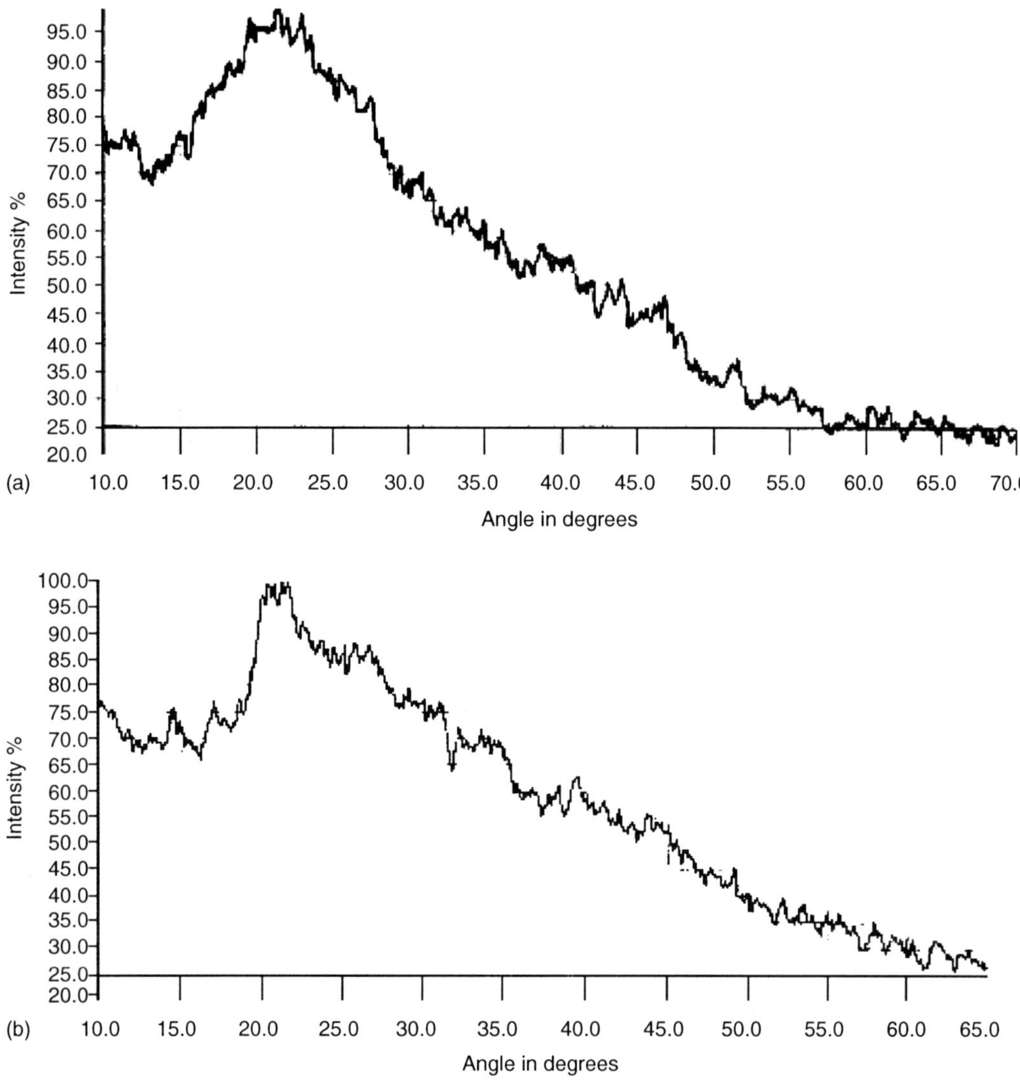

Figure 7.2 XRD spectra of (a) guar gum and (b) PVs grafted guar gum.

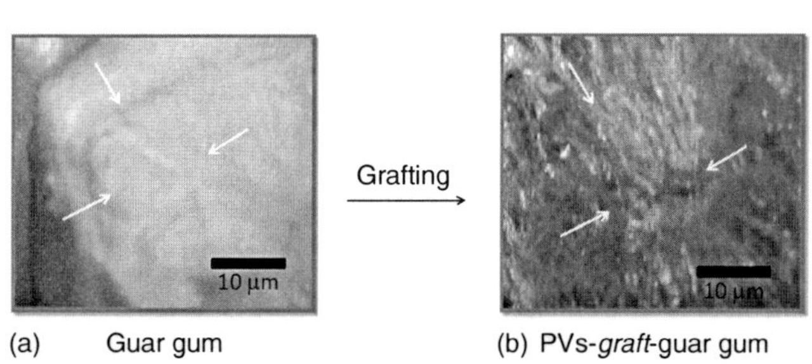

Figure 7.3 SEM images of (a) guar gum and (b) PVs grafted guar gum.

However, the demands of the environment and the solid waste loads encountered by bioplastics restoratives require further property improvements in these materials. Specifically, bioplastics with improved modulus, better efficiency of the free-radical polymerization, low water sorption, improved processability, and low local shrinkage are needed. Selective functionalization of the bioplastics can lead to better interactions at the PVs-guar gum interphase and has been used in studies in which PVs particles and/or matrix were modified to varying extents using two different modifiers and then mixed with a bioplastics.

In another development, the primary use of biopolymers as plastics is in molded products. Biopolymers have been reinforced to increase strength of plastics. PVs content ranges between 20 and 40% (w/w) and sometimes exceeds 50% (w/w) in bioplastics. The automobile body panels, for example, could be composed of poly(ester)-biopolymer graft copolymers. Biopolymers and PVs are not homogeneously mixed on a microscopic level in graft copolymers, especially at very high loadings. The interface between PVs and biopolymer matrix is not large so their interaction is limited. On the other hand, if the PVs are grafted randomly and homogeneously in the guar gum matrix, the Interface area is enormous and a large interaction is expected.

7.5 Conclusions and Future Challenges

Current opportunities for bioplastics from the biopolymers, arise from the multitude of packaging applications. Typically, bioplastics must meet certain design and functionalities, including biocompatibility, biodegradability, mechanical properties, and, in some cases, aesthetic demands. The underlying solution to the use of bioplastics in vastly differing applications is the correct choice of biopolymer chemistry, copolymer type, and biopolymer-PVs attachments and/or interactions. The design of packaging interfaces necessitates a combination of amphiphilicity and amorphous segment. Future efforts in this field are directed at assessing the thermogenic potential of these bioplastics, which would be critical in their application as packaging edges at higher temperature. Interest in these bioplastics varies from application-oriented design to understanding a multitude of structure-property relations. However, the basis of the performance of these bioplastics are interactions between the biopolymer and synthetic polymer, i.e., PVs, which can be tuned and perfected to suit specific packaging needs. We hope that further research into these interactions will prove valuable in contemplating the design of novel bioplastics for packaging applications. On the other hand, acceptable handling of large volumes of acid solutions is an environmental challenge for their producers.

Acknowledgements

The authors owe their heartfelt gratitude to the National Institute for Materials Science, Japan for providing infrastructure facility and to the JSPS, JST-CREST and MEXT Japan for generous financial support to carry out this research.

References

1. Ashutosh Tiwari, Chemical study of plant seed gums, PhD thesis, University of Allahabad, India, 2005.
2. Ashutosh Tiwari, Mani Prabaharan, Santosh Aryal, Songjun Li, "Polysaccharides based colloidal carriers for drug delivery applications," *Current Focus on Colloids and Surfaces*, Ed. Songjun Li, Chapter-1, 1–21, Research Signpost, Trivandrum, Kerala, India, 2009.
3. Mathur, N.K., Mathur, V., Nagori, B.P.; "Scope for the development of modified and derivatised guar gum products in India," In *Trends in Carbohydrate Chemistry*, Ed.; Soni, P. L., Surya International Publication, Dehradun, India, 2, pp. 45–57, 1996.
4. Hallagen, J.B., La Du, B.N., Pariza, M.W., Putnam, J.M., Bozelleca, *J. F., Food Chem. Toxicol.*, Vol. 35, 625–632, 1997.
5. Kapoor, V.P.; "Commercial seed galactomannans: New sources and structural chemistry," In *Current Concepts in Seed Biology;* Eds., Mukherji, K.G.,Bhatnagar, A.K., Tripathi, S.C., Bansai, M., Saxena, M., Naya Prakash, Naya Prokash, Calcutta, India, I[st] Ed., pp. 87–114, 1992.
6. Gerard. Friedlander, *Anal. Chem.*, 1950, Vol. 22 (12), pp 1545–1551.
7. Tomonori Ishigaki, Yasunori Kawagoshi, Michihiko Ike and Masanori Fujita, *World Journal of Microbiology and Biotechnology*, Vol. 15, 3, 321–327, 1999.
8. Hoffmann J.; Reznckova I.; Kozakova J.; Ruzicka J.; Alexy P.; Bakos D.; Precnerova L., *Polymer Degradation and Stability*, Vol. 79, 3, March 2003 , pp. 511–519(9).
9. Ferry, J.D., *Viscoelastic Properties of Polymers*, Wiley International, New York, 3rd Ed., 1980.
10. Singh, Vandana, Tiwari, Ashutosh, Tripathi, D.N., Sanghi, R.; *Carbohydr. Polym.*, Vol. 58, 1–6, 2004.
11. Elfak, A.M., Geoffrey, P., Glynn, P.O., Robert, M.G., *J. Sci. Food Agri.*, Vol., 28, 895, 1977.
12. Deshmukh, S.R., Chaturvedi, P.N., Singh, R.P., *J. Appl. Polym. Sci.*, Vol. 30, 2013, 1985.
13. Vandana Singh, Ashutosh Tiwari, D.N. Tripathi, R. Sanghi, *Biomacromolecules*, Vol. 6, 453–456, 2005.
14. V. Singh, P. L. Kumari, Ashutosh Tiwari, S. Pandey, *J Appl Polym Sci*, Vol. 117, 3630–3638, 2010.
15. Maier, H., Anderson, M., Karl, C., Whistler, R.L.; "Guar, Locust bean, tara and fenugreek gum: Polysaccharides and their derivatives," In *Industrial Gums*, Eds., Whistler, R.L., BeMiller, J.N., Academic Press, Inc., London, 3rd Ed., pp. 181–221, 1993.
16. Ger., U.S. Pat. Chem. Abstr., 1956.
17. Prakash, A., Kappor, R.C., *Cell. Chem. And. Tech.*, Vol. 18, 207–212, 1984.
18. Kapoor, R.C., Prakash, A., *J. Ind. Chem. Soc.*, Vol. 61, 600–603, 1984.
19. Deshmukh, S.R., Chaturvedi, P.N., Singh, R.P., *J. Appl. Polym. Sci.*, Vol. 30, 2013, 1985.
20. Gulrajani, M., Choudry, L., Roy, A.K., *Colourage*, Vol. 28(10), 3–12, 1981.
21. Nordgren, R., U.S. Pat. Chem. Abstr., 1967.
22. Carroll, J.W., Griffith, *L.G., Chem. Abstr.*, 1979.
23. Guiseley, B.K., *A.C.S. Symp. Ser.*, 1978.
24. Costanza, R.J.O., Ronald, Dementino, N., Arthur, Goldstein, M., U.S. Pat. Chem. Abstr., 1978.
25. Ronald, W.H., James, K.L., U.S. Pat. Chem. Abstr., 1966.
26. Nayak, B.R., Singh, R.P.; *J. Appl. Polym. Sci.*, Vol. 81(7), 1776–1785, 2001.
27. Soppinath, S.K., Kulkarni, A.R., Aminabhavi, T.M.; *J. Controlled Release*, Vol. 75, 331–345, 2001.
28. Lokhande, H.T., Vardarajan, P.V., Iyer, V., *J. Appl. Polym. Sci.*, Vol. 48, 495, 1993.
29. Naidoo, S., Joff, R., Neuse, E. W., Angew, *Makromol. Chem.*, Vol. 59, 156, 1988.
30. Mehrotra, U. S.; Mushran, S. P., *J Ind Chem Soc.*, 1970, 47, 41.
31. Vandana Singh, Ashutosh Tiwari, Devendra Narayan Tripathi, Rashmi Sanghi, *J Appl Polym Sci.*, 92: 1569–1575, 2004.
32. Zheng, X.-X.; Luo, Y.-B.; Cheng, Z.-F.; Zheng, C.-Y. *Shiyou Huagong* 2000, 29, 19.
33. Vandana Singh, Ashutosh Tiwari, Premlata, Ajit K. Sharma, *Polymers for Advanced Technologies*, Vol. 18, 379–385, 2007.
34. Gabriel, C., Gabriel, S., Grant, E.H., Halstead, S.J., Mingos, D.M.P., *Chemical Society Reviews*, Vol. 27, 213–223, 1998.

35. Zhenping, C., Xiulin, Z., Chen, G., Xu, W., Lu, J.; *J. Polym., Sci., Polym. Chem. Part-A,* Vol. 40, 3823–3834, 2002.
36. Vandana Singh, Ashutosh Tiwari, S. P. Singh, P.K. Shukla, R. Sanghi, *Reactive and Functional Polymers,* Vol. 66, 1306–1318, 2006.
37. Chowdhury, P., Samui, S., Kundu, T., Nandi, M.M.; *J. Appl. Polym. Sci.,* Vol. 82(14), 3520–3525, 2001.

8

Starch Based Composites for Packaging Applications

K. M. Gupta

Professor (Department of Applied Mechanics) and ex-Dean (Research and Consultancy), Motilal Nehru National Institute of Technology, Allahabad, India

Abstract

Environmental problems of synthetic polymeric composites are explained. Importance of starch based composites is discussed in favor of the clean and green environment. Different kinds of starches suitable as matrix component of composites are highlighted along with their characteristics and mechanical properties. Vivid kinds of biopolymers and reinforcing agents such as natural fibers etc. are described along with their suitability and usefulness. The concept of various kinds of composites viz. flaked composite, fiber composite, sandwich composite and hybrid composites are included in. Mechanics of the composites is discussed and derivation is done for a better understanding of the strength and stiffness behavior of starch based composite. The requirements of packaging, their applications, and suitability of starch based composites for such applications are elaborated. Flexible packaging, active and intelligent packaging, testing standards and norms for packaging are given in. Characteristics, properties and behavior of a large varieties of starch based composites, nanocomposites, films and coatings are discussed. Effects of various parameters on the behavior of packaging purpose starch based biocomposites are included. Manufacturing methods and futuristic research outlook are other features of this chapter.

Keywords: Different sources of starches as matrix material, mechanical properties of starches, biopolymers, plasticization, different types of biocomposites, natural fibers, mechanics of composites, packaging requirements, vivid packaging applications, testing standards of packaging

8.1 Introduction

Starch is used as a starting material for a wide range of green biomaterials. Different routes are used to modify starch to improve the product properties and to extend the application range. 75% of all organic material on earth is present in the form of polysaccharides. An important polysaccharide is starch. Plants synthesize and store starch in their structure as an energy reserve. It is generally deposited in the form of small granules or cells with diameters between 1–100 μm. Starch is found

Srikanth Pilla (ed.) Handbook of Bioplastics and Biocomposites Engineering Applications, (189–266)

in seeds (i.e. corn, maize, wheat, rice, sorghum, barley, or peas) and in tubers or roots (i.e. potato or cassava) of the plants. Most of the starch produced worldwide is derived from corn. The worldwide production of starch in 2008 is estimated to be around 66 million tons [1]. Starch is generally extracted from the plant by wet milling processes. Starch is polymers consisting of two types of anhydroglucose (AHG) Polymers are amylose and amylopectin. Amylose is essentially a linear polymer in which AHG units are predominantly connected through α-D-(1,4) glucosidic bonds. Amylopectin is a branched polymer, containing periodic branches linked with the backbones through α-D-(1,6) glucosidic bonds. The content of amylose and amylopectine in starch varies and largely depends on the starch source. Typically, the amylose content is between 18–28%.

Utilization of polymer products at the end of their service life is an environmental problem. The traditional way of treating them viz. burning, recycling, waste disposal and pyrolysis cannot improve the ecological situation. Environmental pollution caused due to accumulation of non-destructible solid waste is a matter of serious concern today. It has created stir and has attracted the attention of researchers to develop bio-based biodegradable polymers and composite materials. The biodegradable materials are the appropriate answers to this menace. Such materials are composed quickly under the action of natural environment such as soil micro-organism, light, water, and other factors. In developing such materials, one has to meet a mutually contradictory environments namely, high parameters of mechanical properties on one hand and the accelerated biodegradability on the other hand. In this connection, the material based on biodegradable polymer obtained from renewable vegetable as raw materials are of increasing interest. Special among them is starch-based plastics. Starch is a typical natural polymer. It is biodegradable, and after plasticization can be processed into items on processing equipment.

Basic drawback of materials based on starch is its low mechanical properties. A promising method to improve the mechanical properties of starch is to introduce fibrous or lamellar particles of filler. By doing so, the effect of reinforcement (strengthening) can be reached not only because of the considerably higher values of strength and rigidity of the filler but also because of the particle geometry (the characteristics ratio of their sizes, known as the aspect ratio). Good results are obtained on using cellulose microfibers as the filler. Therefore, during the past years, an interest in starch-based composites containing natural fibers, layered silicate etc. as filler has increased. These composites consist of a completely biodegradable matrix and ecologically safe filler.

8.1.1 Starch: History, Characteristics and Structure

Starch, also known as amylum, is an important carbohydrate that remains present in human diet. It is composed of a large number of glucose units (Figure 8.1). These units are joined together by glycosidic bonds. Its molecular formula is $(C_6H_{10}O_5)_n$. All green plants produce it as store of energy. Starch molecules arrange themselves in the plant in semi-crystalline granules. Each plant species has a unique starch granular size. For example, the rice starch is relatively small (about 2 μm)

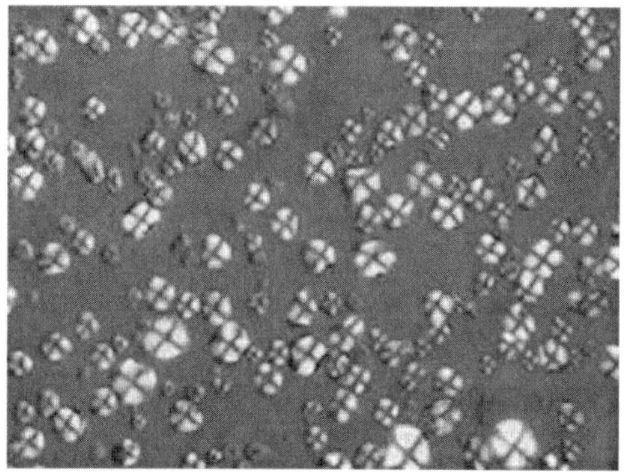

Figure 8.1 Starch, 800× magnified, under polarized light.
Source: Ref [2].

while potato starches have larger granules (up to 100 μm). Although in absolute mass, only about one quarter of the starch granules in plants consists of amylose, there are about 150 times more amylose molecules than amylopectin molecules. Amylose is a much smaller molecule than amylopectin.

The word starch is derived from "structure," an English word, which means "stiffness" [2]. It is named as "amylum" in Latin, and "amulon" is Greek language. The history of usage of starch is very old. Egyptians were using wheat starch as paste for stiffening the cloth and weaving of linen. Romans used wheat starch pastes in cosmetic creams, for powdering the hair and thickening sources. Persians used wheat starch to make dishes. Indian used it to make wheat halva. Chinese used the rice starch for surface treatment of paper.

Starch in pure form is a powder of white appearance. It is odorless and tasteless, and is insoluble in alcohol and cold water. It is heavier than water having the density of 1500 kg/m^3. It does not possess a sharp melting point, rather decomposes over a temperature range. It auto-ignites at 410°C. Starch is composed of two different kinds of molecules viz. linear and helical amylase structure, and branched amylopectin structure. These molecules vary plant to plant. Generally 20 to 25% amylase and 75 to 80% amylopectin is found in starch. The structure of an amylase molecule is shown in Figure 8.2.

Starch becomes soluble in water when heated. The granules swell and burst, the semi-crystalline structure is lost and the smaller amylose molecules start leaching out of the granule, forming a network that holds water and increasing the mixture's viscosity. This process is called starch gelatinization. During cooking, the starch becomes a paste and increases further in viscosity. During cooling or prolonged storage of the paste, the semi-crystalline structure partially recovers and the starch paste thickens, expelling water. This is mainly caused by the retrogradation of the amylose. This process is responsible for the hardening of bread or staling, and for the water layer on top of a starch gel (syneresis).

Some cultivated plant varieties have pure amylopectin starch without amylose, known as waxy starches. The most used is waxy maize, others are glutinous rice, waxy potato starch. Waxy starches have less retrogradation, resulting in a more stable paste. High amylose starch, amylomaize, is cultivated for the use of its gel

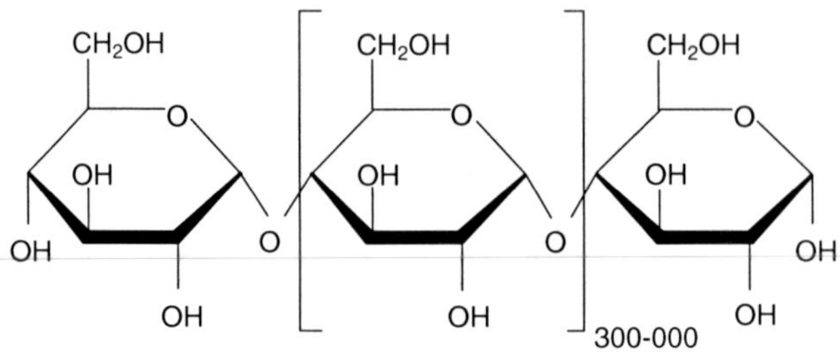

Figure 8.2 Structure of the amylose molecule.

strength. Major source of starch is found in foods such as wheat, rice, corn (maize), potatoes, and cassava.

8.1.2 Different Sources of Starch and Modified Starches

Various sources of starch that are used for making the biocomposites are the following:

- Commercial corn
- Sugarymutant corn
- Buckwheat
- Alderman pea
- Steadfast pea
- Water chestnut
- Waxy corn
- Sweet potato
- White potato
- Mung bean
- Colacasia
- Sorghum
- Oat
- Arracacha
- Rye
- Kudzu
- Yarns
- Wheat
- Arrowroot
- Tapioca
- Cannas
- Oca
- Sago
- Chestnuts
- Lentils
- Barely
- Banana

- Millet
- Breadfruit
- Kataburi
- Taro
- Malaga
- Favas

Starches are modified to make them suitable for specific applications. They are generally chemically treated to function properly during processing and storage. The treatments involve high heat, cooling, freezing, high shear, low pH etc. Various modified starches are listed as under.

- Hydroxypropyl starch
- Cationic starch
- Oxidized starch
- Carboxymethylated starch
- Alkaline treated starch
- Acid treated starch
- Bleached starch

These starches find use in different applications such as food, food additives, paper making, bio-fuel, corrugated board adhesives, clothing/laundry, textile industry, printing industry , to produce bio-plastics that are bio-degradable e.g. polylactic acid, body powder, in oil exploration to adjust the viscosity of drilling fluid.

Foamed starch: Starch can be blown by environmentally friendly means into a foamed material using water steam. Foamed starch is antistatic, insulating and shock absorbing, therefore constituting a good replacement for polystyrene foam.

8.1.3 Processing of Starch before Using as Matrix in Composite

In making the starch suitable for industrial use, first of all the starch is extracted from roots, tubers and seeds of the plants. It is then refined by conducting the processes of wet grinding, washing, sieving and drying. Mainly the following refined starches are used commercially.

- Wheat starch
- Tapioca
- Rice starch
- Mung bean starch
- Potato starch
- Corn starch
- Sweet potato starch
- Sago starch

Untreated starches are heated for thickenings (i.e. gelatinizing). Starch are also pre-coated and thickened in cold water instantly.

The starch is used for sizing and coating of paper. It is preferable because of potential to reduce total cost of composite. Proper cooking and complete hydration of starch molecule is essentially desired for all applications. Potato starch is very easy to cook. Complete dispersion of starch granule depends on cooking temperature and shear ability. High shear is a desired property.

Modified starches such as acid-modified, oxidized, and ethylated are used in textile industry. These starches are used to treat the cloth as it comes from the loom [3]. The starch removes the impurities of various applied dyes, chemicals, softeners etc. It is also used for de-sizing.

8.1.4 Improving the Properties of Starch

To improve the properties of starch, various physical and chemical modifications such as blending, derivation and graft copolymerization are done. It can be done in following ways.

- **By blending with synthetic degradable polymers**

To prepare completely biodegradable starch-based composites the components to blend with starch are aliphatic polyesters, polyvinyl alcohol (PVA) and biopolymers [4]. The commonly used polyesters are poly(β-hydroxyalkanoates) (PHA) obtained by microbial synthesis, and poly-lactide (PLA) or poly(ε-caprolactone) (PCL), derived from chemical polymerization. The goal of blending completely degradable polyester with low cost starch is to improve its cost competitiveness whilst maintaining other properties at an acceptable level.

PLA is one of the most important biodegradable polyesters with many excellent properties. It possesses good biocompatibility and processability as well as high strength and modulus. However, PLA is very brittle under tension and bending loads, and develops serious physical aging during application. Moreover, PLA is a much more expensive material than the industrial polymers.

To improve the compatibility between PLA and starch, suitable compatibilizer is added. Besides, gelatinization of starch is also a good method to enhance the interfacial affinity. Starch is gelatinized to disintegrate granules and overcome the strong interaction of starch molecules in the presence of water and other plasticizers, which leads to well dispersion. The glass transition temperature and mechanical properties of TPS/PLA blend depend on its composition and the content of plasticizer.

PCL is another important member of synthetic biodegradable polymer family. It is linear, hydrophobic, partially crystalline polyester, and can be slowly degraded by microbes. Blends between starch and PCL have been well used. The weakness of pure starch materials including low resilience, high moisture sensitivity and high shrinkage is overcome by adding PCL to starch matrix even at low PCL concentration. Blending with PCL, the impact resistance and the dimensional stability of native starch is improved significantly.

PCL/starch blends can be further reinforced with fiber and nano-clay respectively. Moreover, the other properties of the blends such as hydrolytic stability, degradation rate, and compatibilization between PCL and starch are also improved.

PVA is a synthetic water-soluble and biodegradable polymer. PVA has excellent mechanical properties and compatibility with starch. PVA/ starch blend is assumed to be biodegradable since both components are biodegradable in various microbial environments.

- **By blending with biopolymers**

Natural polymers such as chitosan and cellulose and their derivatives are inherently biodegradable, and exhibit unique properties. Starch and chitosan are abundant naturally occurring polysaccharide. Both of them are cheap, renewable, non-toxic, and biodegradable. The starch/chitosan blend exhibits good film forming property, which is attributed to the inter- and intra-molecular hydrogen bonding that formed between amino groups and hydroxyl groups on the back-bone of two components. The mechanical properties, water barrier properties, and miscibility of biodegradable blend films are affected by the ratio of starch and chitosan.

- **By chemical derivatives**

One problem for starch-based blends is that starch and many polymers are non-miscible, which leads to the mechanical properties of the starch/polymer blends generally become poor. Thus, chemical strategies are taken into consideration. Chemical modifications of starch are generally carried out via the reaction with hydroxyl groups in the starch molecule [4]. The derivatives have physico-chemical properties that differ significantly from the parent starch but the biodegradability is still maintained. Consequently, substituting the hydroxyl groups with some groups or chains is an effective means to prepare starch-based materials for various needs.

8.2 Composite Materials

Composite materials are the material systems that are composed of two or more dissimilar constituents, differing in forms, insoluble in each other, physically distinct and chemically inhomogeneous. The resulting product possesses properties much different from the properties of constituent materials. Reinforced composites are made-up of two basic constituents viz. (i) Matrix or body constituent, and (ii) Reinforcing constituent. The matrix constituent comprises of different kinds of starches as mentioned earlier, while the commonly used reinforcing agents are the following.

1. Synthetic fibers such as carbon, graphite, kevlar, glass, ceramic etc.
2. Natural fibers such as polymra, kenaf, jute, hemp, ramie, flax, sisal etc.

The two constituents, therefore, make a large variety of composites.

8.2.1 Advantages and Limitations of Composites

Besides having favorable effects, the composites suffer from some shortcomings and limitations also. These are given as follows.

Advantages: Main among them are as follows.

1. They possess a combination of excellent mechanical, structural and other desired properties.
2. They are lightweight materials possessing higher specific strength and specific modulus than the conventional materials.
3. Alternatively it helps in weight reduction.
4. Composites can be molded to any shape and size, and according to any desired specification.
5. They possess excellent anti-chemical and anti-corrosion properties.
6. Making, repairing and fabricating of composites are easier than the metals and RCC.
7. Assembling and disassembling of components is easy and quick.
8. Efficient utilization of material may be done. The fibers may be oriented in such a way so as to provide greatest strength and stiffness in the desired direction.
9. Seepage and weathering problems are negligible.
10. Composites may be desired to obtain aesthetic appearance.

Limitations: Main among them are as follows.

1. They have low flash and fire points because of biopolymeric matrix.
2. They may develop undesired biological effects as seen in many polymers.
3. Polymeric biocomposites are not suitable for high temperature applications.
4. Cost of composites is still higher than many conventional materials.
5. On prolong exposure to sunlight, the colors of composites generally fade-out.

8.2.2 Classification of Starch-Based Biocomposites

The starch based biocomposites may be classified in many ways such as given below.

I. On the Basis of the Nature of Reinforcing Material Used
   - Particulate composite
   - Fiber composite
   - Flake composite
   - Sandwich composite
   - Hybrid composite
   - Whisker reinforced composite

II. Natural Fiber-reinforced Biocomposites [5–20]such as:
   - PLA-cellulose composite
   - PLA-kenaf composite
   - PLA-flax composite
   - PLA-hemp composite

- PLA-bamboo composite
- PLA-jute composite
- PLA-wood fibers composite
- PLA-cotton fibers composite
- PLA-silkworm silk fibers composite for tissue engineering applications
- PLA-agriculture waste fibers composite (water bamboo husk, rice straw)
- PLA-rice straw
- PHBV-rice straw
- PLA-nanocomposites
- PLA-chicken feather fibers composite

III. Synthetic Fiber-reinforced Biocomposites such as:
- PLA-rayon composite
- PLA-polypropylene composite

IV. On the Basis of Type of Aspect Ratio of Fibers Used
- Short-fiber biocomposite
- Long-fiber biocomposite

V. Based on the Arrangement of Fiber Lay
- Unidirectional (U/D) composite
- Bidirectional or cross-plied composite
- Angle-plied composite
- Off –axis composite
- Randomly oriented composite

VI. On the Basis of Type of Surface Modification of Fiber Biocomposites
- Modification by silane-treatment
- Modification by alkaly silane-treatment
- Modification by sodium hydroxide
- Modification by hybridization
- Modification by steam explosion

VII. On the Basis of Method of Manufacturing/Fabrication
- Hot press composite
- Film stacking composite
- Injection molded composite
- Micro-braining technique composite
- Melting compound method
- Coupling agent

VIII. Starch Based Nanocomposites such as:
- Starch nano-clay composite
- MMT- potato starch nanocomposite
- Cellulose-Starch nanocomposite
- Sweet potato-OMMT nanocomposite

IX. Starch Based Flexible Films and Coatings such as:
- Chitosan starch foam
- Cassava starch foam

8.2.3 Particulate Biocomposites

Particulate composites have one or two dimensional macroscopic additive constituents randomly embedded in the matrix. The size, nature and function of these particles vary widely. The particles are 1μm or more in size and having volume concentrations of 20 to 40 percent. Small particles of uniform size with proper orientation exhibit a more strengthening effect. Plasticized biopolymer matrix with potato starch and thermocell (a wood product) [21] is an example of this kind.

8.2.4 Flake Biocomposites

These are composites of two-dimensional nature, and are preferred when planer isotropy is desired in components. It should be noted that the composites generally possess orthotropy or anisotropy, and not the isotropy. Moreover, flakes of two-dimensional geometry can be more closely packed than the fibers. These qualities make them suitable for various packaging applications. Silver flakes are used where good conductivity is desired.

8.2.5 Hybrid Biocomposites

Fiber composites are made up of only a single type of fiber such as glass, carbon, Kevlar sisal etc.

Each type of fiber has its own limitations in terms of strength, cost and other material properties. These limitations can be overcommed by using a combination of two or more types of fibers in the same matrix. Mixing of two or more different types of continuous fibers in the same matrix is called *hybridization* and the resulting product is called *hybrid composite.* Improvement in properties of such composites is due to the *hybrid effect.*

Figure 8.3 shows an idealized stress-strain characteristic curve of two fibers system. The initial slope OA gives initial modulus and slope BC gives the final modulus. The fall AB in the curve is due to failure of first fiber system at strain ε_A. The second fiber system continues to take the load until final fracture occur at strain ε_c. In a typical case, the first and second fiber systems may be those of flax and jute.

Hybrid composites are subdivided into following four types:

a. Interply hybrid composites
b. Intraply hybrid composites
c. Inter-Intraply hybrid composites
d. Super hybrids.

The *interply hybrid* composite consists of alternate lamina (layer or ply) of the same matrix but different fibers. *Intraply hybrid* composite contains each lamina having two or more kinds of fiber system. The *inter-intraply* hybrid composite is a combination of the above two types.

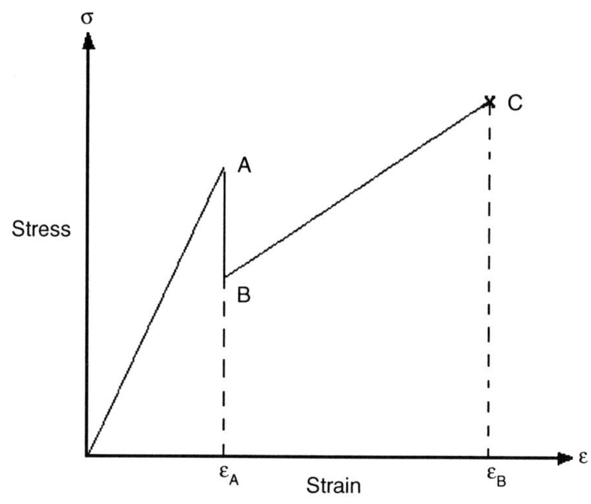

Figure 8.3 Idealized stress- strain curve of a hybrid composite showing hybrid effect.

8.2.6 Sandwich Biocomposites

A sandwich composite is constructed by sandwiching foam core between two skins of FRP laminates as in Figure 8.4. A lightweight cellular structure (typically hexagonal nested cells) is used as core in composite sandwich structures. It may be made from either metallic (e.g., aluminum) or nonmetallic (e.g., resin-impregnated paper or woven fabric) sheet materials. Rectangular sheets are adhesively bonded together in stacks, by means of parallel stripes of adhesive placed at regular intervals along one axis. Stacks are sliced across the transverse axis, and each sliced stack is expanded to form a honeycomb grid.

The thickness of the skin t_s is kept up to 3 mm, and the thickness of core t_c is kept deeper. The core is either foamed or made of honeycomb material so that its density is very less. As t_c is deep ($t_c >> t_s$), the area moment of inertias of the cross-section is enhanced too much; due to which the flexural rigidity of sandwich beam becomes more. This higher flexural rigidity construction along with

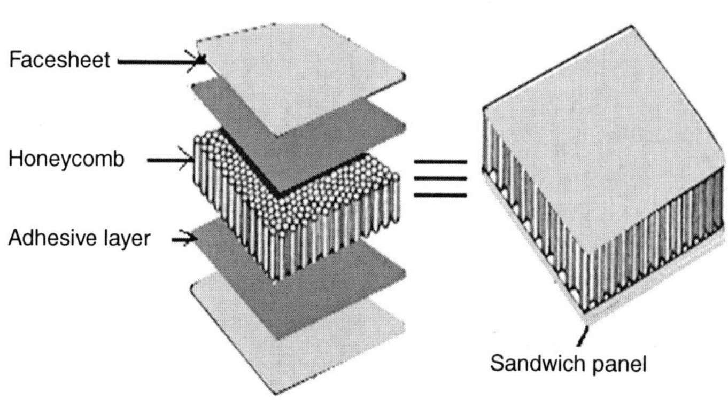

Figure 8.4 Sandwich composite structure.
Source: Ref. [20].

very light weight makes the sandwich composite most suitable as a beam. The sandwich core materials are much lighter in weight. For example the density of graphite-polyimide (Gr-PI) is 30 kg/m^3, of aluminum honeycomb is 32 kg/m^3 and glass-polyimide (GI-PI) is 40 kg/m^3 only [22].

8.3 Biopolymers/Biodegradable Polymers for use as Matrix of the Composite

Biopolymers are obtained in different ways viz. (1) From renewable resources, (2) By synthesizing microbially, and (3) By synthesizing from petroleum-based chemicals. The biopolymers based on renewable resources are the following.

1. Agro-polymers such as starch and cellulose plastics
2. Polyhydroxyalkanoates (PHA) e.g. polyhydroxybutyrate (PHB), etc.
3. Polylactides, etc.

Biodegradable polymers may be used to constitute the matrix part of composite materials. But, they generally have high cost and performance limitations. This deficiency, however, may be overcomed by using the biodegradable films such as starch. This is an attractive way to get cost-effective bio-based polymer. Such matrix material will be less expensive, biodegradable and with ample availability. Due to hygroscopic nature of starch and lack of affinity, the adhesion between starch and hydrophobic biopolymers can be increased by preparing the multi-layered biocomposites.

Different types of biodegradable polymers are:

1. Petroleum/fossil fuel-based such as
   - Aliphatic polyesters
   - Aliphatic-aromatic polyesters
   - Poly(ester amide)
   - Poly(alkyene succinates)s
   - Poly(vinyl alcohol

2. Renewable resource based such as
   - Polyactides (PLA)
   - Cellulose esters
   - Starch plastics
   - Polyhydroxy-alkanoates (PHAs)

3. Mixed resource based : renewable resources + petroleum resources such as blendings of
   - Two/more biodegradable polymers(example: starch plastic + PLA)
   - One biodegradable + one fossil fuel-made polymer(example: starch plastic + polyethene)
   - Epoxidized soyabean oil + petro- based epoxy resin

8.3.1 Important Bio-Polymers

Modified PET: Polyethylene Tetraphalate (PET) is a rigid polymer to which alphatic monomers can be added to enhance biodegradability, such as PBAT (polybutyleneadipate/terephthalate) and PTMAT (polytetramethyleneadipate/terephthalate).

Polyhydroxyalkanoates (PHA): PHAs are linear aliphatic polyesters produced in nature by bacterial fermentation of sugar or lipids. More than 100 different monomers can be combined within this family to produce materials with extremely different properties. They can be either thermoplastic or elastomeric materials, with melting-points ranging from 40 to 180°C. The most common type of PHA is PHB (polybeta-hydroxybutyrate).

Polyhydroxybutyrate (PHB): PHB has properties similar to those of polypropylene; however it is stiffer and more brittle.

Polyhydroxybutyrate-valerate copolymer (PHBV): Polyhydroxybutyrate-valerate is a PHB copolymer which is less stiff and tougher, and it is used as packaging material.

Polylactic acid (PLA): It is a biodegradable polymer derived from lactic acid. PLA resembles clear polystyrene. It provides good aesthetics (gloss and clarity), but it is stiff and brittle and needs modification for most practical applications (e.g. plasticizers increase its flexibility).

Polylactic acid aliphatic copolymer (CPLA): Biodegradable CPLA is a mixture of polylactic acid and other aliphatic polyesters. It can be either a hard plastic (similar to PS) or a soft flexible one (similar to PP) depending on the amount of aliphatic polyester present in the mixture.

10 % Starch composites: Starch can be used as a biodegradable additive or replacement material in traditional oil-based commodity plastics. If starch is added to petroleum derived polymers (e.g. PE), it facilitates disintegration of the blend, but not necessarily biodegradation of the polyethylene component. Starch accelerates the disintegration or fragmentation of the synthetic polymer structure. Microbial action consumes the starch, thereby creating pores in the material which weaken it and enable it to break apart.

50 % Starch composites: It is also called plastified starch materials. Such materials exhibit mechanical properties similar to conventional plastics such as PP, and are generally resistant to oils and alcohols. However, they degrade when exposed to hot water. Their basic content (40–80%) is corn starch, a renewable natural material. The balance is performance-enhancing additives and other biodegradable materials.

90 % Starch composites or Thermoplastic starch: Usually referred to as thermoplastic starch, they are stable in oils and fats. However depending on the type, they can vary from stable to unstable in hot/cold water. They can be processed by traditional techniques for plastics.

8.3.2 Biodegradable Polymers from Starch and Cellulose

In order to produce useful plastics from biopolymers, biopolymers have to be modified. The best known renewable resources capable of making biodegradable

Table 8.1 Properties of some traditional polymers and biodegradable polymers.

S. No.	Property	PBS[a]	PS	LDPE	PP	PLA (Eco-PLA)	PHB[b] (P226[c])	PHBV	PCL (Tone 787)
1.	Tensile stress at break (MPa)	57	35–64	8–10	34	45	24–27	25	41
2.	Tensile Modulus (MPa)	–	2800–3500	100–200	–	2800	1700–2000	1000	386
3.	Elongation at break (%)	700	1–2.5	150–600	12	3	6–9	25	900
4.	Density (g/cm³)	1.26	1.04–1.09	0.92	0.90	1.21	1.25	1.25	1.145
5.	Melting Point (⁰C)	115	–	124	164	177–180	168–172	135	60

[a] Bionolle 1001b Film Grade
[b] Plasticizes PHB, (*Source:* Biomer, Germany, http://www.biomer.de/)
[c] From technical data sheet, Showa Highpolymer Co. Ltd (Adopted from Ref. [4] and modified)

plastics are starch and cellulose. Starch and cellulose are not plastics in their native form, but are converted into plastics through various approaches such as extrusion cooking, fictionalization, plasticization. Starch is one of the least expensive biodegradable materials. It is a versatile biopolymer with immense potential for use in non-food industries. Starch-based polymers can be produced from corn, rice, wheat, or potatoes. Starch can be made thermoplastic through destructurization in presence of specific amounts of plasticizers (water and/or polyalcohols) under specific extrusion conditions. Three phenomena (i.e., fragmentation of starch granules, hydrogen-bond cleavage between starch molecules leading to loss of crystallinity, and partial depolymerization of the virgin starch polymers) generally occur during conversion of starch to starch plastic under extrusion conditions. Comparison of the properties of biodegradable polymers is shown in Table 8.1 as compared to that of the traditional polymers [4].

8.3.3 Biodegradable Thermoplastic Polymer: Polylactic Acid (PLA)

Polylactic acid [23] is a class of crystalline biodegradable thermoplastic polymer with relatively high melting point and excellent mechanical properties. Recently it has been highlighted because of its availability from renewable resources such as corn and sugar beets. PLA is synthesized by the condensation polymerization of D- or L-lactic acid or ring-opening polymerization of the corresponding lactide. Under specific environmental conditions, pure PLA can degrade to carbon dioxide, water and methane over a period of several months to two years, a distinct advantage compared to other petroleum plastics that need much longer periods. The final properties of PLA strictly depend on its molecular weight and crystallinity. PLA has been extensively studied as a biomaterial in medicine, but only recently it has been used as a polymer matrix in composites.

8.4 Starch as a Source of Bio-Polymer (Agro-Polymer)

Potato. The potato (Figure 8.5) is a starchy, tuberous crop of solanaceae family. Potatoes are the world's fourth-largest food crop; following rice, wheat, and maize. Potatoes yield abundantly with little effort, and adapt readily to diverse climates as long as the climate is cool and moist enough for the plants to gather sufficient water from the soil to form the starchy tubers. In terms of nutrition, the potato is best known for its carbohydrate content (approximately 26 grams in a medium potato). The predominant form of this carbohydrate is starch. Its starch content along with other nutritional values is given in Table 8.2.

Sweet Potato. The sweet potato (Figure 8.5) is a dicotyledonous plant that belongs to the family of convolvulaceae. Its large, starchy, sweet tasting tuberous roots are important root vegetable. Its starch content along with other nutritional values is given in Table 8.2.

Rice. Rice (Figure 8.5) is the seed of the monocot plant. As a cereal grain, it is the most important staple food for a large part of the world's human population. It is the grain with the second highest worldwide production, after maize (corn). Its starch content along with other nutritional values is given in Table 8.2.

Wheat. Wheat (Figure 8.5) is a grass. In 2007, world production of wheat was 607 million tons, making it the third most-produced cereal after maize (784 million

Figure 8.5 Starch as a source of bio-polymer (agro-polymer).
Source: Ref. [2].

Table 8.2 Starch content and nutritional values of bio-polymers per 100g.

Item	Potato	Sweet Potato	Rice	Wheat	Maize	Banana
Energy	321 kJ	360 kJ	1527 kJ	1506 kJ	360 kJ	371 kJ
Carbohydrate	19 g	20.1 g	79 g	51.8 g	19.02 g	22.84 g
Starch	15 g	12.7 g	75.8 g	71.9 g	79.5 g	–
Sugar	–	12.7 g	0.12 g	–	3.22 g	12.23 g
Fat	2.2 g	0.1 g	0.5 g	0.5 g	4.6 g	0.016–0.4 g
Dietary fiber	0.1–2 g	3.0 g	1.3 g	1.3 g	2.7 g	2.6 g
Protein	75 g	1.6 g	712 g	7.12 g	10.2 g	1.09 g
Water	1.8 mg (14%)	–	11.62 g	11 g	–	68.6–78.1 g

Percentages are relative to US recommendations for adults. *Source*: Ref [2,24,25], nutritiondata.com and USDA Nutrient database.

tons) and rice (651 million tons) [22]. Much of the carbohydrate fraction of wheat is starch. Wheat starch is an important commercial product of wheat. The principal parts of wheat flour are gluten and starch. These can be separated in a kind of home experiment, by mixing flour and water to form a small ball of dough, and kneading it gently while rinsing it in a bowl of water. The starch falls out of the dough and sinks to the bottom of the bowl, leaving behind a ball of gluten. Its starch content along with other nutritional values is given in Table 8.2.

Maize. Maize is also known as corn. Maize is one of the most widely grown crop. There are many maize varieties. Sweet corn is usually shorter than field-corn varieties. Many forms of maize (Figure 8.5) are used for food. They are also classified as following subspecies depending upon the amount of starch each had.

- Flour corn
- Waxy corn
- Dent corn
- Pod corn
- Sweet corn
- Popcorn
- Amylomaize
- Flint corn
- Striped maize

Maize is a major source of starch. Starch from maize can also be made into other chemical products. The corn steep liquor, a plentiful watery byproduct of maize wet milling process, is widely used in the biochemical industry and research as a culture medium to grow many kinds of microorganisms. Maize (Corn) contains about 70% starch, other components being protein, fibers and fat. The basis of the maize milling process is the separation of the maize kernel into its different parts. Maize starch is produced by the wet milling process, which involve grinding of softened maize and separation of corn oil seeds(germs) gluten (proteins), fibers (husk) and finally pure starch. Its starch content along with other nutritional values is given in Table 8.2.

Cassava. Cassava (Figure 8.6), also called yuca or manioc, is a woody shrub of spurge family. It is extensively cultivated as an annual crop in tropical and subtropical regions for its starchy tuberous root, a major source of carbohydrates. Cassava is the third largest source of carbohydrates for meals in the world. It is classified as "sweet" or "bitter" depending on the level of toxic cyanogenic glucosides. Commercial varieties can be 5 to 10 cm in diameter at the top, and around 15 cm to 30 cm long. Cassava roots are very rich in starch, and contain significant amounts of calcium (50 mg/100g), phosphorus (40 mg/100g) and vitamin C (25 mg/100g). However, they are poor in protein and other nutrients. Its tuberous roots have innumerable industrial uses also, particularly for starch extraction.

Cassava starch has very good properties that are highly desirable for the paper manufacturer. Cassava starch possesses a strong film, clear paste, good water holding properties, and stable viscosity. Properties of the starch used are abrasion resistance, flexibility, ability to form a bond to the fiber, to penetrate the fiber bundle to some extent and to have enough water holding capacity so that the fiber itself does not rob the size of its hydration.

Starch is a popular base for adhesives, particularly those designed to bond paper in some form to itself or to other materials such as glass, mineral wool, and clay. Starch can also be used as a binder or adhesive for non paper substances such as charcoal in charcoal briquettes, mineral wool in ceiling tiles and ceramics before firing. The starches most commonly used for the manufacture of adhesive pastes are maize, potato, and cassava; of these cassavas starch appears more suitable in several respects.

Cassava starch adhesives are more viscous and smoother working. They are fluid, stable glues of neutral pH that can be easily prepared and can be combined with many synthetic resin emulsions. Corn and rice starches take a much longer time to prepare and a higher temperature to reach the same level of conversion. For top-quality work, cassava starch is thought to be ideal, because it is slightly stronger than a potato starch adhesive while being odorless and tasteless, excellent as an adhesive for postage stamps, envelope flaps, and labels. Certain potato pastes have bitter tasting properties while cereal starches exhibit a cereal flavor.

Banana. The banana plant has long been a source of fiber for high quality textiles. Harvested shoots are first boiled in lye to prepare fibers for yarn-making. These banana shoots produce fibers of varying degrees of softness, yielding yarns and textiles with differing qualities for specific uses. For example, the outermost fibers of the shoots are the coarsest, and are suitable for tablecloths, while the softest innermost fibers are desirable for cloth-making process that requires many steps, all performed by hand.

Banana fiber is used in the production of banana paper. Banana paper is used in two different senses: to refer to a paper made from the bark of the banana plant, mainly used for artistic purposes, or paper made from banana fiber, obtained with an industrialized process from the stem and the non-usable fruits. The paper itself can be either hand-made or in industrial processes. Distribution of the world banana production average on the 2003–2007 period is shown in Figure 8.6. There is abundance of starch available in agro-polymer crops, which is indicated in Table 8.3. Worldwide production of various starch yielding agro- polymer crops in (million metric ton) are displayed in Table 8.5.

Barley. Barley (Figure 8.7) has a wider ecological range than any other cereal because it is more adaptable than other cereals. Barley can be grown on soils unsuitable for wheat, and at altitudes unsuitable for wheat or oats. Because of its salt and drought tolerance, it can be grown near desert areas also.

Buckwheat. Buckwheat (Figure 8.7) refers to variety of plants in dicot family. The name "buckwheat" or "beech wheat" comes from its triangular seeds, which

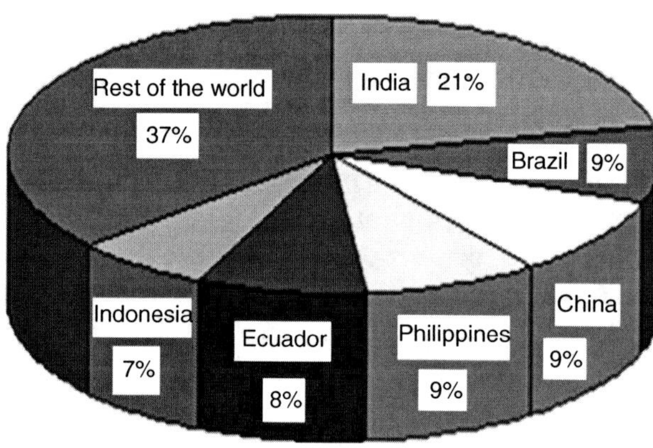

Figure 8.6 Distribution of the world banana production.
Source: NNCTAD, UN Food & Agriculture Organization (FAO)/fao.org from FAO statistics) average on the 2003–2007 period.

Table 8.3 Worldwide production of various starch yielding agro-polymer crops in (million metric ton).

Producer Countries	Potato in 2006	Rice in 2007	Wheat in 2008	Cassava in 2005
World total	315	–	690	202.58
People's Republic of China	70	187	112	4.20
Russia	39	–	64	–
India	24	144	79	6.7
USA	20	–	68	–
Indonesia	–	57	–	19.26

Figure 8.7 Starch as a source of bio-polymer (Agro-polymer).
Source: Ref [2] and Encyclopedia Britannica.

resemble the much larger seeds of the beech nut from the beech tree, and the fact that it is used like wheat. The starch content in seeds of buckwheat is as follows.

> 71–78% in groats, 70–91% in different types of flour. Starch is 25% amylose and 75% amylopectin. Depending on hydrothermal treatment buckwheat groats contain 7–37% of resistant starch.

Rye. Rye (Figure 8.7) is a cereal crop that is grown extensively for its grains. Scientifically known as "secale cereale," it belongs to the wheat tribe "Triticeae." Rye bears a lot of resemblance to wheat and produces kernels in the same manner as wheat. However, the kernels of rye are much smaller as compared to those of wheat. Rye contains many healthy nutrients like dietary fibers and proteins.

Taro. Taro (Figure 8.7), is a tropical plant grown primarily as a vegetable food for its edible corm, and secondarily as a leaf vegetable. Taro is loosely called elephant ear. Taro leaves are rich in vitamins and minerals. Taro corms are very high in starch, and are a good source of dietary fiber. In North India, it is called "Arbi."

8.4.1 Aliphatic Polyester-Grafted Starch

Natural polysaccharides such as starch, are often used in polymer blende and composite to produce the widely sought biodegradable property. However, conventional melt-processing usually provides the starch-based materials with very poor mechanical properties, mainly due to thermal decomposition of starch before melting, strong water absorption and poor interfacial adhesion. An alternative consists of the chemical modification of starch, such as non-degradative substitution of the hydroxyl groups with functional groups like ester, ethers, isocyanates etc. From various observations, it emerged that the best results in terms of grafting level and polymerization kinetics were obtained by activating the starch hydroxyl groups with $AIET_3$.

Polyester-grafted starch [26] composites have been synthesized by in-situ ring opening polymerization of the corresponding lactose monomer in the presence of granular starch. Polymerization is carried out either in bulk or in toluene dispelsion after adequate modification of the hydroxyl groups, available at the starch surface, into aluminum alkoxides. This process proves to be very effective in promoting fast polymerization and covalent grafting of the polyester chains into the starch surface. The grafting of the polymeric chains have been evidenced by Differential Scanning Calorimetry, Laser Scattering Granulometry, Secondary Ion Mass Spectrometry. The very good interfacial adhesion between the polyester coating and starch surface has been found by SEM observations.

8.5 Fibers

Fibers are classified in different ways such as given below.

1. Natural and synthetic fibers
2. Continuous and short fibers
3. Organic and inorganic fibers

Natural fibers such as jute, hump, silk, felt, cotton, flax etc. are obtained from natural sources such as plants, animals and minerals. Synthetic fibers are produced in industries. They are cheaper and more uniform in cross-section than the natural fibers. Their diameters vary between 10 µm to 100 µm. Bio fibers such as carbon and graphite fibers are light in weight, flexible, elastic and heat sensitive. Inorganic glass, tungsten and ceramic fibers have high strength, low fatigue resistance and good heat resistant. The strength of composites increases when it is made of long continuous fibers. A smaller diameter of fibers also enhances the overall strength of composite.

8.5.1 Natural Fibers

Most natural leaf, stalk (bast) and seed fibers can be used in filling or reinforcing the thermoplastics. Typically, the greater is the aspect ratio (length: diameter) of fiber, the greater the improvement in properties over the pure thermoplastic. Bast fibers are typically the best for improvements in tensile and bending strength and modulus. For toughness, the coarse fibers such as sisal, and coir (coconut husk fiber) are best. In the addition to fibers, fines from processing of wood, coir, agave, hemp, jute etc. as well as rice and nut hulls, cereal straws (oat, rye, wheat, etc), and corn cobs can be used to improve dimensional stability and stiffness of thermoplastics. Some examples of natural fibers and their category are listed below.

Leaf: Pineapple
Bast: Kenaf, Hemp, Jute, Ramie, Flax, Sugar cane, Banana, Wood, Bamboo, Sisal
Seed: Cotton, Coconut, Milkweed, Rice hulls

Ramie Fibers. Ramie [27] is a flowering plant of the nettle family. The true ramie (or china grass) also called Chinese plant or white ramie is the Chinese cultivated plant. A second type, known as green ramie or rhea, has smaller leaves which are green on the underside. Ramie plant is shown in Figure 8.8. Ramie is one of the oldest fiber crops. It is a bast fiber, and the part used is the bark (phloem) of the vegetative stalks. Unlike other bast crops, ramie requires chemical processing to de-gum the fiber. The extraction of the fiber occurs in three stages.

Ramie is one of the strongest natural fibers. It exhibits greater strength when wet. It is not as durable as other fibers, and so is usually used as a blend with other fibers such as cotton or wool. It is similar to flax in absorbency, density and microscopic appearance. Because of its high molecular crystallinity, ramie is stiff and brittle, and breaks if folded repeatedly in the same place. It lacks resiliency and is low in elasticity and elongation potential. Ramie fiber has a moisture content of 8.0 wt% and fracture strain as 0.025% [25]. Other mechanical and physical properties are given in Table 8.4.

Sisal Fibers. Sisal (i.e. Agave sisalana), is a plant of the agave family. Sisal fibers are made of the leaves of the plant. The fiber is usually obtained by machine decortications in which the leaf is crushed between rollers. The resulting pulp is scraped from the fiber, and the fiber is washed and then dried by mechanical or natural means. The lustrous fiber strands are usually creamy white, average 100 to 125 cm in length and 0.2 to 0.4 cm in diameter. Sisal plant and fibers are shown in Figure 8.9a,b.

Sisal fiber is fairly coarse and inflexible. It is valued for its strength, durability, ability to stretch, affinity for certain dyestuffs, and resistance to deterioration in saltwater. Sisal ropes are employed for marine, agricultural, shipping, and general industrial uses. Water absorption capacity of sisal fiber is 5.8%–6.1% and elongation at break is 4.3% [25]. Other Mechanical and physical properties are given in Table 8.4.

Banana Fibers. The banana fiber products are popular for their household utility. These utility items are like laundry basket, office waste paper basket, and fruit or egg trays. Banana fiber products also serve as house deco. They can also be used as

Figure 8.8 Ramie plant.
Source: http://en.wikipedia.org/wiki/File:Boehmeria_nivea_1.jpg.

paper, for both industrial grade and stationery products. The paper is made of 20% natural banana fiber and 80% post consumer waste. By using this method instead of wood fiber, the burden of deforestation is assuaged. A banana tree and banana fibers are shown in Figure 8.9c,d. Other mechanical and physical properties are given in Table 8.4.

Coir Fibers. Coir fiber is produced in India, Sri Lanka and Thailand. Fibers are gained from coconut husks. Approximately 40–50% of a matured husk consists of fibers. After harvesting, the retting process takes place for gaining white fibers. This fruit fiber is contained in the husk of coconuts. Fiber length ranges from 100–300 m. Coir fibers are light in weight, strong and elastic and have a low light resistance. They have a high durability (because of the fiber composition; 35–45% cellulose, 40–45% lignine and 2.7–4% pectins and 0.15–0.25% hemicelluloses). A coir tree and coir fibers are shown in Figure 8.10a-b respectively. Thermal conductivity of coir fiber is 0.047 W/mK and value of elongation at break is 15–17.3% [25]. Other mechanical and physical properties are given in Table 8.4.

Flax Fibers. Flax is obtained from flax fiber plant. Its botanical name is "linum usitatissimum." It is commonly known as "patsan" which is a substitute of "jute." It is produced from the external to xylem. The flax fiber is strong and wiry, longer and finer in nature. It lies in the category of bast(or soft) fibers.

Flax fiber has a length of 75–120 cm and is valued for its strength, luster, durability and moisture absorbency. It absorbs little dirt, is free of bacteria, does not cause fluffs, and has good resistance against bases. The flax tree and fibers are shown in Figure 8.10 c,d. Thermal conductivity of flax fiber is 0.055W/mK and value of

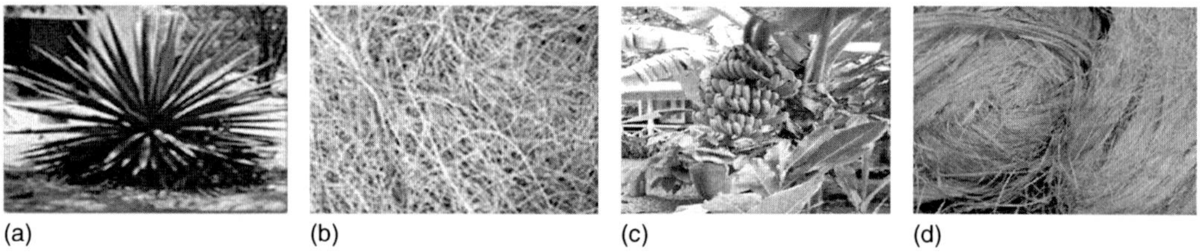

Figure 8.9 a) Sisal plant. b) Sisal fiber. c) Banana tree. d) Banana fiber.
Source: a) Wigglesworth & Co. Limited, London SE1 2NY. b) http://www.matbase.com/material/fibers/natural/sisal/properties. c) http://askpari.files.wordpress.com/2009/06/100_4807_banana_tree.jpg). d) http://ropeinternational.com/images/uploaded_ images/banana%20silky%20fiber.jpg.

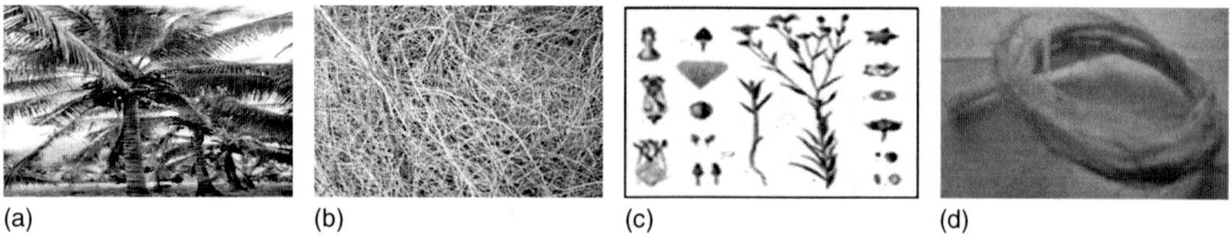

Figure 8.10 a) Coir tree. b) Coir-fiber. c) Flax fiber plant. d) Natural flax fiber.
Source: a) http://fida.da.gov.ph/Coco%20tree.jpg. b) http://product-image.tradeindia.com/00281409/b/0/Coir-Fiber.jpg. c) Ref [28]. d) Ref [28]).

elongation at break is 1.5–4%. The water absorption capacity is 8–10% [25]. Other mechanical and physical properties are given in Table 8.4.

Jute Fibers. It is a bast fiber and one of the cheapest natural fibers. It possesses a poor resistance against moisture, brittles under influence of light, absorbs paint easily. Jute consists of very short elementary fibers (length 0.7–6 mm) which are stuck together by lignin to form long brittle fibers (length of 300–400 mm). The jute tree and fibers are shown in Figure 8.11 a,b. For jute fiber, value of elongation is 0.8–2% and the water absorption capacity is 2–35% [25]. Other mechanical and physical properties are given in Table 8.4.

Hemp Fibers. It is a bast fiber, and is yellow-brown in color. It resembles flax in appearance, but is coarser and harsher. It is strong, lightweight and has very little elongation. The hemp tree and fibers are shown in Figure 8.11 c,d respectively. Thermal conductivity of hemp fiber is 0.048W/mK and value of elongation at break is 1–6%. The water absorption capacity is 8–30% [25]. Other mechanical and physical properties are given in Table 8.4.

Cotton Fibers. It is a seed fiber. Length of its fibers are 8–50 mm, diameter 12–20 μm and width 16–40 μm. Cotton fabrics are usually available in 1.2 m to 1.4 m width. Well-known varieties are furniture fabric, velours, gobelin and velvet. A cotton plant and cotton fibers are shown in Figure 8.12 a,b respectively. For cotton fiber, value of elongation at break is 3–8%. And the water absorption capacity is 20–100% [25]. Other mechanical and physical properties are given in Table 8.4.

Palmyra Fibers. Palmyra fiber is also known as toddy, wine siwalana, lontar, or taluuria baha palm. It is obtained from palm (toddy) plant. It is a bast fiber. Its botanical name is "boralessus flabillifer." Palmyra fibers are produced from the

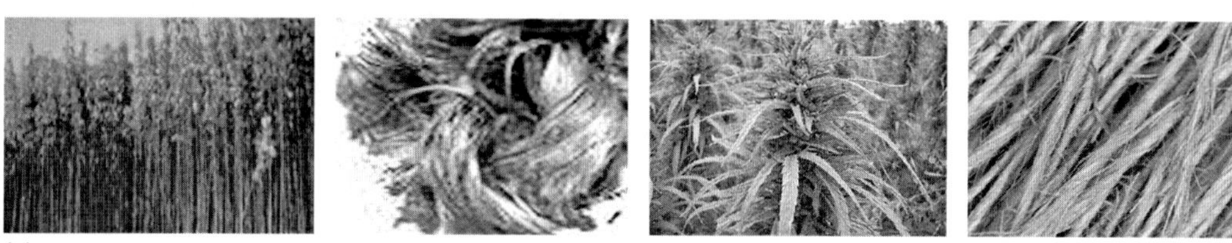

(a) (b) (c) (d)

Figure 8.11 a) Jute fiber plant. b) Natural jute fiber. c) Hemp in field. d) Hemp fiber.
Source: a) Ref [29]. b) Ref [29] c) http://keetsa.com/blog/wp-content/uploads/2008/09/hemp_field2.jpg. d) http://www.hempsa.co.za/images/Fiber/HempFiberRaw.jpg.

(a) (b) (c) (d)

Figure 8.12 a) Cotton plant. b) Cotton fiber. c) Palmyra tree. d) Palmyra fiber.
Source: a) http://www.djc.com/blogs/BuildingGreen/wp- content/uploads/2009/03/cotton-plant.jpg. b) http://www.diplomatie.gouv.fr/en/IMG/ jpg/trans2_150.jpg. c) Ref. [30]. d) Ref. [30].

Table 8.4 Characteristic values for the density, diameter, and mechanical properties of vegetable fibers.

Fiber	Density (g cm^{-3})	Diameter (μm)	Tensile Strength (MPa)	Young's Modulus (GPa)	Elongation at Break(%)
Ramie	1.55	–	400–938	61.4–128	1.2–3.8
Sisal	1.45	50–200	468–700	9.4–22	3–7
Banana	–	–	540–600	–	2.82–3
Coir	1.15–1.46	100–460	131–220	4–6	15–40
Flax	1.5	40–600	345–1500	27.6	2.7–3.2
Jute	1.3–1.49	25–200	393–800	13–26.5	1.16–1.5
Hemp	1.47	25–500	690	70	1.6
Cotton	1.5–1.6	12–38	287–800	5.5–12.6	7–8
Kenaf	–	–	930	53	1.6
Oil Palm EFB	0.7–1.55	150–500	248	3.2	25

Source: Ref. [31] Page 41, Table 8.1 and modified.

leaf sheath (petioles) of palmyra tree. The fiber is strong and wiry, shorter and finer in nature. Its lies in the category of hard fibers. Palmyra tree and palmyra fiber are shown in Figure 8.12c and 8.12d. For palmyra fiber, the value of elongation at break is 3–8%. Its specific strength is 70–270 MPa. Other mechanical and physical properties are given in Table 8.4.

8.6 Mechanics of Fiber Composite Laminates

The strength and moduli of composites laminates are greatly influenced by their fiber orientation. Most laminates are made of several distinct layers of unidirectional (U/D) laminate. The mechanics of a structural laminate, therefore, requires adequate knowledge of the properties of individual lamina. The study of properties of U/D lamina has been carried-out in the article that follows [22].

8.6.1 Rule of Mixture for Unidirectional Biocomposites Lamina

Volume and Weight fraction. A typical U/D lamina (Figure 8.13), has parallel fibers embedded in a matrix. The longitudinal, transverse, and normal directions are shown by L, T and T′ respectively. We consider the volumes v_c, v_f and v_m for

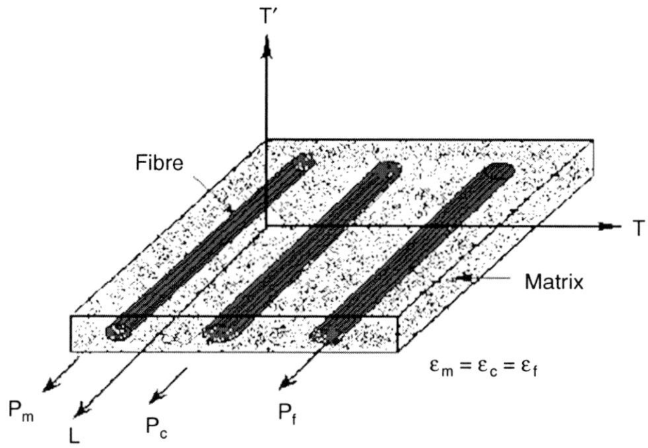

Figure 8.13 Representation of an unidirectional lamina.

composite, fiber and matrix respectively. Let w_c, w_f and w_m be the corresponding weight of composite, fiber and the matrix respectively. The volume fractions V_f and V_m, and weight fractions W_f and W_m of fibers and matrix respectively are defined as

$$V_f = v_f / v_c \text{ and } V_m = v_m / v_c \tag{8.1}$$

$$\text{where} \quad V_f + V_m = 1 \tag{8.2}$$

$$\text{and} \quad W_f = w_f / w_c, \quad W_m = w_m / w_c \tag{8.3}$$

$$\text{where} \quad W_f + W_m = 1 \tag{8.4}$$

In these equations the volume of voids v_v is assumed to be zero.

Determination of density of biocomposite. The density of composite ρ_c may be obtained in terms of densities of its constituents by simple rule of mixture as

$$\rho_c = \rho_f V_f + \rho_m V_m \tag{8.5}$$

Here ρ_f and ρ_m are densities of fiber and matrix respectively. The conversion between weight fraction and volume fraction may be obtained from

$$W_f = (\rho_f / \rho_m) V_f, \quad W_m = (\rho_m / \rho_f) V_m \tag{8.6}$$

Load carried by biocomposite. The resultant load P_c carried by composite along longitudinal direction is the sum of P_f and P_m shared by the fibers and matrix respectively. It is expressed by

$$P_c = P_f + P_m \tag{8.7a}$$

This load P_c induces an average stress σ_c in the composite of cross-sectional area A_c. Thus

$$P_c = \sigma_c A_c = \sigma_f A_f + \sigma_m A_m \tag{8.7b}$$

where A_f and A_m are the cross-sectional areas of fibers and matrix, σ_f and σ_m are the stresses produced in fiber and matrix respectively.

Longitudinal Strength and Modulus of biocomposites. The longitudinal strength is obtained from Eq. 8.7b as

$$\sigma_c = \sigma_f \left(A_f / A_c\right) + \sigma_m \left(A_m / A_c\right)$$

But for U/D composite, the volume fractions may be written as

$$V_f = A_f / A_c \text{ and } V_m = A_m / A_c$$

Therefore,

$$\sigma_c = \sigma_f V_f + \sigma_m V_m \tag{8.8}$$

Now differentiating Eq. 8.8 with respect to strain, and knowing that $\varepsilon_f = \varepsilon_m = \varepsilon$ (say), we get

$$(d\sigma_c / d\varepsilon) = (d\sigma_c / d\varepsilon)\, V_f + (d\sigma_m / d\varepsilon)\, V_m$$

Assuming linear stress-strain curve of the materials, the slope $(d\sigma/d\varepsilon)$ is a constant and corresponds to elastic modulus. Hence

$$(d\sigma_c / d\varepsilon) = E_c, \quad (d\sigma_c / d\varepsilon) = E_f \text{ and } (d\sigma_m / d\varepsilon) = E_m$$

Thus

$$E_c = E_f\, V_f + E_m\, V_m \tag{8.9}$$

where E_f and E_m are the Young's modulus of fiber and matrix respectively.

Transverse Strength and Modulus of biocomposites: Tensile properties of composite are obtained using a mathematical model shown Figure 8.14. Here, the composite is stressed in direction T. Let the elongations in composites, fiber and matrix be δ_c, δ_f and δ_m; and the thickness t_c, t_f and t_m respectively. In this case

$$\delta_c = \delta_f + \delta_m \tag{8.10a}$$

therefore,

$$\delta_c = \delta_c\, t_c, \quad \delta_f = \varepsilon_f\, t_f \text{ and } \delta_m = \varepsilon_m\, t_m \tag{8.10b}$$

On putting Eq. 8.10b in Eq. 8.10a, we get

$$\varepsilon_c\, t_c = \varepsilon_f\, t_f + \varepsilon_m\, t_m$$

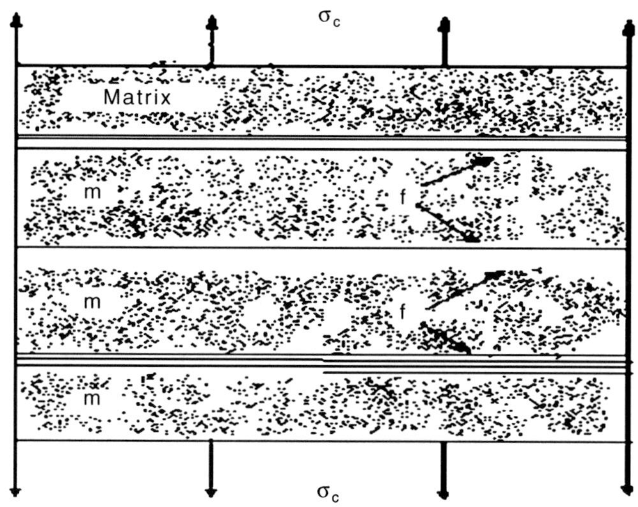

Figure 8.14 Mathematical model for predicting the transverse properties in U/D composite.

therefore,

$$\varepsilon_c = \varepsilon_f\,(t_f/t_c) + \varepsilon_m\,(t_m/t_c)$$
$$\text{or}\quad \varepsilon_c = \varepsilon_f V_f + \varepsilon_m V_m \tag{8.11}$$

where suffices c, f and m represent composite, fiber and the matrix respectively. Knowing that $\varepsilon_c = \sigma_c/E_c$, $\varepsilon_f = \sigma_f/E_f$ and $\varepsilon_m = \sigma_m/E_m$ in elastically deformed composite, the Eq. 8.11 can be written as

$$(1/E_c) = (V_f/E_f) + (V_m/E_m) \tag{8.12}$$

where E_c is transverse modulus.

Poisson's Ratio. Based on the rule of mixture of the constituents of U/D composite, the Poisson's ratio in longitudinal-transverse (LT) plane may be obtained from

$$\nu_{LT} = \nu_f V_f + \nu_m V_m \tag{8.13a}$$

where ν_f and ν_m are the Poisson's ratio of fiber and matrix respectively.
Poisson's ratio in LT´ plane can be determined by

$$\nu_{LT}' = \nu_f V_f + [\{1 + \nu_m - \nu_{LT}'(E_m/E_c)\}/\{1 - \nu^2_m + \nu_m\,\nu_{LT}'\,(E_m/E_c)\}]\,\nu_m V_m \tag{8.13b}$$

where E_c is longitudinal modulus. Poisson's ratio in TT´ plane may be obtained from above equation by replacing TT´ for LT´.

Shear Moduli. The fibers and matrix experience the same shear stress τ_{LT}, hence the shear strain γ_{LT} in composite is given by

$$\gamma_{LT} = \gamma_f V_f + \gamma_m V_m \tag{8.14}$$

where $\gamma_f = (\tau_{LT}/G_f)$, $\gamma_m = (\tau_{LT}/G_m)$ and $\gamma_{LT} = (\tau_{LT}/G_{LT})$.

Now the shear modulus G_{LT} of composite may be obtained from

$$(1/G_{LT}) = (V_f/G_f) + (V_m/G_m) \tag{8.15}$$

where G_f and G_m are shear moduli of fibers and matrix respectively. The same equations may be used to approximate G_{LT}.

8.6.2 Generalized Hooke's Law and Elastic Constants

Generalized Hooke's law is different from Hooke's law. Where the isotropic materials such as metals obey Hooke's law, the non-isotropic materials (such as composites) follow generalized Hooke's law. The generalized Hooke's law may be expressed mathematically as follows [20].

$$\begin{aligned}
\sigma_L &= A_{11}\,\varepsilon_L + A_{12}\,\varepsilon_T + A_{13}\,\varepsilon_T{'} + A_{14}\,\gamma_{LT}{'} + A_{15}\,\gamma_{TT}{'} + A_{16}\,\gamma_{LT} \\
\sigma_T &= A_{21}\,\varepsilon_L + A_{22}\,\varepsilon_T + A_{23}\,\varepsilon_T{'} + A_{24}\,\gamma_{LT}{'} + A_{25}\,\gamma_{TT}{'} + A_{26}\,\gamma_{LT} \\
\sigma_T{'} &= A_{31}\,\varepsilon_L + A_{32}\,\varepsilon_T + A_{33}\,\varepsilon_T{'} + A_{34}\,\gamma_{LT}{'} + A_{35}\,\gamma_{TT}{'} + A_{36}\,\gamma_{LT} \\
\tau_{LT}{'} &= A_{41}\,\varepsilon_L + A_{42}\,\varepsilon_T + A_{43}\,\varepsilon_T{'} + A_{44}\,\gamma_{LT}{'} + A_{45}\,\gamma_{TT}{'} + A_{46}\,\gamma_{LT} \\
\tau_{TT}{'} &= A_{51}\,\varepsilon_L + A_{52}\,\varepsilon_T + A_{53}\,\varepsilon_T{'} + A_{54}\,\gamma_{LT}{'} + A_{55}\,\gamma_{TT}{'} + A_{56}\,\gamma_{LT} \\
\tau_{LT} &= A_{61}\,\varepsilon_L + A_{62}\,\varepsilon_T + A_{63}\,\varepsilon_T{'} + A_{64}\,\gamma_{LT}{'} + A_{65}\,\gamma_{TT}{'} + A_{66}\,\gamma_{LT}
\end{aligned} \tag{8.16}$$

where the coefficients A_{11}, A_{12}, ...A_{23}, ...A_{44}, ...A_{55}, ..., A_{66} are elastic constants. Some of these elastic constants are Young's moduli, some are shear moduli, many are Possion's ratios, and others are coupling constants.

Considering first row and first column of Eq. 8.16, we can see that that A_{11} is *Young's modulus in longitudinal direction* as $\sigma_L/\varepsilon_L = A_{11}$ is Young's modulus for uniaxial case. Considering fourth row and fourth column, we notice that A_{44} is *shear modulus* for LT′ plane because $\tau_{LT}{'}/\gamma_{LT}{'} = A_{44}$ which is the modulus in uniaxial case. Now, we consider third row and sixth column which shows that, a shear strain γ_{LT} is produced in LT plane on application of direct tensile stress $\sigma_T{'}$ in T′-direction. Hence, the effect of direction T′ has reached to plane LT due to coupling effect. Thus A_{36} is termed as a coupling coefficient.

From first row and second column, we observe that a longitudinal stress σ_L causes a transverse strain ε_T. Similarly, the first column of second row reveals that the stress in transverse direction σ_T produces a longitudinal strain ε_L. Thus A_{12} and A_{21} are Poisson's ratios. Of these A_{12} is called *major Poisson's ratio* and A_{21} is known as *minor Poisson's ratio*. Numerical value of major Poisson's ratio may be less than that of minor Poisson's ratio.

8.7 Introduction to Packaging and its Functions

Currently, the raw materials used for packaging are petroleum-based, such as polystyrene and polyethylene etc. Disposal of used packaging products has become an ecological problem owing to their non-degradability. The utilization of biodegradable packaging materials has greater potential as eco-friendly material. The starch is such an alternative raw material for packaging because it is a biodegradable polymer with low cost.

The primary function of any packaging is to protect the product against the environment, to enhance the products life time and conserve its content. It also means that it can imply the protection of the environment against the packaged product (e.g. for harmful or toxic content). In either case, packaging provides possibilities to facilitate the transport of the packaged content. Packaging innovation seeks to increase resource efficiency, eliminate the production of waste and reduce environmental impact through improved design and use of alternative materials.

Packaging innovation is all about trying to gain differentiation through truly novel packs that enhance products and therefore make them more appealing to the consumer. For example, with the availability of technologies allowing the tailoring of material structure at the nanoscale, the next generation of packs will be "intelligent" ones. Materials science will also become more and more important to packaging as we continue to look for ways of making packaging materials sustainable yet functional. A packaging serves the following purposes.

i. Containment of item/articles/goods
ii. Convenience
iii. Protection of item/articles/goods
iv. Communication

8.7.1 Characteristics of a Good Packaging Material

A good packaging material should have the following characteristics.

- Excellent tensile and compressive strength to weight ratio, i.e. σ/w ratio
- Cost effectiveness
- Good visual and aesthetic appeal
- High puncture resistance
- It should be re-usable/recyclable
- It should be light in weight and easy to use
- It should be versatile for multipurpose wrapping
- It should be flame and fire resistant
- It should be sealable
- It should be flexible to protect all shapes of products
- Resistance to moisture absorption
- High tear resistance

8.7.2 Vivid Kinds of Packaging Materials and their Applications

The packaging industry has grown immensely now-a-days. It has become an important arena in the scientific and industrial world. Packaging is widely used for a large variety of applications, the main among them are food packaging, medicine packaging, electronics device packaging, and opto-electronic packaging etc.

A large number of food packaging applications are in use, and many more are in process of development.

A large variety of packagings are designed for vivid uses [32, 33]. These are:

- Multi-material packaging
- Transparent packaging
- Non-transparent packaging
- Corrugated packaging of paper grades, of fluting type, for lining purpose

The packaging materials may be of wide ranges, such as:

- Traditional packaging materials
- Environmental-friendly packaging materials
- Anti-corrosion packaging materials
- Anti-moisture packaging materials
- Anti-shock and anti-vibration packaging materials
- Anti-impact and anti-abrasion packaging materials

The packaging may be meant for the following services

- Export packaging services
- Hazardous goods services, etc.

The traditional/conventional materials used in packaging applications are the following.

1. Low density polyethylene (LDPE)
2. Foam and foam padding
3. High density polyethylene (HDPE)
4. Aluminum foils and rolls
5. Polystyrene
6. Bubble wrap and film cushioning
7. Paper and recycled waste paper
8. Sealed barrier foil bags
9. Timber
10. Desiccants such as silica gel
11. Polypropylene, PVC and Vinyl tapes
12. Void fill airbags, paper packaging, loose fill chips etc.
13. Activated clay which protects from water vapor damage, and provides protection in varying climates and temperatures
14. Bubble films, which are transparent cellular air packaging material that uses the air trapped inside the polyethylene film as a cushion for protecting products from abrasion, shock and vibration

8.7.3 Necessity of Biodegradable Packaging in Food Industry

Food packaging and edible films are two major applications of the starch-based biodegradable polymers in food industry. The requirements for food packaging include reducing the food losses, keeping food fresh, enhancing organoleptic characteristics of food such as appearance, odor, and flavor, and providing food safety. Traditional food packaging materials such as LDPE have the problem of environmental pollution and disposal problems. The starch-based biodegradable polymers can be a possible alternative for food packaging to overcome these disadvantages and keep the advantages of traditional packaging materials. However, the components in the conventional starch-based polymer packaging materials are not completely inert. The migration of substances into the food possibly hap-pens, and the component that migrates into food may cause harm for the human body. In view of this, new starch-based packaging materials are being developed. For instance, a starch/clay nanocomposite food packaging material is developed, which can offer better mechanical property and lower migration of polymer and additives [34]. Starch-based edible films are odorless, tasteless, colorless, non-toxic, and biodegradable. They display very low permeability to oxygen at low relative humidity and are proposed for food product protection to improve quality and shelf life without impairing consumer acceptability. In addition, starch can be transformed into a foamed material by using water steam to replace the polystyrene foam as packaging material. It can be pressed into trays or disposable dishes, which are able to dissolve in water and leave a non-toxic solution, then can be consumed by microbic environment.

8.8 Starch Based Packaging Materials

Starch based packaging materials are an environmentally-aware packaging materials with "green" credentials. These are biodegradable loose fill materials produced from starch-based raw materials. The biodegradable bubble film cushionings are another eco-friendly products whose appearance and performance are very high. They are free from toxic residues, safe to deposit in landfill where the film is oxo-biodegradable into bio-mass, CO_2 and water. This bubble film can reduce the volume of landfill and pollution also. Starch based packaging materials may be made of different materials and by different processes. Some of them are given below.

Currently, biodegradable packaging offers best for applications of shorter shelf life, high WVTR to prevent condensation, and high oxygen barrier. The proteins and carbohydrates offer many sites for potential improvement through chemistry. Also to replace the wide variety of petroleum-based packaging (PS, PP, PE, EVA, etc.), one will need a wide variety of starting materials. The commodity like fruits: orange, mango, sugarcane etc.; food like: wheat, maize etc. may be used to produce bio-based packaging as follows.

Commodity	Product	Waste
Fruits and grains	Juice, cheese, flour, vegetable oil, ethanol, bio-diesel etc.	Pulp, peel, glycerol, whey, feather, sugar beet fibers.

From these wastes, the different kinds of bio-based packagings obtained are composites, film, PHA, PLA, jars, monomers, biopolymers, coatings foils, bottles and natural fibers, boxes molded cups from whey proteins, PHA from fermented vegetable, animal oils and bio-glycerol.

8.8.1 Bio-degradable Packaging from Agricultural Feed Stocks

These are: pectin based film, edible film and coatings, whey-based packaging, soy-based packaging, straw-based packaging, edible film from dairy proteins, corn-film from extruded films, wheat-based starch molded packaging, keratin-based packaging.

The French AGRIPACK company produces packing material from maize starch [35]. The beads obtained are spherical and calibrated at approximately 15–20 mm in diameter. This is shown in Figure 8.15.

Active edible coatings for food packaging [36] are useful in many respects. These can reduce or replace barrier layers in traditional packaging systems, and can makes recycling of plastics easier. The carriers like antimicrobials-sorbates, benzoates, bacteriocins can prevent post-processing contamination of nutrients—vitamins, minerals, and phenolic compounds. Details of edible film and coating is given in Figure 8.16. These packing materials are also made by molding. Various molded packagings are shown in Figure 8.17 a–d, [36, 37].

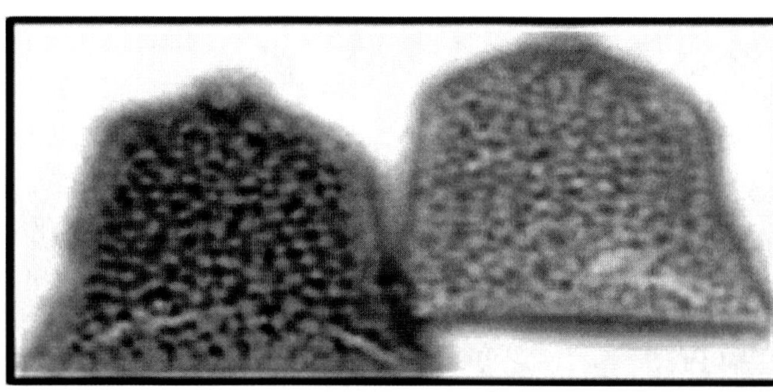

Figure 8.15 AGRIPACK packing material from maize starch.
Source: Ref. [35].

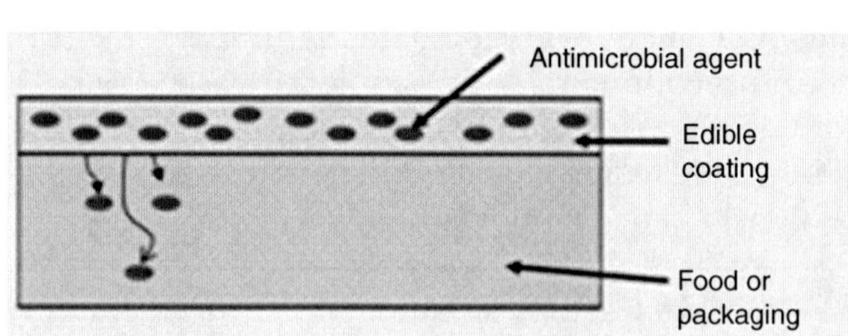

Figure 8.16 Edible film and coatings.
Source: Ref. [36, 37] and redrawn.

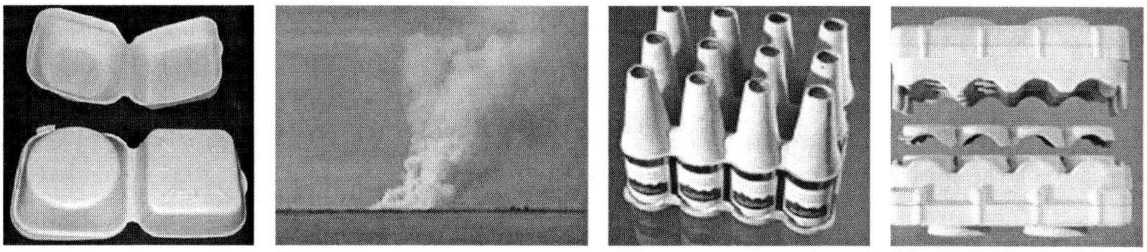

(a) (b) (c) (d)

Figure 8.17 Various molded packaging materials (a) starch based; (b), (c), (d) straw based. *Source:* Ref. [36, 37].

8.9 Flexible, Active and Passive, and Intelligent Packagings

Flexible Packaging. Flexible packaging covers a wide range of packaging that can be single and multi-layered, and is supplied in reels or bags. It can be paper/ poly/foil / nylon/ or a combination of materials which are supplied either plain/ printed/coated and/or laminated to provide long shelf life properties. End products packaging include confectionery, snack foods, frozen foods, soups and pharmaceuticals.

Flexible packages can take a variety of physical forms. *Wraps* consist of a layer of a material surrounding the product. It can be classified as intimate wraps, which make direct contact with the product as building wraps and is used to join together two or more products; so they can be handled as a single unit. *Stretch wrap* functions by utilizing the elasticity of the plastic film. It causes the film to return to its original dimensions when it has been stretched. Over time, this force decreases due to relaxation within the plastic. *Shrink wrap* also uses elasticity. Shrink wrap is commonly used for small wraps, and stretch wrap for longer ones, although both can be used in a variety of sizes.

The sides of the flexible packages can be formed by folding the film using a tube, or by joining two edges together. Joining is most commonly done by heat-sealing, which requires use of plastic. *Stand-up pouches*, one of the most rapidly growing forms of flexible packaging are pouches which, when filled with product, are capable of standing upright on a shelf and can serve as replacements for rigid containers.

Intelligent and Active Food Packaging. Intelligent packaging is emerging as a new branch of packaging technology that offers exciting opportunities for enhancing food safety, quality, and convenience. A new concept is emerging in which the packaging science, food science, biotechnology, sensor science, information technology, nanotechnology, and other disciplines are coming together to develop abreakthrough in packaging technology. The advancement in this technology will require researchers to continue to think out new challenges. It monitors the condition of packaged foods to give information about the quality of the packaged food during transport and storage. Active packaging actively changes the condition of the packaged food to extend life or to improve safety and sensory properties, while maintaining the quality of the packaged products.

Passive Food Packaging. Passive packaging provides protection from external elements such as air and moisture. This is a traditional packaging that involves the use of a covering material, characterized by some inherent insulating, protective and ease-of-handling qualities. The most common example of this type of packaging is a simple bio plastic bag.

8.9.1 Necessity of Active and Intelligent Packaging

With changes in the way the food products are produced, distributed, stored and retailed; and reflecting the continuing increase in consumers demand for improved quality and extended shelf life for packaged foods, the consumer are placing greater and greater demands on the performance of food packaging. Considering these aspects, innovative active and intelligent packaging concepts are being developed to:

- retain integrity and actively preventing the food spoilage i.e. enhanced shelf-life
- enhance the product attributes such as look, taste, flavor, aroma etc.
- respond actively to changes in product or package
- communicate product information, product history or condition to user
- assist with opening and indicate seal integrity
- confirm product authenticity and act to counter theft

Opto-Electronics Packaging. Optoelectronics packaging involves maintaining the functionalities of active and passive devices by providing optical and electrical interconnection, mechanical support, and protection from the operating environment to maintain the integrity of the performance. Epoxies are widely used for optoelectronic packaging [38], but there is a need for biocomposite optoelectronic packaging.

8.10 Testing Standards/Norms for Packaging

Packaging testing involves the measurement of characteristics involved with packaging. This includes packaging materials, packaging components, primary packages as well as the associated processes. Testing measures the effects and interactions of the levels of packaging, the package contents, external forces, and end-use. It can involve controlled laboratory experiments, or field testing. With some types of packaging such as food and pharmaceuticals, chemical tests are conducted to determine suitability of food contact materials. Package testing can extend for the full life cycle. Packages can be tested for their ability to be recycled and their ability to degrade as surface litter, in a sealed landfill or under composting conditions [39, 40].

Packaging testing might have a variety of purposes to serve. Main purposes among these are: to provide standard data for other scientific, engineering, and quality assurance functions; to validate suitability for end-use, to provide a basis for technical communication, to provide a technical means of comparison of several options, to help solve problems with current packaging , to predict the performance of a package during distribution and use.

With distribution packaging, one vital packaging development consideration is to determine that a product will not be damaged throughout the entire process of getting to the customer from the manufacturer. Primary purpose of a package is to ensure the safety of a product during transportation and storage. If a product is damaged during this process, then the package has failed to accomplish its primary objective Packages are usually tested when there is a new packaging design, a revision to a current design, a change in packaging material, and various other reasons. Testing a new packaging design before full scale manufacturing can save time and money.

Several standards organizations publish test methods for package testing. These are:

- International Organization for Standardization, ISO ASTM International
- European Committee for Standardization, CEN TAPPI
- International Safe Transit Association
- Governments and also many corporate test standards in use

Materials testing: The basis of packaging design and performance is the component materials. Packaging materials testing is often needed to identify the critical material characteristics and engineering tolerances. These are used to prepare and enforce specifications. For example, shrink film data might include: tensile strength, elongation, elastic modulus, surface energy, thickness, moisture vapor transmission rate, oxygen transmission rate, heat seal strength, heat sealing conditions, heat shrinking conditions, etc. Average and process capability are often provided. The chemical properties related for use as food contact materials may be necessary.

Thermal testing: Many packages are used for products that are sensitive to temperature. Exposure to high temperatures is also for shelf life testing of products: temperature (and other factors such as relative humidity) can accelerate the degradation of many products. The ability of packaging to control product degradation is frequently a subject of laboratory and field evaluations. Temperature is also one of the factors that can accelerate the natural aging of packaging. Accelerated aging of packaging can be based on time and the exposure to temperature, relative humidity, light, the contents of the package, and several other environmental factors. An Arrhenius equation ($\Phi = A\ e^{(-Ea/kT)}$) is often used to correlate chemical reactions at different temperatures. Some of the relevant tests in this regard are the following.

ASTM D3103- Standard Test Method for Thermal Insulation Performance of Distribution Packages

ASTM F1980- Standard Guide for Accelerated Aging of Sterile Medical Device Packages

Vacuum testing: Vacuum chambers are used to test the ability of a package to withstand low pressures which might be encountered during distribution. They are also used to stress the package to test the strength of seals and the tendency for leakage. Some of the relevant tests in this regard are the following.

ASTM D3078- Standard Test Method for Determination of Leaks in Flexible Packaging by Bubble Emission
ASTM D4991- Standard Test Method for Leakage Testing of Empty Rigid Containers by Vacuum Method
ASTM D6653- Standard Test Methods for Determining the Effects of High Altitude on Packaging Systems by Vacuum Method
ASTM F2391- Standard Test Method for Measuring Package and Seal Integrity Using Helium as the Tracer Gas

Barrier properties: The ability of a package to control the permeation and penetration of gasses is vital for many types of products. Tests are often conducted on the packaging materials but also on the completed packages. These tests are: moisture vapor transmission rate, oxygen transmission rate, and carbon dioxide transmission rate. Some of the relevant tests in this regard are the following.

ASTM D 3201-94: To Measure Water Vapor Absorption of Biocomposite Films
ASTM E 96-94 (desiccant method): To Measure Water Vapor Permeability of Biocomposite Films

Shock and impact testing: Both primary (consumer) packages and shipping containers have a risk of being dropped or being impacted by other items. Package integrity and product protection are important packaging functions. Tests are conducted to measure the effectiveness of package cushioning to isolate fragile products from shock. Some of the relevant tests in this regard are the following.

ASTM D1596- Standard Test Method for Dynamic Shock Cushioning Characteristics of Packaging Materials
ASTM D3332- Standard Test Methods for Mechanical-Shock Fragility of Products, Using Shock Machines
ASTM D5265- Standard Test Method for Bridge Impact Testing
ASTM D5276- Standard Test Method for Drop Test of Loaded Containers by Free Fall
ASTM D6537- Standard Practice for Instrumented Package Shock Testing For Determination of Package Performance

Vibration testing: Vibration is encountered during shipping (vehicle vibration, rough roads, etc.) and movement on conveyors. The ability of a package to withstand these vibrations and to protect the contents can be measured by several laboratory test procedures. Some allow searching for the particular frequencies of vibration that have potential for damage. Others use specified bands of random vibration to better represent vibrations measured in field studies of distribution environments. Some of the relevant tests in this regard are the following.

ASTM D999- Standard Test Methods for Vibration Testing of Shipping Containers
ASTM D3580- Standard Test Methods for Vibration (Vertical Linear Motion) Test of Products
ASTM D4728- Standard Test Method for Random Vibration Testing of Shipping Containers
ASTM D5112- Standard Test Method for Vibration (Horizontal Linear Sinusoidal Motion) Test of Products

Compression testing: Compression testing relates to stacking or crushing of packages, particularly shipping containers. It usually measures the force required to crush a package, stack of packages. A force deflection curve used to obtain the peak load or other desired points. Dynamic compression is sometimes tested by shock or impact testing with an additional load to crush the test package. Dynamic compression also takes place in stacked vibration testing. Some of the relevant tests in this regard are the following.

ASTM D5331- Standard Test Method for Evaluation of Mechanical Handling of Unitized Loads Secured with Stretch Wrap Films
ASTM D5414- Standard Test Method for Evaluation of Horizontal Impact Performance of Load Unitizing Stretch Wrap Films
ASTM D5416- Standard Test Method for Evaluating Abrasion Resistance of Stretch Wrap Films by Vibration Testing

Dangerous goods testing: Hazardous materials, dangerous goods are highly regulated. There are some material and construction requirements but also performance testing is required. The testing is based on the packing group (hazard level) of the contents, the quantity of material, and the type of container. Some of the relevant tests in this regard are the followings.

ASTM D4919- Standard Specification for Testing of Hazardous Materials Packaging
ASTM D7387- Standard Test Method for Vibration Testing of Intermediate Bulk Containers (IBCs) Used for Shipping Liquid Hazardous Materials (Dangerous Goods)
ISO 16104 - 2003 Packaging - Transport packaging for dangerous goods - Test methods

Medical packaging testing: Medical packaging is highly regulated. Often medical devices and products are sterilized in the package. The sterility must be maintained throughout distribution to allow immediate use by physicians. A series of special packaging tests is used to measure the ability of the package to maintain sterility. Some of the relevant tests in this regard are the following.

ASTM D1585- Guide for Integrity Testing of Porous Medical Packages
ASTM F2097- Standard Guide for Design and Evaluation of Primary Flexible Packaging for Medical Products
EN 868-1- Packaging materials and systems for medical devices which are to be sterilized. General requirements and test methods

8.11 Recent Advances in Starch Based Composites for Packaging Applications

The researches being carried out for the development of starch based composites for packaging applications are focused on different areas. They may be grouped under following categories.

1. Plasticized starch and fiber reinforced composites for packaging applications.
2. Starch based nanocomposites for packaging applications.
3. Starch foam, film and coated composites for packaging applications.
4. Starch based smart (or intelligent) composites for packaging applications.

Their details are elaborated in subsequent sections.

8.12 Plasticized Starch and Fiber Reinforced Composites for Packaging Applications

Starch is considered as a polymer with high potential for packaging applications because of low cost, renewability and biodegradability. During the extrusion of starch; the combination of shear, temperature and plasticizers allows to produce a molten thermoplastic material by disruption of the native crystalline granular structure and plasticization. This plasticized starch could be suitable for injection molding or thermoforming. In fact, different weaknesses limit the utilization of plasticized starch in packaging applications. Major drawbacks are water sensitivity, change of mechanical properties with time, crystallization due to ageing, plasticization by water adsorption, and low impact strength resistance.

8.12.1 Plasticized Wheat Starch (PWS) and Cellulose Fibers Composites for Packaging Applications

Luc Avérous, Cristophe Frigant and Laurence Maro [41] have investigated the plasticized wheat starch (PWS) and cellulose fibers composites for packaging applications. The materials and methods used by them are: wheat starch cellulose fibers (fiber length = 60 to 90 μm, diameter d = 20 μm, and l/d ratio = 3 to 45) and glycerol of 99% purity grade as plasticizer. Starch, fibers and glycerol were mixed in a turbo batch-mixer. After dehydration for 45 min in a vented oven at 170°C, the mixture was stirred and the desired amount of water was slowly added. The resulting "dry blend" was then extruded at 150°C in a single screw extruder equipped with a conical shaped element and granulated. The granules were equilibrated at 65% relative humidity (RH) for 8 days before injection molding into dumbbell specimen used for mechanical testing. An injection molding machine was used with a clamping force of 50 tones. The screw barrel was regulated from 100 to 130°C. Injection pressure was 1500 bar. Holding pressure and time were 1000 bar and 23s, respectively. Injected parts were moisture equilibrated during several weeks at 50% RH, prior testing. The results indicated a significant improvement in stiffness, obtained by blending the PWS with cellulose fibers. Young's modulus (E modulus) increased by a 5 to 9 order of magnitude. Thermo-mechanical behavior has been studied by DMTA. The evolution of the thermo-mechanical properties for different materials is presented in Figure 8.18 and mechanical properties in Table 8.5.

8.12.2 Biodegradable Packing Materials based on Waste Collagen Hydrolysate Cured with Dialdehyde Starch

Longmaier, Maladek, Mokers and Kolomaznic [42] have studied the biodegradable packing materials based on waste collagen hydrolysate cured with

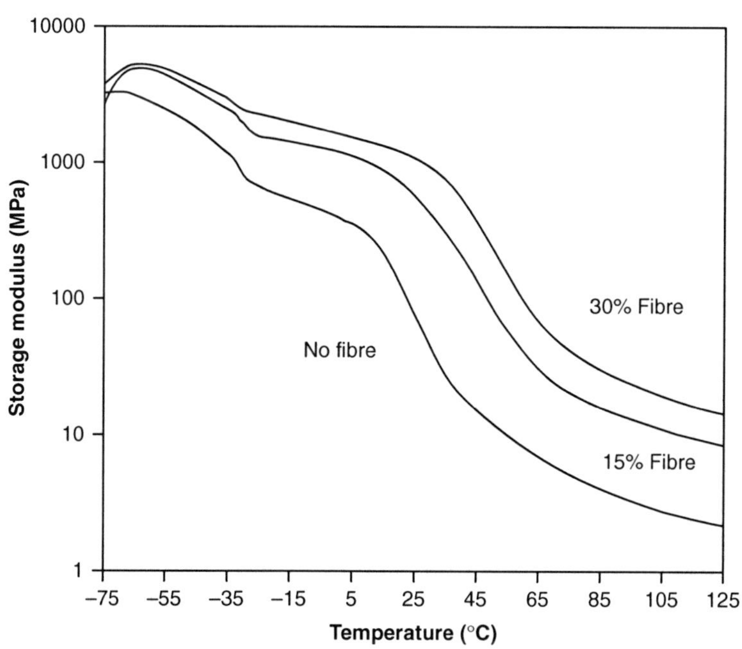

Figure 8.18 Thermo-mechanical behavior of thermoplastic starch-based materials reinforced with cellulose fibers.
Source: Ref. [41].

Table 8.5 Mechanical properties of PWS-cellulose fiber composites: influence of relative humidity during storage.

Material	% RH	Modulus (MPa)	Impact Strength (kJ/m²)	Strain at Break (%)	Max. Strength (MPa)
PWS 1 (reference)	50	1110	2	3	25
PWS 1 (reference)	75	100	>100	60	3
PWS 2 + 20% fibers	50	1150	3	8	26
PWS 2 + 20% fibers	75	n.d.*	n.d.*	n.d.*	n.d.*
PWS 3 + 30% fibers	50	1280	17	6	24
PWS 3 + 30% fibers	75	200	22	15	7

n.d.* = not determined. (*Source*: Ref. [41]).

dialdehyde starch. The collagen waste has so far finished in landfills of industrial waste. Environmental aspects of suitably utilized collagen waste have lately been accentuated by the ever growing production of packaging materials based on synthetic polymers. Biodegradability of packaging materials has become a very closely watched field, promoting application of collagen hydrolysates in industrial practice. Hydrolysate of chrome-tanned leather waste does not meet the requirements for packages of pharmaceutical and food products. The biodegradable character of such packages does not resolve the issue of used packages disposal. When suitably processed, such materials may provide time-controlled release of active substances. Hydrolysate of abattoir collagen waste, or of waste from manufacture of edible meat product casings, meets the criteria of edible (and biodegradable) packages for pharmaceutical and food products, and possibilities of its industrial application are wider.

Processing collagen hydrolysates into edible packages has to take into account their lower molecular mass. That usually attains values around 15–30 kDa which, when compared with gelatin currently used for edible packages, is a level of 5–10 times lower. Results of preliminary studies lead to concluding the reaction of collagen hydrolysates with dialdehyde starch allows to prepare thermo-reversible gels unless hydrolysate concentration in reaction mixture exceeds 30% and dialdehyde starch concentration attains at most 15% based on hydrolysate.

8.12.3 Novel Starch Thermoplastic/Bioglass® Composite

Leonor *et al.* [43] have studied the mechanical properties, degradation behavior and in-vitro bioactivity of starch thermoplastic/Bioglass® composites and have evaluated the possibility of creating new degradable, stiff and highly bioactive composites. Combination of a biodegradable thermoplastic starch-based polymeric blend and Bioglass® filler should allow for the development of bioactive and degradable composites with a great potential for a range of temporary applications. In their work, a blend of starch with ethylene-vinyl alcohol copolymer (SEVA-C) is reinforced with a 45S5 Bioglass® powder presenting a granulometric distribution between 38 and 53 μm. Composites with 10 and 40 wt% of 45S5 Bioglass® are compounded by twin-screw extrusion (TSE) and subsequently injection molded under optimized conditions. Its mechanical properties are evaluated by tensile testing, and their bioactivity assessed by immersion in a simulated body fluid (SBF) for different periods of time. The stiffness and strength of SEVA-C/ Bioglass® composites are 3.8 GPa and 38.6 MPa respectively.

8.12.4 Bio-Based Polymer Composites Using Poly-Lactic Acid

Maurizio Avella *et al.* [44] have studied the eco-challenges of bio-based polymer composites using poly-lactic acid. Poly-lactic acid (PLA) is a class of crystalline biodegradable thermoplastic polymer with relatively high melting point and excellent mechanical properties. Source of PLA is renewable resources such as corn and sugar beets. Under specific environmental conditions, pure PLA can degrade to carbon dioxide, water and methane over a period of several months to two years, a distinct advantage compared to other petroleum plastics that need much longer periods. Natural fiber reinforcements could also considerably lower the price of bio-based composites. The final properties of PLA strictly depend on its molecular weight and crystallinity. Recently, PLA has been used as a polymer matrix in composites. The tensile and flexural modulus results could be improved by increasing the content of cellulose or cellulose based reinforcements in PLA based composites, whereas tensile and flexural strength remain practically unchanged. With regards to impact properties, the toughness results are impaired for PLA composites reinforced with cellulose, whereas small improvements are recorded with the addition of cotton or kenaf fibers. Different natural fibers have been employed to modify the properties of PLA. Generally in the most studied, natural fiber reinforcements for PLA are kenaf, flax, hemp, bamboo, jute and wood fibers. Besides conventional natural fibers, recently reed fibers have been tested in appropriate PLA composites in order to improve the tensile modulus and strength. Generally, the mechanical properties of natural fiber reinforced composites were improved by using surface modified fibers.

8.12.5 Protein-Starch Based Plastic Produced by Extrusion and Injection Molding

Huang, Chang and Jane[45] have investigated the mechanical and physical properties of protein-starch based plastic produced by extrusion and injection molding They have used soy protein. Their properties and high-amylose corn starches are chemically modified to improve the viscosity, texture, and stability of pastes for coating, binding, and sizing applications. A twin-screw extruder and an injection molding machine have been used. The processing conditions, ingredient ratio, and moisture content are optimized for making soy protein/starch based plastics. Ingredient mixing is done by using a metering pump and a high speed mixer. The optimal processing temperature for injection molding is 130°C (at this temperature tensile strength of soy protein/starch based plastics is 3.9 MPa), and the moisture content of extruded pellets is 10–14%. Processing effects are investigated by measuring the tensile properties and water resistance of specimens. Reduction of water and glycerol in mixtures increased the barrel pressure of the extruder. Mold release is improved by incorporating 0.25 parts tallow per 100 parts of solid material (soy protein and starch). The water absorption of the specimens is reduced by adding acids to adjust the pH to the isoelectric point of soy proteins (pH 4.5). The processibility of pellets is stable after a 4-week-storage period, despite some moisture loss. Injection-molded specimens, after being stored for up to 6 months at dry conditions [50 and 11% relative humidity (RH)] at room temperature and for 4 week in a 50°C oven, show no surface crack. However, humid (93% RH) storage at room temperature promoted fungal growth after storage for 3 months, indicating that preservatives such as potassium sorbate and propionic acid are needed.

8.12.6 Mechanical Properties of Starch Modified by Ophiostoma SPP for Food Packaging Industry

Chen Bei Huang *et al.* [46] have studied the production, characterization, and mechanical properties of starch modified by ophiostoma spp. for food packaging industry. Since the starch is a biodegradable polysaccharide, produced in abundance at low cost, and exhibits thermoplastic behavior, therefore, it is used as an alternative material to replace the traditional plastics in certain segments such as the food packaging industry. Microbial modification of starch with *Ophiostoma* spp. is investigated with the purpose of developing a novel packaging material for the food or pharmaceutical industries. The yield is about 85%. The biopolymer derived from modification of different starch sources (corn, potato, tapioca and rice) by *O.* spp., and ultimately, develop a commercially viable process for large scale production of a biopolymer that can be used as packaging material for food or even medical applications, are characterized by gel-filtration chromatography (GFC) and Fourier Transform Infrared (FT-IR) to determine the changes in the biopolymer following fermentation. Films are also cast to test the mechanical properties of the polymer.

Fungus could be used to modify starch and improve its attributes as a bioplastic. *Ophiostoma* spp. (*O.* spp.) is able to successfully modify starch into a biopolymer with improved mechanical properties in about 72 hours. Results from GFC demonstrated a substantial increase in molecular weight during the modification process. The increasing molecular weight also contributed to the improved mechanical properties of the starch films. The pyranose ring is maintained after the modification, but the hydrogen bonds between molecules intensified. Peak shifts and ratio changes suggested the fixation of new chemical functional groups or new linkages between starch molecules. One pathway involves the fungus producing a polymer that can bond starch molecules together and form new crosslinked structures.

The second possible pathway involves the fungus attaching one or more functional groups that help to strengthen the starch polymer.

8.12.7 Functional Properties of Extruded Starch Acetate Blends

Guan, Fang and Hanna studied the functional properties of extruded starch acetate blends [47]. They stated that since the Starch is difficult to process into functional end products because of its hydrophilic characteristic of native starch, therefore it is necessary to modify starch to improve the hydrophobic properties of the end products. The hydrophobicities of the extruded foams are significantly higher than that of native starch-based extruded foams, whereas other functional properties are same or even better. Modified starches via chemical, physical, or enzymatic treatments have improved functional properties as compared to native starches. One common hydrophobic (modified) starch is starch acetate. Starch acetate, with degree of substitution of 2, is blended with 0, 7.5 and 15% polylactic acid (PLA), Eastar Bio Copolyester 14766 (EBC) or Mater-Bi ZF03U (MBI) and 10%, 13%, or 16% (d.b.) ethanol and twin-screw extruded at 160°C barrel temperature. Physical characteristics of the extrudates such as radial expansion ratio, unit and bulk densities, and of the mechanical properties, including unit spring index and bulk spring index, are measured. Too much polymer hindered the radial expansion ratio. Type of polymer, polymer content, and ethanol content significantly affected the physical characteristics and mechanical properties. The sample extruded with 7.5% PLA and 13% ethanol has the highest expansion ratio and bulk spring index. The sample with 15% MBI and 16% ethanol has the lowest unit density, while the sample with 7.5% PLA and 16% ethanol had lowest bulk density. The highest unit spring index is expressed in the sample containing 7.5% PLA and 10% ethanol. Unit spring index and bulk spring index increased at lower ethanol.

8.12.8 Thermoplastic Starch and Bacterial Cellulose Based Biocomposite

Ivo M.G. Martins *et al* have studied the thermoplastic starch and bacterial cellulose based biocomposite [48]. Starch, a promising raw material for the development of novel materials, can be converted into a thermoplastic material, known as Thermoplastic Starch (TPS), through the disruption of the molecular chain interactions under specific conditions, in the presence of plasticizer. In addition to the enhanced mechanical properties of these reinforced TPS materials, a significant improvement in water and thermal resistance is also obtained by adding cellulose crystallites or microfibrillated cellulose. Owing to unique properties such as high mechanical strength, high crystallinity and a high pure nano-fibrillar network structure, it is becoming a promising material for several applications.

The Young modulus and the tensile strength, determined from the typical stress-strain curve, shows at a fiber content of 5%, the Young modulus was 30 times higher than that of the unfilled TPS, for bacterial cellulose composites [48]. Accordingly, the presence of cellulose fibers caused a considerable decrease in the elongation of cellulose fiber-based composite was also higher than that of the unfilled TPS matrix. The introduction of BC fibers into the TPS matrix resulted in a shift in the main relaxation to higher temperatures, which further confirmed the strong fiber-matrix interfacial interactions. The only drawback is the strong sensitivity to higher relative humidity. Still, due to other improved mechanical properties, these materials are promising candidate in applications like food packaging and biodegradable artifacts.

8.12.9 Starch/Rubber Composites

The coagulum obtained by coagulating starch paste and rubber latex by aqueous solution, is dried in an oven at 80°C for 18h to a moisture content of about 10%. Then starch/rubber blend is prepared [49]. The above preparation technique employs compounding rubber latex with natural starch paste and directly coagulating the mixture using an electrolyte without cross-linking starch. This strategy originated from the fact that water is a very good chemical medium to dissociate the hydrogen bond in starch, and most rubbers have a latex form. In an aqueous system, the rubber latex particles, which are generally smaller than 100 nm, can be mixed with starch paste uniformly, giving starch/rubber composites with a fine dispersion.

The starch/rubber composites [49] prepared by co-coagulation exhibit higher hardness, stress at 100%, tensile strengths and tear strengths. This difference is assumed to result from the fine dispersion of the starch in these rubber/starch composites. The crystalline nature of starch was reduced and sometimes even disappeared in starch/rubber compounds produced by co-coagulation, and the size of the starch particles decreased dramatically; more specifically, to smaller than 1μm. this caused a large improvement in the mechanical properties of the starch/rubber composites. This technique also has an advantage of being suitable for rubber with a latex form.

8.12.10 Fiber-Reinforced PLA Composites

Interest in biodegradable polymers and natural fiber-reinforced polymers has recently grown because of increasing environmental concerns. Natural fiber reinforcements can lower the price of bio-based composites considerably. It has been reported that [50] tensile and flexural modulus results could be improved by increasing the content of cellulose or cellulose based reinforcements in PLA based composites, whereas tensile and flexural strength remain practically unchanged or even worsened. With regards to impact properties, the toughness results are impaired for PLA composites reinforced with cellulose, whereas small improvements are recorded with the addition of cotton or kenaf fibers.

Different natural fibers have been employed in order to modify the properties of PLA. The most studied natural fiber reinforcements for PLA are kenaf, flax, hemp, bamboo, jute and wood fibers. Besides conventional natural fibers, recently reed fibers have been tested in appropriate PLA composites in order to improve the tensile modulus and strength. Generally the mechanical properties of natural fiber

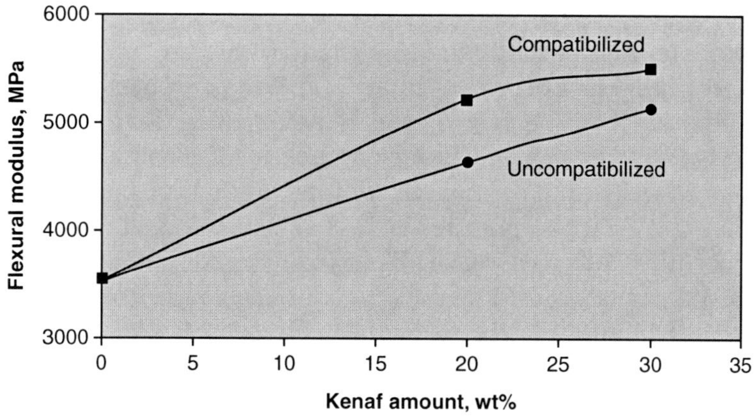

Figure 8.19 Flexural modulus of PLA based composites as a function of the kenaf content and the compatibilization.
Source: Ref. [50].

reinforced composites were improved by using surface modified fibers. Flexural modulus of PLA based composites as a function of the kenaf content and the compatibilization is shown in Figure 8.19.

8.12.11 Biodegradation of Starch and Polulactic Acid-Based Materials

A biopolymer-based material may be expected to exhibit a partial intrinsic biodegradability because its constituents are not completely chemically modified throughout the manufacturing process. Biodegradability of an insoluble polymer is based on the percentage of mineralization of the materials carbon content. Richard Gattin *et al.* [51] have presented the comparative study of biodegradability of starch and polylactic acid-based material. Experiment was conducted on about 2cm × 2cm pieces of film, each weighing between 0.4 and 0.5 gm.

In this material, the starch is an easily biodegradable material while the polylactic acid-based material is not easily biodegradable. Experiments conducted according to ASTM and ISO/CEN standards showed that the final mineralization percentage was always greater than 60%, which is the minimum assigned value for biodegradable material. Moreover, the percentage of biodegradation including the microbial bio-assimilation of the material carbon was between 82% and 90%.

8.12.12 Bacterial Cellulose Fiber-Reinforced Starch Biocomposites

Cellulose, the most abundant natural homopolymer is considered to be one of the most promising renewable resources and an environmentally friendly alternative to products derived from the petrochemical industry. Besides cellulose from plants, cellulose is also secreted extracellularly as synthesized cellulose fibers by some bacterial species, which is called bacterial cellulose (BC). Compared with cellulose from plants, BC has high mechanical properties like tensile strength and modulus, possesses higher water holding capacity, higher crystallinity and a finer web-like network. BC has been widely used in foods, in acoustic diaphragm for audio speakers, and has medical applications as wound dressing and artificial skins, artificial blood vessels, and tissue engineering scaffolds.

The tensile strength of the BC/Starch composites [52] is 2.03, 2.18 and 2.37 times of the pure starch when the fiber loading is 7.8, 15.1 and 22.0 wt % respectively. The tensile modulus increases by 111.7%, 116.7% and 132.4% respectively, at 7.8, 15.1 and 22.0 wt% fiber loadings. However, a decrease in elongation at break is observed for all BC/Starch composites. In the present BC-Starch system [38] both starch and BC are hydrophilic which may render the composites high moisture absorption. It is accepted that moisture absorption degrades polymer composites. The kinetics of water absorption of BC/Starch biocomposites follows Fick's Law of Diffusion. The M_{∞}, K, and D values of the BC/Starch biocomposites decrease with BC fiber loading and a lower than those of the starch. The tensile strength deteriorates after water absorption.

Soil burial experiments shows the average degradation rate of about 1%/day and 0.9%/day, respectively, for the starch and the 15.1 wt% BC fibers. The tensile strength of the BC/Starch biocomposite containing 15.1 wt% BC nanofibers and the starch decreases drastically upon exposure to micro organism attacks. The BC/Starch biocomposite shows higher tensile strength retention when compared to the starch owing to the higher resistance to microorganism attacks of the BC fibers in comparison to starch.

8.12.13 Starch-based Completely Biodegradable Polymer Materials

Synthetic polymers and natural polymers that contain hydrolytically or enzymatically liable bonds or groups are degradable. The advantages of synthetic polymers are obvious, including predictable properties, batch to batch uniformity, and can be tailored easily, but they are quite expensive. Thus we focus on natural polymers which are inherently biodegradable and can be promising candidates to melt different requirements.

Among natural polymers, starch is of interest. It is completely biodegradable, has low cost and is renewable. Starch is mainly composed of two homopolymers of D-glucose [53]: amylase, a mostly linear α-D (1, 4')-glucon and branched amylopectin, having the same backbone structure as amylose but with many α-1,6'-linked branch points. There are a lot of hydroxyl groups on starch chains, two secondary hydroxyl groups at C-2 and C-3 of each glucose residue, as well as one primary hydroxyl group at C-6 when it is not linked.

Starch based biodegradable polymers find many applications. In food industry, food packaging and edible films are two major applications. They find three major applications in agriculture: the covering of greenhouse, mulch film and fertilizer controlled release materials. Due to their good biocompatibility, biodegradability, proper mechanical properties and degradation products being non-toxic, find major applications in medical field.

8.12.14 Maleated-Polycaprolactone/Starch Composite

Among the commercially available biodegradable plastics, polycaprolactane (PCL) has received much attention due to its high flexibility and biodegradability and because of its hydrophobic nature. PCL is, however, too expensive to be used widely. Thus many attempts have been focused on blending plastic materials with cheap and biodegradable natural polymers, such as starch. But physical properties of PCL become significantly worst when blended with starch due to poor compatibility between the two phases. Such an operation therefore requires a compatibilizer or a toughness to enhance the compatibility between the two immiscible phases and to improve the mechanical properties.

Thus, use of PCL-g-MAH (maleic anhydride-grafted-PCL) in place of pure PCL has been investigated in [54]. Compatibility and mechanical properties of a PCL/Starch composite was improved by using PCL-g-MAH. The blending of PCL-g-MAH with starch leads to the formation of an ester carboxyl group not present in PCL/Starch. This group is responsible for changed mechanical properties. PCL-g-MAH/Starch is more easily processed due to its lower melt temperature and torque requirement. The fusion heat (ΔH_f) of PCL/Starch decreased with increasing starch content while that of PCL-g-MAH /Starch increased. The improved compatibility between the constituents of PCL-g-MAH/Starch is illustrated by the noticeable reduction in the starch phase size. Tensile strength of PCL-g-MAH/Starch was enhanced, and elongation at break improved, though these properties worsened in both composites as starch content improved. Although water resistance of PCL-g-MAH/Starch was higher than that of PCL/Starch, in a soil environment the compatibilized blend showed only a slightly lower biodegradation rate than the uncompatibilized one.

8.13 Starch Based Nanocomposites for Packaging Applications

The development of newer kinds of nano biocomposites for packaging applications demand for improved properties and enhanced strength in them. These can be accomplished by various means such as coating of starch sheet by various means such as coating of starch sheet by silicone and wax emulsion, mixing of clay to improve barrier properties, by mixing oil to improve resistance to water and chemical attack etc. The details of such methodologies are given in subsequent section.

8.13.1 Biodegradable Starch-based Nano-clay Composites

The biodegradable starch-based nano-clay composites have been developed by School of Engineering and Design, Bernel University, London [55]. In it, the thermoplastic starch (TPS) material is based on wheat starch along with nano-clay composite. The starch sheet has been coated with silicone and wax emulsion. It results in enhanced moisture resistance and reduction in water vapor permeability. The rate of biodegradation is high i.e. its compostability in home environment is superior as compared to bio-polymers and other paper based materials. The TPS sheet is produced by direct extrusion process. Whereas the rigid sheet is thermoformed into packaging lids and trays, the flexible and transport films are used as window materials for food packaging.

8.13.2 MMT-Filled Potato Starch Based Nanocomposites

Robert D. Maksimov *et al.* [56] have investigated the mechanical properties and water vapor permeability of starch/montmorillonite nanocomposites. Glycerin, which is usually used in combination with water, exerts a plasticizing effect on starch. The basic drawback of materials based on starch and a softener is the low mechanical properties. A promising direction in mechanical property of plasticize starch (PS) is the introduction of anisomeric (fibrous or lamellar) particles of a filler. In this case, the effect of reinforcement (strengthening) can be reached not only because of the considerably higher values of strength and rigidity of the filler but also because the particle geometry (the characteristics ratio of their sizes, known as the aspect ratio). Good results are obtained on using cellulose microfibers as the filler.

They have shown that the glycerin, which is usually used in combination with water, exerts a plasticizing effect on starch. The polymer composites containing plane nanoparticles of layered silicates as the filler increases rapidly. The clay mineral montmorillonite (MMT) is the silicate used most widely in polymer nanocomposites. The introduction of a relatively small amount of MMT makes it possible to improve the mechanical properties of different synthetic polymers (polyamides, polypropylene, polycaprolactam, polyimides, epoxy resins, elastomers, etc.). From the ecological viewpoint, the drawback of these materials is that, after the useful life, practically they do not degrade under the influence of natural environment. Along with mechanical properties, other operational characteristics of nanocomposites are also of importance. The limits of their application depend much on their barrier properties, i.e. the permeability to gases, vapors, and liquids. Results of the investigations show that the barrier properties of many polymers can be improved considerably by doping them with a rather small amount of MMT. The decrease in permeability is achieved mainly because of the increased path of diffusing sorbate molecules caused by shielding effect of the plate like filler particles. In this case, the degree of exfoliation of the layered particles and their aspect ratio and orientation in the material are of great importance.

In this study, the filler of nanocomposites was purified, unmodified clay whose basic rock-forming mineral was MMT, potato starch of density $1.5\,g/cm^3$ and moisture content of 9.7%, were used. Glycerol with a molar mass of 92.09 g/mol and density of $1.26\,g/cm^3$ was taken as a plasticizer.

8.13.3 Sweet Potato Starch/OMMT Nanocomposite for Packaging Application

A biodegradable nanocomposite has been prepared from carbamide/ethanolamine plasticized thermoplastic starch and dodecyl benzyl-di-methyl ammonium bromide (12-OREC)- activated montmorillonite (MMT) [57]. The starch used is sweet potato (11.6%). The thermoplastic starch (TPS) and nanocomposite (TPS/OMMT) was prepared with 15% carbamide, 15% ethanolamine and different contents of organic activated montmorillonite (OMMT) by twin-screw extruder with the 130° C barrel temperature. Mechanical properties investigation indicated that the tensile strength and modulus of TPS/OMMT nanocomposites were better than those of TPS, while elongation at break was descended with the increasing of OMMT contents. When the content of OMMT was 4%, the tensile strength and modulus of TPS was improved from 4.2 and 42 MPa to 6.0 and 76 MPa, respectively. Improved mechanical properties are due to introduction of carbamide and ethanolamine which forms hydrogen bonds with starch under the high temperature and high shear stress effect produced by twin- screw extruder. Thus the crystallization structure of starch is destroyed. As a result, the starch mechanical and processing properties were improved.

The hydrogen bonding formed in TPS becomes stronger with the increase of OMMT content, leading to the increase of Young's modulus and stress. However, the impermeability of OMMT would decrease the flexibility of starch molecules and the strain of materials.

8.13.4 Biocomposites from Wheat Straw Nanofibers

In a starch-based thermoplastic polymer (STP) the reinforcing potential of cellulose nanofibers is agro-residues. Cellulose-based biofibers including cotton, flax, hemp, jute and sisal, and wood fibers are used to reinforce plastics due to their relative high-strength, high stiffness and low density. They are biodegradable and offer potential advantages over synthetic plastics in disposable applications but it has poor mechanical properties and brittle in nature, therefore, it is a poor alternative for any synthetic thermoplastic. The thermal and mechanical properties of starch are improved by blending them with other polymers such as polylactic acid, polyethylene and polyvinyl alcohol and the addition of natural fillers. Recent advances in producing bio-fibers, micro-fibrillated or nano-size fibers with high-strength and surface area, offer manufacturing of high-performance composites from these bio-fibers. Alemdar and Mohini Sain [58] have studied the morphology, thermal and mechanical properties of biocomposites from wheat straw nanofibers.

A chemi-mechanical technique is used for isolation of cellulose nanofibers from wheat straw and determined to have diameters in the range of 10–80 nm and lengths of several thousand nanometers. Thermo-gravimetric analysis is used to find thermal stability of the fibers. The solution casting method is used to prepare the nanocomposites from the wheat straw nanofibers and the thermoplastic starch. Thermo-gravimetric analysis (TGA) and dynamic mechanical analysis (DMA), and tensile testing are used to find their thermal and mechanical performance, and it's evaluated and compared with the pure thermoplastic starch. A uniform dispersion of the nanofibers in the polymer matrix of the nanocomposites is shown by the scanning electron microscopy (SEM) images. The tensile strength and modulus of the nanocomposites films revealed significantly enhanced properties compared to the pure thermoplastic starch and its glass transition is shifted to higher temperatures with respect to the pure thermoplastic starch.

8.13.5 Cellulose Nanocomposites with Starch Matrix

Nanocomposites, i.e. composites containing fillers that have at least one nano-sized dimension, represent a new class of materials that have interesting penetrant diffusion properties. Because of their small size, the surface to volume ratio of the nanoparticles is significantly greater than that of the corresponding micro sized particles at the same mass content. Cellulose nanocomposites with starch matrix [59] have interesting mechanical properties and mimics biological plant structures in several ways. Processing is possible at room temperature in a water medium.

The moisture uptake decreases with increasing content of cellulose. Cellulose nanofibers are less hygroscopic than starch due to the higher degree of molecular order. In the cellulose, the disordered regions are likely to be preferred sorphin sites. A well dispersed nano-fiber network (70 wt% cellulose nanofibers) reduced the moisture uptake of the composite to half the value of the pure plasticized starch film. The moisture sorption kinetics in the biomimetic nanocomposites (cellulose nanofiber) must be described by a moisture concentration –dependent, rather than a constant diffusivity in several cases. The observed reduction in moisture diffusivity could be due to cellulose characteristics and geometrical impedance, swelling constraints due to a high-modulus/hydrogen bonded fiber network, and strong molecular interactions between cellulose nanofibers& with the amylopectin matrix.

8.14 Starch Foam, Film, and Coated Composites for Packaging Applications

8.14.1 Blended Composite Film of Chitosan and Starch

Chitosan is well known for its industrial use to make blend films with polymers such as polyethylene, polyvinyl alcohol, cellulose etc. Starch is another abundantly available low cost natural polymer, which is used to produce biodegradable plastics. A blended film of chitosan and starch has been prepared by Du Yumin *et al* [60] in which the starch forms the film and chitosan acts as supporting (matrix) material.

The materials used were chitin, twice by 47% NaOH under 100% for about 2h; and soluble starch which was chemically pure, dissolvable in 70–80°C hot water with 1% concentration. The blended films were produced in weight ratio of chitosan: starch as 10:0, 9:1, 8:2, 7:3, and 6:4 respectively. The mixtures were kept in water at 95°C for about 30 min and stirred. 18% glycerol was added as plasticizer.

Tensile strength of blended films is obtained as a function of starch content, Figure 8.20. Tensile strength is variable for different percentage of starch. It is maximum 781 kg/cm^2 (766.16 kPa) at 30% starch content. The blended films may be used in food packaging, biodegradable plastics, medical uses etc.

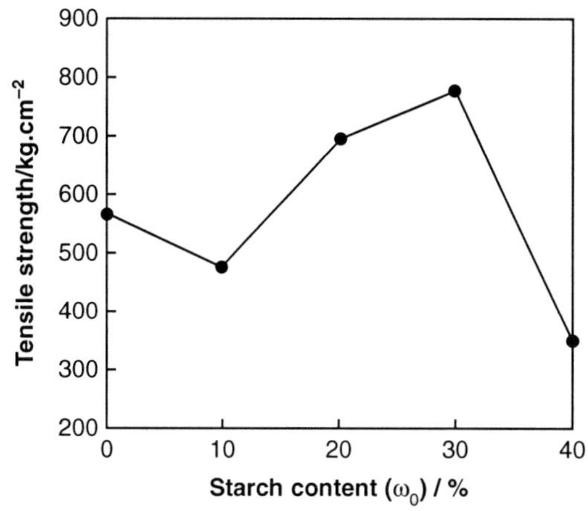

Figure 8.20 Effect of starch content on tensile strength of the films.
Source: Ref. [60] with permission from Du Yumin *et al.*, *J. of Natural Sciences*, 2, 2, 220–224, 1997.

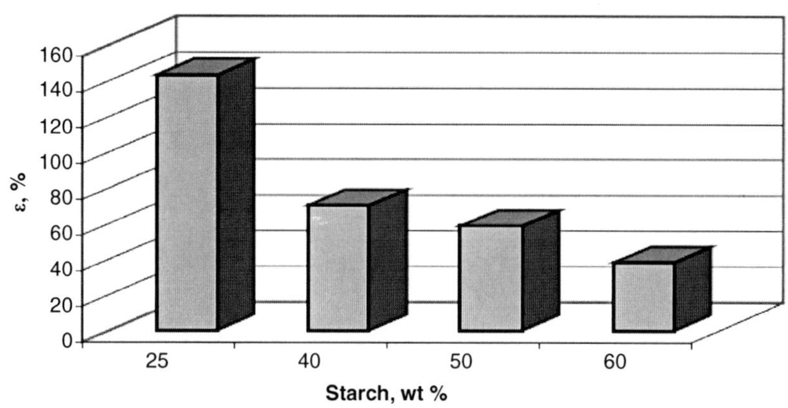

Figure 8.21 Elongation at break (ε) of plasticized PHB–starch films.
Source: Ref. [21] with permission.

8.14.2 PHB Matrix with Potato Starch and Thermo-cell Filled Biocomposites for Films and Coatings

A heterogeneous multi-components biocomposites films and coatings have been formed using modified matrices of a bio-polymer "polyhyroxybutyrate (PHB)" incorporating the additives of destructive chemical nature (disperse renewable fillers, plasticizers, copatibilizers) [21]. Three grades of microbiologically synthesized biodegradable polymers (PHB) are used as matrix polymers for Biocomposites formation. PHB is recovered from the biomass of Azotobacter chrococcum, Bisoflex DOP (B P chemicals) is used as plasticizer, and potato starch and thermocell are used as fillers for PHB composites.

Solutions of plasticizers PHB as a matrix is used to develop potato starch + thermo-cell filled biocomposites. The effect of starch content 25 to 60% by wt. on mechanical characteristics has given the results as shown in Figure 8.21. It shows that the effect of increasing starch content is to decrease the tensile strength σ_T and elongation at break ε. The values of σ_T is in the range of 4–7 MPa and $\varepsilon = 42–120\%$.

8.14.3 Jute and Flax-Reinforced Starch Based Composite Foams

N. Soykeabkaekaew *et al.* [61] have studied the preparation and characterization of jute- and flax-reinforced starch based composite foams. Starch-based composite foams (SCFs) are successfully prepared by baking starch-based batters. In this process, the jute or flax fibers are mixed inside a hot mold. The flexural strength and the flexural modulus of elasticity increase with increasing aspect ratio of the fibers. The improvement in the mechanical properties of SCFs was attributable to the strong interaction between fibers and the starch matrix. As compared to flax-fibers, jute fibers have a greater reinforcing effect. Starch can be used as an alternate material for making foams. Foams made from pure starch are brittle and sensitive to moisture and water. Therefore, they are further treated to attain reasonable strength, good flexibility and water resistance. For this, mineral filler and wood fibers are added to improve strength and whereas coating the surface with wax improve water resistance of starch from products. However, the major drawbacks for these composite foams are lower tensile strength and lower elongation.

8.14.4 Egg Albumen-Cassava Starch Composite Films Containing Sunflower-Oil Droplets

Blending of biopolymers is one of the most effective methods to develop biopolymeric materials with new and improved properties. Nowadays different types of starch-protein composites have been developed to produce effective and biocompatible micro- and nano-capsules for delivery of flavors, enzymes and antimicrobials. Wongasulak *et al.* [62] have studied the thermo- mechanical properties of egg albumen-cassava starch composite films containing sunflower-oil droplets. Dynamic mechanical analysis (DMA), differential scanning calorimetry (DSC) and scanning electron microscopy (SEM) are used for analysis of the effect of moisture content on the thermo-mechanical and structural properties of egg albumen–cassava starch composite films containing sunflower oil droplets [49]. Composite films are prepared by cold gelation, dried in a moisture controlled incubator (83.5%RH) at 25°C for 8 days and aged at different relative humidity at room temperature (21 ± 1 °C) for 7 days to obtain composite films with moisture contents of 4%, 7%, 11%, 17% and 46% (dry weight basis).

The effect of moisture content and heating temperature on the microstructure of the composite matrix is analyzed by SEM images. In binary mixtures, the compatibility between two biopolymers is of key importance. Drying a mixed biopolymer system may lead to phase separation and the formation of new microstructures. The release of functional components entrapped in a composite biopolymer matrix depends on the properties of the matrix, the characteristics of the release agent, solution composition, and environmental conditions. Matrix properties can be changed by varying the concentration and type of protein-based and starch used the moisture content, the pH and the ionic strength. The rate of release of sunflower oil from a composite matrix of egg albumen and cassava starch depends on solvent type and the relative humidity and temperature under which the film is dried.

8.14.5 Starch Based Loose-Fill Packaging Foams

H.A. Pushpadass *et al.* [63] have studied the effects of temperature, moisture and talc on physical properties of starch based loose-fill packaging foams. A single-screw laboratory-scale extruder is used to make starch-based loose-fill packaging foams. Corn starch is blended with polystyrene in the ratio of 70:30 and extruded into foams using talc and polycarbonate as additives. Extrusions are done at moisture contents of 16, 18 and 20% (dry basis), and at barrel temperatures of 140 and 160°C. The effects of moisture and talc contents on the radial expansion of foams are found to be critical, while the role of temperature is close to significant. The expansion ratio increased when the moisture content is to be increased from 16 to 18%, and then decreased when moisture content is increased to 20%. The expansion ratios of foams are generally higher at 160°C as compared to 140°C. Although polycarbonate mixed well with the starch-polystyrene melt, it is not effective as a structural and anti-shrinking agent, and it does not contribute to the radial expansion. The bulk densities and unit densities of the starch foams generally decreases as the moisture content and extrusion temperature increases. Scanning electron microscope (SEM) images show that the addition of talc yielded foams with smaller-sized cells, with less expansion of the foam melt, and thus a higher density. The crystallinity of starch foams increases post-extrusion (shows by X-ray diffractograms), and there is adequate dispersion of the starch and polystyrene polymers to make the foam water-resistant.

8.14.6 Chemically Modified Starch (RS4)/PVA Blend Films

Synthetic or biosynthetic polymers as well as plant-based polymers such as starch and pectin which are biodegradable and recyclable and, therefore, may help satisfy the increasing consumer and regulatory demands for materials with these properties. Starch, renewable and biodegradable polysaccharides, generally it contains about 30% amylose, 70% amylopectin and less than 1% lipids and proteins from plant. Biodegradable starch-based plastics such as starch/cellulose, starch/PVA (polyvinyl alcohol), etc., have great potential marketability in agricultural foils, garbage or composting bags, food packaging, fast food industry as well as biomedical fields. Starch based films have been synthesized from corn starch, wheat starch, rice starch, potato starch, and cassava starch, and the investigations of their mechanical properties revealed that amylose content of starch affected the properties of films.

Starch-based films have an effect on physical properties because of linear structure of amylose and branch structure of amylopectin of starch. Especially, these films are soluble in water because of the amylopectin of branch structure. The physical properties of films are synthesized by using native corn starch (NS) and chemically modified starch (RS4) [64]. The mixing process and the casting method are used to synthesize the NS or RS4/PVA blend films. The additives are Glycerol (GL), sorbitol (SO), and citric acid (CA). The chemically modified starch (RS4) is synthesized by using sodium trimetaphosphate (STMP) and sodium tripolyphosphate

(STPP) as a cross-linker. Then, the RS4 thus synthesized was confirmed by using the pancreatin-gravimetric method, swelling power and an X-ray diffractometer (XRD).

8.14.7 Starch/Polycaprolactone Films

The increase in environmental awareness and strict waste management policies created a need for natural-based biodegradable materials instead of nonbiodegradable petrochemical-based plastics. Because of the solid waste problem, the use of biodegradable plastics is highly encouraged in environmental standards especially in the packaging sector. Biodegradable plastics are suitable for composting. Therefore, their production and consumption rates are expected to increase in the near future. Starch, the natural biodegradable polymers, has been considered as one of the most promising candidates primarily because of its attractive combination of availability and price for the manufacturing of plastic-like materials. But, starch-based plastics have poor long-term stability caused by water absorption, poor mechanical properties, and processibility. To overcome these disadvantages, starch is usually mixed with a biodegradable synthetic polymer such as polycaprolactone (PCL). PCL or starch granules have been modified with various chemical agents to facilitate mixing of hydrophilic starch granules with hydrophobic PCL by using cross-linking [65].

Cross-linking reinforces the hydrogen bonds in the granule with chemical bonds that act as bridges between the starch molecules. In this paper, starch granules are modified with trisodium trimetaphosphate (TSTP) and characterized by P^{31}-NMR, FTIR and DSC. By using solvent casting technique 70μ films are prepared from modified starch and polycaprolactone blends. Three different types of films—PCL (100% polycaprolactone), MOD-ST/PCL (50% modified starch and 50% polycaprolactone blend) and NONMOD-ST/PCL (50% nonmodified starch and 50% polycaprolactone blends)—are prepared, and their thermal, mechanical, and morphologic properties are show the increased performance of PCL with the addition of starch and also the effect of modification. The Young's modulus of polycaprolactone increases and became less ductile, whereas tensile strength and elongation at break values decreases with the addition of starch. In a compost environment, degradation is faster, and all polymer films are broken into pieces within first 7 days of degradation and no film remained after 15 days. Surface modification of starch granules with TSTP increased the interconnection between two phases and eliminated the need of a compatibilizer and Modification of starch granules reduced its degradation rate.

8.15 Effects of Various Parameters on Behavior of Packaging Purpose Biocomposites

8.15.1 Influence of Fibers on Mechanical Properties of Cassava Starch Foam

Influence of fibers on mechanical properties of cassava starch foam has been studied by Laura G. Carr *et al.* [66]. The starch may be formed into foam by the processes of swelling, gelatinization and thermo pressing. The foam generally has a dense outer skin and a less dense interior with large, mostly open cells molded into many shapes and used for several applications, such as food packaging until automotive industry. Starch alone is brittle and sensitive to water; therefore further treatments are necessary to obtain the strength, flexibility and water resistance for commercial applications of foams. Water resistance can be improved by coating with polyester, while the mechanical properties of starch foams can be improved by reinforcing with natural fibers. The mechanical properties of a fiber reinforced polymer composite depend on many factors, like fiber-matrix adhesion, volume fraction of fiber, fiber aspect ratio (l/d) and fiber orientation.

Matsui *et al.* [67] added long fibers in cardboard made with cassava bagasse and improved the mechanical properties and water resistance of material. Curvelo *et al.* [68] used cellulosic fiber from eucalyptus urograndis pulp as reinforcement for thermoplastic starch. This composite shows an increase of 100% in tensile strength and more than 50% in modulus with respect to non-reinforced thermoplastic starch. Lawton *et al.* [69] made starch foam with 2.5–45% fiber content and showed that the strength of the foam increased with increase in the fiber content of the foam, until fiber content reached about 15%. Laura G. Carr *et al.* have studied the influence of fiber on mechanical properties of cassava starch foam whose results are depicted in Figure 8.22 a–c and Figure 8.23 a,b.

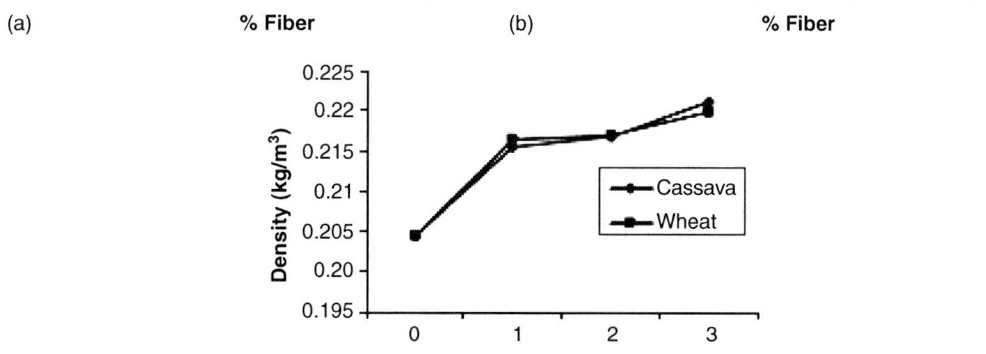

Figure 8.22 a) Strength of foam with no fiber comparing with the foam added with 1, 2 and 3% of cassava and wheat fiber. b) Flexibility foam with no fiber compared with foam added with 1, 2 and 3% of cassava and wheat fiber. c) Density of foam with no fiber comparing with foam added with 1, 2 and 3% of cassava and wheat fiber.
Source: a) Ref. [66]. b) Ref. [66]. c) Ref. [66].

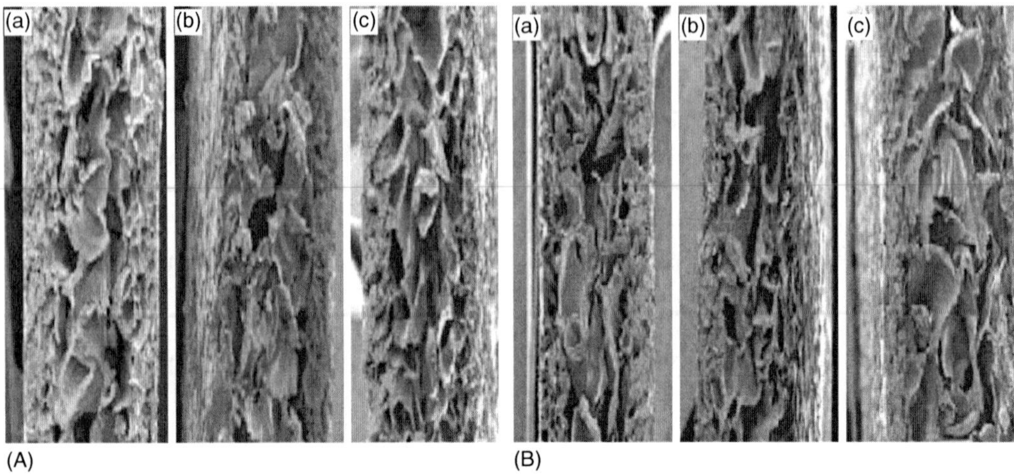

Figure 8.23 a) Starch foam with different content of wheat fiber a = 1%, b = 2% and c = 3%. b) Starch foam with different content of cassava fiber a = 1%, b = 2%, and c = 3%.
Source: a) Ref. [66]. b) Ref. [66].

8.15.2 Water Absorption Behavior of Oil Palm Fiber-Low Density Polyethylene Packaging Purpose Composites

Shinoj *et al.* [70] have studied the water absorption pattern and dimensional stability of oil palm fiber-linear low density polyethylene composites. Alkali treatment of fibers is done to reduce the hydrophilic nature of the composites. It is to be found that the water absorption in most of the combinations followed typical Fickian behavior. The rate of water absorption and swelling increased with fiber loading. Alkali treatment of the fibers resulted in a reduction of water absorption at higher fiber loadings only, and composites with higher fiber sizes exhibited higher water absorption. The thickness swelling is also increased with fiber; however, a constant trend is not obtained for the 75–175 μm fiber size. Linear expansion in composites also happens during water absorption, but less than thickness swelling. In addition to thickness swelling, composites also expanded linearly during water absorption; however, linear expansion was considerably less than thickness swelling. Linear expansion is more during higher fiber loading and alkali treatment. Maximum solid loss on water immersion occurred with small-sized and also alkali-treated fiber composites. An increase in thickness and a decrease in linear dimension are observed after one sorption-desorption cycle. This irreversible change is to be proportional to fiber loading and alkali treatment.

8.15.3 Hygroscopic Effect on PHB Matrix with Potato Starch Biocomposites for Food Packaging

In the hygroscopic properties of filled PHB biocomposites (Figure 8.24), starch content of 25–40 by wt % assured 9–11% water vapor absorption in comparison with 5–6% for the same content of thermo-cell. A significant advantage of PHB-based

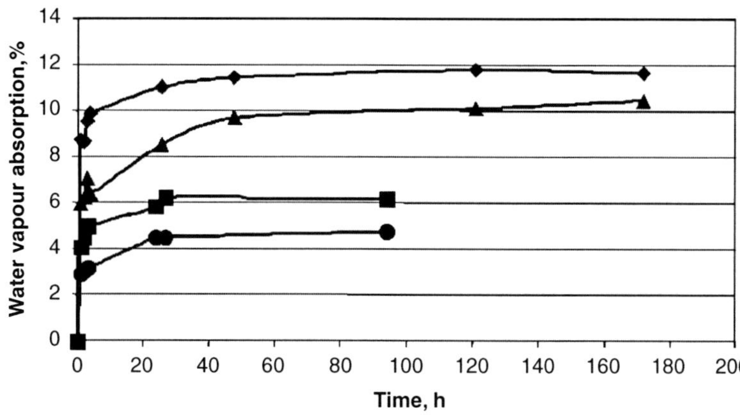

Figure 8.24 Kinetics of water vapor absorption of PHB based composites: ◆ – starch 40 wt %, ▲ – starch 25 wt %, ■ – thermo-cell 40 wt %, ● – thermo-cell 25 wt %.
Source: Ref. [21] with permission from Dace Erkske *et al.*, *Proc. Estonian Acad. Sci. Chem.*, Vol. 55, No.2, p. 70–77, 2006.

Table 8.6 Water vapor permeability of PHB-based films and coatings.

Content (wt. %)	**Thickness (mm)**	**Water Vapor Permeability (g.m/Pa.s.m²)10⁻¹¹**
Paper (grammage 45 g/m²)	0.051	4.713
PHB film	0.061	0.245
PHB+23% Bisoflex (film)	0.071	0.307
Paper-PHB	0.063	3.674
Paper-v PHB	0.066	1.437
Paper-Br. PHB	0.073	0.737
Paper-v PHB + 23% Bisoflex	0.068	3.120
Paper-Br. PHB + 23% Bisoflex	0.065	2.914
Paper-v PHB + 23% Bisoflex +25% starch	0.083	5.668

Source: Ref. [21] with permission from Dace Erkske *et al.*, *Proc. Estonian Acad. Sci. Chem.*, Vol. 55, No.2, p. 70–77, 2006.

biocomposites is for short life packaging applications [21]. The results testified that a weight loss of 50–60% of the initial mass (which is considered as the criterion of the biodegradation process) was observed during two to three weeks for all systems tested. It is ascertained that the plasticized PHB-based films can be used for food biopackaging in terms of some barrier properties and possibility of using PHB-based biocomposites as paper coatings. Experimental value of results of water vapor permeability of the obtained laminated systems for paper-biocomposite systems are summarized in Table 8.6.

The water vapor permeability of plane PHB films was found to be about 20 times lower than the corresponding value of untreated paper. Use of higher molecular weight PHB resulted in increased values of water vapor permeability. A pronounced increase of water vapor permeability was observed for starch containing coating due to hydrophilic character of starch and the formation of a heterogeneous structure.

Plasticized PHB employed as continuous polymer matrices for the elaboration of PHB-starch-based films and paper coatings. It is demonstrated that the use of high molecular weight PHB- and PHB-starch-based coatings for paper significantly increases the water vapor permeability of the material. The results testify the that combination of inexpensive water sensitive starch with hydrophobic PHB offers a potential for creating ecologically sound biocomposites for special kinds of application.

8.15.4 Effect of Degradation and Mineralization of Starch in Different Media

The waste disposal is a major environmental concern, and programs to recycle, incinerate or convert plastic wastes have been developed because synthetic plastic materials resist environmental degradation. Cause of the unsuitability of collecting and recycling some plastic wastes (food packaging, disposable diapers, and hospital wastes) have stimulated the development of new biodegradable plastics. Natural polymer-based (biodegradable) materials such as starch, have consequently received much research attention. Richard Gattin *et al.* [71] have made comparison of mineralization of starch in liquid, inert solid and compost media according to ASTM and CET norms for the composting of packaging materials. Biodegradation is defined in these norms as that percentage of the carbon of the polymer converted in CO_2 during aerobic degradation.

8.15.5 Effect of Blending of Chitosan and Starch

Chitosan has received worldwide interest for its industrial use. Yumin, Z uyong, and Rong [60] have studied the effect of blending of chitosan/starch. Chitosan-starch blend films are prepared, and their structure and properties are studied by FT-IR, X-ray diffraction, SEM and measurement of tensile strength. Starch is mixed with synthetic polymers to develop biodegradable plastics. When the starch is less than 30% by weight, the two polysaccharides are compatible. This result obtained from IR spectra and SEM analysis. But the hydrophilic starch is incompatible with hydrophobic synthetic polymers, and simple mixing blends tend to phase separation. The blends are compatible and films of suitable mechanical properties are obtained. However, the blend films are water soluble, thus their utilization is limited. Crystallization of starch was inhibited, and recrystallization of chitosan is also affected by starch. These are observed by X-ray diffraction (XRD) patterns of blend films. Crystal form I, one of the main two crystal forms of chitosan, drastically increased in 30% starch content films. These results indicated that the interactions between chitosan and starch molecules exist in the blend films. The tensile strength of the film are improved when chitosan and starch are blended by weight ratios of 8:2 and 7:3, in which the highest tensile strength (781 kg/cm^2) is achieved.

8.15.6 Effect of Starch Composition on Structure of Foams

The use and disposal of oil-based products causes emissions of CO_2 to the atmosphere, while biopolymer based products emit the same amount of CO_2 as is used during the plants growing period. Starch, the natural biodegradable polymers, has been considered as one of the most promising candidates primarily because of its attractive combination of availability and price. Sjoqvist and Gatenholm [72] have studied the effect of starch composition on structure of foams prepared by microwave treatment. Blends of polyethylene and starch, using starch as filler in a synthetic polymer matrix, and esterified starch has been prepared. Foams can be prepared by extrusion, baking in a hot mold, compression molding, and freeze drying. In this paper microwave technique is used to prepare foams from different potato starches in granular form, with varying amounts of amylose content, and water. In addition to native potato starch (PN), high amylose potato starch (HAP) and potato amylopectin (PAP) are used, as well as mixtures thereof. The native crystallinity of starch granules is lost upon microwave treatment and an amorphous material is created.

An increased concentration of starch in the initial water dispersion resulted in a less dense foam structure. The potato amylopectin formed open cell foams, whereas increased amylose content, as in native potato starch, yielded a more compact structure with irregular pore shapes. The high amylose potato starch yielded a structure with hardly any porosity. Water evaporates more rapidly from a high amylose starch solution than native potato starch and potato amylopectin solutions. The amylose solution had much lower viscosity than starch and amylopectin. This confirms that polymer–water interaction, such as in amylopectin solution, favors stabilization of bubbles formed upon boiling and evaporation of water, which yields high porosity materials. A better water holding capacity is amylopectin, based on its ability to swell in water. Swelling is the key to bubble formation since it gives the granule an elastic behavior, which allows it to hold steam bubbles and create pores during the process. Gelatinization should take place before expansion gelatinization creates the gel which should trap the steam bubbles and create the foam. High amylose starch is not completely gelatinized, when bubbles are introduced in the gel, HAP does not have sufficient strength to trap the bubbles and create pores.

8.16 Characterization of Biocomposites

The biocomposites are characterized by different means to determine their characteristics and properties. For it, mainly the following means are adopted.

1. Fourier Transforms Infrared (FT-IR) Spectroscopy
2. X-ray differactometry
3. Scanning electron microscope (SEM)
4. Mechanical testing
5. Thermal analysis

8.16.1 Characterization of Starch/OMMT Nanocomposites for Packaging Applications

In this regard the characterization of starch/OMMT nanocomposites for packaging applications [57] is illustrated below. The thermoplastic starch (TPS) and nanocomposite (TPS/OMMT) was prepared with 15% carbamide, 15% ethanolamine and different contents of organic activated montmorillonite (OMMT). Fourier transforms infrared spectroscopy and wide angle X-ray diffraction have shown that the alkylamine in dodecyl benzyl dimethyl ammonium bromide could react with MMT via cation exchange reaction. After treated, the *d*(001)space distance of MMT increased from 1.5 to 1.7 nm. Scanning electron microscope revealed that the lower contents of OMMT could disperse well in the matrixes of TPS. The carbamide, ethanolamine and the OMMT could destroy the crystallization behavior of starch, but only the OMMT restrained this behavior for long-term storing. Mechanical properties investigation indicated that the tensile strength and modulus of TPS/OMMT nanocomposites were better than those of TPS, while the elongation at break was descended with the increasing of OMMT contents. When the content of OMMT was 4%, the tensile strength and modulus of TPS was improved from 4.2 and 42 MPa to 6.0 and 76 MPa, respectively.

FT-IR absorption spectra. The FT-IR absorption spectra of MMT and OMMT are shown in Figure 8.25. A comparison of spectra displays that the absorption peaks of bands at 2927 and 2835 cm^1 appeared when MMT was activated by dodecyl benzyl dimethyl ammonium bromide(12-OREC). The band at $^{-1}$ is related to –CH3 and –CH2– 2927 and 2835 cm stretching vibration, respectively. It indicates that alkylamine can react with MMT via cation exchange reaction and produced OMMT.

XRD pattern. The X-ray diffraction pattern of pure sweet starch (ST) and thermoplastics starch (plasticized by 15 wt% carbamide and 15 wt% ethanolamine) (TPS) were shown in Figure 8.26. From the pattern, it can be found that there are three crystallization peaks in the ST sample. The peaks at 15.2 and 18.2° were attributed to A type crystallization. The peak at 23.1° was the V_H type crystallization. These crystallization structures made the starch brittle and process difficult. Compare with the ST, the height of crystallization peaks of TPS descended obviously. It indicated that the introduced carbamide and ethanolamine, as plasticizers, could form hydrogen bonds with starch under the high temperature and high shear stress effect produced by twin screw extruder. So the crystallization structure of starch was destroyed. As a result, the starch mechanical and processing properties were improved, Figures 8.25 and 8.26.

SEM photographs. The scanning electron microscope (SEM) photographs of starch and TPS/OMMT nanocomposites are shown in Figure 8.28. A comparison of photographs demonstrated that the crystallization behavior of starch was restrained for the existing of OMMT. When the content of OMMT was little the crystallization behavior of starch can hardly be observed. Though the crystallization behavior of starch can dimension of global crystal in starch descend obviously. Its reason may be attributed to the small contents of OMMT can dispersed well in the matrixes of TPS while the high contents of OMMT tend to agglomerated in starch, Figure 8.27.

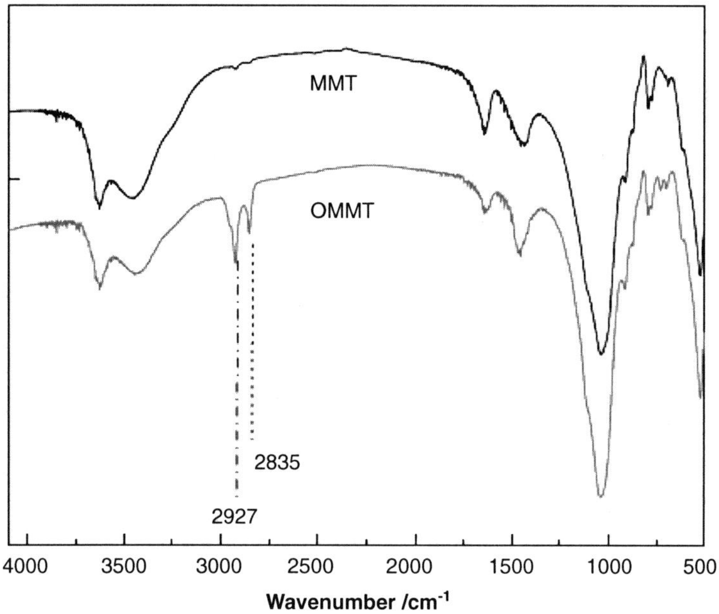

Figure 8.25 FT-IR spectra of MMT and OMMT.
Source: Ref.[57] with permission from Penggang Ren *et al.*, *J Polym Environ*, 17, 203–207, 2009.

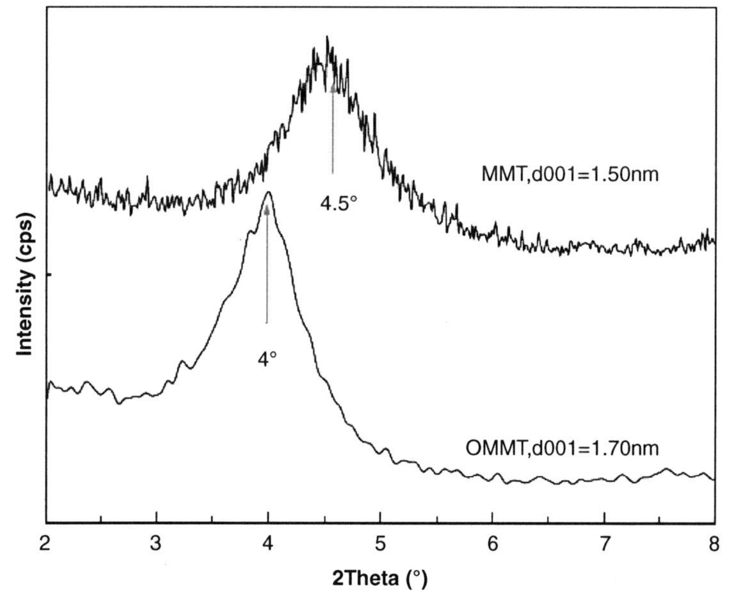

Figure 8.26 WXRD patterns of ST and TPS.
Source: Ref.[57] with permission from Penggang Ren *et al.*, *J Polym Environ*, 17, 203–207, 2009.

Mechanical Properties. The stress–strain curves of TPS and TPS/OMMT are shown in Figure 8.28. It was found that the tensile strength and modulus of TPS/OMMT nanocomposites increased with the content of OMMT increasing.

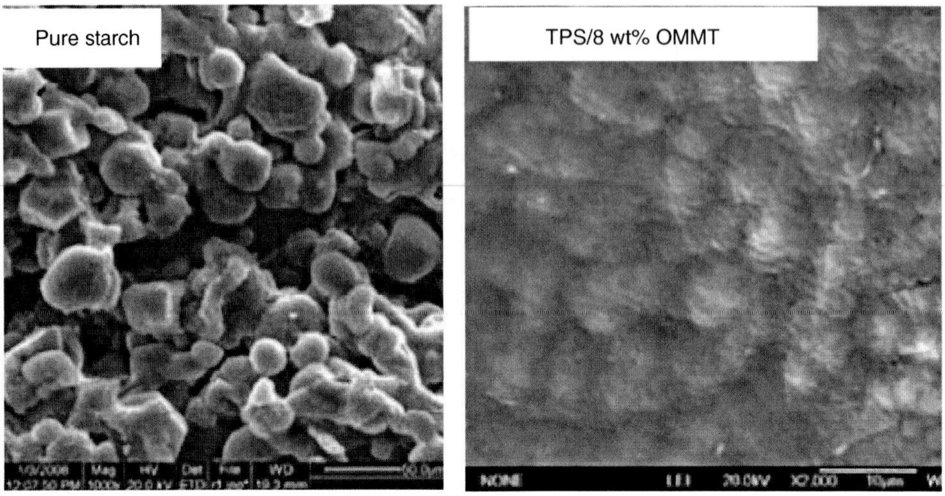

Figure 8.27 SEM photographs of starch and TPS/OMMT nano-composites.
Source: Ref. [57] with permission from Penggang Ren *et al.*, *J Polym Environ*, 17, 203–207, 2009.

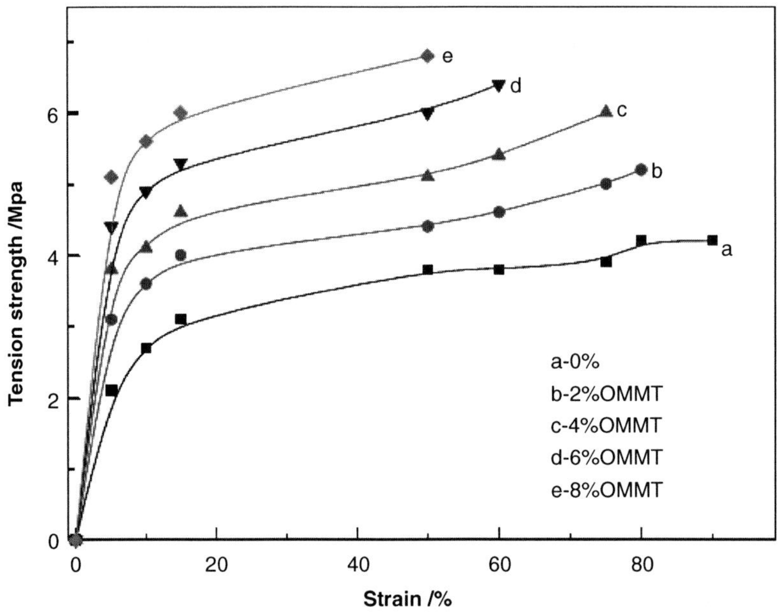

Figure 8.28 Stress–strain curves of TPS/OMMT nano-composites with different mass contents.
Source: Ref. [57] with permission from Penggang Ren *et al.*, *J Polym Environ*, 17, 203–207, 2009.

The tensile strength increased from 4.2 MPa for TPS, to 5.2 MPa for 2% OMMT, 6 MPa for 4% OMMT, 6.4 MPa for 6% OMMT and 6.8 MPa for 8% OMMT introduced, respectively. The tensile modulus of TPS was increased from 42 to 62°MPa for 2% OMMT, 76°MPa for 4% OMMT, 88°MPa for 6% OMMT and 102°MPa for 8% OMMT introduced, respectively. When OMMT content is in the range of 2–4%, the mechanical properties of TPS/OMMT nanocomposites are obviously ameliorated.

While the elongation at break descended with the increasing of OMMT contents. It descended from 90% for TPS to 50% for the TPS nanocomposites with 8% OMMT introduced. It indicated that the mechanical properties of nanocomposites were better than those of TPS except the elongation at break, Figure 8.28.

8.16.2 Characterization of Blend Film of Chitosan Starch

Chitosan-starch blend films were prepared, and their structure and properties were studied by FT-IR, X-ray diffraction, SEM and measurement of tensile strength [60]. IR spectra and SEM analysis showed that the two polysaccharides were compatible, when the starch was less than 30% by weight. From X-ray diffraction patterns of blend fIlms,it was observed that crystallization of starch was inhibited, and recrystallization of chitosan was also affected by starch. Crystal form I , one of the main two crystal forms of chitosan,drastically increased in 30% starch content films. These results indicated that the interactions between chitosan and starch molecules exist in the blend films. The tensile strength of the film were improved when chitosan and starch were blended by weight ratios of 8 ' 2 and 7 : 3, in which the highest tensile strength (781 kg/cm') was achieved.

FT-IR analysis of the films. Figure 8.29 shows IR spectra of the films. In this blend film, it is shown that with starch content increasing, absorbance of chitosan molecules about at 898 cm^{-1} characteristic absorbance of starch molecules appeared enhanced.

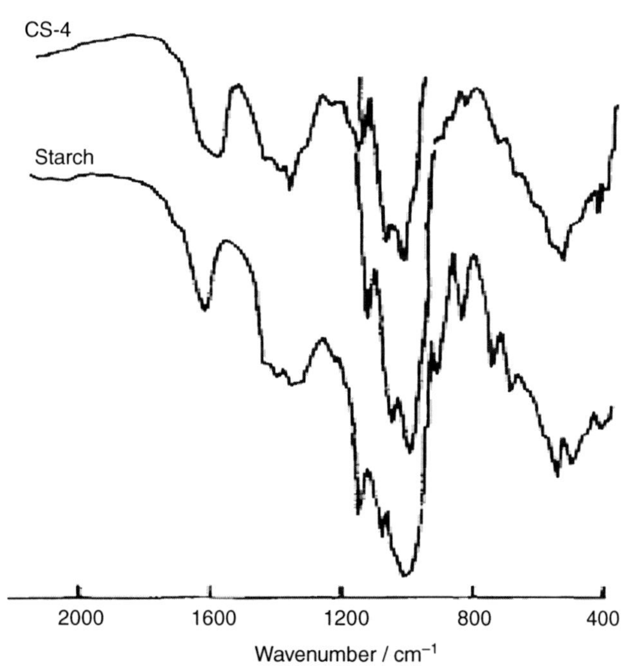

Figure 8.29 IR spectra of the films.
Source: Ref. [60] with permission from Du Yumin *et al.*, *J. of Natural Sciences*, 2, 2, 220–224, 1997.

X-ray diffraction analysis. X-ray diffraction patterns of the blend films are shown in Figure 8.30. From the pattern of chitosan films can observe a mixture of crystal form I and I [13] in the crystal structure of crystalline chitosan. It is obvious that patterns of the blends is quite different from that of either chitosan or starch. Crystal peaks attributed to starch can't be found in patterns of the blend films.

Scanning electron microscopy analysis. Figures 8.31a,b show SEM of the films. In pure chitosan films (Figure 8.31a) chitosan molecules or molecule groups gathered

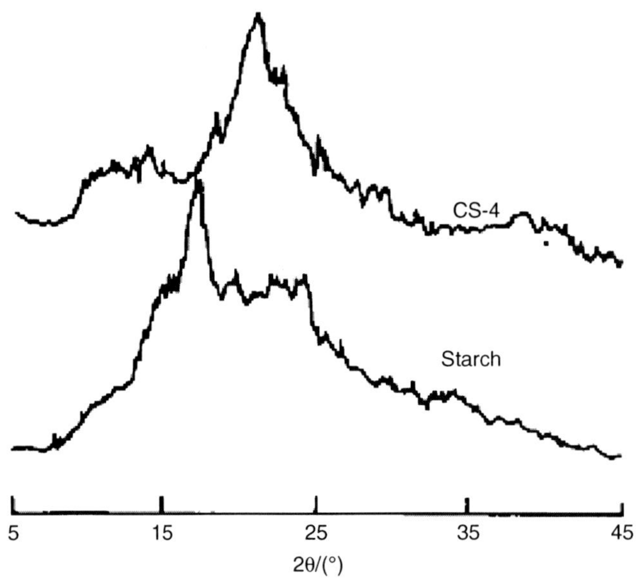

Figure 8.30 X-ray diffractograghs of the films.
Source: Ref. [60] with permission from Du Yumin *et al.*, *J. of Natural Sciences*, 2, 2, 220–224, 1997.

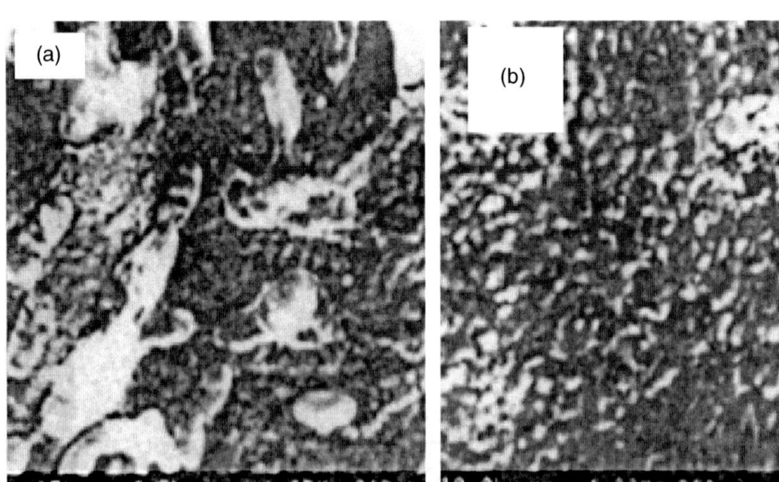

Figure 8.31 Scanning electron microgragh s of (a) CS-O and (b) CS-2.
Source: Ref. [60] with permission from Du Yumin *et al.*, *J. of Natural Sciences*, 2, 2, 220–224, 1997.

together and distributed follow random directions. There were no clear linkages among the particles, and media buffer spaces existed in the microstructure. But when starch molecules were incorporated into the films (Figure 8.31b) the structure changed and big chitosan particles were not find. However, much more little size particles arranged in the films.

Tensile strength of blend films. The effect of starch content on the mechanical properties of the blend films are already described in Figure 8.20 earlier. 20% and 30% starch had higher tensile strength than others, and the maximum value of 781 kg/cm^2 was observed at 30% starch content. The improved tensile strength suggest further the occurrence of interactions between chitosan and starch molecules in the blend films, Figure 8.20

8.16.3 Morphological and Thermomechanical Characterization of Thermoplastic Starch/Monomorillonate Nanocomposites

Starch is a semi-crystalline polymer composed of two polymers, amylase and amylopectin, a linear and a highly branched polysaccharide, respectively with repeating -D-glucopyranosyl units. Various plasticizers have been used with starch to convert it into thermoplastic starch (TPS), mainly water and glycerol. Recently, the vegetable oils have been used as additives for polymeric materials. The presence of oil/fatty acid chain in the polymer structure improves some physical properties of polymer in terms of adhesion and resistance to water and chemical attack. Oil, mixed to petrochemical polymers, provides flexible and photoluminescent materials, with excellent thermal stability. A combination of starch and clay for the preparation of nanocomposite materials is proposed. A significant advance has occurred with the preparation of nanocomposites, where the structural order within the material can be controlled in nanoscale. The reinforcement with filler is particularly important for polymers from renewable resources, since most of them have the disadvantages of lower softening temperatures and lower modulus. Additionally, the hydrophilic behavior of most natural polymers offers a significant advantage, since it provides a compatible interface with the nanoclay.

Deniela Schlemmer *et al.* [73] have studied the morphological and thermomechanical characterization of thermoplastic starch/monomorillonate nanocomposites. In this work starch is plasticized by pequi (caryocar brasiliense) oil, and, thermoplastic starch (TPS)/montmorillonite (MMT) nanocomposites are analyzed by X-ray diffraction (XRD), thermogravimetry (TG), and thermomechanical analyses (TMA) and scanning electron microscopy (SEM). Exfoliated and intercalated nanocomposites are found to be dependent on MMT content. Exfoliation is the predominant mechanism of clay dispersion for low filler loading. Increase of the clay loading (>5 wt. %) causes intercalation. The introduction of low content (<5 wt. %) of MMT improves the thermal stability and the stiffness of the materials. There is a limit content of clay that can be added to improve the thermal and thermomechanical properties of the composites. Beyond that value the composite presents properties below the original polymer.

The films obtained from the cassava starch and vegetable oils, are homogeneous and with good properties. At low filler, exfoliation is the predominant mechanism of clay dispersion. Increase of the clay loading impedes complete exfoliation due to the restricted area remaining available in the polymer matrix and enhances the degree of intercalation. The introduction of low content (65 wt. %) of MMT improves the thermal stability and the stiffness of the materials. Almost all materials showed a higher temperature of decomposition in the second stage of degradation (referring to starch), except the 90:10 TPS/MMT material.

8.17 Composite Manufacturing Methods

Polymeric composites can be manufactured by various methods. These methods are different for thermoset and thermoplast composites, and are listed below.

Manufacturing processes for thermoset composites:

1. Prepreg lay-up processes or autoclave processing or vacuum bagging process
2. Hand lay-up (or wet lay-up) process
3. Spray-up process
4. Filament winding process
5. Pultrusion process
6. Resin transfer molding process
7. Structural reaction injection molding (SRIM) process
8. Compression molding process
9. Roll wrapping process
10. Injection
11. Molding process

Manufacturing processes for thermoplastic composites:

1. Thermoplastic tape winding process
2. Thermoplastic pultrusion process
3. Compression molding of glass mat thermoplastic (GMT)
4. Hot press technique
5. Autoclave processing
6. Diaphragm forming process
7. Injection molding process

Other methods of preparation of biocomposites:

1. Conventional solution casting technique to prepare films
2. Wet process of preparing the film forming suspension with the fillers (e.g. starch)
3. Manual molding

A brief description of some of the above processes is given in subsequent sections. Other processes can be referred to in any related text/reference [74].

8.17.1 Prepreg Lay-up Process

This is a very common process in aerospace industry. Complicated shapes having very high volume fractions V_f, can be produced by this process. This is an open mold process with low volume capability. During manufacturing the prepreg are cut, laid down in desired fiber orientation and then vacuum bagged. Thereafter, the composite is kept inside an autoclave or oven along with the mold. After that, the pressure and heat are applied for curing and consolidation of the part to be manufactured.

It an intensive laborious process in which the labor cost is about 50 to 100 times greater than the pultrusion, filament winding and other high volume processes. However, it is advantageous for small quantity manufacturing of huge parts. Its major applications are in the manufacturing of following items.

- Radomes
- Wing structures
- Yacht parts
- Sports goods
- Landing gear door

8.17.2 Wet Lay-up (or Hand Lay-up) Process

This process is employed in fabricating the components of marine industry and for other prototype parts. Similar to prepreg lay-up process, this is also an open mold type process. In its working, the resin in liquid form is applied on to the mold and the reinforcement is placed over it. A roller is pressed and rolled over the resin reinforcement mix to impregnate the fibers with resin. Another layer of resin and reinforcement is applied on the first layer. In this way, the stacking of layers is continued until the laminate of suitable thickness is built-up. Different types of fibers, mats and fabric reinforcement can be employed in this process. This is an easier and less expensive, but laborious process. Since the entire manufacturing process is performed manually, therefore this is also called "hand lay-up process." Major use of this process may be found in fabrication of following items.

- Windmill blades
- Storage tanks
- Swimming pools
- Sports boat

8.17.3 Thermoplastic Pultrusion Process

This process is similar to the pultrusion process used for thermoset composites. In it, the thermoplastic prepregs or the fibers are pulled through a die to get the final product. Since the thermoplastic resins are of high viscosity, the processing demands for a greater pulling force. Thermoplastic pultrusion suffers from a drawback that it provides an inferior surface quality than the thermoset pultrusion. Major uses of this process are in the fabrication of following items.

- Angles and channels section
- Circular tubes
- Rods and squares
- Rectangular bars
- Strips
- Beams
- Flooring
- Handrails
- Ladders, etc.
- Light poles
- Electrical enclosures
- Walkways

8.17.4 Starch Wet Milling Process

In this case, the starch granules are wrapped in protein matrix. The protein matrix has to be opened to free the starch granules from protein matrix. Typical procedures for starch extraction include germ removal, fiber removal, protein matrix removal, and final starch purification (Figure 8.32). Starch extraction requires large amounts of water, that is why the starch extraction is referred to as wet milling. Commercial corn starch milling is an example of this process. The corn based packaging materials [37] and the process to make packaging material from them are schematically given in Figure 8.32.

8.17.5 Comparison of Various Manufacturing Processes

The raw material used, shape and size of the products that can be manufactured; strength, production speed and cost etc. for various processes are compared in Table 8.7 [74].

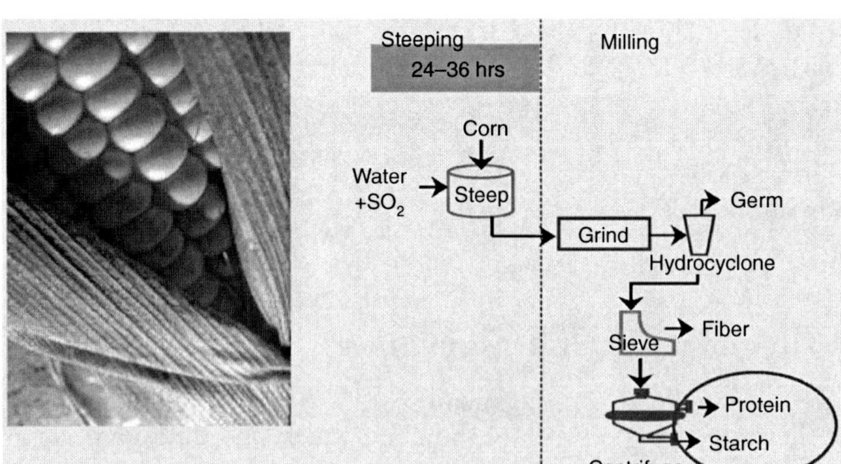

Figure 8.32 Process of wet milling to use corn.
Source: [Ref. 37].

Table 8.7 Comparison of various composite manufacturing processes from different view points.

S.No.	Process	Raw Material	Shape	Size	Strength	Cost	Production Speed
1.	Hand lay up	Prepreg and fabric with epoxy resin	Simple to complex	Small to large	High	High	Slow
2.	Wet lay up	Fabric/mat with polyester and epoxy resins	Simple to complex	Medium to large	Medium to high	Medium	Slow
3.	Spray-up	Short fiber with catalyzed resin	Simple to complex	Small to medium	Low	Low	Medium to fast
4.	Pultrusion	Continuous fibers, usually with polyester and vinyleser resin	Constant Cross-section	No restriction on length, small to medium size cross-section	High (along longitudinal direction)	Low to medium	Fast
5.	Filament winding	Continuous fiber with epoxy and polyester resins	Cylindrical and axis-ymmetric	Small to large	High	Low to high	Slow to fast
6.	Roll wrapping	Prepreg	Tubular	Small to medium	High	Low to medium	Medium to fast
7.	Compression molding	Molded compound e.g. SMC, BMC	Simple to complex	Small to medium	Medium	Low	Fast
8.	Injection molding	Pallets (short fibers with thermoplastic)	Complex	Small	Low to medium	Low	Fast

Table 8.7 (cont.) Comparison of various composite manufacturing processes from different view points.

S.No.	Process	Raw Material	Shape	Size	Strength	Cost	Production Speed
9.	Stamping	Fabric impregnated with thermoplastic tape	Simple to contoured	Medium	Medium	Medium	Fast
10.	SRIM*	Fabric or perform with polyisocyanurate resin	Simple to complex	Small to medium	Medium	Low	Fast
11.	RTM**	Perform and fabric with vinyl ester and epoxy	Simple to complex	Small to medium	Medium	Low to medium	Medium

* SRIM means Structural Reaction Injection Molding.
** RTM means Resin Transfer Molding

8.18 Futuristic Research Outlook

Bio-based polymer composites have been the subject of many scientific and research projects, as well as many commercial programs because of global environmental and social problems. The high rate of depletion of petroleum resources and present environmental regulations has forced the search for biodegradable composites and green materials, compatible with the environment. Industrial progress in packaging technology in future, will depend upon the ability to produce newerbreed of bio-materials. For that, researches have to be made to develop starch based smart materials, functionally graded materials (FMGs), whiskers etc. These will become essential for newer generation active and intelligent packaging. And also for opto–electronic packaging, packaging of MEMs and NEMs devices, ferroelectric parts, optical and magnetic components etc; the starch based photonic materials, photo-refractive materials, piezoelectric and magnetostrictive materials are to be studied and investigated.

8.19 Glossary of Terminology

Accelerator: A chemical additive that hastens cure or chemical reaction.

Additive: An ingredient mixed into resin to improve properties (e.g., plasticizers, initiators, light stabilizers and flame retardants).

Amylose: A component of starch consisting of a chain polymer of linked D-glucopyranosyl structures. Thermoplastic starch polymers consist largely of amylose.

Anaerobic degradation: Degradation in the absence of air (oxygen) as in the case of landfills. Anaerobic degradation is also called biomethanization. Anaerobic degradation of plastics can be determined by measuring the amount of biogas released as described in ASTM 5210-91.

Bioassimilation: Chemical assimilation of a substance into the natural environment.

Biodegradability is defined (as per ASTM and CEN norms) as the percentage of carbon of the polymer counted in CO_2 during aerobic degradation.

Biodegradable: The ASTM defines biodegradable as "capable of undergoing decomposition into carbon dioxide, methane, water, inorganic compounds, or biomass in which the predominant mechanism is the enzymatic action of microorganisms that can be measured by standardized tests, in a specified period of time, reflecting available disposal condition."

Biomass: The weight of all the organisms in a given population, trophic level or region.

Bulk molding compounds (BMCs) are the premixed material of short fibers, preimpregnated with starch and various additives.

Cellulose: The main fibrous material in paper.

Compostable: Compostable is defined as "capable of undergoing biological decomposition in a compost site as part of an available program, such that the plastic is not visually distinguishable andbreaks down to carbon dioxide, water, inorganic compounds, and biomass, at a rate consistent with known compostable materials (e.g. cellulose)."

Composting: Breaking down of plant and animal material using micro-organisms under aerobic conditions.

Crimp: Degree of waviness of a fiber, which determines its capacity to cohere.

Degradability: Ability of materials to break down, by bacterial (biodegradable), thermal (oxidative) or ultraviolet (photodegradable) action. When degradation is caused by biological activity, especially by enzymatic action, it is called biodegradation.

Functional coatings: The lamination of polyethylene and/or plastic or foil films to paper substrates, providing a water or greaseproof barrier. Typically used in high humidity applications.

Gel: To enter an initial jelly-like, semi-solid phase during a resin curing process.

Gel coat: An unreinforced, clear or pigmented coating resin applied to the surface of a mold or part to provide a smooth, more impervious finish on the part exterior.

Gel time: The period of time from initial mixing of liquid reactants in a resin to the point when gelation occurs as defined by a specific test method.

Hand layup: A fabrication method in which reinforcement layers, preimpregnated or coated afterwards, are placed and arranged in a mold manually.

Homogeneity and heterogeneity of composites: When the material properties do not change from point to point in a certain direction, the nature is termed as *homogeneity*. If the properties change from point to point in a certain direction, the nature is called *heterogeneity*.

Lamina: Subunit of a laminate consisting of one or more adjacent plies of the same material with identical orientation. (Plural: laminae.)

Laminate: A laminate is the stack of lamina having different orientations of reinforcing materials in the lamina.

Peel ply: A layer of material that, when applied to a layup surface, can be removed from the cured laminate prior to bonding operations, leaving a clean, resin-rich surface suitable for bonding.

Peel strength: Strength of an adhesive bond between sheet materials; determined by applying parting stress at a right angle (perpendicular) to the plane of the adhesive interface.

Polysaccharide is starch, being produced by green plants as an energy store.

Porosity: The presence of voids open to the surface of a solid material into which air or liquids may pass.

Pot life: Length of time in which a catalyzed thermosetting resin retains sufficiently low viscosity for processing.

Ramping: A programed gradual increase/ decrease in temperature and/or pressure to control cure or cooling of composite parts.

Sheet molding compounds (SMCs) are starch-reinforced-composite laminates, made-up by pressing together many unidirectional (U/D) laminate, one over the other.

Shelf life: Length of time a material can be stored and continue to meet specification requirements, remaining suitable for its intended use. (Also see storage life.)

Sizing: A chemical solution used to coat fiber filaments, facilitating operations such as weaving or braiding. Sizing protects the filament from water absorption and abrasion (to minimize fiber wear) and also can be used to bind together and stiffen warp yarns during weaving. Sizing is usually removed and replaced with finish before matrix application. Also called size.

Storage life: The length of time a material can be stored and retain specific properties.

Substrate: Material that provides the surface on which an adhesive-containing substance is applied for any purpose, such as bonding or coating.

Thermoforming: The process of shaping a plastic sheet of styrene or PVC under heat and pressure.

Volatiles: Materials such as water and alcohol, in a sizing or resin formulation that can be vaporized at ambient or slightly elevated temperatures.

Woven roving: Heavy, coarse fabric produced by weaving continuous roving bundles.

Yarn: A continuous, ordered assembly of essentially parallel, collimated filaments, usually with a twist.

Zero bleed: Laminate fabrication procedure that does not allow loss of resin during cure.

Acknowledgements

The author would like to acknowledge the help of my students from Motilal Nehru National Institute of Technology, Allahabad: Mr. Brahma Yadav (M.Tech. III sem., Materials Science and Engineering), Mr. Mayank Bhayana (B.Tech V sem., Mechanical Engineering), and Mr. Nishu Gupta (M.Tech. III sem; Nanoscience and Technology, Delhi Technological University, Delhi) without whose help, the completion of this text would not have been possible. They provided assistance in collecting the information from Internet and Research Journals, in their proper compilation, computer typing and page setting, editing of jist of many articles/papers, drawing some figures and sending the emails to authors and publishers of various papers.

The author also wishes to offer many thanks to Prof. Velta Tupureina, Editor of eXPRESS Polymer Letters, Prof. Ferdinand Langmaier, Dr. Sc. (Thomas Bata University, Faculty of Technology, Department of Polymeric Engineering, the Czech Republic), Penggang Ren (Institute of Printing and Packaging Engineering, Xi'an University of Technology, People's Republic of China), Ülo Niine (Director, Estonian Academy Publishers, Kohtu 6, 10130 Tallinn, Estonia), Monique Ritchie (Copyright and Digital Resources Officer, Library, Brunel University, U.K), Nikoletta Schalbert (Publisher, Akademiai Kiado, Prielle K.u. 19/d, H-1117 Budapest, Hungary), Charles Onwulata (USDA, ARS, Eastern Research Center, Wyndmoor, PA 19038), Maurizio Avella (ICTP-CNR, Pozzuoli, NA, Italy) for granting the kind permission to use figures, photographs, tables etc. from their works to include in this text.

Special thanks are due to Dr. Srikanth Pilla for inviting and guiding me to write this chapter. Last but not the least, I shall like to thank Prof. Arun B. Samaddar (Director, Motilal Nehru National Institute of Technology, Allahabad, India) for encouraging and allowing me to write this chapter, and to my respected teachers Prof. Ashok K. Govil and Prof. V. K. Bindal (both Visiting Professor, Applied Mechanics Deptt., MNNIT, Allahabad, India) for encouragement and helpful comments, and Mrs. Shefalika Ghosh Samaddar (Assistant Professor in ISEA Project, Computer Science and Engg. Deptt., MNNIT, Allahabad, India) for informing about IPR and Copyright norms.

I shall not forget to remember the blessings of my mother Smt. Bela Devi, brother-in-law Mr. Jawahar Lal, Sister Smt. Savitri Lal; for extracurricular assistance of my friends Mr. Ranjeet Singh Virmani (Asstt. General Manager, PNB, Kolkata) and Mr. K.R.D. Tewari (Construction Consultant, Allahabad); for emotional encouragement of my wife Smt. Rita Rani Gupta (Advocate), my son-in-law Mr. Ritesh Shankar Gupta (Accounts Officer, BHEL, India), daughter Smt. Nidhi Gupta, and grand-son RAM-Akarsh.

References

1. Asaf Kleopas Sugih, *Synthesis and Properties of Starch Based Bio-materials*, University of Groningen, December 2008.
2. Starch-Wikipedia, the free Encyclopedia, http://en.wikipedia.org/wiki/Starch
3. Fungulani, T. and E. Maseko, 2001, "Starch in textile Industry," *International Starch Trading*, Science Park, Aarhus, Denmark (©2004–2009).
4. Richard P. Wool, and Xiuzhi Susan Sun, *Bio-based Polymer and Composites*, Elsevier Academic Press, 2005., UK.
5. M.S. Huda, A.K Mohanty, L.T. Drzal, E. Schut, and M. Misra, "Green Composites from Recycled Cellulose and Poly (Lactic Acid): Physico-mechanical and Morphological Properties Evaluation," *J. Mater. Sci.*, Vol. 40, p. 4221–4229, 2005.
6. Graupner, N. "Application of Lignin as Natural Adhesion Promoter in Cotton Fiber-Reinforced Poly(Lactic Acid) (PLA) Composites," *J. Mater. Sci.*, 2008, Vol. 43, 5222–5229.
7. M.S. Huda, L.T. Drzal, A.K. Mohanty, and M. Misra, "Effect of Fiber Surface-Treatments on the Properties of Laminated Biocomposites from Poly(Lactic Acid) (PLA) and Kenaf Fibers", *Compos. Sci. Technol.*, Vol. 68, p. 424–43 2,2008.
8. M. Avella, G. Bogoeva-Gaceva, A. Buzarovska, M.E. Errico, G. Gentile, and A. Grozdanov, "Poly(Lactic Acid)-Based Biocomposites Reinforced with Kenaf Fibers," *J. Appl. Polym. Sci.*, Vol. 108, p. 3542–3551. 2008.

9. M. García, I. Garmendia, and J. García, "Influence of Natural Fiber Type in Eco-Composites," *J Appl. Polym. Sci.*, Vol. 107, p. 2994–3004, 2008.
10. B. Bax, and J. Mu ssig, "Impact and Tensile Properties of PLA/Cordenka and PLA/Flax Composites," *Compos. Sci. Technol.*, Vol. 68, p. 1601–1607, 2008.
11. E.Bodros, I. Pillin,N. Montrelay, and C. Baley, "Could Biopolymers Reinforced by Randomly Scattered Flax Fibers be Used in Structural Applications," *Compos. Sci. Technol.* Vol. 67, p. 462–470.2007.
12. R.Hu, and J. Lim, "Fabrication and Mechanical Properties of Completely Biodegradable Hemp Reinforced PLA Composites," *J. Compos. Mater.*, Vol. 41, p. 1655–1669, 2007.
13. R.Tokoro, D.M.Vu, K. Okubo, T. Tanaka, T. Fujii, and T. Fujiura, "How to Improve Mechanical Properties of PolyLactic Acid with Bamboo Fibers," *J. Mater. Sci.*, Vol. 43, p. 775–787, 2008.
14. N. Shikamoto, A. Ohtani, Y.W. Leong, and A. Nakai, Fabrication and Mechanical Properties of Jute/PLA Composites. In 22nd Technical Conference of the American Society for Composites 2007, Composites, Enabling a New Era in Civil Aviation, Curran Associates, Inc: Red Hook, NY, USA, p. 151:1–151:10, 2007.
15. M.S. Huda, L.T. Drzal, A.K. Mohanty, and M. Misra, "Wood-Fiber-Reinforced Poly(Lactic Acid) Composites, Evaluation of the Psycomechanical and Morphological Properties," *J. Appl. Polym. Sci.*, Vol. 102, p. 4856–4869, 2006.
16. M.S. Huda, L.T. Drzal, A.K. Mohanty, and M. Misra, "Effect of Chemical Modifications of the Pineapple Leaf Fiber Surfaces on the Interfacial and Mechanical Properties of Laminated Biocomposite," *Compos. Interface*, Vol. 15, p. 169–191, 2008.
17. Y.Q. Zhao, K.T. Lau, T. Liu, S. Cheng, P.M. Lam, and H.L. Li, "Production of a Green Composite, Mixture of Poly(Lactic Acid) and Keratin Fibers from Chicken Feathers," *Adv. Mat. Res.*, Vol. 47–50, p. 1225–1228, 2008.
18. K.H. Wang, T.M. Wu, Y.F. Shih, and C.M. Huang, "Water Bamboo Husk Reinforced Poly (Lactic Acid) Green Composites," *Polym. Eng. Sci.*, Vol. 48, p. 1833–1839, 2008.
19. A. Buzarovska, G. Bogoeva-Gaceva, A. Grozdanov, M. Avella, G. Gentile, and M.E. Errico, "Potential Use of Rice Straw as a Filler in Eco-Composite Materials," *Aust. J. Crop. Sci.*, Vol. 1, p. 37–42, 2008.
20. A. Grozdanov, A. Buzarovska, G. Bogoeva-Gaceva, M. Avella, M.E. Errico, and G. Gentile, "Rice Straw as an Alternative Reinforcement in Polypropylene Composites," *Agron. Sustain. Dev.*, Vol. 26, p. 25 1–255, 2006.
21. Dace Erkske, Ilze Viskere, Anda Dzene, Velta Tupureina, and Ludmila Savenkova, "Bio-Based Polymer Composite for Films and Coatings," *Proc. Estonian Acad. Sci. Chem.*, Vol. 55, 2, p. 70–77, 2006.
22. K. M. Gupta, *Material Science and Engineering*, Umesh Publications, New Delhi, 2008.
23. Starch papers, *International Starch Trading*, Science Park, Aarhus, Denmark, © 2003–2007, www.starch.dk
24. FAO (2007)."FAOSTAT", http://faostat.fao.org/site/526/default.aspx, 2007.
25. http://www.matbase.com/material/fibers/natural/
26. Delphine Rutot, Philippe Degee, Ramani Narayan and Philippe Dubois, "Aliphatic Polyester Grafted Starch Composites by in situ Ring Opening Polymerization," *Composites Interfaces*, Vol. 7, 3, p. 215–225, 2000.
27. http://en.wikipedia.org/wiki/Ramie#Properties
28. Dual Matrix –Single Fiber Hybrid Composite, M. Tech. Thesis, Applied Mechanics Department, Moti Lal Nehru National Institute of Technology, Allahabad, U.P., India, Submitted by Asheesh Kumar under supervision of Dr. K. M. Gupta, 2009.
29. Mechanical Characterization of Jute Epoxy Hybrid Composite, M. Tech. Thesis, Applied Mechanics Department, Moti Lal Nehru National Institute of Technology, Allahabad, U.P., India, Submitted by Anshuman Srivastava under supervision of Dr. K. M. Gupta, 2009.
30. K. M. Gupta, and A. Srivastava, Tensile Characterization of Individual Palmyra Fibers, p. 424–29. International conference on recent advances in composite materials (ICRACM-2007), sponsored by: Air Force Office of Scientific Research—Asian Office of Aerospace Research & Development, U.S.A, and the Department of the Navy, Allied Publishers Private Ltd., New Delhi, eds: Vijaya K Srivastava et al., Feb 2007.

31. Amar K. Mohanty, Manjusari Misra, and Lawrence T. Drazal, *Natural Fibers, Biopolymers, and Biocomposites*, Taylor & Francis, CRC Press, 2005.
32. "Packaging development is more than "putting a box around a product" http://www.packagingconsultancy.com/package-development.html.
33. Rosewood Packaging (Scotland), 9–15 Napier Place Cumbernauld, Glasgow, U. K., www.rosewoodpackaging.co.uk *2008*.
34. M. Avella, J.J. de Vlieger, M.E. Errico, S. Fischer, P. Vacca, and M.G. Vope, "Biodegradable Starch/Clay Nanocomposite Films for Food Packaging Applications," *Food Chemistry*, Vol. 93, p. 548–558, 2009.
35. *Magali Rocher, ADEME, France, Prof. Gyula Marton, University of Veszprém, Hungary, Interactive European Network for Industrial Corps and their Applications (IENICA), Newsletter*, Number 19, March 2003.
36. Biodegradable Packaging Research in the Agricultural Research Service, Eastern Regional Research Center Agricultural Research Service, U.S. Department of Agriculture, Wyndmoor, Pennsylvania.
37. Kirsten Dangaran, Charles Onwulata, and John Cherry (Center Director), *Packaging, Films and Coatings: Research Technologies and Applications*, Eastern Regional Research Center Agricultural Research Service, USDA, Wyndmoor, PA.
38. Michael J. Hodgin, Epoxies for Opto Electronic Packaging; Applications and Material Properties, Proceeding of the 36th Annual IMAPS Conference, Boston MA, pp. 26–30, Nov 17–20, 2003.
39. "Packaging development is more than 'putting a box around a product'" http://www.packagingconsultancy.com/package-development.html
40. Package Testing-Wikipedia, the free Encyclopedia, http://en.wikipedia.org/wiki/Package_testing
41. Luc Averous, Cristophe Fringant, and Laurence Moro, "Starch-Based Biodegradable Materials Suitable for Thermodynamics Packaging," *Starch/Starke*, Vol. 53, p. 368–371, 2001.
42. F. Langmaier, M. Mladek, P. Mokerjs, and K. Kolomaznic, "Biodegradable Packaging Materials Based on Waste Collagen Hydrolysate Cured with Dialdehyde Starch," *Journal of Thermal. Analysis and Calorimetry*, Vol. 93–2, p. 547–552, 2008.
43. I.B. Leonor, R. A. Sousa, A. M. Cunha, R. L. Reis, Z. P. Zhong, and D. Greenspan, "Novel Starch Thermoplastic/Bioglass® Composite, Mechanical Properties, Degradation Behavior and In-Vitro Bioactivity," *Journal of Materials Science, Materials in Medicine*, Vol. 13, p. 9 3 9–9 4 5, 2002.
44. Maurizio Avella, Aleksandra Buzarovska, Maria Emanuela Errico, Gennaro Gentile, and Anita Grozdanov, "Eco-Challenges of Bio-Based Polymer Composites," *Materials*, Vol. 2, p. 911–925, 2009.
45. H.C. Huang, T.C. Chang and J. Jane, "Mechanical and Physical Properties of Protein-Starch Based Plastic Produced by Extrusion and Injection Molding," *JAOCS*, Vol. 76, 9, p. 1101–1108, 1999.
46. Chen Bei Huang, Robert Jeng, Mohini Sain, Bradely A. Saville and Martin Hubbles, "Production, Characterization, and Mechanical Properties of Starch Modified by Ophiostoma SPP.," *Bio-Resources*, Vol. 1(2), p. 257–269, 2006.
47. J. Guan, Q. Fang and M. A. Hanna, "Functional Properties of Extruded Starch Acetate Blends," *Journal of Polymers and the Environment*, Vol. 12, No. 2, 2004. [33]
48. M.G. Ivo, Martins, Sandra P. Magina, Lucia Oliveira, Carmen S.R. Freire, Armando J.D. Silvestre, Carlos Pascoal Neto, and Alessandro Gandini, "New Biocomposites Based on Thermoplastic Starch and Bacterial Cellulose," *Composites Science and Technology*, Vol. 69, p. 2163–2168, 2009.
49. You-Ping Wu, Mei-Qin Ji, Qing Qi, Yi-Qing Wang and Li-Qun Zhang, "Preparation, Structure, and Properties of Starch/Rubber Composites Prepared by Co-Coagulating Rubber Latex and Starch Paste," *Macromolecular Rapid Communications*, Vol. 25, p. 565–570, 2004.
50. Maurizio Avella, Aleksandra Buzarovska, Maria Emanuela Errico, Gennaro Gentile, and Anita Grozdanov, "Eco-Challenges of Bio-Based Polymer Composites," *Materials*, Vol. 2, p. 911–925, 2009.
51. Richard Gattin, Alain Capinet, Celne Bertrand, and Yves Couturier, "Comparative Biodegradation Study of Starch and Polylactic Acid-based Materials," *Journal of Polymers and the Environment*, Vol. 9, No. 1, 2001.

52. Y.Z. WanHonglin Luo, F. He, H. Liang, Y. Huang, and X.L. Li, "Mechanical, Moisture Absorption, and Biodegradation Behaviours of Bacterial Cellulose Fiber-Reinforced Starch Biocomposites," *Composites Science and Technology*, Vol. 79, p. 1212–1217, 2009.
53. D. R. Lu, C.M. Xiiao, and S.J. Xu, "Starch-Based Completely Biodegradable Polymer Materials," *Express Polymer Letters*, Vol. 3, 6, p. 366–375, 2009.
54. Chin-San Wu, "Physical Properties and Biodegrability of Maleated-Polycaprolactone/Starch Composite," *Polymer Degradation and Stability*, Vol. 80, p. 127–134, 2003.
55. Biodegradable Starch-based Nano-Clay Composites for Food Packaging Applications, School of Engineering and Design, Bernel University, West London, DEFRA Food Link, 2004–2007.
56. Robert D. Maksimov, Aivars Lagzddins, Nadezda Lilichenko, and Egils Plume, "Mechanical Properties and Water Vapor Permeability of Starch/Montmorillonite Nanocomposites," *Polymer Engineering and Science*, Dec. 2009.
57. Penggang Ren, Tingting Shen, Fang Wang, and Zhengwei Zhang, "Study of Biodegradable Starch/OMMT Nanocomposites for Packaging Applications," *J Polym. Environ.*, Vol. 17, p. 203–207, 2009.
58. Ayse Alemdar, and Mohini Sain, "Biocomposites from Wheat Straw Nanofibers: Morphology, Thermal and Mechanical Properties," *Composites Science and Technology*, Vol. 68, p. 557–565, 2008.
59. Anna J. Svagan, Mikael S. Hedenqvist, and Lars Berglund, "Reduced Water Vapor Sorption in Cellulose Nanocomposites with Starch Matrix," *Composites Science and Technology*, Vol. 69, p. 500–506, 2009.
60. Du Yumin, Xia Zuyong, and Lu Rong, "Blends Films of Chitosan/Starch," *Wuhan University Journals of Natural Sciences*, Vol. 2, No. 2, p. 220–224, 1997.
61. Nattakan Soykeabkaekaew, Pitt Supaphol, and Ratana Rujiravanit, "Preparation and Characterization of Jute- and Flax-Reinforced Starch-Based Composite Foams," *Carbohydrate Polymers*, Vol. 58, p. 53–63, 2004.
62. S. Wongasulak, T. Yoodihya, S. Bhumiratna, P. Hongsprabhas, D.J. McClements, and J. Weiss, "Thermo-Mechanical Properties of Egg Albumen-Cassava Starch Composite Films Containing Sunflower-Oil Droplets as Influenced by Moisture Content," *Journal of Food Research International*, Vol. 39 (3), 2006.
63. Hearwin Amaladhas Pushpadass, Govindranjan Suresh Babu, Robert W. Weber, and Milford A. Hanna, "Extrusion of Starch-Based Loose-Fill Packaging Foams: Effects of Temperature, Moisture and Talc on Physical Properties," *Packaging Technology and Science*, Vol. 21, No. 3, p.171–183, 2008.
64. Wan-Jin Lee, Young-Nam, Yeon-Hum Yun, and Soon-Do Yoon, "Physical Properties of Chemically Modified Starch (RS4)/PVA Blend Films—Part 1," *J. Polym. Environ.*, Vol. 17, p. 35–42, 2007.
65. Hulya Yavuz, and Ceyhun Babac, "Preparation and Biodegradation of Starch/Polycaprolactone Films," *Journal of Polymers and the Environment*, Vol. 11, No. 3, p. 107–113, 2003.
66. Laura G. Carr, Duclerc F. Parra, Patricia Ponce, Ademar B. Lungao, and Pedro M. Buchler, "Influence of Fibers on the Mechanical Properties of Cassava Starch Foams," *J. Polym. Environ.*, Vol. 14, p. 179–183, 2006.
67. K.N. Matsui, F.d.S. Larotonda, S.S. Paes, D.B. Luiz, A.T.N. Pires, and J.B. Laurindo, *Carbohydrate Polym*er, Vol. 55, p. 237, 2004.
68. A.A.S. Curvelo, A.J.F. Carvalho, and J.A.M. Agnelli, *Carbohydrate Polymers*, Vol. 45:183, 2001.
69. J.W. Lawton, R.L. Shogren, and K.F. Tiefenbacher, *Ind. Crops Prod.*, Vol. 19:41, 2004.
70. S. Shinoj, S. Panigrahi, and R. Visvanathan, "Water Absorption Pattern and Dimensional Stability of Oil Palm Fiber-Linear Low Density Polyethylene Composites," *Journal of Applied Polymer Science*, Vol. 117, p. 1064–1075, 2010.
71. Richard Gattin, Christophe Poulet, Alain Copinet, and Yves Couturier, "Comparison of Mineralization of Starch in Liquid, Inert Solid and Compost Media According to ASTM and CET Norms for the Composting of Packaging Materials," *Biotechnology Letter*, Vol. 22, p. 1471–1475, 2000.

72. Mia Sjoqvist, and Paul Gatenholm, "The Effect of Starch Composition on Structure of Foams Prepared by Microwave Treatment," *Journal of Polymers and the Environment*, Vol. 13, No. 1, 2005.
73. Deniela Schlemmer, Romilo S. Angelica, and Maria Jose A. Sales, "Morphological and Thermomechanical Characterization of Thermoplastic Starch/Monomorillonate Nanocomposites," *Composite Structures*, Vol. 92, p. 2066–2070, 2010.
74. Sanjay K. Mazumdar, *Composite Manufacturing: Materials Product and Process Engineering*, CRS Press, NY, 2002.

PART 3
CIVIL ENGINEERING APPLICATIONS

9

Vegetable Oil Based Rigid Foam Composites

Venkata Chevali, Michael Fuqua and Chad A. Ulven

Mechanical Engineering Department
North Dakota State University
Fargo, North Dakota, USA

Abstract
Rigid polymeric foams reinforced with natural or synthetic fibers constitute a major class of semi-structural composite materials. With the existing emphasis shifting from petroleum-based resources towards renewable resources, and sustainable engineering and materials, the usage of vegetable oils for producing biopolymers and resins through chemical syntheses is ever increasing. For rigid-foam composites, the usage of vegetable-oil-based, polyether- and polyester-urethane resins bring about a predominant renewable fraction. The degree of renewability in rigid foam composites is further increased by the addition of a natural fiber/filler, such as flax fiber or hemp fiber, which with exceptions, always cause increased mechanical performance over their unreinforced foam counterparts. Production of these rigid foam composites is dependent upon the fiber content and the specific foam type, with reaction injection molding (RIM) and mold casting positioned as leading production methods. Many industrial sectors have been penetrated by rigid foam biocomposites, chiefly marine and construction, but a major market for these materials is the transportation industry, which harnesses the superior mechanical performance, cost-effectiveness, and biorenewability of rigid-foam biocomposites in many underbody applications.

Keywords: Rigid foam, vegetable oil, fiber reinforced composite, reaction injection molding, rigid foam applications

9.1 Rigid Foam Composites

Rigid polymeric foams have found use in a wide variety of engineering applications (i.e., construction, transportation, military, etc.) due to their lightweight, thermal insulation, and moderate physical properties. In addition to their use as non-structural stand-alone materials, many rigid polymeric foams have been laminated between facesheets of monolithic metals or polymer matrix composites to create sandwich structures or have been reinforced throughout the foam structure to create polymer matrix composites. Sandwich composite structures have a rich history and well documented past in many different industries such as aerospace

Srikanth Pilla (ed.) Handbook of Bioplastics and Biocomposites Engineering Applications, (269–284)

and construction, however, rigid foam composites as addressed in this chapter are a relatively niche area of materials development. Reinforcing rigid foams has been accomplished using fillers, short or long fibers, and most recently, promise has been shown with nanoparticulate reinforcement.

Specifically, this chapter focuses on the development of rigid polymeric foam composites based on renewable agricultural resources to produce both the foam matrix and the reinforcement phase. As concerns for the environment and world dependence on petroleum continue to be on the forefront of global debate, industrial and academic researchers continue to develop new ways to synthesize "greener" material options for non-structural, semi-structural, and structural applications. Flexible and rigid foams have been shown to be produced from a variety of renewable natural resources (i.e., vegetable oils, proteins, starches, etc.) but the focus of this chapter has been limited to those of vegetable oil origin, being the most widely studied and used precursor for rigid biobased foams to-date. In addition, many of the traditional foam processing techniques with and without reinforcement are reviewed as they are now being proven as the means to create biobased foams and biocomposites based on natural fiber reinforcement. With the advent of new biobased foams, advances in natural fiber processing, new composites processing technologies and/or modifications to traditional processing technologies, biocomposites based on biobased foams and natural fibers will be entering the marketplace in growing numbers over the next couple of decades.

This chapter explores numerous varieties of vegetable oil based rigid foams which can be made as a result of 1) the variety of renewable resources they can be derived from, 2) their flexible chemistry and different ways to be synthesized, and 3) ways reinforcing materials can be added to them in order to create polymer matrix composites. In addition to the development of these vegetable oil based rigid foams and their composites, the potential environmental impacts of these materials are discussed. Finally, as evidence of this growing class of biobased material, an industrial application is highlighted and described. The biocomposite materials developed in this case study are compared and contrasted against the traditional materials that they were designed to replace. The emphasis of this review chapter is that a class of biobased materials based on vegetable oil-derived rigid foams and natural fiber reinforcement to create biocomposites is on the rise and has been shown as a viable option to replace traditionally used engineering materials and/or their synthetic counterparts.

9.2 Biofoams

Biofoams are obtained from renewable sources, primarily vegetable oils with triglycerides as reactive sites for chemical reactions. The polyol or triol component is derived from the vegetable oil by modifying the oils. For example, –OH groups are added to an unsaturated triglyceride through (a) hydroxylation of C=C or (b) triglyceride alcoholysis or (c) esterification of fatty acids and glycerol in presence of a catalyst, to produce a monoglyceride. The three largest production volumes of commodity oils in descending order are soybean oil, palm oil and rapeseed oil [1].

However, commercial and academic research is most concentrated on bio-polyols derived from soybean oil and acrylated epoxidized soybean oil [2–13]. Apart from soybean oil derivatives, several other sources have been successfully harnessed for synthesizing polyol, including palm oil [14–16], castor oil [17–19], sunflower oil [20], corn oil and other corn derivatives [20]. Several other resources, such as rapeseed oil [21], rubber seed oil [22], tung oil [23, 24], linseed oil [20], and canola oil [20] have also been explored. Table 9.1 summarizes the specific focus of these works.

Table 9.1 A list of leading authors and research on rigid foams.

Author	Vegetable Oil Source	Filler/Reinforcement
Dumont [25]	Canola oil	–
Zlatanić [20]	Canola, Soybean, Sunflower, Corn, Linseed	–
Aranguren [17]	Castor Oil	Pinewood, Hemp
Manjula [18]	Castor-Oil	Silk fibers
Javni [9]	Epoxidized Soy-Oil	–
Chian [14]	Palm Oil	–
Chuayjuljit [15]	Palm Oil	Montmorrilonite (MMT)
Yaakob [16]	Palm Oil	Saw Dust
Hu [21]	Rape-seed Oil	–
Bakare [22]	Rubber-seed-oil	–
Bandhopadhyay-Ghosh [2]	Soy Oil	–
Banik [3, 4]	Soy Oil	Paper
Banik [3]	Soy Oil	–
Chang [5]	Soy Oil	Soy Flour
Chang [6]	Soy Oil	Soy Flour
Fuqua [7]	Soy Oil	Flax
Javni [8]	Soy Oil	–
Latere Dwan'isa [13]	Soy Oil	Glass
Petrović [10]	Soy Oil	–
Petrović [26]	Soy Oil	–

Table 9.1 (cont.) A list of leading authors and research on rigid foams.

Author	Vegetable Oil Source	Filler/Reinforcement
Singh [11]	Soy Oil	–
Song [12]	Soy Oil	MMT
Tu [19]	Soy Oil, Epoxidized Soy Oil, Castor Oil	–
Mosewicki [23, 27]	Tung Oil	Pinewood Flour

The properties of a biofoam are dictated by cell density, cell size and distribution, and edge connectivity [28, 29]. These properties are dictated by the foaming technologies that are tailored specifically for each polymer to optimize its final properties as foams. Additional modifications may be necessary with the introduction of a fibrous component in the foam. The mechanical properties of these foams are a strong function of the cellular structure, especially the ratio of foam density and the solid portion of the foam, also known as the relative density. Relative density is a parameter in the formulation for several scaling laws that describe the mechanical behavior of the foams. The cell size is also critical, as equiaxed cells cause isotropic properties, and non-equiaxed cells lead to strong directionality in properties. Also, with a smaller cell size, insulation capability is strengthened, concomitant with an improved energy absorption. Nonetheless, the edge and face connectivity dictate certain mechanical properties, but the mode (open versus closed) and degree of connectivity depends on structure and dispersion of the foam. For example, closed cell foams are ideal for higher compressive strength and impact toughness, whereas open cell foams are suitable for noise insulation. Hence, the formation of cell structure during foaming affects the mechanical, thermal, and insulative properties of the foams.

As an example, foaming of polyol with a diisocyanate causes the liberation of the carbon dioxide within the free-rising foam. Following evacuation of such gases, cells form (Figure 9.1a), whose cell walls may break because of low mechanical integrity, forming an open cell in lieu of the closed cell structure expected. In case of higher cell integrity, an even distribution and standard geometry, such as a hexagonal structure (Figure 9.1b) are obtained. With the addition of a reinforcement for producing a composite, two scenarios are encountered: 1) Fibers may be well distributed and contrived within the foam bulk (Figure 9.2a), or 2) Fibers may be dispersed in a random fashion, in which case the cell growth may be partially affected by their presence (Figure 9.2b). For optimal mechanical properties, a uniform distribution of the fibers across the foam bulk is beneficial, along with a low cell size for high energy absorption.

9.2.1 Reactant Chemistry

Vegetable oil-based rigid foam properties are a function of the reactant chemistry used in the foaming reaction. For polyurethane foams, with an increasing fraction

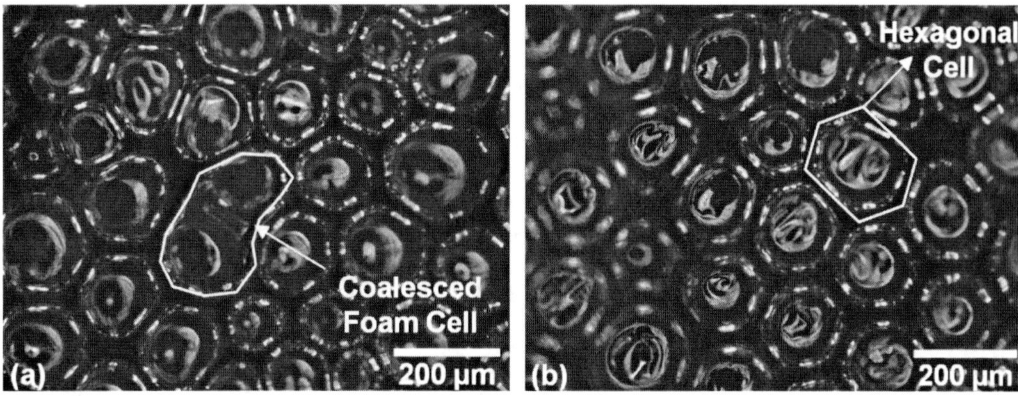

Figure 9.1 The foam cell structure of a polyurethane foam during foaming, showing a) coalescence of adjacent cells, and b) formation of near equivalent hexagonal cells.

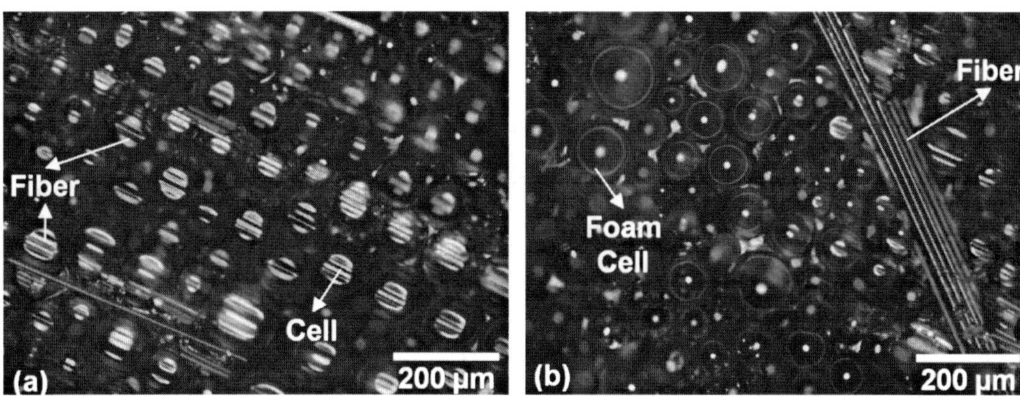

Figure 9.2 The foam cell structure of a polyurethane/glass composite, showing a) uniform distribution of glass fibers across the foam cells, and b) a cross section of the molded foam panel with glass fibers constricting the cell growth.

of isocyanate, foams of higher strength can be obtained as higher –OH conversion can be obtained [8]. Aromatic triisocyanates cause higher density and glass transition temperature (T_g), along with high impact resistance and dimensional stability in solvents, i.e., glassy behavior. Aliphatic triisocyanates and diisocyanates produce low crosslink densities and rubbery foams with high ductility and low dimensional stability in solvents, i.e., elastomeric/rubbery behavior. Foams that are produced with aromatic and cycloaliphatic diisocyanates show properties which are intermediate to those of the triisocyanates [8]. The change in molecular weight of diamines causes low degree of variation in T_g, and a high degree of variation in the hardness and tensile strength, as degree of crosslinking and hydrogen bonding vary accordingly [9]. Maintaining the stoichiometric ratio of carbonate/amine functional group is another parameter that aids in obtaining improved properties over non-stoichiometric ratios. In addition, the functionality of the polyol component also dictates the formation of the type of foam, wherein the usage of a polyol with a –OH functionality greater than 200 is likely to form glassy and brittle foams, and a functionality of less than 200 is likely to cause elastomeric foams [10]. In case the polyol is functionalized, the densities, rheological properties, and molecular weights of the mixtures decrease with the decreasing complexity of the added group [26]. For example, polyurethane properties decrease in this polyol functionalization sequence-brominated, chlorinated, methoxylated, and hydrogenated [26].

A chemical blowing agent having the chemical compatibility with –OH groups of the vegetable oil-derived polyol governs the mechanical strength of the foams. A blowing agent is selected for a specific resin by also considering its thermodynamic properties, processing ease, and the targeted foam application [28]. For polyurethane systems, the amount of water in the reaction is a contributor to strength as it reacts with isocyanate to form rigid poly(urea) structures, thereby causing higher foam stiffening [11]. In these stiffening mechanisms of the foams, increasing the water content causes faster stiffening of domains with a final increased modulus. The effect of addition of a synthetic polyol in soy-based polyol is lowering the liquid foam plateau and a corresponding reduction in the overall reaction time [11]. In most cases, the usage of an efficient surfactant for proper mixing of reactants is necessary to obtain a well-distributed cellular structure. Albeit the chief parameter in the reaction is the foaming reaction temperature that controls the rate of modulus development in a four stage mechanism, i.e., 1) evolution and growth of bubbles, 2) filling of bubbles and liquefied foam, 3) separation of urea phase and opening of cells, and 4) final curing.

9.2.2 Environmental Impact

The environmental and toxicity impact from the processing of vegetable oil-derived rigid foams is mainly from the chemical blowing agents that may be used in addition to the volatile gases that are expelled during the foaming [28]. However, the most commonly yielded chemical vapor during rigid foam production using vegetable oil-derived resin systems is carbon dioxide, a resultant of isocyanate and water blowing process. It is important to recognize that a Material Data Safety Sheet (MSDS) should always be consulted regarding the potential health effects from the various chemicals used in the reaction process. Despite being relatively benign compared to many reactive polymer systems used in composite productions, chemicals such as polyols, isocyanates, and epoxidized vegetable oils have been known to cause irritation of the eyes, nose, throat, lungs and skin, as well as potentially cause allergic reactions of the skin and lungs. Basic precautions when working with these chemicals, such as using proper material handling equipment and clothing, as well as maintaining a good ventilated working environment, are important.

The environmental impact of vegetable oil-derived rigid foams is also tied with its eventual end-of-life service retirement and subsequent recycling or scrapping. While typically derived from 40% to 70% biorenewable material, currently 100% biodegradable rigid foam systems do not exist for composite applications. As such, the same end-of-life design and planning concerns that are in the polymer and polymer composite world are prevalent for foams. Being a thermosetting system, simple recycling is not possible. However, companies [30] are starting to incorporate up to 10% reground foam into their products. The purpose of this approach is to take an end-of-life product, and ground for re-introduction as a non-intrusive filler for new foam or composite production, thereby keeping the older parts from being land-filled or incinerated.

9.3 Production Methods

Vegetable oil based foams can be processed in a variety of methods that are standardized for traditional polymer foam production. Most vegetable oil based foams are created by crosslink foam processing, whether in the processing of derived vegetable oil polyols for polyurethane or polyisocyanurate foams, or acrylated epoxidized vegetable oils for polyester foaming. Crosslinked foams are produced using quick reaction molding techniques, due to the rapid reaction times for the initial foaming. This time constraint limits the subset of processing methods that may be used for crosslinking foam processing to only those catered to quick final shape formation. When coupled with the incorporation of fibers or fillers, processing methods can be classified into a number of distinct types, including mold casting, reaction injection molding (RIM), and slabstock molding.

9.3.1 Mold Casting

Mold casting represents a range of molding techniques in which a premixed crosslinking foams system is poured into a mold for final shaping as the mixture foams and cures. Mold casting techniques are most commonly used for lab scale research of material properties and manufacturing capabilities. Due to the difficulty of mixing a foaming system and pouring it into a cast before the reaction begins foaming, this method is most viable for parts of small dimensions.

Casting methods can be divided into two major classes, open mold casting and closed mold casting. In open mold casting, the individual liquid reactants of a system are mixed and poured into an open-faced mold. The foaming reaction is allowed to progress unrestricted, resulting in rising foam with density that is controlled by the chemical foaming process (i.e. free-rise). However, in closed mold casting, the foaming resin system is poured into a mold, which is closed with a set compressive pressure applied to the mold. By constricting the growth/rise of the foam, parts manufactured in this manner result in a higher foam density than open mold casting produces, as well as a high density skin due to the contact and restriction occurring at the mold surface.

For composite production, open mold casting is only viable for short fiber or nanofiller reinforcement because of the free expansion of the resin during the foaming process. The use of continuous fibers in an open mold process is not advantageous because of the inability of a free rising foam to achieve proper impregnation while maintaining the control of fiber placement. However, as demonstrated with short agricultural fiber such as sisal [31], agricultural fillers [3, 4, 5, 6], and mineral fillers such as nanoclays [15], open mold casting is applicable for producing composite foam systems using a vegetable oil-derived resin matrix. Short fibers and fillers are added to one part of the reacting chemistry (such as the vegetable oil-derived polyol in a polyurethane system) and mixed until a homogenous distribution is achieved. This fiber-dispersed component is mixed with the other components of the reaction and poured into mold, and allowed to expand without restriction. A random fiber orientation is achieved, yielding relatively isotropic foam properties.

Closed mold casting for composite production can be used with both short and long fiber reinforcements [17]. In the case of short fiber or filler addition, the fibers are premixed in a manner similar to that used for open mold casting. The reacting mixture is poured into a mold, which is closed with pressure being applied simultaneously and maintained to constrict foam expansion. As a result, tighter cell packing and higher fiber volume fraction are obtained. In the case of continuous fiber reinforcement, the fibers are pre-placed in the mold, and the reaction resin system is poured into the mold and closed. The restriction of foam expansion allows the pressurized reaction mixture to achieve high foam impregnation of the fiber system. Through fiber placement in the mold, anisotropic properties can be achieved.

9.3.2 Reaction Injection Molding

Reaction injection molding (RIM) is a process in which a reactive thermosetting system such as polyurethane is injected into a closed mold through an impinging mixer. The liquid components of the resin system are metered separately at high pressures until the impingement mixing of the liquid begins the foaming reaction as the pressurized resin system is injected into a mold cavity. Because the mixture enters the mold at relatively high pressures, reaction injection molding allows for the creation of larger dimension parts compared to those produced by an open or closed mold casting process.

Two approaches can be adopted with RIM for composite production. The first is reinforced reaction injection molding (RRIM), in which reinforcing agents are preblended into one component of the liquid system prior to impinging the mixture and injecting into the mold cavity. Much in the same manner as open cast molding, this method is limited to short fibers or fillers, and results in parts with low degree of fiber orientation and thus relatively isotropic foam properties. The second method is structural reaction injection molding (SRIM), in which continuous fiber matting is placed in the mold prior to injection, and the foaming resin system is then infused into the mold, resulting in impregnation.

9.3.3 Slabstock Molding

Slabstock molding is a process for producing high volumes of thermosetting foam by a free-rising process. In slabstock molding, the liquid components of a foam resin are metered continuously to a mixing head, which moves across a conveyer in a pattern of ribbons. The foam is moved continuously by the conveyer, resulting in uniform distribution of the reactive thermosetting system to produce a continuous output foam slab. The foam is free to expand upwards, however width of the produced slabs is controlled with adjustable sideboards. Foam cell density is controlled by both the chemical formulation of the foam as well as temperature controls throughout the slab output conveyance. From a composite standpoint, slabstock molding is normally limited to the production of short fiber or filled composites because of its inability to attain a proper impregnation of continuous fiber mats.

9.4 Reinforcement Effects

Vast arrays of fillers and fibers have been explored as reinforcements for rigid foam composites. Two approaches are adopted for rigid foam composite production: long fiber reinforcement, which are aimed mainly at improving or altering the mechanical performance of a system, or short fiber and filler reinforcements, which hold several advantages in both mechanical and thermal properties. In short fiber and long fiber forms, a wide variety of reinforcements are available. These reinforcements include both conventional synthetic and inorganic fibers and fillers, as well as agriculturally-derived reinforcements that add an additional "green" component to the potentially bio-renewable vegetable oil-derived foams.

9.4.1 Short Fiber/Fillers

The utilization of short fibers and fillers within vegetable oil-derived rigid foams has been proven a very advantageous pursuit. Fillers and short fibers in two general classes are currently being explored in vegetable oil-derived rigid foams. The first group comprises of inorganic filler agents, many of which (such as layered silicates) are heavily explored for utilization in synthetic polymer matrices. The other group is the agriculturally-derived lignocellulosic fillers and fibers, including vegetable flours from soy or wood source, cellulose fillers derived from paper, or short, chopped-bast fibers such as sisal or hemp fibers. Agriculturally-derived fillers hold a strong potential for advancing environmentally-friendly and sustainable development from the utilization of vegetable oil-derived rigid foams.

Short fibers and fillers have a major influence on rigid foam density and cell structure. An increasing inorganic or agricultural fiber or filler loading has shown to cause increases in cell density. Based on observations from infrared spectroscopy and thermal analyses, density changes in rigid foam are inferred to be from the modification of the foaming behavior caused by intermolecular interactions of the fiber or filler [3]. For inorganic fillers such as layered silicates, this variation is caused mainly because of the viscosity increase of the reacting mixture, which restricts the expansion of the foam cells [15]. In a rigid foam composite with agriculturally-derived fillers, the effect on foam and cell density can also be explained by secondary reactions occurring due to the addition of the fillers. Since agricultural fillers are hydrophilic, they inherently contain moisture content. This water content serves as a blowing agent within isocyanate based foam systems, where the reaction between isocyanate and water produces carbon dioxide [6]. Hence, the presence of water within natural fillers will lead to production of higher amount of carbon dioxide (and polyurea), which lead to a higher cell density, characterized by a large number of cells and small individual cell size [3, 4, 6, 15, 17, 31]. Nevertheless, at high fiber loadings, agglomeration effects hinder the reactivity of the fiber and affect the overall properties [3].

The increases in cell density that are caused by the addition of fillers and short fibers show multiple effects on several performance characteristics of the reinforced composites. The mechanical properties of rigid foams are a strong function of the foam cell density [32]. Clear influence of filler and short fiber addition on compressive performance of vegetable-oil-derived rigid foam composites is expected. Specifically, fiber loading increases the compressive strength, independent of the filler type [6, 15, 17, 31]. In a study of the addition of soy flour to polyurethane foam that uses a soy-based polyol, the filled composites of similar densities as the neat soy-based foam samples showed higher compressive strengths [6]. However, short fiber length is beneficial for improving compressive strength initially, but may ultimately be disadvantageous when exceedingly long fiber lengths are used. This behavior is attributed to the probability of fibers in contact with gas cells that increases if fiber length increases at a constant cell density, detrimental for maintaining high compressive strength and cell integrity [31]. Similarly, filler addition affects the modulus of vegetable oil-derived rigid foam systems. Compressive modulus, for example, increases with filler and short fiber loading, a product of the cell density improvements gained through filler or fiber addition [6, 31].

Unique dynamic mechanical behavior is seen with the addition of fillers and short fibers into vegetable oil-derived rigid foam systems, particularly in conjunction with agriculturally-derived reinforcements. A combination of the hydrodynamic effects of the fillers within the viscoelastic matrix, coupled with the inherent mechanical strength obtained through the presence of fillers, leads to large variations in the behavior of the reinforced composites against neat foam systems. For example, low degree of variation between glassy modulus and rubbery storage modulus are observed for filled composite systems versus unfilled vegetable oil-derived foams. The width of the tan δ peaks for filled composites widen compared to unfilled foams, suggesting that the addition of a filler or short fiber improves damping capabilities significantly over the unreinforced rigid foams [17].

Thermally, the addition of fillers or short fibers to vegetable oil-derived rigid foams has a number of effects. Work with both soy flour and wood flour have shown that both the addition of filler, as well as increasing the initial water content in the foam composites, contribute to a higher glass transition temperature [5, 6, 17]. The increase in number of urea bonds with increased moisture in the water/isocyanate reaction partially contributes to this effect. Polyurea is more thermally stable than polyurethane, and thus increasing the number of urea groups (domains) lead to increases in thermal stability [6]. The thermal stability of the foam composites is also increased by the addition of filler or short fiber loading, both for agriculturally-derived fillers as well as inorganic fillers such as layered silicates [3, 4, 15]. As fillers restrict curing and increase cell density, urethane formation is decreased vis-à-vis urea production, which is increased. The higher thermal stability of the urea, coupled with the general resistance to thermal expansion by the fillers and fibers causes overall thermal stability improvements. Fire resistance is achieved in some cases by the addition of boric and/or phosphoric acid that also induce modifications in the structure.

9.4.2 Long Fiber

Long fiber reinforcements for use in rigid vegetable oil-derived foams are classified into two groups: inorganic and organic. For rigid foam composite applications, the inorganic fibers utilization is concentrated within glass fibers, but carbon, aramid, and ultrahigh-molecular weight polyethylene fibers are also used. The usage of glass fibers is justified because of their low cost and high applicability in a vast number of applications where the use of higher performance fibers, e.g., carbon or aramid, in relatively low-mechanical-performance rigid foams is not practically feasible. Long organic fibers, on the other hand, encompass a much greater array of sources and types, including agriculturally-derived bast fibers such as hemp and flax, but also span into other naturally-derived fibers such as protein-based fibers like silk.

Glass fiber is a well-understood long fiber reinforcement that is commercially utilized with traditional petroleum-derived (synthetic) rigid foams composites. As such, they have been successfully engineered for application within vegetable oil-derived rigid foam composites. Glass fiber loadings bring significant improvements in both strength and modulus. Strength and stiffness of these composites also is highly dependent on fiber orientation. Improvements are achieved with a suitable fiber orientation, leading to potential improvements. For example, with randomly-oriented long fiber E-glass reinforcement at 50 wt. % fiber loading, an approximately 260% increase in flexural strength and 480% increase in flexural stiffness over unreinforced foam is obtained with glass fiber usage in a soy phosphate ester polyol-derived rigid polyurethane foam matrix [13]. In addition, about a 1600% increase in notched Izod impact performance was also seen, because the glass fibers introduce dissipation mechanisms and crack propagation, such as fiber breakage, that are not present in the unreinforced foam. Thermally, long fiber reinforcements such as glass fibers show similar effects as short fibers and fillers. A clear increase in the T_g is realized, compared to the unreinforced foam along with improvements in thermal stability because of reduction in the volume fraction of the vegetable oil-based polyurethane component.

Agriculturally derived organic long fibers possess benefits within rigid foam composites, beyond their advantages of potential renewability and biodegradability. In particular, moisture content of these fibers allows for improved interfacial interaction and improvements of foaming kinetics of vegetable oil-derived rigid foams causes increases in mechanical properties. Just as with natural fillers or short fibers, the moisture in long agricultural fibers assists in the blowing process, yielding slightly higher cell densities and greater yield of polyurea that acts as an improved means of initiating surface interaction between the fibers and foam. For example, randomly-oriented bast fibers such as hemp and flax cause significant improvements in flexural strength and modulus over unreinforced vegetable oil-derived rigid foams [7, 17]. Other organic fiber types, such as protein-based silk fiber, also show yield increases in mechanical properties over their neat unfilled foams [18]. Similar to glass fibers, silk fibers also increase the T_g of the resultant biocomposites and improve their thermal stability. Hence, the incorporation of fibers, regardless of the source holds promise and potential towards improvements in a variety of properties simultaneously retaining or improving the renewability and biodegradability of the resultant composite.

9.5 Applications/Case Study

9.5.1 Potential Industry Utilization

Many industrial applications require lightweight, insulative, vibrational damping, and semi-structural materials. As transportation costs continue to increase, the need to produce mass transit vehicles, personal vehicles, construction materials, etc., which are lighter in weight is paramount towards sustaining the current way of life. Therefore, many of the foams and their composites discussed in this chapter are gaining interest because of their process and design flexibility as well as potential integration of other structures such that a multifunctional material system can be manufactured. An example of this philosophy would be the development of structural insulated panels (SIPs) for building construction [33]. The advantage of SIPs is the ability to transport final structures on site, thereby reducing the amount of excess material and improving the bulk density of building materials being transported to the site. Other growing applications of rigid foams and their composites include lightweight flooring in buildings, boat hull stringers and transoms, bulkheads of boats, recreational vehicles, tractors/trailers, etc. An example of reducing weight in mass transit vehicle applications would be in the replacement of structural metallic frame supports and plywood covering with integrated sandwich composites composed of multifunctional cores. The following section details the development of a commercially viable vegetable oil-based rigid foam with natural bast fiber reinforcement to create a biocomposite capable of replacing plywood.

9.5.2 Mass Transit Application Case Study

The potential of vegetable-oil-derived rigid foam composites was determined as a replacement for plywood paneling in mass transit applications. Currently, most mass transit vehicle flooring is produced with 12.7 mm thick plywood floorboards supported by steel frames. However, plywood is extremely susceptible to decay over the lifecycle of a floor, even with chemical treatments to mitigate potential decay, and requires replacement that brings substantial vehicle downtime and lost profits [34]. Synthetic polyurethane composites reinforced with glass fibers have been proven a potential alternative, but they are environmentally detrimental in terms of their carbon footprint. As a solution, long agricultural fiber reinforcements such as flax and hemp are being used as fibers within soy-polyol-based rigid polyurethane foams [7] to produce panels that can replace plywood. A natural bast fiber/soy-based rigid foam system is composed of about 55% to 70% renewable materials by weight, thus constituting an environmentally-friendly biocomposite alternative to plywood.

Increases in both strength and stiffness of a rigid foam have been established previously with the addition of long fibers, such as flax, even with low quality and inexpensive random-oriented long-fiber mats. In vegetable-oil-derived- or synthetically-derived-rigid foams, this is a breakthrough since improvements in performance can be brought with cost-effective fibers, often comparable to plywood. To determine the viability of replacing plywood flooring with a natural fiber reinforced, vegetable oil-derived rigid foam, randomly oriented linseed derived flax mat was used. The fiber mat was composed of 55% to 60 % fiber, with the remainding 40% to 45% consisting of a light polymer binder and shive, which are thin fragments of the woody core of the flax stem having little mechanical benefit.

The results of this study indicated that the effect of both fiber volume fraction and cell density was critical for composite strength. As fiber volume content is increased and the corresponding void content is decreased, flexural strength increases, but at the cost of a higher part density. Overall, the specific flexural strength of the composites remains near constant for composites ranging from 6% to 16% volume fraction fiber. This specific flexural strength is approximately 30% lower than plywood paneling [7]. While not equivalent to the currently used material, the properties fall in an acceptable tolerance range for effective utilization in semi-structural mass transit applications.

This work also highlighted that as fiber volume content is increased and void content is decreased, the flexural modulus of the composite undergoes improvements in stiffness versus the unreinforced foam. However, the specific flexural modulus varies little with fiber volume fraction, and results in a composite that is approximately 60% less rigid than plywood [7]. While this challenge can be potentially addressed through molding with a higher quality flax or other agricultural fibers with less shive content, the difference in performance between the composite system and plywood does not necessarily limit the foam system from being used as a replacement. Other modifications of hybridizing reinforcements or manipulating fiber architecture could also potentially develop specific mechanical properties approaching those of traditional plywood.

A unique mechanical property which affects the suitability of a foam composite to be used as a replacement for plywood in mass transit flooring application is the ability to secure a given panel to the vehicle frame. Plywood is traditionally mechanically-fastened using wood screws, which allows the removal of panels for maintenance, repair, or replacement. Foam systems, conversely, are generally adhesively bonded to their mounting points, preventing easy replacement or repair. However, rigid foam composites with random oriented flax fiber are an exception to this rule. An internally developed test to determine screw retention during normal pullout showed that increased fiber loading and decreased cell content, both help in improving screw retention. When compared against material density, a positive linear correlation was observed between the force required for screw pullout and composite density. Especially at an equivalent density as plywood, a random-oriented-flax-mat-reinforced, soy-based-polyurethane-foam-composite possesses a similar screw retention strength [7]. Hence, the applicability of these composites to be mechanically fastened using screws rather than adhesives was established.

The current mechanical methods of fastening plywood planks in mass transit vehicles have been shown to be valid for highly-renewable composite foam flooring. In addition, part replacement concerns have also been addressed as adhesive bonding can be used as an alternative method. Although the specific mechanical properties of these rigid foam composites were still lower than those of plywood, strong potential of utilization was demonstrated with the incorporation of a cost-effective agricultural bast fibers and byproducts as reinforcement. These developments will likely strengthen vegetable-oil-derived-rigid position for new markets. These advancements in the engineering of fiber and filler reinforced rigid foams highlight the potential of vegetable oil-derived rigid foams in the future development of semi-structural and structural composite materials for a wide variety of applications.

References

1. Y. Lu and R. Larock, "Novel Polymeric Materials from Vegetable Oils and Vinyl Monomers: Preparation, Properties, and Applications." *ChemSusChem*, Vol. 2, p. 136, 2009.
2. S. Bandyopadhyay-Ghosh, S. Ghosh, and M. Sain, "Synthesis of Soy-Polyol by Two Step Continuous Route and Development of Soy-Based Polyurethane Foam." *Journal of Polymers and the Environment*, Vol. 18, p. 1, 2010.
3. I. Banik and M. Sain, "Water Blown Soy Polyol-Based Polyurethane Foams of Different Rigidities." *Journal of Reinforced Plastics and Composites*, Vol. 27, p. 357, 2007.
4. I. Banik and M. Sain , "Role of Refined Paper Fiber on Structure of Water Blown Soy Polyol Based Polyurethane Foams." *Journal of Reinforced Plastics and Composites*, Vol. 27, p. 1515, 2008.
5. L. Chang, Y. Xue, and F. Hsieh, "Comparative Study of Physical Properties of Water-Blown Rigid Polyurethane Foams Extended with Commercial Soy Flours." *Journal of Applied Polymer Science*, Vol. 80, p. 10, 2001.
6. L. Chang, Y. Xue, and F. Hsieh, "Dynamic-Mechanical Study of Water-Blown Rigid Polyurethane Foams with and without Soy Flour." *Journal of Applied Polymer Science*, Vol. 81, p. 2027, 2001.
7. M.A. Fuqua et al., "Development of Flax Fiber/Soy-Based Polyurethane Composites for Mass Transit Flooring Application." *SAE International Journal of Materials & Manufacturing*, Vol. 3, p. 230, 2010.
8. I. Javni, and Z.S. Petrović, "Effect of Different Isocyanates on The Properties of Soy-Based Polyurethanes." *Journal of Applied Polymer Science*, Vol. 88, p. 2912, 2003.
9. I. Javni and Z.S. Petrović, "Soy-Based Polyurethanes by Nonisocyanate Route." *Journal of Applied Polymer Science*, Vol. 108, p. 3867, 2008.
10. Z.S. Petrović, et al., "Polyurethane Networks from Polyols Obtained by Hydroformylation of Soybean Oil." *Polymer International*, Vol. 57, p. 275, 2008.
11. A. Singh, and M. Bhattacharya, "Viscoelastic Changes and Cell Opening of Reacting Polyurethane Foams from Soy Oil." *Polymer Engineering and Science*, Vol. 44, p. 1977, 2004.
12. B. Song, et al., "Compressive Properties of Epoxidized Soybean Oil/Clay Nanocomposites." *International Journal of Plasticity*, Vol. 22, p. 1549, 2006.
13. J.P. Latere Dwan'isa et al., "Biobased Polyurethane and its Composite with Glass Fiber." *Journal of Materials Science*,Vol. 39, p. 2081, 2004.
14. K.S. Chian and L.H. Gan, "Development of a Rigid Polyurethane Foam from Palm Oil." *Journal of Applied Polymer Science*, Vol. 68, p. 509, 1998.

15. S. Chuayjuljit, T. Sangpakdee, and O. Saravari, "Processing and Properties of Palm Oil-Based Rigid Polyurethane Foam." *Journal of Metals, Materials and Minerals*, Vol. 17, p. 17, 2007.
16. Z. Yaakob et al., "Oleic Acid-based Polyurethane and its Biocomposites with Oil Palm Trunk Fiber Dust." *Journal of Thermoplastic Composite Materials*, Vol. 23, p. 447, 2010.
17. M. Aranguren, I. Rácz, and N. Marcovich, "Microfoams Based on Castor Oil Polyurethanes and Vegetable Fibers." *Journal of Applied Polymer Science*, Vol. 105, p. 2791, 2007.
18. K. Manjula et al., "Biobased Chain Extended Polyurethane and its Composites with Silk Fiber." *Polymer Engineering and Science*, Vol. 50, p. 851, 2010.
19. Y. Tu et al., "Physical Properties of Water-Blown Rigid Polyurethane Foams from Vegetable Oil-Based Polyols." *Journal of Applied Polymer Science*, Vol. 105, p. 453, 2007.
20. A. Zlatanić et al., "Effect of Structure on Properties of Polyols and Polyurethanes Based on Different Vegetable Oils." *Journal of Polymer Science Part B: Polymer Physics*, Vol. 42, p. 809, 2004.
21. Y. Hu et al., "Rigid Polyurethane Foam Prepared from a Rape Seed Oil Based Polyol." *Journal of Applied Polymer Science*, Vol. 84, p. 591, 2002.
22. I. Bakare et al., "Synthesis And Characterization of Rubber-Seed-Oil-Based Polyurethanes." *Journal of Applied Polymer Science*, Vol. 109, p. 3292, 2008.
23. M.A. Mosiewicki et al., "Polyurethanes from Tung Oil: Polymer Characterization and Composites." *Polymer Engineering and Science*, Vol. 489, p. 685, 2009.
24. U. Casado et al., "High-Strength Composites Based on Tung Oil Polyurethane And Wood Flour: Effect of the Filler Concentration on the Mechanical Properties." *Polymer Engineering and Science*, Vol. 49, p. 713, 2009.
25. M. Dumont, X. Kong, and S. Narine, "Polyurethanes from Benzene Polyols Synthesized from Vegetable Oils: Dependence of Physical Properties on Structure." *Journal of Applied Polymer Science*, Vol. 117, p. 3196, 2010.
26. Z.S. Petrovic et al., "Structure and Properties of Polyurethanes Based on Halogenated and Nonhalogenated Soy-Polyols." *Journal of Polymer Science Part A: Polymer Chemistry*, Vol. 38, p. 4062, 2000.
27. M.A. Mosiewicki et al., "Vegetable Oil Based-Polymers Reinforced with Wood Flour." *Molecular Crystals and Liquid Crystals*, Vol. 484, p.143, 2008.
28. S.T. Lee, C.B. Park, and N.S. Ramesh, *Polymeric Foams : Science and Technology*, Boca Raton, CRC/Taylor & Francis, 2007.
29. L. Gibson and M. Ashby, *Cellular Solids: Structure and Properties*, New York, Cambridge University Press, 1999.
30. General Plastics Manufacturing Company, *www.generalplastics.com*, 2010.
31. S. Wu, et al., "Plant Oil-Based Biofoam Composites with Balanced Performance." *Polymer International*, Vol. 58, p. 403, 2009.
32. G. Oertel, *Polyurethane Handbook*, Munich, Carl Hanser, GmbH & Co., 1993.
33. Structural Insulated Panel Association, *www.sips.org*, 2010.
34. U.K. Vaidya et al., "Design And Manufacture of Woven Reinforced Glass/Polypropylene Composites for Mass Transit Floor Structure." *Journal of Composite Materials*, Vol. 38, p. 1949, 2004.

10

Sustainable Biocomposites Based for Construction Applications

Hazizan Md Akil and Adlan Akram Mohamad Mazuki

School of Materials & Mineral Resources Engineering, Universiti Sains, Malaysia

10.1 Introduction

10.1.1 Polymer Matrix Composites (PMC's)

An increase in the environmental awareness has led the scientist to produce biodegradable composites known as bio-composites. Recently, a large portion of composites industries are mainly producing Polymer Matrix Composites (PMC's) with the vast majority of the matrix polymers used are synthetic and non biodegradable. The most significant advantages of using polymer are, ease of processing, high productivity and low cost, in combination with their versatility [1]. However, the usage of synthetic polymer as a matrix in PMC poses several environmental concerns particularly when land filling is the mean of disposal. These synthetic plastic materials are non-degradable in nature resulting in continuous accumulation in the environment causing severe pollution. Degradation of plastic requires a long time and most of them end up over burdening on landfill [2]. Therefore, the productions of PMC with natural and biodegradable polymers are of primary concern recently [3–5]. Environmentally friendly composites are today keenly required by utilizing natural fibers as reinforcements combined with biodegradable polymer as matrices. Depending on the natural fiber used, the profiles exhibited specific properties equivalent to the properties of glass fiber reinforced composites. This makes the natural fibers such as sisal, coir, jute, ramie, pineapple leaf (PALF), and kenaf are appropriate alternative candidates to replace glass or other traditional reinforcement materials in composites [5]. The natural fiber can be applied for a broad area of applications such as in construction, electrical and industry. The adoption of natural fiber composites in this industry is lead by motives such as availability in large amounts, renewable, biodegradable, low cost, low density, less equipment abrasion and less skin and respiratory irritation. The natural fiber also has good, high specific properties such as stiffness, impact resistance and flexibility modulus. Considering the ability of natural fibre reinforced polymer composites to withstand the above mentioned environments, this chapter will be concentrating on looking at the potential of kenaf fibre reinforced

Srikanth Pilla (ed.) Handbook of Bioplastics and Biocomposites Engineering Applications, (285–316)

polymer composite produced via pultrusion method to replace either steel-based or synthetic components in these applications.

10.2 Problem Statement

10.2.1 Minimum Environmental Impact

Manufacturing high performance composites from natural fibers resources is one of an ambitious goal currently being pursued by researchers across the globe. The ecological benefits of this material are clearly saved valuable resources which are environmentally sound and do not cause health problems [6–9]. The natural fiber is appropriate alternative candidates to replace the synthetic and other reinforcement because of their renewable resources properties for long term solution to the problem.

Reconstitution of natural fibers such as wood, stone/ceramic, minerals, chitin and collagen involves processing to enable the materials to be converted into the desired shape, thus allowing the molecular structure to form. In the case of wood, the structure can be reconstituted as particleboard, plywood or wooden flour [10]. Stone can be reconstituted as a ceramic or as a composite with concrete. Cellulose can be derivatized and/or dissolved directly to form new fibers and films. Cellulosic fibers are the most prevalent components of natural composites and they have been used in many semi-synthetic composites [5]. Materials generally must be melted or dissolved in order to mould them into the desired shaped and product. Waste sugar cane bagasse can be formed into moderately strong panels with good insulation properties by binding with synthetic polymer [11].

Therefore, the intention in this project is to utilise the locally available kenaf fiber which is much cheaper than synthetic fibers such as glass, aramid and carbon fibers in an attempt to reduce the cost of producing composites and hence become more competitive than conventional materials such as galvanized steel and etc. As motivation, National Kenaf Research and Development Program have been formed in the effort to develop kenaf plants as a possible new industrial crop for Malaysia. The government has allocated RM12 million for research and further development of the kenaf –based industry under the 9th Malaysia Plan (2006-2010) in recognition of kenaf as a commercially viable crop.

10.2.2 Water and Humidity Issues

Most natural materials are hydrophilic. Composites made from natural materials will be susceptible to changes in strength, chemistry or dimensions with water. Loss of water through excessive heat will cause friability of cellulosic fibers. Degradation will be a continuing and accelerating process. Desirably, composites will have an application period during which no change occurs, followed by onset of degradation after the required lifetime. Since water can act as a plasticiser, absorbed moisture in composites can influence both the dimension stability and the mechanical properties [12]. This problem is the factor that makes natural composites less successful than synthetic ones in many applications.

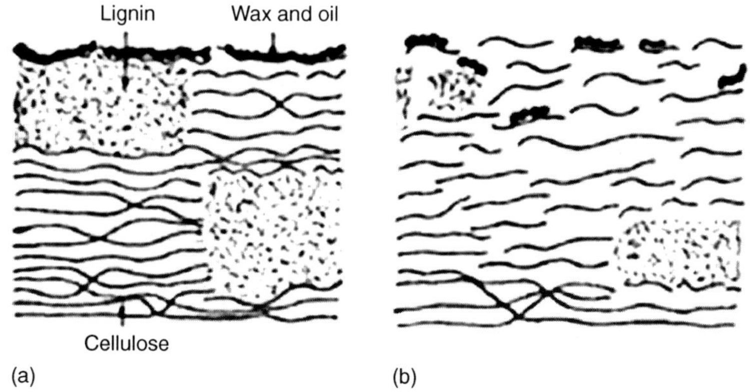

Figure 10.1 Typical structure of a) untreated and b) alkalized cellulosic fiber [14].

In fact, surface modification of natural fibers can potentially influence moisture uptake from the environment [13]. Improved interfacial adhesion can potentially be achieved either by better wetting during processing or by chemically bonding between fiber and matrix. Natural fiber surfaces are rich in hydroxyl groups and therefore have poor compatibility with hydrophobic polymer matrices. However, the presence of hydroxyl groups also ensures a reactive fiber surface that is highly suitable to the chemical modification. Scientifically, natural fibers are chemically treated to remove lignin, pectin, waxy substances, and natural oils covering the external surface of the fiber cell wall (Figure 10.1[a]). This reveals the fibrils, and gives a rough surface topography to the fiber (Figure 10.1[b]).

Sodium hydroxide (NaOH) is the most commonly used chemical for bleaching and/or cleaning the surface of plant fibers. It also changes the fine structure of the native cellulose I to cellulose II by a process known as alkalization [15–17]. The reaction of sodium hydroxide with cellulose is show in eq. (10.1).

$$\text{Cell -OH} + \text{NaOH} = \text{Cell} - \text{O}^- \text{NA}^+ \text{ - } \text{H}_2\text{O} + [\text{surface impurities}] \qquad (10.1)$$

It is worth pointing out that alkalization depolymerises the native cellulose I molecular structure producing short length crystallites (Figure 1[b]).

10.2.3 Processing of Fiber Reinforced Polymer Composites (FRP)

The rapidly expanding usage of composite components in automotive, construction, sports and leisure and other mass production industries has focused attention on continuous production techniques with the optimum properties. One of the techniques for the manufacturing of structural profiles from composites on a continuous basis definitely is pultrusion. Pultruded composites are traditionally manufactured using thermosetting resin systems. These profiles are produced by pulling a carefully specified mass of wetted-out reinforcement material through a heated metal die containing a cavity of the desired cross-section [19].

Pultrusion method is chosen among many other methods of producing fiber reinforced polymer composites due to the main advantage such as the ability to produce continuous long constant cross section profile with various shapes by only changing the die design, which is not possible with other production method. Besides that capability, it also offers an automization process with limited number of labour for operating at tremendous production rate and often quoted as the most cost-effective method of producing fibre reinforced polymer composites. Pultuded fiber reinforced polymer composites have also proven to be a better choice of composite under the corrosive environment due to its durability and capability as compared to the hand layup and spray layout composites. With long and continuous profile and percentages of fiber loading which is greater than 60% in composites, it is the most suitable composite to be used in the applications such as in building and construction industries.

Currently, the world pultrusion technology is still in the infancy state. So far, most of kenaf is utilized for animal feeding. Elsewhere, this technology has been utilized to the extent that some of the bridges and architect monuments in USA and European countries are built partly using pultruded composites. This chapter should bring along the expansion of pultrusion technology in producing structural composites for engineering application.

10.3 Case study: Fabrication, Characterization and Properties of Pultruded Kenaf Reinforced Composites

In order to demonstrate the various factors describe before, which influence the composites properties, kenaf bast fibers composites were manufactured and analysed properly in terms of the morphological, mechanical and thermal properties of composites.

10.3.1 Raw Materials

Kenaf raw fibers used in this work are supplied by locally supplied by Lembaga Tembakau Negara (LTN) Malaysia) and came in straight long fibers. The fibers have been separated from their stalks by water retting for about 20 days in LTN. After the water retting process is completed, the fibers were then cleaned with water and dried under the sunlight before they were delivered to us. Kenaf fiber was further processed into yarn by Institute of Natural Fibers, Poznan, Poland with Tex unit 2200 g/km. Unsaturated polyester resin (Reversol P-9941) for pultrusion grade was obtained from Revertex (Malaysia) Sdn. Bhd.

10.3.2 Fiber Chemical Treatment

Kenaf fibers were immersed in NaOH solution with different concentrations (3%, 6% and 9% of NaOH) for 48 h at room temperature. After treatment, the fibers were thoroughly washed with running water and allowed to dry at room temperature before being placed in an oven for 5 hours at 100°C.

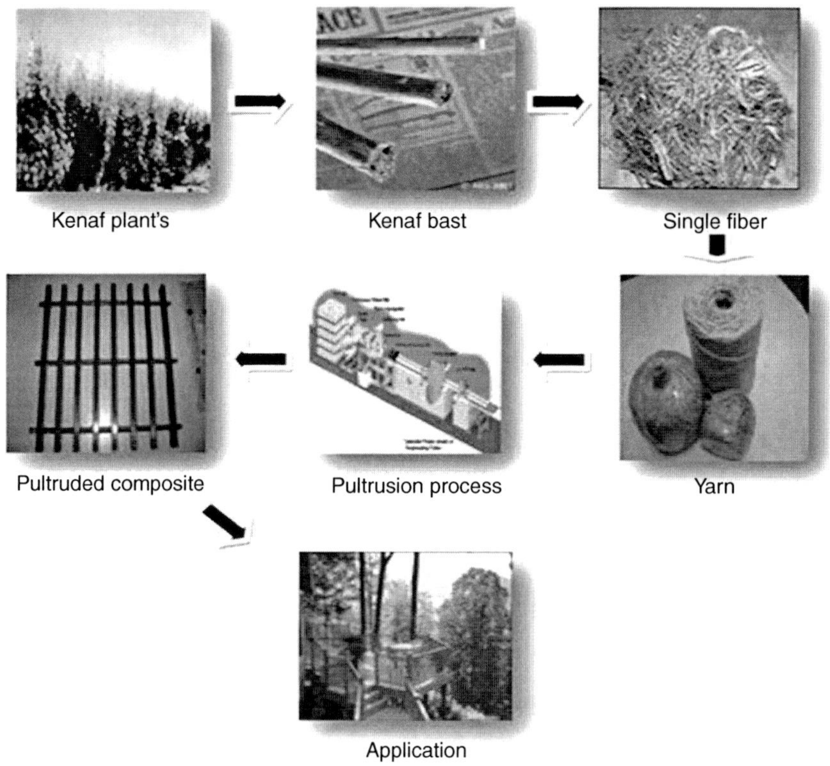

Figure 10.2 Process/product of pultruded kenaf reinforced composites.

10.3.3 Preparation of Pultruded Composites

The composites were prepared using a SVS Pultrusion machine at School of Materials and Mineral Resources Engineering, University Sains Malaysia, Malaysia. During manufacturing of composites, the optimizing processing parameter was recorded. The pulling speed used was 180 mm/min with temperature of the die, 90°C. Pultruded composites were prepared an average diameter of all composite rods is 12.7 mm of kenaf fiber composites with fiber and matrix respectively with the (50, 60, 65, 70 and 75)%.v/v. The composites were produced and classified as untreated pultruded kenaf reinforced composites (UTPKRC) and treated pultruded kenaf reinforced composites (TPKRC). The detailed process of composites is given in Figure 10.2.

10.3.4 Testings

10.3.4.1 Fiber Bundle Tensile Test

Fiber bundle tensile strength tests were performed using a computer controlled Instron machine with a gauge length of 40 mm and a crosshead speed of 5 mm/min. For every set of chemical treatment, 5 specimens were tested to determine the average fiber bundle strength. The tests were conducted at a standard laboratory atmosphere of 23 °C and 50% relative humidity. The maximum breaking load was determined directly from the stress–strain curve and the unit break (UB) is calculated as follows [13]:

$$UB = F/d \tag{10.2}$$

Where
F = Maximum breaking load (N)
d = Cross-sectional area of the fiber (mm2)

10.3.4.2 Flexural Testing

Flexural test was carried out using Instron 8802 according to the standard ASTM D4476–03. Specimens (pultruded rod with diameter of 12.7 mm) were cut into two parts with the cross section of each part is smaller than a half-round section. The total specimen length is 125 mm with overhang length of 12.5 mm at both supports. The crosshead speed for flexural test was set at 5 mm/min. Three specimens for each condition were used to obtain a satisfactory result. The testing was performed to study the effect of chemical treatment to the mechanical properties of composites in different percentages of fiber loading, % v/v.

10.3.4.3 Dynamic Mechanical Analysis (DMA)

The dynamic mechanical properties of the resin and composites were measured using a Mettler Toledo Model DMA 861 under the flexural mode of testing. The dimensions of the specimens were cut to 50 m×12mm×3 mm. The heating rate was set at 2 °C per min.

10.3.4.4 Degradation Test

The water absorption study was performed in accordance to ASTM D 570-98. Specimens were cut similar to flexural test dimension. The percentage of water absorption in the composites was calculated by weight difference between samples immersed in water and dry samples using the Equation 10.3.

$$M_t\,(\%) = \frac{w_t - w_0}{w_t} \times 100 \tag{10.3}$$

Where

M_t (%) is the moisture content in percentage;
W_t is the weight of the wet sample at the time;
W_o is the initial weight of the sample.

Diffusion coefficient, D is calculated from the slope of moisture content versus the square root of time using the Equation 10.4.

$$D = \pi\left(\frac{h}{4M_m}\right)^2\left(\frac{M_2 - M_1}{\sqrt{t_2} - \sqrt{t_1}}\right)^2 \tag{10.4}$$

Assuming the absorption process is linear at an early stage of immersion, time is taken at the beginning of absorption process, so that the weight change is expected to vary linearly with the square root of time. The permeability of water molecules through the composite sample depends on the sorption of water by the fibre. The permeability coefficient (P) which implies the net effect of sorption and diffusion is given by the relation [20–21].

$$P = D\,x\,M_t \tag{10.5}$$

10.3.4.5 Scanning Electron Microscopy (SEM)

Scanning electron microscopy (SEM) Leo Supra 35VP was used to identify the tensile fracture morphology of the composite samples. The samples surfaces were sputter coated with gold to avoid charging.

10.4 Result and Discussions

10.4.1 Single Kenaf Fiber

10.4.1.1 Morphological Study of Kenaf Fiber

Scanning electron microscopy (SEM) provides an excellent technique for examining the surface morphology of untreated and treated kenaf fibers at different concentration of NaOH (3% M, 6% M and 9% M). Studies of the fiber surface topography could provide vital information on the level of interfacial adhesion that would exist between the fiber and the matrix later when they are used as reinforcement fiber with and without treatment. All micrographs in this work are taken with 4000 times (4.00 KX) magnification.

The SEM micrograph of untreated fiber (3a) showed the presence of waxy substances on the untreated fiber surface. According to Mohanty *et al.*, 2003 [22], such waxy substance contributed to ineffective fiber–matrix bonding and poor surface wet out. On the other hand, Figure 10.3(b) shows similar fiber after 3%NaOH treatment. In both figures, there are still a lot of impurities that remain on the fiber surface. It indicates that 3% NaOH was not good enough to effectively remove the impurities from kenaf fiber surfaces. Figure 10.3(c) shows the SEM micrograph of 6% NaOH treated kenaf fiber. It can be observed that almost all impurities have been removed from the fiber surface. Figure 10.3(d) shows the absence of impurities on the fiber surface treated with 9% NaOH. As compared to the untreated fiber, the 9% NaOH treated fiber has a cleaner surface but looks jagged and feels rougher when touched. According to Cao *et al.*, 2006 [23], the fibers in the untreated fibers were packed together but split up after the treatments. The phenomenon is called as fibrillation, which breaks the untreated fiber bundle down into smaller ones by dissolution of the hemicellulose. The fibrillation is reported to increase the effective surface area available for contact with the matrix and hence the interfacial was improved [24].

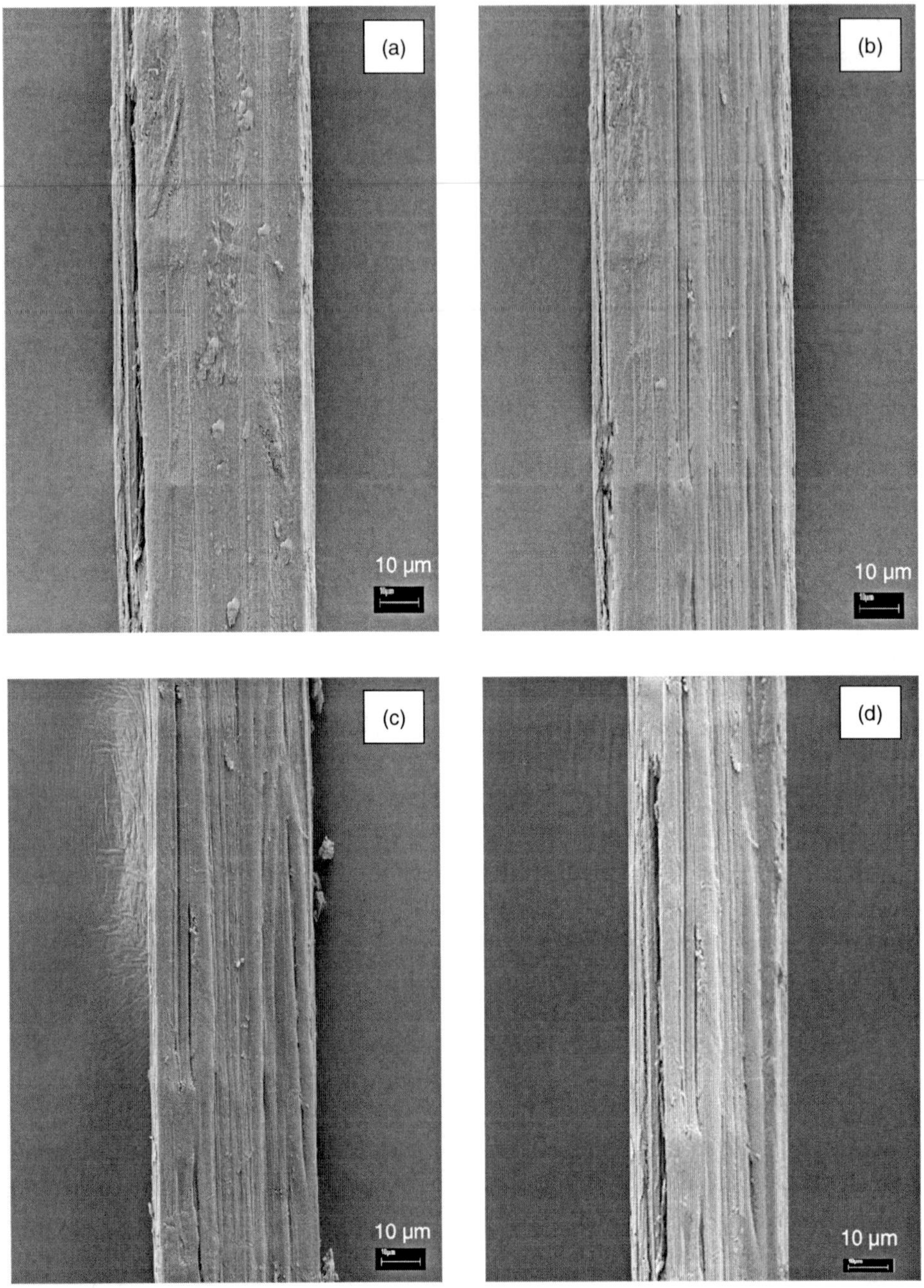

Figure 10.3 SEM micrograph with 4000 times (4.00 KX) magnification of a) an UTPKRC, b) 3% NaOH of TPKRC, c) 6% NaOH of TPKRC and d) 9% NaOH of TPKRC.

10.4.1.2 Fourier Transmission Infrared (FTIR) Analysis

The absorbance peaks of interest in this study have been identified and shown in Figure 10.4. Alkali treatment reduces hydrogen bonding due to removal of the hydroxyl groups by reacting with sodium hydroxide [25]. This result in the increase

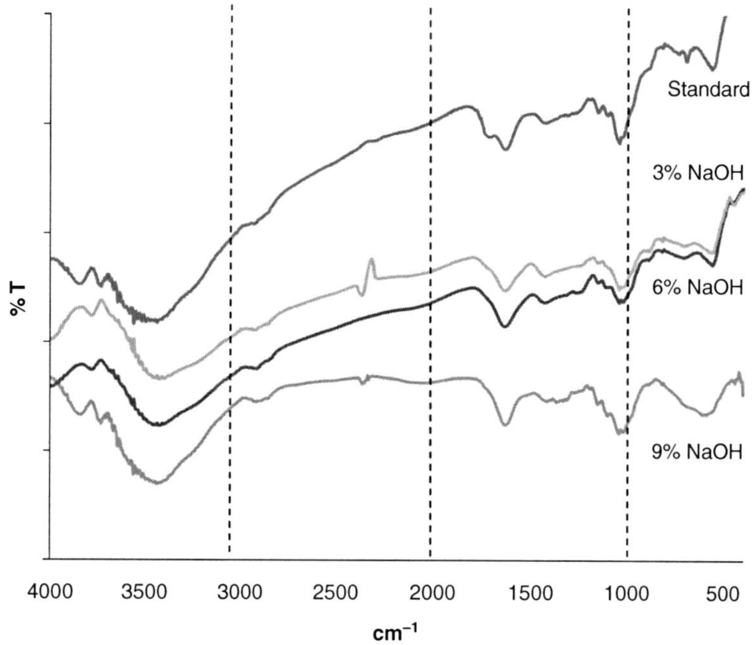

Figure 10.4 Infrared spectra of kenaf fiber immersed in various concentration of NaOH solutions.

of the -OH concentration which is evident with the increase in intensity of the peak between 1000 and 1500 cm^{-1} bands compared to the untreated fiber. Absorbance between this ranges are indicative of the hemicelluloses. The hydroxyl groups are also involved in hydrogen bonding with the carboxyl groups, perhaps of the fatty acids, available on the fiber surface of natural fibers. This is indicated by the reduction of the peaks between 3200 cm^{-1} to 3600 cm^{-1}.

The peak between 1736 and 1740 cm $^{-1}$seen in untreated fibers disappears upon treated by alkali. This is due to the removal of the carboxylic group by alkali treatment by a process called deesterification [25]. The carboxylic group may also be present in the fiber as traces of fatty acids present in oils. The FTIR spectra of untreated fiber indicates that it contains more fatty acids than the treated fibers studied due to the intensity of the 1639 cm^{-1} peak (C=C stretching) followed by 3%, 9%, and 6% NaoH treated fibers. The observed peak at 1437 cm^{-1} and between 1245–1259 cm^{-1} indicate the presence of lignin and hemicellulose, respectively [14,25]. The peak at 1437 cm^{-1} shows diminishing intensity as the fibers is subjected to higher concentration of NaOH.

The disappearance of the peak between 1245–1259 cm^{-1} after alkalization indicates the complete removal of hemicellulose materials rather than lignin. This implies that hemicelluloses are easily removed by alkalization but not lignin. The peak observed at range 895–898 cm^{-1} indicates the presence of the β-glycosidic linkages between the monosaccharides. The COOH bending peak is observed between ranges 560 cm^{-1} to 668 cm^{-1}. From these results it is clear that several reactions take

Table 10.1 Summarized of infrared transmittance peaks (cm^{-1}) kenaf fiber immersed in different solutions.

Bond Type	Treatment			
	Standard (cm^{-1})	3% NaOH (cm^{-1})	6% NaOH (cm^{-1})	9% NaOH (cm^{-1})
-OH stretching	3448.23	3435.79	3452.16	3468.36
C-H vibration	2929.43	2927.21	2926.71	2926.12
C=O stretching	1737.2	–	–	–
C=C stretching	1639.12	1643.3	1642.82	1643.39
C-H bending	1437.34	1437.04	1437.09	1437.09
C-H bending	1259.9	1257.62	–	–
C-C stretching	1000–1162	1000–1162	1053.11	1000–1162
C-H stretching	898.1	897.2	895.7	896.8
-OH	592.55	564.10	563.91	568.55

place during alkalization. Table 10.1 summarized the FTIR spectra for the alkaline treatment on kenaf fiber at different concentration of NaOH.

10.4.1.3 Fiber Bundle Tensile Test

Fiber bundle tensile test of untreated, 3%, 6% and 9% NaOH treated kenaf fiber bundles has been measured and the results was shown in Figure 10.5. Five specimens were tested using the Instron machine, and their break unit was calculated using Equation 10.2. From the figure, it is observed that the treatment has improved the tensile properties of the fibers. The average unit break of the bundle of 3% NaOH treated kenaf fibers is higher than the untreated kenaf fiber bundle. When the NaOH concentration is increased to 6%, a further increase of the average unit break is noticed. Kenaf fibers treated with 6% NaOH at high temperature show the highest average unit break over all. This is explained by the increase of uniformity that contributes to the increase in strength, due to the removal of the impurities. However, when NaOH concentration further increased up to 9%, the fiber bundle tensile strength was suddenly decreased. The value recorded was even lower than that of the untreated fibers. The observation was similarly shown by Mwaikambo *et al.*, 2002 [14] in their experiment on treated hemp, jute, sisal and kapok fibers with various concentrations of NaOH and found that 6% was the optimum concentration in terms of cleaning the fiber bundle surfaces and retaining a high index of crystalline.

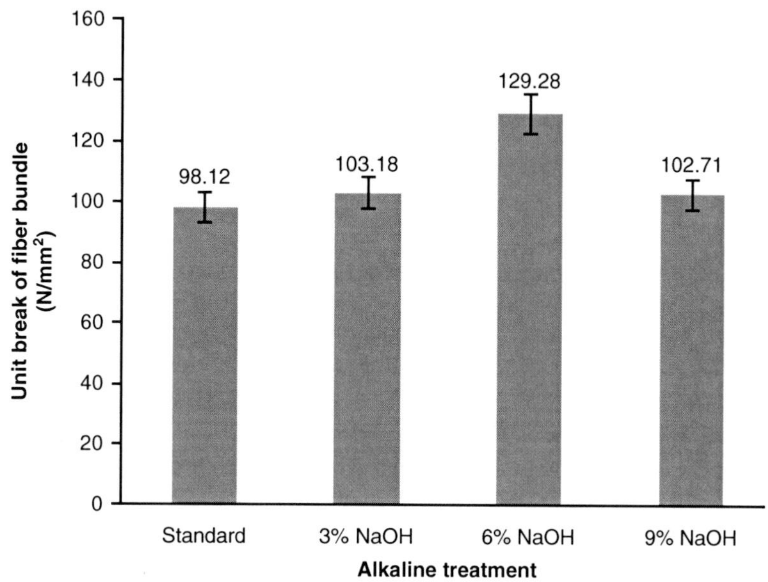

Figure 10.5 Average unit break of kenaf fiber bundles.

10.4.2 Pultruded Composites

10.4.2.1 Apparent Density of Composite and Void Content

Figure 10.6 shows the density of composite at different content of treated and untreated fibre. It was found that composites produced with treated fiber have higher values of density compared with composites produced with untreated fibre. A negative change would signify cell wall damage leading to de-polymerization of the cellulose molecule. Higher concentrations of NaOH are likely to damage the cell wall and reduced the bulk density. Therefore, results obtained for treated kenaf indicate that 6% NaOH treatment did not cause cell wall damage. The appearance of tougher surface topography indicates a better fibre-matrix interfacial adhesion which resulted in increasing mechanical properties. This may be due to the fact that during harsher alkali treatment, the alkali will penetrate into the yarn more effectively and produced more sides for polymer impregnation.

Table 10.2 shows the effect of varying fiber loading upon composites density compared to the theoretical density calculated from measured density of UP ($1.12\,g/cm^3$) and kenaf fiber ($1.44\,g/cm^3$). It is apparent from Table 10.2 that theoretical and experimental density of PKRC increase with increasing fiber loading. This is anticipated as the density of PKRC is higher than neat UP itself. It is also notable that experimental density values are slightly lower than theoretical density values. This is due to presence of voids in the fiber-matrix interface [26-27]. However, the percentages of void content (%) of TPKRC are lower than UTPKRC. This might be due to efficiency of alkaline treatment in improving fiber-matrix adhesion and fiber dispersion in the composites.

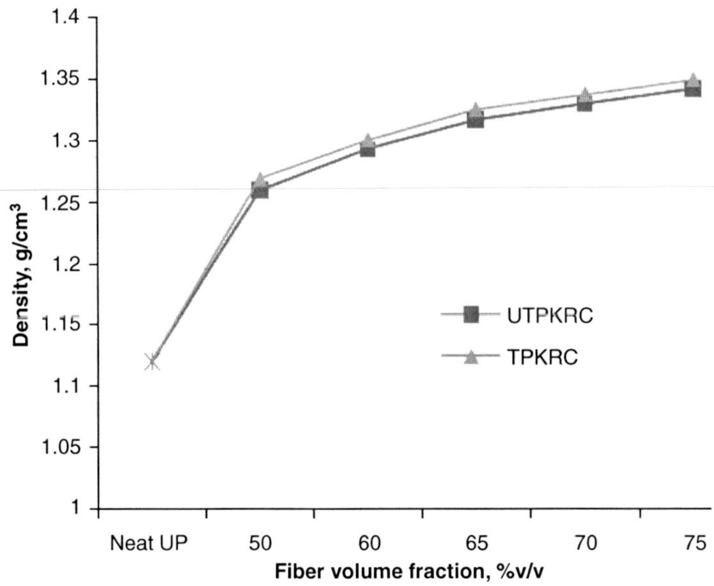

Figure 10.6 Density comparisons of neat UP, UTPKRC, and TPKRC at different fiber (%.v/v).

Table 10.2 Measured densities of UTPKRC and TPKRC at different fiber loading.

Type of Composites	Theoretical Density (g/cm^3)	Actual Density (g/cm^3)	Void Content (%)
Neat UP		1.12	–
50 % UTPKRC	1.28	1.26	1.56
60 % UTPKRC	1.31	1.293	1.29
65 % UTPKRC	1.33	1.317	0.97
70 % UTPKRC	1.34	1.33	0.74
75 % UTPKRC	1.36	1.342	1.32
50 % TPKRC	1.28	1.269	0.86
60 % TPKRC	1.31	1.301	0.68
65 % TPKRC	1.33	1.326	0.3
70 % TPKRC	1.34	1.338	0.14
75 % TPKRC	1.36	1.35	0.73

10.4.2.2 Flexural Test

Figure 10.7 reflects the effect of treated and untreated kenaf content at 50%, 60%, 65%, 70% and 75% of volume fiber on flexural properties of the composites, respectively. The flexural properties of neat unsaturated polyester resin (UP) were used as references. Each value represents an average data of three specimens. It is observed that, all of treated kenaf fiber had improved the flexural properties of the polyester composites. It is expected that NaOH reacts with hydroxyl groups of cementing material hemicelluloses and brings on the destruction of cellular structure and thereby the fiber split into filaments. The untreated fibre bundle breaks down into smaller fibril by the dissolution of the hemicelluloses due to the fibrillation process [28]. The fibrillation increased on surface area of the biocomposites. Thus, the contact area and interfacial bonding between fibre and matrix was improved.

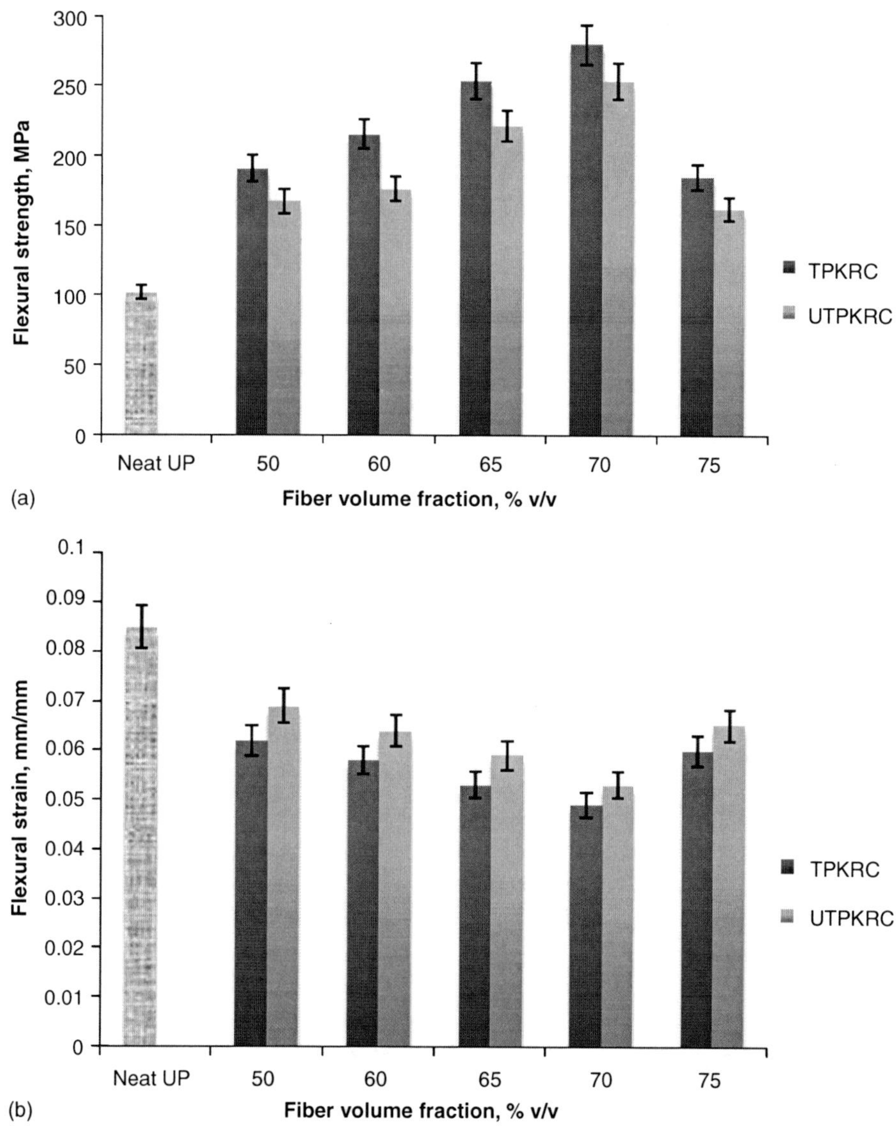

Figure 10.7 Comparison of flexural properties of treated and untreated PKRC with a) flexural strength, b) Flexural strain at break.

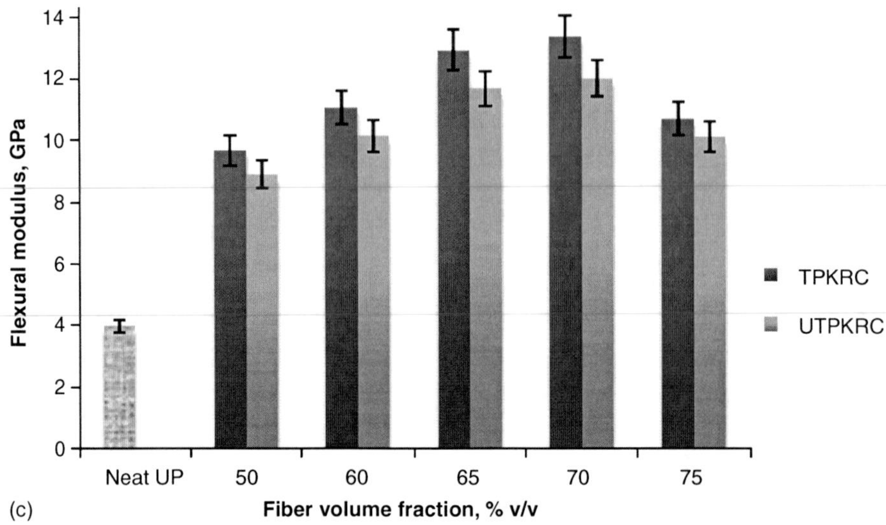

Figure 10.7 (cont.) Comparison of flexural properties of treated and untreated PKRC with c) flexural modulus at various fiber content. Neat UP is used as a reference.

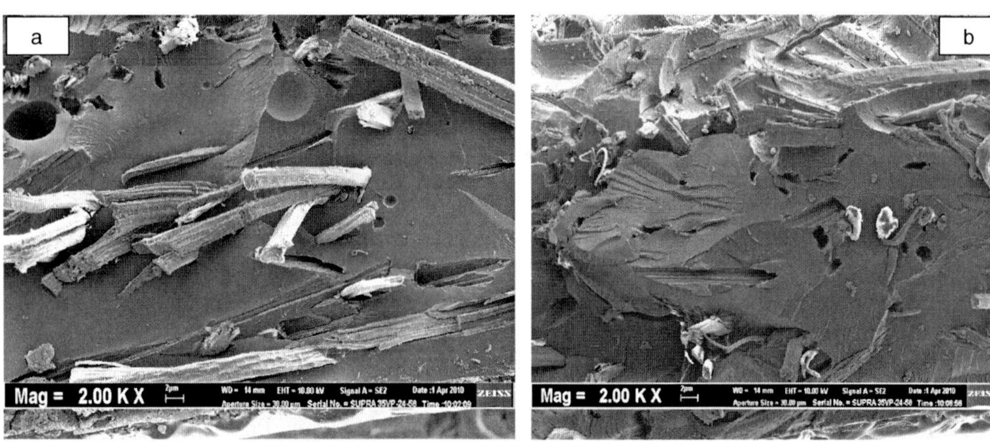

Figure 10.8 SEM photomicrographs of flexural fracture surface of a) untreated fiber based composites; b) NaOH fiber based composites 2.00K magnification.

Mechanical properties of composite increased with the increase of fibre content. The maximum value of flexural properties was exhibited at the fibre content of 70% v/v. Pothana *et al.*, 2003 [29] states that,when the fiber concentration is lower, the packing of the fibers will not be efficient in the composite. This leads to matrix rich regions and thereby easier failure of the bonding at the interfacial region. When there is closer packing of the fibers crack propagation will be prevented by the neighboring fibers. The decreasing of mechanical properties for the composite with the fibre content above 70% is due to the insufficient filling of the matrix resin and it was represented by composites with 75% of fiber volume content.

SEM photomicrographs of (a) untreated fiber based composites; (b) NaOH fiber based composites 2.00K× magnification Figure 10.8 shows the fracture photomicrographs of the specimen subjected to the flexural load. It can be seen in Figure 10.8(a) for untreated kenaf fiber based composites, the phenomenon of pull-out fibers occurred in greater extent than those of treated fiber based composites. It could be observed that in NaOH treated based composites (shown in Figure 10.8[b]), the fibers are still embedded in the resin together with some cavities left by pulled-out fibers, which indirectly indicating better adhesion exists at the interphase.

10.4.2.3 Dynamic Mechanical Analysis (DMA)

10.4.2.3.1 Storage Modulus

Figure 10.9 shows the variation storage modulus (E′) for neat UP, UTPKRC and TPKRC, plotted against temperature. From the DMA curves, the E′ values fall steeply around the glass transition temperature (T_g) of the polyester (91.3 °C). The E′ values were higher for treated fiber based composites compared with those of untreated fiber composite. For example at 40 °C, the E ′value of NaOH treated fibers composites exhibit higher values compare to untreated fiber composites for all cases of fiber volume content. These results indicate that the fiber modified NaOH treatments exhibited better compatibility with the polyester resin than the untreated fibers. This is in accordance with the explanation by previous reseachers where higher E′ value of treated kenaf–polyester composite is due to greater interfacial adhesion and bond strength between resin and fiber [30–32]. From the DMA curves, incorporation of treated fibers imparts stiffness to the composite material, the similar trend was observed in the static flexural modulus shown earlier.

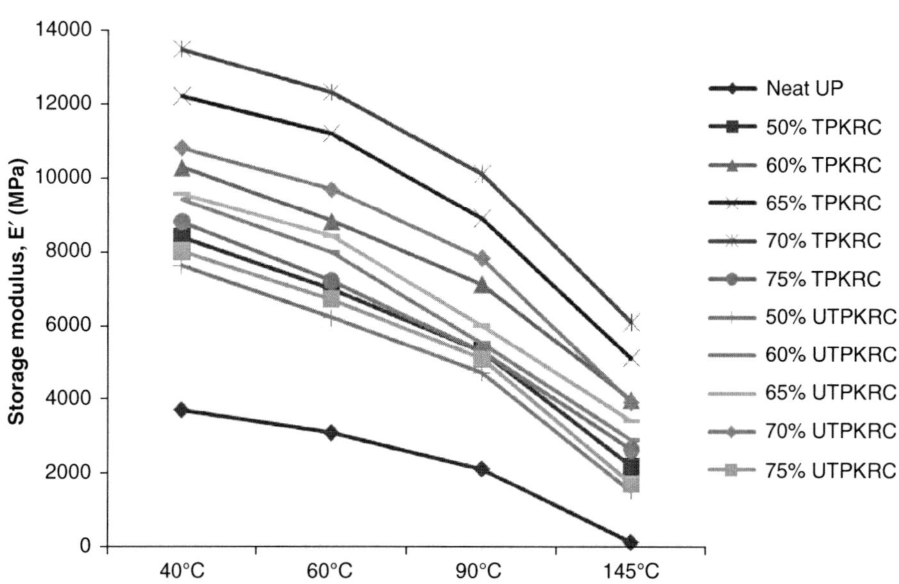

Figure 10.9 Variation storage modulus (E′) for neat UP, UTPKRC and TPKRC, plotted against temperature.

Table 10.3 Calculated constants effectiveness of fillers on the moduli, C of the composites different fiber (%.v/v).

Sample	Coefficient C	
	UTPKRC	TPKRC
50 %	1.312	1.2741
60 %	1.012	0.98
65 %	0.979	0.969
70 %	0.92	0.89
75 %	1.142	1.0331

Variation in modulus occurs due to the effect of the incorporated fibers to the matrix. The increase of E′ in the rubbery plateau is maximum for the composites with 70% of fiber loading (v/v).The effectiveness of fillers (kenaf fibers) on the moduli of the composites can be represented by a coefficient C such as [33]:

$$C=\frac{\left(\dfrac{E'g}{E'r}\right)comp}{\left(\dfrac{E'g}{E'r}\right)resin} \tag{10.6}$$

where E′g and E′r are the storage modulus values in the glassy and rubbery region respectively. The higher value of the constant C shows lower the effectiveness of the filler [29]. The measured E′ values at 45 °C and 140 °C for polyester were employed as E′g and E′r respectively. The values obtained for the different systems at frequency 1 Hz are given in Table 10.3. In this case the lowest value has been obtained for 70% fiber loading and the highest value for 50% fiber loading. Therefore, the effectiveness of the filler is extremely being précised at 70% fiber loading (v/v).

It is important to mention that modulus in the glassy state is determined primarily by the strength of the intermolecular forces and the way the polymer chains is packed. In all the cases, the E′ value is highest to a TPKRC loading of 70%.v/v. This could be attributed to the combination of the hydrodynamic effects of the fibers embedded in a viscoelastic medium and to the mechanical restraint introduced by the higher volume of fibers, which reduce the mobility and deformability of the matrix.

10.4.2.3.2 Damping Energy (Tan δ)

From the graph plotted of tan δ in Figure 10.10(a), it is observed that for the TPKRC, the tan δ peak assigned as the glass transition of polyester, and slightly shifted to higher temperature. According to Nishino *et al.*, 2003 [31] this phenomenon is due to a strong interaction between the treated fiber with the polymer resin, resulting decreases the polymer chain mobility. The similar report by Aziz *et al.* 2005 [32] which indicates that the incorporation of stiff fibers reduces the tan δ peak by restricting the movement of polymer molecule.

Figure 10.10(b) delineates the effect of temperature on tan δ. Improvement in interfacial bonding in composites occurs as lower tan δ values is observed. The high damping at the interfaces, have influenced the interface adhesion. The variation of tan δ with temperature of the composites has been analysed with respect to fiber loading (%.v/v). Incorporation of fibers reduces the tan δ peak height by

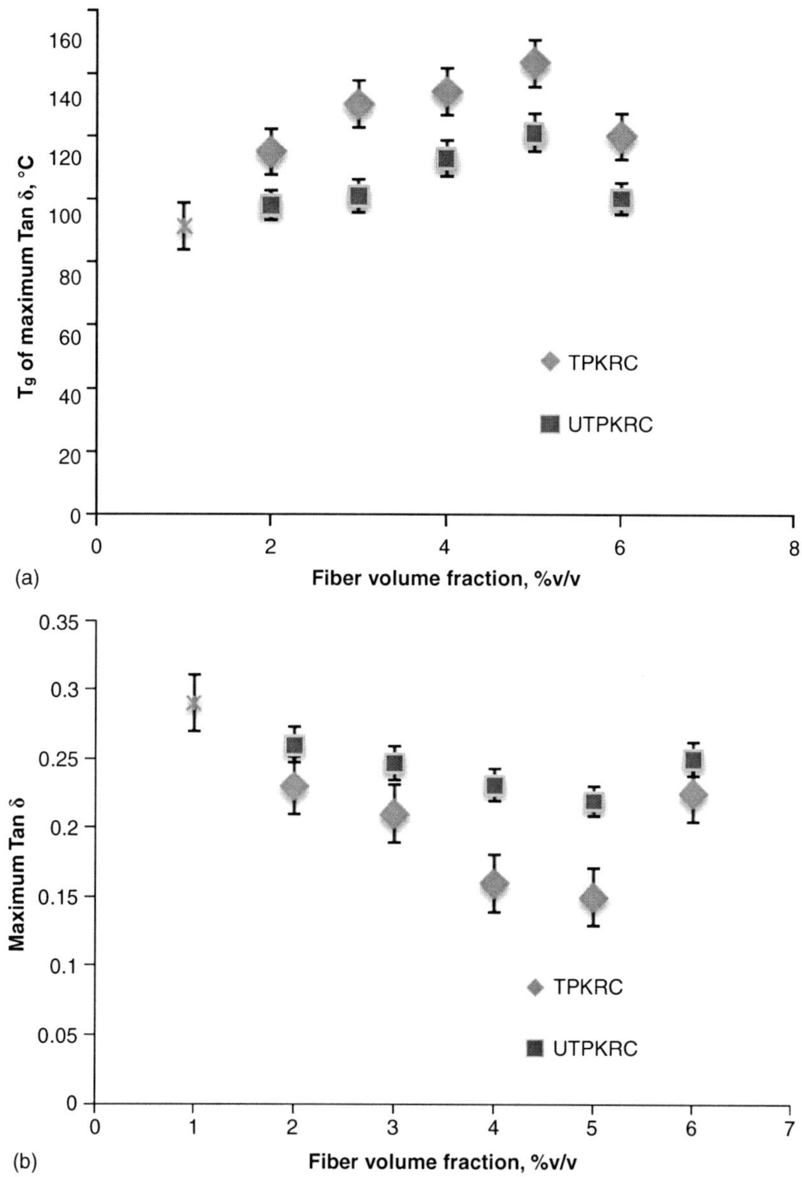

Figure 10.10 a) Effect of temperature on the tan δ value of the composite at different fiber (%.v/v). b) Maximum peak of tan δ value of the composite at different fiber (%.v/v).

restricting the movement of the polymer molecules. Magnitude of the tan δ peak is indicative of the nature of the polymer system. In an unfilled system, the chain segments are free from restraints. Results show that the increase in percentages of fiber loading (%.v/v), the T_g values show a positive shift, stressing the effectiveness of the fiber as a reinforcing agent due to the high polymer- filler interaction. The result is consistent in terms of the E′ values obtained. The introduction of filler reduces the magnitude of the tan δ peak, and thus shifting the temperature. The shifting of T_g to higher temperatures can be associated with the decrease in mobility of the chains due to the addition of fibers. Elevation of T_g is taken as a measure of the interfacial interaction. In addition, the stress field surrounding the particles induces the shift in T_g. It has been reported before that composite with poor interface bonding tends to dissipate more energy than that with good interface bonding [30].

At high fiber loading (%.v/v), when strain is applied to the composite, the strain is controlled mainly by the fiber in such a way that the interface, which is assumed to be the more dissipative component of the composite, is strained to a lesser degree [31]. The width of the tan δ peak also becomes broader than that of the matrix. The behaviour suggests that there are molecular relaxations in the composite that are not present in the pure matrix. The molecular motions at the interfacial region generally contribute to the damping of the material apart from those of the constituents [32]. Hence the width of the tan δ peak is indicative of the increase in volume of the interface.

Table 10.4 shows the peak width at half height of the samples from the damping curve. The peak width is found to be optimum for treated composites with 70% fiber loading. Increase in the concentration of the filler increases the interface [34]. Stress induced motions may also occur in the composite. The hydrogen bonds

Table 10.4 Peak width at half height of the tan δ curves.

Type of Composites	Peak Width (cm)
Neat UP	6.52
50 % UTPKRC	6.87
60 % UTPKRC	6.95
65 % UTPKRC	6.99
70 % UTPKRC	7.22
75 % UTPKRC	6.91
50 % TPKRC	7.12
60 % TPKRC	7.16
65 % TPKRC	7.17
70 % TPKRC	7.39
75 % TPKRC	7.14

between segments may rupture under stress. After such a bond is broken, a small amount of molecular motion can occur which makes it difficult for the same bond to reform. As a result, a new hydrogen bond forms, which initially carries little, if any, stress. The net result is energy dissipation and mechanical damping [29].

10.4.2.3.3 Frequency studies

The storage modulus, E' and damping peaks (tan δ) have been found to be affected by frequency [35]. The variation of E' with frequency of neat polyester as a function of temperature is shown in Figure 10.11(a). Increase of frequency has been found

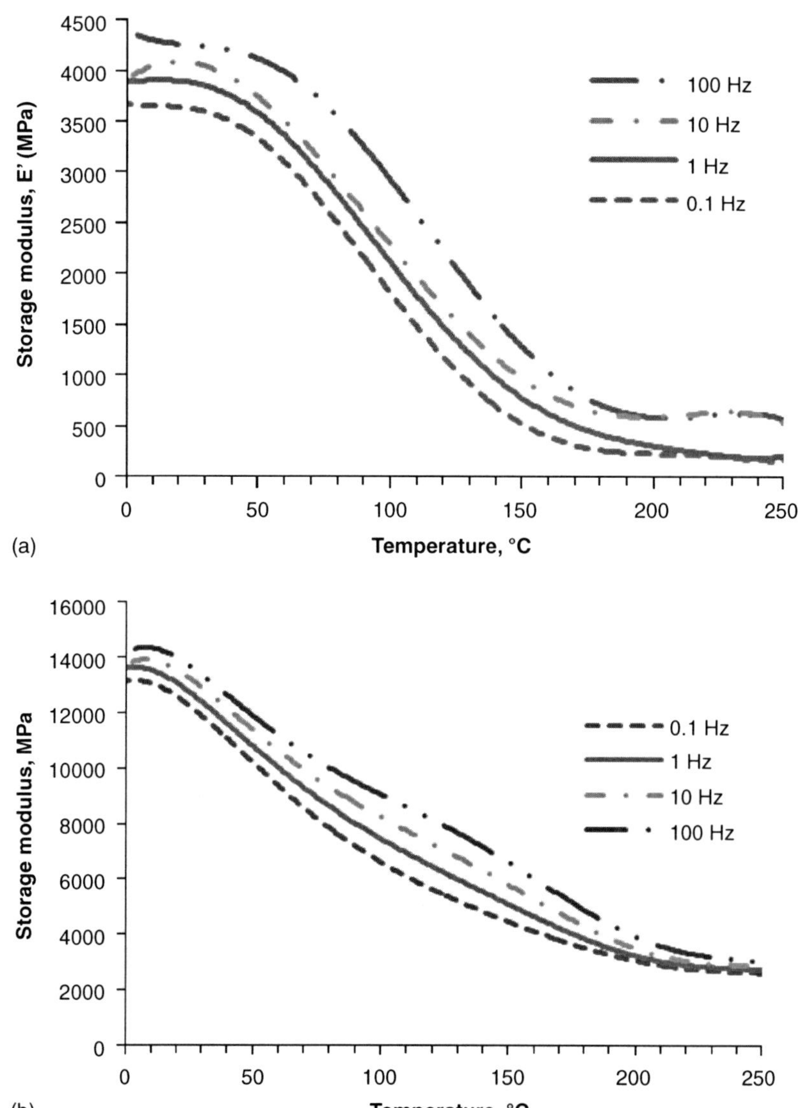

Figure 10.11 a) The variation of storage modulus, E' of neat unsaturated polyester with different frequency (Hz). b) The effect of frequency on the dynamic modulus of samples with 70%.v/v fiber loading.

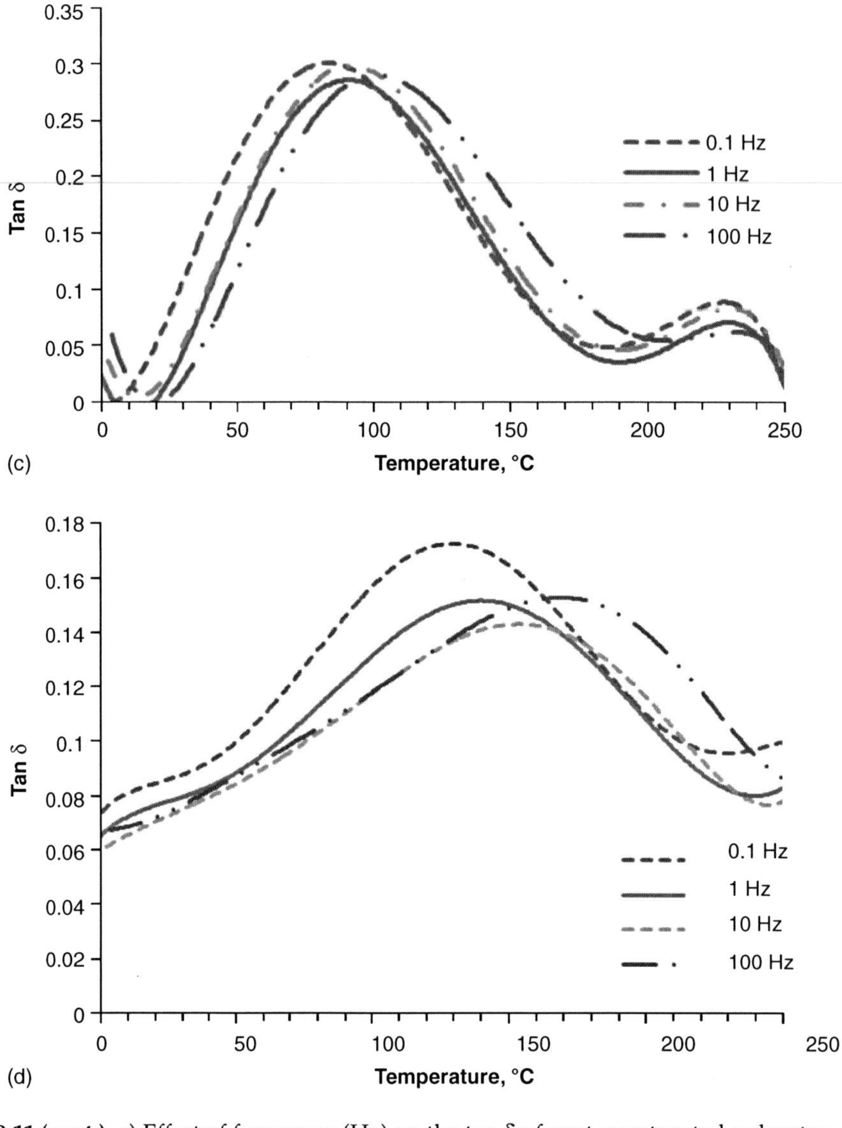

Figure 10.11 (cont.) c) Effect of frequency (Hz) on the tan δ of neat unsaturated polyester. d) Effect of frequency (Hz) on the tan δ of composites with 70%.v/v fiber loading.

to increase the modulus values. Figure 10.11 (b) shows the effect of frequency on the dynamic modulus of samples with 70% fiber loading (v/v). Frequency has a direct impact on the dynamic modulus especially at high temperatures. The modulus values are found to drop at a temperature around 45°C. The drop in modulus value continues steadily until a temperature of 140°C is reached. The molecular motion is believed to be set at 45°C. The change in dynamic properties is also associated with crazing and formation of microscopic cracks and voids. At high temperature breaking up of the fiber agglomerates and breaking up of the bond between the fiber and polymer phases may also occur [29]. Frequency is seen to have a direct impact on the tan δ values as well.

The viscoelastic properties of a material are dependent on temperature, time and frequency. If a material is subjected to a constant stress, its elastic modulus will decrease over a period of time. This is due to the fact that the material undergoes molecular rearrangement in an attempt to minimize the localized stresses. Modulus measurements performed over a short time (high frequency) result in high values whereas measurements taken over a long period of time (low frequency) result in lower values [29]. In this system also, the modulus measurements over a range of frequencies have been studied. Higher values were observed for measurements made over a short time. The tan δ values measured over a range of frequencies for the neat polyester samples are shown in Figure 10.11 (c). The tan δ peak is found to shift to higher temperature with the increase of frequency. The damping peak is associated with the partial loosening of the polymer structure so that groups and small chain segments can move. The tan δ curve peak, which is indicative of the glass transition temperature, is also indicative of the degree of cross-linking of the system.

Figure 10.11(d) shows the effect of frequency on the tan δ curve of samples with 70% loading (v/v) of treated PKRC. Increase of frequency is found to have a broadening effect on the tan δ curve. Broadening of the curve is due to some kind of heterogeneity in the network structure. This broadening is more prominent in composites with high fiber content. Addition of fiber increases the free volume between monomeric units. The introduction of fibers, which in turn affects the curing reaction, will also affect the molecular motions and diffusion.

Table 10.5 shows the tan δ max and the corresponding T_g values for the different composites. The values of T_g obtained positively shift due to plasticization results from the addition of fiber within the polyester matrix. With increase in frequency, the tan δ peak, which corresponds to the glass transition temperature, is also found to be shifted to higher temperature.

Table 10.5 Values of tan δ maximum and T_g values of neat polyester and kenaf fiber reinforced polyester composites at different fiber loading.

TPKRC								
Fibre Loading	**Tan ∂ max**				**T_g from Tan ∂ max (°C)**			
	Frequency (Hz)				**Frequency (Hz)**			
	0.1	**1**	**10**	**100**	**0.1**	**1**	**10**	**100**
UP	0.3	0.29	0.27	0.28	85.2	91.3	99.2	105.7
50%	0.24	0.23	0.21	0.22	110	115	121	127
60%	0.22	0.21	0.17	0.19	125	130	135	140
65%	0.19	0.16	0.15	0.16	130	134	138	143
70%	0.17	0.15	0.14	0.15	140.5	143.3	147	151.4
75%	0.23	0.225	0.22	0.21	114	120	125	130.5

Table 10.5 (cont.) Values of tan Δ maximum and T_g values of neat polyester and kenaf fiber reinforced polyester composites at different fiber loading.

UTPKRC								
Fibre Loading	**Tan ∂max**				**T_g from Tan ∂ max (°C)**			
	Frequency (Hz)				**Frequency (Hz)**			
	0.1	**1**	**10**	**100**	**0.1**	**1**	**10**	**100**
50%	0.28	0.26	0.23	0.22	91.3	98	103.6	109.3
60%	0.25	0.247	0.21	0.21	97.2	101	106.2	112.2
65%	0.24	0.231	0.2	0.19	108.5	113	116.8	121.1
70%	0.23	0.219	0.18	0.16	118	121.1	125.4	128.6
75%	0.26	0.25	0.2	0.22	95	100.4	106.7	112.7

10.4.2.3.4 Activation Energy

The activation energy of the glass transition, ΔH, was obtained by applying the Arrhenius law [35]. In dynamic mechanical experiments, ΔH was estimated by using the time-temperature superposition principle, to superimpose the tan δ peaks determined at different test frequencies [36]. Accordingly, individual tan δ peaks can be shifted for superposition along the logarithmic time axis by the shift factor, $\log_{aT}$. The temperature dependence of the test frequency may then be expressed as:

$$f = f_0 \exp\left(-\Delta H/RT\right) \tag{10.7}$$

where f and f_o are analogous to the rate constant and pre-exponential factor of the Arrhenius equation and R is the gas constant. The shift of the glass transition temperatures, T_{g1} and T_{g2}, due to change in the test frequencies f_1 and f_2 allows the determination of the activation energy of T_g;

$$\Delta \mathrm{H} = -\mathrm{R}\left[\frac{d(\ln f)}{d(1/Tg)}\right] \tag{10.8}$$

Equation (10.8) describes the temperature-dependence of polymer relaxations, where ΔH is the activation enthalpy of the glass transition relaxation, f and T_g are the measuring frequency and the glass transition temperature for the dispersion peak respectively, and R is a gas constant (8.314 × 10.3 kJ $mol^{-1}K^{-1}$). Table 10.2 shows the values of ln$f$ and $(1/T_g)$ for T_g determined from the tan δ peaks. They are plotted as $(1/T_g)$ *vs.* ln$f$ in Figure 10.12(a) and 10.12(b), respectively.

Activation energy of the different composite samples was calculated from the Arrhenius relationships by using linear regression analysis [35]. The activation energy values are given in Tables 10.6 and 10.7. The correlation coefficients,

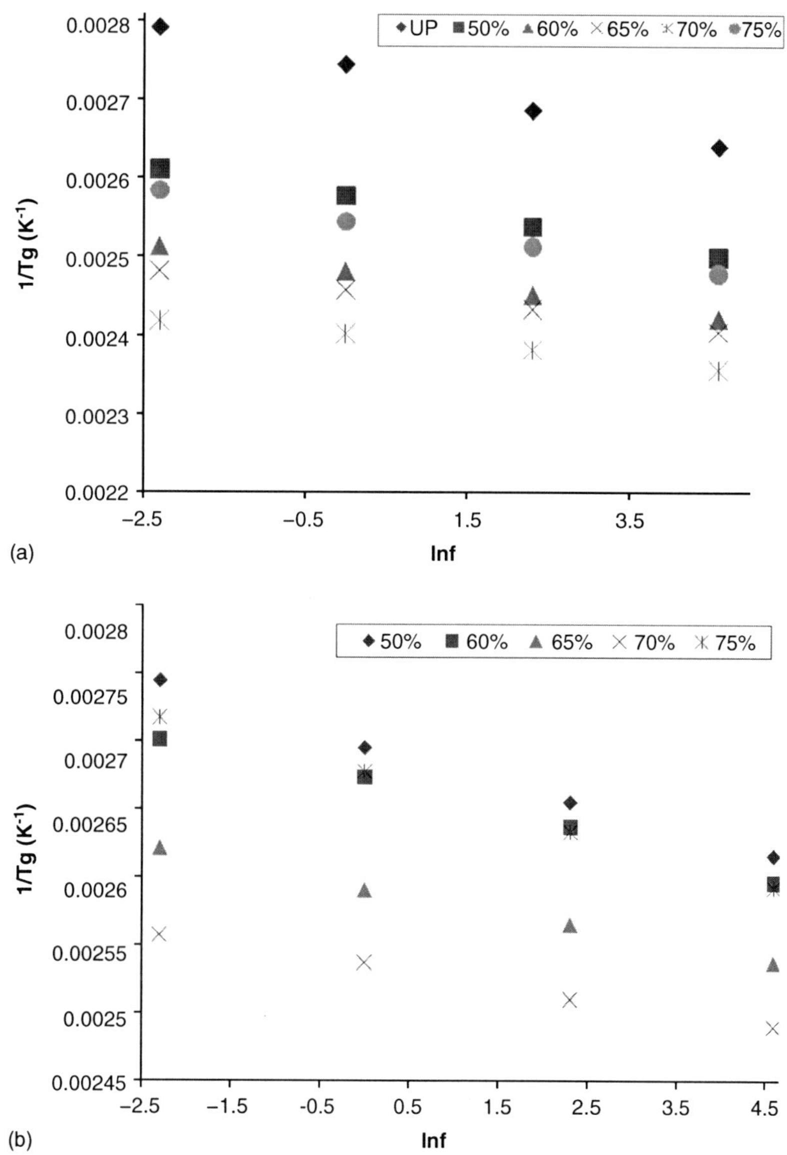

Figure 10.12 Plot of $(1/Tg)$ *vs.* ln$f$ based on tan δ peaks of a) TPKRC and b) UTPKRC at different fiber loading.

(R^2), have also been included. The activation energy values of the composites with70%.v/v of treated PKRC are the maximum, 926.4 kJ/mol. The activation energy values for neat polyester (UP) samples are 380.37 kJ/mol. At low fiber loading, the fiber/matrix adhesion is low and the activation energy is also low. It is interesting to note that the treated composites also give higher activation energy as compared to untreated composites. This might due to the high interfacial interaction and effective stress transfer in composites system. This increases the activation energy value [35].

Table 10.6 Values of ln*f* and (1/Tg) based on Tan δ peaks.

TPKRC							
Frequency, f (Hz)	**ln f**	**Fiber Loading (%)**					
		UP 1/T tan δ	**50% 1/T tan δ**	**60% 1/T tan δ**	**65% 1/T tan δ**	**70% 1/T tan δ**	**75% 1/T tan δ**
0.1	–2.303	0.002792	0.002611	0.002513	0.002481	0.002418	0.002584
1	0	0.002745	0.002577	0.002481	0.002457	0.002402	0.002545
10	2.303	0.002687	0.002538	0.002451	0.002433	0.002381	0.002513
100	4.605	0.002641	0.0025	0.002421	0.002404	0.002356	0.002478
UTPKRC							
Frequency, f (Hz)	**ln f**	**Fiber Loading (%)**					
		UP 1/T tan δ	**50% 1/T tan δ**	**60% 1/T tan δ**	**65% 1/T tan δ**	**70% 1/T tan δ**	**75% 1/T tan δ**
0.1	–2.303	–	0.002745	0.002701	0.002621	0.002558	0.002717
1	0	–	0.002695	0.002674	0.002591	0.002537	0.002678
10	2.303	–	0.002655	0.002637	0.002565	0.00251	0.002634
100	4.605	–	0.002616	0.002596	0.002537	0.00249	0.002593

Table 10.7 Activation energies Δ*H* calculated from t and peaks and correlation coefficient (*R*2)

Fiber loading (%)	**TPKRC**		**UTPKRC**	
	(0.1, 1, 10, 100) Hz		**(0.1, 1, 10, 100) Hz**	
	(ΔH)tan δ/kJ mol^{-1}	**R^2**	**(ΔH)tan δ/kJ mol^{-1}**	**R^2**
Neat UP	380.37	0.9979	–	–
50%	517.44	0.999	445.24	0.9969
60%	624.31	0.9999	547.01	0.9924
65%	745.92	0.9977	683.77	0.9987
70%	926.4	0.9918	844.65	0.996
75%	541.85	0.9953	463.2	0.9995

10.4.2.4 Thermogravimetric Analysis (TGA)

Thermogravimetric curves of neat UP, UTPKRC and T PKRC containing 70%.v/v fiber are shown in Figure 10.13. The temperature range used for the analysis is 30–900 °C. From the graph presents, neat UP almost decomposes at the temperature of 400 °C, while for the untreated composites (UTPKRC), the dehydration as well as degradation of lignin occurs in the temperature range 100–450 °C and most of the cellulose is decomposed at a temperature of 700 °C. Between, it is interesting to note that 6% NaOH treated of PKRC shows the highest thermal stability with degradation start in range 130–510 °C and almost decomposed at 780 °C. It reveals the fact that fiber filled system degrades later than the neat UP matrix, i.e. the thermal stability of the composite is higher than that of the neat matrix and highest shown by treated composites. This increased stability of composite compared to neat UP and untreated composites is due to improved fibre–matrix interaction [37].

Step analysis of neat UP thermogravimetric scan from 30 to 100 °C shows a percentage mass drop of 3.43% whereas UTPKRC with 2.65% and TPKRC show a lowest mass drop with 2.14%. At 200 °C, the mass drop of UTPKRC is about 8.76% and TPKRC with 7.22%, respectively. This may be attributed to the degradation of lignin in the kenaf fibre. At 350 °C the weight loses for neat UP, UTPKRC and TPKRC are 77.7%, 35.12% and 31.2% respectively. However, at a temperature of around 400 °C, neat UP is completely decomposed. Weight loses of neat UP, UTPKRC and PKRC at different temperatures are summarized in Table 10.8.

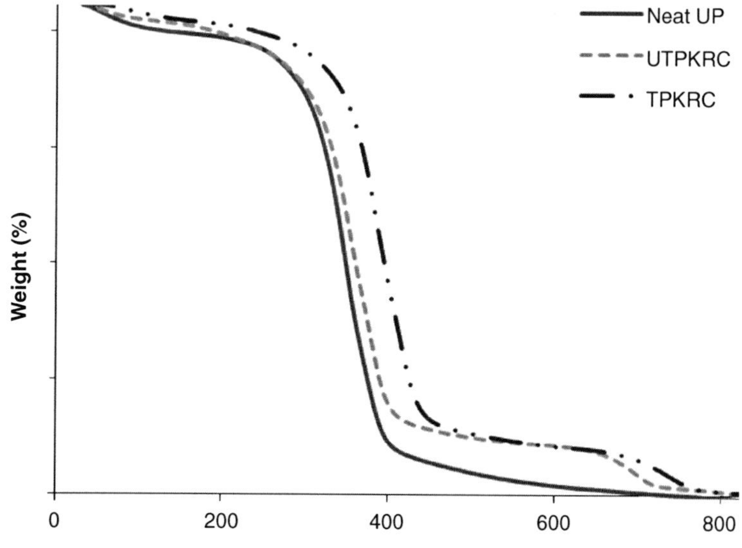

Figure 10.13 Thermogravimetric curves of neat UP, UTPKRC and TPKRC containing 70%.v/v fiber.

Table 10.8 Detailed weights loses of neat UP, UTPKRC and PKRC at different temperatures.

Temperature (°C)	Weight losses (%)		
	Neat UP	UTPKRC	TPKRC
100	3.43	2.65	2.14
200	10.7	8.76	7.62
350	77.7	35.12	31.2
400	99.4	75.56	71.2
700	100	80.43	78.6

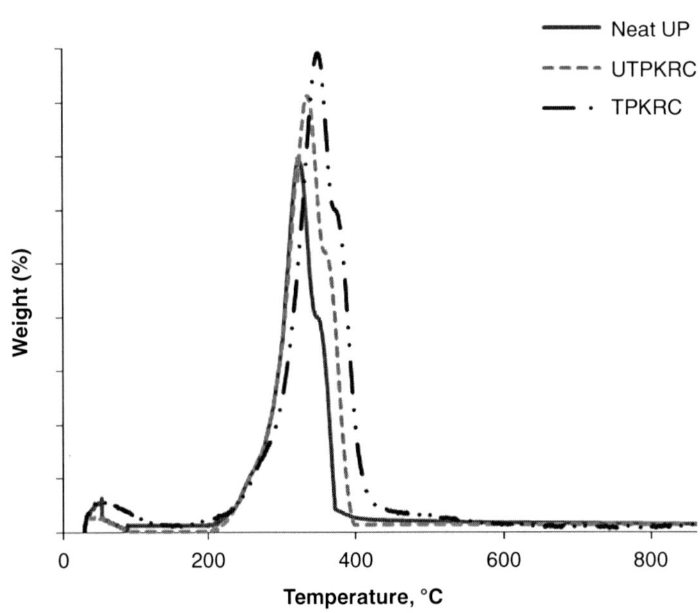

Figure 10.14 DTG curves of neat UP, UTPKRC and PKRC at different temperatures.

DTG curves also give evidence for this pattern (Figure 10.14). The major peak of the DTG curve of neat UP is observed at 325 °C, which indicates the degradation of saturated and unsaturated carbon atoms in polyester. In the case of TPKRC, the peak is shifted to higher temperature region compared to UTPKRC, suggesting that the thermal stability of the composite is higher than those of the untreated fibre and neat UP due to fibre/matrix interactions.

Figure 10.15 summarized the percentages of weight loss (%) versus temperature (°C) for various types of composites. From the figure, the increase of temperature has increased the percentages of weight loss. This is due to thermal stability

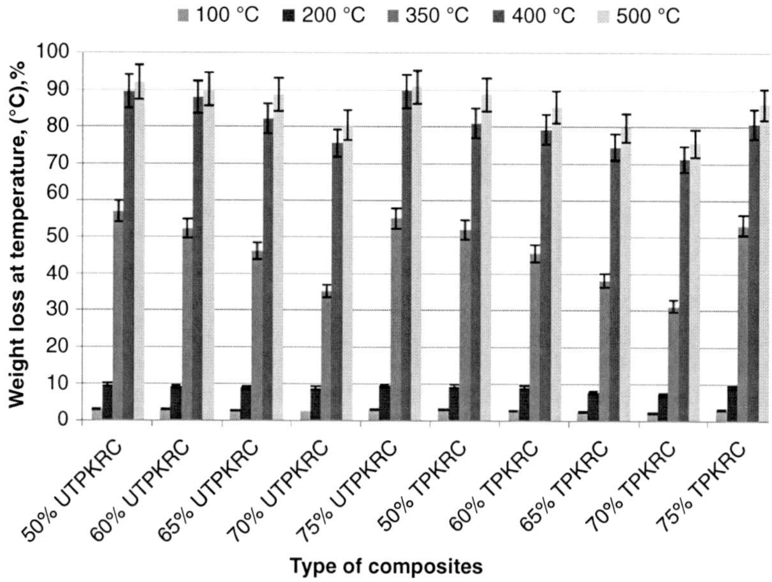

Figure 10.15 Summarized thermogravimetric analyses of UTPKRC and TPKRC at different fiber loading.

Table 10.9 Summarized weights loses analysis of UTPKRC and PKRC at different fiber loading

Type of Composites	Weight Loss at Temperature, (°C), %				
	100	200	350	400	500
50 % UTPKRC	3.11	9.68	56.9	89.6	92.1
60 % UTPKRC	3.02	9.23	52.2	88.1	90.1
65 % UTPKRC	2.79	8.98	46.2	82.2	88.7
70 % UTPKRC	2.65	8.76	35.1	75.6	80.4
75 % UTPKRC	3.15	9.56	55.1	89.7	90.9
50 % TPKRC	2.98	9.43	52.1	81.1	88.8
60 % TPKRC	2.82	9.1	45.7	79.3	85.4
65 % TPKRC	2.39	7.79	38.2	74.5	79.8
70 % TPKRC	2.14	7.22	31.2	71.2	75.6
75 % TPKRC	3.08	9.24	53.3	80.9	86.2

properties of composites againts temperature. However, it is interesting to note that, the increase of fiber loading, decreases the amount of fiber decomposed. This statement proved that high content of fiber will improve the fiber-matrix interaction. Table 10.9 gives evident of quantitive data of Figure 10.15. From the table, it can be deduced that the most stable composite is towards 70% TPKRC as compared to others.

10.4.3 Degradation Test

10.4.3.1 Water Absorption Behavior

Hydrophylicity is an important characteristic of biomaterials [38–40]. One of the parameters used to determined the hydophylicity of materials is via the the water absoprtion testing. Water absorption of the compsoites is an important characteristic that determines the applications in which this materials can be used. Water absorption could lead to a decrease in some of the properties and should be considered when selecting the application of composites.

Therefore, to determine the hydrophilicity of the Pultruded kenaf reinforced composites (PKRC), their water absoprtion abilities were investigated.Water absorption curves for immersed specimens of Neat UP, UTPKRC and TKRC in relation to exposure time in distilled water at room temperature are shown Figure 10.16.

In all cases, the water absorption processes are sharp at the beginning and level off for some length of time as they approach equilibrium. It is considered that the change of weight gain for all samples is a typical Fickian diffusion behavior. The composites show lower water absorption compared to neat UP resin. It is

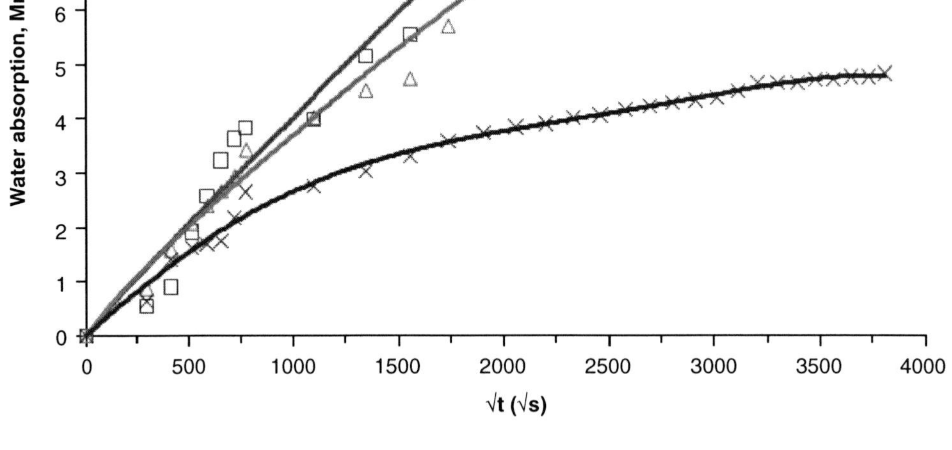

Figure 10.16 Water absorption curve for immersed specimens of Neat UP, TPKRC and UTPKRC for 24 weeks.

Table 10.10 Diffusion coefficient, D and maximum of the moisture content, M_m and permeability coefficient, P of Neat UP, TPKRC and UTPKRC for 24 weeks.

Specimens	Result		
	D (m^2/s)	**M_m (%)**	**P (m^3/s)**
Neat UP	7.558×10^{-13}	4.732	3.576×10^{-14}
UTPKRC	1.261×10^{-12}	9.017	1.093×10^{-13}
TPKRC	9.675×10^{-13}	8.121	7.857×10^{-14}

apparent that the NaOH treatments of kenaf fibers had reduced the water uptake of the composite system. As shown earlier, chemical treatment can reduce the hydroxyl group in the cell wall of the natural fiber molecule, thus decreasing the water absorption of the composites. However, it should be reiterated that water absorption in a fibrous composites is dependent on temperature, fiber loading, orientation, permeability of fiber, surface protection, area of the exposed surfaces, diffusivity, etc [38–39].

According to Das *et al.*, 2000 [41], initially, water saturates the cell wall of the kenaf fiber, and next, water occupies void spaces. Table 10.10 summarizes the diffusion coefficient (D), moisture content at maximum of the moisture content, M_m and permeability coefficient, P of Neat UP, TPKRC and UTPKRC. From the D values of the composite samples, it was found that the NaOH treated fiber polyester composites has better resistance towards water absorption than those of untreated fiber composites. Untreated fiber polyester composites show the highest D values. Higher D value might also indicate higher void content in the system where void generates more pathways for water to start diffusing into the composite. With better adhesion between matrix and fibers, the velocity of the diffusion processes decreases since there are fewer gaps in the interfacial region. According to Edeerozey *et al.*, 2007 [13] chemical treatments of natural fibers can reduce the hydroxyl group in the cell wall of natural fibers, thus decreasing the water absorption of the composites. The M_m values of untreated fiber composites are higher compared to treated fiber composites. Poor adhesion between fiber and matrix can cause an increase value of M_m [20, 38–39].

10.4.3.2 *Morphological Assessment*

Figure 10.17a and 10.17b represents a microstructure of UTPKRC and TPKRC after immersion in distilled solutions for 24 weeks. Clearly, significant damage was observed on the surface of composites due to the degradation of fiber-matrix. In terms of microstructure, the UTPKRC shows severe damage with the fiber/matrix debonding and fiber pull out from composites, while the TPKRC presents a better interface reflected a good fiber-matrix interface.

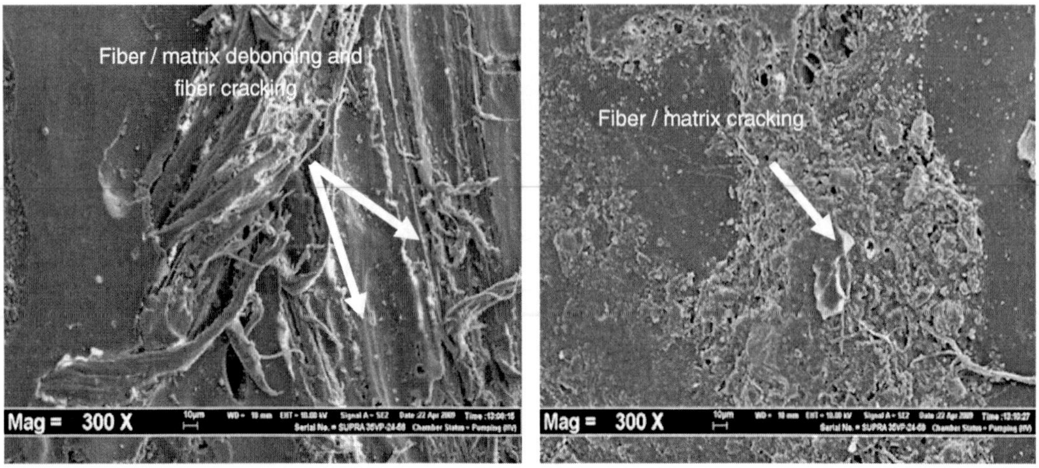

Figure 10.17 SEM of the a) UTPKRC and b) TPKRC immersed in pH 7 solutions.

10.5 Conclusions

The development of high performance composite structures using pultrusion process has been greatly prepared based on locally kenaf fibers. From the results and discussion above, the conclusion of this project can be summarized as follows:

1. Higher flexural properties and dynamic mechanical analysis were obtained for 6% NaOH treated fiber composites compared to those of untreated fiber based composites.
2. Addition of higher amount of fiber loading results in higher flexural properties of the kenaf fiber–reinforced polyester composites, with significant trend shown by 70% of volume fraction
3. Chemical treatments via NaOH decrease the water absorption of composites with good contact between fiber and matrix.

Acknowledgement

The authors wish to thank Universiti Sains Malaysia (USM) for their assistant and supportive grant RU814013, RU 8032027, Ministry of Science, Technology and Innovation (MOSTI) Malaysia, Malaysian Agricultural Research and Development Institute (MARDI), National Kenaf & Tobacco Board (NKTB), Malaysia for their assistances that have resulted in this article.

References

1. Georgopoulos S.T., Tarantili P.A., Avgerinos E., Andreopoulos A.G. and Koukios E.G., "Thermoplastic polymers reinforced with fibrous agricultural residue," *Polymer Degradation and Stability*, Vol. 90 (2005) 303–312.

2. Xu Y.X., Lim K.M., Hanna M.A. and Nag D., "Chitosan-starch Composites Film: Preparation and Characterization," *Industrial Crops and Products*, Vol. 21 (2005) 185–192.
3. Srivinasa P.C., Ramesh M.N., Kumar K.R. and Tharanathan R.N., "Properties and Sorption Studies of Chitosan-Polyvinyl Alcohol Blend Films," *Carbohydrate Polymers*, Vol. 53 (2003) 431–438.
4. Gallstedth M. and Hedenquist M.S., "Packaging-related Mechanical and Barrier Properties of Pulp-Fiber-Chitosan Sheets," *Polymers*, Vol. 63 (2006) 46–53.
5. A.K. Bledzki, J. Gassan. "Composites Reinforced with Cellulose Based Fibres," *Progress in Polymer Science*, Vol. 24 (1999) 221–274.
6. Joseph K., Thomas S., Pavithran C., "Effect of Chemical Treatment on the Tensile Properties of Short Sisal Fiber-Reinforced Polyethylene Composites," *Polymer*, Vol. 37 (1996), 5139–45.
7. Varma I.K., Ananthakrishnan S.R., Krishnamoorthi S., "Comp. of Glass/Modified Jute Fabric and Unsaturated Polyester," *Composites*, Vol. 20 (1989) 383.
8. Geethamma V.G., Thomas Mathew K., Lakshminarayanan R., Thomas S., "Composite of Short Coir Fibers and Natural Rubber: Effect of Chemical Modification, Loading and Orientation of Fiber," *Polymer*, Vol. 39 (1998) 1483.
9. Sreekala M.S., Kumaran M.G., Thomas S., "Oil Palm Fibers: Morphology, Chemical Composition, Surface Modification and Mechanical Properties," *J Appl Poly Sci*, Vol. 66 (1997) 8–821.
10. Nunez, A.J., Sturm, P.C., Kenny, J.M., Aranguren, M.I., MArcovich, N.E. and Reboredo, M.M. (2003), "Mechanical Characterization of Polypropylene-Wood Flour Composites," *J. Appl. Polym. Sci.*, Vol. 58, 527–35.
11. Stael, G.C., Tavares, M.I.B., and d'Almeida, J.R.M, (2001), "Impact Behavior of Sugarcane Bagasse Waste-EVA Composites," *Polym. Test.*, Vol. 20, 869–72.
12. V.A. Alvarez, A.N. Fraga, A. Vázquez, "Effects of the Moisture and Fiber Content on the Mechanical Properties of Biodegradable Polymer–Sisal Fiber Bio-Composites," *Journal of Applied Polymer Science*, Vol. 91, Issue 6, (2004), pages 4007–4016.
13. A.M. Mohd Edeerozey, Hazizan Md Akil, A.B. Bakar, M.I. Zainal Ariffin, "Chemical Modification of Kenaf Fibers," *Materials Letters*, Vol. 61 (2007), 2023–2025.
14. L. Mwaikambo, M. Ansell, "Chemical Modification of Hemp, Sisal, Jute and Kapok Fibers by Alkalization," *Journal of Applied Polymer Science*, Vol. 84 (12) (2002) 2222–2234.
15. Shenouda, S.G., *Applied Fibre Science*; Happey, F., ed.; Academic Press: London, 1979, vol. 3, Chap. 7.
16. Zeronian, S.H., *Cellulose Chemistry and Its Applications;* Nevell, T. P.; Zeronian, S. H. eds.; Ellis Horwood: Chichester, 1985, p. 159.
17. Atkins, E., *Applied Fibre Science;* Happey, F., ed.; Academic Press: London, 1979, vol. 3, chap. 8.
18. Nguyen, T., Zavarin, E., Barrall, E.M.,"Thermal analysis of lignocellulosic materials,"*J Macromol Sci. Rev. Macromol. Chem.* 1981, chap. 20, 1.
19. Meyer R.W., *Handbook of Pultrusion Technology*, Chapman and Hall Ltd., 1985, USA.
20. Adlan Akram Mohamad Mazuki, Hazizan Md Akil, Sahnizam Safiee, Zainal Arifin Mohd Ishak, Azhar Abu Bakar, Degradation of dynamic mechanical properties of pultruded kenaf fibre reinforced composites after immersion in various solutions. *Composites Part B: Engineering.* Volume 42 (1), (2011), 71–76.
21. Kim H.J., Seo D.W., "Effect of Water Absorption Fatigue on Mechanical Properties of Sisal Textile-Reinforced Composites," *International Journal of Fatigue*, 2006; Vol. 28: 1307–1314.
22. Mohanty A.K., Mishra S., Drzal L.T., Misra M., Parija S., Nayak S.K., Tripathy S.S., "Studies on the Mechanical Performance of Biofiber/Glass Reinforced Polyester Hybrid Composites," *Compos. Sci. Technol.*, 2003; Vol. 63:1377–85.
23. Cao Y., Shibata S., Fukumoto I., "Press Forming of Short Natural Fiber– Reinforced Biodegradable Resin: Effect of Fiber Volume and Length on Flexural Properties," *Polym Test*, 2005; Vol. 24:1005–11.
24. A.C. Albuquerque, K. Joseph, L.H. Carvalho, "Effect of Wettability and Ageing Conditions on the Physical and Mechanical Properties of Uniaxially Oriented Jute-Roving-Reinforced Polyester Composite," *Composites Science and Technology*, Vol. 60 (Nov. 1999) 833–844.
25. Sun, R. C.; Fang, J. M.; Bolton, *J. Holzforschung*, 1999, Vol. 53, 253.
26. C.A.S. Hill, H.P.S. Abdul Khalil, "Effect of Fiber Treatments on Mechanical Properties of Coir or Oil Palm Fiber Reinforced Polyester Composites," *Journal of Applied Polymer Science*, Vol. 78, 1685–1697.

27. Facaa, A.G., Korschort, M. & Yan, N., "Predicting the Elastic Modulus of Natural Fibers Reinforced Thermoplastic," *Composite: Part A: Applied Science and Manufacturing*, Vol. 13, 278–292.
28. Cao Y, Shibata S, Fukumoto I., "Mechanical Properties of Biodegradable Composites Reinforce with Bagasse Fiber Before and After Alkali Treatments," *Composites Part A*, 2006; Vol. 37:423–9.
29. Laly A. Pothana, Zachariah Oommen, Sabu Thomas, "Dynamic Mechanical Analysis Of Banana Fiber; Reinforced Polyester Composites," *Composites Science and Technology*, 2003; Vol. 63; 283–293.
30. Aziz Sharifah H, Ansell Martin P., "The Effect of Alkalization and Fiber Alignment on the Mechanical and Thermal Properties of Kenaf and Hemp Bast Fiber Composites: Part 1 – Polyester Resin Matrix," *Compos. Sci. Technol.*, 2004; Vol. 64:1219–30.
31. Takashi Nishino, Hirao Koichi, Kotera Masaru, Nakamae Katsuhiko, Inagaki Hiroshi, "Kenaf Reinforced Biodegradable Composites," *Compos. Sci. Technol.*, 2003; Vol. 63:1281–6.
32. Aziz Sharifah H, Ansell Martin P, Clarke Simon J., Panteny Simon R., "Modified Polyester Resins Natural Fiber Composites," *Compos Sci Technol.*, 2005; Vol. 65:525–35.
33. Tan J.K., Kitano T, Hatakeyama T.J., "Crystallisation of Carbon Fiber Reinforced Polypropylene," *Mater. Sci.*, 1990; 25:3380–4.
34. Landel R.F. *Mechanical Properties of Polymers and Composite*, New York: Marcel Dekker, Inc., 1994.
35. G. Li, P. Lee-Sullivan and R.W. Thring, "Determination of Activation Energy for Glass Transition of an Epoxy Adhesive Using Dynamic Mechanical Analysis," *Journal of Thermal Analysis and Calorimetry*, Vol. 60 (2000) 377–390.
36. L. Barral, J. Cano, A. López, P. Nogueira, C. Ramírez, "Determination of the Activation Energies for α and β Transitions of a System Containing a Diglycidyl Ether of Bisphenol a (DGEBA) and 1,3-bisaminomethylcyclohexane (1,3-BAC)," *J. Thermal Anal.*, 1994; Vol. 41; 1463–1467.
37. Seena Joseph, M.S. Sreekala, Sabu Thomas, "Effect of Chemical Modifications on the Thermal Stability and Degradation of Banana Fiber and Banana Fiber-Reinforced Phenol Formaldehyde Composites," *Journal of Applied Polymer Science*, 2008; Vol. 110 (4); pages 2305–2314.
38. Hazizan Md Akil, Leong Wei Cheng, Z.A. Mohd Ishak, A. Abu Bakar, M.A. Abd Rahman, "Water Absorption Study on Pultruded Jute Fibre Reinforce Unsaturated Polyester Composites," *Composites Science and Technology*, 2009; Vol. 69 (11–12):1942–1948.
39. Norlin Nosbi, Hazizan Md Akil, Z.A. Mohd Ishak and A. Abu Bakar, "Degradation of Compressive Properties of Pultruded Kenaf Fiber Reinforced Composites after Immersion in Various Solutions," *Materials & Design*, 2010; Vol. 31(10), 4960–4964.
40. Hazizan Md Akil, Leong Wei Cheng, M.H. Affzan, A. Abu Bakar, Z.A. Mohd Ishak, "Water Absorption Study on Pultruded E-glass Fibre Reinforce Unsaturated Polyester Composites," *Advance Composites Letters*, 2010; Vol. 19 (3); 67–73.
41. Das S, Sara A.K., Choudhury P.K., Basak R.K., Mitra B.C., Todd T., et al, "Effect of Steam Pretreatment of Jute Fiber on Dimensional Stability of Jute Composite," *J. Appl. Polym. Sci.*, 2000; Vol. 76:1652–61.

11

Starch as a Biopolymer in Construction and Civil Engineering

Chandan Datta

Department of Polymer Engineering, Birla Institute of Technology, Mesra, Ranchi

Abstract

One of the great challenges that we face in the 21st century is to build up new manufacturing industries based on renewable resources. The construction industry has become a major field of use for biopolymers. In the construction industry, starch and starch derivatives, usually starch ethers, based on a variety of raw materials, are used as additives for hydraulic binders (e.g., cement, lime and gypsum). Value-addition can be simple as sterilizing products required for the pharmaceutical industry to highly complex chemical modification to confer properties totally different from the native starch. Fields of application in the construction industry are plaster (machine plaster and hand plaster), adhesives for tiles, fillers, plaster boards, concrete applications (shotcrete, self-compacting concrete, concrete goods etc.) polymers of natural origin (e.g., starch and cellulose) must be modified either physically or chemically in order to make them suitable for processing as thermoplastic resins. For example, the structure of starch can be made thermoplastic by using adjuvant such as glycerol and water. A method of making a substitute wood product includes the steps of combining ingredients including from 20–80% by weight pre-dried starch, from 20–78% by weight of a synthetic resin, from 0.5–4% by weight of a compatibilizer, said compatibilizer having a melt index of 2–150, and from 0–15% by weight of a fiber; processing the ingredients to achieve a melt temperature of 260–400 degree F; and extrusion shaping the substitute wood product. The detailed process of starch modified biologically degradable polymer foam and its uses are discussed. Such biodegradable polymer foams have the most varied uses in, among other things, packaging, thermal insulation, acoustic insulation, and construction use of daily life.

Keywords: Starch, construction, joint composition, thermoplastic foam

11.1 Introduction

One of the great challenges that we face in the 21st century is to build up new manufacturing industries based on renewable resources [1]. Traditional organic chemicals, materials and pharmaceutical companies rely on fossil resources, notably

Srikanth Pilla (ed.) Handbook of Bioplastics and Biocomposites Engineering Applications, (317–344)

petrochemicals, from which they make the products that go into such everyday substances as fabrics, dyes, packaging, drugs, construction materials and electronic goods. But fossil resources are finite, and oil—the source of petrochemicals—will become increasingly scarce as the century progresses. Biomass—in the form of starch—represents a real long-term solution.

Nature produces about 170,000 million tones per annum of renewable carbon as 'biomass', though we only harvest about 3 per cent of this for food and non-food applications [2]. Less than 1 per cent of this would be sufficient to meet the foreseeable demands of the chemicals and related industries that currently consume fossil carbon [3]. The challenge for us will be to convert the complex chemical structures that Nature produces into the chemical structures that we need (Figure 11.1).

Starch is one of Nature's three biggest products, the other two being cellulose and chitin. All three are rich in carbon but this carbon is trapped in macromolecular networks. We can break these large molecules down into the smaller molecules more commonly encountered in organic chemistry—ethanol (with huge potential as a biofuel), lactic acid, succinic acid etc. So can the chemical industry use Nature's renewable organic macromolecules, in particular starch, for some of its applications, and perhaps some new applications?

Major starch sources include potatoes, corn, rice and wheat. Starch is a combination of two polymeric carbohydrates (polysaccharides)—amylose (1), a linear structure, and amylopectin [2], a branched structure. The relative amounts of these polymers vary between species with high amylose cornstarch, for example, having about 85 per cent amylose while waxy cornstarch comprises 99 per cent amylopectin (Figure 11.2).

While starch is used predominantly for food, it is also used as a thickening agent, as an adhesive—the glue on the back of stamps—as a stabilizer, and as a binder. Starch is also often used as a carrier for drugs and as a viscosity modifier in paints. From the chemists' point of view starch has many appealing properties—it is abundant and sustainable, non-hazardous, and biodegradable—properties

Figure 11.1 SEM picture of potato starch.

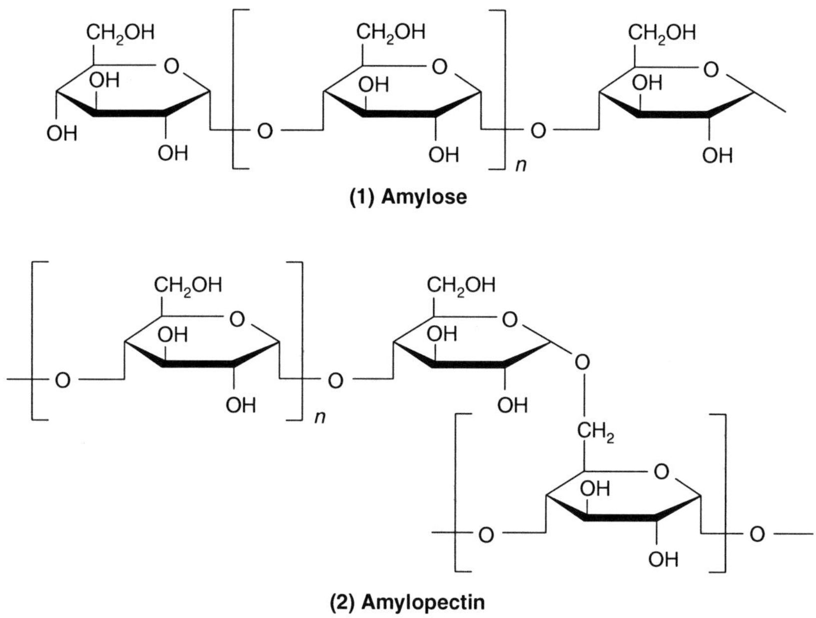

Figure 11.2 Chemical structure of Amylose and Amtlopectin.

that are becoming increasingly important in these environmentally conscious and sustainability-driven days [4–7]. Furthermore, legislation affecting chemicals is increasing at an exponential rate and the use of hazardous substances is becoming increasingly difficult and restricted. Starch also has potentially useful functionality, notably hydroxyl groups to assist adsorption and chemical modification.

The construction industry is becoming a major field of use for biopolymers. In 2000, an estimated $1±1.5 bn in sales was made at the manufacturer's level, and this growth is expected to continue. Applications of biopolymers in construction are widespread and diverse. In some cases, biopolymers offer distinct advantages in performance and/or cost over synthetic polymers, while in other areas biopolymers may be the only product available that can provide certain properties for building materials. Biopolymers also bear the image of being environmentally more acceptable than synthetic polymers produced in a chemical plant, and although this point can be argued it does influence the choice of materials used, especially for interior home building. This chapter begins with a brief description of the construction industry and its usage of chemicals, in order to introduce the market. The technology of building materials using biopolymers is then presented to enable the reader to understand the functionality and benefits of biopolymers, after which the main applications of biopolymers in various segments of the construction industry are described. Because of limited space, only biopolymers with a significant usage volume are discussed here. Although many more biopolymers are in current use, their volume is often very limited, and so they were omitted from this discussion. Rather, an attempt was made to present details of the major biopolymer-like starch, to highlight their advantages over synthetic materials, and to identify their overall contribution to modern construction technology.

11.1.1 Chemicals used in Concrete

Concrete uses almost exclusively liquid chemical admixtures, the main reason being ease of dosing and mixing. Major chemical admixtures for concrete include: dispersants based on lignosulfonates, naphthalenesulfonate resins (BNS), melamineformaldehyde sulfite resins (MFS), or polycarboxylates [PC; e.g., methacrylic acid poly(ethyleneglycol) methacrylate ester copolymers]; retarders based on sodium gluconate or sugar-rich lignosulfonate; accelerators based on calcium nitrate or calcium formate; air-entraining agents based on root resin extracts, alkylsulfates of phenol ethoxylates; foamers based on protein hydrolysates; anti-segregation admixtures based on welan gum or starch; anti-washout admixtures based on hydroxypropyl cellulose; shotcrete accelerators based on sodium aluminate or fine, amorphous aluminum oxide; and shrinkage-reducing admixtures based on neopentyl glycol. Comprehensive overviews on chemical admixtures used in concrete have been produced by Ramachandran (1995) and Rixom and Mailvaganam (1999). Clearly, the concrete industry uses a great diversity of admixtures, some important members of which belong to the group of biopolymers.

US demand to grow 6% annually through 2012 Natural Polymer demand is expected to grow 6.0 percent annually to $5 billion in 2012, reaching 2.3 billion pounds. Increased levels of food production, and opportunities in packaging, oilfield, medical, cosmetic, toiletry and other areas will stimulate gains. Average natural polymer prices are expected to decrease slightly based on declining prices for starch and fermentation products. Prices of other natural polymers will be moderated by the commodity nature of most materials and the dominance of price over other considerations. Threats to further growth include mature applications and variable suppliers for products such as guar gum due to climatic and political uncertainties. With many natural polymers harvested offshore, such as carrageenan and gum Arabic, imports will constitute a growing share of domestic demands. Starch, vegetable gums among best prospects best opportunities are anticipated for starch and fermentation products, followed by exudates and vegetable gums, and marine and protein polymers. Starch and fermentation product demand will grow at a double digit pace to nearly $1.1 billion in 2012 based on increased capacity and declining prices for polylactic acid and starch blend polymers in packaging and textile fiber uses.

11.2 Starch as a Biopolymer

Few can deny that the indigenous starch crops of the tropics are true wonders of nature. With sun and rain, and little or no artificial inputs, they are able to grow in great abundance. Whether it is cassava, arrowroot, sago, taro, sweet potato or yam, for centuries tropical starches have served as staple foods for millions of people, throughout the hot and humid regions of the world. Indeed, these starch crops are so proficient at supplying essential calories to even the very poorest peoples of the world that they are considered to be the quintessential subsistence crop.

But, what is considered to be a blessing in one situation can turn out to be a burden under another set of circumstances. In the majority of tropical developing countries, the only foreseeable route to economic development is through agricultural development. The irony is that the very crops that have proven to be most suited for tropical agro-climatic conditions and upon which economic development will depend have been relegated to the role of subsistence crops. Although these crops have been the subjects of much investigation in the area of basic production, they have not benefited from the kind of value-added research required for economic competitiveness on an international scale.

It is extremely difficult to break out of this subsistence crop mode and compete with mainstream starch products such as corn, wheat or potato starches, particularly when it is not the commodities themselves that are the competition, but rather the functional characteristics of the value-added products. Consequently, for many indigenous tropical starch crops, the lack of competitive market access has become the major obstacle to their contribution to agricultural development. Efforts to improve production and yields often result in excess supplies of basic commodities for the existing market demand that, in turn, discourages future production. On the other hand, modern value-added products are generally very application-specific and are thus far less susceptible to the sort of market fluctuations that cause chaos to developing countries whose economies are built upon basic commodities.

Until recently, the starch markets of the world were virtually closed to foreign countries. Import duties were so high that it was practically impossible to sell anything but the most basic commodities, at a price dictated by the buyer. All talk of value-addition to starches of developing countries was considered absurd. However, on April 12, 1994 the GATT Uruguay Round was signed in Marrakesh, paving the way for new trade opportunities.

As far as starch is concerned, what are some of the possible consequences of the Uruguay Round? There is tremendous potential for the profitable commercial use of tropical starches, but considerable research and product development of a new type is necessary to properly exploit these materials. The model for product quality and reliability has already been set by the international starch industry. That is who the competition is. If locally produced tropical starches cannot reach an equivalent level of quality, functionality or reliability, then these products will never survive in the competitive market. There is only so much that a more equitable trade environment can offer.

A review of the sort of research that has been done by both international institutes show that extensive work has been carried out on agronomic and phenotypic properties for most tropical crops, but relatively little study has been carried out on the sort of functional properties which are of direct technical and economic interest to competitive food and non-food industry. There is little purpose increasing the yield potential of crops that are unsuitable for processors or of limited acceptability to consumers. Far more work must be carried out on those characteristics, which will result in products that are more convenient to distribute, easier to process, and have the physical, chemical and organoleptic properties required by the target markets. For those starches that do not have the native functional characteristics that are desired by the target market, an additional effort must be made to value-add or modify them so that they can compete internationally. According to the opinion of expert marketers, large markets require a consistent supply and a reliable price and quality. They do not like to be pioneers and it is extremely difficult to interest markets in new products unless these criteria can be assured with some confidence. Another factor which large markets require is time—time to test and re-test new products until they are absolutely certain that they are suitable.

Once these basic factors are accounted for, the next most critical consideration is product performance, which, in turn, depends upon the functional characteristics. In fact, that shows starch should be viewed—as a set of functional characteristics suited to a particular application. The functional characteristics we are after are the same ones that have firmly established other starches and natural polymers in specific markets. These functional characteristics follow on from the basic physico-chemical properties of the starch granules and can often be enhanced through value-addition of one type or another. The most basic of the physical properties of starch granules are their size as exemplified in Table 11.1.

The size and distribution of starch granules can be very important for specific applications and even this very basic physical characteristic can be value-added (Table 11.1). For example, the small granule size of rice starch makes it very suitable for applications laundry sizing of fine fabrics and for skin cosmetics. Carbonless paper requires the use of starch as a stilt material to protect ink capsule from premature rupturing, as can be seen in Figure 11.3.

This application requires a starch that is of a particular size and uniformity and arrowroot was the product of choice for many years. A starch such as wheat could

Table 11.1 Granule size distribution of various starches.

Starch Species	Granule Size Range (μm)	Average Size (μm)
Waxy Rice	2–13	5.5
High Amylose Corn	4–22	9.8
Corn	5–25	14.3
Cassava	3–28	14
Sorghum	3–27	16
Wheat	3–34	6.5, 19.5
Sweet Potato	4–40	18.5
Arrowroot	9–40	23
Sago	15–50	33
Potato	10–70	36
Canna (Aust. Arrowroot)	22–85	53

not be used because its bimodal distribution of starch granules made it unsuitable (Figure 11.4 and Figure 11.5).

However, the variation in supply and cost of arrowroot prompted one company to develop a process of separating the population of small granules from the large ones through centrifugation which resulted in an immediate take over of this market from arrowroot. It was a case of value addition to the very basic physical characteristic of the starch. Here is another interesting example of unique size, with all granules in the 1µm range (Figure 11.6).

Other simple physical characteristics, which have an impact on functionality, are starch granule shape and surface. This is often a critical factor for applications requiring starch to be a surface carrier of materials such as colors, flavors, seasonings and even pesticides.

Figure 11.3 Carbonless paper.

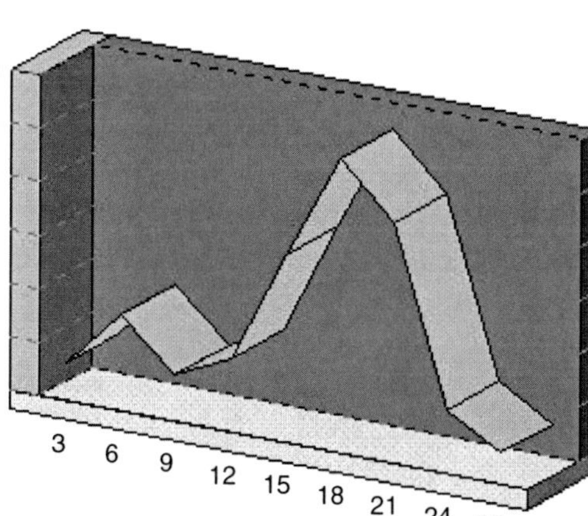

Figure 11.4 Wheat starch granule distribution µm.

Figure 11.5 Air-classified wheat starch.

Figure 11.6 Cow cockle starch.

Starch has two major components: amylose and amylopectin. These polymers are very different structurally; amylose being linear and amylopectin highly branched—each structure playing a critical role in the ultimate functionality of the native starch and its derivatives. The amylose/amylopectin ratios of starches can be genetically manipulated and offer a significant opportunity for the researcher with certain crops. Viscosity, shear resistance, gelatinization, textures, solubility, tackiness, gel stability, cold swelling and retrogradation are all functions of their amylose/amylopectin ratio (Table 11.2).

When aiming at functional properties in starch, most commercial companies examine the characteristics of competitive starches in particular applications. This sets the target to shoot for. For those characteristics, which are unattainable with native starches, the only alternative is to look towards some form of

Table 11.2 Amylose content of various starches.

StarchSource	% Amylose
Waxy Rice	0
High Amylose Corn	70
Corn	28
Cassava	17
Waxy Sorghum	0
Wheat	26
Sweet Potato	18
Arrowroot	21
Sago	26
Potato	20

value-addition to achieve the desired results. Value-addition can be as simple as sterilizing products required for the pharmaceutical industry to highly complex chemical modification to confer properties totally different from the native starch.

Simple value-addition is represented by washing, air classification, centrifugation and pre-gelatinization. The latter process can be done in many from boiling in crude pots to drum dryers to modern multi-screw extruders, each method having its particular advantages and disadvantages. The wide range of chemically modified starches found in the food, paper and textile industries represents complex value-addition.

The most common non-food applications for native and value added starches are as follows: (Table 11.3)

As can be seen, there are a great variety of value-added applications for starch in the non-food area, and each application requires very particular functional characteristics. Even in the most basic non-food applications of starch, a great deal of value-addition is employed. Adhesives starches are acid or alkali treated; they are modified with oxidizing agents, salts and different alcohols. Textiles starches are esterified, oxidized and are subject to various cross-linking agents.

The use of sophisticated, value added starches in paper products is even more noticeable, when one considers the wide range of applications in that industry. Starches are used to provide greater strength to tissues and paper towels, and they allow a greater use of recycled paper in linerboard and cardboard. The growing demand for biodegradability promises to provide additional volumes as starch is used in plastic films and sheets as well as in natural fiber formulations that will eventually replace plastic foams. The volume of starch going into non-food uses is enormous and it is all based upon the functional characteristics of the individual products.

Table 11.3 Non-Food applications of starches.

Adhesives	**Explosives Industry**	**Paper Industry**
• Hot-melt glues • Stamps, bookbinding, envelopes • Labels (regular and waterproof) • Wood adhesives, laminations • Automotive, engineering • Pressure sensitive adhesives corrugation paper	• Wide range binding agent • Match-head binder	• Internal sizing • Filler retention • Surface sizing • Paper coating (regular and color) • Carbonless paper stilt material • Disposable diapers, • Feminine products sacks
Construction Industry	**Cosmetic and Pharmaceutical Industry**	**Mining Industry**
• Concrete block binder • Asbestos, clay/limestone binder • Fire-resistant wallboard • Plywood/chipboard adhesive • Gypsum board binder • Paint filler	• Dusting powder • Make-up • Soap filler/extender • Face creams • Pill coating, dusting agent tablet binder/dispersing agent	• Ore flotation • Ore sedimentation • Oil well drilling mud
Miscellaneous		
• Biodegradable plastic • Film • Dry cell batteries • Printed circuit boards • Leather finishing		

The non-food uses of starch are a prime indicator of a country's economy. During recessions, the volume of starch going into non-food use drops considerably. On the other hand, an active economy needs construction materials for buildings, industrial plants and housing; it needs paper for the bureaucracy, for packaging and wrapping various products, for corrugated boxes and it need adhesives to stick all this economic activity together. As the economy booms, so does the volume of starches going into non-food uses. As countries develop, so does their demand for high quality, highly functional, value-added starches.

11.2.1 Thermoplastic Starch Products

Thermoplastic starch biodegradable plastics (TPS) have a starch (amylose) content greater than 70% and are based on gelatinized vegetable starch, and with the use of specific plasticizing solvents, can produce thermoplastic materials with good performance properties and inherent biodegradability. Starch is typically plasticized, destructured, and/or blended with other materials to form useful mechanical properties. Importantly, such TPS compounds can be processed on excisting plastics fabrication equipment.

High starch content plastics are highly hydrophilic and readily disintegrate on contact with water. This can be overcome through blending, as the starch has free hydroxyl groups which readily undergo a number of reactions such as acetylation, esterification and etherification.

Foam loose fill packaging and injected moulded products such as take-away containers are also potential applications. Foamed polystyrene can be substituted by starch foams that are readily biodegradable in some loose-fill packaging and foam tray applications.

Foamed starch loose-fills are rather easy products to produce and this area has become an early market for biodegradable plastics. During its preparation, raw starch is premixed with 25 to 50 weight percent water and fed into an extruder capable of imparting intensive shear and operating at high temperature (higher than the boiling point of water, i.e., 150–180°C). Under these conditions of shear and temperature, starch breaks down, loses its crystallinity, and gets plasticized with water, resulting in a homogenous amorphous mass. When this gelatinized starch/water mixture exits the extruder, the water that is present in the mass at a temperature higher than its boiling point expands into steam due to a sudden drop in pressure, and the foam is formed. Generally a plasticiser (such as glycerol) and another polymer (such as polyvinyl alcohol) impart more reproducible properties to starch foam.

11.2.2 Starch Synthetic Aliphatic Polyester Blends

Blends of biodegradable synthetic aliphatic polyesters and starch are often used to produce highquality sheets and films for packaging by flat-film extrusion using chill-roll casting or by blown film methods since it is difficult to cast films from 100% starch in a melted state. Approximately 50% of the synthetic polyester (at approximately $4.00/kg) can be replaced with natural polymers such as starch (at approximately $1.50/kg), leading to a significant reduction in cost. Furthermore, the polyesters can be modified by incorporating a functional group capable of reacting with natural starch polymers.

Lim *et al.* (1999) studied the properties of aliphatic polyester blended with wheat starch. The polyester was synthesized from the poly-condensation of 1,4-butanediol and a mixture of adipic and succinic acids. The wheat starch-polyester blends were found to have melting points near that of the polyester alone. A plasticiser was added to the starch, making the blends more flexible and processable than the polyester itself. Plasticized blends were found to retain a high tensile strength and elongation at the break point, even at high concentrations of starch.

Blending starch with degradable synthetic aliphatic polyesters such as PLA and PCL has recently become a focus of biodegradable plastic development. Biodegradable plastics can be prepared by blending up to 45% starch with

Table 11.4 PCL polymers (commercially available).

Polymer Type	Trade-name	Supplier	Origin
Starch-polycaprolactone (PCL) blends	Mater-Bi™	Novamont	Italy
	Bioflex™	Biotech	Germany

degradable PCL . This new material is not strong enough for most applications, as the melting temperature is only 60°C and it gets soft at temperatures above 40°C. These drawbacks greatly limit the applications of the starch-PCL blends. Table 11.4 details some starch-PCL polymers that are commercially available.The applications for starch-synthetic aliphatic polyester blends include high-quality sheets and films.

11.2.3 Starch and PBS/PBSA Polyester Blends

Other polyesters that are blended with starch to improve material mechanical properties are polybutylene succinate (PBS) or polybutylene succinate adipate (PBSA). A small amount (5% by weight) of compatibiliser (maleic anhydride functionalized polyester) can be added to impart phase stability to these starch based polymer blends. At higher starch content (>60%), such sheets can become brittle. For this reason, plasticisers are often added to reduce the brittleness and improve flexibility.

Ratto *et al.* (1999) investigated the properties of PBSA and corn starch blends of varied compositions. PBSA is biodegradable, and exhibits excellent thermoplastic properties. The objective of the study was to obtain a mixture that maximized these properties while minimizing cost. Corn starch is an inexpensive polysaccharide that was blended with PBSA at concentrations of 5–30% by weight for analysis. Tensile strength of the blends was lower than that of the polyester alone, but there was not a significant drop in strength with increasing starch content. In addition, melt temperature and processing properties were not appreciably affected by the starch content.

11.3 Starch-plastic Composite Resins and Profiles made by Extrusion

A method of making a substitute wood product includes the steps of combining ingredients including from 20–80% by weight pre-dried starch, from 20–78% by weight of a synthetic resin, from 0.5–4% by weight of a compatibilizer, said compatibilizer having a melt index of 2–150, and from 0–15% by weight of a fiber; processing the ingredients to achieve a melt temperature of 260–400° F.; and extrusion shaping the substitute wood product. The starch source is selected from the group consisting of wheat, corn, rice, tapioca, potato and mixtures thereof. The compatibilizer is selected from the group consisting of maleated polyethylene, maleated polypropylene, and mixtures thereof. The synthetic resin is selected from the group consisting of polyolefines, polyethylene, polypropylene, polyurethane, polystyrene, polyamides, polyesters, and combinations thereof. Fiber is selected from the group consisting of glass fibers, cotton, hardwood fibers, softwood fibers, flax, abaca, sisal, ramie, hemp, bagasse, recycled paper fibers, cellulose fibers, polymer fibers, and mixtures thereof.

We also blended polypropylene (PP) with amylose (AM)and/or dodecanoyl ester of amylose (DODAM) in an effort to make it biodegradable [8]. The content of AM/ DODAM was varied from 0 to 40% in the blends. The Biodegradability, mechanical properties, melt flow indexes MFIs], and morphologies of the blends were studied. Biodegradability increased with increase in AM/DODAM content. It was found to be dependent on DODAM content and was at a maximum in blends containing 40% AM+DODAM. Blends with no DODAM or 2.5% DODAM showed almost no adherence of the phases. Dispersion of AM improved in blends with 5% DODAM, and it showed satisfactory adherence to PP also. The tensile strength, elongation at break, and Izod impact strength decreased with increasing AM content. However, in blends with both AM and DODAM, all these properties, especially the elongation at break, showed improvements. The same trend was observed for MFI.

Polycaprolactone (PCL) was blended in a twin-screw extruder with chemically modified thermoplastic starch (CMPS) to provide biobased and biodegradable resin composition. Reacting starch with maleic anhydride (MA) in the presence of a plasticizer and a free radical initiator provided the CMPS. The starch modification improved interfacial adhesion and processability in blending with other thermoplastic polyesters. The rheological, mechanical, thermal, and morphological properties of the blends were examined. Differential scanning calorimetry (DSC), scanning electron microscopy (SEM), and Fourier transform infrared (FTIR) studies revealed that the PCL/CMPS blends are thermodynamically immiscible. However, they formed compatible blends due to the reaction of the carboxyl groups on starch backbone with hydroxyl groups of the PCL chain ends. The tensile strength and elongation decreased with increasing CMPS content, whereas the modulus increased. Dynamic viscoelastic measurements showed that the flow behavior of PCL was that of Newtonian fluid within the tested frequencies, whereas the CMPS exhibited strong shear thinning characteristics. The flow behavior of the blends varied with the CMPS content. The complex viscosity, storage, and loss moduli of the blends containing more than 40% of CMPS were higher than those of pure CMPS and PCL. In addition, the properties of CMPS to those of chemically unmodified thermoplastic starch (TPS) were compared [9].

11.4 Construction Industry – Starch and its Derivatives as Construction Material

With a long history supplying biomaterials for construction applications (Figure 11.7), our technologies can be used to influence hydration and adhesion in dry mix mortars, pastes and cements.

Figure 11.7 Biomaterials for construction applications.

Concrete block binder, asbestos, clay/limestone binder, fire resistant wall board, plywood/ chipboard adhesive, gypsum board binder paint filler, ad-mixer retarders are some of examples.

In the construction industry, starch and starch derivatives, usually starch ethers, based on a variety of raw materials, are used as additives for hydraulic binders (e.g., cement, lime, and gypsum). Starch or starch derivatives have a strong effect on the rheology of aqueous systems. In particular, they act as efficient thickening agents, rheology enhancers as a means of improving water retention, and as processing additives. Substantial properties of the used starch derivatives – commonly, cold water soluble products are applied – are their rapid swelling ability and therefore their rapid thickening and rheology-enhancing behavior. Moreover, starch ethers have a gluing effect and stabilize the systems (gypsum, cement, or lime basis) to which they are added Fields of application in the construction industry are:

Summary of the invention

- Plasters (machine plaster and hand plaster)
- Adhesives for tiles
- Fillers
- Plaster boards
- Concrete applications (shortcrete, self-compacting concrete, concrete goods, etc.)
- Emulsion paints and synthetic

In plaster applications the consistency of the mortar is an important criterion. It is determined via the "Flow Table Test." With this method it is also possible to determine the water demand of starch derivatives. Water retentivity examination according to standard DIN 18555 is another criterion for plasters. The air pore distribution can also be determined with this method.

The following figures demonstrate the consistency of mortars to which different additives were admixed. While cellulose ether has a thickening effect, starch ether, added at a similar concentration level, already has a plastifying effect on the mortar. Zero mortar, to which the same share of water is added, appears liquid. (Figure 11.8)

A. Zero mortar without additive
B. Mortar with 0.04 % cellulose ether
C. Mortar with 0.04 % starch ether

For these applications special application examinations are conducted.

Concerning adhesives for tiles, the steadiness, which is described by the slip resistance, is gaining importance. This special behavior cannot be achieved by viscosity-increasing cellulose derivatives. Only by means of rheology-enhancing starch derivatives it is possible to adjust a sufficiently high yield point to ensure steadiness of tiles (Figure 11.9). This special property of these tile adhesives is determined by the slip of the tiles according to standard method EN 1308.

Viscosity properties are determined via Brookfield measurement. The open time is used as an important criterion for the workability time. This test procedure is conducted according to EN 1346. The setting performance of mortars containing starch derivatives is determined by means of an automatic Vicat® unit.

Zuckerforschung Tulln has been engaged for some time with the development of **starch thickeners** for use in construction paints. A marketable product has already been developed and is being marketed by Agrana Stärke GmbH under the name of **Amitropaint**. The new product is intended for use as a thickening agent

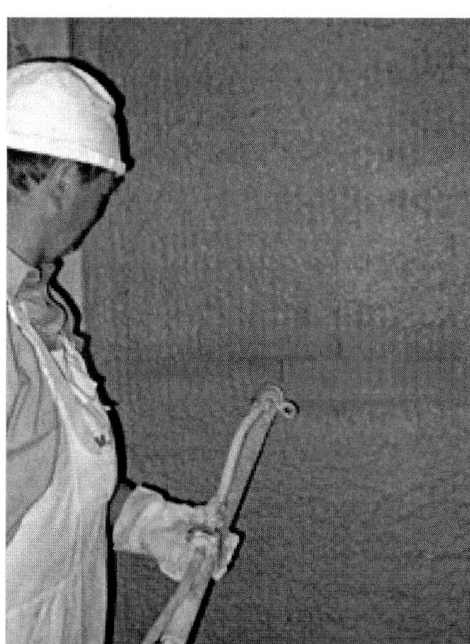

Figure 11.8 Photograph of resin plaster application.

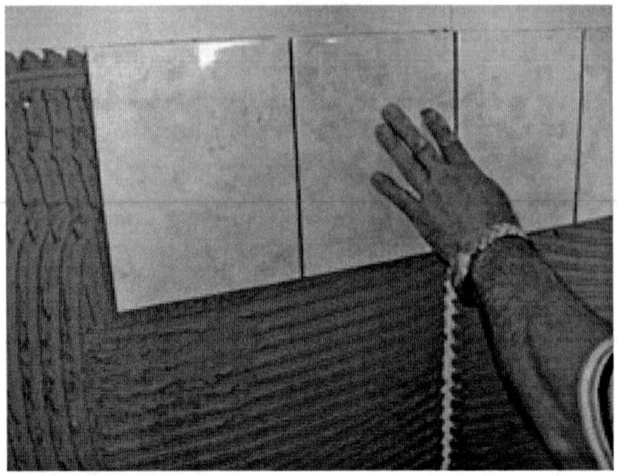

Figure 11.9 Adhesives for tiles.

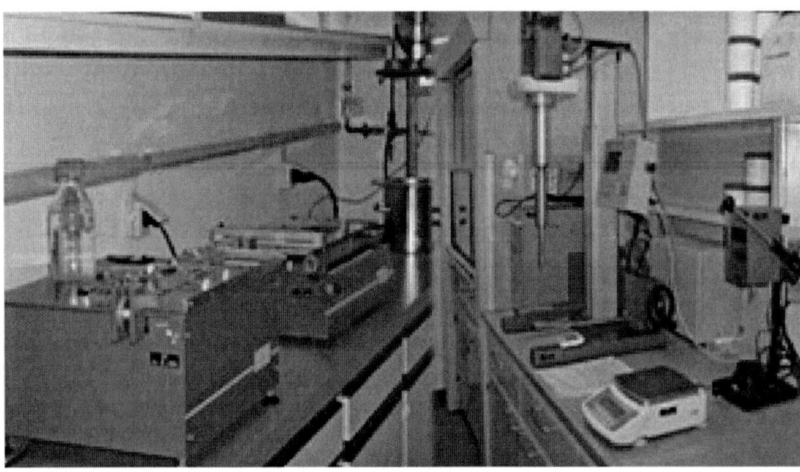

Figure 11.10 The automatic vicat-needle-device.

in combination with high-viscosity celluloses, particularly in dispersion binder-based paint systems. This combined use offers cost advantages and at the same time ensures the good properties of high quality paints. In addition to excellent scrub resistance, flow properties, and brilliance, it also features improved resistance to sagging and good rolling behavior.

Besides Paints the products can also be used in **synthetic resin plaster**. In adequate recipes this product is able to fully substitute celluloses. Further developments should provide the alone utilization of this starch product as a thickener in paints. This product "Amitropaint Plus" has already approved its readiness for marketing in the laboratory. (Figure 11.10)

The following **tests** can be conducted at Zuckerforschung Tulln:

- Paint manufacture with a dispersion system
- Viscosity measurement (Bohlin, Stormer, Brookfield)

- Test of film build with standard test blades
- For leveling
- And sagging
- Scrub resistance in accordance with ISO, DIN, and ASTM
- Coverage (wet and dry)
- Color and degree of whiteness measurement, gloss measurement
- Roll properties with spatter tendencies

11.5 Setting Behavior

The setting behavior is measured by the penetration of a needle (Vicat-Needle) into a tile adhesive. The automatic Vicat-Needle-Device enables us to watch the setting performance of cement, plaster, and mortar-systems continuously.

The start of the setting of a certain tile adhesive is defined by the time when the needle penetrates the tile adhesive only 36 mm deep. The end-point of the setting is arrived when the needle penetrates only 4 mm deep. The picture below displays the zero-sample as well as two different starch-types, which were investigated as adhesive additives. By optimization of the derivatization it is possible to shorten the delay of the setting (Figure 11.11).

Starch ethers as an additive for the reduction of the rebound were developed within a publicly funded project. The resulting product for the dry shotcrete application was named Amitrolit 8865. Even at very low dosages of 0,1–0,2% based on the spray cement it showed a significant reduction of the rebound by approximately 20% (absolute). The mode of operation can be explained in that way that the starch ether causes a change in the rheology of the mortar and that a "softer" concrete bed is formed. This extraordinary product shows in contrast to other

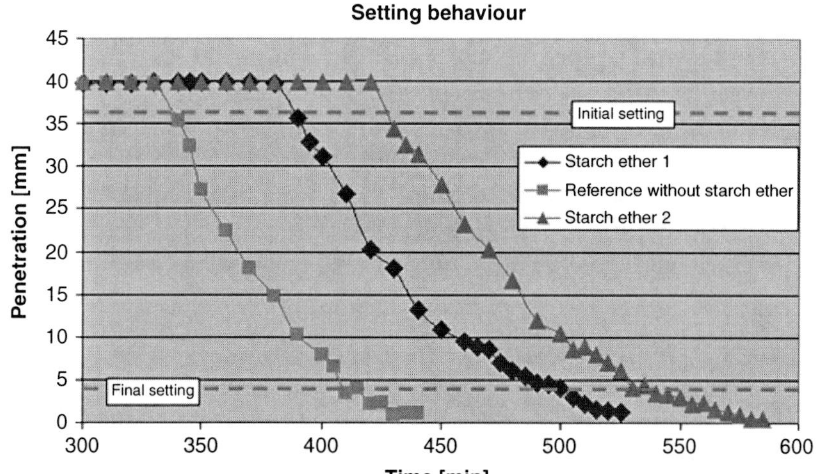

Figure 11.11 The setting behavior of starch ether-1, Starch ether-2, reference without starch ether.

Figure 11.12 Use of starch in shotcrete.

products no negative influences on the quality of the concrete. Ecological and economical advantages are gained by reducing the rebound, because less time for the disposal of the rebounded material is necessary, lower costs for this disposal arise, less concrete is consumed altogether and the somatic stress for the workers is minimized. (Figure 11.12)

An additional product for the wet shotcrete application emerged from the development too. Its potential will be investigated in a follow up project. Preliminary application test at an experimental rig showed a rebound reduction by 50% based on the spray concrete reference.

11.6 Rheological Measurement of Cements

Rheological attributes of mortars and cement plasters can be determined directly in the respective system by means of a building material rheometer. Apart from the relative "yield point" it is also possible to determine the shear viscosity.

The following figures display two types of starch ethers with different extent of derivatization. These starch ether types show a higher yield point than zero mortar and can be differentiated by their yield point and shear viscosity. Starch ethers have a typical shear reducing behavior, which is highly important for their pumping properties. Moreover, thixotrope behavior is visible very well.

11.6.1 Other Specific Applications

11.6.1.1 Joint Composition Including Starch

A filler composition of the type used to hide the joint of adjoining wallboard panels and comprising filler, binder, bulking agent and improved starch-bodying agent, and optionally a water retention agent. various types of starches to joint compositions in which the starch, depending on the type used, functions as a binder The improved starch bodying agent is a high water-carrying starch and it is used typically in an amount at which its binding properties are not exhibited to a significant degree. Accordingly, compositions within the scope of the present invention should include another material which functions as the principal binder constituent. Such a binder can be another type of starch, that is, a low water-carrying starch [10].

Any material capable of binding the other constituents of the composition in the manner desired of a joint composition can be used. Examples of materials which can be used as a binder in the compositions of the present invention include starch, ethylene vinyl acetate copolymer, poly (vinyl alcohol), poly(vinyl acetate), and butadiene-styrene copolymer. Preferred binders are ethylene vinyl acetate copolymer, poly(vinyl alcohol) and the type of starches used as binders in the compositions of the examples.

The composition includes a water-retention agent which functions to retard evaporation of water and to keep the water constituent from being absorbed in a blotter-like effect by the paper facing of the wallboard core. The water retention agent can function also to thicken the composition. Examples of materials which function as water retention agents are methyl cellulose, hydroxy ethyl cellulose, guar gum derivatives, alginates and certain starches.

If the improved starch-bodying agent of the present invention is used in admixture with one or more other bodying agents, it is preferred that it comprise at least about 20% of the proportion of bodying agents used.

The starch-bodying agent used in the compositions is a hydroxyl propylated waxy starch containing 6–7% substitution of propylene oxide on the starch molecule and containing virtually 100% amylopectin. dried joint compositions of the examples to bond to reinforcing tape of the type used in the joint of adjoining wallboard panels. the use of a mixture of bodying agents, that is, the propylated waxy starch of the present invention and attapulgus clay which has been used heretofore as an asbestos substitute. Asbestos is another example of a bodying agent that can be used in combination with the improved starch-bodying agent, but it appears that for the present at least, this would not be advisable because of governmental regulations regarding its use. starch sold under the trademark Sta-Gel 136 by A. E. Staley Manufacturing Co. The use of this starch results in a composition which has properties substantially equivalent to one containing aforementioned Gelatinized Dura-Gel starch.

11.6.1.2 Starch Ether

- An additive combination containing a) a water-soluble cellulose ether or a derivative, b) polyacrylamide, c) superabsorbent polymer (alkali metal salt or ammonium salt of a crosslinked polyacrylate which may have been grafted with starch), d) starch ether and e) a water-soluble alkali metal salt, alkaline earth metal salt or ammonium salt of arylsulphonic acid/ formaldehyde condensation products or of a sulphonic-acid-modified polycondensation product of melamine and formaldehyde. This additive combination reduces the stickiness of mortars which is caused by cellulose ethers. [11]

- Synergistic interactions between crosslinked hydroxyethyl starch and carboxyalkyl cellulose and/or xanthan gum. [12]
- Synergistic interactions between crosslinked HE-starch and carboxyalkyl cellulose and/or xanthan gum. [13]
- Synergistic interactions between crosslinked hydroxyethyl starch and HEC (hydroxyethyl cellulose). Named products: Natrosol 250 HHR, Bohramyl CR. [14]
- Synergistic interactions between crosslinked HE-starch and HEC (hydroxyethyl cellulose). Named products: Natrosol 250 HHR, Bohramyl CR. [15]
- Synergistic interactions between CM-starch ("Solvitose C5"), guar gum (or a derivative), and xanthan gum. [16]

Synergistic interactions between starch ethers and superabsorbents based on crosslinked polyacrylates (if desired, additionally grafted with starch) and [3] superplasticizers based on the formaldehyde condensation products of naphthalenesulfonate, phenolsulfonate, or melaminsulfonate. Hydroxypropyl starch, crosslinked with eBerolan ST 902 is a special starch ether for construction products. It effects a pseudo-plasticity and increases the consistency of the building material. A low adhesiveness to the tools allows a easy handling of the product. Water retention is controlled by combination with cellulose ether. Berolan ST 902 is dosed depending on the application in the range from 0,005 to 0,3%.

Berolan ST 902 is potato starch ether, soluble in cold water. [17]

- Appearance: white–yellowish powder
- Odor: neutral
- Loss of drying: max. 10,0 %
- Ash: max. 10,0 %
- pH value: approx. 8 (1%, 20°C)
- Propox content: 19,0–24,0 in TM
- Solubility: Soluble in cold and hot water
- Viscosity: 300–500 mPas (5%; standard)
- epichlorhydrin, was used.

11.6.2 Plasters

1–4% VINNAPAS RE 5010N added to cement and lime-cement finishing plasters improves their adhesion, abrasion resistance and flexibility. In addition, 0.2–0.4% cellulose ether or starch ether or a combination of the two should be incorporated [18].

11.6.2.1 Acoustic Construction Panel

An acoustic construction panel for use in the construction of walls, floors, or ceiling structures to improve the acoustical properties thereof, and a method of making that panel. [19] The panel comprises a composition of natural wood fibers, paper and starch, and is absent of any chemical toxic products. The panel has a minimum thickness of about 3/4-inch, and an average density in the range of from about 15-lb/ft^3 to 17-lb/ft^3. A plurality of cavities are perforated on one surface of the panel to increase the acoustical surface properties of the panel. In the construction of the panel the wood pulp is directed into a holding tank for a predetermined period of time in order to expand the wood fibers, and further in which a composite mixture is produced by introducing into the wood pulp predetermined quantitites of starch and wax.

A flexible, waterproof material that will solidify in water has long been desired in civil engineering. They have developed a new class of material, called Aquaphalt, which has these and other desirable properties. Aquaphalt is composed of an asphalt emulsion, cement and a water-absorbing polymer [20]. The components are liquid at ambient temperature and can therefore be pumped, but they form a gel almost instantly when mixed. The hardened mixture is similar to hard bitumen, and has very low water permeability, high ductility and good adhesion to other materials. Here they described the characteristics of Aquaphalt, with particular emphasis on those properties that give it potential as a shock-absorbing, waterproof backfill material for tunnels and dams, and as an anti-liquefaction agent for protection of buildings exposed to earthquake hazards.

The production of foamed polymers is a known process and is effected either by mechanically without pressure or by means of foam-forming agents or else by sudden expansion of gases, expansion agents, or solvents which, at higher pressures, produce an inflation pressure in the plastic or liquid polymer composition. Such polymer foams [21–23] have the most varied uses in, among other things, packaging, thermal insulation, acoustic insulation, construction and many fields of use of daily life. As is generally the case with polymers or plastics, disposal or degradability constitutes an important factor also in the case of foamed materials, particularly if the foamed materials have a high strength and compressing is not readily possible.

For this reason, a number of foamed substances of so-called biologically degradable polymers are known such as, for instance, starch foamed materials, in connection with which, starting from, for instance, native or so-called disaggregated starch, such a foamed material is produced by means of an expansion agent. Thus for almost a century a sponge made from starch has been known which is produced in the manner that a boiled starch paste is cooled to temperatures below the freezing point and the water then removed from the sponge mass by thawing. In Federal Republic of Germany 23 04 736, a process for the production of a foamed material is described in which carbohydrates or polysaccharides in granulated, compacted or coarsely crystalline form are heated until dry in a tunnel furnace for 10–40 minutes at 200° to 400° C. with the addition of small amounts of organic or inorganic acids or acid salts. Due to the pyrolysis of the carbohydrate material which takes place, inflation occurs, whereby a carbonated foam material is obtained. In Federal Republic of Germany 32 06 751, a relatively rigid foam is obtained by an extrusion process in which heating of the starch material is effected already in an extruder due to the shearing forces and the pressure, inflation and foaming of the emerging gelating starch taking place due to the development of gas upon the reaction of expansion agent additives with simultaneous solidification of the starch paste. As expansion agent additives, calcium carbonate and phosphoric acid are described, whereby carbon dioxide is produced.

In WO 91/18048, a nucleating agent is applied to granules of starch. The nucleating agent is decomposed by heat, whereupon the development of the foam commences. As nucleating agent, carbonates enter into consideration, so that once again carbon dioxide is the responsible expansion agent. Starch foam can however also be produced in the manner, for instance, that powdered starch is mixed with water, this mass is extruded and upon the extrusion the starch is inflated by the steam which is produced.

All such starch foams of this type which have been described are, as a rule, partially or completely biologically degradable, in which connection, of course, the degradability can be negatively influenced by the addition of synthetic additives or plastic additives. Furthermore, it has been found that by the introduction of water as expansion agent or by the use of carbon dioxide as expansion gas, a foaming of the starch can be effected, but a non-uniform cell structure is established which, in its turn, requires additional additives. Furthermore, the required percentage of expansion agent or water necessary to produce the corresponding steam is very high and amounts up to about 20%.

The use of water in combination with the starch furthermore has other disadvantages which can be recognized in particular in connection with the development of so-called thermoplastic starch, which disadvantages are described in detail in a number of publications such as international patent application WO 90/05161 as well as the article "Sorption Behavior of Native and Thermoplastic Starch" by R. M. Sala and I. A. Tomka, in *Die angewandte makromolekulare Chemie* 199: 45–63, 1992; as well as ETH Dissertation No. 9917 by R. M. Sala, 1992, ETH Zurich.

As a result thereof, it would be advantageous to use thermoplastic starch or polymer blends containing thermoplastic starch and, for instance, polycaprolactone as basis for the production of a starch foam. Since water which is bound in the starch does not enter into consideration for the production of such a foam, ordinary physical or chemical expansion agents are necessary, which make the advantage of the biological degradability of pure thermoplastic starch and its blends questionable or do not represent naturally growing resources. Furthermore, a number of naturally occurring expansion agents are compatible with the thermoplastic starch or are thermally unstable or increase the thermal degradation of the thermoplastic starch.

A method [21] is proposed for the production of substantially biologically degradable polymer foam, starting from thermoplastic or disaggregated starch or from a polymer mixture consisting of thermoplastic or disaggregated starch with at least one other biologically degradable hydrophobic polymer. The starch or starch mixture is first of all mixed with a biologically degradable fibrous or capsular material which has the ability to bind water by capillary action and which is at least partially treated or substantially fully saturated with water. The biologically degradable/material mixture thus produced can be either isolated and, for instance, granulated so as to be subsequently processed stepwise in a separate process, or else, be directly processed, with pressure and temperature control, in such a manner that the water bound by capillary action in the material is released so as to effect a foaming of the polymer. The great advantage of introducing water as expansion agent by means of the fibrous or capsular materials is that, due to the binding of water by capillary action, undesired interactions with the polymer matrix can be avoided. Furthermore, the ratio of total-water to amount of expansion agent can in this way be kept very low, namely at less than 0.1 wt. %, referred to the total weight of the starch/material mixture.

In order to achieve the excellent physical and/or mechanical properties of thermoplastic starch in a starch foam of the invention, one preferably starts from thermoplastic starch or a polymer mixture containing thermoplastic starch in which the content of water in the thermoplastic starch or polymer mixture is less than 5 wt. % and preferably less than 1 wt. %. The thermoplastic starch or polymer mixture can be mixed with up to 30 wt. % of fibrous or capsular material which is treated or saturated with water. It is essential, upon the mixing of the polymer or the polymer mixture with the material treated with water that the process parameters such as pressure and temperature do not reach values which lead to the liberation of the capillarily bound water and/or the expansion agent. As a rule therefore, the mixing of the thermoplastic starch or the polymer mixture with the material which has been treated with water takes place, for instance in an extruder, in a temperature range of about 100° C. to 200° C., this temperature range or the optimal temperatures to be selected depending on the plasticizing or swelling agent in the starch and thus on the melt viscosity of the starch. As fibrous or capsular materials the following fibers enter into consideration: hemp, jute, sisal, cotton, flax/linen, natural silk or abaca. So-called ramie fibers also known, for instance, as fibers of so-called China grass have been found to be particular advantageous. Ramie fiber is frequently also referred to as so-called high-performance fiber directly from nature, since it represents a true alternative to the synthetic industrial fibers. Ramie fibers are therefore frequently also used as reinforcing fibers in view of their high tear strength, low elongation upon rupture, as well as their high adherence. Uses include the strengthening of rubber bands, as reinforcing fiber for building materials such as cement and plaster, as reinforcing fibers for thermosetting polymers, as well as reinforcing fiber for geotextiles. But ramie fiber has the particular advantage that it is completely decomposable or biologically degradable due to its natural origin. As fibrous or capsular materials, however, substances such as expanded clay aggregate, silica gel, agarose gel, cephatex gel and ceolith are also suitable.

The thermoplastic starch preferably contains as plasticizing agent or swelling agent one of the following substances: glycerol, sorbitol, pentaerythritol, trimethyl propane, polyvinyl alcohol, amino alcohol, other polyhydric alcohols, mixtures of these components, ethoxylated polyalcohols such as glycerol and ethoxylate or sorbitol and ethoxylate. This list is not limitative and the use of other plasticizing or swelling agents which are suitable for the production of thermoplastic starch is possible, in which connection, as already stated above, water is not suitable.

In accordance with a preferred embodiment of the process, thermoplastic starch or the polymer mixture containing thermoplastic starch which has a water content of less than 1 wt. % is mixed with 2–20 wt. %, and preferably about 4–8 wt. %, of water-saturated ramie fiber having a fiber length of about 0.3–4.0 mm in a temperature range of about 130° C. to 170° C. The mixing can take place, for instance, in a single-shaft or co-rotating or counter-rotating twin-shaft extruder or a Kokneader® or else in a batchwise unit such as an internal mixer or "Banbury® mixer." It is important in this mixing process that the water which is bound by capillary action in the ramie fiber not be liberated. On the other hand, however, temperature and pressure as well as the mechanical work introduced must be such that no degradation of the thermoplastic starch takes place. The temperature for the incorporation is furthermore dependent on the swelling agent of plasticizing agent used in the thermoplastic starch, which can greatly affect the melt viscosity of the starch. Thus, for instance in the case of a thermoplastic starch which contains glycerol, lower temperatures are to be used upon the incorporating of the ramie fibers than in the case of thermoplastic starch which contains sorbitol for instance.

Now, it is possible, in principle, to isolate the thermoplastic starch to which ramie fiber has been added or the polymer blend containing thermoplastic starch and to store it as a so-called polymer raw material for the further production of starch foam in a separate operation or at a later time. The other possibility consists of directly processing the soft compound further, for instance by injection into an injection mold in which case also the molding produced in this manner is, as before, not defoamed.

For the production of the foam, it is now important that the starch compound be processed at elevated temperature and pressure, for instance 200° C. to 210° C., whereby the capillary active water in the ramie fiber is released so as to foam the starch. In this connection, it is possible to process the starch melt containing the ramie fibers in an extruder at the said temperature of about 200° C., or to extrude or injection mold it, the thermoplastic starch or the polymer mixture foaming upon leaving the die. Or, however, the injection molding which has already been produced can be introduced into a mold and be foamed at elevated temperature and pressure.

In contradistinction to the various starch foams known from the prior art, the foam produced in accordance with the invention, consisting of thermoplastic starch or the polymer mixture containing thermoplastic starch, has an extremely uniform cell structure, a low density, and excellent mechanical properties. The mechanical properties are, of course, also decisively influenced by the presence of the ramie fibers, since, as is known, ramie fibers are capable of substantially improving the mechanical properties of polymers or plastics.

Upon the production of the biologically degradable polymers suitable for foaming such as, for instance, the thermoplastic starch, it is, of course, also possible, and at times also advantageous, to operate with additives such as, for instance additional plasticizing agents, lubricants, softeners, etc. Furthermore, it may be advantageous if the material intended for incorporation which has the water bound therein by capillary action, such as for instance the ramie fibers, be treated on its surface before the incorporation, for instance gummed or degummed, in order to permit better wettability by the polymer. Additives such as fire-proofing agents, coloring substances, etc. can also be used upon the compounding of the polymer or the polymer mixture.

Upon the production of the polymer foam, it must be seen to it, however, when adding additives such as plasticizing agents, lubricants and, in particular, softeners, that the viscosity is not too low. A low viscosity has a negative effect on the production of the foam and in case of too low a viscosity, there is the danger of collapse. Low viscosity on the part of the material to be foamed can, however, also result if the further polymer component used in the polymer mixture in addition to the thermoplastic starch has a very low viscosity and the overall viscosity is low due to too high a proportion of this further component. In general, upon producing a polymer blend for use for the production of foam in accordance with the invention, it should be seen to it that the viscosity of the thermoplastic starch contained therein is not substantially reduced. Too low a viscosity, however, also results if the water content in the material to be foamed is too high. Finally, too low a viscosity can also be a sign that the thermoplastic starch used in the material to be foamed is too strongly degraded.

The properties of the polymer foam can, however, also be controlled by, for instance, the length of the ramie fibers used, or else by admixing different materials such as, for instance, ramie fibers with cotton fibers in order, for instance, to produce a greater flexibility of the foamed material. The mixing of ramie fibers with, for instance expanded clay aggregate or silica gel, etc. is, for instance, also possible. The present invention accordingly provides biologically degradable polymer foam consisting essentially of foamed thermoplastic starch or a foamed thermoplastic polymer mixture which is compounded or treated with a material which is capable of binding water by capillary action. This material can either be a fibrous material such as hemp, jute, sisal, cotton, flax/linen, natural silk, abaca, or preferably ramie fibers, or else a capsular material such as, for instance, expanded clay aggregate, silica gel, agarose gel, Sephatex gel or Ceolith.

The polymer foam defined in accordance with the invention is suitable, inter alia, as packaging material, as thermal or acoustic insulation, or, in general, as absorbing material and for various uses in construction.

The invention has explained in further detail with reference to one example.

One starts from thermoplastic starch which has been prepared by digesting 65% starch with 35% sorbitol. The operation is carried out in a Theysohn TSK 045 Compounder (twin-shaft extruder with shafts rotating in the same direction) with different liquid/solid ratios. The following temperature profile is selected in the extruder: Zone 1, 25° C.; Zone 2, 130° C.; Zone 3, 150° C.; Zone 4, 170° C.; Zone 5, 150° C.; Zone 6, 140° C.; Zone 7, 140° C.; Zone 8, 170° C. 10 kg/hr of thermoplastic starch granulates are introduced into Zone 1 and melted. In Zone 5, 1500 g/hr of thermoplastic starch, 840 g/hr of ramie fibers having a fiber length of 0.5 mm, and 200 g/hr of stearic acid are furthermore added. The ramie fiber had been pretreated by moistening or substantially saturating with water before its admixture. This was followed by mixing and the removal of the melt and cooling. It should be seen to it in this connection that the material does not foam already upon the compounding, which can be obtained by temperatures which are definitely below 200° C. The following extruder values were selected: Speed of rotation of extruder: 200 rpm, Torque: 65% of the maximum torque, Mass pressure (die): 4–8 bar.

As an alternative to the procedure used, one can also start from native starch, in which case the thermoplastic starch is first of all digested by the addition of sorbitol. It should be seen to it in this connection that any moisture present in the native starch is removed by the application of a vacuum. It is essential that the thermoplastic starch have only a low moisture content upon the processing or the incorporating and compounding with the ramie fiber, i.e. that the moisture content is preferably less than 1% by weight.

From the foaming tests carried out, it resulted, furthermore, that it is preferable to operate with a fiber length of about 0.5 to 0.6 mm of the ramie fibers. The proportion of ramie fibers was preferably 4–8 wt. % referred to the total weight of the foam since in the case of saturated ramie fibers and a higher percentage, the amount of water thus resulting in the material to be foamed may be too high, as a result of which the viscosity is greatly reduced due to too high a content of water. Low viscosity upon the production of the foam is, however, as already stated, not desired.

A cylinder of a height of 15 mm was used. The foam material which was introduced into it was pushed in 3 mm, held in inward-pushed condition for 1 minute, and then released, a further minute being waited before measuring the restoration. As comparison with this, a thermoplastic starch containing 31.5 wt. % sorbitol without fibers was used for reference purposes, this thermoplastic starch being foamed with 3.5% water and 0.15 wt. % microtalc (as nucleation agent). Both the four foamed materials used in accordance with Examples 20–23 and the reference foam were conditioned at 70% humidity.

The foamed materials containing ramie fibers gave a restoration of 82–91%, the material in accordance with Example 22 gave a restoration of 87–91%, and the foam in accordance with Example 23 a restoration of 88–91%. As compared with this, the reference foam without ramie fibers gave a restoration of 81–89%.

It is thus shown that the foamed materials produced in accordance with the invention have a somewhat higher compressive strength than the reference foam. Finally, it may also be pointed out that the foamed materials produced showed a higher resistance to humidity than the reference foam. This effect is probably due, in particular, to the proportion of ramie fibers in the foam.

Finally, mention should also be made of polyvinyl alcohol, known for instance under the brand name Noviol, in which case the polyvinyl acetate used for its production is preferably 88% hydrolyzed. Since the thermoplastic starch is hydrophilic and the above-mentioned partners for the production of a polymer mixture are of a hydrophobic character, it is necessary or advantageous as a rule to use for the production thereof a so-called phase mediator which is compatible both with thermoplastic starch and at the same time with the hydrophobic polymer. Due to the different cohesion energy densities of starch and the hydrophobic polymers, block copolymers enter into consideration, namely ones which consist of a block which is soluble in starch and a block which is soluble in the hydrophobic polymer phase. It is, of course, essential in this connection that the phase mediator also be biologically degradable and that it can be suitably processed thermoplastically. As an example thereof, a polycaprolactone/polyvinylalcohol copolymer may be mentioned. As phase mediator, however, there also enter into consideration reaction products between a hydrophobic biologically degradable polymer and the thermoplastic starch which are compatible with the hydrophobic polymer phase. In this connection, for instance, the biologically degradable polymer can, for instance, have reactive groups such as, for instance, epoxy groups, or else acid-anhydride groups, which react with at least a part of the thermoplastic starch.

The phase mediator to be used or the quantity thereof to be employed is, in the final analysis, a question of optimalization; it is essential in the case of the polymer mixtures which are to be used for the production of the foam that it be as homogeneous or uniform as possible in order to be able to produce a foam which is also as uniform as possible.

References

1. H. Clark, Green chemistry: today (and tomorrow), *Green Chem.*, RSC Publishing, Cambridge, UK.,2006, 8, 17.
2. M. Eggersdorfer, J. Meijer and P. Eckes, *FEMS Microbiol. Rev.*, 1992, Vol. 103, 355.
3. C. Okkerse and H. van Bekkum, *Green Chem.*, RSC Publishing, Cambridge, UK, 1999, Vol. 1, 107.
4. Vitaly Budarin, James H. Clark, Jeffrey J.E. Hary, Rafael Luque Dr.Krzysztof Milkowski, Stewart J. Tavene, Ashley J. Wilson, Starbons: New Starch-Derived Mesoporous Carbonaceous Materials with Tunable Properties, *Chem. Comm.*, RSC Publishing, Cambridge, UK, 2005, 2903.
5. S. Doi, James H. Clark *Chem. Comm.*, RSC Publishing, Cambridge, UK, 2002, 632.
6. Gronnow M J, Luque R, Macquarrie D J and Clark H J . *Green Chem.*, RSC Publishing, Cambridge, UK, 2005, 7, 552.
7. Budarin, V., Clark J.H., Hardy, J.J.E, Luque, R., *Angew. Chem.*, John Wiley & Sons, Inc, USA, 23th ed., 45, 3782, (2006).
8. Diya Basu, Chandan Datta, Amarnath Banerjee "Biodegradability, mechanical properties, melt flow index, and morphology of polypropylene/ amylose/ amylose-ester blends: *J Appl. Polym. Sci.*, Vol. 85: 1434–1442, 2002.
9. Boo Young Shin, Ramani Narayan, Sang Il Lee, Tae Jin Lee : Morphology and rheological properties of blends of chemically modified thermoplastic starch and polycaprolactone; *Polym Eng Sci* 48(11):2126–2133 (2008).
10. Ptasienski, Mitchell P. Gill, Joseph W., Dry cement composition comprising cellulosic thickener gelled starch, polyvinyl alcohol and polyvinyl acetate, United States Patents: 3003979, assigned to UNITED STATES GYPSUM CO., October 10, 1961.
11. Additive combination for improving the workability of water containing building material mixtures, European Patent: 0530768-B1, September, 1992.
12. Nevins M, Reid K. NL Industries Inc. Aqueous well servicing fluids, GB 2110698-A, Publisher: NL Industries In (1983-06-22).
13. Thickening system for building material mixtures (German) The thickener for tile adhesives and gypsum-based smoothing mortars contains cellulose ether, starch ether, and phyllosilicates. Publisher: Clariant GmbH, 1997, EP0773198.
14. House RF, Hoover LD. NL Industries Inc. Aqueous well servicing fluids, GB 2110699-A (1983-06-22).
15. Additive mixtures for gypsum materials based on cellulose ethers (German)Additive for gypsum-based mortars, containing cellulose ether, cationic polyacrylamide, hydroxypropyl starch and superplasticzer (i.e. Ca-lignosulfonate).Publisher: Aqualon GmbH, DE3920025-A1, 1991.
16. Racciato, Joseph S. (San Diego, CA), Thickening compositions containing xanthan gum, guar gum and starch, United States Patent: 4105461, Merck & Co., Inc. (Rahway, NJ), August 8, 1978.

17. Girg, Friedrich (Idstein, DE) Bohme-kovac, Jozef (Dexheim, DE), Building material products containing organic polymers as thickeners, United States Patent: 5432215, July 11,1995.
18. http://www.wacker.com/cms/en/products-markets/trademarks/vinnapas/vinnapas.jsp; Wacker Polymers Manual, V 2.00/18-04-05/KB VINNAPAS RE 5010N.
19. Ducharme, Robert (Ste-Anne-de-Bellevue, CA) Boisvert, Andre (St-Leon, CA) Zinkewich, Johanne (Verdun, CA) Laroche, Lucie (Danville, CA), Acoustic construction panel, United States Patent: 5125475, June 30,1992 .
20. Akihiro Moriyoshi, Ichiro Fukai and Mikio Takeuchi, A composite construction material that solidifies in water, *Nature* 344, 230–232, Nature Publishing Group, 15 March 1990.

PART 4
BIOMEDICAL APPLICATIONS

12

Cellulose Based Green Bioplastics for Biomedical Engineering

A.K. Mishra[1] and S.B. Mishra[2]

[1]*UJ Nanomaterials Science Research Group, Department of Chemical Technology, University of Johannesburg, Doornfontein, South Africa*
[2]*Department of Chemical Technology, University of Johannesburg, Doornfontein, South Africa*

Abstract
Cellulose is the most abundant green bioplastic, found in nature. This natural biopolymer is the main component of plants cell walls and also found in few bacterial, algal and fungal species. The non toxicity, non-mutagenicity and biocompatibility of the polysaccharide is well known and has been explored for various biomedical engineering categories especially tissue, neural, pharmaceutical and fabrications of implants. The occurrence, morphology, structure and applications of cellulose have been mentioned. The cellulose as a green plastic and its recent research in the field of biomedical engineering has been discussed in this chapter.

Keywords: Cellulose, tissue engineering, pharmaceutical engineering, implant, biomedical engineering

12.1 Green Bio plastics

The climate change, natural disasters, global warming and depletion of ground water table and few more such factor, which living species in all forms are facing today is the gift of technology based industrialization. To normalize the present situation, it becomes highly important to search the similar or equally competitive products to carry on with the lifestyle which we are accustomed to. Large group of researchers and environmental agencies such as US EPA & UNEP, are particularly interested in investigating and working towards commercializing the bio-based products. Eatables, utensil, clothes, drugs, electronics, aerospace and many more which are part of our life today are primarily polymer or plastic based essentials.

Recovery of a plastic from natural, renewable resources especially the biomass refers to green bioplastics or organic plastics (Figure 12.1). The bioplastics can be degradable or non degradable based on the type of origin i.e if it is derived from

Srikanth Pilla (ed.) Handbook of Bioplastics and Biocomposites Engineering Applications, (347–356)

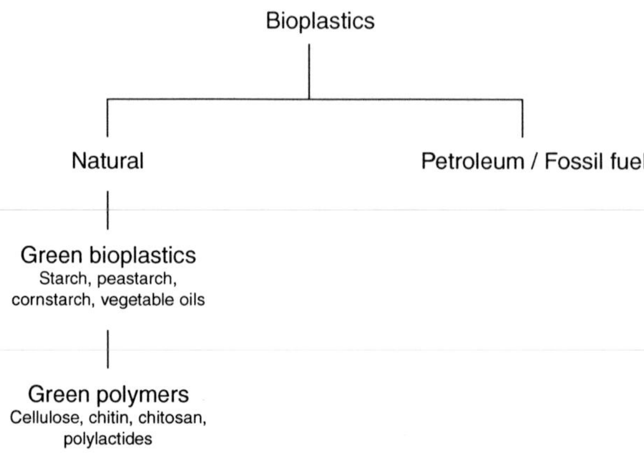

Figure 12.1 Classification of bioplastics.

agricultural source or from fossil fuel. The most important sources of the green bioplastics are starch, corn starch, pea starch and vegetable oil. Few applications of the green bioplastics include packaging, shopping bags, drug capsules, car interiors, pipes & cell phone housing. Before proceeding into details, it is necessary to understand the fact that the nature & the properties associated with the parent polymer decides if the product developed from it is biodegradable or non biodegradable, disposable or non disposable. These natural polymers or green bioplastics have also contributed to a large extent in the field of biomedical sciences and engineering. Among these bioplastics that are widely used or being investigated are, cellulose, starch, chitin, chitosan, polylactides.

12.2 Biomedical Engineering

Medical science is a discipline which has benefited millions of people across the world and improving the mortality rate. In a larger perspective, medical science is an interdisciplinary subject which is a culmination of aspects with respect to diagnostics, prevention and cure which cannot be met without the help of other scientific and engineering domains. Biomedical engineering is one of those fields where medical science is related to engineering aspects. To better understand with the biomedical engineering or what one should infer from this complicated term is that it includes various type of engineering such as tissue, genetic, neural, medical imaging and implants. Figure 12.2 shows some of the outcomes of this interdisciplinary area.

As shown in the above Figure 12.2, some of the other fields of biomedical engineering are bionics, medical imaging, biomechanics, bio-nanotechnology, clinical engineering and so forth. Since it is covering diverse fields from medical science to engineering and technology, we have to actually focus on the niche areas where the green bioplastics are playing a crucial role. In this chapter, therefore we will focus on cellulose and it utility as green plastic in the biomedical engineering.

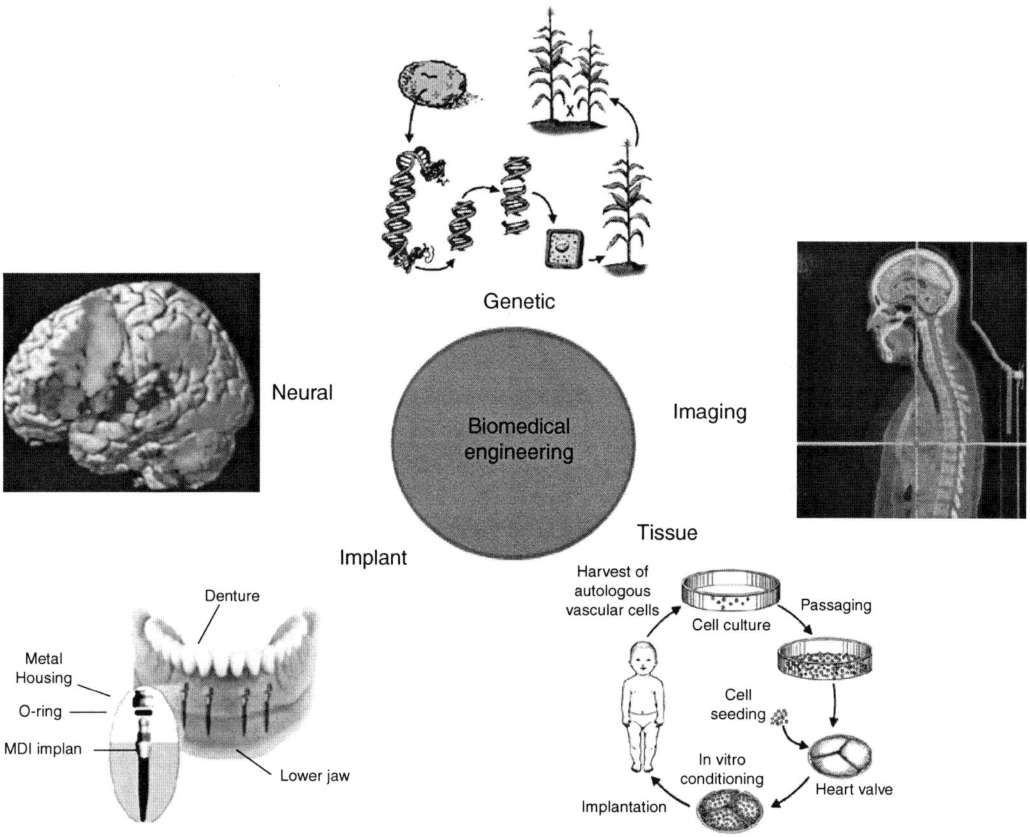

Figure 12.2 Various fields of biomedical engineering.

12.3 Cellulose

In 1838, Payen identified the prominent polymer in nature and coined the term "Cellulose". Cellulose is a natural polymer or the biopolymer abundantly found in nature and its composition varies from 30% to 95% depending upon the type of plant. It is estimated that 10^{11} – 10^{12} tons of cellulose is generated in pure form annually with the help of photosynthesis. It is principle biopolymer matrix along with hemicelluloses, lignin and other extracts to form wood composite. The chemical structure of this polysaccharide is composed of β (1-4) glucose units that combines together to form a linear chain. The linear chain can be made up of hundreds to thousands of these units having glycosidic bonds. The basic structure of this polysaccharide is shown below in Figure 12.3. The functionalization or modification of cellulose was reported to have carried out was as early as 1920s. The first modification was acetylation and deacetylation of this green plastic.

Some of the other sources of cellulose are bacteria and algae. Various bacterial species have been investigated and successfully used for deriving cellulosic structures. Among the bacteria are Gluconacetobacter **xylinium** & Acanthamoeba castellani, where as for algae Valonia ventricosa and Chaetamorpha melagonicum.

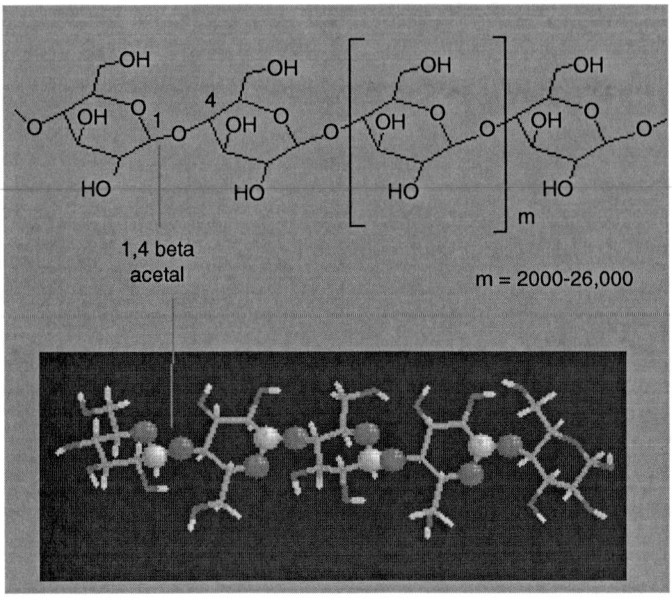

Figure 12.3 Structure of cellulose.

The morphology of cellulose is shown in Figure 12.4 which shows well organized fibrillar structure. Each elementary fibril is therefore the simplest morphological unit of cellulose and is composed of primary cell wall of diameter 10 nm, a secondary cell wall with two layers of thickness ranging from 100 nm–300 nm and finally tertiary that touches the fiber lumen.

The properties of cellulose are well known in today's scientific world. Some these that make it a unique valuable green bioplastic is its hydrophilic behaviour, insolubility in water and most of the organic solvents, crystalline nature, high tensile strength due to hydrogen bonding among the linear polymeric chains and last but not the least is ability to biodegrade. These properties thus renders this bioplastic either in its pristine state or modified state to be used for various applications ranging from water treatment, paper production, displays to medical or health care. Here, we are discussing some of the research work carried out using cellulose or modified cellulose in the field of biomedical engineering applications.

12.4 Cellulose Based Bioplastics for Biomedical Engineering

12.4.1 Tissue and Neural Engineering

When cells, materials and engineering are combined together, these give rise to an interdisciplinary domain of tissue engineering. Tissue engineering is tailor made approach to design or replace biochemical functions such as bone, cartilage, skin and bladder etc. With the help of tissue engineering, the scientists are also trying to develop artificial organs. Here, we are discussing some recent work done using cellulose bioplastic for tissue and neural engineering.

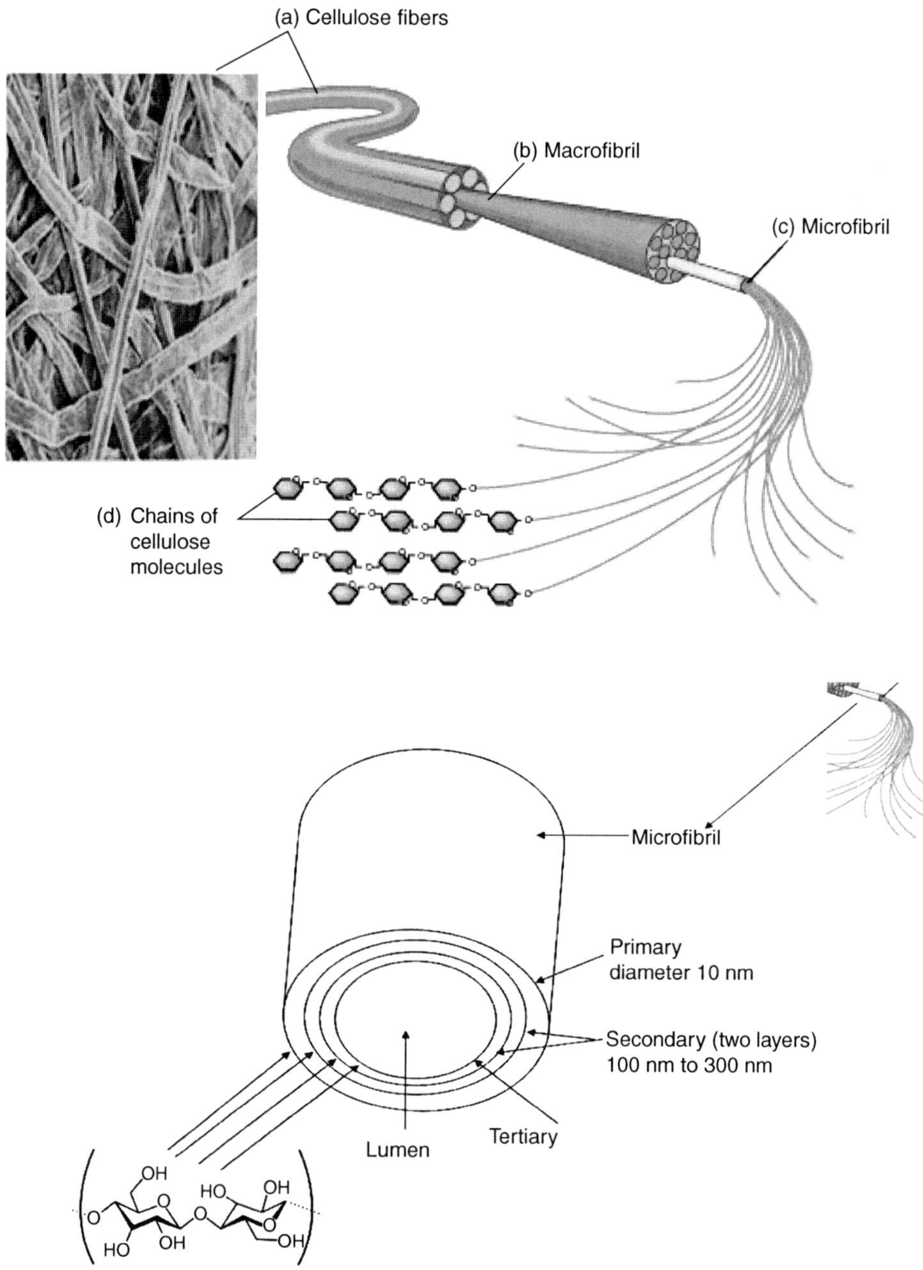

Figure 12.4 Morphology of cellulose.

Cellulose matrix was modified to 2,3-dihydrazone cellulose *via* 2,3-dialdehyde cellulose formation. The modified cellulose was tested for cytocompatibilty using mice fibroblast cell and thus was recommended for scaffold tissue engineering [1]. Acetobacter xylinum was elsewhere used for developing bacterial cellulosic membranes with tiny pores of 60 to 300 μm that did not show the crack formation or border failure. These micorporous membranes were therefore expected to be used for tissue repair [2]. 2,3-dialdehyde bacterial cellulose nano-network was synthesized and was found to be successfully degraded into a porous scaffold having microsized pores in water and simulated body fluid [3].

In a different study, for in vitro cartilage tissue engineering and chondrocyte cell response, non woven cellulose was chosen. The researchers were able to show the homogenous distribution of seeded cells followed by the development of cartilage tissue [4]. In another investigation Gluconacetobacter **xylinus** was used for studying its secretion of cellulose which was also sulphonated and phosphorylated to be used for novel scaffold material. A comparative study of the natural and modified bacterial cellulose was done with respect to tissue culture plastic and alginated. It was concluded that the natural bacterial cellulose had the highest in growth of chondrocytes onto this scaffold [5]. For urinary diversion, tissue engineered conduit was developed using bacterial cellulose scaffold that was seeded with human urine derived stem cells.

The scaffold was reported to be three dimensional and highly porous that successfully was able to form multilayer urothelium and cell-matrix infiltration [6]. In another study on bacterial cellulose scaffold, the researchers incorporated paraffin wax during the fermentation process. They, thus, used the scaffolds for studying the cell growth of osteoprogenitor cells [7]. Growth of functional cardiac cell constructs was studied using cellulose acetate and regenerated cellulose. It was shown that the biodegradability of these bioplastics could be controlled with the help of deacetylation, hydrolysis and cytocompatibilty [8]. Interconnected macroporous hydrogels in aqueous environment were produced by hydroxypropylcellulose modified by allyl isocynate. Minimal inflammatory response was shown by in vivo cytocompatibilty tests when implanted subcutaneously in mice [9]. Cellulose based bioplastics especially the gel forming derivatives have also been investigated for wound dressing. Cellulose crosslinked with hyaluronic acid was found to be effectively proliferating the keraticnocytes [10]. Growth of chondrocytes and in vivo formation of cartilaginous tissue with in mice was observed in silanized hydroxyl propyl cellulose scaffolds [11]. In another study, fibroblast of L929 strain was incorporated into hydrogels derived from bacteria *G. hansenii* ATCC 23769 and was therefore recommended for guided tissue regeneration [12].

12.4.2 Pharmaceutical Engineering

Pharmaceutical engineering is a hybrid of biomedical and chemical engineering. And is one of the important subdivisions of biomedical engineering where cellulose and its derivatives play a crucial role in formulation properties. The below Figure 12.5 show concept of control release of the drug from a formulation which is supposedly the result of pharmaceutical engineering.

In this section we will discuss the cellulose and its derivatives that have been investigated for pharmaceutical engineering. Colloidal dispersion of microcrystalline cellulose with chemically gelatinized maize starch was prepared to be used for multifunctional pharmaceutical recipients with enhanced disintegration abilities [13]. It has been shown by a group of researchers that the manufacturing factors have an influence on the material properties and functionalities of

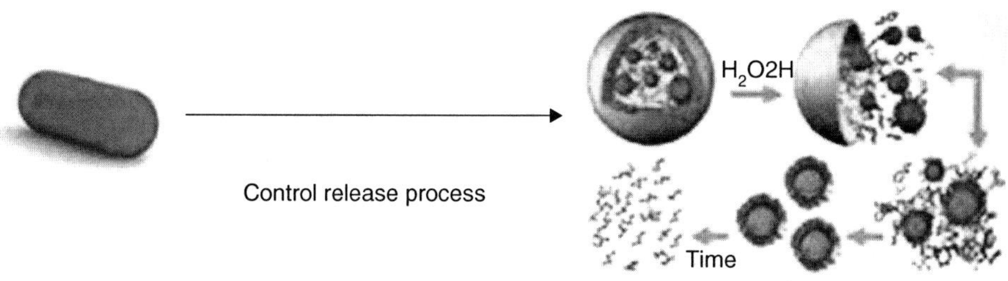

Figure 12.5 Control release of the drug from a formulation in a capsule.

microcrystalline cellulose to be used for pharmaceutical engineering application [14]. Moisture induced degradation of acetyl salicylic acid with the structural effect of cellulose was studied. The authors reported to have observed higher stability of the acetyl salicylic acid with cellulose of low crystalline index [15]. Cellulose derivatives are the most popular for film forming bioplastics for coated drug release forms. The film forming properties of this green bioplastic was investigated for an improvement by incorporating alkenyl succinic anhydrides as plasticizers with excellent mechanical strength [16].

In a different study, the effect of variation of interlot and inter supplier onto the properties of cellulose and modified cellulose for the control release of theophylline was studied. The researchers concluded that the molecular weight distribution and molecular size of cellulose were the important factors that were affecting drug release [17]. Chitosan and ethyl cellulose microspheres were blended and were used to load ciprofloxacin hydrochloride. The drug was reported to be stable in the blend and a significant improvement of the release time was observed [18]. Microcrystalline cellulose paste was investigated for extrusion-spheronization and it was found that either water or dimethylsuphoxide was suitable solvent that enhances the mechanical property of cellulose [19]. The size and shape of paracetamol particles affects the flow and compression behavior when blended with cellulose. In this study cellulose was blended with three different type of paracetamol partiles *viz* untreated paracetamol and two small particles sized, micronized and SAXD processed. Out of these three blends, cellulose-SAXD blend was reported to be the best combination to have shown enhanced properties [20].

Comparative study of powdered and microcrystalline cellulose, as sole excipients cellulose was carried using extrusion spheronization, forming furosemide pallets. Hydrophobic drug release rate was significantly higher in the powdered cellulose [21]. Non-Fickian trend of controlled release of ketrolac tromethamine was studied from the semi interpenetrating polymer network microspheres of gelatin and carboxymethyl cellulose crosslinked by glutaraldehye [22]. Chitosan/cellulose multimicrosphere were loaded with ranitidine hydrochloride, acetaminophen and 6-mercaptopurine. It was concluded the loading was directly proportional to the hydrophobicity of the drugs [23].The cellulose and cyclodextrin was co-dried to study the effect of wet granulation and lubrication on the tablet properties. Avicel pH 101 and 301 were taken for comparison and were found to be sensitive to lubrication with decrease disintegration [24].

12.4.3 Implants

Implants are often considered as medical devices which are able to replace natural organs such as pacemakers. Implants, so far developed are basically composed of titanium, silicon or appetite. However, in recent years, cellulose and functionalized cellulose are also being investigated and recommended for developing various biocompatible implants. Implants are of various types such as dental, orthopedic or biomedical. Some of the implants are shown below in Figure 12.6.

The cellulose and its derivatives are the widely investigated green plastic that is used for fabricating for various types of implants. Carboxy methyl cellulose hydrogels have been recommended green plastic to develop breast implants due to non toxic and non mutagenic behavior [25]. Hydroxyapetite cellulose sponges were used for the induction of granulation tissue. These sponges were later implanted subcutaneously in rats and were able to attract macrophages and fibroblast promoting angiogenesis [26]. In another study, free fat graft was compared with cellulose membrane to prevent laminectomy membrane in dogs. Although cellulose coverage membrane were found to be better, also had neurological deficits [27].

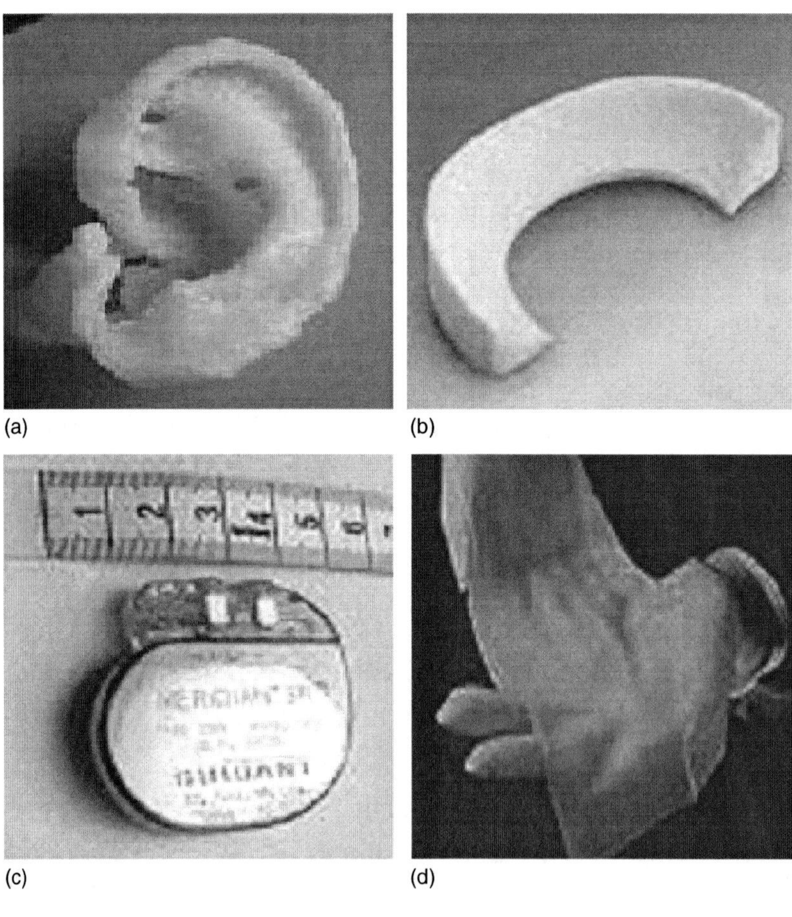

Figure 12.6 (a) Cartilage implant (b) Menisucal implant (c) Artificial pacemaker (d) Artificial skin.

Bacterial cellulose, on the other hand was also recommended for meniscus implants as it is inexpensive, can be developed into any shape of meniscus and favors cell migration [28]. It is important to note that the choice of cellulosic structures for building implants. Some researcher worked in this regards where they found that the viscous cellulosic implants were suitable for short period whereas cellulosic implants recommended for long term [29]. Bacterial cellulose further was investigated for developing artificial vascular implants. The results related to this investigation showed that the bacterial cellulose scaffold was promising candidate for small diameter artery [30].

12.5 Concluding Remarks

Cellulose and its derivatives investigated for various biomedical applications have been found to be one of the greenest bioplastic. The tissue, neural, pharmaceutical engineering and implants fabrication are the categories of biomedical engineering that are widely studied for the cellulose. Cellulose in various forms such as microcrystalline, powder, sponges or nano-structure defines the area of its application and affects the properties related to it to a large extent. It is worthwhile to mention here this green plastic has made a significant contribution in this multidisciplinary domain of science.

References

1. V. Verma, P. Verma, P. Ray and A.R. Ray, *Mater. Sci. & Eng. C*, Vol. 28, p. 1441, 2008.
2. C.R. Rambo, D.O.S. Recouvreux, C.A. Carminatti, A.K. Pitlovanciv, R.V. Antônio and L.M. Porto, *Mater. Sci. & Eng. C*, Vol. 28, p. 549, 2008.
3. J. Li, Y. Wan, L. Li, H. Liang and J. Wang, *Mater. Sci. & Eng. C*, Vol. 29, p. 1635, 2009.
4. F.A. Muller, L. Muller, I. Hofmann, P. Greil, M.M. Wenzel and R. Staudenmaier, *Biomaterials*, Vol. 27, p. 3955, 2006.
5. A. Svenssona, E. Nicklassonb, T. Harrah, B. Panilaitis, D.L. Kaplan, M. Brittberg and P. Gatenholm, *Biomaterials*, Vol. 26, p. 419, 2005.
6. A. Bodin, S. Bharadwaj, S. Wu, P. Gatenholm, A. Atala and Y. Zhang, *Biomaterials*, 2010 (In Press).
7. M. Zaborowska, A. Bodin, H. Bäckdahl, J. Popp, A. Goldstein and P. Gatenholm, *Acta Biomaterialia*, Vol. 6, p. 2540, 2010.
8. E. Entcheva, H. Biena, L. Yina, C-Y. Chunga, M. Farrella and Y. Kostovc, *Biomaterials*, Vol. 25, p. 5753, 2004.
9. Z. Yue, F. Wen, S. Gao, M.Y. Ang, P.K. Pallathadka, L. Liu and H. Yu, *Biomaterials*, Vol. 31, p. 8141, 2010.
10. A. Sannino, S. Pappadà, M. Madaghiele, A. Maffezzoli, L. Ambrosio and L. Nicolais, *Polymer*, Vol. 46, p. 11206, 2005.
11. C. Vinatier, J. Guicheux, G. Daculsi, P. Layrolle and P. Weiss, *Biomed. Mater. Eng.*, Vol. 16, p. 107, 2006.
12. F.V. Berti, D.O.S. Recouvreux, C.R. Rambo, R.M. Ribeiro-do-Valle, P.F. Dias and L.M. Porto, 11th International conference on advanced materials, ICAM 2009, Brazil, 2009.
13. P.F. Buildersa, A.M. Bonaventurea, A. Tiwaladeb, L.C. Okpakoc and A.A. Attamad, *International Journal of Pharmaceutics*, Vol. 388, p. 159, 2010.
14. J-S. Wu, H-O. Ho and M-T. Sheu, *European Journal of Pharmaceutical Sciences*, Vol. 12, p. 417, 2001.

15. A. Mihranyana, M. Strømmeb and R. Eka, *European journal of pharmaceutical sciences*, Vol. 27, p. 220, 2006.
16. M. Tarvainen, R. Sutinen, S. Peltonenc, H. Mikkonenc, J. Maunusa, K. Vaha-Heikkila, V-P. Lehto and P. Paronen, *European Journal of Pharmaceutical Sciences*, Vol. 19, p. 363, 2003.
17. C. Alvarez-Lorenzo, E. Castro, J.L. Go´mez-Amoza, R. Martínez-Pacheco, C. Souto and A. Concheiro, *Pharmaceutica Acta Helvetiae*, Vol. 73, p. 113, 1998.
18. P. Shi, Y. Zuo, Q. Zou, J. Shen, L. Zhang, Y. Li and Y.S. Morsi, *International Journal of Pharmaceutics*, Vol. 375, p. 67, 2009.
19. S. Mascia, C. Seiler, S. Fitzpatrick and D.I. Wilsona, *International Journal of Pharmaceutics*, Vol. 389, p. 1, 2010.
20. J.S. Kaerger, S. Edge and R. Price, *European Journal of Pharmaceutical Sciences*, Vol. 22, p. 173, 2004.
21. L. Alvarez, A. Concheiro, J.L. Go´mez-Amoza, C. Souto and R. Martınez-Pacheco, *European Journal of Pharmaceutics and Biopharmaceutics*, Vol. 55, p. 291, 2003.
22. A.P. Rokhade, S.A. Agnihotri, S.A. Patil, N.N. Mallikarjuna, P.V. Kulkarni and T.M. Aminabhavi, *Carbohydrate Polymers*, Vol. 65, p. 243, 2006.
23. H.Y. Zhou, X.G. Chen, C.S. Liu, X.H. Meng, C.G. Liu and L.J. Yu, *Biochemical Engineering Journal*, Vol. 31, p. 228, 2006.
24. J.-S. Wu, H.-O. Ho and M.-T. Sheu, *European Journal of Pharmaceutics and Biopharmaceutics*, Vol. 51, p. 63, 2001.
25. C.A. Brunner and R.W. Gröner, *Can. J. Plast. Surg.*, Vol. 14, p. 151, 2006.
26. M. Tommila, J. Jokinen, T. Wilson, A.-P. Forsback, P. Saukko, R. Penttinen and E. Ekholm, *Acta Biomaterialia*, Vol. 4, p. 354, 2008.
27. R.C.Da Costa, N.L. Pippi, D.L. Grac, S.A. Fialho, A. Alves, A.C. Groff and Ubirata Rezler, *The Veterinary Journal*, Vol. 171, p. 491, 2006.
28. A. Bodin, S. Concaro, M. Brittberg, P. Gatenholm, *Journal of Tissue Engineering and Regenerative Medicine*, Vol. 1, p. 406, 2007.
29. M. Martson, J. Viljanto, T. Hurme, P. Laippal and P. Saukko, *Biomaterials*, Vol. 20, p. 1989, 1999.
30. D.A. Schumann, J. Wippermann, D.O. Klemm, F. Kramer, D. Koth, H. Kosmehl, T. Wahlers and S. Salehi-Gelani, *Cellulose*, Vol. 16, p. 877, 2009.

13

Chitin and Chitosan Polymer Nanofibrous Membranes and Their Biological Applications

Ahsanulhaq Qurashi [1*]

Center of Excellence in Nanotechnology and Department of Chemistry, King Fahd University of Petroleum and Minerals, Dhahran, Kingdom of Saudi Arabia

Abstract

A variety of shapes of polymer nanostructures, nanocomposites, nanofibrous membranes were intentionally studied to discern their possible applications. This chapter highlights the chitin and chitosan nanofiber structures, nanofibrous membranes and their biocompatible nanocomposites. These chitin/chitosan nanofibrous membranes and their nanocomposites were found to be sustainable, biodegradable, antimicrobial, non-toxic and exhibited tremendous other biological activities compared to their micro or bulk membranes. Chitin and chitosan possesses interesting inimitable structures, multidimensional properties, highly complicated functionalities and possibility of engineering into nanofibrous membranes. Chitin and chitosan is one of the important polymers investigated so far and yield potential applications in areas such as filtrations, recovery of metal ions, drug release, dental, bone tissue engineering, catalyst and enzyme carriers, wound healing, protective clothing, skin regeneration, biosensors, medical implants and liver functioning respectively. Nanofibers matrices so far showed fascinating results in tissue engineering scaffolds due to their ultrafine, continuous fibers high porosity, variable pore-size distribution, high surface to volume ratio.

Keywords: Polymer nanostructures, chitin, chitosan, nanofibrous membranes, nanocomposites and biomedical applications

13.1 Introduction

Polymeric materials possess many attractive properties such as high toughness and recyclability. Some possess exceptional biocompatibility, biodegradability, and can offer various biofunctionalities [1–7]. An appropriate combination of functional polymers and biomolecules can offer tailored properties for various biomedical applications, but the ability to process them at the nanoscale to form well-defined functional structures is largely underdeveloped. Nanofabrication techniques for

---

Srikanth Pilla (ed.) Handbook of Bioplastics and Biocomposites Engineering Applications, (357–370)

feature sizes less than 100 nm are highly available for silicon-based or inorganic nanomaterials using high cost methods.

In recent times conducting polymer-based one-dimensional (1D) nanostructured materials (nanowires) less than 100 nm have acknowledged much attention transversely in scientific and engineering disciplines, due to their light weight, large surface area, adjustable transport properties, chemical specificity, low cost, ease of processing and readily scalable production [8–22]. Furthermore, biological modifications can be made prior to polymerization, during polymerization or after polymerization. This makes them very attractive and adaptable materials because optimal conditions can be used for each step to obtain optimum polymer conductivity and orientation of the biofunctionalized moiety.

Chitin and chitosan are natural aminopolysaccharides polymers with inimitable structures, multidimensional properties, highly complicated functionalities and extensive applications in biomedical and other industrial areas [23–27]. Also these chitin and chitosan polymers are sustainable, biocompatible, biodegradable, antimicrobial and non-toxic polysaccharides of great importance in many fields of application [28–32]. Among the many types, the cationic polysaccharide chitosan is a very promising polymer for producing functional nanofibers. Although the material has good physicochemical properties, the electrospinning of the polymer is far from easy. In this chapter we intend to study the formation of different types of polymer nanostructures. Also we will focus on the biocompatibility of chitosan and chitin polymer. Nanostructures of chitin or chitosan yield potential applications in areas such as filtrations, recovery of metal ions, drug release, dental, tissue engineering, catalyst and enzyme carriers, wound healing, protective clothing, cosmetics, biosensors, medical implants and energy storage [32–37].

In this book chapter we present the current research activities regarding the chitin and chitosan polymer nanofibrous membranes and their potential and promising applications in various biomedical field for instance tissue engineering, skin regeneration, liver functioning etc.

13.2 Shape of Polymer Nanostructures

2.1: Polymer nanostructures have attracted tremendous interest in optoelectronic devices and biomedical applications and many other related fields. These useful and attractive applications fascinated the researchers to develop various shapes of polymer nanostructures. Few of interested morphologies of polymer nanostructures, their nanofibrous membranes and their possible biomedical applications are summarized in this chapter.

Figure 13.1 showed the optical and SEM images of the poly (pyrrolepropylic acid [PPA]) nanowires [42]. The average diameter of nanowires was about 175–250 nm. These nanowires were deposited by electrochemical deposition template directing method. Before the electrochemical deposition process, one side of alumina template was sputtered with a thin film of gold metal and acted as the seed layer. The length of the poly (PPA) nanowires was controlled by the amount of charge passed.

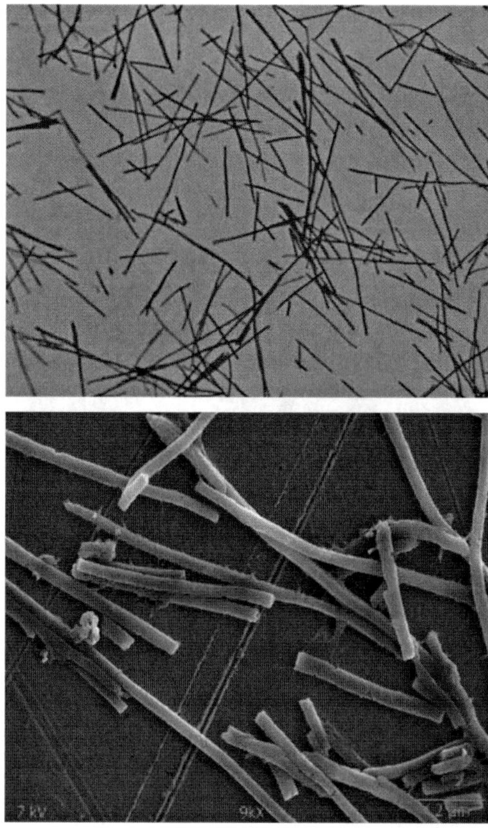

Figure 13.1 Optical (top) and SEM (bottom) images of bundles of poly (PPA) nanowires. Reprinted from Ref. [42], copyright permission from Springer 2010.

High aspect-ratio nanofibers of cationic polysaccharide, chitosan derivative such as N-[(2-hydroxy-3- trimethylammonium)propyl] chitosan chloride (HTCC), have been formed by electrospinning aqueous solution of poly(vinyl alcohol) (PVA)–HTCC blends [43]. Electrospinning is an effective method for preparing ultrafine polymer nanofibers with diameter ranging from micrometers to nanoscale. SEM image reveals the surface morphology and average diameter of the electrospun nanofibers of PVA-HTCC as shown in Figure 13.2. The results showed that diameters of the electrospun fibers of PVA–HTCC blends were in the range of 200–600 nm, depending on the electrospinning conditions. The SEM results indicated that weight and applied voltage considerably affected the final morphology of the nanofibers. It was found that by increasing HTCC contents in the polymer reduced the average diameter of nanofibers.

Dumbbell shaped polymer nanoparticles were synthesized by step seeded emulsion polymerization. It is important to note that polymer nanoparticles have tremendous applications in drug discovery, biosensors and lithography. In order to achieve minitured electronic devices, fine substrate patterning is an important and remains serious issue. However polymer nanoparticles showed tremendous performance for nanoscale patterning of substrates for small scale

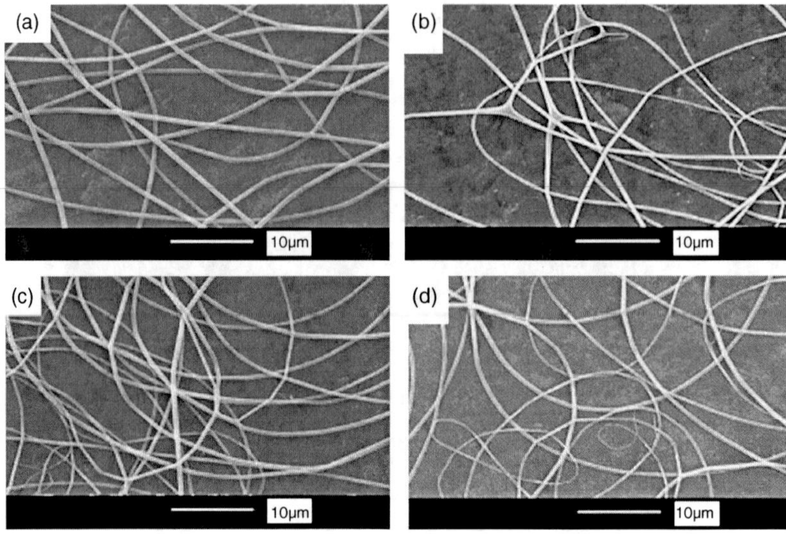

Figure 13.2 SEM-micrographs and average diameter distribution of the fibers prepared from PVA–HTCC blend solutions. Weight ratios PVA–HTCC = 100:0 (a), 95:5 (b) 90:10 (c) and 75:25 (d), applied voltage 10 kV, spinning distance 15 cm. Total polymer concentration 12 wt %. Ref. [43], copyright permission from Elsevier 2010.

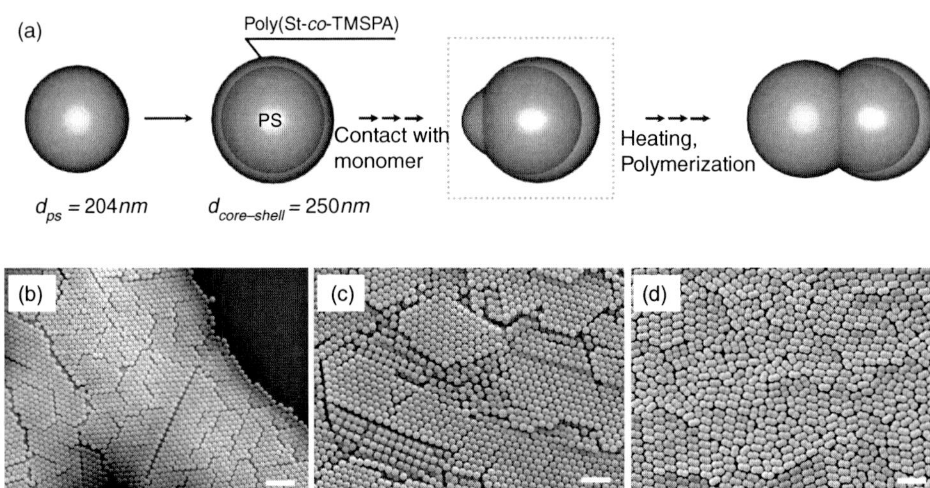

Figure 13.3 SEM images of (a) a polypyrrole-based CPNEJ grown on one of the ten 2 μm wide electrode gaps between electrode groups A and B and (b) a PEDOT-based CPNEJ grown between electrode groups C and D. SEM images of (c) polypyrrole CPNWs (ca. 80–150 nm in diameter) and (d) PEDOT CPNWs (ca. 60–120 nm in diameter) that were grown on the working electrode surfaces (ca. 4000 μm^2) in the electrode groups A and C, respectively. Ref. [13], copyright permission from institute of physics 2010.

optoelectronic devices. Figure 13.3 (a) shows schematic illustration for two step seeded polymerization. Figure 13.3 (b) shows polystyrene spheres (PS) polymer nanoparticles [44]. Figure 13.3 (c) shows polystyrene and trimethoxysilylpropylacrylate (TMSPA) core-shell polymer nanoparticles. Figure 13.3 (d) shows FESEM image of symmetric dumbbell shaped polymers.

Generally nanostructures were prepared by two general methods top-down and bottom-up. However by combination of both the techniques highly ordered and complex micro and nanostructures can be obtained. By simple and versatile techniques various types of complex polymer nanostructures were obtained which include spin coating, multibeam interference, developing and drying respectively. These complex nanostructures are highly applicable in nanobiosystem engineering [45]. Figure 13.4 (a–d) shows plane and tilted FESEM images of complex periodic polymer nanostructures formed at different exposure time of

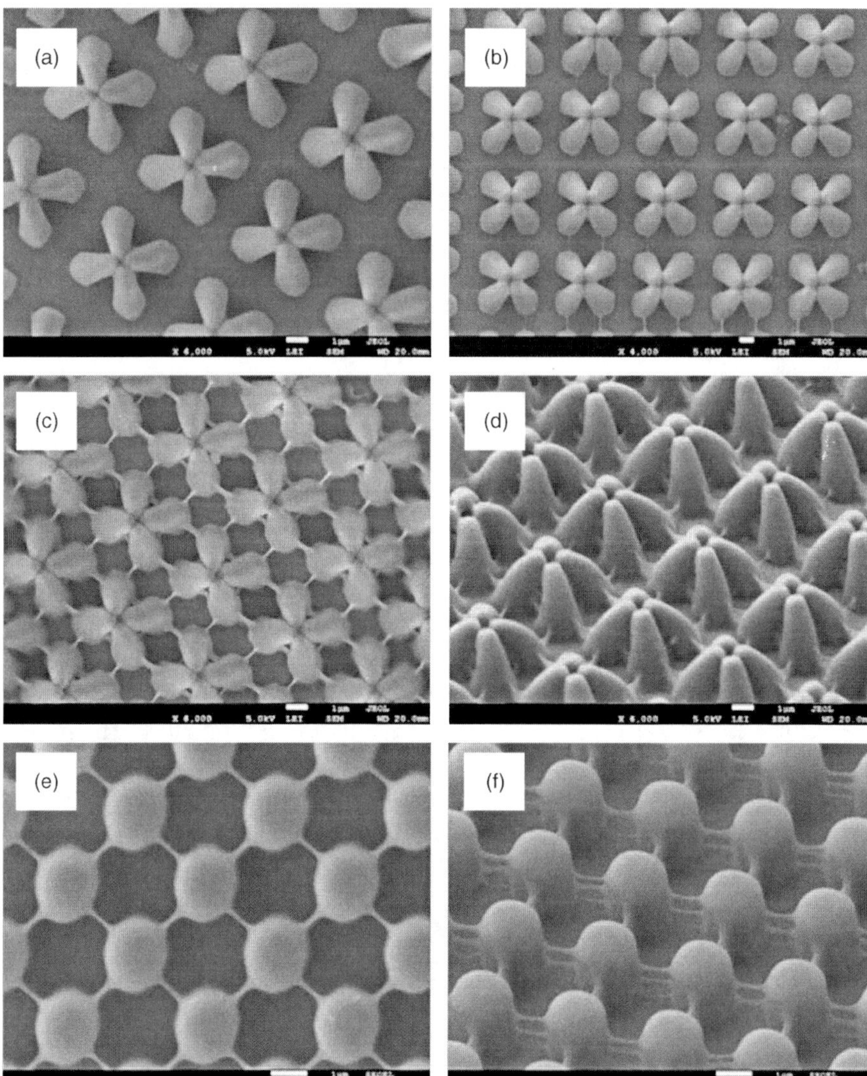

Figure 13.4 Synthesis of dumbbell-shaped polymer nanoparticles: (a) Schematic representation of two-step seeded emulsion polymerization. (b-d) Scanning electron micrographs of (b) PS nanoparticles, (c) PS/poly(St-*co*- TMSPA) core-shell nanoparticles, and (d) symmetric dumbbell-shaped nanoparticles. For the synthesis of (d), V_m/V_{core}-shell, the volume ratio of the monomer solution to the core-shell seed particles, was 0.9. The scale bars in the micrographs represent 1.0 μm. Ref. [44], copyright permission from American chemical society 2010.

3, 4 and 5 s respectively. These nanostructures were prepared by combination of top-down laser microfabrication and bottom-up self organization.

13.3 Application of Chitosan Nanofibers

13.3.1 Lipase Immobilization

Nanomaterials attracted attentions of researchers due to their high-surface to volume ratio, high surface area, enhanced enzyme activities, catalytic efficiency etc [46–58]. Thus, it is necessary to use nanostructures as supporting material for enzyme immobilization. Jun *et al.* developed a nanofibrous membrane with a fiber diameter of 80–150 nm from mixed chitosan/poly(vinyl alcohol) (PVA) solution via an electrospinning technique [59]. It was found that lipase loading on nanofibrous membrane (chitosan) was reached upto 63.6 mg/g and retention activity of enzyme was about 49.8% at optimum conditions. However the pH and thermal stabilities of lipase was highly improved after immobilization on chitosan nanofibrous membranes. From the storage ability and reusability it was observed that the residual activities of lipase were enhanced enormously.

13.3.2 Antibacterial Activities of Quarternay Chitosan Nanofibers

Polymers with bactericidal activities represent an important class of materials due to their ability to wound healing and wound dressings respectively. Chitosan (polysaccharide) is reported to possess fabulous biological properties like non toxicity, intrinsic antibacterial properties, and haemostatic activity etc. Ignatova *et al.* recently reported microbiological screening of quarternay chitosan [60]. They found the antibacterial activity of photo-cross-linked electrospun mats against *Staphylococcus aureus* and *Escherichia coli* respectively.

The cross-linked quarternay chitosan (QCh) poly vinyl alcohol (PVA) electrospun mat containing 2845 μg/mL QCh killed *S aureus* bactria with in 60 min. of contact time.

However photo-cross-linked PVA did not affect the bacterial growth (Fig. 13.5A).

These results indicated that antibacterial activities of QCh/PVA nanofibers mat resulted from QCh which effected against S *aureus* as bactericidal. Similarly the electrospun QCh/PVA mats were also exposed to Gram-positive bacteria (E. coli). E. Coli bacteria and showed the reduction for cross-linked QCh/PVA nanofibers containing 2885 μg QCh was 98% after 120 min. contact time (Fig 13.5B).

13.3.3 Wound Dressing

Nanofibrous membranes (NFM) present a variety of advantages over the conventional wound dressing processes. Due to their huge surface area and porosity, these NFM can start signaling pathway and draw fibroblasts to the derma layer. This derma layer can easily secrete necessary extra cellular matrix components to repair the damaged tissues.

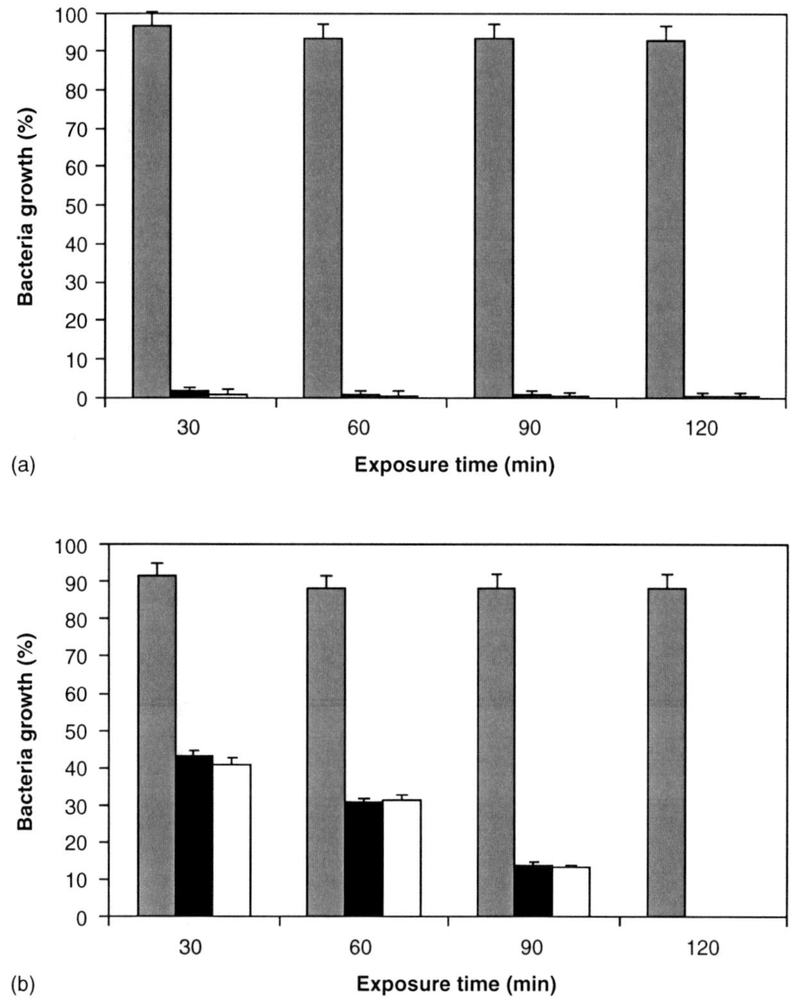

Figure 13.5 Plot of the bacteria growth, in percent of the control, versus the exposure time: for electrospun cross-linked QCh/PVA mats (□), for cross-linked QCh/PVA films prepared by solvent casting method (■) and for electrospun cross-linked PVA mats (▒). The test was carried out against S. aureus (A) and against E. coli (B). The error bars are the standard deviations for triplicated experiments. Ref. [60], copyright permission from Elsevier 2010.

Chin *at al.* investigated electrospun collagen/chitosan NFM as wound dressing material[61].

NFM doesn't showed cytotoxicity towards growth of 3T3 fibroblasts and demonstrated excellent *in vitro* biocompatibility. When compared with animal studies, NFM proved better than the gauze and commercial collagen sponge wound dressing in healing rate. Thus, it was accomplished that this NFM will have great potential as a wound dressing for skin regeneration.

Furthermore chitin NFM showed interesting results for biodegradability and cellular test. Recently Noh *et al* studied the chitin nanofibrous (Ch-N) and commercial Chitin microfibrous (Ch-M) for biodegrability and cellular response to normal human keratinocytes and fibroblasts [62]. Compared to Chi-M, high cell

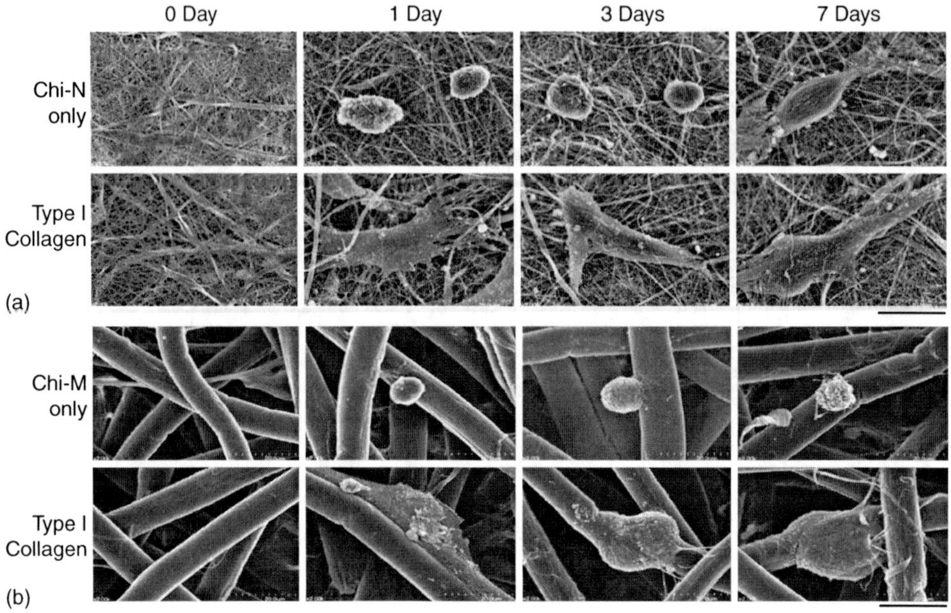

Figure 13.6 SEM micrographs of the interaction between NHGF and chitin nanofibrous (A) or microfibrous (B) structures coated with type I collagen after 0, 1, 3, and 7 days of culture. Chi-N only, uncoated chitin nanofibers; Type I collagen, type I collagen-coated chitin nanofibers (A); Chi-M only, uncoated chitin microfibers; Type I collagen, type I collagen-coated chitin microfibers (B). Magnification 2000 X; Bar, 20 μm. Ref. [62], copyright permission from Elsevier 2010.

attachment and uniform spreading of cells were observed in Chi-N. Figure 13.6A shows primary normal human gingival fibroblasts (NHGF) adhered and spread on the surface of chitin nanofibrous and microfibrous networks. NHGF started growing deeply within the matrix. However the cells could not grow properly under layer of chitin microfibrous as shown in Figure 13.6B.

13.3.4 Cellular Compatibility

It is highly significant to know the reliability of nanofibrous materials in water and celluar biocompatibility to serve as potential scaffolding material for tissue engineering purposes. Recently Bhattarai *et al.* developed chitosan/polyethylene oxide based nanofibrous materials to investigate the celluar biocompatibility and reliability in water [38]. They found that the matrix with a chitosan/PEO ratio of 90/10 maintained exceptional integrity of the fibrous structure in water. Figure 13.7 (A, B) shows the SEM images of osteoblast (MG-63) grown on the chitosan nanofibers after 5 day cell culture. The cells were attached suitably and showed numerous and long microvilli on their surfaces. Figure 13.7 (C and D) shows SEM images of chondrocyte (HTB-94) cells grown on chitosan nanofibers after 5 days of culture at low and high magnification. There was proper adherence of cells and demonstrated round shaped chondrocytes, representing that the nanofibers continuing phonotype of chondrocytes. These results indicated that nanofibrous maintained distinguishing cell morphology and viability during the study period.

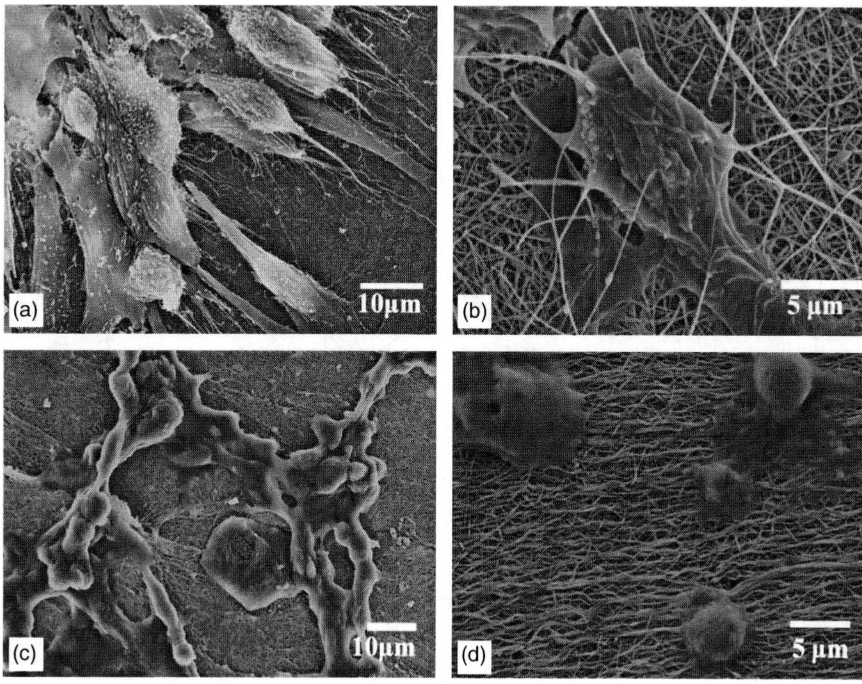

Figure 13.7 SEM images of osteoblasts (MG-63) (A and B) and chondrocytes (HTB-94) (C and D) seeded on nanofibrous membranes of chitosan/PEO (90/10) after 5 day culture; (A) 800 x original magnification, (B) 3500 x original magnification, (C) 800 x original magnification and (D) 2500 x original magnification. Ref. [38], copyright permission from Elsevier 2010.

13.3.5 Bone Tissue Engineering

New materials remain important issue for bone tissue engineering and attained considerable attention of biologists and material scientists. Nanocomposites based on hydroxyapetite ($Ca_{10}(PO_4)_6(OH)_2$) is more frequently used for bone tissue engineering due to their structural and compositional properties to engineer functional native bone-like substitutes by employing tissue engineering approach. Zhang *et al.* studied nanocomposite nanofibers hydroxyapetite/chitosan for bone tissue engineering [63]. By playing with the chemistry of HAp nanoparticles loading into chitosan scaffold resulted into significant bone formation oriented outcomes, contrast to that of pure electrospun CTS (chitosan) scaffolds. The compositional and structural features of HAp/CTS nanocomposite fibers were found to be close to natural mineralized nanofibril complements and have prospective attention for bone tissue engineering applications.

Figure 13.8 shows FESEM images of cell-scaffold constructs after 10 and 15 days of culture.

The nanofibrous nanocomposite scoffold surfaces found completely covered with multi-layers of cells. The cells secreted extra cellular matrix (ECM) representing convergence of human fetal osteoblast (hFOB) cell growth. HAp/CTS based scaffolds produced exclusively more mineral deposits and aggregated into coalesce and larger mineral bunchs compared to the electrospun CTS. Figure 13.8 (E and F)

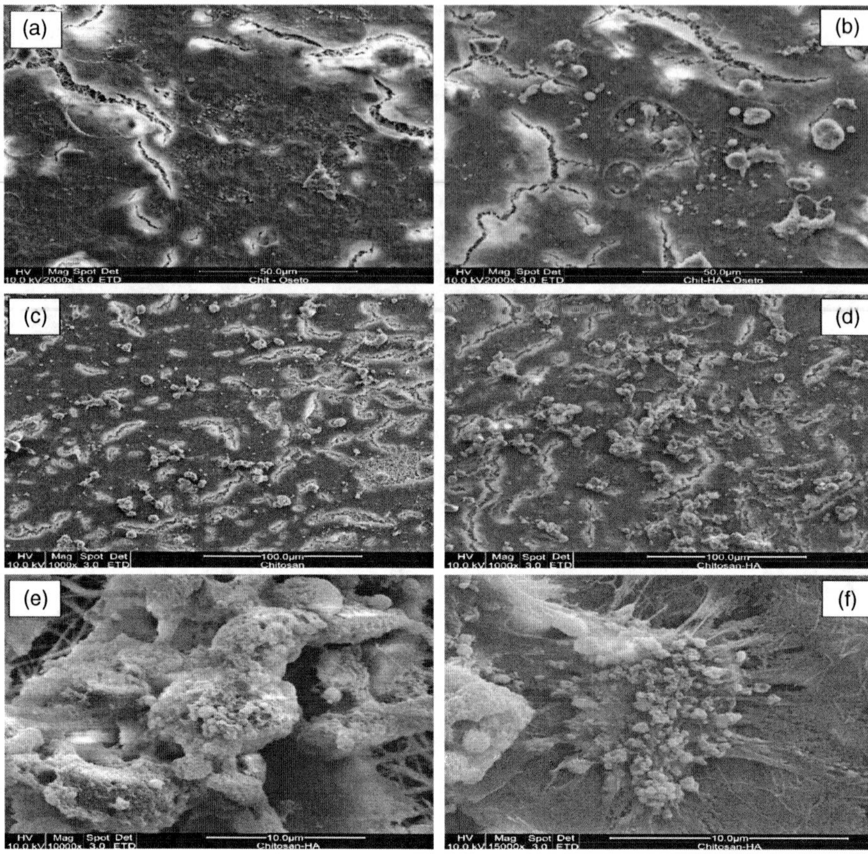

Figure 13.8 Mineral depositions of hFOB on the electrospun nanofibrous scaffolds: CTS, day 10 (A) and day 15 (C); HAp/CTS, day 10 (B) and day 15 (D); apatite-like morphology of deposit at higher magnification (E); visible tiny globular minerals and collagen bundles associated with a single hFOB cell viewed at higher magnification (F). Ref. [63], copyright permission from Elsevier 2010.

also showed minerals deposits apetite-like porous morphology deposited on basal surface of hFOB cells.

13.3.6 Skin Regeneration

Skin is one of the largest organ of human body and comparatively composed of soft tissue, covering the complete external surface and forming about 8% of the total body mass. Millions of people's suffers every year due to injuries and require skin graft. The conventional skin graft techniques have demerits like high cost, the limited availability of skin grafts in severely burned patients etc. One of the best ways to deal with damaged skin is to develop effective tissue engineering substitutes. Nanofibers matrices so far showed fascinating results in tissue engineering scaffold due to their ultrafine, continuous fibers are oxygen-permeable high porosity, variable pore-size distribution, high surface to volume ratio, and most importantly and morphological similarity to natural extracellular matrix (ECM) in skin. Zhou *et al.* studied biocompatible carboxyethyl chitosan/poly(vinyl alcohol)

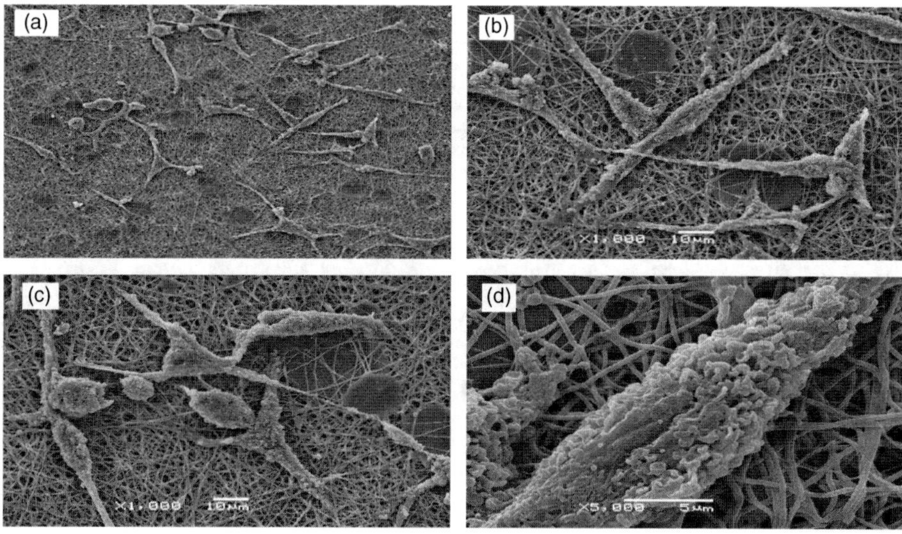

Figure 13.9 SEM images of L929 cell seeded on nanofibrous membrane of CECS/PVA (50/50) after 48 h culture. Ref. [64], copyright permission from American chemical society 2010.

(CECS/PVA) nanofibers for skin generation applications [64]. The possible utilization of nanofibrous mat of CECS/PVA as scaffolding materials for skin regeneration was examined *in vitro* using mouse fibroblasts (L929) as reference cell lines. Their cell cultured results demonstrated that nanofibrous mat was excellent in promoting cell growth attachment and proliferation. FESEM images (Figure 13.9) of L929 grown on the cross-linked nanofibers CECS/PVA after 48 hr cell culture showed well adherence and revealed normal morphology on the surface because of large surface area for cell attachment. Figure 13.9 (b, c and d) showed cells attachment to the surface by filopodia and revealed smaller and numerous microvilli on their surfaces. Microvilli seem to grow along the polymer nanofibers.

13.3.7 Liver Functioning

Primary hepatocyte culture plays crucial role in the clinical treatment of hepatic failure patients. This is carried out in order to avoid natural human immune response and to provide enough replacement of synthetic and metabolic functions of liver. Particularly bioengineers devised tissue engineering, regeneration and bioartificial liver assisted device (BLA). To obtain the higher level of liver-specific functions and mechanical stability these heapocytes were cultured on different types of substrates with diverse biomaterials and structures. Recently Feng *et al.* developed nanofibrous galactosylated chitosan which showed slow degradation and highly attractive mechanical properties [65]. Figure 13.10 shows confocal microscope images of double-stained fluorescence demonstrated bioactivity of hepatocyte aggregates. In this investigation the hepatocyte aggregates seeded on GC films and GC nanofibers demonstrated tremendous cell bioactivity without death of inner hapotocytes in course of 5 days culture period. Conversly on 5th day of culture some of hepatocyte on GC film started to loose their activity

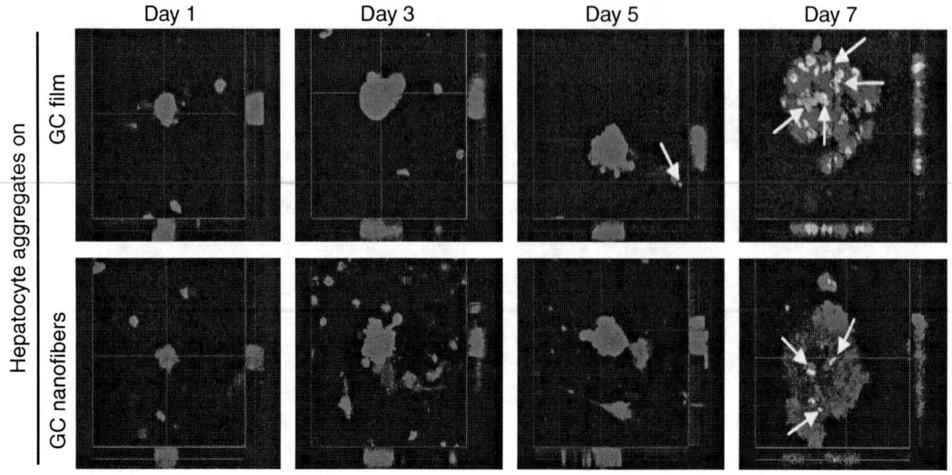

Figure 13.10 Confocal microscopy images of double-staining fluorescence (Living – green, Calcein-AM; Dead – red, Cytox) of hepatocyte aggregates cultured on GC substrates at various time points: Days 1, 3 and 5 (Scan bar: 250 μm); Day 7 (Scan bar: 100 μm). Ref. [65], copyright permission from Elsevier 2010.

(yellow arrow). Later on 7th day various dead hepatocytes (red) were found over the aggregates on GC films. However hepatocyte aggregates on GC nanofibers retained their bioactivity during the 7 days of culture period. In conclusion the GC nanofibrous scaffold showed good bioactivity and high levels of liver functioning for longer period of time compared to the hepatocytes on GC films.

13.4 Conclusion

In this chapter, we have emphasized the evolution of different shapes of polymer nanostructures like nanowires, nanospheres and nanoflowers and chitin and chitosan polymer nanofibers, their nanostructured membranes and nanocompsoites. By employing different techniques it is possible to prepare various kinds of polymer nanostructures similar to the metals and metal chalcogenides and carbon nanostructures. In addition detailed study was performed on chitin, chitosan and their nanocomposite membranes for different biological applications for instance enzyme immobilization, antibacterial activities, wound dressing, cellular biocompatibility, bone tissue engineering, skin regeneration and liver functioning. It was found that NFMs highly useful for biological application due to their nanoscale morphology, high surface-to-volume ratio, high surface area compared to the commercial or their bulk structured membranes.

References

1. C. Montserrat, J.S. Antonio, V.-R. Mar´ıa, *Chem. Mater.*, Vol. 18, p. 5676, 2006.
2. E. Chiellini, A. Corti, S. D'antone, R. Solaro, *Prog. Polym. Sci.*, Vol. 28, p. 963, 2003.

3. A.P.V. Pereira, L.V. Wander, R.L. Orefice, *J. Non-Cryst. Solids*, Vol. 273, p. 180 2000.
4. P. Willi, P.S. Chandra, *Am. J. Biochem. Biotechnol.*, Vol. 2, p. 41, 2006.
5. T. Yamaoka, Y. Tabata, Y. Ikada, *J. Pharmaceut. Pharmacol.*, Vol. 47, p. 479, 1995.
6. Y.F. Tang, Y.M. Du, X.W. Hu, X.W. Shi, J.F. Kennedy, *Carbohydr. Polym.*, Vol. 67, p. 491, 2007.
7. L.C. Wang, X.G. Chen, D.Y. Zhong, Q.C. Xu, *J. Mater. Sci.: Mater. Med.*, Vol.18, p. 1125, 2007.
8. B.W. Maynor, S.F. Filocamo, M. W. Grinstaff, J. Liu, *J. Am. Chem. Soc.*, Vol. 124, p 522, 2002.
9. M. Reza Abidian, D. H. Kim, D.C. Martin, *Adv. Mater.*, Vol. 18, p. 405, 2006.
10. K. Ramanathan, M. A. Bangar, M. Yun, W. Chen, Nosang V. Myung, and A. Mulchandani, *J. Am. Chem. Soc.*, Vol. 127, p. 496, 2005.
11. K. Ramanathan, M. A. Bangar, M. Yun, W. Chen, A. Mulchandani, N.V. Myung, *Nano Lett.*, Vol. 4, p. 1237, 2004.
12. Y. Dan, Y. Cao, T. E Mallouk, S. Evoy T.C. Johnson, *Nanotechnology*, Vol. 20, p. 434014, 2009.
13. Y. Wang, K.K Coti, J Wang, M. M Alam, J.J. Shyue, W. Lu, N.P Padture and H. R. Tseng, *Nanotechnology*, Vol. 18, p. 424021, 2007.
14. M. A. Bangar, D.J. Shirale, W. Chen, N.V. Myung, A. Mulchandani, *Analytical Chemistry*, Vol. 81, 2168, 2009.
15. C.M. Hangarter, M. Bangar, A. Mulchandani N.V. Myung. *Mater. Chem.*, Vol. 20, p. 3131, 2010.
16. S. B. Tolani, M. Craig, R.K. DeLong, K. Ghosh, A.K. Wanekaya, *Anal Bioanal Chem.*, Vol. 393, p. 1225, 2009.
17. A. Rahman, M. K. Sanyal, *Adv. Mater.*, Vol. 19, p. 3956, 2007.
18. C. Huang, N. Lu, Y Wang, L. Tian, B. Yang, B. Dong, L Chi. *Langmuir*, Vol. 26, p. 9142, 2010.
19. S.Y. Lee, H. Lim, G.R. Choi, J.D. Kim, E. K. Suh, S. K. Lee, *J. Phys. Chem. C.*,Vol. 114, p. 11936, 2010.
20. Y.S. Jung, W. C. Jung, H.L. Tuller, C. A. Ross *Nano Lett.*, Vol. 8, p. 3776, 2008.
21. Y. Cao, A. E. Kovalev, R. Xiao, J. Kim, T.S. Mayer, T.E. Mallouk, *Nano Lett.*, Vol. 8, p. 4653, 2008.
22. J. Xu, Y. Jiang, Y. Yang, *Materials Science and Engineering: B*, Vol. 157, 2009, p. 87.
23. K. Aoi, A. Takasu, and M. Okada, *Macromolecules*, Vol. 30, p. 6134. 1997.
24. K. Aoi, A. Takasu, M. Tsuchiya, M. Okada, *Macromol. Chem. Phys.*, Vol. 199, p. 2805, 1998.
25. T. Chandy, P.C. Sharma, *Biomater Artif. Cells Artif. Organs*, Vol.18, p. 1, 1990.
26. W. Paul, P.C. Sharma, *STP Pharma Sci.*, Vol.10, p. 5 2000.
27. R.A.AMuzzarelli RAA,Muzzarelli, *Adv. Polym. Sci.*, 2005, Vol. 186, p. 151–209.
28 K.V.H. Prashanth, R.N. Tharanathan, *Food Sci. Technol.*, Vol. 18, p. 117, 2007.
29. E. Khor. Curr Opin, *Solid State Mater. Sci.*, Vol. 6, p. 313, 2002.
30. M. Rinaudo, *Prog. Polym. Sci.*, Vol. 31, p. 603, 2006.
31. R. Jayakumar N. Nwe S. Tokura H. Tamura, *Int. J. Biol. Macromol.*, Vol. 40, p. 175, 2007.
32. M. Rinaudo, *Main Polym. Int.*, Vol. 57, p. 397, 2008.
33. B.M. Min, S.W. Lee, J.N. Lim, Y. You, T.S. Lee, P.H. Kang, W. H. Park, *Polymer*, Vol. 45, p. 7137, 2004.
34. K. Ohkawa, K. Minato, G. Kumagai, S. Hayashi, H. Yamamoto *Biomacromolecules*, Vol. 7, p. 3291, 2006.
35. J. Lyons, F. K. Ko, *Encycl. Nanosci. Nanotechnol.*, Vol. 6, p. 727, 2004.
36. K. Ohkawa,D. Cha, H. Kim, A. Nishida, H. Yamamoto, *Macromol. Rapid Commun.*, Vol. 25, p. 1600,. 2004.
37. V.S. Fridrikh, J.H. Yu, M.P. Brenner, C.G. Rutledge, *Phys. Rev. Lett.*,Vol 90, p. 144502/1. 2003.
38. N. Bhattarai, D. Edmondson, O. Veiseh, A.F. Matsen, M. Zhang, *Biomaterials*, Vol. 26, p. 6176, 2005.
39. L. Li, Y.L. Hsieh, *Carbohydr. Res.*, Vol. 341 (3), 374, 2006.
40. M.B. Min, W.S. Lee, N.J. Lim, Y. You, S.T. Lee, P.H. Kang, H.W. Park, *Polymer*, Vol. 45, p. 7137, 2004.
41. Y.S. Shin, N.H. Park, Y.H. Kim, H.M. Lee, S.Y. Choi, J.Y. Park, M.Y. Lee, Y. Ku, C.I. Rhyu, B.S. Han, S.J. Lee, P.C. Chung, *J. Periodontol.*, Vol. 76, p. 1778, 2005.
42. B.S. Tolani, M. Craig, R. K. DeLong, K. Ghosh A.K. Wanekaya, *Anal. Bioanal. Chem.*, Vol. 393, p. 1225, 2009.
43. S.M. Alipour, M. Nouri, J. Mokhtari, S.H. Bahrami, *Carbohydrate Research*, Vol. 344, p. 2496, 2009.

44. J.G. Park, J.D. Forster, and E.R. Dufresne, *J. Am. Chem. Soc.*, p. 132, p. 5960, 2010.
45. D. Wu, Q. Chen, B.B. Xu, J. Jiao, Y. Xu, H. Xia, and H.B. Sun, *Appl. Phys. Lett.*, Vol. 95, p. 091902, 2009
46. Q. Ahsanulhaq, A. Umar, and Y. B. Hahn, *Nanotechnology*, Vol. 18, p. 115603, 2007.
47. Q. Ahsanulhaq, J. H. Kim, and Y. B. Hahn, *Nanotechnology*, Vol. 18, p. 485, 2007
48. Q. Ahsanulhaq, S. H. Kim, and Y. B. Hahn, *J. Alloys. Comp.*, Vol. 484, p. 17, 2009.
49. A. Qurashi, N. Tabet, M. Faiz, and T. Yamazaki, *Nanoscale Res. Lett.*, Vol. 4, p. 948, 2009.
50. N. K. Reddy, Q. Ahsanulhaq, J. H. Kim, M. Devika, and Y. B. Hahn, *Nanotechnology*, Vol. 18, p. 445710, 2007.
51. Q. Ahsanulhaq, J. H. Kim, Y. H. Im, and Y. B. Hahn, *J. Nanosci. Nanotech.*, Vol. 10, p. 4950, 2010.
52. A. Qurashi, E. M. El-Maghraby, T. Yamazaki, and T. Kikuta, *J. Alloys Compd.*, Vol. 48, p. L9 2009.
53. A. Qurashi, E. M. El-Maghraby, T. Yamazaki, Y. Shen, and T. Kikuta, *J. Alloys Compd.*, Vol. 481, p. L35, 2009.
54. A. Qurashi, El Maghraby, T. Yamazaki, and T. Kikuta, *Appl. Phys. Lett.*, Vol. 95, p. 153109, 2009.
55. A. Qurashi, E. M. El-Maghraby, T. Yamazaki, and T. Kikuta, *Sensors and Actuators B*, 147, p. 482010.
56. E. M El-Maghraby, A. Qurashi, and T. Yamazaki, "*J. Nanosci. Nanotec.*, Vol.10, p. 4950, 2010.
57 Q. Ahsanulhaq, J.H. Kim, J.S. Lee, Y.B. Hahn, *Electrochem. Communi.*, Vol. 12, p. 475, 2010.
58. Q. Ahsanulhaq J.H. Kim J.H. Kim Y.B. Hahn, *Nanoscale Res. Lett.*, Vol. 5, p. 669, 2010.
59. X.J. Huang, D. Ge, Z.K. Xu *Eur. Polymer Journal*, Vol. 43, p. 3710, 2007.
60. M. Ignatova, K. Starbova, N. Markova, N. Manolova, I. Rashkov, *Carbohydrate Research*, Vol. 341, p. 2098, 2006.
61. J.P. Chen, G. Yun Chang, J.Chen, Colloids and Surfaces A: Physicochem. Eng. Aspects, Vol. 313–314, p. 183, 2008.
62. H. K.Noh, S.W. Lee, J. Kim, J. Oh, K.H. Kim, C.P. Chung, S.C. Choi, W. H. Park, B.M. Min, *Biomaterials*, Vol. 27, p. 3934, 2006.
63. Y. Zhang, J. R. Venugopal, A. Turki, S. Ramakrishna, B. Su, C.T. Lim *Biomaterials*, Vol. 29, p. 4314, 2008.
64. Y. Zhou, D. Yang, X. Chen, Q. Xu, F. Lu, J. Nie, *Biomacromolecules*, Vol. 9, p. 349, 2008.
65. Z. Feng, X. Chu, N.P. Huang, T. Wang, Y. Wang, X. Shi, Y. Ding, Z.Z. Gu *Biomaterials*, Vol. 30, p. 2753, 2009.

PART 5
AUTOMOTIVE APPLICATIONS

14

Biobased and Biodegradable PHBV-Based Polymer Blends and Biocomposites: Properties and Applications

Alireza Javadi[1,2], Srikanth Pilla[2], Shaoqin Gong[1,2] and Lih-Sheng Turng[2,3]

[1]*Department of Biomedical Engineering, University of Wisconsin–Madison, WI, USA*
[2]*Wisconsin Institute for Discovery,University of Wisconsin-Madison, WI, USA*
[3]*Department of Mechanical Engineering, University of Wisconsin–Madison, WI, USA*

Abstract
Petroleum-based polymers have made a significant contribution to the human society due to their extraordinary adaptability and processability. However, over the past few decades, the widespread application of plastics in various sectors has led to growing concerns over the undesirable environmental impact of plastics. Many strategies including more efficient plastics waste management and employment of biodegradable materials obtained from renewable resources have been investigated. Plastics waste management is at the beginning stages of development and has proven more expensive than expected. Thus, there is a growing interest in developing sustainable biobased and biodegradable plastics produced from renewable resources, which can offer a comparable performance while providing additional advantages such as biodegradability, biocompatibility, and a reduced carbon footprint. Poly(3-hydroxybutyrate-co-3-hydroxyvalerate) (PHBV) is one of the most promising biobased and biodegradable polymers. In fact, many petroleum-based polymers such as poly(propylene) (PP) and polystyrene (PS) can be potentially replaced by PHBV due to its unique material properties. Despite PHBV's attractive properties, there are several drawbacks including high cost, brittleness, and thermal instability, which hamper the widespread usage of this specific polymer. Several strategies (such as forming blends or composites with biodegradable polymers, natural fibers, or inorganic fillers, as well as developing novel processing techniques) have been investigated to overcome the aforementioned shortcomings, which will be discussed in this chapter.

Keywords: Biobased, biodegradable, microcellular injection molding, mechanical properties, viscoelastic properties, PHBV, polymer, crystallinity, thermal properties, biomedical applications

Srikanth Pilla (ed.) Handbook of Bioplastics and Biocomposites Engineering Applications, (373–396)

14.1 Introduction

Owing to growing environmental concerns over the use of synthetic, non-biodegradable polymers, there is a strong drive to replace some of the non-degradable petroleum-based polymers with biobased and biodegradable polymers made from renewable resources [1]. Biobased and biodegradable polymers not only offer a promising solution to growing environmental issues by conserving limited non-renewable resources (petroleum) and reducing CO_2 emissions but also by stimulating the growth of agricultural industries around the world as they produce the raw materials and feedstock needed for this thriving industry [1]. In the past few years, extensive research on biobased and biodegradable materials has led to a better understanding of their properties and morphologies, as well as their structure–property relationship. Poly(hydroxyalkanoates) (PHAs), a family of linear polyesters produced in nature by bacterial fermentation, are among the most promising biobased and biodegradable materials currently being investigated for potential industrial applications in various sectors such as packaging, biomedical, civil, and automotive [1].

PHAs can be synthesized through the bacterial fermentation of various renewable sources such as sugars, lipids, and alkanoic acids [2]. These biobased and biodegradable polymers can be broken down into CO_2 and water in the presence of appropriate biological conditions [3]. In addition, PHAs' mechanical and thermal properties are similar to those of polyolefins (synthesized from non-renewable resources), which makes them a promising candidate for replacing commodity polymers in diverse applications such as packaging, civil and construction, agricultural, automotive, and biomedical industries [4].

Among PHAs, poly(3-hydroxybutyrate) (PHB) and its copolymers have attracted a lot of attention in the past two decades due to their unique properties. PHB is a fully biobased and biodegradable polymer that can be processed via injection molding and extrusion. It has a high modulus and a tensile strength similar to that of isotactic polypropylene [4]. However, PHB possesses low impact strength/toughness and poor thermal stability. Degradation via chain scission caused by hydrolysis at high temperatures also makes it difficult to process [4]. Several approaches, such as annealing and recrystalization [5] and copolymerization of PHB with HV, [3, 4, 6, 7] have been explored to overcome the aforementioned disadvantages. Poly(3-hydroxybutyrate-co-3-hydroxyvalerate) (PHBV) is synthesized by microorganisms by consuming sugars in the presence of propionic acid or produced directly from plants [8]. PHBV has drawn considerable attention because of its biodegradability and improved properties as compared to PHB. By controlling bioreaction conditions, PHBV can be produced with HV units up to 30%. Mechanical and thermal properties such as toughness, Young's modulus, and crystallization rate can be engineered by optimizing the HV molar ratio in the PHBV [8]. PHBV is available commercially under various names including Tianan Biologic's ENMAT Y1000P™, Biomer's Biomer L™, and Metabolix's Biopol™. The schematic structure of repeating PHBV units is illustrated in Figure 14.1.

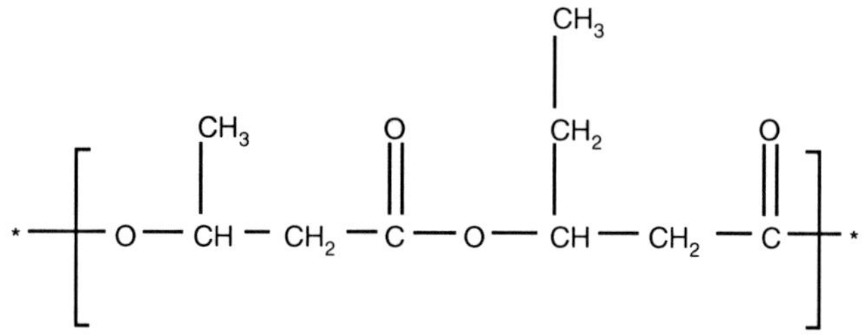

Figure 14.1 Schematic chemical structure of Poly(3-hydroxybutyrate-co-3-hydroxyvalerate) (PHBV).

In spite of improved mechanical (e.g., toughness) and thermal properties compared to PHB, PHBV still exhibits some disadvantages including low strain-at-break, a narrow processing window, a slow crystallization rate, and a higher cost as compared to petroleum-based synthetic polymers [9]. In order to tailor its properties and decrease its total cost, several approaches have been proposed such as forming blends or composites with biodegradable polymers, natural fillers, or inorganic fillers.

PHBV-based polymer blends have been extensively studied in order to reduce their material cost, improve their processability, tailor their biodegradability, and engineer their mechanical (e.g., toughness) and thermal properties (e.g., degree of crystallinity) [15]. Thus, biodegradable polymers such as poly(caprolactone) (PCL) [10], poly(butylene succinate) (PBS) [11], poly(ethylene succinate) (PES) [12], poly(butylene adipate-co-terephthalate) (PBAT) [13], and poly(hydroxyethyl methacrylate) (PHEMA) [14] have been blended with PHBV. In addition to PHBV-based polymer blends, PHBV/natural fiber composites have also been actively studied by many research groups. Due to their low cost, availability, unique mechanical properties, and biodegradability, natural fibers can effectively lower the total cost of PHBV-based materials while providing some improvements to their mechanical and physical properties [16]. The effects of adding various types of natural fibers such as recycled cellulose fiber [17], lignocellulosic flour [18], pineapple fiber [19], recycled wood fiber [16], kenaf fiber [20], bamboo fiber [21], wheat straw [22], flax [23], abaca [24], jute [24], and coir fiber [25], on PHBV have been investigated. These studies have shown that natural fibers can be embedded in the PHBV matrix as an excellent reinforcer to improve mechanical properties.

Besides natural fibers, the effect of adding various inorganic fillers into the PHBV matrix were also investigated by various research groups. In a study reported by Choi *et al.*, Cloisite® 30B acted as a nucleating agent and enhanced PHBV's crystallization rate. Also, the incorporation of nanoclay increased the thermal stability of PHBV composites [9, 26–28]. Moreover, bioactive fillers such as hydroxyapatite (HA) [29–31], wollastonite [32, 33], tricalcium phosphate (TCP) [34, 35], and sol-gel-bioactive glass (SGBG) [36], have been incorporated into the PHBV matrix to improve its biocompatibility, biodegradability, and mechanical properties.

As mentioned previously, PHBV can potentially replace polyolefins, especially polypropylene, due to its similar mechanical and thermal properties [4].

PHBV has the potential to be widely used in automotive, packaging, civil and construction, and biomedical industries. In order to fully utilize PHBV in these diverse industries, improving its thermal and mechanical properties (such as brittleness and low strain-at-break) and employing economic processing techniques (such as microcellular injection molding [13]) is important. Microcellular injection molding is an environmentally friendly polymer processing method capable of mass-producing components with minimally compromised material properties while consuming less energy and materials, as compared to components produced by the conventional injection molding process [37].

This book chapter presents an overview of the synthesis, processing, properties, and applications of biobased and biodegradable PHBV and its blends and composites.

14.2 Synthesis of PHBV

PHBV is a linear, aliphatic co-polyester produced either by (1) bacterial fermentation or (2) directly from plants. These synthesis routes are presented schematically in Figure 14.2 and discussed below. In addition to being biobased, it is a compostable and biodegradable polymer [3].

For the fermentation synthesis method, several bacteria including *Paracoccus denitrificans, Ralstonia eutropha, Escherichia coli,* and *Alcaligenes latus* have been reported to be able to efficiently produce PHBV by up to 85% of its dry cell weight using glucose and propionic acid, as substrate and co-substrate, respectively [38]. PHBV forms inside the bacterial cells in two distinct steps. In the first step, monomers of 3-hydroxybutyrate (HB) and 3-hydroxyvalerate (HV) are produced intracellularly [39]. Then, in the second step, PHBV is formed as cytoplasmic inclusions inside of the bacterial cell as the result of polymerization of the HB and HV monomers [39]. The biochemical pathway for PHBV production is initiated by the condensation of acetyl-CoA with propionyl-CoA which yields 3-ketoacyl-CoA. Then, the D-specific acetoacetyl-CoA reductase produces 3-hydroxy products by reducing the 3-ketoacyl-CoA and consequently, PHBV is synthesized [40]. Despite the fact that bacterial fermentation is an effective method which yields high production of PHBV, large-scale synthesis of PHBV using this method suffers from high cost [41, 42].

Through genetic engineering, PHBV can also be synthesized directly from green plants and is expected to be economically favorable as compared to bacterial fermentation [43, 44]. Synthesizing PHBV in plants requires two different metabolic precursors; i.e., acetyl-CoA and propionyl-CoA. Unlike acetyl-CoA, which is found plentifully and naturally in plants as the precursor of fatty acids, propionyl-CoA is scarcely present as an intermediate of amino acid degradation in the peroxisomes [45]. Gruys *et al.* improved the production of propionyl-CoA by modifying the isoleucine biosynthetic pathway and synthesized PHBV in *Arabidopsis thaliana* leaves and *Brassica napus* seeds (oilseed rape) [46]. Using this method, they produced propionyl-CoA derived from threonine via threonine deaminase and the pyruvate dehydrogenase complex (PDC) [46]. Thereafter, the additional

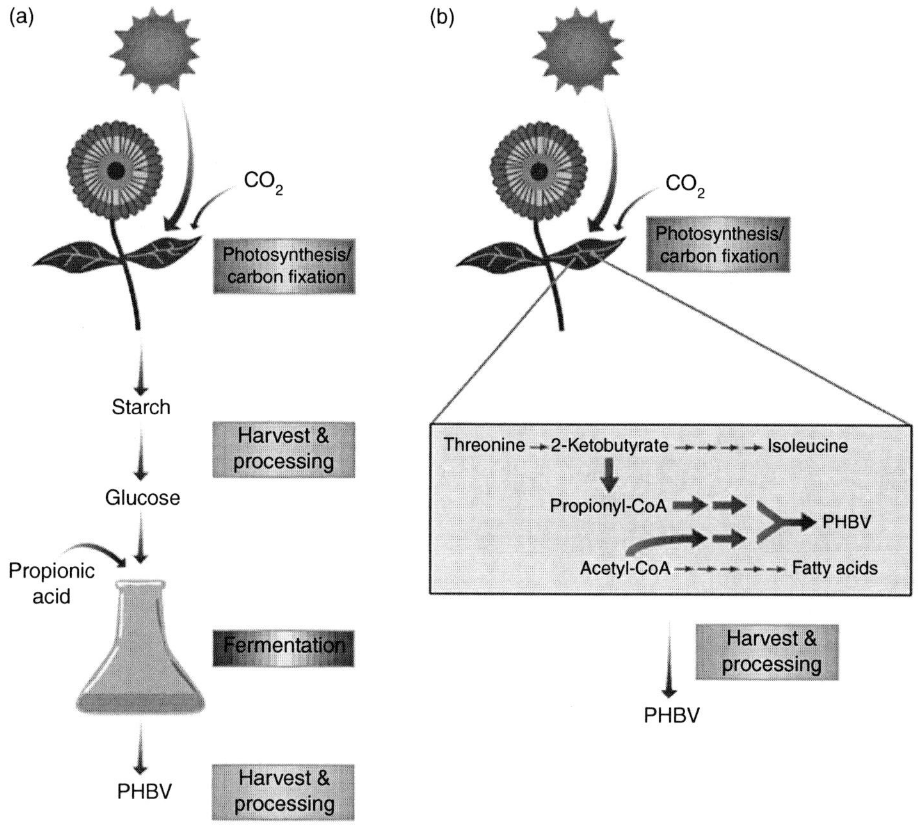

Figure 14.2 (a) Synthesis of PHBV by bacterial fermentation process; (b) Direct synthesis of PHBV in crop plants. Reprinted with permission from Y. Poirier, *Nature Biotechnology*, Vol. 17, p. 960, 1999, ©1999 Nature America Inc.

condensation of propionyl-CoA and acetyl-CoA leads to the production of the PHBV copolymer directly from plants. Despite being a cost effective method, direct synthesis of PHBV from plants yields lower production levels of PHBV, compared to the fermentation method [45]. Enhancing the production of PHBV via direct synthesis from plants can stimulate the growth of the agriculture industry; however, this method requires the modification of the plant's genome over the expression of four genes [45], which is still challenging.

14.3 Microcellular Injection Molding

Efforts on producing polymeric foams using supercritical fluids (SCF) were first reported during the 1970s and 1980s by Martini [47] and Okonishnikov [48]. The fundamental principle of microcellular foaming is to create a large number of microcells smaller than the preexisting flaws in the component to reduce the amount of polymer used without compromising the mechanical properties [49].

Several processing techniques can be used to produce microcellular components such as extrusion, injection molding, blow molding, etc. However, in this chapter, the focus will be on PHBV-based foams produced by microcellular injection molding. A continuous microcellular injection molding technique was commercialized by Trexel Inc. (under the trade name Mucell®) [50]. The microcellular injection molding process encompasses three major steps: gas dissolution, cell nucleation, and cell growth; all of which are briefly described below [51, 52].

1. Gas Dissolution: Supercritical fluid (SCF) such as N_2 or CO_2 is injected into the barrel and mixed with the polymer melt to form a single-phase polymer–gas solution.
2. Nucleation: Nucleation occurs at the nozzle or gate and is triggered by a rapid pressure drop.
3. Cell Growth: Cell growth occurs during the molding stage and within the mold cavity. Cell growth is thermodynamically favorable only when the size of the nucleated cell is greater than the critical size. Nucleated cells smaller than the critical size dissolve back into the polymer–gas single-phase solution. The growth rate is controlled by the gas diffusion rate and the melt strength of the polymer–gas solution [37].

There are other factors that affect the cell nucleation and growth rate such as time allowed for cells to grow, state of saturation, hydrostatic pressure applied to the polymer, temperature of the system, nucleating agents, blending with other polymers, and viscoelastic properties of the single-phase polymer–gas solution [53].

Microcellular injection molding has many benefits compared to conventional injection molding. The microcells can potentially improve mechanical properties by serving as crack arrestors due to blunting of the crack tip, thereby enhancing the material's impact strength, fracture toughness, and fatigue life [Bledzki *et al.*, 2006]. Also, components processed by the microcellular foaming technique possess an increased strength/density ratio and toughness, higher thermal and acoustic insulation properties, low dielectric constant, low thermal conductivity, and use less material due to smaller and more uniform cells (generally less than 100 μm) as compared to conventional foaming techniques [49, 54–57].

Due to their unique properties, microcellular plastics are particularly attractive for applications such as food packaging, the automotive industry, sporting equipments, roof sheet insulators, microelectronic circuit board insulators, electronic wire insulation, and molecular-grade filters [37]. Figure 14.3 depicts an SEM image of a tensile fractured surface of a plastic component processed by microcellular injection molding.

14.4 Thermal Properties

The thermal properties, including the glass transition temperature (T_g), melting temperature (T_m), and crystallinity of biobased and biodegradable PHBV-based polymer blends and composites are generally studied by differential scanning

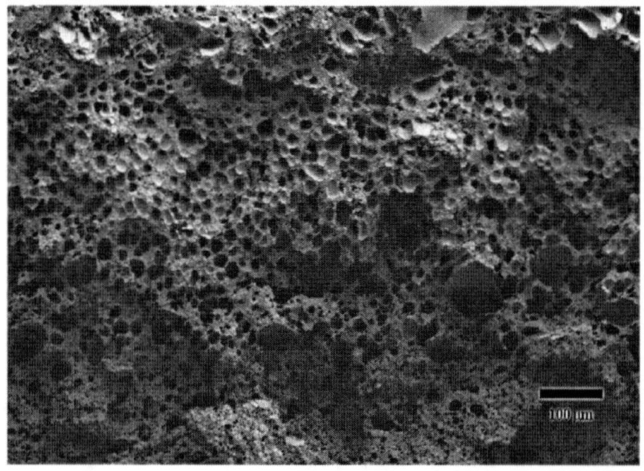

Figure 14.3 Representative scanning electron microscopy (SEM) image of the tensile fractured surface of a component processed by microcellular injection molding.

calorimetry (DSC) as discussed in this section. With an increasing HV molar ratio (from 0% to 34%) in PHBV, the T_m decreased from 175 °C to 97 °C and the T_g decreased from 9 °C to –9 °C as reported by Mark *et al.* and Brandup *et al.* [58, 59]. Scandola *et al.* reported that the degree of crystallinity of PHBV slightly changed with the HV monomer ratio and was consistently higher than 50% over the entire composition range of the HV monomer ratio (i.e. from 0% to 95%) [60]. However, with an increasing HV monomer molar ratio in PHBV (up to 55%), the crystallization rate became significantly slower (four orders of magnitude lower as compared to PHB) and at an HV molar ratio higher than 55%, the crystallization rate increased [60]. This behavior can be attributed to the fact that at an HV molar ratio around 55%, the crystalline phase shifts from a PHB to a PHV lattice, which leads to a much slower crystallization rate at this transition range [60]. A number of research groups have studied the effects of incorporating different polymers, natural fibers, and inorganic nanofillers on the crystallinity and thermal properties of PHBV.

The phase behavior and crystallinity of biodegradable PHBV/PES blends studied by Miao *et al.* showed that the PHBV and PES were immiscible and the degree of PHBV crystallinity decreased with the addition of PES whereas the degree of PES crystallinity remained unchanged at various blend compositions [12]. Chun and Kim investigated the thermal properties of biodegradable PHBV/PCL blends and reported that with the addition of PCL, the PHBV crystallization rate in PHBV/PCL blend decreased compared to that of neat PHBV indicating that PCL suppressed the nucleation of PHBV in the PHBV/PCL blend [61]. Similar observations for biodegradable PHBV/poly(ethylene oxide) (PEO) blends were reported by Tan *et al.* [62]. The thermal properties of solid and microcellular biodegradable PHBV/PBAT blends investigated by Javadi *et al.* revealed that the degree of PHBV crystallinity decreased consistently with an increasing PBAT content in the PHBV matrix for both solid and microcellular components and no significant difference was observed in the degree of PHBV crystallinity between solid and microcellular components [13].

Different natural fibers have been incorporated into the PHBV matrix in order to study the thermal properties of the resulting green composites. Dufresne *et al.* reported that the addition of lignocellulosic flour into the PHBV matrix increased the PHBV enthalpy of fusion and the rate of crystallization whereas no significant change was observed in T_g and T_m [18]. They concluded that lignocellulosic flour acted as heterogeneous nucleation sites for PHBV crystallization [18]. Reinsch *et al.* [63] and Avella *et al.* [22] reported similar results on wood fiber and wheat straw fiber filled PHBV composites. The thermal properties of the solid and microcellular PHBV/coir fiber composites studied by Javadi *et al.* showed an enhancement in the degree of PHBV crystallinity with the addition of coir fiber while no obvious change was observed in the degree of PHBV crystallinity between solid and microcellular components [25]. On the contrary, Gatenholm *et al.* [64], Avella *et al.* [65], Luo *et al.* [66], and Buzarovska *et al.* [67] all reported that incorporating wood cellulose fiber, steam exploded straw fiber, pineapple fiber, and kenaf fiber into the PHBV matrix did not cause any significant change to the degree of PHBV crystallinity or the crystallization kinetics.

The thermal properties of PHBV-based nanocomposites were also investigated by different groups. Choi *et al.* [28] studied the effect of incorporating organically modified nanoclay (Cloisite 30B) into the PHBV matrix and demonstrated that the intercalated organically modified nanoclay acted as a nucleating agent thereby enhancing the PHBV crystallization temperature and crystallization rate. Also, their study revealed that the melting temperature (T_m) of PHBV shifted to lower temperatures [28].

Similar findings were reported by Wang *et al.* [9] and Chen *et al.* [26, 27], who presented a detailed study of the thermal properties of the PHBV/organically modified montmorillonite (OMMT) nanocomposites. It was demonstrated that with an increasing OMMT content, the overall crystallization rate of PHBV was enhanced [26, 27]. They also concluded that OMMT played two opposite roles in the PHBV crystallization process. Firstly, acting as nucleating agents, OMMT nanoplatelets enhanced the nucleation and crystallization rate of PHBV. Secondly, OMMT nanoplatelets hampered the movement of PHBV chains and subsequently decreased PHBV's degree of crystallinity as a result of strong interactions between PHBV chains and OMMT nanoplatelets [27]. Figure 14.4 depicts the spherulitic crystalline structure developed in the PHBV and PHBV/OMMT nanocomposites (at different OMMT loading levels) obtained under a polarized optical microscope [9]. As shown in Figure 14.4, with the addition of OMMT to the PHBV matrix, the number of nucleation sites increased compared to those of neat PHBV [9]. Wang *et al.* reported that with the addition of OMMT, the formation and growth of spherulites occurred at a faster rate compared to pure PHBV. Also, with the addition of OMMT, more uniform spherulites were formed.

14.5 Thermal Degradation Properties

One of the major drawbacks of PHBV is its poor thermal stability [68]. This co-polyester, similar to other types of polyesters, undergoes thermal degradation and hydrolysis which can lead to a reduction in molecular weight at temperatures

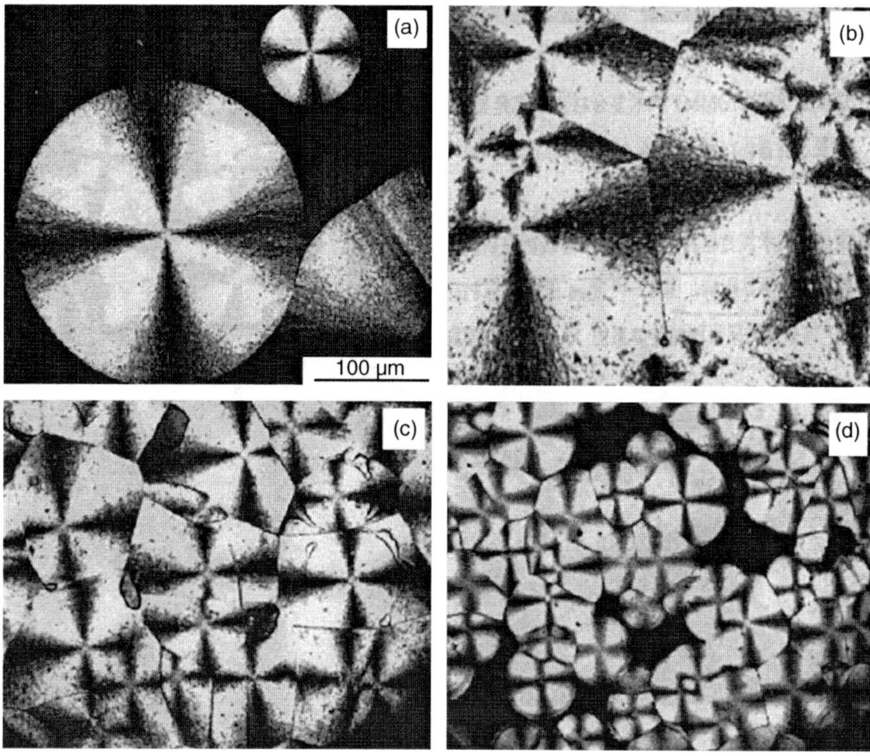

Figure 14.4 Polarized optical microscope photographs of PHBV/OMMT nanocomposites. (a) PHBV; (b) PHBV/ 3% OMMT; (c) PHBV/ 7% OMMT; and (d) PHBV/10% OMMT. Reprinted with permission from S. F. Wang *et al.*, *Polymer Degradation and Stability*, Vol. 87, p. 69, 2005, © 2004 Elsevier Ltd.

above 170 °C. Thermal degradation occurs through a random chain scission at the esteric group by β-hydrogen elimination as shown in Figure 14.5 [69].

Thermal degradation studies of PHBV were conducted using a thermogravimetric analyzer (TGA). Several methods such as incorporation of supercritical fluids, natural fibers, and inorganic nanofillers into the PHBV matrix have been proposed to improve the thermal stability of PHBV. Jenkins *et al.* demonstrated that incorporating supercritical fluid (CO_2) into the PHBV matrix during the melt blending process minimized the thermal degradation of PHBV due to a significant reduction of the PHBV melting point [70]. Several research groups have studied the thermal degradation of PHBV in the presence of natural fibers such as kenaf fiber [20], pineapple fiber [66], and bamboo fiber [21]. As reported in the aforementioned cases, the addition of natural fibers into the PHBV matrix improved the thermal stability of PHBV and yielded a higher ash content. The effect of incorporating inorganic nanofillers, such as organically modified nanoclay, on the thermal degradation of PHBV was investigated by Bruzaud and Bourmaud [71]. They reported that PHBV-based nanocomposites degraded at higher temperatures as compared to that of neat PHBV. They also reported the thermal stability

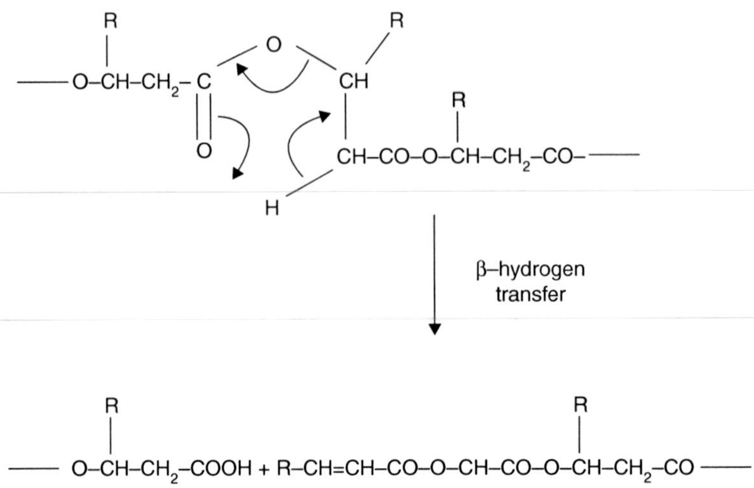

Figure 14.5 Chain scission at the esteric group of biobased and biodegradable polyesters as a result of thermal degradation. Reprinted with permission from M. Avella *et al.*, *Journal of Materials Science*, Vol. 35, p. 523, 2000, © 2000 Kluwer Academic Publishers.

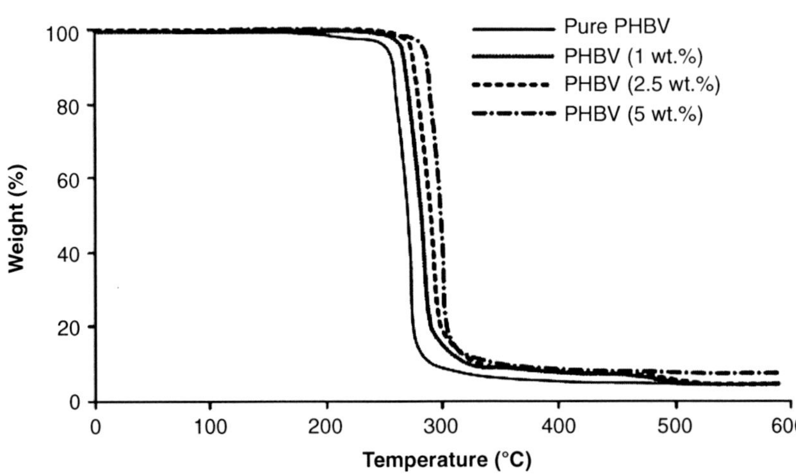

Figure 14.6 TGA curves of pure PHBV and PHBV-based nanocomposites containing various amounts of Cloisite 15A. Reprinted with permission from S. Bruzaud and A. Bourmaud, *Polymer Testing*, Vol. 26, p. 652, 2007, © 2007 Elsevier Ltd.

of PHBV-based nanocomposites increased with an increasing nanoclay loading content (Figure 14.6) [71]. This might be due to the fact that nanoclays create a more tortuous diffusion path for oxygen and other volatile products [72]. Also, Javadi *et al.* [73] investigated solid and microcellular PHBV/PBAT/recycled wood fiber (RWF)/nanoclay hybrid composites and found that adding RWF and nanoclay improved the thermal stability of PHBV.

14.6 Mechanical Properties

As stated in the introduction section, the wide-spread application of PHBV has been hindered due to its high cost, narrow processing window, and especially inferior mechanical properties such as brittleness, low impact resistance, and low elongation at break [15]. The brittleness of PHBV has been attributed to several factors: (1) low nucleation density in addition to a high degree of crystallinity and a slow crystallization rate which leads to the formation of large spherulites [5, 74]; (2) a logarithmic increase in the degree of PHBV crystallinity during storage time when more amorphous regions integrate into the crystalline regions, which will result in physical aging and a significant reduction in the impact strength [75, 76]; and (3) circular and radial cracks inside the large spherulites which can act as stress concentration spots and promote the brittleness of PHBV [77–79]. To improve the mechanical properties of PHBV, several approaches such as blending with tough polymers, natural fibers, and organic/inorganic nanofillers have been investigated [77]. Tensile testing is the most employed method to measure the static mechanical properties of polymers such as tensile modulus, tensile strength, elongation at break, and toughness. Li *et al.* incorporated poly(propylene carbonate) (PPC) into the PHBV matrix by reactive and mechanical blending and observed an increase in elongation at break up to 1300% and 74%, respectively. Also, the toughness (amount of energy absorbed by the component during tensile testing, which is equal to the area under the stress–strain curve) of the biodegradable PHBV/PPC blends produced by reactive and mechanical blending was 167 and 9.9 times higher than that of PHBV, respectively [15]. However, this was accompanied by a reduction in tensile modulus and tensile strength. The authors ascribed the improved elongation at break and toughness of the blends to a reduction in both spherulite sizes and degree of PHBV crystallinity [15]. Willett *et al.* investigated the mechanical properties of biodegradable PHBV/starch-graft-poly(glycidyl methacrylate) (PGMA) blends and demonstrated a significant improvement in tensile strength, fracture toughness, and flexural yield strength compared to those of neat PHBV [80]. The authors claimed this observation was due to the reaction of epoxide groups in PGMA with the hydroxylic or carboxylic functional end groups present in PHBV [80]. The resulting covalent bond between the PHBV chain and starch granules may improve the mechanical properties by enhancing stress transfer across the polymer–filler interface [80]. Similar results were reported by Shogren for PHBV/PEO-coated starch blends [81]. Wang *et al.* studied ternary blends of PHBV/poly(d,l-lactide) (PDLLA)/poly(ethylene glycol) (PEG) and reported an increase in elongation at break and impact strength at the expense of reduction of tensile modulus and tensile strength with the addition of PEG to the PHBV/PDLLA blend [82]. The mechanical properties of solid and microcellular biodegradable PHBV/PBAT blends were investigated by Javadi *et al.* [13]. They demonstrated an increase in elongation at break and toughness with the addition of tough PBAT to the brittle PHBV matrix. On the other hand, the tensile modulus and tensile strength were inferior as compared to those of neat PHBV. Microcellular components had comparatively lower values of elongation at break, tensile strength, and toughness, which were ascribed to the presence of certain large voids in the microcellular samples due to the dynamic nature of microcellular processing [13].

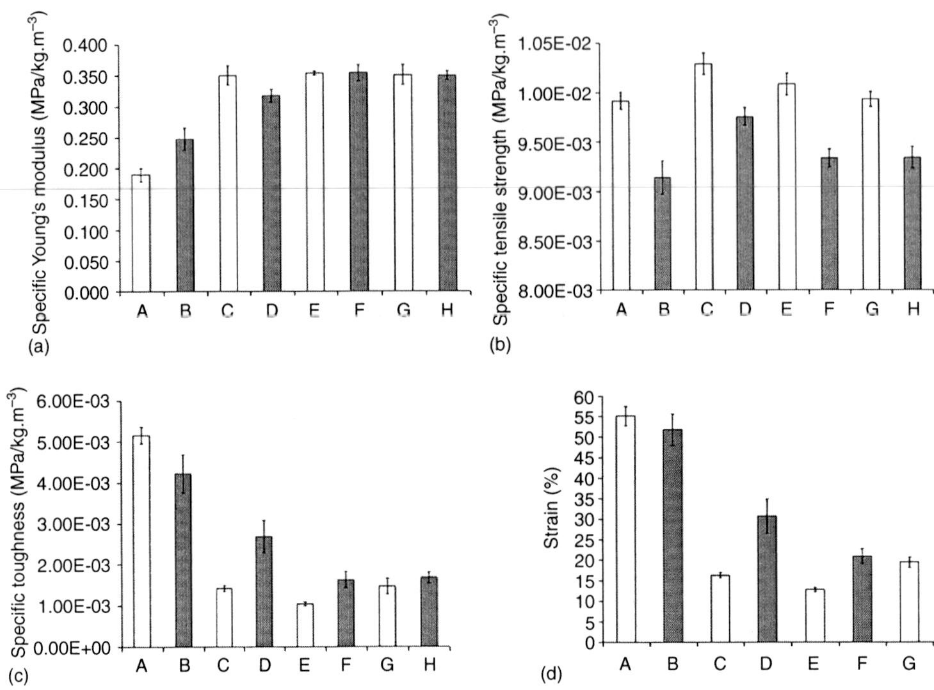

Figure 14.7 Specific mechanical properties of solid and microcellular PHBV/PBAT and their RWF/nanoclay composites; (a) specific Young's modulus (MPa/kg m^3); (b) specific tensile strength (MPa/kg m^3); (c) specific toughness (MPa/kg m^3); and (d) strain-at-break (%). (A) PHBV/PBAT blend (solid); (B) PHBV/PBAT blend (microcellular); (C) PHBV/PBAT/untreated-RWF composite (solid); (D) PHBV/PBAT/PHBV/PBAT/untreated-RWF composite (microcellular); (E) PHBV/PBAT/silane-treated-RWF composite (solid); (F) PHBV/PBAT/silane-treated-RWF composite (microcellular); (G) PHBV/PBAT/silane-treated-RWF/nanoclay composite (solid); and (H) PHBV/PBAT/silane-treated-RWF/nanoclay composite (microcellular). Reprinted with permission from A. Javadi *et al.*, *Composites Part A: Applied Science and Manufacturing*, Vol. 41, p. 982, 2010, © 2010 Elsevier Ltd.

Several research groups have studied the mechanical properties of PHBV/natural fiber composites. The addition of natural fibers such as wood fiber [16], bamboo fiber [21], wheat straw [22], and flax [23] to the PHBV matrix generally decreased the elongation at break and toughness of the resulting composites. However, an increase in tensile modulus was generally observed. Javadi *et al.* studied the mechanical properties of solid and microcellular biodegradable PHBV/coir fiber composites and reported an improvement in elongation at break and toughness of the composites for both solid and microcellular components when the coir fibers were treated with silane [25]. Significant improvements in tensile strength and tensile modulus were observed by incorporating nanoclay nanoparticles into the PHBV matrix as reported by Choi *et al.* [28] and Wang *et al.* [9]. Javadi *et al.* investigated the biodegradable PHBV/PBAT/RWF/nanoclay hybrid composites. They showed an improvement in tensile strength and tensile modulus with the addition of RWF and nanoclay into the PHBV/PBAT blend as shown in Figure 14.7. Also, they demonstrated that with the addition of RWF and nanoclay, microcellular components had a higher elongation at break and toughness as

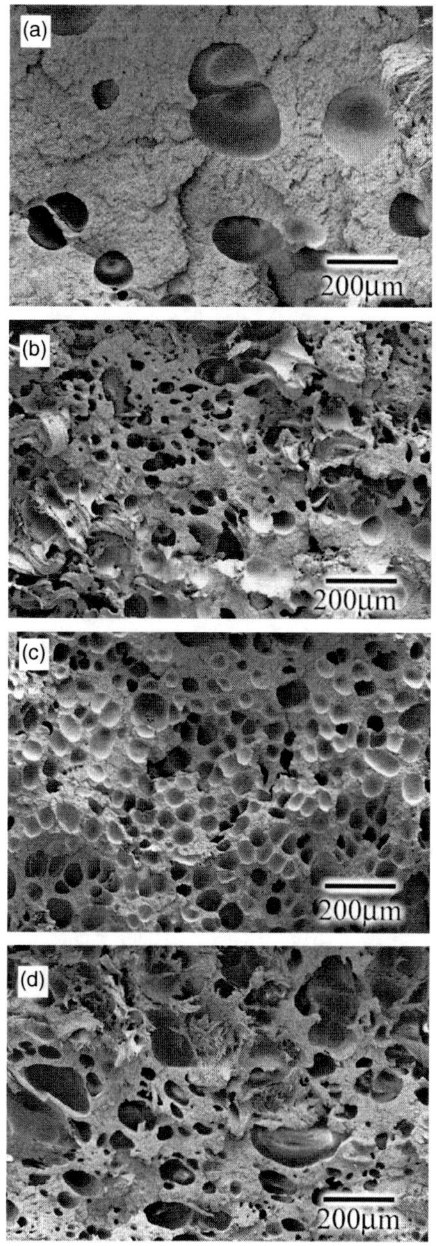

Figure 14.8 Tensile fractured surfaces of microcellular components made of (a) PHBV/PBAT blend; (b) PHBV/PBAT/untreated-RWF composite; (c) PHBV/PBAT/silane treated-RWF composite; and (d) PHBV/PBAT/silane-treated-RWF/nanoclay composite. Reprinted with permission from A. Javadi *et al.*, *Composites Part A: Applied Science and Manufacturing*, Vol. 41, p. 982, 2010, ©2010 Elsevier Ltd.

compared to their solid counterparts, whereas microcellular components of neat PHBV/PBAT blends showed a lower elongation at break and toughness as compared to their solid counterparts (Figure 14.7). These observations were attributed to the formation of much smaller cells and a much higher cell density (Figure 14.8) due to the addition of RWF and nanoclay. These small cells may have acted as crack arrestors which subsequently improved the toughness and elongation at break [73].

14.7 Viscoelastic Properties

Polymers exhibit viscoelastic properties. In general, the effects of temperature or frequency on the viscoelastic properties of polymers are studied by a dynamic mechanical analyzer (DMA) and/or a rheometer by applying a sinusoidal or periodic deformation [16, 21]. The viscoelastic characteristics of polymers are complex quantities consisting of an elastic component (storage modulus: G′ (tension–torsion mode) or E′ (bending mode)) and a viscous component (loss modulus: G″ (tension–torsion mode) or E″ (bending mode)) [16, 21]. The ratio of loss modulus to storage modulus, known as Tan δ , is usually used to determine the occurrence of first order transitions such as glass transition temperatures. Also, the area under Tan δ is a measure of the damping ability of the polymeric material [16, 21]. The viscoelastic characteristics of biobased and biodegradable PHBV-based polymer blends and composites were investigated by several groups using a DMA and/or a rheometer. Buchanan *et al.* investigated PHBV/cellulose acetate butyrate (CAB) blends and found that the PHBV/CAB blends with a 20% to 50% PHBV content were miscible as demonstrated by a single Tan δ peak [83]. At higher CAB contents (greater than 50%), the PHBV/CAB blend became amorphous and the storage modulus (E′) of the blend decreased significantly due to the hampered crystallization of PHBV in the presence of CAB. Jenkins *et al.* investigated the viscoelastic properties of biodegradable PHBV/PCL blends and found that addition of PCL to PHBV did not induce any significant shift in the T_g's of PHBV and PCL, thus indicating the immiscibility of the two components [70]. The viscoelastic properties of biodegradable PHBV/poly(l-lactic acid) (PLLA) blends were studied by Ferreira *et al.* [84]. They observed two separate values of T_g for all blends, thus indicating that PHBV and PLLA were immiscible [84]. Javadi *et al.* studied the viscoelastic properties of solid and microcellular biodegradable PHBV/PBAT blends and reported that with an increasing PBAT content in the PHBV/PBAT blend, the storage modulus decreased and the area under the Tan δ curve increased for both solid and microcellular components as shown in Figure 14.9 [73]. A higher area integration underneath the Tan δ curve indicates a better damping ability attributed to the addition of a tougher phase; i.e., PBAT.

The viscoelastic properties of various natural fiber/PHBV composites have also been studied. The storage modulus of biodegradable PHBV/bamboo fiber composites increased with an increasing bamboo fiber content as reported by Singh *et al.* [16, 21]. Moreover, the area under the Tan δ curve decreased with an increasing fiber content, thus indicating a reduced energy loss of the composite as compared to neat PHBV (Figure 14.10 (a), (b)). This might be attributed to the restriction of polymer chain movement in the presence of the fiber [16, 21].

Dufresne *et al.* investigated the viscoelastic properties of biodegradable PHBV/lignocellulosic flour composites [18] and observed an increase in the storage

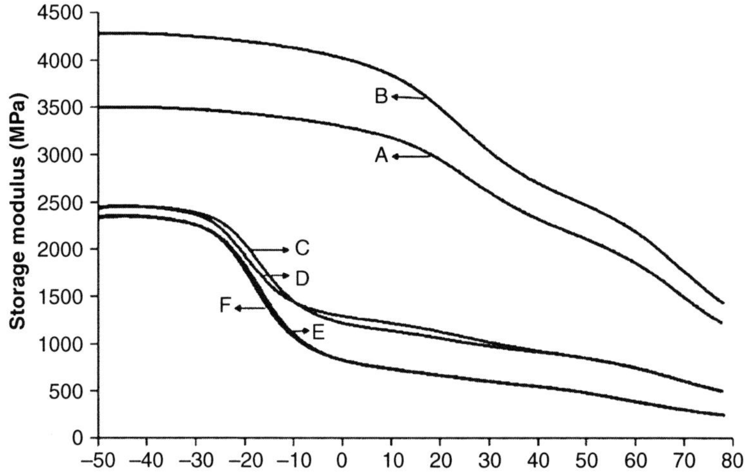

(a)

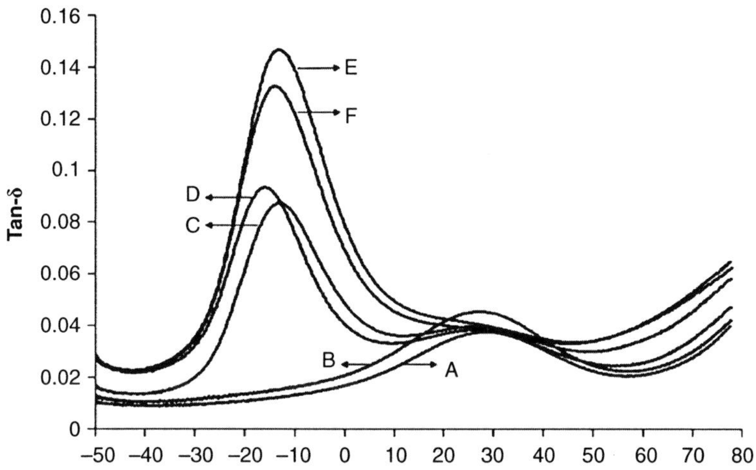

(b)

Figure 14.9 Viscoelastic properties of solid and microcellular PHBV/PBAT blends. (a) Storage modulus as a function of temperature; (b) Loss factor (Tan δ) as a function of temperature. (A) PHBV (solid); (B) PHBV (microcellular); (C) PHBV/PBAT (weight ratio: 45/55) (solid); (D) PHBV/PBAT (weight ratio: 45/55) (microcellular); (E) PHBV/PBAT (weight ratio: 30/70) (solid); and (F) PHBV/PBAT (weight ratio: 30/70) (microcellular). Reprinted with permission from A. Javadi *et al.*, *Polymer Engineering and Science*, Vol. 50, p. 1440, 2010, © 2010 Society of Plastics Engineers.

modulus with the addition of lignocellulosic flour fibers. This enhancement was ascribed to both the reinforcing effect of the lignocellulosic flour and the increase of PHBV crystallinity with the addition of lignocellulosic flour fibers [18]. Gatenholm *et al.* reported an increase in storage modulus and a decrease in loss modulus in biodegradable PHBV/cellulose fiber composites [64, 85]. They attributed this observation to the restriction of chain movements in amorphous regions as a result of interactions with cellulose fibers [64, 85]. Javadi *et al.* studied the viscoelastic properties of solid and microcellular biodegradable PHBV/coir fiber composites

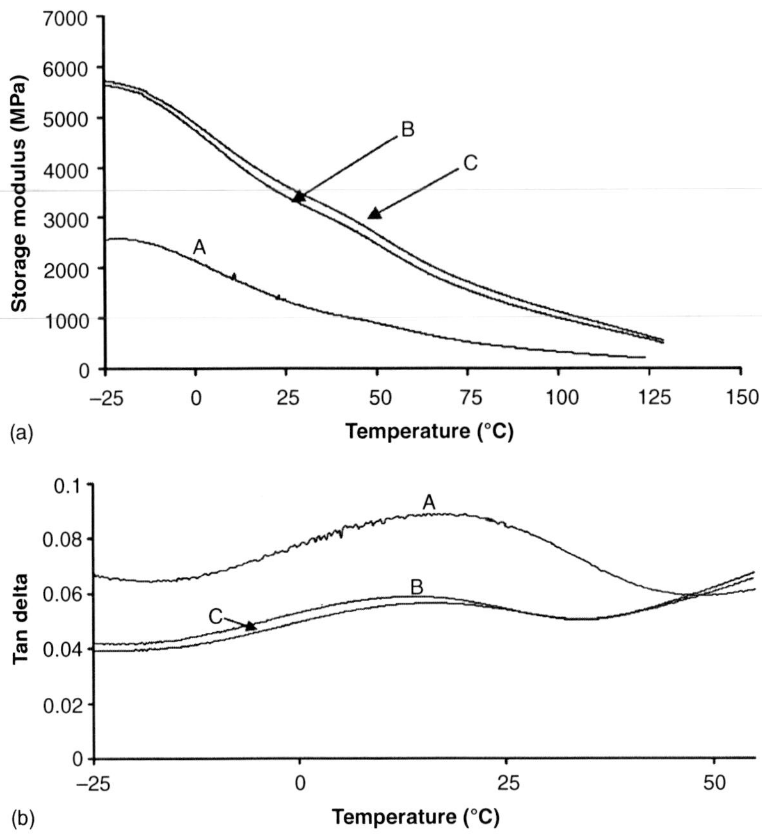

Figure 14.10 (a) Storage modulus versus temperature of PHBV with varying loading levels of fiber; (b) Tan δ versus temperature of PHBV with varying loading levels of fiber. (A) PHBV, (B) PHBV/bamboo fiber (70:30) and (C) PHBV/bamboo fiber (60:40). Reprinted with permission from S. Singh, *et al.*, *Composites: Part A*, Vol. 39, p. 875, 2008, © 2008 Elsevier Ltd.

and reported that with the addition of coir fiber, microcellular samples showed a higher storage modulus compared to their solid counterparts [25].

Viscoelastic properties of the PHBV/nanoclay nanocomposites were studied by different research groups [9, 26–28]. Chen *et al.* observed an increase in storage modulus and T_g with an increasing nanoclay content as shown in Figure 14.11 (a). The authors ascribed this observation to the fact that with an increasing nanoclay loading level and further intercalation and exfoliation, clay nanoplatelets restricted the motion of the polymer chains [27]. Moreover, as can be seen in Figure 14.11(b), two peaks were observed in the Tan δ curves. The first peak appeared at around –110 °C and was attributed to the β-relaxation temperature which remained unchanged with the addition of nanoclay [27]. The β-relaxation temperature is generally associated with the local crankshaft motion of the $(CH_2)_n$ segments of the polymer chains. The second peak appeared at around 15 °C to 17 °C and was associated with the glass transition temperature which shifted to a higher temperature as previously mentioned [27].

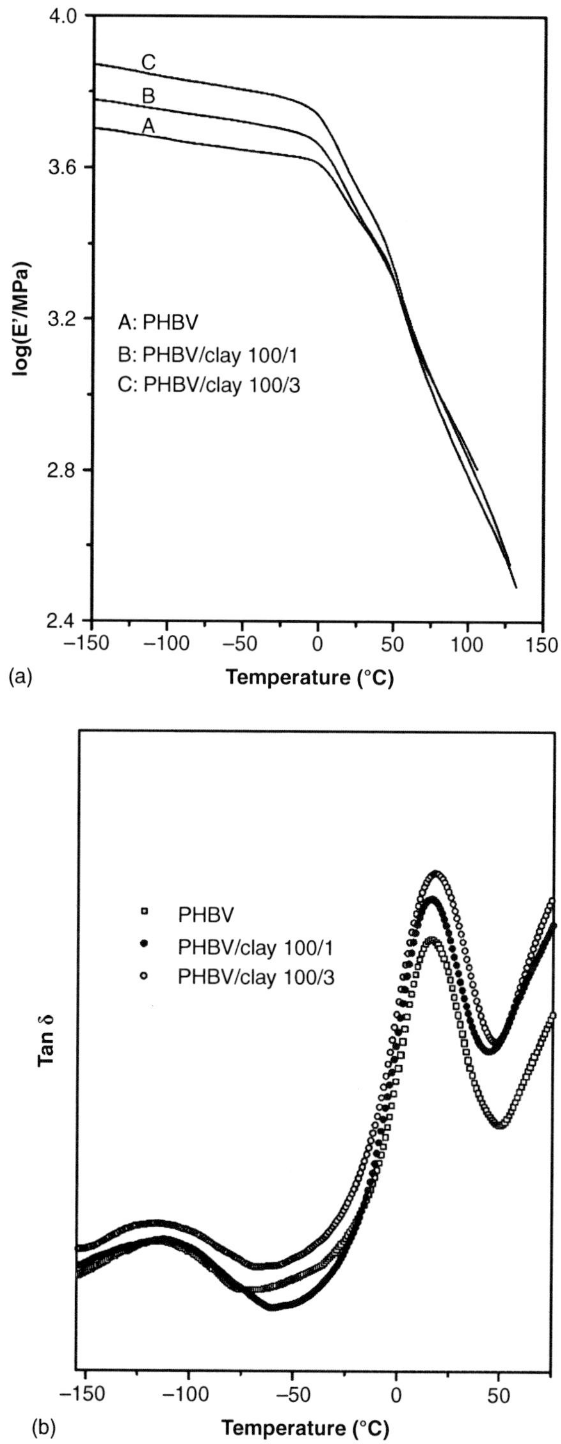

Figure 14.11 (a) Dynamic mechanical properties of PHBV and PHBV/clay nanocomposites with various OMMT content; (b) Tan δ of PHBV and PHBV/clay nanocomposites with various OMMT content. Reprinted with permission from G.X. Chen *et al.*, *Journal of Materials Science Letters*, Vol. 21, p. 1587, 2002, © 2002 Kluwer Academic Publishers.

14.8 Biocompatibility

Biodegradable PHBV and its blends and composites have been extensively studied as potential candidates in various biomedical applications such as surgical sutures, drug carriers, implant patches, biodegradable nerve guidance channels, and bone replacements because of their biocompatibility, biodegradability, and biologically favorable properties including low immunogenicity, low cytotoxicity, and controllable biodegradation rate [42, 86–95]. The in-vivo study of PHBV sutures did not show any significant acute vascular reactions, inflammation, malignant tumor formation, and/or tissue necrosis, at the site of implantation [96]. Porous PHBV materials were found to be promising substrates for in-vitro proliferous cells [86]. Various types of cells including fibroblasts, endothelium cells, retinal epithelium, and isolated hepatocytes demonstrated high levels of adhesion and proliferation on the surface of porous PHBV [97]. The biocompatibility of PHBV can be further improved by surface modification via various methods. Tesema *et al.* modified the surface of PHBV films with immobilized collagens and demonstrated an enhancement in cell proliferation compared to untreated PHBV films [98]. Ke *et al.* modified the inner surface of porous PHBV by UV-polymerization of polyacrylamide and observed an improvement in the mechanical properties and initial cell adhesion of bone marrow-derived stromal cells (BMSCs) without any adverse effects on the viability and/or proliferation of BMCSs [99].

Santos *et al.* investigated the cell adhesion and growth of Vero cells on pure PHBV and PHBV/PLLA blends [100]. They observed a superior cell adhesion and more efficient cell growth on PHBV/PLLA blends compared to neat PHBV [100]. Luklinska and Schluckwerder reported that PHBV/hydroxyapatite (HA) composites used as bone implants had desirable mechanical properties, as their compressive strength was similar to that of several human bones [101]. Also, the in-vivo study of PHBV/HA composites showed strong osteocytes and osteoblast regeneration at the interface of the bone implant [101]. Li *et al.* studied the mechanical properties and biocompatibility of three-dimensional, porous PHBV/wollastonite composite scaffolds. They showed enhanced bioactivity, tailored biodegradability, and engineered mechanical properties when the concentration of wollastonite in the PHBV matrix was optimized [32, 102]. Further study on PHBV/wollastonite composite scaffolds showed an enhancement in the adhesion of BMSCs and an improvement in the stimulation and differentiation of hBMSCs towards osteoblasts [103].

14.9 Biodegradability

Biodegradation of polymeric materials encompasses the physical and/or chemical alteration in their structure as a result of the synergistic effects of abiotic degradation (such as mechanical degradation, photo-degradation, thermo-oxidative degradation, or chemical degradation) and biotic degradation which involves the biological activity of the microorganisms such as bacteria and fungi [104]. Two major factors—i.e., environmental conditions and polymer characteristics—affect the rate of biodegradation. Environmental factors include temperature, moisture, pH, microbial population, and enzyme specificity. Polymer characteristics include chemical structure, chain flexibility, molecular weight, crystallinity, and copolymer composition [105].

The process of biodegradation consists of four major steps as described below:

1. Biodeterioration: In the first step, biodegradable polymer components will break down into smaller segments due to the combined action of decomposing organisms and/or abiotic degradation [106, 107].
2. Depolymerization: Depolymerization involves the dissociation of biodegradable polymer chains and generation of smaller segments such as monomers, dimmers, and oligomers by means of catalytic reactions in the presence of enzymes such as extracellular and intracellular depolymerases. This will result in an extensive reduction of molecular weight in biodegradable polymers by means of the catalytic reactions of enzymes [108].
3. Assimilation: Assimilation involves the transportation of the generated molecules (recognized by the receptors of the microbial cells) across the cytoplasmic membrane followed by the production of metabolites, storage vesicles, biomass, and/or energy inside the microbial cell [108].
4. Mineralization: In the final step, some simple and complex metabolites and their products such as CO_2, N_2, CH_4, and H_2O are discharged into the extracellular environment [108].

Several characterization methods such as visual observation, weight loss measurement, mechanical testing, CO_2 evaluation, radiolabeling, clear-zone testing, and controlled composting testing are used to measure the biodegradation of the biodegradable polymers [104]. In general, PHAs, and specifically PHBV, have the advantage of biodegradation in both aerobic and anaerobic environments [3]. The extracellular degradation of PHBV will result in the formation of 3-hydroxybutyrate and 3-hydroxyvalerate which are soluble in water and can be absorbed by microbial cells wherein CO_2, H_2O, and CH_4 are produced as the result of their metabolism under aerobic and anaerobic conditions [109, 110]. As discussed previously, PHBV has a lower degree of crystallinity as compared to PHB, therefore its biodegradability is higher than PHB [111]. In addition, PHBVs with low molecular weights exhibit better biodegradability than those with high molecular weights [111]. PHBV can be degraded by microbial depolymerases, enzymes, and hydrolysis [111]. Several research groups have studied the biodegradability of PHBV blends. They demonstrated that blending PHBV with other biodegradable polymers such as PHB [112, 113], starch [114, 115], and PBA [116], resulted in an increase in the biodegradation rate. This observation was ascribed to a lower degree of PHBV crystallinity as a result of blending. Also, Wang *et al.* investigated the biodegradability of PHBV/OMMT nanocomposites and found that the biodegradability of the nanocomposites decreased with an increasing OMMT content in the PHBV matrix which was in agreement with previous biodegradability studies of aliphatic polyester/OMMT nanocomposites [9]. This was attributed to the slower diffusion rate of water, oxygen, and microorganisms into the bulk polymer nanocomposites due to the presence of OMMT.

14.10 Applications

PHBV can be potentially used in a wide variety of applications such as automotive, construction, packaging, agricultural, and biomedical industries because of its unique mechanical and thermal properties in addition to its biocompatibility, biodegradability, and sustainability [4]. Owing to the fact that it has similar mechanical and thermal properties to polyolefins, PHBV is considered a promising alternative for synthetic-based polymers in the automotive, construction, agricultural, and packaging industries [4]. PHBV exhibits excellent barrier properties; thus, in packaging industries, PHBV film and latex can replace aluminum as the inner linings of packaging cardboard [117]. Moreover, PHBV can be vastly utilized in agricultural industries. There are several manufacturers which produce mulch films, composting bags, and bacterial inoculants (used to improve N_2 fixation in plants) [118]. In the agricultural industry, PHBV is also used as a carrier for pesticides in order to achieve the controlled release of pesticides via PHBV biodegradation [118]. The biodegradation of PHBV is affected by the presence of pests in the environment [118]. PHBV is fully biodegradable and biocompatible because its in-vivo biodegradation yields 3-hydroxybutyrate and 3-hydroxyvalerate which are typical components that can be found in blood [119]. Additionally, due to its natural origin and microbial polymerization process, PHBV does not contain any catalytic or solvent residues, which makes it suitable for biomedical applications such as bone tissue engineering, cartilage tissue engineering, nerve guidance channels, intestinal patches, wound dressings, surgical sutures, and drug carrier systems [120].

Several research groups have blended PHBV with other biodegradable polymers such as PCL [10], PBS [11], PES [12], PBAT [13], and PHEMA [14], to modify its mechanical, biodegradation, and morphological properties and to broaden its applicability in various industries. Also, natural fibers such as wood fiber [16], bamboo fiber [21], wheat straw [22], flax [23], abaca [23], jute [24], and coir fiber [25], which are cheap, lightweight, and abundantly available, have been incorporated into the PHBV matrix to tailor its mechanical properties and to reduce its weight and also its production cost. Moreover, inorganic nanofillers such as nanoclays have been incorporated into the PHBV matrix to modify the mechanical and thermal properties of PHBV [26]. Bioactive fillers such as HA [29–31], wollastonite [32, 33], TCP [34, 35], and SGBG [36] have been added to the PHBV matrix to tailor its mechanical properties (especially compression strength), biocompatibility, and biodegradation rate for several biomedical applications such as porous scaffolds for bone and cartilage tissue engineering, nerve guidance channels, implant patches, and drug delivery devices. With the continuous development of new PHBV-based blends and composites and new processing technologies, an even broader range of applications are anticipated for biobased and biodegradable PHBV.

14.11 Conclusion

During the past two decades, many scientists have focused on developing new classes of biobased and biodegradable polymeric materials made from renewable resources which can potentially replace synthetic polymeric materials made from non-renewable resources such as petroleum and its derivatives. One of the most promising biobased and biodegradable polymeric materials are the polyhydroxyalkanoates (PHAs), a family of aliphatic polyesters. The most extensively studied polymer in the PHA family is poly(hydroxyl butyrate-co-hydroxyvalerate) (PHBV). PHBV can be produced by either a bacterial fermentation process or directly from plants. Biobased and biodegradable PHBV has garnered a great deal of attention in the academic and scientific world as well as in industry. Owing to its unique material properties coupled with its biocompatibility, biodegradability, and renewability, PHBV is a promising alternative for synthetic polymers in a wide variety of applications in the automotive, packaging, construction, agricultural, and biomedical industries. In order to widen its application range, reduce production costs, tailor its mechanical, thermal, and morphological properties, and modify its biocompatibility and biodegradability, PHBV has been blended with various biodegradable polymers, natural fibers, and inorganic fillers and has also been processed using novel techniques such as microcellular injection molding. In this chapter, the methods of synthesizing PHBV have been presented along with a brief introduction to a microcellular injection molding technique which is a novel processing technique used to process polymeric materials with lower cost, lower component weight, and minimally compromised material properties. The crystallinity, thermal degradation, mechanical properties, viscoelastic properties, biocompatibility, and biodegradability of PHBV-based polymer blends and composites have also been discussed. Finally, various applications of the PHBV-based polymer blends and composites in automotive, packaging, agricultural, and biomedical industries have been presented briefly.

Acknowledgements

The authors would like to acknowledge the partial financial support from the National Science Foundation (CMMI 1032186) for this work.

References

1. K.G. Satyanarayana, G.G.C. Arizaga, F. Wypych, *Progress in Polymer Science*, Vol. 34, p. 982, 2009.
2. C. Scholz, "Poly(β-hydroxyalkanoates) as Potential Biomedical Materials: An Overview," in: C. Scholz, R.A. Gross, eds., *Polymers From Renewable Resources*, Washington, American Chemical Society, pp. 328–334, 2000.
3. S. Philip, T. Keshavarz, I. Roy, *Journal of Chemical Technology and Biotechnology*, Vol. 82, p. 233, 2007.
4. L. Jiang, E. Morelius, J. Zhang, M. Wolcott, J. Holbery, *Journal of Composite Materials*, Vol. 42, p. 2629, 2008.
5. G. J. M. Koning, P. J. Lemstra, *Polymer*, Vol. 34, p. 4089, 1993.

6. R.A. Shanks, A. Hodzic, S. Wong, *Journal of Applied Polymer Science*, Vol. 91, p. 2114, 2004.
7. L. Yu, K. Dean, L. Lin, *Progress in Polymer Science*, Vol. 3, p. 576, 2006.
8. A. K. Mohanty, M. Misra, G. Hinrichsen, *Macromolecular Materials and Engineering*, Vol. 276, p. 1, 2000.
9. S. F. Wang, C. J. Song, G. X. Chen, T.Y. Guo, J. Liu, B.H. Zhang, S. Takeuchi, *Polymer Degradation and Stability*, Vol. 87, p. 69, 2005.
10. Z. Qiu, W. Yang, T. Ikehara, T. Nishi, *Polymer*, Vol. 46, p. 11814, 2005.
11. Z. Qiu, T. Ikehara, T. Nishi, *Polymer*, Vol. 44, p. 7519, 2003.
12. L. Miao, Z. Qiu, W. Yang, T. Ikehara, *Reactive & Functional Polymers*, Vol. 68, p. 446, 2008.
13. A. Javadi, A. J. Kramschuster, S. Pilla, J. Lee, S. Gong, L. S.Turng, *Polymer Engineering and Science*, Vol. 50, p. 1440, 2010.
14. G. Mabilleau, M.F. Moreau, R. Filmon, M.F. Basle, D. Chappard, *Biomaterials*, Vol. 25, p. 5155, 2004.
15. J. Li, M.F. Lai, J.J. Liu, *Journal of Applied Polymer Science*, Vol. 98, p. 1427, 2005.
16. S. Singh, A.K. Mohanty, *Composites Science and Technology*, Vol. 67, p. 1753, 2007.
17. R. Bhardwaj, A.K. Mohanty, L.T. Drzal, F. Pourboghrat, M. Misra, *Biomacromolecules*, Vol. 7, p. 2044, 2006.
18. A. Dufresne, D. Dupeyre, M. Paillet, *Journal of Applied Polymer Science*, Vol. 87, p. 1302, 2003.
19. S. Luo, A.N. Netravali, *Journal of Materials Science*, Vol. 34, p. 3709, 1999.
20. M. Avella, G.B. Gaceva, A. Buzarovska, M.E. Errico, G. Gentile, *Journal of Applied Polymer Science*, Vol. 104, p. 3192, 2007.
21. S. Singh, A. K. Mohanty, T. Sugie, Y. Takai, H. Hamada, *Composites: Part A*, Vol. 39, p. 875, 2008.
22. M. 26, G. Rota, E. Martuscelli, M. Raimo, P. Sadocco, G. Elegir, *Journal of Materials Science*, Vol. 35, p. 829, 2000.
23. N.M. Barkoula, S.K. Garkhail, T. Peijs, *Industrial Crops and Products*, Vol. 31, p. 34, 2010.
24. A.K. Bledzki, A. Jaszkiewicz, *Composites Science and Technology*, Vol. 70, p. 1687, 2010.
25. A. Javadi, Y. Srithep, S. Pilla, J. Lee, S. Gong, L. S. Turng, *Materials Science and Engineering: C*, Vol. 30, p. 749, 2010.
26. G.X. Chen, G.J. Hao, T.Y. Guo, M.D. Song, B.H. Zhang, *Journal of Applied Polymer Science*, Vol. 93, p. 655, 2004.
27. G.X. Chen, G.J. Hao, T.Y. Guo, M.D. Song, B.H. Zhang, *Journal of Materials Science Letters*, Vol. 21, p. 1587, 2002.
28. W.M. Choi, T.W. Kim, O.O. Park, Y.K. Chang, J.W. Lee, *Journal of Applied Polymer Science*, Vol. 90, p. 525, 2003.
29. N. Galego, C. Rozsa, R. Sanchez, J. Fung, A. Vazquez, J. S. Tomas, *Polymer Testing*, Vol. 19, p. 485, 2000.
30. E. I. Shishatskaya, I.A. Khlusov, T. G. Volova, *Journal of Biomaterials Science-Polymer Edition*, Vol. 17, p. 481, 2006.
31. S. M. Cool, B. Kenny, A. Wu, V. Nurcombe, M. Trau, A.I. Cassady, L. Grondahl, *Journal of Biomedical Materials Research*, Vol. 82, p. 599, 2007.
32. H. Y. Li, J. Chang, *Biomaterials*, Vol. 25, p. 5473, 2004.
33. H. Y. Li, W. Y. Zhai, J. Chang, *Journal of Materials Science-Materials in Medicine*, Vol. 19, p. 67, 2008.
34. Y. Liu, M. Wang, *Key Engineering Materials*, Vol. 334–335, p. 1217, 2007.
35. Y. D. Zheng, Y. J. Wang, H. A. Yang, X. F. Chen, Z. H. Chen, *Journal of Biomedical Materials Research Part B-Applied Biomaterials*, Vol. 80, p. 236, 2007.
36. Y. D. Zheng, Y. J. Wang, X. F. Chen, Y. B. Ren, G. Wu, *Chemical Journal of Chinese Universities-Chinese*, Vol. 24, p. 1325, 2003.
37. S. Gong, L.S. Turng, C. Park, L. Liao, "Microcellular Polymer Nanocomposites for Packaging and other Applications," in: A. Mohanty, M. Misra, H.S. Nalwa, eds., *Packaging Nanotechnology*, American Scientific Publishers, pp.144, 2008.
38. J. Choi, S. Y. Lee, *Applied Microbiology and Biotechnology*, Vol. 53, 646, 2000.
39. S. Iadevaia, N.V. Mantzaris, *Journal of Biotechnology*, Vol. 128, p. 615, 2007.
40. A.J. Anderson, E.A. Dawes, *Microbiological Reviews*, Vol. 54, p. 450, 1990.
41. D. Byrum, *Microbiological Reviews*, Vol. 102, p. 247, 1992.
42. P.A. Holmes, S.H. Collins, F. Wright, 3-Hydroxybutyrate Polymers. US Patent 4477654, 1984.
43. Y. Poirier, C. Nawrath, C. Somerville, *Biotechnology*, Vol. 13, p. 142, 1995.

44. F.R. van der Leij, B. Witholt, *Canadian Journal of Microbiology*, Vol. 41, p. 222 1995.
45. Y. Poirier, *Nature Biotechnology*, Vol. 17, p. 960, 1999.
46. S. Slater, T. A. Mitsky, K. L. Houmiel, M. Hao, S. E. Reiser, N. B. Taylor, M. Tran, H. E. Valentin, D. J. Rodriguez, D. A. Stone, S. R. Padgette, G. Kishore, K. J. Gruys, *Nature Biotechnology*, Vol. 17, p. 1011, 1999.
47. V. Martini, N.P. Suh, F.A. Waldman, Microcellular Closed Cell Foams and Their Method of Manufacture, U.S. Patent, 4473665, 1984.
48. G. B. Okonishnikov, E. I. Blednykh, V. P. Skripov, *Polimerov*, Vol. 2, p. 370, 1973.
49. D.F. Baldwin, N.P. Suh, ANTEC Technical Paper, p. 1503, 1992.
50. D. Pierick, D. Szczurko, Trexel Technical Papers, Trexel Inc., 1998.
51. C. B. Park, N. P. Suh, *Cellular Polymers*, Vol. 38, p. 69, 1992.
52. J. Xu, D. Pierick, *Journal of Injection Molding Technology*, Vol. 5, p. 152, 2001.
53. H.E. Naguib, C.B. Park, P.C. Lee, *Journal of Cellular Plastics*, Vol. 39, p. 499, 2003.
54. A. K. Bledzki, O. Faruk, H. Kirschling, J. Kuhn, A. Jaszkiewicz, *Polimery*, Vol. 51, p. 697, 2006.
55. K.A. Seeler, V. Kumar, *Journal of Reinforced Plastics and Composites*, Vol. 12, p. 359, 1993.
56. D.I. Collias, D.G. Baird, *Polymer*, Vol. 35, p. 3978, 1994.
57. M. Shimbo, D.F. Baldwin, N.P. Suh, *Polymer Engineering and Science*, Vol. 35, p. 1387, 1995.
58. H. F. Mark, N. G. Gaylord, N. M. Bikales , *Encyclopedia of Polymer Science and Technology*, New York, London, Sydney, Inter science Publishers (Wiley & Sons), 1964.
59. J. Brandup, E. H. Immergut, *Polymer Handbook*, New York, Inter science Publishers (Wiley & Sons), 1975.
60. M. Scandola, G. Ceccorulli, M. Pizzoli, M. Gazzano, *Macromolecules*, Vol. 25, p. 1405, 1992.
61. Y.S. Chun, W.N. Kim, *Polymer*, Vol. 41, p. 2305, 2000.
62. S. M. Tan, J. Ismail, C. Kummerlowe, H. W. Kammer, *Journal of Applied Polymer Science*, Vol. 101, p. 2776, 2006.
63. V.E. Reinsch, S.S. Kelley, *Journal of Applied Polymer Science*, Vol. 64, p. 1785, 1997.
64. P. Gatenholm, J. Kubat, A. Mathiasson, *Journal of Applied Polymer Science*, Vol. 45, p. 1667, 1992.
65. M. Avella, E. Martuscelli, B. Pascucci, M. Raimo, B. Focher, A. Marzetti, *Journal of Applied Polymer Science*, Vol. 49, p. 2091, 1993.
66. S. Luo, A.N. Netravali, *Polymer Composites*, Vol. 20, p. 367, 1999.
67. A. Buzarovska, G. Bogoeva-Gaceva, A. Grozdanov, M. Avella, G. Gentile, M. Errico, *Journal of Materials Science*, Vol. 42, p. 6501, 2007.
68. M. Avella, E. Martuscelli, M. Raimo, *Journal of Materials Science*, Vol. 35, p. 523, 2000.
69. M. Kunioka, Y. Doi, *Macromolecules*, Vol. 23, p. 1933, 1990.
70. M.J. Jenkins, Y. Cao, L. Howell, G.A. Leeke, *Polymer*, Vol. 48, p. 6304, 2007.
71. S. Bruzaud, A. Bourmaud, *Polymer Testing*, Vol. 26, p. 652, 2007.
72. P. Bordes, E. Pollet, L. Avérous, *Progress in Polymer Science*, Vol. 34, p. 125, 2009.
73. A. Javadi, Y. Srithep, S. Pilla, J. Lee, C. Clemons, S. Gong, L. S. Turng, *Composites Part A: Applied Science and Manufacturing*, Vol. 41, p. 982, 2010.
74. F. Biddlestone, A. Harris, J. N. Hay, *Polymer International*, Vol. 39, p. 221, 1996.
75. P. A. Holmes, "In Developments in Crystalline Polymer," in D. C. Bassett, ed., Cambridge University Press: Cambridge, England, 1981.
76. G. J. M. Koning, A. H. C. Scheeren, P. J. Lemstra, M. Peeters, H. Reynaers, *Polymer* Vol. 35, p. 4598, 1994.
77. P. J. Barham, A. Keller, *Journal of Polymer Science Part B: Polymer Physics*, Vol. 24, p. 69, 1986.
78. A. El-Hadi, R. Schnabel, E. Straube, G. Mueller, S. Henning, *Polymer Testing*, Vol. 21, p. 665, 2002.
79. J. K. Hobbs, T. J. McMaaster, M. J. Miles, P. J. Barham, *Polymer*, Vol. 37, p. 3241, 1996.
80. J. L. Willett, M. A. Kotnis, G. S. O'Brien, G. F. Fanta, S. H. Gordon, *Journal of Applied Polymer Science*, Vol. 70, p. 1121, 1998.
81. R. L. Shogren, *Journal of Environmental Polymer Degradation*, Vol. 3, p. 75, 1995.
82. S. Wang, P. Ma, R. Wang, S. Wang, Y. Zhang, Y. Zhang, *Polymer Degradation and Stability*, Vol. 93, p.1364, 2008.
83. C. M. Buchanan, S. C. Gedon, A. W. White, M. D. Wood, *Macromolecules*, Vol. 25, p. 7373, 1992.
84. B. M. P. Ferreira, C. A. C. Zavaglia, E. A. R. Duek, *Journal of Applied Polymer Science*, Vol. 86, p. 2898, 2002.

85. P. Gatenholm, A. Mathiasson, *Journal of Applied Polymer Science*, Vol. 51, p. 1231, 1994.
86. G.-Q. Chen, Q. Wu, *Biomaterials*, Vol. 26, 6565, 2005.
87. C. Kunze, T. Freier, S. Kramer, K.P. Schmitz, *Journal of Materials Science-Materials in Medicine*, Vol. 13, p. 1051, 2002.
88. P.J. Hocking, R.H. Marchessault, in: G.J.L. Griffin, editor. Chemistry and Technology of Biodegradable: Polymers, London, Chapman & Hall, pp. 48, 1994.
89. C. Doyle, E.T. Tanner, W. Boneld, *Biomaterials*, Vol. 12, p. 841, 1991.
90. J.C. Knowles, G.W. Hastings, H. Ohta, S. Niwa, N. Boeree, *Biomaterials*, Vol. 13, p. 491, 1992.
91. L.J. Chen, M. Wang, *Biomaterials*, Vol. 23, p. 2631, 2002.
92. R.C. Young, M. Wiberg, G. Terenghi, *British Journal of Plastic Surgery*, Vol. 55, p. 235, 2002.
93. B. Kramp, H.E. Bernd, W.A. Schumacher, M. Blynow, W. Schmidt, C. Kunze, D. Behrend, K.P. Schmitz, *Laryngo-Rhino-Otologie*, Vol. 81, p. 351, 2002.
94. L.N. Novikov, L.N. Novikova, A. Mosahebi, M. Wiberg, G. Terenghi, J.O.Kellerth, *Biomaterials*, Vol. 23, p. 3369, 2002.
95. J. Ni, M. Wang, *Materials Science and Engineering: C*, Vol. 20, p. 101, 2002.
96. E.I. Shishatskaya, T.G. Volova, A.P. Puzyr, O.A. Mogilnaya, S.N. Efremov, *Journal of Materials Science-Materials in Medicine*, Vol. 15, p. 719, 2004.
97. S. Gogolewski, M. Jovanovic, S.M. Perren, J.G. Dillon, M.K. Hughes, *Journal of Biomedical Materials Research*, Vol. 27, p. 1135, 1993.
98. Y. Tesema, D. Raghavan, J. Stubbs, *Journal of Applied Polymer Science*, Vol. 93, 2445, 2004.
99. Y. Ke, Y.J. Wang, L. Ren, Q.C. Zhao, W. Huang, *Acta Biomaterialia*, Vol. 6, p. 1329, 2010.
100. A.R. Santos, B.M.P. Ferreira, E.A.R. Duek, H. Dolder, R.S. Wada, M.L.F. Wada, *Artificial Organs*, Vol. 28, p. 381, 2004.
101. Z.B. Luklinska, H. Schluckwerder, *Journal of Microscopy-Oxford*, Vol. 211, p. 211, 2003.
102. H. Li, J. Chang, *Polymer Degradation and Stability*, Vol. 87, p. 301, 2005.
103. H. Y. Li, W. Y. Zhai, J. Chang, *Journal of Biomaterials Applications*, Vol. 24, p. 231, 2009.
104. A. A. Shah, F. Hasan, A. Hameed, S. Ahmed, *Biotechnology Advances*, Vol. 26, p. 246, 2008.
105. G. Kale, T. Kijchavengkul, R. Auras, M. Rubino, S. E. Selke, S. P. Singh, *Macromolecular Bioscience*, Vol. 7, p. 255, 2007.
106. H.O.W. Eggins, T.A. Oxley, *International Biodeterioration and Biodegradation*, Vol. 48, p. 12, 2001.
107. J.H. Walsh, *International Biodeterioration and Biodegradation*, Vol. 48, p. 16, 2001.
108. N. Lucas, C. Bienaime, C. Belloy, M. Queneudec, F. Silvestre, J. E. Nava-Saucedo, *Chemosphere*, Vol. 73, p. 429, 2008.
109. G. Scott, *Polymers in Modern Life: Polymers and the Environment*, Cambridge, UK, The Royal Society of Chemistry, 1999.
110. W.D. Luzier, *Proceedings of the National Academy of Sciences of the United States of America*, Vol. 89, p. 839, 1992.
111. S. Gogolewski, M. Jovanovic, S.M. Perren, M.K. Hughes, *Journal of Biomedical Materials Research*, Vol. 27, p. 1135, 1993.
112. Y. Kumagai, Y. Doi, *Polymer Degradation and Stability*, Vol. 36, p. 241, 1992.
113. H. Satoh, N. Yoshie, Y. Inoue, *Polymer*, Vol. 35, 286, 1994.
114. H. Verhoogt, N. St Pierre, F. S. Truchon, B. A. Ramsay, B. D. Favis, J. A. Ramsay, *Canadian Journal of Microbiology*, Vol. 41, p. 323, 1995.
115. B. A. Ramsay, V. Langlade, P. J. Carreau, J. A. Ramsay, *Journal of Applied Environmental Microbiology*, Vol. 59, p. 1242, 1993.
116. M. Avella, L. Calandrelli, B. Immirizi, M. Malinconico, E. Martuscelli, B. Pascucci, P. Sadcco, *Journal of Environmental Polymer Degradation*, Vol. 3, p. 49, 1995.
117. C.A. Lauzier, C.J. Monasterios, I. Saracovan, R.H. Marchessault, B.A. Ramsay, *Tappi Journal*, Vol. 76, p. 71, 1993.
118. P. A. Holmes, UK Patent Application, Great Britain, 2160208, 1985.
119. R.N. Reusch, *Canadian Journal of Microbiology*, Vol. 41, p. 50, 1995.
120. C.W. Pouton, S. Akhtar, *Advanced Drug Delivery Review*, Vol. 18, p. 133, 1996.

15

Bioplastics and Vegetal Fiber Reinforced Bioplastics for Automotive Applications

Daniela Rusu[1,2,*], Séverine A.E. Boyer[1,2], Marie-France Lacrampe[1,2] and Patricia Krawczak[1,2]

[1]*Univ. Lille Nord de France, F-59000 Lille, France*
[2]*Ecole des Mines de Douai, Department of Polymers and Composites Technology & Mechanical Engineering, Douai, France*

Abstract
Evergrowing concerns related to sustainability and ecology have been the key driving forces for developing bio-based plastics, especially for single-use packaging applications where biodegradability is an advantage. Automotive applications are however much more challenging, since durable bioplastics are expected to meet very demanding requirements, such as high thermo-mechanical performance at both very short term (e.g. impact) and very long term (e.g. creep, fatigue) often coupled to chemical resistance to aggressive automotive fluids. The present chapter focuses on the main classes of thermoplastic and thermosetting bioplastics and natural fiber-reinforced bioplastics, also called biocomposites, with current or emerging interest for the modern car industry. It points out the great potential of these renewable materials and their expected future evolution, without forgetting to mention their present drawbacks and the necessary improvements for enhancing their durable applications in automotive and related sectors.

Keywords: Bioplastics, biocomposites, automotive applications, polylactic acid, thermoplastic starch, polyamides, polyolefins, polyurethanes, unsaturated polyesters, epoxy, natural/vegetal fibers, processing, recycling

15.1 Introduction

15.1.1 Plastics and Automotive Applications

Nowadays, polymer materials represent approximately 20 weight percent of the total car, in other words 100 to 150 kg [1–4]. The major driving forces for the growing demand of plastics and composites in automotive applications include the need of car weight reduction for limiting the fuel consumption, the production

* Corresponding author: Daniela Rusu. E-mail: daniela.rusu@mines-douai.fr. Phone +33 327712464, Fax: +33 327712981

Srikanth Pilla (ed.) Handbook of Bioplastics and Biocomposites Engineering Applications, (397–450)

gains and increasing design flexibility for easier assembling/dismantling and integration of parts and systems. On the other hand, the permanent improvements in properties of polymers and composites lead to materials which are more fitting to the automakers' specific requirements.

Typical modern vehicles are made up to 15.000 parts, including approximately 600 plastic parts. Each one of these parts is subjected to specific constraints, depending on its role and position in the automobile. Generally, one can distinguish four main classes of automotive applications for polymers [2,4–5]:

- interior trims, which consume 47–50% of the plastics found in automobiles;
- external parts, which use 29–35% of the plastics;
- structural parts and fuel systems, which may use up to 13% of the plastics;
- under-the-hood applications, which consume 12–15% of the plastics.

Yet, they are also polymer-based materials present on "hidden" parts, including electrical components and electronics, coatings, varnishes, paintings, sealers, adhesives.

Within the polymers used in automotive industry, a main role is played by polyolefins (polypropylene - 43%, polyethylene and copolymers), polyurethanes (14%), and engineering thermoplastics such as polyamides (12%), acrylonitrile-butadiene-styrene, polycarbonate, etc. [2]. Until recently, these polymer materials were mainly issued from petrochemical feedstocks. Nowadays, the increasing crude oil price, the growing environmental concerns about fossil resources depletion, greenhouse gas emissions, together with the regulation-imposed end-of-life products management and high "white" pollution with durable plastics, have encouraged the automotive industry to develop, adapt or recall more eco-friendly bio-based plastic materials and composites for their modern cars. The recent international regulations for the greenhouse gas emissions from the passenger cars [6–8] and the strict standards concerning the recyclability of end-of-life vehicles [9] are also very effective economical drivers stimulating the automakers, Original Equipment Manufacturers (OEMs) and suppliers to develop lighter materials with tailored properties and better ecological footprint, allowing additional functional utilities, good durability and ageing behavior, multi-material assembly abilities and better recycling and waste management [3,4,10]. Therefore, in the last decades, two classes of polymer materials are regaining a special interest for automotive applications and they will form the scope of this chapter: the bioplastics and biocomposites.

Obtaining plastic materials from renewable sources and using them in automotive applications is not a new idea. For instance, casein plastics (resins from casein-milk protein and formaldehyde) had been first proposed around 1897 [11], and by 1913, patents had been issued for preparing plastics from soy-protein [12]. And soon, several automotive applications of plastic materials from renewable resources were proposed by Henry Ford. For instance, coil cases for his 1915 Model T Ford were made from a wheat gluten-resin reinforced with asbestos fibers [13].

In the 1920s, Ford's researchers developed a thermoplastic resin by reacting soybean meal with phenol and formaldehyde, to make nonstructural car parts, such as plastic horn buttons, gearshift knobs, coil cases, distributor heads, accelerator foot pedals, glove compartment doors, and tractor seats [14]. And in early 1940s, Henry Ford developed the first prototype composite car made from hemp fibers and soy-phenol-formaldehyde resin, but unfortunately this car was never transferred to mass production [13,15].

Since 1950s–1960s, the exponential development of petrochemical products, cheaper and better performing had diminished the industrial interest in materials from renewable feedstocks. But the present economical and ecological context is calling back the renewably materials to substitute synthetic polymers and possibly some metal parts in specific automotive applications [13,16–18].

15.1.2 Definitions of Bioplastics and Biocomposites

Bioplastics, or plastics based on renewable resources, concern a large family of polymer materials. It is therefore important to clearly define them.

According to European Commission reports [19–20], bio-based products refer to non-food products derived from biomass (plants, algae, crops, trees, marine organisms and biological waste from households, animals and food production). Bio-based products may range from high-value added fine chemicals such as pharmaceuticals, cosmetics, food additives, etc., to high volume materials such as general bio-polymers or chemical feedstocks (i.e. building blocks). The concept excludes traditional bio-based products such as pulp, paper, and wood products, and biomass as an energy source.

Japan BioPlastics Association (JBPA) proposed since 2006 a certification program, called BiomassPla, for plastic products with at least 25 wt% biomass-based plastic content. The JBPA defines biomass-based plastic ratio as the proportion of the total weight consisting of components derived from biomass in the composition of the biomass-based plastic contained in the raw material and product (percentage by weight) [21].

ASTM International recommends a test method [22] providing precise measure of the bio-based content of a product. The bio-based (renewable) content of a product is considered to be the amount of bio-based carbon (via the radiocarbon ^{14}C) as fraction weight or percent weight of the total organic carbon in the product. The European Committee for Standardization (CEN/TC 249/WG 17) is also preparing technical specification for calculating the bio-based carbon content in monomers, polymers, plastic materials and products, based on the ^{14}C content measurements. The bio-based carbon content will be expressed by a fraction of sample mass, as a fraction of the total carbon content or as a fraction of the total organic carbon content (FprCEN/TS 16137 under approval – expected for 2011–02) [23].

Despite their origin from renewable feedstocks, the bioplastics are not all biodegradable or compostables. We briefly recall that biodegradable plastics are plastics able to degrade under the action of naturally occurring microorganisms such as bacteria, fungi, and algae. The compostability notion is more restrictive, asking certain specified criteria to the plastic biodegradation, such as degradation rate, maximum residue of material left after a specific period of time, and the requirement for the material to have no harmful impact on the final compost or the composting process. The European standard EN 13432/2000 for compostable plastics asks for a compostable product to degrade up to 90% under commercial composting conditions, within 180 days. The ASTM 6400 standard sets a less stringent threshold of 60% biodegradation within 180 days, in commercial composting conditions [24].

Some bioplastics are biodegradable/compostable and they were the first driving force for developing biodegradable/compostable materials for disposal applications, i.e. packaging, mulch film, disposable cutlery. In the meantime, the R&D achievements in this field allowed to better understand the potential of bioplastics for durable applications (e.g. electronics, automotive applications), and they extend the industrial interest to not biodegradable bioplastics such as bio-based polyethylene or polypropylene. Finally, in the last few years, the durable applications of bio-based plastics tend to become the motivating force for the use of renewable resources in the plastics industry.

The biocomposites are the second class of greener polymer materials of clear interest for automotive applications. Biocomposites generally concern the materials formed by a matrix (resin) and a natural fibrous reinforcement/fillers. The polymer matrix can be either thermoplastic or thermosetting, from petrochemical or bio-based resources. This chapter will focus only on the bioplastics-based biocomposites for car applications.

Another group of greener polymer materials with interest for automotive industry concerns the bioplastic nanocomposites, also called bionanocomposites, which are multiphase systems having one minor phase uniformly dispersed into the bioplastic on nanoscale level (10^{-9} m) [dedicated Chapter 4 in the present handbook].

15.2 Bioplastics for Automotive Applications

Within the bioplastics existing today on the market, some are already validated for different automotive applications: it is the case for polylactic acid formulations and fabrics, bio-based polyamides and bio-based polyurethane foams. Other current and emerging bioplastics with potential/validated use in automotive industry are belonging to the class of bio-based polyesters and copolyesters, starch plastics, bio-based polyolefins and bio-based thermosetting polymers such as unsaturated polyester resins or bio-based epoxies. All these classes of bioplastics will be presented in the following sections of this chapter. And for giving to the reader a first overview on the bioplastics for automotive applications, Table 15.1 presents the most important ones, together with their thermal and mechanical characteristics, for the sake of comparison.

As one can see, the current and emerging bioplastics already covers the range from commodity thermoplastics up to engineering materials. And even if their present properties are not always optimal for durable automotive applications, their development is exponential and the bioplastics could offer in the future real alternatives for petrochemical plastics in modern cars.

Table 15.1 Main characteristics of bio-based thermoplastics for automotive applications and comparison with some petrochemical polymers.

Polymer		Density (g/cm^3)	Tensile Properties				Flexural Modulus (GPa)	Impact Strength @ 23°C			T_g † (°C)	T_m (°C)	Renewable Content (wt %)
			Yield Stress (MPa)	Ultimate Strength (MPa)	Young's Modulus (GPa)	Elongation at Yield/Break (%)		Unnotched (Charpy) (kJ/m^2)	Notched Charpy (kJ/m^2)	Notched Izod (J/m)			
PA 11[a] [25]		1.05	36[1]	47–50[a]		22 / 360[1]	1.17	–	30–40[5]	99[1]	45	180–189[1]	100
PA 6,10[b] [25]		1.08	55; 59[1]		2.4[1]	10 / 70–100[1]	2.00[1]	–	4–10[1]	50[1]	45–70[1]	225	≈60
PLA	PLA[c] [25,26]	1.24–1.33	53–100		2.98–3.45	2.5–8[3,c]	2.85–3.83	–	–	12.8–144	45–60	120–170	100
	L-PLA [27]	1.24–1.29	28–50		1.2–3.0	3–10[3]	1.4–3.25	–	–	26[2]	53–64	145–185	
	D.L-PLA [27]	1.24–1.29	29–35		1.9–2.4	2–10[3]	1.95–2.35	–	–	18[2]	50–57	–	
	sc-PLA (50/50) [25,28]	–	–	–	–	–	–	–	–	–	60–70	210–240	
Bio-based PTT[e,f] [25*,29,30]		1.35*; 1.4[e]	66*		6.2[e]	10 / 15[f,1]	2.7*; 5.7[e]	–	–	2.7[f]	45–55*	228*	35
Bio-based PBS[d] [24,28]		1.2–1.3	33[3,4]		2.6	– / 560[3]	2.3	–	–	294	–32	144	
Durable starch plastics[g] [31]		1.03–1.05	–	17.4	0.96–1.1	– / 10	0.69–0.75	–	–	37–55	HDT at 0.45 MPa: 74°C		≈50
Bio-based LDPE [25]*		0.92	10–18		0.15–0.20	12 / 100–800[1]	0.24–0.33[1]	no break[5]	no break[5]	no break[1]	–30	105–130	100
Bio-based PP homopolymer [25]*		0.91	33–35		0.8–2.0	10 / 700[1]	1.45–1.55	no break[5]	3–17[5]	27[i,1]	–13 to 0[h] –18 to –1[i]	160–170	100

Table 15.1 (cont.) Main characteristics of bio-based thermoplastics for automotive applications and comparison with some petrochemical polymers.

Polymer		Density (g/cm³)	Tensile Properties				Flexural Modulus (GPa)	Impact Strength @ 23°C			$T_g^{\dagger}$ (°C)	T_m (°C)	Renewable Content (wt %)
			Yield Stress (MPa)	Ultimate Strength (MPa)	Young's Modulus (GPa)	Elongation at Yield/Break (%)		Unnotched (Charpy) (kJ/m²)	Notched				
									Charpy (kJ/m²)	Izod (J/m)			
Petrochemical polymers - for sake of comparison													
PET[5]	Partly crystalline	1.38	81	42	2.8	4 / 70	2.0[1]	no break	3.0	90[1]	98	255	–
	Amorphous	1.34	55	38	2.2	3.5 / 300		no break	2.5		86	255	
PBT[5]		1.30	52	50	2.6	4 / 200	2.3[1]	no break	3.5	53[1]	60	223	–
PS (General purpose)[5]		1.05	–	45–65	3.2–3.25	– / 3–4	3.1[1]	5–20	2–2.5	20[1]	80–100	240[1,‡]	–

*Bio-based plastics supposed to be chemically identical with their petrochemical counterparts. So, the physical and thermal properties are also identical.
[a]Rilsan (Arkema); [b]Amilan™ CM 2001 (Toray); [c]Ingeo™ series (NatureWorks LLC); [d]Bionolle 1000 (Showa Highpolymer); [e]DuPont™ Sorona EP; [f]DuPont™ Biomax®PTT 1100; [g]Cereplast Hybrid Resins; [h]isotactic PP; [i]atactic PP.
†Tg of the amorphous polymer or respectively, of the amorphous phase of a semi-crystalline polymer; ‡pseudo-melting temperature.
[1][27]; [2][32]; [3][24]; [4][33] – data estimated from figure; [5][34].

15.2.1 Bio-based Polyamides (PAs) and Copolyamides

The polyamides are engineering thermoplastics with repeating amide linkage [-CO-NH-] in the polymer backbone. Generally, PAs are synthesized either via diamines and dicarboxylic acids reaction (PAs with two basic repeating units), or amino acids or lactams polymerization (PAs with single repeating unit). The PAs with a single basic unit are abbreviated as ***PA x***, where *x* represents the number of carbon atoms in the basic unit. The PAs with two repeating units are abbreviated as ***PA x,y***, where *x* indicates the number of carbon from diamine part and *y*, the number of carbon atoms from the dicarboxylic acid.

The PAs-key properties combine excellent mechanical properties, such as high mechanical strength and stiffness, wear properties, good heat resistance, together with chemical resistance to oils and solvents, dielectric properties, fire resistance, good appearance, and good processability [35,36]. These nice mechanical and thermal performances design them for high-end automotive applications and especially in their "undisputed domain," the under-the-hood car compartments. In fact, approx. 10% of the plastics from the modern cars are PAs and the automotive industry is using about 41% of the PAs market [37,38]. The PA 6 and PA 6,6 are the most used in car applications, but also PA 11, PA 12, PA 6,10, PA 4,6 and PA 6,12. The typical ratio consumption between PA 6, PA 6,6 and the other PAs was appreciated to about 5/4/1 in Western Europe, Japan and the United States [35]. The main applications of PAs in modern cars concerns [38]:

- exterior parts, that need thermal performance, mechanical strength, design flexibility;
- interior parts, generally components and upholstery with safety standards (i.e., brake lines and fittings, accelerator pedal module);
- under-the-hood components, requiring high temperature resistance and inertia to corrosive substances (i.e., air cleaner, throttle valve guards and throttle valves in automobile engines);
- technical parts involving fluids transport and aggressive fuel contact (i.e., multilayer fuel lines and fittings);
- 'invisible' products (i.e., anti-corrosion coating treatment, gasket).

Until recently, the polyamides for automotive applications were petro-based, except the Rilsan®PA 11 from Arkema, derived from castor oil. Today, several other new bio-based PAs appeared on the market, derived from renewable feedstocks such as castor beans and sugar cane. They are promising new solutions for replacing the petrochemical PAs, but also for extending the metal substitution in car applications, improving automotive comfort, design and insulation, and enriching the performances with fuel economy and reduced CO_2 emissions [38–45]. The most interesting current and emerging bio-based PAs for car applications, their chemical structure and origin, trademarks and producers are presented in Table 15.2. Examples of mechanical and thermal properties of these classes of bio-based PAs, plus their current/potential use in automotive applications and a comparison with some petrochemical counterparts are shown in Table 15.3.

Table 15.2 Current and emerging bio-based PAs and derivates for automotive applications.

Bio-based PAs and Derivates		Producer	Chemical Structure/Origin
Commercial Products			
PA 11	Rilsan®PA 11	Arkema [46]	$-[NH-(CH_2)_{10}-CO]_n-$ • bio-based part: ω-amino-undecanoic acid (amino 11) from castor oil
PA 6,10	Ultramid® Balance	BASF [47]	$-[NH-(CH_2)_6-NH-CO-(CH_2)_8-CO]_n-$ • 1,6-hexamethylenediamine • bio-based sebacic acid from castor oil
	Amilan® CM2001	Toray [48]	
	Technyl eXten®	Rhodia [49]	
	Zytel® RS PA 6.10	DuPont [50]	
	Vestamid® Terra HS	Evonik Industries [51]	
	Grilamid 2S	EMS Grivory [52]	
	Akromid S	Akro Plastik [53]	
	Uni-Rez®	Arizona Chemical [53]	
PA 10,10	Vestamid® Terra DS	Evonik Industries [51]	$-[NH-(CH_2)_{10}-NH-CO-(CH_2)_8-CO]_n-$ • bio-based 1,10-decamethylenediamine from castor oil • bio-based sebacic acid from castor oil
	Zytel® RS PA 10.10	DuPont [50,54]	
	Grilamid 1S	EMS Grivory [52]	
PA 4,10	EcoPaxx™	DSM [55]	$-[NH-(CH_2)_4-NH-CO-(CH_2)_8-CO]_n-$ • 1.4-tetramethylenediamine • bio-based sebacic acid from castor oil
PA 10,12	Vestamid® Terra DD	Evonik Industries [51]	$-[NH-(CH_2)_{10}-NH-CO-(CH_2)_{10}-CO]_n-$ • bio-based 1,10-decamethylenediamine from castor oil • fossil or bio-based dodecanedioic acid
PEBA	Pebax® Rnew	Arkema [46]	• polyether-block-amide (PEBA) copolymer consisting of segments of bio-based amino-11 and polyether

Table 15.2 (cont.) Current and emerging bio-based PAs and derivates for automotive applications.

Bio-based PAs and Derivates		Producer	Chemical Structure/Origin
Commercial Products			
PPAs	Rilsan® HT	Arkema [56,57]	• copolymer using the Amino 11 chemistry
	Vestamid® HT*plus* M3000	Evonik Industries [58]	PA 10T-copolyamides. from which the bio-based part is 1,10-decamethylenediamine from castor oil (T = terephthalic acid)
	Grivory HT3	EMS Grivory [52]	
R&D products			
PA 6	Potential automotive applications	DuPont [50]	$-[NH-(CH_2)_5-CO]_n-$ • ε-caprolactam from glucose (sugar fermentation)
PA 6,6		[28]	$-[NH-(CH_2)_6-NH-CO-(CH_2)_4-CO]_n-$ • 1,6-hexamethylenediamine • bio-based adipic acid by glucose (sugar fermentation)
PA 5,10		BASF [47]	$-[NH-(CH_2)_5-NH-CO-(CH_2)_8-CO]_n-$ • bio-based 1,5-diaminopentane via lysine from glucose fermentation • bio-based 1,10-decamethylenediamine from castor oil
PA 4,6		DSM [55]	$-[NH-(CH_2)_4-NH-CO-(CH_2)_4-CO]_n-$ • 1,4-tetramethylenediamine • bio-based adipic acid from glucose
PA 6,9		[28]	$-[NH-(CH_2)_6-NH-CO-(CH_2)_7-CO]_n-$ • 1,6-hexamethylenediamine • bio-based azelaic acid (oleic acid: olive oil. palm berry oil. animal fats)

15.2.1.1 PA 11

Rilsan®PA 11, from Arkema, is a high performance aliphatic bioPA derived from castor oil feedstocks (see Tables 15.2 and 15.3). Its key-properties concern low modulus and superior impact properties at both ambient and sub-ambient temperatures, abrasion resistance twice as much as PA 6,6, chemical resistance to hydrolytic reagents, and lower moisture absorption than other PAs. Its use temperature ranges from –40°C to 130 °C [46,56,63,64].

Table 15.3 Mechanical and thermal characteristics of petrochemical and bio-based PAs* for automotive applications.

	Density (g/cm^3)	Tensile Strength (MPa)	Young's Modulus (GPa)	Flexural Modulus (GPa)	Charpy Impact Strength (kJ/m^2)	T_g[†] (°C)	T_m (°C)	Water Absorption (%)	Processing	Current and Emerging Automotive Applications
Neat Petrochemical Polyamides										
PA 6[a]	1.13	79	2.30–2.50	2.8		47–57	220	2.7	• processed by all conventional thermoplastics processing techniques plus casting from monomer and reaction injection molding	• under-the-hood components: engine and rocker valve covers, air injectors, automobile air intake manifolds • structural components: front end modules • exterior parts: door and tailgate handles, front-end grilles, exterior mirror housings, fuel caps and lids, wheel covers • interior parts: driver side airbags, seat adjuster handles, pedals
PA 6,6[a]	1.14	82.7–90.3	3,3	1.207–1.301		78	264	3–4		• under-the-hood: power-steering reservoir, engine and rocker cover, air intake manifold, radiator end tank, thermostat housing, shifter module, release handle

										• chassis: shifter detent, air piston, carbon canister • exterior parts: fan and shroud, headlamp bezels, mirror bracket, wheel covers, fuel filler door • interior parts: seat levers & seat belt components, airbag bolts
PA 6,10[a]	1.07–1.09	59	2.4	2.24		42–67	225	1.4–1.5		• housings and transmission components • connectors, tubing and reservoirs in coolant circuits, wheel speed sensors
PA 12[a,b,c]	1.02[a]	49[c]	1.27–2.60[a]	1.41[c]		40[b], 55[a]	170–179[b]	0.25[c], 0.8[a]		• mono- and multilayer fuel lines, connectors • tank filler necks
Neat Bio-Based Polyamides										
PA 11[b,c]	1.03[b], 1.05[c]	57[c]	1.10–1.40[a]	1.17[c]	No break[b] un-notched	45[b]	180–189[b]	0.25[c]	• extrusion • extrusion-blow molding • injection molding • rotomolding • fiber spinning • powder coating formulations	**1.** *PA 11 resin*: • flexible tubing, mono-wall fuel lines and Rilperm® multi-layer fuel lines [59], such in ESD-Flex conductive fuel-pump module for General Motor car models [60]

Table 15.3 (cont.) Mechanical and thermal characteristics of petrochemical and bio-based PAs* for automotive applications.

	Density (g/cm³)	Tensile Strength (MPa)	Young's Modulus (GPa)	Flexural Modulus (GPa)	Charpy Impact Strength (kJ/m²)	T_g† (°C)	T_m (°C)	Water Absorption (%)	Processing	Current and Emerging Automotive Applications
										• fluid transfer lines (brake, cooling, clutch), friction parts, quick connectors, pneumatic brake noses **2.** *Rilsan® PA 11 fibers*: potential 'like-vegetal' fiber reinforced-bioplastics for automotive applications [46]. **3.** *Rilsan® PA FinePowders* as coatings for door handles, oil and fuel filters, engine blocks, wheels, coil springs, steering shafts, interior small parts, seat rails [61].
PA 6,10[d,e]	1.08[d]	55[d]	2.10[d]	2.00[d]	notched 40[d]	48[e]	225[d], 222[e]	3[e]	• extrusion • injection molding	• fuel contact line • air filter system
PA 10,10[f]			1.70			37	206	2		• fuel lines and special cables
										• charge air coolers, turbo air ducts, engine mounts, cylinder head covers, oil pans, transmission parts

PA 5,10[g]	1.07					50	215	1.8		• air filter housing (Daimler) [62] • accelerator pedal module, cogwheel for the steering angle sensor, cooling fan (Mercedes) [62]
Glass-Reinforced Bio-Based Polyamides										
PA 6,10–30% GF[h]	1.33	150			75		222	4.0	• extrusion • injection molding	• technical automotive parts
PA 10,10–50% GF[i]	1.51	200			100		200	2.0		
PA 10,10–65% GF[j]	1.76	230			52		200	2.0		

*Dried PAs.

†T_g of the amorphous phase of a semi-crystalline polymer.

[a][27], [b]Rilsan®PA 11 [46], [c][35], [d]Amilan™ CM2001 [48], [e]Vestamid®Terra HS [51], [f]Vestamid®Terra DS [51], [g][62], [h]Grilamid XE3976-black9992 [52], [i]Grilamid 1SVX-50H [52], [j]Grilamid 1SVX-65H [52].

The PA 11 already found its role as engineering polymer in high-end automotive applications, and in Rilsan®PA FinePowders formulations for coating large automotive parts. Additionally, Rilsan®PA 11 fibers are expected to be used in 'like-vegetal' fiber reinforced-bioplastics for automotive applications. The extruded bio-fibers are reported as 40% lighter than polyesters, and bacteriostatic. They are already used in commercial products such as footwear, clothing and luggage market companies [65].

Additionally, an effective way to extend PA 11 automotive applications by improving some of its use properties is offered by the nanoparticles technology. In fact, the automotive industry has pioneered the use of polymer nanocomposite, as they provide strength and stiffness, enhance thermal stability, and possibly flame retardancy, improve dimensional stability, gas barrier, and electrical conductivity (for detailed data see the pertinent reviews of Ray *et al.* [66] and Pavlidou *et al.* [67]). The concept was introduced in 1985 by Toyota research group, and the first commercial nanocomposite was a petro-based PA 6/nanostructured silicate clay (4 wt%) used in timing-belt cover [68–70]. Presently, academic and industrial attention is paid on PA 11/clay nanocomposites and focuses on the importance of nanoclays dispersion within the PA 11 matrix, for improving the use properties of nanocomposites [69,71–72]. For instance, the thermal stability and mechanical properties of the exfoliated PA 11/clay nanocomposites were found to be superior to those of the intercalated ones, due to the finer dispersion of organoclay among the matrix. Additionally, the hardness and modulus of the PA 11-nanocomposites gradually increase with increasing clay content. Adding 5 wt% clay into PA 11 matrix was found to improve its hardness and elastic modulus by about 30% [72].

15.2.1.2 *Other Commercial Bio-based PAs*

Today, several new bio-sourced polyamides are present on the market, derived from renewably castor oil (see Table 15.2).

Bio-based PA 4,10: Recently, DSM launched EcoPaXX™, a bio-based PA 4,10 with about 70% biocontent [55,73,74]. It exhibits high melting point (approx. 250 °C), low moisture absorption, hydrolysis resistance. It also has excellent resistance to various chemical substances, including road salt, and high crystallization kinetics (typical for petro-based PA 6,6 and PA 4,6). Due to its low moisture absorption, which assure good maintain of strength and stiffness after conditioning, this PA 4,10 is suitable for demanding automotive applications such as under-the-hood components (engine covers, air injectors, airducts), structural components (front end modules), exterior parts (door handles, exterior grills, exterior mirror housings), and interior parts (driver side airbags, seat adjuster handles, pedals) [74,75].

Bio-based PA 6,10: Several bio-based PA 6,10 commercial grades are now available, with about 60% biocontent (see Table 15.2). Typically, the PA 6,10 has comparable mechanical and thermal performances to PA 6, while chemical resistance is comparable to PA 12. The PA 6,10 has lower density than PA 6 and better low-temperature toughness. At low temperature, it shows good impact strength. Bio-based PA 6,10 has the potential to replace classical PA 6 in automotive applications, especially where low water absorption is required, as for instance in flexible tubes for the power-assisted control system, fittings and adapters for the engine fuel. A recent example of an under-the-hood application is the new automotive radiator end tank proposed by Toyota, Denso and DuPont Automotive consortium [76,77], based on DuPont's Zytel® RS, and used in some 2009 Toyota Camrys vehicles.

Another example is the air filter housing for Daimler, made from BASF bioPA 6,10 reinforced with 10 wt% glass fiber and 20 wt% mineral substances [62].

Bio-based PA 10,10: Several commercial grades of the entirely bio-sourced PA 10,10, derived from castor oil feedstocks, are present on the market (see Table 15.2). PA 10,10 offers properties between PA6 and PA12, and is reported as suitable for glass-fiber-reinforced molding compounds. Despite its crystallinity, it is translucent. Its automotive applications include fuel lines and special cables, charge air coolers, turbo air ducts, engine mounts, cylinder head covers, oil pans, transmission parts [50,51,54].

Bio-based PA 10,12: Evonik Industries recently launched a PA 10,12 with up to 45% biocontent, the Vestamid®Terra DD (see Table 15.2). It is a transparent PA and stands out for its high impact resistance [78].

15.2.1.3 Bio-based PAs—in R&D State

Some other bio-based PAs with potential automotive interest are in R&D state (Table 15.2) [28,35], and the most promising are:

Bio-based PA 5,10: BASF has developed a 100% bio-based PA 5,10 with technical performances suitable for automotive applications. However, the bioPA 5,10 is a rather expensive PA, and this limits today broad automotive applications. PA 5,10 will be most probably found in small parts, such as accelerator pedal module [47,53].

Bio-based PA 6 and **PA 6,6**: Bio-route exists to obtain these two polyamides from renewable feedstocks, bur they are not yet economically interesting [28]. Once commercially available, one can imagine that they will gradually replace their petrochemical counterparts in automotive applications and especially in the severe under-the-hood applications.

Bio-based PA 4,6: DSM Engineering works on PA 4,6 grade expected to favor metal parts replacement under-the-hood such as turbo diesel systems components [44,55].

15.2.1.4 Bio-based Polyether-block-amides (PEBAs)

The polyether-block-amides belong to the class of thermoplastic elastomers, and similar to other thermoplastic elastomers (polyurethanes and polyesters), their backbone chains contain alternating hard and soft segments. The PEBA soft segments are composed of polyether diols responsible with the flexibility properties over a broad temperature range, while the hard segments are polyamides. These hard segments of different chains are interconnected by strong hydrogen bonds (typical for PAs), which act as physical cross-linking points and are responsible for the high tensile strength, good abrasion and friction characteristics, plus high heat and weathering resistance. Upon heating, the hard phase melts and these hydrogen bonds disappear, letting the material to be processed as a classical thermoplastic.

Bio-based PEBAs are already on the market, with either the polyamide block or both types of block polymers derived from renewable feedstocks. Thus, Arkema proposes the Pebax®Rnew elastomers, with Rilsan®PA 11 as polyamide block [79,80]. They have 20 to 95% biocontent; an entirely bio-sourced grade, the Pebax®Rnew100, combines a bio-sourced polyol with the bio-based PA 11 [81]. EMS-Grivory proposes a bio-sourced PA 12-Grilflex PEBA line, which includes bio-based materials with 10% to 100% renewable content derived from castor and canola oil [82]. Similar to their petrochemical counterparts, the bio-based PEBAs are intended to replace the conventional elastomers in different fields, as for instance in high-end automotive applications asking for good chemical resistance and preservation of mechanical properties in severe use conditions.

15.2.1.5 *Polyphtalamides (PPAs)*

Polyphthalamides are semi-aromatic polyamides and copolyamides in which the aromatic part comes in general from terephthalic (T) (Figure 15.1) or isophthalic acid (I).

Generally, the PPAs performances have been shown to be superior to that of aliphatic PA 6 and PA 6,6 [83]. Indeed, the presence of an aromatic ring structure into PA backbone provides several advantages to the polymer, i.e., higher T_g, higher melting point and reduced absorption of moisture and solvents. From practical viewpoint, this means improved dimensional stability, improved solvent and hydrolysis resistance, and better conservation of their mechanical properties in high temperature applications. Their mechanical and thermal properties design the PPAs for extending the replacement of metal car parts in compartments submitted to high use temperature or to direct contact with fuel or water.

Nowadays, several bio-based PPAs are commercially available, with 50–70% biocontent (see Table 15.2). Automotive applications of these commercial bioPPAs and their glass-fiber composites include automotive powertrain [80,84,85], charge air duct for automotive engines [58], connectors and connection elements for use in contact with fuel, and parts involving direct contact with cooling water [52]. Rilsan®HT from Arkema is used in the exhaust gas recirculation of PSA Peugeot Citroen cars, as replacement of aluminum tubes [3], and for replacing under-the-hood metallic flexible tubes in Volkswagen car models [57].

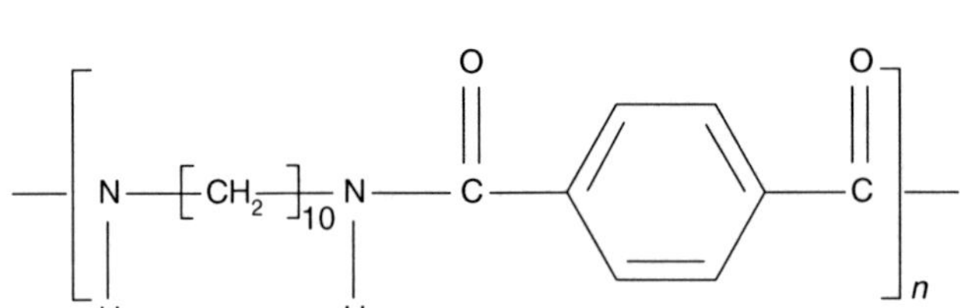

Figure 15.1 Example of a PPA chemical formula, namely PA 10T.

15.2.1.6 Conclusion

The current and emerging bio-based polyamides are green materials with confirmed technical potential to replace the petrochemical PAs in the automotive applications, and additionally to increase the part of plastics in the modern cars by extending the substitution of metal parts. Therefore, they are offering new opportunities for increasing the functionality and the reliability of different car parts submitted to severe use conditions and for reducing the total car weight, which brings fuel economy and less CO_2 emissions.

15.2.2 Polylactic Acid (PLA)

15.2.2.1 PLA and PLA-based Compounds

PLA (polylactic acid) is an aliphatic biodegradable polyester (Figure 15.2) obtained by polymerization of lactic acid, which is a sugar fermentation product from corn, sugar beets, sugar cane, potatoes and other plants. In the future, cellulosic non-food biomass is indented to be used to produce PLA [25].

While the biopolyamides naturally come as engineering polymers for high-end automotive applications, the PLA is a rather new polymer in durable applications and from some aspects, still in development. For long time, PLA applications have concerned only biomedical and packaging fields, but in last years new PLA-improved materials were proposed in durable applications, for transportation and E&E (electrical appliances and electronics) [25].

PLA exists in several isomeric forms, which exhibits different end-use properties: two semi-crystalline homopolymers, poly(L-lactide), L-PLA, and poly(D-lactide), D-PLA, and a racemic poly(DL-lactide), DL-PLA, which is amorphous. The mechanical and thermal properties of PLA and its copolymers are highly related to the ratio between the two isomers, L and D, of the lactic acid monomer used in the polymerization (see Table 15.1). It is also possible to obtain stereo-block PLA, which exhibits better heat resistance than the previous forms. And finally, the most promising from the mechanical and thermal viewpoint is the stereocomplex PLA, sc-PLA, obtained by stereocomplex crystallization via blending of lactides monomers or blending of L-PLA and D-PLA in the melt state. Additionally, this sc-PLA has a reported crystallization rate comparable to that of PA 6 and PE, and largely superior to that of racemic DL-PLA [28,86-PURAC].

Today, most of the commercial PLA grades are copolymers of L-PLA and DL-PLA [26,87], and exhibit mechanical properties close to polystyrene. These

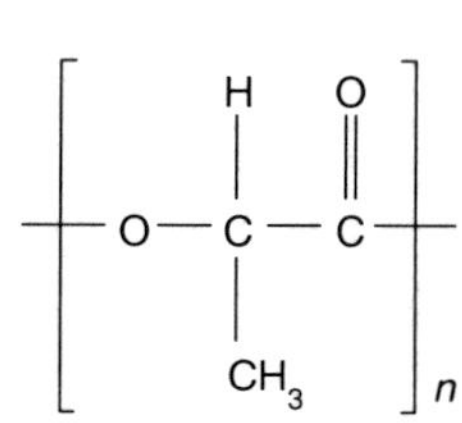

Figure 15.2 PLA chemical structure.

unmodified PLAs are yet not directly suitable for durable automotive applications because of their low heat resistance, low impact resistance, low crystallinity/crystallization kinetics, and also because of long-term durability questions by reason of hydrolysis tendency. As a consequence, in the last decade the R&D activities have looked for answers to these practical issues limiting PLA-durable applications. The mechanical and thermal drawbacks can generally be overcame by using different types of additives (fillers, impact modifiers, etc.) or by blending the PLA with other petro- or bio-based polymers, or via fiber-reinforcement (e.g. biocomposites). The processing issues related to the low crystallization rate of PLA in the mold need also to be solved for automotive applications, as the PLA-based compounds require longer processing cycles and, moreover, increase the risk of slow continuing crystallization after the processing step, which means dimensional/geometrical evolution in time for the parts, and also changes in mechanical and thermal properties. To overcome this specific aspect, the use of appropriate nucleating agents, such as talc or sc-PLA, makes it possible to improve both processing cycle times and heat resistance [88].

In the following, we will briefly present the interest of each one of these solutions. Table 15.4 presents a non-exhaustive list of these improved PLA-materials, together with their key-properties and current/emerging automotive applications. The PLA-formulations are organized within four categories: typical inorganic additives, PLA blends with petro-based and bio-based polymer, and finally PLA biocomposites.

15.2.2.1.1 Filled and Reinforced PLA

Generally, microfillers are added to PLA to improve its modulus at use temperature and to increase the heat resistance, or to reduce the final cost of the product. From this category one can cite the talc, mica, calcium carbonate or calcium sulfate microfillers. Table 15.4 presents the main improvements reported for this kind of additives. A second effective way to improve mechanical and thermal PLA properties is to use nanofillers, either pre-formed nanoparticles, in-situ formed nanoparticles, or by reactive extrusion of nanoparticles/lactide monomer premixture [105]. Table 15.5 gives three examples of PLA/Montmorillonite (MMT) nanocomposites, with emphasis on their improved mechanical and thermal properties. The promising results for PLA-nanocomposites obtained under appropriate preparation conditions demonstrate that nanoparticles do improve technical performances of neat PLA, by making it more competitive with petro-chemical polymers and conventional composites.

A third way to obtain interesting PLA-composite materials for automotive applications is offered by the natural fibrous reinforcements, which will be presented in the third section of this chapter.

15.2.2.1.2 PLA-based Blends

Another method to overcome PLA brittleness is to blend it with other polymers. Literature data indicate than effective toughened PLA-based materials can be obtain by blending PLA with thermoplastics such as PE or PP, or by using impact modifiers such as ABS, MBS or polyurethane elastomers Tables 15.4–15.6. Yet, as can be seen in Table 15.6, a major challenge for the toughening strategies still

Table 15.4 Non-exhaustive list of commercial and emerging PLA and PLA-based materials for current/emerging automotive applications.

PLA–Based Materials	Properties	Current or Potential Automotive Applications
Standard commercial PLAs	• mechanical properties ≈ PS • clarity • classical processing techniques BUT • brittleness • low HDT	• not yet enough performance for durable automotive applications • immediate possibilities: using PLA for non-durable auto applications such as protective wrappings for vehicle manufacturing and transit [89]
PLA fibers and fabrics	• comfort/feel of natural fibers • performance and easy care [26,90,91]	• canvas roof and carpet mats in Ford Model U (2003) [92] • floor mats – in Toyota Raum and Prius cars (2003) [93]
Biofront™ stereocomplex PLA, co-developed Teijin & Mazda	• higher crystallinity than PLA • heat-resistant bioplastic • $HDT_{80\text{–}100\%\ sc\text{-}PLA} > 150°C$ [86-PURAC] • higher strength [94,95] • resistant to abrasion and damage from sunlight • flame retardant • hydrolytic and hydrothermal resistance ≈ PBT and PET	• Biofront™ is intended for fibers and automotive applications • automotive interior components • car seat fabric made 100% from Biofront™ - in Mazda Premacy Hydrogen RE Hybrid vehicle [94,95] • door trim in Mazda Premacy Hydrogen RE Hybrid [95]

Table 15.4 (cont.) Non-exhaustive list of commercial and emerging PLA and PLA-based materials for current/emerging automotive applications.

PLA –Based Materials	Properties	Current or Potential Automotive Applications
1. Additives		
+ talc as a nucleating agent	• speeds PLA crystallization and reduces molding cycle time, when added 10–30%	[96,97]
+ very fine-particle (0.05-micron) silica	• increases toughness while maintaining clarity, even at lower levels than 10–30%	[96,97]
+ high-aspect-ratio precipitated calcium carbonate from Specialty Minerals, called EMforce™ Bio as reinforcing agent	• reduces PLA brittleness • adding 30% of EMforce™ Bio into PLA => surprisingly ductile failure with high energy adsorption	[96]
2. PLA/Petrochemical Polymer blends		
+ ABS	• reduce brittleness (see Table 15.6) • easy to mix • opaque two-phase blend	• durable applications like cell phones (Samsung Cheil Industries in Korea) [97]
+ PE, PP	• toughening effect, i.e. Izod impact resistance of PLA/LLDPE 80/20 is 30–38 times higher than for neat PLA [98]	• Toyota Motor uses PLA/PP for interior vehicle parts: scruff plates, cowl side trim, floor finish plate. toolbox [99]

+ PC (GL-1401 alloy from Cheil Industries. part of Samsung Group)	• better fatigue resistance than PC/ABS • higher Izod impact strength than PC/ABS • better flexural modulus than PC/ABS	• compounds for durable applications like cell phones • Cheil's biomaterials are now being tested by General Motors and Ford. [100]
+ engineering thermoplastic resins + bio-based polymers (polyhydroxy-alkanoates and other bio-polyesters) => PolyOne's resound™ biopolymer compounds	• up to 50%wt bio-derived content • good impact resistance (Notched Izod impact: 534–641 J/m vs. 16–144 J/m of neat PLA) • medium heat resistance (HDT: 108–115°C) • cleanable, good chemical resistance • good color stability • specific gravity 1.21	• automotive interior parts [101,102]
+ epoxidized natural rubber with high epoxy functionality (50%)	• increases toughness • 20% of added rubber increases by 5 times the impact strength	[96]
+ polyurethane thermoplastic elastomers	• very compatible with PLA • increases toughness when added 5–10%	[96]
+ aliphatic/aromatic polyesters such as Hytrel™ (DuPont)	• very compatible with PLA • increases toughness when added 5–30%	[96]
+ epoxy-based chain-extender	• enhanced processability and improved properties of solid and microcellular PLA • increases PLA thermal stability	• potential applications for structural automotive parts [103]

Table 15.4 (cont.) Non-exhaustive list of commercial and emerging PLA and PLA-based materials for current/emerging automotive applications.

PLA –Based Materials	Properties	Current or Potential Automotive Applications
3. PLA/Bioplastics Blends		
+ sc-PLA as nucleating agent	• 1–10% of sc-PLA in injection molded commercial PLA acts as nucleating agent and bring its HDT at more than 100°C	[86-PURAC]
+ sc-PLA nucleating agent + shock-absorbing flexible ingredients + compatibilizer → Mazda PLA-based bioplastic (over 80% plant-derived content)	• high-strength compounds • 3 times the shock impact resistance and 25% higher HDT than classical bioplastics used for items such as electrical appliances • high quality finish • injection and extrusion-moldable	• Mazda biotechmaterial for vehicle's instrument panel and other interior fittings needing thermal and shock resistance and beautiful finish in Mazda's Premacy Hydrogen RE Hybrid [95]
4. Biocomposites		
+ (chemically coupled) kenaf fiber → Eco-Plastic PLA compounds, from Toray	• improved strength and HDT than standard PLA (see Table 15.9)	• cover spare wheel on Toyota Prius and Toyota Raum (2003) [104] • translucent roof PLA/kenaf and ramie biocomposite on Toyota 1/X plug-in hybrid concept vehicle [93] • Toyota plans to use Eco-Plastic in around 60% of interior components, starting with the next-generation Prius hybrid

Table 15.5 PLA-nanocomposites and their improved properties [adapted from 105].

Nanocomposite	Mechanical Properties, % as Reported to Neat PLA	HDT, % as Reported to Neat PLA	Reference
PLA + 4 wt% organically modified C_{18}-MMT	• flexural modulus (117%) • flexural strength (154%) • distortion at break (168%)	HDT (123%)	[106]
PLA + 2–8 wt% Cloisite® 25A* (organically modified MMT)	• ultimate strength (168%) • tensile modulus (143%) • elongation at break (141%)		[107]
PLA + 4–7 wt% organically modified $C^2C^2_{18}$-MMT	• flexural modulus (112%) • flexural strength (119%) • distortion at break (205%) • for an optimum of 4 wt% MMT	HDT (140%)	[108]

*Cloisite® 25A – Trademark from Southern Clay Products, Inc.

remains achieving durable toughening of the PLA without compromising its tensile properties or compostability [109].

15.2.2.2 Durability Issues of PLA Components

One important aspect related to the PLA automotive applications is its durability over time, under real use conditions of temperature and humidity, throughout the lifetime of the vehicle (>10 years).

Previous studies on PLA for biomedical or packaging short-time applications had already reported the degradation mechanisms of PLA when exposed to heat and moisture. It was shown that the degradation rate highly depends on the PLA grade and microstructure, on the geometry/thickness of the investigated part, on the environmental conditions such as moisture and heat exposure, presence of microbes, and so on [32,110,111].

More recently, a study initiated by Ford Motor Company [111] assesses about the long-term durability of injection-molded PLA, exposed to elevated temperature and humidity over several weeks, for simulating the automotive interior environment. The experimental results indicate that after 12 weeks in this aggressive atmosphere, the PLA materials could no longer be tested mechanically. In other words, the injection-molding PLA grades commercially available today are not suitable for use in applications that require long-term durability in environments subject to elevated temperature and humidity. In the meantime, these results do not disqualify the commercial PLA for durable car applications, but remind the

Table 15.6 Properties of some PLA-blends with commercial polymers [adapted from 97 and 104].

Additive	Type	% of Added Polymer into PLA	In % (Relative Values Compared to the Neat PLA)				HDT@ 0.45MPa (°C)
			Tensile Yield	Elong.	Tensile modulus	Notched Izod	
Neat PLA	biodegradable polyester	–	100	100	100	100	55–59
Blendex™ 415 Crompton Corporation	ABS with 65% rubber	15	70	2300		180	
Blendex™ 360 Crompton Corporation	ABS with 50% rubber	20	76	2800		400	
Blendex™ 338 Crompton Corporation	ABS impact modifier resin, based on 70% PB	20	70	2810		1940	
Magnum™ 555 Dow Chemical	ABS	20	106	114	106	83	58
		80	116	146	104	137	73
Paraloid™ BTA 753 Rohm and Haas	MBS impact modifier	20	57	3000		420	
Paroloid™ EXL 3691A Rohm and Haas	MBS core-shell impact modifier	20	69	2800		240	
Paroloid™ EXL 2314 Rohm and Haas	acrylic impact modifier	20	62	2500		200	
Bionolle™ 3001 Showa High Polymer	biodegr. copolymer of succinic and adipic dimethyl esters with BDO	15	94	2300		180	

Pellethane™ 2102-75A. Dow Chemical Company	TPU- PCL Polyester	30	68	4100		2880	
Kraton™ FG Kraton Polymers LLC	SEBS (30% PS) grafted with ≈ 2%w MA	20	53.9	1000			
Hytrel™ 3078, DuPont	TPE	30	55.6	4300		740	
Caliber™ 200-22, Dow Chemical	PC	20	100	86	96	94	59

PC = polycarbonate; ABS = acrylonitrile butadiene styrene; MBS = methacrylate butadiene styrene; BDO = 1,4-butanediol; TPU = thermoplastic polyurethane elastomer; TPU-PCL = polycaprolactone-TPU; SEBS = styrene-ethylene/butylene-styrene copolymer; TPE = thermoplastic copolyester elastomer.

importance of stabilizing it either by blending with other polymers able to reduce the hydrolysis to an acceptable level, or by eliminating the hydrolysis reaction itself via scavengers or equivalents. These improvements are vital before imagining extending the PLA automotive applications [111].

15.2.2.3 Conclusion

Today, PLA is one of the most commercialized polymers on the bioplastics market, and in the last decade, improved PLA-based materials have started to be tested with some success in nonstructural automotive interior and hidden parts. However, some more improvements are still necessary as far as its heat resistance, impact behavior, and durability aspects are concerned, but one can already imagine the potential of green PLA-based materials to replace common petrochemical thermoplastics such as PP and PE in their long-time applications.

15.2.3 Bio-based Polyesters and Copolyesters – other than PLA

15.2.3.1 PTT from Bio-based 1,3-Propanediol

Polytrimethylene terephthalate (PTT, Figure 15.3) is a non-biodegradable aromatic polyester obtained by polycondensation of 1,3-propanediol (PDO) with either terephthalic acid (TPA) or dimethyl terephthalate (DMT).

The presence on the market of bio-based-PDO (DuPont Tate & Lyle Susterra™ renewably sourced from corn sugar), made possible to obtain bio-based PTT: DuPont Sorona® polymers (37 wt% biocontent) for fibers and fabrics, and DuPont Sorona® EP thermoplastic polymers (20–37wt% biocontent), for engineering plastics applications. Similar to its petrochemical counterpart, the bio-based PTT is reported to combine the good strength, stiffness, toughness and heat resistance of polyethylene terephthalate (PET), with the good processability of polybutylene terephthalate (PBT) (see Table 15.1), additionally offering improved surface appearance and gloss [28,29,44]. Therefore, the PPT resins are considered to be adequate for automotive parts and components, such as critical electrical and electronics [29,112]. The fibers from bio-based PTT offer the softness and stain resistance suitable for automotive carpet, fabrics and plastic parts. The PTT fabrics are design for seat covers, door trim and headliners. The lower moisture content of PTT allows the fabrics to dry faster, thus reducing the risk of odor and mildew [113]. Honda introduced PTT seat biofabrics in its "FCX" model car, and Mitsubishi uses PTT floor mat in its "i-MiEV" model, an electric fleet test vehicle [114].

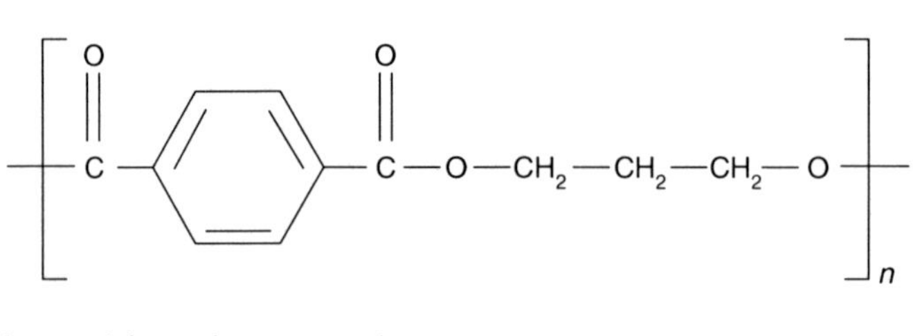

Figure 15.3 PTT chemical formula.

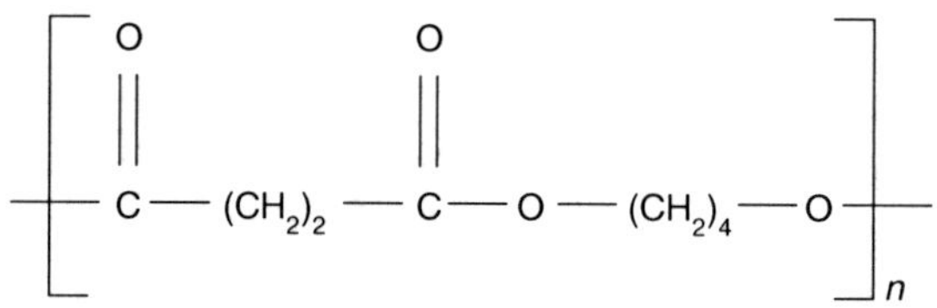

Figure 15.4 PBS chemical formula.

15.2.3.2 PBS from Bio-based Succinic Acid

Polybutylene succinate (PBS, Figure 15.4) is a biodegradable thermoplastic aliphatic polyester, obtained at the moment by polycondensation of petro-based 1,4-butanediol (BDO) and succinic acid. The PBS has comparable mechanical properties with general-purpose thermoplastics such as polyethylene and polypropylene.

Recently, the possibility of large scale production of bio-based succinic acid opens the road to produce bio-based PBS. Indeed, several industrial projects intend to produce succinic acid from renewable feedstocks: Bioamber (ARD/DNP) [115], Myriant Technologies (USA) [116], DSM and Roquette Frères (France) [117], BASF and PURAC [118].

The current petrochemical PBS is mainly used for disposal products such as mulch film, packaging film, bags and 'flushable' hygiene products. In the meantime, possible durable applications do emerge for the PBS-reinforced biocomposites, as shown by Mitsubishi Motors Corp. who proposes a bio-PBS/bamboo fiber biocomposite with good rigidity and strength, suitable for interior-trim automotive applications [119]. Mitsubishi Concept cX car is already using this Mitsubishi's Green Plastic biocomposite for door trim, tailgate trim and seat back panels [120,121].

15.2.3.3 Bio-based Thermoplastic Copolyesters and Copolyetheresters

DSM Engineering Plastics announced in 2010 its bio-based series of high performance thermoplastic copolyesters, the Arnitel® Eco, with 20%-50% biocontent. These new engineering bioplastics are reported to combine performance with a reduced carbon footprint, and are intended for automotive interior applications. They are currently not suitable for high temperatures, but future generations of the product are envisaged [122].

DuPont recently launched the bio-based thermoplastic elastomers Hytrel®RS, which are high-performance copolyetheresters with 20%-60% biocontent. The soft segment of these copolymers is based on the 100% bio-based polyether diol DuPont™ Cerenol™, obtained by direct polymerization of bio-based 1,3-propanediol [123]. They are reported as ideal for injection-molded parts for demanding applications such as air bag doors and energy dampeners [123].

15.2.3.4 Conclusion

Within the family of bio-based polyesters and copolyesters, PLA remains by far the most largely produced and commercialized. Other current and emerging bio-based polyesters and copolyesters have first to consolidate their markets, before broadening their automotive applications. It is also the case for the polyhydroxyalkanoates (PHAs), which are interesting biodegradable aliphatic polyesters produced in nature by bacterial fermentation of sugar or lipids, rather too expensive today for automotive applications.

All in all, the bio-based polyesters and copolyesters have good opportunities for automotive applications, especially for interior-trim parts.

15.2.4 Thermoplastic Starch (TPS) and its Non-biodegradable Blends

Starch is a natural storage material accumulated by green plants. It is a mixture of linear polysaccharide molecules (amylose) and branched molecules (amylopectin). Starch from corn, tapioca, potatoes, rice represents an inexpensive renewable polymer with important food and non-food fields of applications. The native starch cannot be used as plastic, but its plasticized form – the thermoplastic starch (TPS) – is already used in packaging and short-time live consumer goods. In the meantime, recent R&D achievements have been made to propose durable materials based on starch as attractive alternatives to petro-based thermoplastics. For instance, blending TPS with other natural or synthetic polymers such as polyolefins was identified as an effective way to adjust starch plastics characteristics (e.g. regarding rigidity vs. flexibility, water sensitivity and relatively poor mechanical properties) for durable applications [124]. As a result, the potential of TPS blends to substitute commodity plastics, namely low density polyethylene (LDPE), high density polyethylene (HDPE) and polypropylene (PP) is actually highly increased [28].

Today, Cereplast is proposing bio-based Cereplast Hybrid Resins® products, which are replacing 50% or more of the petroleum content in traditional plastic products with starches from corn, tapioca, wheat and potatoes. The first product from this family of Bio-polyolefins® is Biopropylene 50™, a 50% starch-based resin exhibiting similar characteristics (HDT, modulus and impact strength) as traditional polypropylene (see Table 15.1). Biopropylene™ resin can be used in conventional manufacturing processing (i.e., injection molding, profile extrusion, etc.) and require less energy in the production process by using lower process temperatures. Moreover, the price per pound of Cereplast Hybrid Resins® products is similar to the price of traditional polyolefin [31]. Since these blends from starch and PP are design to approach traditional PP characteristics, they will potentially allow replacement of petrochemical PP from some of its current automotive applications.

Furthermore, Cereplast have in mind to introduce Bio-PS and Bio-PE, which will combine 50/50 starch and high-impact polystyrene (HIPS) or HDPE. Knowing that starch bonds readily to PP, but not to PS and PE, the latter two compounds will require compatibilizing agents [31].

These starch/polyolefin hybrids are not biodegradable, but represent an effective way to increase the use of renewable resources. Because they are made with up to 50% conventional thermoplastics, Hybrids overcome what has been a limitation of most bio-based resins to date-limited heat resistance in use and heat stability in processing.

In France, Roquette Frères also presented their intention to produce, starting with 2011, a starch-based bioplastic with potential automotive applications [125].

Mazda, PURAC and Nishikawa Rubber of Japan are collaborating on the development of heat-resistant automotive parts based on combination of starch with stereocomplex PLA [25- PURAC private communication].

15.2.5 Bio-based Polyolefins: BioPE and BioPP

The idea of obtaining polyethylene from bio-sourced ethylene is not new, but the low crude oil prices did not encouraged these products since recently. Now, the increasing prices of fossil resources make attractive the production of bio-based polyolefins, and the possibility of making use of bio-based PE and PP greatly interests the automakers, as their modern cars are already largely using petrochemical polyolefins (about 43% of PP and 8% PE) [2].

Bio-based PE (Figure 15.5a): From technical point of view, the bio-based polyethylene can be synthesized from bio-based ethylene monomer, such as the monomer prepared by Braskem (Brazil) via bio-ethanol made from sugar cane [28].

Bio-based PEs are expected to have exactly the same chemical, physical and mechanical properties as their petrochemical counterparts. By now, Braskem already polymerized some grades of HDPE, LLDPE and UHMWPE entirely made from renewable material. Similar to their petrochemical counterparts, the PE applications will most probably include varied automotive parts, from plastic fuel tanks, filler pipes and air ducts (mainly from HDPE) [126] to noise reduction tapes for dashboard (for squeak and rattle abatement, from UHMW-PE, as approved by General Motors, Ford, Chrysler) [127].

Bio-based PP (Figure 15.5b): Several attempts are made today for obtaining bio-sourced PP via bio-ethanol from different renewably feedstocks. For instance, Braskem and Novozymes recently announced a research partnership to develop large-scale production of green polypropylene from sugarcane, a resin that Braskem has already obtained on laboratory scale and was certified as 100% renewable [128]. In the same time, Mazda is actively developing a bio-route for obtaining various PP and ethylene-propylene copolymers from cellulosic biomass [95]. These new bio-based materials are intended in future to replace their petrochemical counterparts automotive applications (i) in exterior: car bumpers and bumper spoilers, lateral siding, roof/boot spoilers, rocker panels, body panels, wheel arch liners; (ii) in interior: dashboards and dashboard carriers, door pockets and panels, consoles, pillar claddings and seats; (iii) under-the-hood: heating ventilation air conditioning, battery covers, electronic housing, air ducts, pressure vessels, splash shields,

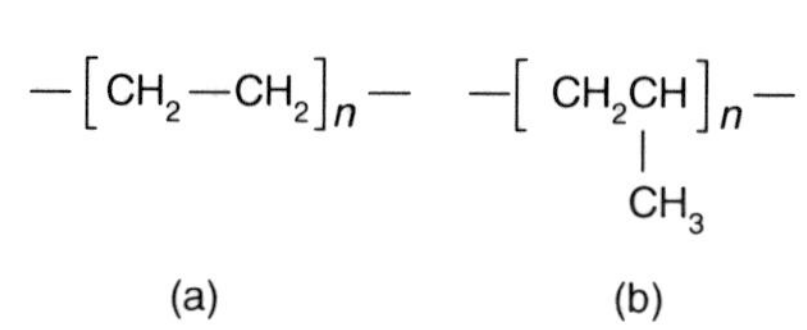

Figure 15.5 a) PE-building block; b) PP-building block.

reservoirs. Already, Ford is using sugarcane-based PP for the side shields of his EnviroSeat [99 – Ford sources].

In future, the new bio-based PP could also gradually shift the petrochemical PP from its biocomposites with natural fibers, in trim parts applications in dashboards, door panels, parcel shelves, seat cushions, backrests and cabin linings, car disk brakes and even for exterior applications, such as the engine/transmission covers in Mercedes-Benz Travego Coach [16].

15.2.6 Bio-based Polyurethanes (PURs)

Nowadays, 15–20% of PURs are used in automotive applications (20–40 kg PURs/ per car), and a growing part from these are starting to use bio-based feedstocks. Some general features of PURs and more specifically their emerging bio-based versions will be presented hereafter.

Commonly, polyurethanes are polymers with repeating urethane linkages (-NHCOO-) in their backbone. Their chemical and physical properties vary over a wide range, as a function of the constituent monomers and reaction conditions. PURs are generally prepared by polyaddition of an isocyanate, di- or polyfunctional, with a diol and/or polyol [129]. Commonly used isocyanates can be aromatic, such as toluene diisocyanate (TDI), methylene diphenyl isocyanate (MDI), polymeric MDI, or aliphatic such as hexamethylene diisocyanate (HDI). While the isocyanate component is generally petro-based, the polyol and/or diol component can be derived now from renewable resources such as sorbitol and isosorbide (from starch) or vegetable oil-based polyols derived from soybean oil, castor oil, sunflower oil or rapeseed oil. Consequently, the bio-based PURs can have from 8 to 70% biocontent, according to the chosen polyol and its own biocontent that can vary from 30 to 100% [28,130].

Due to their large range of chemical structures, PURs are highly versatile and can cover applications from flexible foams in car seats, to rigid foams as insulation, to thermoplastic polyurethanes, to coatings, adhesives and sealants [131–133]. The automotive industry consumes mainly flexible polyurethane foams and thermoplastic polyurethanes and within these two categories, the bio-based PURs take a growing place, as described in the forthcoming part.

15.2.6.1 *Bio-based Thermoplastic Elastomeric Polyurethanes (TPUs)*

The TPUs (Figure 15.6) are linear segmented block copolymers composed of hard and soft segments, obtained by reacting diisocyanates, long-chain diols (called polyols) and short-chain diols (called chain extenders). This particular

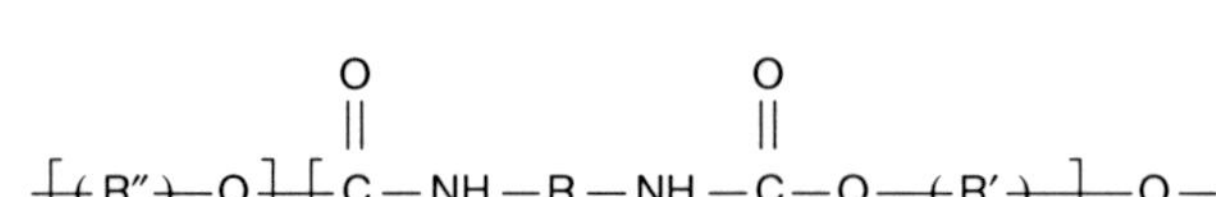

Figure 15.6 TPUs general chemical structure (R = from the isocyanate component, R′ = polyol segment, R″= diol segment).

segmented structure allows the TPUs to behave as elastomers in a broad range of use temperatures, and to act as thermoplastics when in molten state.

Some bio-based TPUs are already commercially available and proposed for automotive applications, such as Pearlthane® Eco and Pearlbond®Eco (from Merquinsa), based on polyols from vegetable oils and fatty acids and having 40 to 95% biocontent [134,135]. In parallel, GLS/PolyOne proposes OnFlex™ BIO series of soft TPUs with at least 20% renewable material from soybean oil, obtained by using Merquinsa's patent-pending Pearlthane®Eco TPU technology [136]. The performances of bio-based TPUs are said to be similar to standard TPUs. Additionally, their glass fiber-composites and alloys with styrene-ethylene/butylene-styrene triblock copolymers (SEBS) offer interesting solutions for automotive applications such as covers, soft-touch applications, interior trim, skins [137].

15.2.6.2 Bio-based Thermosetting Polyurethane Foams

We saw previously that bifonctional monomers reactions form linear TPUs. When the monomers functionality is greater than 2, a tridimensional cross-linked network is formed, and the polyurethane behaves as a thermoset polymer. Within this class of PURs, the flexible and semi-flexible polyurethane foams are extensively used for manufacturing car cushions for seats, headrests, armrests, door and roof liners, dashboards and instrument panels [138].

Until recently, these polyurethane foams were from petrochemical feedstocks, but now several commercial bio-based versions exist, based on different bio-sources polyols. Table 15.7 presents a non-exhaustive list of bio-based polyurethane foams for car applications.

Table 15.7 Examples of commercial bio-based polyurethane foams for automotive applications presented according to their bio-based raw materials.

Raw Materials	Trade Name of PUR/Polyols	Type	Company for PUR/ Polyols	Automotive Applications
Soybean oil	SoyOyl®-based PUR foams, such as Baydur® polyurethanes	flexible foams	• Baydur® PUR: Bayer [28] • SoyOyl® polyols: Urethane Soy System [138]	• seat-cushion and seat-backs in Ford Mustang, Ford Expedition, Focus, Escape, Escape Hybrid, Mercury Mariner... • soy-foam headliner in Ford Escape and Mercury Mariner [89,92]
	BioFoam™ based on BiOH™ polyols	flexible foams	• BioFoam™ PUR: Woodbridge [139] • BiOH™ polyols: Cargill [140]	• seat-cushion and seat-backs in Ford Escape 2009

Table 15.7 (cont.) Examples of commercial bio-based polyurethane foams for automotive applications, presented according to their bio-based raw materials.

Raw Materials	Trade Name of PUR/Polyols	Type	Company for PUR/ Polyols	Automotive Applications
Soybean oil (cont.)	Renuva™-based PUR foams	flexible and rigid foams	Dow Polyurethanes and other PU-product manufacturers [141]	• automobile seats, arm and head rests, instrument panels, door panels and consoles, head liners, impact-absorbing foams, noise, vibration and harshness/under carpet foams • Renuva-based RIM body panels & bumper fascia
	Agrol®-based PUR foams	flexible foams	Bio-foam Insulation Systems [142]	• head and arm rests for Toyota, Honda, Ford and Chrysler 2008 vehicles
Castor oil	Lupranol® BALANCE50-based PUR foams	flexible and rigid foams	BASF [47]	• automobile seats, arm and head rests
Sunflower and/or rapeseed oil	Bio-based PUR foams	flexible and rigid foams	Mitsui Chemicals [28,143]	• automobile seats
	Rubex® Nawaro	flexible foams	Metzeler Schaum [144]	• potential for automobile seats (Toyota) [130]

15.2.6.3 Conclusion

The current and emerging bio-based polyurethanes range from thermoplastic elastomers to thermosetting flexible and rigid foams, and have by now started to replace the petrochemical polyurethanes in automotive applications. One can expect a constantly increasing demand in bio-based PURs, since they are greener solutions fulfilling the OEMs expectations for modern car applications.

15.2.7 Bio-based Thermosetting Resins – Other than Thermosetting Polyurethanes

The thermosets are infusible and insoluble polymer networks obtained via irreversibly curing of a thermosetting polymer (resin), which is initially in a soft solid or liquid state. Curing can be induced by heating, photo-irradiation or electron-beam irradiation, or by mixing with a chemical curing agent [145,146]. A systematic presentation of the bio-based thermosetting materials from renewable resources can be found in the recent review of Raquez *et al.* [147]. Within the different classes of thermosets materials, a special interest for automotive applications is given to the reinforced unsaturated polyesters resins and epoxy-based composites as alternatives to replace metal components, and to the polyurethane foams (see §15.2.6.2).

15.2.7.1 Bio-based Unsaturated Polyesters Resins

The unsaturated polyester resins (UPRs) are one of the most important thermosetting matrices used in composites industry for the preparation of thermoset composites for automotive applications. The UPRs result from the polycondensation of unsaturated and saturated dicarboxylic acids with polyols. To form the final thermoset resin, the UPRs are dissolved in a vinyl monomer (usually styrene) able to react with the unsaturated double bonds of the polyester backbone and to provide the tridimensional cross-linked thermoset network. One of the main advantages of UPR thermosets is the ease of fabrication and low production costs, mainly due to the fast and easily controllable cure process (a free radical polymerization).

Common UPR applications include sheet molding compound (SMC) and bulk molding compound (BMC), already used for automotive vehicle body parts, including exterior panels. Generally, the use of SMC and BMC for automotive applications offers multiple advantages, such as in part consolidation and corrosion resistance, reduced density and lower capital investment for smaller series runs. High heat stability makes SMC vehicle body parts suitable for on-line painting and its excellent class A surface finish a material of choice for exterior panels such as fenders, tailgates, decklids and spoilers [148]. Additionally, SMC composite parts typically reduce by 20–30 wt% of equivalent steel parts, which imply substantial fuel savings and improved performance over the life of the vehicle. SMC composites have high thermal stability, and the molded parts maintain their dimensional in a large range of temperatures (–50°C to 200°C). This is why SMC and BMC have long been used for sunroof frames, headlamp reflector shells and engine bay components.

Traditionally, the standard UPR for thermoset composites are based on 1,2-propanediol, an unsaturated acid component such as maleic or fumaric acid, and a saturated dicarboxylic component, typically phthalic acid. Until recently, all the UPR monomers were petroleum-based products but the development of bio-based polyols opened the door for obtaining bio-based UPRs. Ashland Inc. has been a pioneer in this regard, and its ENVIREZ 1807 polyester resin with a 75/17/8 content ratio of petroleum, soybean oil and corn derived ethanol has been used since 2002 in Class A exterior styling panels on all John Deere combines, as well as in many of the OEM's tractor models [149,150].

Another route to obtain greener UPS is to replace the classical petro-based 1,2-propanediol with the commercially available Susterra™ 1,3-propanediol (DuPont Tate & Lyle BioProducts announced), a 100% renewably sourced glycol made from corn sugar. Using this bio-based glycol, Ashland Inc. prepared since 2008 two other bio-based UPRs from ENVIREZ series [28].

Very recently, DSM launched a UPR with 55% biocontent, Palapreg® ECO P55-01, designed for SMC and BMC applications ranging from under-body shields to Class A exterior body panels. DSM announced that their bioUPR has been already used for the bodywork of a kart competing in Formula Zero, the world's first zero-emission motor racing championship, and has been approved by the Formula One Association. The French company Mixt Composites Recyclables, is being testing Palapreg® ECO on a range of automotive applications and it consistently matched conventional alternatives in terms of performance and functionality [55].

15.2.7.2 Bio-based Epoxy Resins

Epoxy resins are versatile polymers used as adhesives, coatings, and matrices for structural composite parts. Several collaborative R&D projects aiming at precompetitive achievements were carried out during the last decade to develop materials, technologies and design concepts suitable for manufacturing carbon or glass/epoxy composite automotive body structures to be implemented on popular car models in mass production. For example, major European car makers such as Volkswagen (as coordinator), Fiat, Opel, Renault, Volvo, Porsche and Daimler were involved in projects co-funded by the European Community (e.g. TECABS, SLC) investigating the manufacturing of Resin Transfer Molded (RTM) carbon/epoxy floorpans [151–152] or carbon-glass/epoxy hybrids parts [153–154]. Another recent example is the futuristic lightweight concept car presented by Riversimple, a two-seat urban vehicle using epoxy prepregs from Advanced Composites Group Ltd (ACG) for its body shell and bonnet. The versatile prepreg was chosen for its low cost tooling and manufacturing properties, and was processed using out-of-autoclave vacuum bagging [155]. This kind of applications encouraged as to dedicate a short paragraph to the epoxy resins and their bio-based version.

Today, approx. 75% of epoxy resins are derived from diglycidyl ether of bisphenol A (DGEBA), which comes from bisphenol A and epichlorohydrin reaction; both monomers traditionally derived from petrochemical sources. But now, they are new opportunities for obtaining bio-based epoxy-resins, which are supposed to have clear benefits due to their bio-based origin, improved skin-friendliness and low toxicity [156].

One way to produce bio-epoxy-resins is to use bio-based DGEBA, by using bio-based glycerol-derived epichlorohydrin [157]. The bio-based DGEBA is chemically identical with the petrochemical one, so their properties are also assumed to be identical, and so the final cured bio-epoxy resins [28]. Another way to obtain bio-based epoxy resins is to use conventional DGEBA and a bio-based curing agent such as cardanol-based novolac proposed by Cimteclab S.p.A., Italy (cardanol being a phenol derived from cashew nutshell liquid). This cardanol-epoxy resin is already used as matrix in composites reinforced with jute fabrics, presenting potential use for exterior car components [158]. A third way to obtain greener epoxy composites is to blend epoxidized vegetable oils with bio-epoxy-resins, in the presence of suitable curing agents. This was shown to improve not only their biocontent level, but also to reduce their stiffness, which is an important drawback for structural applications [159–161].

15.2.7.3 Other Bio-based Thermosetting Resins

Besides the previous mentioned bio-based thermosetting resins, some other emergent classes are under development but already very promising. It is the case of the new furanic resins proposed by TransFurans Chemicals, Belgium, and evaluated in the recent European project BioComp. The furanic resins BioRez™ and Furolite™ are based on prepolymers of furfuryl alcohol, derived from furfural through conversion of agricultural waste streams (bagasse, corn cobs). They are suitable as matrix for composites reinforced with fiberglass and carbon fibers, as well as natural fibers such as wood, flax, sisal and jute [162–164]. More specifically, the Biorez™ Automotive is a one-component sprayable resin intended for the production of natural-fiber-reinforced biocomposites by hot compression molding. These biocomposites were found to have mechanical properties comparable with epoxy/PUR natural-fiber-reinforced composites, but lower volatile organic compounds (VOCs) and fats, oil and grease (FOG) emission values. A prototype of a BMW door panel was already produced and tested, in cooperation with Polytec Automotive GmbH & Co [163].

15.2.7.4 Conclusion

The market forces that have stimulated the most recent bio-resins development are expected to become even more active for obtaining more eco-friendly thermosetting resins for automotive applications. However, it is important to note that while biocontent is desirable, it is hardly the most important parameter. Mechanical, physical, and liquid (i.e. viscosity) properties of new resins (if completely new formulations, and not only replacement of a petrochemical monomer with a bio-based one) must meet all requirements set by customers before they can even be considered for any application. Simply stated, renewably resourced thermosetting polymers must offer similar or better performance and quality than petroleum-based counterparts at similar price [165].

15.3 Biocomposites Based on Bioplastics for Automotive Applications

When automotive applications need from the bioplastic materials (i.e. mainly PLA up to now), improved mechanical performances, while conserving the eco-friendliness and possibly the compostability/biodegradability features, one efficient and cost-effective way is to use natural fibrous reinforcements. Typically, they are cellulosic or lignocellulosic fibers from wood or non-wood origin. The second category concerns the vegetal fibers coming from bast (i.e., jute, hemp, kenaf, flax), leaf (i.e., sisal, pineapple) or seed (i.e., cotton), each one having distinct mechanical and physical properties (see Table 15.8).

Actually, the vegetal fibers have a real potential as reinforcement systems for automotive applications, as low cost materials, CO_2 neutral, with acceptable specific mechanical properties as compared with the well-known glass fibers [41].

Table 15.8 Comparative properties of natural fibers and man-made fibers [compilation data from 15,41,161,166–173].

Fiber	Fiber Type	Density. ρ (g/cm^3)	Young Modulus. E (GPa)	Specific Young Modulus. E/ρ ($GPa \cdot cm^3/g$)	Tensile Strength. σ (MPa)	Specific Tensile Strength. σ/ρ ($MPa \cdot cm^3/g$)	Elongation at Break (%)	Moisture Absorption (%)
E-glass	mineral	2.55	70–73	27.5–29	2000–3500	785–1373	2.5–5.0	-
Carbon	man-made	1.70	230–260	135–153	3500–5000	2058–2940	0.5–1.8	-
Flax	bast	1.40–1.50	50–110	34–76	345–1500	238–1000	1.2–3.3	7–10
Jute	bast	1.30–1.46	10–30	7–22	350–900	286–650	1.1–1.8	12.6
Kenaf	bast	1.45–1.5	14–53	9.3–36.5	230–930	153–641	1.6	
Hemp	bast	1.40–1.50	35	23.6	310–1834	214–1264	1.6–3.0	10.8
Ramie	bast	1.5	61.4–128	41–85	400–938	265–625	1.2–3.8	12–17
Sisal	leaf	1.33–1.45	9.4–22.0	6.7–16	468–640	340–467	3.0–7.0	11
Abaca	leaf	1.5	31.1–33.6	20.7–22.4	756–813	500–542	2.9	
Curaua	leaf	1.4	11.8	8.4	500–1150	357–821	3.7–4.3	7.9
Coir	seed	1.15–1.25	4–6	3.3–5	130–175	108–145	15–40	8–10
Cotton	seed	1.5–1.6	5.5–12.6	3.5–8.1	400	258–265	7.0–8.0	8–25

Additionally, they offer better acoustic performance, improve passenger safety (via favorable non-brittle fracture on impact), and reduce car global weight, with direct impact in energy/fuel consumption and greenhouse gas emissions [43].

The choice of one or several vegetal fibers in a specific automotive interior or hidden application usually depend on the balance between the cost and (regional) availability, the mechanical features and the possible issues related to the odor. Nowadays, their use in reinforced petrochemical plastic composites has been validated by almost all automakers, for interior or hidden applications in their modern car models [15,16,43,174–176].

As a general trend, an important increase of the use of natural fibers in automotive components is predicted, since they are a solution for European and American car makers to achieve the recent environmental directives [177]. A Nova German report had estimated for 2005 that 19% of the natural fibers were intended for bioplastics-based biocomposites, 30% for injection molding and 8% for press-molding, 30% for modified fibers and fabrics for advanced applications and the rest for other different processes [174 - NOVA Market Survey 2002].

In the meantime, the use of natural fibers for automotive durable composites present some major issues, related to their highly variable quality, depending on unpredictable agricultural conditions, their moisture absorption properties which complicates exterior applications, their restricted maximum processing temperature, and their lower strength properties, despite acceptable specific modulus [178]. Other technical challenges include adhesion between the fiber and matrix, increased viscosity for high fiber contents, which induce shear heating/degradation and affect the ability to fill thin walled parts by injection-molding, as well as appearance problems (poor colorability and opacity) [179]. Last but not least, important aspects have still to be improved concerning durability issues, flame-retardant properties, and emission issues (i.e., fogging, odor) [15,16,168,180].

In the following we present some PLA-based biocomposites, as they seem to be the most advanced for the moment for automotive applications. As seen in the section §15.2.2 and Table 15.4, adding natural fibers into commercial PLAs leads to interesting composite materials with high rigidity and heat resistance. Table 15.9 presents some examples of improvements offered by PLA/natural fiber composites vs. unmodified commercial PLA.

Typically, one can observe that 20–40 wt% of natural fibers into commercial PLA increase its tensile properties and in some cases the impact strength and that normally for lower cost-to-weight ratio as compared to neat PLA. Toyota is already proposing automotive applications for PLA/kenaf biocomposites, such as the cover spare wheel on Toyota Prius and Toyota Raum (2003) [104] or the translucent roof PLA/kenaf and ramie biocomposites on Toyota 1/X plug-in hybrid concept vehicle [93].

However, the long-term properties of renewable materials intended for durable applications are to be validated over different time periods and aggressive environment conditions, before thinking to extend the automotive applications for this kind of eco-friendly materials.

Besides the PLA-based biocomposites, different other attempts have been made for developing biocomposites with potential automotive applications. It is the

Table 15.9 Examples of PLA/natural fiber biocomposites and their mechanical properties.

PLA Biocomposite	PLA Grade	Modified / Unmodified Fibers and/ or Other Treatment	%wt Fibers	In % (Relative Values Compared to the Neat PLA)				Reference/Observations
				Young's Modulus	Tensile Strength	Charpy Impact		
						Un-notched	Notched	
Neat PLA		–	–	100	100	100	100	Neat PLA property = reference as 100%
PLA + Abaca fibers	PLA Lacty 9030, Shimadzu Co. Ltd.	Modified (esterification, alkali treatment or cyanoethylation) & unmodified	20	170*	104*	–	–	• Processing: melt mixing/ injection molding • abaca fiber: improves flexural modulus regardless of the fiber treatment • modified fiber => better interface adhesion [167]
PLA + Cordenka[a]	NatureWorks PLA 2002D	No coupling agent	25	140*	157*	185*	328*	• Processing: pultrusion/ pellets 3–5 mm length/ extrusion/injection molding [181]
	NatureWorks PLA 6202D	Unmodified fibers	30	155	130	447	–	• Processing: carding machine/pressing/ pellets/injection molding • Worse interface adhesion than PLA/flax [182]

PLA + Flax	POLLAIT Fortum	With and without (triacetin) PLA plasticizer	30–40	244	106	63*	–	• Processing: twin-screw extruder (180°C) • PLA/flax adhesion must be improved • Triacetin plasticizer: negative effect on mechanical and impact properties [183]
	NatureWorks PLA 6202D	Modified (MA-PP) & unmodified	30	202	121	69	–	• Processing: carding machine/pressing/pellets/injection molding [182]
PLA + woven Flax Biotex	PLA		40	13.2**	102**	33**	–	• Processing: yarns by novel 'twistless' spinning techniques/fabrics/vacuum consolidation and hot press molding • Already tested for two interior automotive parts from Jaguar and Land Rover [184]
PLA + Jute (non-woven mat of basis weight 300 g/m^2)	Biomer L 5000	–	40	271	182	93	–	• Processing: film-stacking procedure → 2 mm-thick pressed composites [185]

Table 15.9 (cont.) Examples of PLA/natural fiber biocomposites and their mechanical properties.

PLA Biocomposite	PLA Grade	Modified / Unmodified Fibers and/ or Other Treatment	%wt Fibers	In % (Relative Values Compared to the Neat PLA)				Reference/Observations
				Young's Modulus	Tensile Strength	Charpy Impact		
						Un-notched	Notched	
PLA + Kenaf (Kenaf sheet - chemically retted bast fiber)	PLA- LACEA Mitsui Chemicals	-	70	492	286	-	-	• Processing: impregnation/drying method [186]
PLA + Pine Wood Flour (PWF)	NatureWorks PLA 3001D	Modified (silane) & unmodified	20-40	134%-186%	-	-	-	• Processing: K-mixer/ injection molding • Good adhesion PLA/ PWF • Increased modulus but lower toughness and strain-at-break of PLA/ PWF • No silane effect on tensile properties [187]
PLA + Wood fibers (WF)	Biomer L 9000	Modified (MAPP) & modified	20-30-40	improved for 20% WF. decreases for 30–40% WF	233	–	no change	• Processing: micro-compounding/ injection molding (183°C) • Flexural modulus of unmodified PLA/WF 60/40: 309% of neat PLA [188]

PLA + recycled newspaper (RNCF)[b]	Biomer L 9000		30	196%	108	–	no change	• Processing: twin-screw extruder (183°C) • Flexural modulus: 163% of neat PLA • HDT (at 0.46MPa): 64.5°C for neat PLA → 80.2°C for PLA with 30% RNCF (similar to HDT of glass fiber-reinforced PLA composites) [189]

*Values calculated by using information from figures and not given in numeric values in the original articles.

**Absolute values, not reported to the neat PLA data.

[a]Cordenka = rayon fiber (regenerated cellulose spun fiber) usually used to reinforcing tires [182].

[b]The recycled newspaper cellulose fibers (RNCF) were added to this table for comparison sake. The RNCF are reclaimed from newspaper/magazine or Kraft paper stock [189].

case of the bio-polybutylene succinate/bamboo fiber composites proposed by Mitsubishi (see §15.2.3.2), or different possible combinations between the emerging bio-based thermosetting resins based on unsaturated polyesters, epoxy or furanic resins (see §15.2.7) and a broad range of the natural fibers, such as, for example, the cardanol-epoxy resin/jute fabric composites proposed by Cimteclab S.p.A. (see §15.2.7.2), or biocomposites made of jute fabric reinforcement and two-component laminating polyester resin from castor oil proposed by Bioresin Ltd., Brazil [190].

15.4 Specific Issues Concerning Processing and Recycling

15.4.1 Processing

15.4.1.1 Bioplastics

Two families of materials may be distinguished among the list of current and emerging bioplastics for automotive applications: on one hand, bio-based versions of already existent petrochemical polymers (such as bio-based PE or bio-based PP, and most of the biopolyamides), and on other hand, bioplastics such as PLA and durable starch plastics, which are newly used for automotive applications. For the first category of eco-friendly polymer materials, the classical processing and drying (when necessarily) facilities and all the practical know-how from their petrochemical counterparts, can be directly applied. The second category of bioplastics requires special care for processing during injection molding, generally because of their relatively narrow processing window, reduced heat resistance (in case of unmodified grades), together with shear and hydrolytic lower stability than for classical petroleum-thermoplastics. Processing of these bioplastics requires understanding of product and tool design, processing equipment and process parameters [88]. For instance, PLA and thermoplastic starch plastics are hygroscopic and moisture sensitive, and similarly to other bioplastics, such as polyesters or polyamides, they need drying prior to the processing stage and proper handling at all stages to minimize moisture uptake. However, PLA drying requirements are stricter than for ABS, polyamides or polycarbonate [179]. The issues of moisture sensitivity and lack of heat resistance appear to be the biggest issues of unmodified bioplastics based on PLA and durable thermoplastic starch plastics, motivating the bio-based modifiers to solve these limitations, via blending or using appropriate additives or by proposing improved bio- and nanocomposites based on these polymers [88].

To conclude, one should keep in mind that processing some of the newly durable bioplastics based on PLA or starch is not a simple copy from the processing conditions of conventional thermoplastics, but in the mean time, they do not need completely new specific facilities.

15.4.1.2 Biocomposites

Besides the processing issues of bioplastics, adding natural fibers in these materials brings some supplementary constrained physical limits due to: (i) the fiber hydrophilic nature, which causes fiber swelling and possible risk of decomposing via fungi attack; (ii) the poor interfacial adhesion between polar-hydrophilic fiber and non polar-hydrophobic matrix, i.e. polyolefin matrix; and (iii) the upper limiting temperature at which the fiber can be processed.

To reduce the moisture absorption of the natural fibers, chemical and physical (surface) modifications are generally performed. For instance, hydrothermal treatment is one of the approaches to reduce moisture absorption of natural fibers, which can increase the crystallinity of cellulose and therefore contributing to a reduced moisture uptake. Most of the chemical treatments decrease the strength properties because of the breakage of the bond structure and the disintegration of the non-cellulosic materials. Silane and acrylation treatments have been reported to lead to strong covalent bond formation and thereby the strength is enhanced marginally. Acrylation, alkali and silane treatments improved the Young's modulus of the fibers. Pretreatments of natural fibers in fiber-reinforced composites often show improvement in tensile properties upon different modifications owing to the increased fiber-matrix adhesion [170]. For improving the interfacial adhesion between the natural fiber and polymer matrix, which directly reflects on the composite final mechanical properties, it is usually necessary to compatibilize/couple the fibers to the polymer matrix [15,170].

Process temperature is another important limiting factor in natural-fiber applications, as the generally upper limit before fiber degradation is around 150°C for long processing cycles, although fibers may withstand short-term exposures to 220°C. The result of prolonged high-temperature exposure may be discoloration, volatile release, poor interfacial adhesion, or embrittlement of the cellulose components.

15.4.2 Recycling

An effective recycling of end-of-life compounds from vehicles, require the partnership of automakers, OEMs and raw material producers, together with environment and energy management agencies [191]. This is true for both petrochemical and bio-based plastic materials, whatever they are biodegradable/compostable or not. The plastic recycling efficiency from end-of-life vehicles largely depends on the design and assembling techniques of the different car parts, on the proportion of plastic mono-material vs. complex composites, on the dismantling facilities, and of course, on the chemical nature of the plastics.

One can distinguish three different situations:

A. Non-biodegradable bio-based thermoplastics, such as polyamides, aromatic polyesters, polyolefins, are supposed to follow the existing recycling streams for classical petrochemical counterparts. This could be an important advantage, as the recovery and recycling of post-consumer/post-industrial plastic and composites goods from automotive industries are now well structured. Indeed, the end-of-life cars are already dismantled, the plastics and composites are sorted and reground to be further used as fillers or blended with virgin materials to manufacture new goods [192]. This first recycling option generally corresponds to the mechanical recycling (reprocessing) and is applicable to practically all thermoplastic bioplastics. A second option, the chemical recycling, is possible for condensation polymers such as polyamides or polyesters, which can be chemically reacted to form starting materials.

B. Thermosetting resins and composites, both petrochemical and bio-based, are cross-linked networks and their commercially viable recycling at the end-of-life is more difficult. Some thermosetting polymers, such as polyurethanes, can be converted back to their original monomers, but the main part of the thermosetting resins practically does not allow recuperating their initial components, and the presence of glass/carbon fibers inside the thermosetting matrix is not helping for an easy waste management. In the meantime, one has to keep in mind that the substitution of precursors derived from fossil resources with bio-sourced counterparts is completely neutral with respect to the recycling issues of the thermosetting resins and their composites. Only recently these recycling aspects have started to be taken into account and new solutions of valorization are currently under development as alternatives to the classical incineration technique. Some techniques are at preindustrial stage, such as the mechanical recycling involving grinding techniques, which generates recyclates for partial substitution of fillers in new short-fiber molding compounds such as SMC/BMC (see companies as MCR in France [193], ERCOM in Germany and Phoenix Fiberglass in Canada [194]), or for use in the cement industry, as validated by the main European Cement Groups [193,195,196]. Other thermal recycling methods for glass/carbon fiber composites are also under study as alternatives to the classical incineration: the fiber recovery using fluidized bed process or pyrolysis processes to produce potentially useful organic products from the polymer, but none is yet commercially viable [194,197]. In this context, effective recycling of the new thermoset biocomposites appears to be even more challenging, due to the biodegradability/thermal stability of the natural fibrous reinforcements.

In the meantime, it is important to note the recent initiatives for stimulation and financial assistance of the composites recycling activities, such as the creation of an European Composites Recycling Concept (ECRC) proposed by the European composites industry, with a 'Green label' for composites from manufacturers guarantying the appropriate recycling of the components according to the legislative requirements [194–198].

C. Biodegradable or compostable bioplastics and biocomposites, such as PLA and PLA- based biocomposites. First, as thermoplastic aliphatic polyesters, PLA can be recycled either mechanically [199–200] or chemically, to lactic acid [28,90,201]. In parallel, the possibility of mechanical recycling of the PLA biocomposites also seems very promising, as reported recently for PLA/flax biocomposites with 20–30 wt% fiber [202]. A third recycling strategy comes from its biodegradable/compostable[1] nature. From practical view point, NatureWorks PLAs ("Ingeo biopolymers") are shown to be compostable in industrial composting facilities, where appropriate temperature and humidity conditions will cause PLA to lose molecular weight and become biodegradable to naturally occurring microorganisms [203]. However, as far the PLA-automotive applications are concerned, it is rare to find today car

[1] PLA is biodegradable polymer under European standards, and a compostable but not biodegradable polymer in the United States, according to the Federal Trade Commission Green Guide [203].

parts made from completely unmodified polymers. And actually, the compostability/ biodegradability of PLA-biocomposites or PLA-complex formulations highly depend on the nature of the other chemical additives present in these materials. A more favorable case is expected to be the one of PLA-biocomposites car parts, but no practical data were published so far on this topic.

Finally, one has to keep in mind that end-of-life treatment of plastics and biocomposites based on renewable resources does not necessarily mean biodegradability/ compostability, and the compatibility of their different recycling streams with the existing recycling infrastructure has still to be validated at large scale [192].

15.5 General Conclusions

Recent economical and ecological increasing concerns are offering strong motivations to substitute the well-known polymer materials derived from fossil feedstocks and, in some cases, some metal materials with more eco-friendly materials from renewable resources, for a wide range of applications. One of the key driving forces for developing bioplastics and biocomposites for durable applications is the automotive industry, due to its well structured high-level network of automakers, OEMs and raw material producers that combines substantial need in plastics and also an important R&D expertise for tailoring materials able to better fit to the specific car applications and to the new environmental requirements.

The main elements playing in favor of the bioplastics and biocomposites are, generally, their enhanced ecological footprint and more equilibrated CO_2 balance as compared to the petrochemical polymer materials, and their capacity to diminish the high fossil feedstock dependency. Additionally, the use of natural fibers in (bio)plastics allows compensating some of the important drawbacks of the polymer matrix and opens access to alternative lightweight and low-cost materials, which permit reducing fuel consumption/CO_2 emissions with respect to the current restrictive legislative standards over the world.

One can reasonably imagine the great development potential for green high-end polymer materials for car applications in the next years, knowing that the next-generation vehicles will need to show much greater efficiency in material use and most probably will be hybrids of many different materials selected to perform the functional requirements of each subsystem. In addition, the present and emerging green material solutions developed for car applications will certainly encourage the use of bioplastics and biocomposites in other fields of durable applications.

References

1. M. Costes, "Masse des carrosseries automobiles: évolution et tendances," Congrès international Carrosserie et plastiques, SFIP-SIA, Sochaux, June 14–15, 2006.
2. C. Duval, "Plastiques et automobile: D'hier à aujourd'hui," *Techniques de l'ingénieur*, AM 3 590-1, 2007.
3. S. Delalande and J.L. Monnet, "PSA Peugeot Citroen: Les Matériaux Verts - Réduire l'impact environnemental des véhicules," Congrès international Carrosserie et plastiques, SFIP-SIA-EMD, Douai, June 9–10, 2010.

4. A. Giocosa, "Innovation-past and future-provided by Composite Materials in the automotive industry," Congrès international Carrosserie et plastiques, SFIP-SIA-EMD, Douai, June 9–10, 2010.
5. L'industrie de l'équipement automobile en chiffres. Fédération de la Plasturgie et Service des études et des statistiques industrielles (SESSI), 2003.
6. F. An and A. Sauer, Comparison of passenger vehicle fuel economy and greenhouse gas emission standards around the world, Report prepared for Pew Center on Global Climate Change, December 2004, http://www.pewclimate.org/docUploads/Fuel%20Economy%20and%20GHG%20Standards_010605_110719.pdf, 2010.
7. European strategy targets car emissions, http://ec.europa.eu/environment/news/efe/24/article_4119_en.htm, 2010 and Proposal for Regulation of the European Parliament and of the Council COUNCIL - Setting emission performance standards for new passenger cars as part of the Community's integrated approach to reduce CO_2 emissions from light-duty vehicles, 2007/0297 COD, http://eur-lex.europa.eu/LexUriServ/LexUriServ.do?uri=COM:2007:0856:FIN:EN:PDF, 2007.
8. New Federal CAFE standards officially released, 34.1mpg by 2016, *The Detroit News*, http://autogreenmag.com/2010/04/02/new-federal-cafe-standards-officially-released-341-mpg-by-2016, 2010.
9. Directive 2000/53/EC of the European Parliament and of the Council of 18 September 2000 on End-of-Life Vehicles, Official Journal of the European Communities, 21 October 2000.
10. P. Kim, "Renault SAS: Composites-Automotive materials of the future?" Congrès international Carrosserie et plastiques, SFIP-SIA-EMD, Douai, June 9–10, 2010.
11. G.H. Brother, Casein plastics, *Industrial Engineering Chemistry*, Vol. 32, p. 31, 1940.
12. B.E. Ralston and T.A. Osswald, *Plastics Engineering*, Vol. 64, p. 36, 2008.
13. E.S. Stevens, *Green Plastics: An Introduction to the New Science of Biodegradable Plastics*, Princeton, Princeton University Press, 2002.
14. W. Shurtleff and A. Aoyagi, Henry Ford and His Employees: Work with Soy, SoyInfo Center, 2004, http://www.soyinfocenter.com/HSS/henry_ford_and_employees.php, 2010.
15. J. Holbery and D. Houston, *JOM Journal of the Minerals, Metals and Materials Society*, Vol. 58, p. 80, 2006.
16. B.C. Suddell and W.J. Evans, "Natural Fiber Composites in Automotive Applications" in A.K. Mohanty, M. Misra and L.T. Drzal, eds., *Natural Fibers, Biopolymers and Biocomposites*, CRC Press, Boca Raton, FL, USA, pp. 237–266, 2005.
17. C. Crawford, *Canadian Chemical News*, http://www.allbusiness.com/chemicals/plastics-rubber-industry-plastics/12399182-1.html, 2009.
18. Bosch, Plastics...naturally, http://researchinfo.bosch.com/content/language2/html/5729.htm, 2010.
19. Commission report "Accelerating the development of the market for Bio-Based products in Europe," in preparation of the Communication on the Lead Market Initiative (COM(2007) 860 final), http://ec.europa.eu/enterprise/policies/innovation/files/lead-market-initiative/prep_bio_en.pdf, 2010.
20. Taking bio-based from promise to market, A report from the Ad-hoc Advisory Group for Bio-based Products in the framework of the European Commission's Lead Market Initiative, November 2009.
21. BiomassPla Identification System B-1, Japan BioPlastics Association, June 2007 (Revised), http://www.jbpaweb.net/english/b-1.pdf, 2010.
22. Standard Test Methods for Determining the Bio-based Content of Solid, Liquid, and Gaseous Samples Using Radiocarbon Analysis, ASTM Standard D6866, ASTM International, 2008.
23. FprCEN/TS 16137, Plastics—Determination of bio-based carbon content—under approval, expected for 2011–02, CEN/TC 249 - Standards under development, http://www.cen.eu/CEN/Sectors/TechnicalCommitteesWorkshops/CENTechnicalCommittees/Pages/WP.aspx?param=6230&title=CEN%2FTC+249, 2010.
24. E. Rudnik, *Compostable Polymer Materials*, Elsevier Science, Oxford UK, 2008.
25. L. Shen, E. Worrell, and M. Patel, *Biofuels, Bioproducts & Biorefining*, Vol. 4, p. 25, 2010.

26. NatureWorks LLC, Technical Data Sheets by Application, http://www.natureworksllc.com/product-and-applications/ingeo-biopolymer/technical-resources/ingeo-biopolymer-technical-data-sheets.aspx, 2005.
27. J.E. Mark ed., *Handbook Polymer Data*, Oxford University Press, New York, 1999.
28. L. Shen, J. Haufe, and M.K. Patel, Product overview and market projection of emerging bio-based plastics (PROBIP 2009), Commissioned by European Polysaccharide Network of Excellence (EPNOE) and European Bioplastics, Group Science, Technology and Society (STS), Copernicus Institute for Sustainable Development and Innovation, Utrecht University, Utrecht, the Netherlands, Report NWS-E-2009-32, 2009
29. DuPont™ Sorona® EP Thermoplastic Polymer, http://www2.dupont.com/Plastics/en_US/assets/downloads/product/sorona/Sorona_EP_Brochure10_07.pdf, 2007.
30. DuPont™ Biomax® PTT 1100, http://www2.dupont.com/Biomax/en_US/assets/downloads/biomax_ptt_1100.pdf, 2010.
31. Cereplast, Cereplast Hybrid Resins®, http://www.cereplast.com/cmspage.php?pgid=71, 2010.
32. D. Garlotta, *Journal of Polymers and the Environment*, Vol. 9, p. 63, 2001.
33. L. Zheng, C. Li, W. Huang, X. Huang, D. Zhang, G. Guan, Y. Xiao, and D. Wang, *Polymer Advanced Technologies*, Published online in Wiley InterScience, August 2009.
34. H. Domininghaus, *Plastics for Engineers. Materials, Properties, Applications*, Carl Hanser Verlag, Munich, 1993.
35. M.I. Kohan, S.A. Mestemacher, R.U. Pagilagan, and K. Redmond, "Polyamides," *in Ullmann's encyclopedia of Industrial Chemistry 7th Edition*. Wiley-VCH Verlag GmbH & Co. KGaA, 2003.
36. D. Rusu and M. Rusu, "Polyamides-based blends" in C. Vasile and A.K. Kulshereshtha eds., *Handbook of Polymer Blends and Composites*, Vol. 4a, RAPRA Technology Press, pp. 201–245, 2003.
37. B. Rosenau, *Kunststoffe International*, p. 66, October 2007.
38. B. Rosenau and R.F. Rodiles, *Kunststoffe International*, p. 82, October 2008.
39. R. Ryntz and R.A. Bring, "Back the steel? The growth of plastics in automotive applications" in *83rd Annual Meeting of the Federation-of-Societies-for-Coatings-Technology. JCT Research* Vol. 3, p. 14, 2006.
40. L. Florez, *Plastics Engineering*, Vol. 63, p. 17, 2007.
41. A.K. Mohanty, M. Misra, and L.T. Drzal, *Journal of Polymers and the Environment*, Vol. 10, p. 19, 2002.
42. M.P. McCourt, G.M. McNally, W.R. Murphy and T. McNally, *Developments in Chemical Engineering and Mineral Processing*, Vol. 11, p. 67, 2003.
43. A. Ashori, *Bioresource Technology*, Vol. 99, p. 4661, 2008.
44. M. Brady and P. Brady, *Reinforced Plastics*, Vol. 52, p. 37, 2008.
45. R. Stewart, *Reinforced plastics*, p. 22, March/April, 2010.
46. Arkema, Rilsan® Polyamides in Automotive Markets, http://www.arkema.com/pdf/EN/products/technical_polymers/rilsan/rilsan_automotive/rilperm_market_oem.pdf, 2010.
47. BASF, Lupranol® BALANCE: polyols on the basis of renewable materials, http://www.basf.com, 2010.
48. Amilan® Nylon Resin, in Our Business—Plastics & Chemicals, Toray, http://www.toray.com, 2010.
49. Rhodia News release, Rhodia's new Technyl® eXten offer boots the bio-based engineering plastics market, http://www.rhodia.com/en/news_center/news_releases/Polyamide_Technyl_eXten_190410.tcm, 2010, and Rhodia, http://www.rhodia.com, 2010.
50. DuPont, Automotive—DuPont™ Renewable Sourced Materials, http://www2.dupont.com/Renewably_Sourced_Materials/en_US/automotive.html, 2010.
51. Evonik, Vestamid® Terra, http://www.degussa-hpp.com/dl/brochure/vestamid_terra_flyer_eng.pdf, 2010 and Evonik Press release, Vestamid®Terra Bio-Polyamide Protects High-Performance Cables, http://corporate.evonik.com/en/media/press_releases/Pages/detail.aspx?newsid=10372, 2010.
52. EMS-Grivory Products Trade Names, Greenline, http://www.emsgrivory.com/index.cfm?id=121, 2010.
53. M. Thielen, *Bioplastics Magazine*, Vol. 3, p. 50, 2010.
54. DuPont, Zytel®RS renewably sourced nylon resin, http://www2.dupont.com, 2010.

55. DSM Press Release, DSM introduces bio-based performance materials for automotive industry, http://www.greencarcongress.com/2010/04/dsm-20100414.html, 2010.
56. Arkema News Release, Arkema and Toyobo to Start Alliance on High Temperature Bio-based Polyamides, http://www.arkema.com/sites/group/en/products/detailed_sheets/technical_polymers/rilsan_12/news_viewer.page?p_filepath=/templatedata/Content/Press_Release/data/en/2010/100428_arkema_and_toyobo_to_start_alliance_on_high_temperature_bio-based_polyamides.xml, 2010.
57. Arkema, *Arketypes the newsletter*, Vol. 26, http://arketypes.e-letter.fr/en/26/rilsan-ht-under-the-hood, 2010.
58. Evonik Press release, Vestamid® HT*plus* both bio-based and high-performing, http://corporate.evonik.com/en/media/press_release/Pages/detail.aspx?newsid=10893, 2010.
59. Arkema, Rilperm® Technology, http://www.rilperm.com, 2010.
60. Fränkische Industrial Pipes, Automotive, http://www.fraenkische-ip.com/cps/rde/xchg/SID-E688281-A9091021/fraenkische_ip/hs.xsl/12621_ENG_HTML.htm, 2010.
61. Arkema, Rilsan® Fine Powders, http://www.rilsanfinepowders.com, 2010.
62. Bio-Polyamides for Automotive Applications, *Bioplastics Magazine*, Vol. 1, p. 10, 2010, http://www.tu-braunschweig.de/Medien-DB/ibvt/bioplasticsmagazine.pdf, 2010.
63. Arkema, Rilsan® Polyamides 11 & 12 (granules), http://www.arkema.com/sites/group/fr/products/product_viewer.page?p_filepath=/templatedata/Content/Product_Datasheet/data/fr/technical_polymers/060331_polyamides_11_et_12_rilsan_granules_.xml, 2010.
64. Arkema, Rilsan®PA 11: Powerful by Nature, http://www.arkema.com/sites/group/en/products/detailed_sheets/technical_polymers/rilsan_11/home.page, 2010.
65. Arkema News Release, Arkema at Techtextil 2007, June, 2009. http://www.arkema.com/pdf/EN/press_release/2007/Press%20kit%20Techtextil%202007.pdf, 2010.
66. S.S. Ray and M. Okamoto, *Progress in Polymer Science*, Vol. 28, p. 1539, 2003.
67. S. Pavlidou and C.D. Papaspyrides, *Progress in Polymer Science*, Vol. 33, p. 1119, 2008.
68. G. Giornano, and H. Inman, *Plastics Engineering*. January, 2010.
69. T. Liu, K.P. Lim, W.C. Tjiu, K.P. Pramoda, and Z.K. Chen, *Polymer*, Vol. 44, p. 3529, 2003.
70. A. Okada and A. Usuki, *Macromolecular Materials and Engineering*, Vol. 291, p. 1449, 2006.
71. X. He , J. Yang, L. Zhu, B. Wang, G. Sun, P. Lv, I.Y. Phang, and T. Liu, *Journal of Applied Polymer Science*, Vol. 102, p. 542, 2006.
72. Y. Hu, L. Shen, H. Yang, M. Wang, T. Liu, T. Liang, and J. Zhang, *Polymer Testing*, Vol. 25, p. 492, 2006.
73. DSM Press release, EcoPaxx™, Breakthrough in polymers, http://www.dsm.com/, 2009.
74. M. Dreisbach and E. Boissard, "DSM Engineering Plastics: Materials solutions and innovations for value creation," Congrès international Carrosserie et plastiques, SFIP-SIA-EMD, Douai, June 9–10, 2010.
75. Chemelot Newsletter, EcoPaXX™: Serious Efforts Made on Issue of Sustainability, http://www.chemelot.nl/?id=506&template=algemeen.htm&taal=en, 2009.
76. DENSO News Releases, DENSO Develops Plant-derived Resin Radiator Tank, http://www.densomediacenter.com, 2009.
77. DENSO Press Release, Tank New Renewably-Sourced Polymer Debuts in Radiator End Tank Program, Jointly Developed with DuPont Engineering Polymers, http://www.globaldenso.com, 2010.
78. H. Häger and J. Limper, *Evonik Science Newsletter*, http://www.degussa-hpp.com/dl/article/Elements_28_en_VESTAMID%20Terra.pfr, 2010.
79. Arkema Press Kits and presentations, Arkema at Kunststoffe, http://www.arkema.com/sites/group/en/press/press_kits.page, 2007.
80. Arkema, Pebax®Rnew, new high performance elastomer, http://www.pebax.com, 2010.
81. Arkema News Release, Pebax®Rnew100, high performance 100% biosourced thermoplastic elastomer, http://www.arkema.com/sites/group/en/press/pr_detail.page?p_filepath=/templatedata/Content/Press_Release/data/en/2009/091005_pebax_rnew100_high_performance_100_bio_sourced_thermoplastic_elastomers.xml, 2009.
82. EMS-Grivory, Grilflex – PEBA High Performance Elastomers Flexible and Enduring, http://www.grilflex.com, 2010.

83. D.W. Garrett and G.R. Owens, "Polyphthalamide resins for use as automotive engine coolant components," *SAE Technical Paper Series #9501192*, February 01, 1995.
84. Arkema News Release, Rilsan®HT: the first flexible, high-temperature thermoplastic to replace metal in engine-compartment tubing, http://www.arkema.com/group/en/products/detailed_sheets/technical_polymers/rilsan_11/news_viewer.page?p_filepath=/templatedata/Content/Press_Release/data/en/2009/090303_rilsan_ht_the_first_flexible_high_temperature_thermoplastic_to_replace_metal_in_engine_com.xml, 2009.
85. Arkema News Release, Arkema rounds out the Rilsan®HT product line designed to replace metal, with Rilsan®HT injection resins, http://www.arkema.com/sites/group/en/press/pr_detail.page?p_filepath=/templatedata/Content/Press_Release/data/en/2009/090717_arkema_rounds_out_the_rilsan_ht_product_line_designed_to_replace_metal_with_rilsan_ht_inje.xml, 2009.
86. R. Haan, Unique Purac Lactides for improved PLA production and properties, Bioplastic Processing & Properties, April 2009. Loughborough University, UK, http://www.soci.org/News/~/media/Files/Conference%20Downloads/Bioplastic%20Processing%20Apr%2009/Haan.ashx, 2009.
87. L. Avérous, "Polylactic Acid: Synthesis, Properties and Applications" in N. Belgacem and A. Gandini, eds., *Monomers, Polymers and Composites from renewable resources*, Elsevier Limited Publication, p. 433, 2008.
88. Biopolymers require special care for processing during injection molding,, http://www.plastemart.com/upload/Literature/Biopolymers-care-processing-during-injection-molding.asp?LiteratureID=1332, 2010.
89. Ford Motor Company, Press Releases, Ford Researchers look to Mother Nature for clues on how to create greener, lighter plastics, http://www.ford.com/about-ford/news-announcements/press-releases/press-releases-detail/pr-ford-researchers-look-to-mother-31009, 2010.
90. D. Farrington, J. Lunt, S. Davies, and R.S. Blackburn, "Poly(lactic acid) fibers" in R.S. Blackburn ed., *Biodegradable and Sustainable Fibers*, Woodhead Publishing, pp. 191–220, 2005, http://www.natureworksllc.com/media/Files/biodegradable%20and%20sustainable%20fibers%20%20%20chapter%206.pdf, 2010.
91. L.T. Lim, R. Auras, and M. Rubino, *Progress in Polymer Science*, Vol. 33, p. 820, 2008.
92. Ford Model U Concept: A Model for change, http://media.ford.com/article_display.cfm?article_id=14047, 2010.
93. Toyota 1/X Plug-In Concept Makes North American Debut at 2008 Chicago Auto Show, Green Car Congress, http://www.greencarcongress.com/2008/02/toyota-1x-plug.html, 2010.
94. Teijin News release 2009, Car Seat Fabric Made of Biofront™ Bioplastic Used in Mazda's Premacy Hydrogen RE Hybrid *and* Teijin Expands Hygrothermal Resistance of Biofront™ Bioplastic Upgraded Version Now Offers High Durability Comparable to PET, http://www.teijin.co.jp/english/news/2009, 2009.
95. Mazda, Biotechmaterials, http://www.mazda.com/mazdaspirit/env/biotech/, 2010.
96. NatureWorks LLC, Technology Focus Report: Polylactic Acid Containing Fillers and Fibers, http://www.natureworksllc.com/product-and-applications/ingeo-biopolymer/technical-publications/~/media/Files/Fillers-and-Fibers-Technology-Report-pdf.ashx, 2007.
97. J.H. Schut, *Plastics Technology*, http://www.ptonline.com/articles/200901fa2.html, 2009.
98. K.S. Anderson, S.H. Lim, and M.A. Hillmyer, *Journal of Applied Polymer Science*, Vol. 89, p. 3757, 2003.
99. R. Kung, Updated Applications for Bioplastics in Global, BASF Biodegradable Polymer Asia Pacific, Seminar Bangkok (Thailand), April 24, 2009, http://www.tbia.or.th/uploads/attach/seminar-10.pdf, 2010.
100. Plastemart, Blends of PLA being developed to improve viability of PLA, http://www.plastemart.com/Plastic-Technicle-Article.asp?LiteratureID=1359, 2010.
101. PolyOne News Release, PolyOne Presents Biopolymer Advances at NatureWorks Innovation Conference, http://www.polyone.com/enus/news/Press%20Release%20Attachments/PolyOne%20to%20Present%20at%20Natureworks%20Conference.pdf, 2010.
102. PolyOne reSound™ RS-1200 Polylactic Acid Compound, Technical Data Sheet http://catalog.ides.com/datasheet.aspx?I=18425&FMT=PDF&E=116453, 2010.

103. S. Pilla, A. Kramschuster, L. Yang, J. Lee, S. Gong, and L.S. Turng, *Materials Science and Engineering: C*, Vol. 29, p. 1258, 2009.
104. NatureWorks LLC, Technology Focus Report: Blends of PLA with Other Thermoplastics, http://www.natureworksllc.com/product-and-applications/ingeo-biopolymer/technical-publications/~/media/Files/Blends-Technology-Focus-Report-pdf.ashx, 2007.
105. S. Joshi, *Journal of Industrial Ecology*, Vol. 12, p. 474, 2008.
106. S.S. Ray and M. Okamoto, *Macromolecules Rapid Communication*, Vol. 24, p. 815, 2003.
107. J. Chang, Y. Uk-An, and G. Sur, *Journal of Polymer Science Part B: Polymer Physics*, Vol. 41, p. 94, 2003.
108. S.S. Ray, K. Yamada, M. Okamoto, and K. Ueda, *Journal of Nanoscience and Nanotechnology*, Vol. 3, p. 1, 2003.
109. NatureWorks LLC, Technology Focus Report: Toughened PLA, http://www.natureworksllc.com/product-and-applications/ingeo-biopolymer/technical-publications/~/media/Files/Toughened-PLA-Technology-Focus-pdf.ashx, 2005.
110. X. Liu, Y. Zou, W. Li, G. Cao, and W. Chen, *Polymer Degradation and Stability*, Vol. 9, p. 3259, 2006.
111. A.M. Harris and E.C. Lee, *Journal of Applied Polymer Science*, Vol. 115, p. 1380, 2010.
112. J.V. Kurian, Sorona® Polymer: Present status and future perspectives, in A.K. Mohanty, M. Misra and L.T. Drzal eds., *Natural Fibers, Biopolymers and Biocomposites*, CRC Press, Boca Raton, FL, USA, pp. 512–542, 2005.
113. DuPont™ Sorona® Automotive applications, http://www2.dupont.com/Sorona/en_US/uses_apps/carpet/automotive_applications.htm, 2010.
114. M. Matsuda and E. Connell, GPEC® 2009 Updates, Toyota Automotive Applications & Expectations of Bio-based Materials, http://www.sperecycling.org/GPEC/GPEC2009/GPEC%202009%20-%20Keynote%20Speaker%20-%20Eric%20Connell%20-%20Toyota.pdf, 2009.
115. Bioamber Press, Bioamber Commissions World's First Renewable Succinic Acid Plant, http://www.bio-amber.com/release/pdf/4921336d21.pdf, 2010.
116. Myriant Technologies, Biobased succinic acid…., http://www.myriant.com/succinicpage.htm, 2010.
117. Roquette News, DSM and Roquette to start bio-based succinic acid joint venture, http://www.roquette.com/delia-CMS/t1/article_id-6130/topic_id-2047/dsm-and-roquette-to-start-bio-based-succinic-acid-joint-venture.html, 2010.
118. BASF News Release, BASF and CSM announce joint production development of bio-based succinic acid, http://basf.com/group/corporate/en/function/conversions:/publish/content/news-and-media-relations/news-releases/downloads/2009/P395-CSM-e.pdf, 2009.
119. Biopolymers in automotive interiors: a step towards sustainability, http://www.plastemart.com/upload/Literature/Biopolymers-in-Automotive-Interiors.asp, 2010.
120. Mitsubishi Monitor October & November 2006, Greening that Matters, http://www.mitsubishi.com/mpac/e/monitor/back/0610/green.html, 2006.
121. Mitsubishi Concept cX, http://www.carbodydesign.com/archive/2007/08/31-mitsubishi-concept-cx/, 2007.
122. DSM Arnitel®, Arnitel® ECO to be launched at K 2010, http://www.dsm.com/le/en_US/arnitel/html/arnitel_eco.htm, 2010.
123. DuPont™ Hytrel® RS Thermoplastic Elastomers, http://plastics.dupont.com/plastics/pdflit/americas/hytrel/HYTREL_RS_Booklet_07_2009.pdf, 2009.
124. F.J. Rodriguez-Gonzalez, B.A. Ramsay, and B.D. Favis, *Polymer*, Vol. 44, p. 1517, 2003.
125. Roquette, *Bulletin d'informations-ChemSuD Montpellier*, Avril 2010, http://chemsud.enscm.fr/files/BIDD0410.pdf, 2010.
126. Total Petrochemicals, Automotive, http://www.totalpetrochemicalsusa.biz/side1/m.asp?lg=en&biz=b1&ent=M&sid=1&m=m1, 2010.
127. DeWAL, Automotive Squeak and Rattle abatement, http://www.dewal.com/2010/automotive.html, 2010.
128. Braskem Release, Braskem and Novozymes to make green plastic, http://www.braskem.com.br/site/portal_braskem/en/sala_de_imprensa/sala_de_imprensa_detalhes_9529.aspx, 2009.

129. D. Dieterich, "Polyurethanes," *in Ullmann's Encyclopedia of Industrial Chemistry 5th Edition*. Wiley-VCH, 1997.
130. O. Wolf, "Techno-economic feasibility of large-scale production of bio-based polymers in Europe. European Commission—Joint Research Centre," European Science and Technology Observatory, Institute for Prospective Technological Studies, Technical Report EUR 22103EN, 2005, http://ftp.jrc.es/EURdoc/eur22103en.pdf, 2010.
131. J.P.L. Dwan'isa, A.K. Mohanty, M. Misra, L.T. Drzal and M. Kazemizadeh, *Journal of Materials Science*, Vol. 39, p. 1887, 2004.
132. J.P.L. Dwan'isa, A.K. Mohanty, M. Misra, L.T. Drzal and M. Kazemizadeh, *Journal of Materials Science*, Vol. 39, p. 2081, 2004.
133. V. Sharma and P.P. Kundu, *Progress in Polymer Science*, Vol. 31, p. 983, 2006.
134. J.A. Grande, "Bio-based TPEs emerge—Thermoplastic elastomers Move Up the Performance Scale," in *Plastics technology*, http://www.ptonline.com/articles/200806fa1.html, 2010.
135. Merquisa News, Merquinsa to showcase its Green TPUs at Chinaplas 2010, http://www.merquinsa.com, 2010.
136. GLS OnFlex™-U Automotive Series, http://www.glstpes.com/products_onflex.php, 2010.
137. PolyOne OnFlex™-U 5300. Soft Grades, 2007, http://www.polyone.com/en-us/docs/Documents/ENG%20ONFLEX-U%205300GR%20Brochure.pdf, 2007.
138. Urethane Soy Systems, "Bio-Renewable" Polyols: An Exceptional "New Use" Agricultural Product, http://www.soyol.com/raw.htm, 2010.
139. Woodbridge Group, Intier Automotive seating and Woodbridge Co-Develop Bio-based seating for 2009 Ford Escape, http://www.woodbridgegroup.com/media/Pressrelease_BioFoamFordEsc.pdf, 2010.
140. BiOH® polyols, http://www.bioh.com, 2010.
141. Dow Polyurethanes, RENUVA™ Renewable Resource Technology Creates Next-Generation Polyols, http://www.dow.com/polyurethane/feature.htm, 2010 and Dow RENUVA™ Renewable Resource Technology—Automotive, http://www.dow.com/renuva/markets/auto.htm, 2010.
142. Bio-foam Insulation Systems, Agrol®, http://www.bio-foam-insulation.com/agricultural-seal.php, 2010.
143. MitsuiChemicals, Products by Applications: Automotive materials – Business & products, http://www.mitsuichem.com, 2010.
144. Metzeler Schaum GmbH, http://www.metzeler-schaum.de, 2010.
145. D.K. Chattopadhyay and K.V.S.N. Raju. *Progress in Polymer Science*, Vol. 32, p. 352, 2007.
146. D. Åkesson, M. Skrifvars, S. Lv, W. Shi, K. Adekunle, J. Seppälä, and M. Turunen. *Progress in Organic Coatings*, Vol. 67, p. 281, 2010.
147. J.M. Raquez, M. Deleglise, M.F. Lacrampe, and P. Krawczak, *Progress in Polymer Science*, Vol. 35, p. 487, 2010.
148. DSM Composite Resins, Automotive, Why SMC/BMC in automotive, http://www.dsm.com/nl_NL/html/drs/why_smc_bmc.htm, 2010.
149. S. Frattini, *JEC Composites Magazine*, n° 38, p. 32, 2008.
150. V.P. McConnell, *Reinforced Plastics*, September 2008, http://www.reinforcedplastics.com/view/1742/new-recipes-for-smc-innovation, 2008.
151. European project TECABS, "Technologies for Carbon Fiber reinforced modular Automotive Body Structures," Project funded by the European Community and coordinated by Volkswagen, contract n° G3RD-CT-2000-00102, 2000–2004, Final Technical Report PR8 TEC278MR, August 2004, http://ec.europa.eu/research/transport/news/article_1507_en.html, 2010.
152. M. Deleglise, C. Binetruy, and P. Krawczak, *Composites Part A: Applied Science and Manufacturing*, Vol. 37, p. 874, 2006.
153. European Project SLC "SuperLightCar—Sustainable Production Technologies of Emission-reduced Lightweight Car Concepts," Project funded by the European Community and coordinated by Volkswagen, contract n° TIP4-CT-2005-516465, 2005-2009, http://www.superlightcar.com/public/index.php and http://ec.europa.eu/research/transport/projects/article_5088_en.html, 2010.

154. M. Deleglise, P. Le Grognec, B. Claude, D. Wilde, A. Terenzi, J. Kenny, C. Binetruy, and P. Krawczak, "Modeling of the injection of an automotive part with a two component highly reactive resin with on-line mixing in RTM," SAMPE Europe 2008, Paris, March 31–April 2, 2008.
155. Riversimple, Reinforced Plastics News, Epoxy prepreg fuels urban car concept, Sept. 2009, http://www.reinforcedplastics.com/view/3914/epoxy-prepreg-fuels-urban-car-concept/, 2009.
156. A.K. Bledzki and A. Jaszkiewicz, *Bioplastics Magazine*, Vol. 5, p. 48, 2010.
157. Solvay Chemicals, EPICEROL®: An innovative environmental breakthrough in Epichlorohydrin production, Solvay, http://www.solvaychemicals.com/info/0,0,1000574-_EN,00.html, 2010.
158. P. Campaner, D. D'Amico, L. Longo, C. Stifani, and A. Tarzia, *JEC Composites Magazine*, n° 56, p. 49, 2010.
159. H. Miyagawa, A.K. Mohanty, M. Misra, and L.T. Drzal, *Macromolecular Materials and Engineering*, Vol. 289, p. 629, 2004.
160. G. Mehta, A.K. Mohanty, M. Misra, and L.T. Drzal, *Green Chemistry*, Vol. 6, p. 254, 2004.
161. A.K. Mohanty, M. Misra, L.T. Drzal, S.E. Selke, B.R. Harte, and G. Hinrichsen, "Natural Fibers, Biopolymers, and Biocomposites: An Introduction" in A.K. Mohanty, M. Misra and L.T. Drzal, eds., *Natural Fibers, Biopolymers and Biocomposites*, CRC Press, Boca Raton, FL, USA, pp. 1–35, 2005.
162. European project BioComp - New classes of Engineering Composites Materials from Renewable Resources, Project n° NMP2-CT-2005-515769, 2005–2008, http://www.biocomp.eu.com/, 2010 and BioComp Final Summary Report, November 2008, http://www.biocomp.eu.com/uploads/Final_Summary_Report.pdf, 2010.
163. H.E. Hoydonckx and W.M. Van Rhijn, *JEC Composites Magazine*, n° 38, p. 34, 2008.
164. H.E. Hoydonckx, G. Switsers, B. M. Weager and E.L. Arnold, *JEC Composites Magazine*, n° 46, p. 41, 2009.
165. D.D. Andjelkovic, D.A. Culkin, and R. Loza, "Unsaturated Polyester Resins Derived from Renewable Resources," in Composites & Polycon, American Composites Manufacturers Association, Tampa, Florida USA, January 15–17, 2009.
166. A.K. Bledzki and J. Gassan, *Progress in Polymer Science*, Vol. 24, p. 221, 1999.
167. M. Shibata, K. Ozawa, N. Teramoto, R. Yosomiya, and H. Takeishi, *Macromolecular Materials and Engineering*, Vol. 208, p. 35, 2003.
168. R. Zah, R. Hischier, A.L. Leão, and I. Braun, *Journal of Cleaner Production*, Vol. 15, p. 1032–1040, 2007.
169. S. Ochi, *Mechanics of Materials*, Vol. 40, p. 446, 2008.
170. S. Kalia, B.S. Kaith, and I. Kaur, *Polymer Engineering & Science*, Vol. 49, p. 1253, 2009.
171. C. Scarponi, *JEC Composites Magazine*, n° 46, p. 46, 2009.
172. C. Alves, P.M.C. Ferão, A.J. Silva, L.G. Reis, M. Freitas, L.B. Rodrigues, and D.E. Alves, *Journal of Cleaner Production*, Vol. 18, p. 313, 2010.
173. L.Y. Feng, *JEC Composites Magazine*, n° 55, p. 29, 2010.
174 M. Karus, M. Kaup, and S. Ortmann, Use of natural fibers in composites in the German and Austrian automotive industry, Market Survey 2002: Status, Analysis and Trends, NOVA-Institut GmbH, Hürth, 2002, http://www.nova-institut.de/pdf/0303_market_nf_composite_1.pdf, 2010.
175. M. Matsuda and E. Connell, GPEC® 2009 Updates, Toyota Automotive Applications & Expectations of Bio-based Materials, http://www.sperecycling.org/GPEC/GPEC2009/GPEC%202009%20-%20Keynote%20Speaker%20-%20Eric%20Connell%20-%20Toyota.pdf, 2009.
176. L. Dufrancatel and C. Peyrelongue, "Faurecia Interior Systems: Light weight solution-Injection molding of polypropylene reinforced with natural fibers," Congrès international Carrosserie et plastiques, SFIP-SIA-EMD, Douai, June 9–10, 2010.
177. O. Faruk, Cars from Jute and Other Bio-Fibers, http://biggani.com/files_of_biggani/mashiur/interview/omar_faruk.pdf, 2010.
178. S.W. Beckwith, *Composites Fabrication magazine*, Vol. Nov/Dec, p. 12, 2003, http://www.sandy-munro.net/articles/Composites_Fabrication_-_Natural_Fiber.pdf, 2010.

179. NatureWorks LLC, Crystallizing and Drying of PLA, http://www.natureworksllc.com/product-and-applications/ingeo-biopolymer/technical-resources/~/media/product%20and%20applications/ingeo%20biopolymer/technical%20resources/processing%20guides/processing-guides_crystallizinganddryingpla_pdf.ashx, 2005.
180. G.C. Ellison, R McNaught, and EP Eddleston, Research & Development Report, United Kingdom Ministry of Agriculture, Fisheries and Food, The use of natural fibers in nonwonven structures for applications as automotive component substrates, Reference NF0309, 2000, http://www.ienica.net/usefulreports/auto.pdf, 2010.
181. J. Ganster and H.P. Fink, *Cellulose*, Vol. 13, p. 271, 2006.
182. B. Bax and J. Müssig, *Composites Science and Technology*, Vol. 68, p. 1601, 2008.
183. K. Oksman, M. Skrifvars, and J.F. Selin, *Composites Science and Technology*, Vol. 63, p. 1317, 2003.
184. B. Weager, *JEC Composites Magazine*, n° 55, p. 32, 2010.
185. D. Plackett, T.L. Andersen, W.B. Pedersen, and L. Nielsen L, *Composites Science and Technology*, Vol. 63, p. 1287, 2003.
186. T. Nishino, K. Hirao, M. Kotera, K. Nakamae, and H. Inagaki, *Composites Science and Technology*, Vol. 63, p. 1281, 2003.
187. S. Pilla, S. Gong, E. O'Neill, R.M. Rowell, and A.M. Krzysik, *Polymer Engineering and Science*, Vol. 48, p. 578, 2008.
188. M.S. Huda, L.T. Drzal, M. Misra, and A.K. Mohanty, *Journal of Applied Polymer Science*, Vol. 102, p. 4856, 2006.
189. M.S. Huda, L.T. Drzal, A.K. Mohanty, and M. Misra, *Composites Science and Technology*, Vol. 66, p. 1813, 2006.
190. Bioresin Ldt., *JEC Composites Magazine*, n° 56, p. 32, 2010.
191. Rhodia News release, Rhodia and its partners create a polyamide recycling channel for end-of-life vehicles, http://www.rhodia.com/en/news_center/news_releases/Polyamide_recycling_channel_070410.tcm, 2010.
192. P. Krawczak, *eXPRESS Polymer Letters*, Vol. 2, p. 23, 2008.
193. InoPlast—Recycling, http://www.inoplast.com/pages/en/26/sustainable-development-recycling.html, 2010 and InoPlats—A green SMC, http://www.inoplast.com/pages/en/31/sustainable-development-a-green-smc.html, 2010.
194. S.J. Pickering, *Composites Part A: Applied Science and Manufacturing*, Vol. 37, p. 1206, 2006.
195. European Composite Recycling, The Green FRP Recycling Label, http://www.nordic-composite.com/admin/common/getimg.asp?FileID=1131, 2010.
196. J.P. de Lary, *JEC Composites Magazine*, n° 57, p. 22, 2010.
197. S.J. Pickering, *JEC Composites Magazine*, n° 17, p. 27, 2007.
198. Automotive Industry endorses 'European Composite Recycling Concept', *NetComposites*, http://www.net-composites.com/news.asp?1644, 2003.
199. I. Pillin, N. Montrelay, A. Bourmaud and Y. Grohens, *Polymer Degradation and Stability*, Vol. 93, p. 321–328, 2008.
200. P. Sarazin, G. Li, W. Orts and B. Favis B, *Polymer*, Vol. 49, p. 599, 2008.
201. Galactic News, LOOPLA® by Galactic, http://www.lactic.com/index.php/news/item/4, 2009.
202. A. Le Duigou, I. Pillin, A. Bourmaud, P. Davies, and C. Baley, *Composites Part A: Applied Science and Manufacturing*, Vol. 39, p. 1471, 2008.
203. NatureWorks LLC, Fact or Fiction?, http://www.natureworksllc.com/product-and-applications/fact%20or%20fiction.aspx#back, 2010.

PART 6
GENERAL ENGINEERING APPLICATIONS

16

Cellulose Nanofibers Reinforced Bioplastics and Their Applications

Susheel Kalia[1*], B.S. Kaith[2] and Shalu Vashistha[3]

[1]*Department of Chemistry, Bahra University, Shimla Hills, Solan, India*
[2]*Department of Chemistry, Dr. B. R. Ambedkar National Institute of Technology (Deemed University) Jalandhar, India*
[3]*Department of Chemistry, Singhania University, Rajasthan, India*

Abstract
One of the main environmental problems we are facing today is the plastic waste and its disposal. Criteria for cleaner and safer environment have directed great part of the scientific research towards eco-composite materials that can be easily degraded or bio-assimilated. Biodegradable composites made entirely from renewable resources are urgently required to improve material recyclability or be able to divert materials from waste streams. Composites with bioplastics matrices and cellulose nanofibers are increasingly regarded as an alternative to conventional composites. Cellulose nanofibers reinforced polymer composites is a fast growing area of research because of their enhanced mechanical, thermal and biodegradation properties. In the present chapter, we have discussed the synthesis and properties of cellulose nanofibers and their applications as reinforcement in some environment benevolent plastics. Applications of cellulose nanofibers and their biodegradable polymer composites are also described in this chapter.

Keywords: Cellulose nanofibers, bioplastics, nanocomposites, morphology

16.1 Introduction

Cellulose materials accounts for 50% of the dry weight of plant biomass and approximately 50% of the dry weight of secondary sources of biomass such as industrial, agricultural and domestic wastes. However, cellulose in natural substrates is constant in the environment and remains as an environment pollutant [1]. Cellulosic materials such as nanofibers can be utilized in many applications and one of promising application is using them as a reinforcing material for synthesis of biocomposite materials.

Due to the high strength and stiffness, biodegradability and renewability of cellulose nanofibers and their production and application in biocomposite materials has gained increasing attention. Application of cellulose nanofibers for the

Srikanth Pilla (ed.) Handbook of Bioplastics and Biocomposites Engineering Applications, (453–470)

synthesis of biocomposites is a relatively new research field [2]. Due to the high stiffness of cellulose crystals, one can potentially exploit cellulose nanofibres as reinforcement in composite materials. It can be achieved by breaking down the hierarchical structure of the plant into individualized nanofibers of high crystallinity, therefore reducing the amount of amorphous material present in them. Since cellulose fibers are hierarchically fibrous, so it is possible yield a fibrous form of the material (nanowhiskers, nanofibrils), which due to their aspect ratio (length/diameter) and therefore reinforcing capabilities are potentially suitable for composite materials. A high aspect ratio to the fibers is desirable as this enables a critical length for stress transfer from the matrix to the reinforcing phase [3].

The introduction of bacterial cellulose onto natural fibers provides new mean of controlling the interaction between natural fibers and polymer matrices. Coating of natural fibers with bacterial cellulose does not only facilitate good distribution of bacterial cellulose within the matrix, it also results in an improved interfacial adhesion between the fibers and the matrix. This enhances the interaction between the natural fibers and the polymer matrix through mechanical interlocking. Bacterial cellulose coated natural fibers introduced nanocellulose at the interface between the fibers and the matrix, leading to increased stiffness of the matrix around the natural fibers [3, 4].

Cellulose nanofibers could be used as a rheology modifier in foods, paints, cosmetics and pharmaceutical products [5], but the main application we have discussed is their reinforcing ability in bioplastics for synthesis of composite materials. In this chapter, we describe various approaches to the preparation of cellulose nanofibers from plant sources. The main focus is on the extraction and characterization of nanocellulose and their applications in biocomposite materials. Bacterial cellulose production and surface modification of natural fibers using bacterial nanocellulose is also briefly discussed in this chapter.

16.2 Cellulose Fibers

The term cellulose fibers cover a broad range of vegetable, animal, and mineral fibers. However, in the composites industry, it usually refers to wood fiber and agro based bast, leaf, seed, and stem fibers. In this section, we have discussed sources, processing techniques, chemical composition and characterization of cellulose fibers.

16.2.1 Sources and Processing Methods

Cellulose fibers can be classified according to their origin and grouped into leaf: abaca, cantala, curaua, date palm, henequen, pineapple, sisal, banana; seed: cotton; bast: flax, hemp, jute, ramie; fruit: coir, kapok, oil palm; grass: alfa, bagasse, bamboo and stalk: straw (cereal). The bast and leaf (the hard fibers) types are the most commonly used in composite applications [6, 7]. Commonly used plant fibers are cotton, jute, hemp, flax, ramie, sisal, coir, henequen and kapok. The largest producers of sisal in the world are Tanzania and Brazil. Henequen is produced in

Mexico whereas abaca and hemp in Philippines. The largest producers of jute are India, China and Bangladesh. Cellulose fibers have so many advantages such as abundantly available, low weight, biodegradable, cheaper, renewable, low abrasive nature, interesting specific properties, as these are waste biomass and exhibit good mechanical properties [8–15]. Cellulose fibers also have some disadvantages such as moisture absorption, quality variations, low thermal stability and poor compatibility with the hydrophobic polymer matrices [16–18].

Fiber processing technology like microbial deterioration and system explosion plays an important role in improving the quality of fibers. Microbial deterioration of the material depends on the environmental conditions. The condition reached thereby are decisive for the energy necessary for delignification and fibrillation and thus also for the attainable fiber masses. In order to obtain a value gain, the more important is to retain the super molecular structure of the fibers. The traditional microbial deterioration process is one of the most important pre-requisite. However, this deterioration process can be partly replaced by the latest chemico-physical processes [19]. In new steam explosion method, steam and additives under pressure and with increased temperature, penetrate the space between fibers of the bundle, because of which the middle lamella and the fibers adherent substances are elementarized softly and are made water soluble which can be removed by subsequent washing and rinsing [20, 21].

16.2.2 Chemical Composition

Natural fibers are constitutes of cellulose fibers, consisting of helically wound cellulose micro fibrils, bound together by an amorphous lignin matrix. Lignin keeps the water in fibers, acts as a protection against biological attack and as a stiffener to give stem its resistance against gravity forces and wind. Hemicellulose found in the natural fibers is believed to be a compatibilizer between cellulose and lignin [22]. The cell wall in a fiber is not a homogenous membrane [23]. Each fiber has a complex, layered structure consisting of a thin primary wall which is the first layer deposited during cell growth encircling a secondary wall. The secondary wall is made up of three layers and the thick middle layer determines the mechanical properties of the fiber. The middle layer consists of a series of helically wound cellular microfibrils formed from long chain cellulose molecules. The angle between the fiber axis and the microfibrils is called the microfibrillar angle. The characteristic value of microfibrillar angle varies from one fiber to another. These microfibrils have typically a diameter of about 10–30 nm and are made up of 30–100 cellulose molecules in extended chain conformation and provide mechanical strength to the fiber [21].

16.2.3 Properties

The properties of cellulose fibers are affected by many factors such as variety, climate, harvest, maturity, retting degree, decortications, disintegration (mechanical, steam explosion treatment), fiber modification, textile and technical processes (spinning and carding) [24]. In order to understand the properties of natural fiber-reinforced composite materials, it becomes necessary to know the mechanical, physical and chemical properties of natural fibers. Flax fibers are relatively strong fibers as compared to other natural fibers. The tensile strength of elementary fibers is in the region of 1500 MPa and for technical fibers a value of circa 800 MPa was observed at 3 mm clamp length [25]. Baley [26] and Lamy and Baley [27] investigated the modulus of flax fibers. The modulus of elementary fibers is dependent on the diameter of fiber and it ranges from 39 GPa for fibers having diameter approximately 35 μm to 78 GPa for fibers having 5 μm diameter. This variation is related to the variation in relative lumen size between fibers having different diameter. An average Young's modulus of 54 GPa was observed after numerous tensile tests on single flax fiber and the results are within the range of moduli measured on technical fibers. The mechanical, chemical and physical properties of plant fibers are strongly harvest dependent, influenced by climate, location, weather conditions and soil characteristics. These properties are also affected during the processing of fiber such as retting, scotching, bleaching and spinning [28].

Cellulose fibers have relatively high strength, high stiffness and low density [29]. The characteristic value for soft-wood-Kraft-fibers and flax has been found close to the value for E-glass fibers. Different mechanical properties can be incorporated in natural fibers during processing period. The fiber properties and structure are influenced by several conditions and varies with area of growth, its climate and age of the plant [30]. Technical digestion of the fiber is another important factor which determines the structure as well as characteristic value of fiber. The elastic modulus of the bulk natural fibers such as wood is about 10 GPa. Cellulose fibers with moduli up to 40 GPa can be separated from wood by chemical pulping process. Such fibers can be further subdivided into micro fibrils within elastic modulus of 70 GPa. Theoretical calculations of elastic moduli of cellulose chain have been given values up to 250 GPa. However, no technology is available to separate these from microfibrils [31]. The tensile strength of natural fibers depends upon the test length of the specimen which is of main importance with respect to reinforcing efficiency. Kohler *et al* [19], Mieck *et al* [32] and Mukherjee *et al* [33] reported that tensile strength of flax fiber is significantly more dependent on the length of the fiber. In comparison to this, the tensile strength of pineapple fiber is less dependent on the length, while the scatter of the measured values for both is located mainly in the range of the standard deviation. The properties of flax fiber are controlled by the molecular fine structure of the fiber which is affected by growing conditions and the fiber processing techniques used. Flax fibers possess moderately high specific strength and stiffness [21].

16.3 Bioplastics: Synthesis, Properties and Applications

Criteria for cleaner and safer environments have directed a great part of the scientific research towards bioplastic materials, which can easily be degraded or bio-assimilated. Bioplastics are natural biopolymers that are synthesized and catabolized by various organisms and these biopolymers do not cause toxic effects in the host and have certain advantages over petroleum derived plastics.

Bioplastics are plastics that are derived from renewable biomass sources of carbon such as hemp and soy bean oil, corn and potato starch. Sometimes the term bioplastics also refer to plastic polymers produced by microbes or plastics which are likely to be biodegradable. There are three major ways to synthesize bioplastics, use of natural polymer (bioplastics: cellulose acetate, starch polymer), use of bacterial polyester fermentation (PHA, PHB) and chemical polymerization (polybutylene succinate, polyglycolic acid, polycaprolactone) [34].

Polyhydroxyalkanoates, macromolecule polyesters naturally produced by many species of microorganisms, are being considered as a replacement for conventional plastics. PHAs can be completely bio-degraded within a year by a variety of microorganism results in carbon dioxide and water, which return to the environment. Whereas, petroleum derived plastics takes several decades to degrade. Polyhydroxyalkanoates (PHAs) and their derivatives are the most widely produced microbial bioplastics. Beijerinck first observed lucent granules of PHA in bacterial cells in 1888 [35]. Macrae and Wilkinson [36] were the first to report the functions of polyhydroxybutyric acid (PHB) appeared in 1958. They reported the rapid biodegradability of PHB produced by *Bacillus megaterium*, by *B. cereus* and *B. megaterium* itself [37].

The biosynthetic pathway for PHB comprises the three enzymes b-ketoacyl-CoA thiolase (PhbA), acetoacetyl-CoA reductase (PhbB) and PHB-polymerase/synthase (PhbC), which are often clustered in bacterial genomes (Figure 16.1). PhbA catalyzes the condensation of two acetyl coenzyme A (acetyl CoA) molecules into acetoacetyl-CoA. PhbB catalyses the reduction of acetoacetyl-CoA to (R)-3-hydroxybutyryl-CoA. Finally, the (R)-3-hydroxybutyryl-CoA monomers are polymerized into PHB by PhbC [38–40].

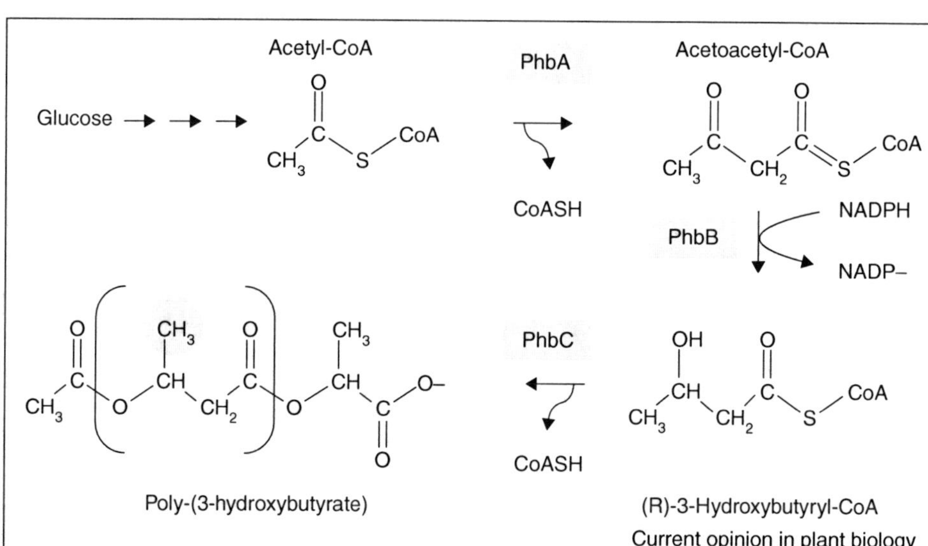

Figure 16.1 Biosynthetic pathway for PHB [40].

Khardenavis *et al* [41] biotechnologically converted agro-industrial wastewaters into biodegradable plastics, PHB. Deproteinized jowar grain-based distillery spentwash yielded 42.3% PHB production (w/w), followed by filtered rice grain-based distillery spentwash (40% PHB) when used as substrates. PHB has properties similar to petroleum derived synthetic plastics like polypropylene (PP) and is completely biodegradable in the environment. However, the production cost of PHB is nine times higher in comparison to synthetic plastics as it involves production of biomass with expensive carbon sources [42]. This has limited the use of PHB to specialized areas like surgery and medicine.

The influence of glutenin-to-gluten ratio on the rheological properties of the glycerol plasticized doughs and the crosslinked bioplastics were investigated by Song *et al* [43]. It has been resulted that the glutenin dough exhibits higher moduli and lower loss factor and equibiaxial deformability in comparison with the gluten dough. Addition of a glutenin-rich fraction to the gluten dough can improve elasticity at small deformation and extensional deformational stress at large deformations but result in reductions in extensibility of the compression molded bioplastics. Song *et al* [44] have prepared bioplastics from a glutenin-rich fraction; that is, the gluten residue insoluble in 70% (v/v) ethanol. They have investigated the influence of reducing agents of sodium bisulfite, sodium sulfite and thioglycolic acid on the properties of the glycerol plasticized doughs and the cross-linked bioplastics. The results showed that reducing agents can be applied to reduce the Young's modulus of the plasticized dough and to improve the Young's modulus of the cross-linked bioplastics. Moisture absorption, weight loss in water, tensile strength, elongation at break and tensile set were also studied to characterize the physical properties of the cross-linked bioplastics.

As with conventional plastics, bioplastics have a very broad application spectrum. Many bioplastics products are still being used in areas where compostability represents a significant benefit. Other applications such as packaging and technical applications are gaining importance. One of the important applications of bioplastics is their use in the synthesis of biocomposite materials using natural fibers as reinforcing materials and bioplastics as binder.

16.4 Cellulose Nanofibers

The name for cellulose fibrils is not consistent, often the terms microfibrillated cellulose, cellulose nanofibrils or cellulose nanofibers are used. A cellulose fiber is made up of bundles of single cellulose fiber, which has a diameter of 25–30 μm. The single cellulose fiber is made up of bundles of microfibers, which have diameters of 0.1–1 μm. The microfiber consists of bundles of nanofibers, which have a diameter in the range 10–70 nm and lengths of thousands of nanometers [45, 46]. Nanofiber is made up of cellulose chains bound by hydrogen bonding.

Many studies have been done on extracting cellulose nanofibers from various sources. Cellulose nanofibers can be extracted from the cell walls by following isolation processes, which may be simple mechanical methods or a combination of both chemical and mechanical methods [47].

16.4.1 Methods of Cellulose Nanofibers Production

16.4.1.1 Electrospinning

Electrostatic fiber spinning or 'electrospinning' is a novel process for forming fibers with submicron scale diameters through the action of electrostatic forces. The effects of solvent system, the degree of polymerization of cellulose, spinning conditions, and post spinning treatment such as coagulation with water on the miscrostructure of electrospun fibers were investigated by Kim *et al* [48]. They have synthesized non-woven mats of submicron-sized cellulose fibers (250–750 nm in diameter) by electrospinning of cellulose solutions. Cellulose are directly dissolved in two solvent systems: lithium chloride (LiCl)/N,N-dimethyl acetamide (DMAc) and N-methylmorpholine oxide (NMMO)/water.

Frenot *et al* [49] have synthesized cellulose nanofibers by electrospinning of cellulose derivatives of carboxymethyl cellulose sodium salt, hydroxypropyl methylcellulose, methylcellulose and enzymatically treated cellulose. Microstructure of the resulting cellulose nanofibers has been studied by scanning electron microscopy. Cellulose nanofibers were directly prepared from cellulose solution using electrospinning techniques. The resulting nanofibers were composed of pure cellulose. The electrospinning procedure can be performed under ambient conditions at room temperature without post-spun treatment. Electrospinning of pre-spun cellulose solution mixed with drug resulted in formation of the drug-loaded nanofiber and the releasing properties were examined with respect to biomedical applications [50].

16.4.1.2 Mechanical & Chemical Defibrillation

Recent advances include producing microfibrillated or nano-size fibers with high-strength and surface area for use in high performance biocomposites. Alemdar and Sain [51] have extracted cellulose nanofibers from wheat straw by a chemical treatment, resulting to purified cellulose. To individualize the nanofibers from the cell walls a mechanical treatment (cryocrushing, disintegration and defibrillation steps) was applied to the chemically treated fibers. Cellulose nanofibers were extracted from the agricultural residues, wheat straw and soy hulls, by a chem-mechanical technique [52]. The wheat straw nanofibers were determined to have diameters in the range of 10–80 nm and lengths of a few thousand nanometers. By comparison, the soy hull nanofibers had diameter 20–120 nm and shorter lengths than the wheat straw nanofibers.

Cellulose is the fibrillar component of plant cells. Bacic *et al* [53] have reported that cellulose chains are packed in an ordered manner to form compact microfibrils, which are stabilized by both inter-molecular and intramolecular hydrogen bonding. Up to 100 glucon chains are grouped together to form long thin crystallites (elementary fibrils) and these crystallites are about 5 nm wide but this varies according to the source of the cellulose. They are organized in groups to form microfibrils that are 8–50 nm in diameter and of lengths of a few microns [54]. These nanofibers give strength to the plant stem because of their crystal structure. Chemical and mechanical treatments of the cellulose fibers result in chemical and structural changes on the fiber surface and the cells, which influence the properties of the fibers in composites. Zimmermann *et al* [55] were separated nanofibrillated cellulose (NFC) at the greatest possible lengths and diameters below 100 nm from different starting cellulose materials by mechanical dispersion and high pressure (up to 1500 bar) homogenization processes. The treatment resulted in nano-scaled fibril networks. Two commercial fibrous celluloses showed bigger cellulose aggregates with micrometer dimensions and a less homogeneous network structure.

The cellulose nanofibers were extracted by Wang and Sain [47] from soybean stock by chemo-mechanical treatments [Figure 16.2]. These are bundles of cellulose nanofibers with a diameter ranging between 50 and 100 nm and lengths of thousands of nanometers.

16.4.1.3 Bacterial Cellulose Nanofibers

Bacterial Cellulose (BC) has gained attention in the research area for the encouraging properties it possesses; such as its significant mechanical properties in both dry and wet states, porosity, water absorbency, moldability, biodegrability and excellent biological affinity [56]. Because of these properties, BC has a wide range of potential applications.

Acetobacter xylinum (or Gluconacetobacter xylinus), a gram-negative strain of acetic-acid-producing bacteria is the most efficient producer of bacterial cellulose (BC) [57, 58]. BC is secreted as a ribbon-shaped fibril, less than 100 nm wide, which is composed of much finer 2–4 nm nanofibrils [59, 60]. In comparison to the

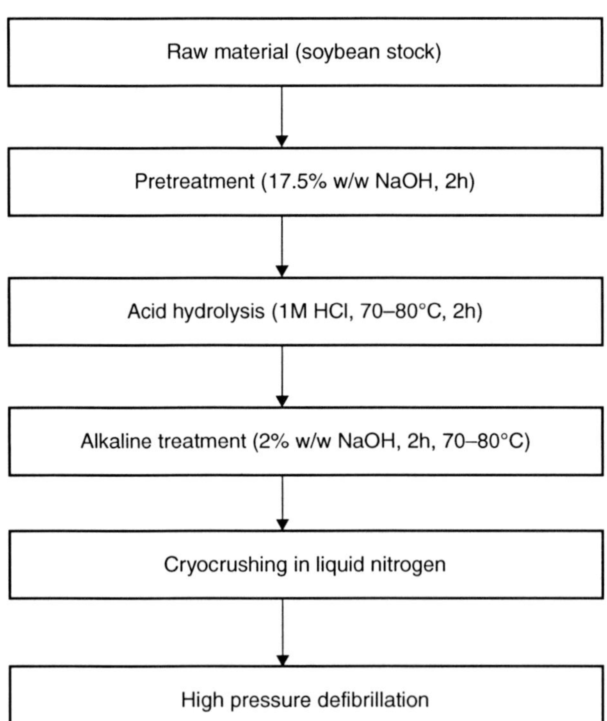

Figure 16.2 Isolation of nanofibers by chemo-mechanical treatment [47].

methods for obtaining nanocellulose through mechanical or chemo-mechanical processes, BC is produced by bacteria through cellulose biosynthesis and the building up of bundles of microfibrils [61].

Pommet *et al* [4] cultivated the cellulose-producing bacteria in the presence of natural fibers (Figure 16.3) such as sisal and hemp, resulting in the coating of natural fiber surfaces by BC. It was resulted that the strong and highly crystalline BC fibrils preferentially attached to the surface of natural fibers thereby creating "hairy fibers" (Figure 16.4) leading to a nanostructured natural fiber surface. Simply weighing the fibers before and after the BC fermentation process confirmed that 5-6 wt% of BC adhered to the fibers after the surface modification. The strength of attachment of the nanocellulose coating to the natural fibers can be attributed to strong hydrogen bonding between the hydroxyl groups present in BC and the lignocellulose in natural fibers [62]. The modification process did not affect the mechanical properties of sisal fibers but it significantly reduced the mechanical properties of hemp fibers [3].

16.4.2 Characterization of Cellulose Nanofibers

Various techniques such as transmission electron microscopy (TEM), scanning electron microscopy (SEM), field-emission scanning electron microscopy (FE-SEM), atomic force microscopy (AFM), wide-angle X-ray scattering (WAXS) and NMR

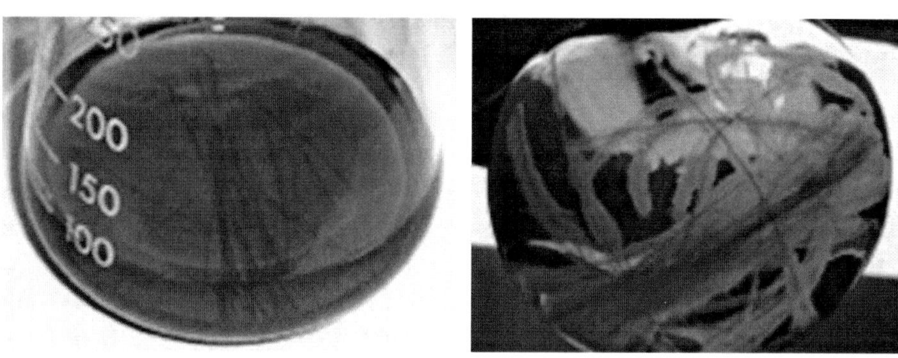

Figure 16.3 Photographs of sisal fibers before and after bacterial culture [4].

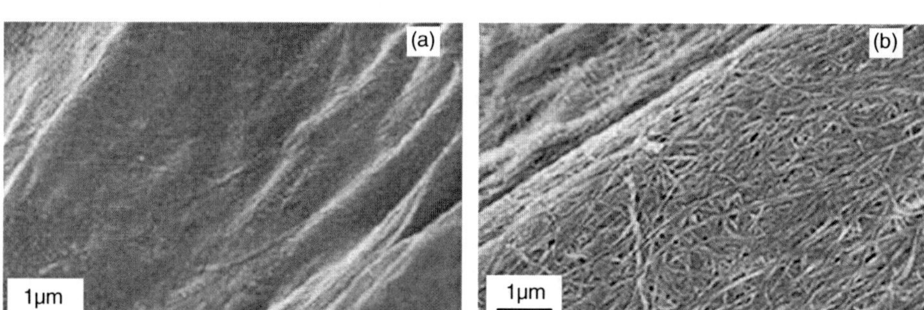

Figure 16.4 SEM micrographs (a) sisal fibre and (b) bacterial cellulose coated sisal fiber [4].

spectroscopy have been used to characterize cellulose nanofiber morphology. A combination of microscopic techniques with image analysis can provide information about widths of cellulose nanofiber but it is very difficult to find out the lengths of nanofiber because of entanglements and difficulties in identifying both ends of individual nanofibers. It is often reported that MFC suspensions are not homogeneous and that they consist of cellulose nanofibers and nanofiber bundles [2].

Teixeira *et al* [63] were obtained the suspensions of white and colored nanofibers by the acid hydrolysis of white and naturally colored cotton fibers. Possible differences among them in morphology and other characteristics were investigated. Morphological study of cotton nanofibers showed a length of 85–225 nm and diameter of 6–18 nm. It was found that there were no significant morphological differences among the nanostructures from different cotton fibers. The main differences found were the slightly higher yield, sulfonation effectiveness and thermal stability under dynamic temperature conditions of the white nanofiber. On the other hand, the colored nanofibers showed a better thermal stability than the white in isothermal conditions at 180 °C.

The structure of the cellulose nanofibers from agricultural residues was investigated by Alemdar and Sain [52]. FT-IR spectroscopic analysis demonstrated that chemical treatment also led to partial removal of hemicelluloses and lignin from the structure of the fibers. PXRD results revealed that this resulted in improved crystallinity of the fibers. Thermal properties of the nanofibers were studied by the TGA technique and found to increase dramatically.

Stelte and Sanadi [64] have studied the mechanical fibrillation process for the preparation of cellulose nanofibers from two commercial hard- and softwood cellulose pulps. The degree of fibrillation was studied using light microscopy (LM), scanning electron microscopy (SEM), and atomic force microscopy (AFM). LM and SEM images (Figure 16.5) of hard- and softwood fibers showed that the hardwood fibers that were fibrillated only on the surface during the refining step are now disintegrated into a network of small fibers. AFM images (Figure 16.6) of the final products after high-pressure homogenization showed that the size distribution of the hard- and softwood nanofibers is in the range of 10–25 nm in diameter.

By chemo-mechanical isolation, soybean stock based nanofibers having a diameter in the range 50–100 nm were produced by Wang and Sain [47]. X-ray crystallography (Figure 16.7) was carried out to investigate the percentage crystallinity after various stages of the chemo-mechanical treatment. It has been found that crystallinity of the samples increased after each stage of nanofiber development.

The network of cellulose nanofibers can be seen in Figure 16.8. The nanofiber suspension obtained after the high pressure defibrillation was analyzed to determine diameters using AFM. The AFM image (Figure 16.8) shows the surface of air-dried soybean stock nanofiber. It is seen that the fibers are indeed nanosized and the diameter of nanofibers is within the range 50–100 nm.

16.4.3 Applications of Cellulose Nanofibers

One of the most important applications of cellulose nanofibers is their use as reinforcement for synthesis of high performance green nanocomposites. This is

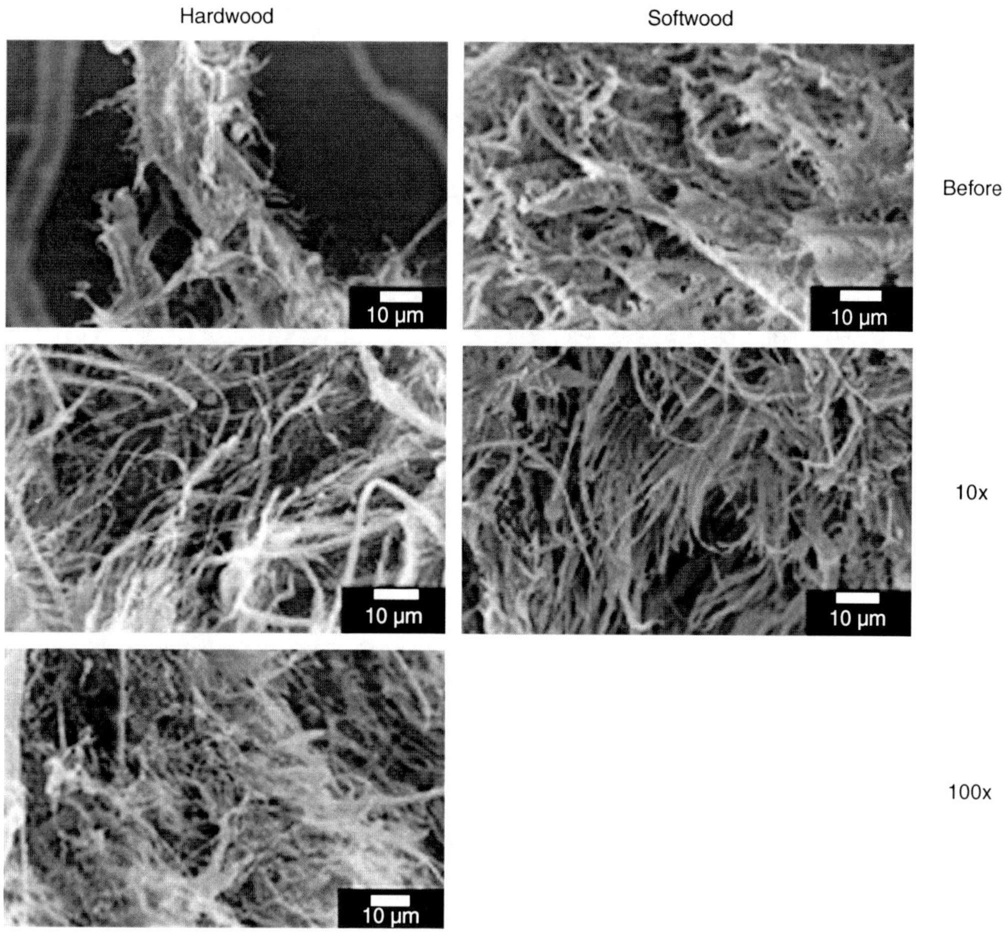

Figure 16.5 Scanning electron micrographs of hard- and softwood cellulose fibers, before and after 10 passes through the homogenizer [62].

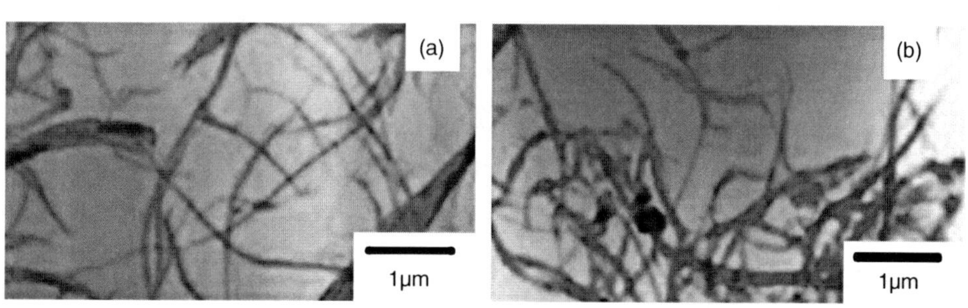

Figure 16.6 AFM images (a) hard- and (b) softwood cellulose nanofibers at process equilibrium [62].

because nanofibers can have even better mechanical properties than micro fibers of the same materials, and hence the superior structural properties of nanocomposites can be expected. Moreover, nanofibers reinforced composite materials may possess some additional merits which cannot be shared by traditional microfiber reinforced composites.

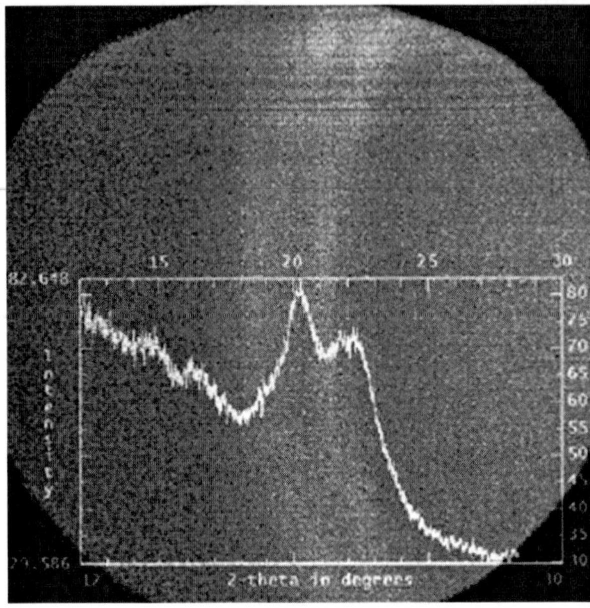

Figure 16.7 X-ray pattern to demonstrate the crystallinity of soybean stock nanofibers [47].

Figure 16.8 Atomic force micrograph of soybean stock nanofibers [47].

Due to high porosity, water absorbance, mechanical properties, formability, and biocompatibility, bacterial cellulose has also recently attracted a great deal of attention for biomedical applications. Bacterial cellulose has long been used in a variety of applications in the paper, food, and electronic industries. For instance, bacterial cellulose has been successfully used for wound dressings and for vascular implants [60].

Bacterial or microbial cellulose is a pure form of cellulose with extensive applications. It is very useful for medical applications especially wound dressings due to its high liquid absorbency and hygienic nature. Bacterial cellulose secreted by *Acetobacter xylinum* has some special properties such as high water retaining capacity, nano fibrils, proper biodegradability, molding ability, and high tensile strength. It has found many industrial applications in high-strength construction materials, food additives and biodegradable products and paper, textiles and biomaterials for cosmetics and medical applications, temporary skin substitute and micro vessel endoprothesis [65]. Meftahi *et al* [65] studied the coating of cotton gauze with microbial cellulose. The modified gauze with microbial cellulose shows high capability of medical liquid absorption and greater dry time, which are the main feature of modern wound dressing. Further, it is also possible to encapsulate various medical agents into this medium according to its unique fibril network.

Cellulose nanofibers could be used as a rheology modifier in foods, paints, cosmetics, pharmaceutical products and biomedical applications [5], but one of the main applications is their reinforcing ability in bioplastics for synthesis of green composite materials.

16.5 Cellulose Nanofibers Reinforced Bioplastics

Application of cellulose nanofibers as reinforcement for synthesis of high performance polymer composites is a relatively new research area. Synthesis of cellulose nanofibers and their application in polymer composite materials has gained increasing attention due to their high strength and stiffness, biodegradation and renewability.

16.5.1 Synthesis and Properties of Nanocomposites

Cellulose nanofibers can be obtained from various natural fibers by chemical treatments followed by innovative mechanical techniques. The nanofibers thus obtained have diameters between 5 and 60 nm. Reinforced composite films comprising 90% polyvinyl alcohol and 10% nanofibers were prepared by Bhatnagar and Sain [66]. Mechanical properties of these composites were studied and it was resulted that reinforcement of composites with nanofibers enhanced the mechanical strength of pure polyvinyl alcohol. Chen and Liu fabricated cellulose nanofibrous mats (CNM) reinforced soybean protein isolate (SPI) composite with high visible light transmittance. The light transmittance, mechanical properties, and swelling ratio of CNM/SPI composite were investigated in terms of CNM content in the composite. It was resulted that strong interfacial interactions occurred at the cellulose nanofiber/SPI interfaces. The reinforcement of 20 wt % cellulose nanofibers in the SPI matrix resulted in great improvement of mechanical strength and Young's modulus by respectively 13 and 6 times more than pure SPI film [67].

Bacterial cellulose nanofibers reinforced starch biocomposites consisting of biodegradable BC and starch are fully biodegradable. The BC nanofibers were incorporated in the starch plasticized with glycerol via a solution impregnation method. Tensile properties of the BC/starch biocomposites were tested and compared with those of the unreinforced starch. The reinforcement of BC nanofibers improves the tensile properties and the resistance to moisture and microorganism attacks [68]. BC has been used as reinforcements for various composites due to its excellent mechanical performance [69, 70]. The introduction of bacterial cellulose onto natural fibers provides new means of controlling the interaction between natural fibers and polymer matrices. Coating of natural fibers with bacterial cellulose does not only facilitate good distribution of bacterial cellulose within the matrix, it also results in an improved interfacial adhesion between the fibers and the matrix. This enhances the interaction between the natural fibers and the polymer matrix through mechanical interlocking. Bacterial cellulose coated natural fibers introduced nanocellulose at the interface between the fibers and the matrix, leading to increased stiffness of the matrix around the natural fibers [3, 4].

The reinforcement of polylactic acid (PLA) using microfibrillated cellulose (MFC, mechanically fibrillated pulp, mostly consisting of nanofibers) is reported by Iwatake *et al.* The MFC increased Young's modulus and tensile strength of PLA by 40% and 25%, respectively, without a reduction of yield strain at a fiber content of 10 wt% [71]. The reinforcing potential of cellulose nanofibers obtained from agro-residues was investigated by Alemdar and Sain [51] in a starch-based biocomposites. Thermal and mechanical performance of these composites were studied and compared with the pure thermoplastic starch using thermogravimetric and dynamic mechanical analysis, and tensile testing. It has been found that tensile strength and modulus of the nanocomposites film were significantly enhanced as compared to the pure thermoplastic starch. The glass transition of the composites was shifted to higher temperatures with respect to the pure thermoplastic starch. Morphological images (Figure 16.9) of the nanocomposites showed a uniform dispersion of the nanofibers in the polymer matrix. Better interfacial adhesion between the nanofibers and the polymer matrix results in an improvement in the mechanical properties of the nanocomposites. Kaushik *et al* [72] have extracted the cellulose nanofibers from wheat straw using steam explosion, acidic treatment and high

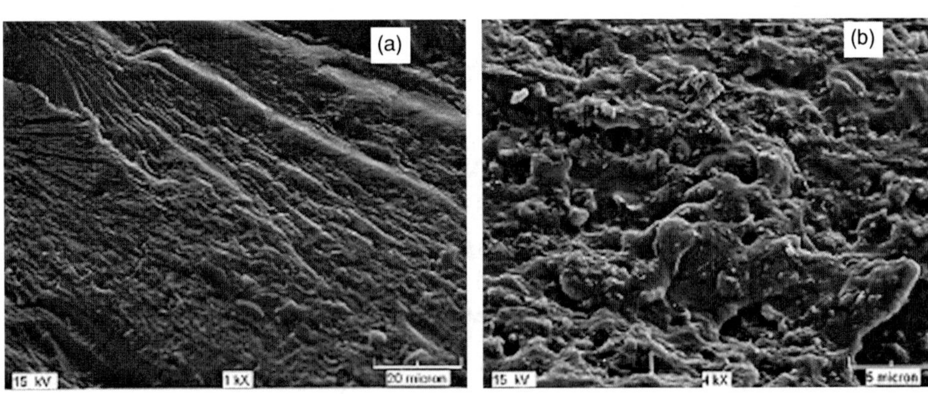

Figure 16.9 SEM images of the cryo-fractured surface of the (a) thermoplastic starch and (b) nanocomposite filled with 10 wt% of cellulose nanofibers [49].

shear mechanical treatment. These nanofibers were dispersed in thermo plastic starch using a Fluko high shear mixer in varying proportions and films were casted out of these nanocomposites. It was found that mechanical properties increased with nanofiber concentration and barrier properties also improved with addition of nanofibers up to 10% but further addition deteriorated properties due to possible fiber agglomeration. Wang and Sain [47] studied the dispersion mechanism of soybean stock-based nanofibers in a plastic matrix. The mechanical properties of nanofiber reinforced PVA film demonstrated a 4– to 5–fold increase in tensile strength, as compared to the untreated fiber-*blend*-PVA film. Ensuring good dispersion of the filler in the composite material was one of the problems encountered in the use of nano reinforcements. The potential use of chemically treated hemp nanofibers as reinforcing agents for biocomposites was studied by Wang and Sain [73]. The cellulose nanofibers were treated using five different chemicals. Bio-nanocomposites were prepared using nanofibers as reinforcement and PLA and PHB as matrices. Nanofibers were only partially dispersed in the polymers and therefore mechanical properties were lower than those predicted by theoretical calculations.

16.5.2 Applications of Nanocomposites

Cellulose nanofiber reinforced nanocomposites materials with superior performance and extensive applications have received great attention. Polymeric biocomposites are biodegradable and offer potential advantages over recalcitrant synthetic plastics in disposable applications.

Cellulose nanofiber reinforced polymer composite materials can be used in medical devices such as biocompatible drug delivery systems, blood bags, cardiac devices, and valves as reinforcing biomaterials. Biological tissues are made from nanosized materials and have led to interest in manufacturing synthetic nanocomposites. Due to their low weight and high strength, they also can be utilized as high strength components in the aerospace and automotive sector [47, 74].

Use of cellulose nanofibers improves polymer mechanical properties such as tensile strength and modulus in a more efficient way than is achieved in conventional micro- or macro- polymer composite materials. Packaging is one area in which nanocellulose reinforced polymer films could be of interest. It is possible to produce nanocellulose reinforced polymer films with high transparency and with improved oxygen barrier properties. High oxygen barrier is often a requirement for food and pharmaceutical packaging applications and such improvement may be a key for capturing new markets. In addition to packaging, cellulose nanofibers could also be used in the electronic device industry in the future [2]. Applications of cellulose nanocomposites in detail were reported by Hubbe *et al.* [75].

16.6 Conclusion

Cellulose nanofibers can be extracted from the cell walls by electrospinning method and simple mechanical methods or a combination of both chemical and mechanical methods. Application of cellulose nanofibers for the synthesis of biocomposites is a relatively new research field due to the high strength and stiffness, biodegradability and renewability of cellulose nanofibers. Introduction of bacterial cellulose onto natural fibers provides new mean of controlling the interaction between natural fibers and polymer matrices. Coatings of natural fibers with bacterial nanocellulose results in an improved interfacial adhesion between the fibers and the matrix. Due to low weight and high strength, cellulose nanocomposites can be utilized in packaging, medical devices and aerospace and automotive sector.

References

1. S. Haruta, Z. Cui, Z. Huang, M. Li, M. Ishii, Y. Igarashi, *Appl. Microbiol. Biotechnol.*, Vol. 59, p. 529, 2002.
2. I. Siro and D. Plackett, *Cellulose*, Vol. 17, p. 459, 2010.
3. S. J. Eichhorn, A. Dufresne, M. Aranguren, N. E. Marcovich, J. R. Capadona, S. J. Rowan, C. Weder, W. Thielemans, M. Roman, S. Renneckar, W. Gindl, S. Veigel, J. Keckes, H. Yano, K. Abe, M. Nogi, A. N. Nakagaito, A. Mangalam, J. Simonsen, A. S. Benight, A. Bismarck, L. A. Berglund, T. Peijs, *J. Mater. Sci.*, Vol. 45, p. 1, 2010.
4. M. Pommet, J. Juntaro, J.Y.Y. Heng, A. Mantalaris, A.F. Lee, K. Wilson, G. Kalinka, M.S.P. Shaffer, A. Bismarck, *Biomacromolecules*, Vol. 9, p. 1643, 2008.
5. A.F. Turbak, F.W. Snyder, and K.R. Sandberg, *J. Appl. Polym. Sci.: Appl. Polym. Symp.*, Vol. 37, p. 815, 1983.
6. G.I. Williams, and R.P. Wool, *Applied Composite Materials*, Vol. 7, p. 421, 2000.
7. F.G. Torres, and R.M. Diaz, *Polymers & Polymer Composites*, Vol. 12, p. 705, 2004.
8. P.R. Hornsby, E. Hinrichsen, and K. Tarverdi, *J. Mater. Sci.*, Vol. 32, p. 1009, 1997.
9. K. Oksman, L. Wallstrom, L.A. Berglund, and R.D.T. Filho, *J. Appl. Polym. Sci.*, Vol. 84, p. 2358, 2002.
10. K. Oksman, *Journal of Reinforced Plastics and Composites*, Vol. 20, p. 621, 2001.
11. R. Heijenrath, and T. Pejis, *Ad. Comp. Let.*, Vol. 5, p. 81, 1996.
12. K. Oksman, *Appl. Comp. Mat.*, Vol. 7, p. 403, 2000.
13. K.P. Mieck, A. Nechwatal, and C. Knobeldorf, *Melliand Textilberichte*, Vol. 11, p. 228, 1994.
14. A.R. Sanadi, D.F. Cauldfield, and R.M. Rowell, *Plastic Eng.*, Vol.4, p. 27, 1994.
15. A.K. Bledzki, S. Reihmane, and J. Gassan, *J. Appl. Polym. Sci.*, Vol. 5, p. 1329, 1996.
16. D. Nabi Saheb, and J.P. Jog, *Adv. Polym. Technol.*, Vol. 18, p. 351, 1999.
17. I.S. Wichman, A.B. Oladipo, and I. Hermann, The influence of moisture on fibre/matrix adhesion for wood/HDPE-composites. In: Proceedings of *9th Annual ASM/ESD Advanced Composites Conference*, ESD. The Engineering Society; 265, 1993.
18. S.Th. Georgopoulos, P.A. Tarantili, E. Avgerinos, A.G. Andreopoulos, and E.G. Koukios, *Polymer Degradation and Stability*, Vol. 90, p. 303, 2005.
19. R. Kohler, and M. Wedler, *Landinfo*, Vol. 3, p. 33, 1995.
20. S.J. Eichhorn, "Regenerated Cellulose Reinforced Plastics," *Natural Fibers, Plastics and composites*, Edited by Frederick T. Wallenberger and Norman Weston, Kluwer Academic Publisher, New York, p. 287, 2004.
21. S. Kalia, B.S. Kaith, and I. Kaur, *Polym. Eng. Sci.*, Vol. 49, p. 1253, 2009.
22. C.M. Hansen, and A. Bjorkman, *Holzforschung*, Vol. 52, p. 335, 1998.
23. M.Z. Rong, M.Q. Zhang, Y. Liu, G.C. Yang, and H.M. Zeng, *Compos. Sci. Technol.*, Vol. 61, 1437 (2001).
24. K. Van de Velde, and P. Kiekens, *Composite Structures*, Vol. 54, p. 355, 2001.
25. H.L. Boss, M.J.A. van den Oever, and O.C.J.J Peters, *J. Mater. Sci.*, Vol. 37, p. 1683, 2002.
26. C. Baley, *Composites Part: A*, Vol. 33, p. 939, 2000.
27. B. Lamy, and C. Baley, *J. Mater. Sci. Lett.*, Vol. 19, p. 979, 2000.

28. A. Jähn, M.W. Schröder, M. Füting, K. Schenzel, and W. Diepenbrock, "Spectrochimica Acta, Part A:" *Molecular and Biomolecular Spectroscopy*, Vol. 58, p. 2271, 2002.
29. J. Gassan, and A.K. Bledzki, *Die Angew Makromaol. Chem.*, Vol. 236, p. 129, 1996.
30. A.J. Michell, *Composite Asia Pacific, Adelaide*, Vol. 89, p. 19, 1989.
31. T.M. Moloney, In: S.M. Lee and R.M. Rowell Editors, *International Encyclopediaof Composites*, VCH Publishers, New York, 656, 1995.
32. K.P. Mieck, A. Nechwatal, and C. Knobeldorf, *Melliand Texilberichte*, Vol. 11, p. 892, 1994.
33. P.S. Mukherjee, and K.G. Satyanarayana, *J. Mater. Sci.*, Vol. 21, p. 51, 1986.
34. E. Matsuura, Y. Ye, and X. He, Sustainability opportunities and challenges of bioplastics, Master Thesis, Blekinge Institute of Technology Karlskrona, Sweden, 2008.
35. A.A. Chowdhury, *Arch. Mikrobiol.*, Vol. 47, p. 167, 1963.
36. R.M. Macrae, and J.F. Wilkinson, *J. Gen. Microbiol.*, Vol. 19, p. 210, 1958.
37. P. Suriyamongkol, R. Weselake, S. Narine, M. Moloney, and S. Shah, *Biotechnology Advances*, Vol. 25, p. 148, 2007.
38. B.H. Rehm, *Biochem. J.*, Vol. 376, p. 15, 2003.
39. L.L. Madison, and G.W. Huisman, *Microbiol. Mol. Biol. Rev.*, Vol. 63, p. 21, 1999.
40. J. Scheller, and U. Conrad, *Current Opinion in Plant Biology*, Vol. 8, p. 188, 2005.
41. A. A. Khardenavis, M.S. Kumar, S.N. Mudliar, and T. Chakrabarti, *Bioresource Technology*, Vol. 98, p. 3579, 2007.
42. L.S. Serafim, P.C. Lemos, R. Oliveira, and M.A.M. Reis, *Biotechnol. Bioeng.*, Vol. 87, p. 145, 2004.
43. Y. Song, Q. Zheng, and Q. Zhang, *Journal of Cereal Science*,Vol. 50, p. 376, 2009.
44. Y. Song, and Q. Zheng, *Journal of Cereal Science*,Vol. 48, p. 77, 2008.
45. M.C. McCann, B. Wells, and K. Roberts, *J. Cell. Sci.*, Vol. 96, p. 323, 1990.
46. M.C. McCann, B. Wells, and K. Roberts, *J. Microsc., Vol.* 166, p. 123, 1992.
47. B. Wang, and M. Sain, *Polym. Int., Vol.* 56, p. 538, 2007.
48. C.W. Kim, D.S. Kim, S.Y. Kang, M. Marquez, and Y.L. Joo, *Polymer*, Vol. 47, p. 5097, 2006.
49. A. Frenot, M.W. Henriksson, and P. Walkenström, *J. Appl. Polym. Sci.*, Vol. 103, p. 1473, 2007.
50. K. Ohkawa, S. Hayashi, A. Nishida, H. Yamamoto, and J. Ducreux, *Textile Research Journal*, Vol. 79, p. 1396, 2009.
51. A. Alemdar, and M. Sain, *Composites Science and Technology*, Vol. 68, p. 557, 2008.
52. A. Alemdar, and M. Sain, *Bioresource Technology*, Vol. 99, p. 1664, 2008.
53. A. Bacic, P.J. Harris, and B.A. Stone, "Structure and function of plant cell walls," in: J. Preiss, P.K. Stumpf, E.E. Conn, (Eds.), The Biochemistry of Plants, Academic Press Inc., CA, Vol. 14, p. 297, 1988.
54. F.A.L. Clowes, and B.E. Juniper, *Plant Cells*, Blackwell Scientific Publications Ltd, Oxford, p. 203, 1968.
55. T. Zimmermann, N. Bordeanu, and E. Strub, *Carbohydrate Polymers*, Vol. 79, p. 1986, 2010.
56. M. Shoda, and Y. Sugano, *Biotechnology and Bioprocess Engineering*, Vol. 10, p. 1, 2005.
57. R.M. Brown, *Abstr. Pap. Am. Chem. Soc.*, Vol. 227, p. U303, 2004.
58. D. Klemm, D. Schumann, F. Kramer, N. Hessler, M. Hornung, H.P. Schmauder, and S. Marsch, *Polysaccharides*, Vol. 205, p. 49, 2006.
59. M. Iguchi, S. Yamanaka, and A. Budhiono, *J Mater Sci*, Vol. 35, p. 261, 2000.
60. E.E. Brown, and M.P.G. Laborie, *Biomacromolecules*, Vol. 8, p. 3074, 2007.
61. A.N. Nakagaito, and H. Yano, *Appl. Phys. A-Mater. Sci. Process*, Vol. 80, p. 155, 2005.
62. D.J. Gardner, G.S. Oporto, R. Mills, and M. Samir, *J Adhesion Sci Technol*, Vol. 22, p. 545, 2008.
63. E.M. Teixeira, A.C. Corre^a, C.R. de Oliveira, and L.H.C. Mattoso, *Cellulose*, Vol. 17, p. 595, 2010.
64. W. Stelte, and A.R. Sanadi, *Ind. Eng. Chem. Res.*, Vol. 48, p. 11211, 2009.
65. A. Meftahi, R. Khajavi, A. Rashidi, M. Sattari, M. E. Yazdanshenas, and M. Torabi, *Cellulose*, Vol. 17, p. 199, 2010.
66. A. Bhatnagar, and M. Sain, *Journal of Reinforced Plastics and Composites*, Vol. 24, 1259, 2005.
67. G. Chen, and H. Liu, *Journal of Applied Polymer Science*, Vol. 110, p. 641, 2008.
68. Y.Z. Wan, H. Luo, F. He, H. Liang, Y. Huang, and X.L. Li, *Composites Science and Technology*, Vol. 69, p. 1212, 2009.

69. H. Yano, J. Sugiyama, A.N. Nakagaito, M. Nogi, T. Matsuura, M. Hikita et al, *Adv. Mater.*, Vol. 17, p. 153, 2005.
70. W. Gindl, and J. Keckes, *Compos. Sci. Technol.*, Vol. 64, p. 2407, 2004.
71. A. Iwatake, M. Nogi, and H. Yano, *Composites Science and Technology*, Vol. 68, p. 2103, 2008.
72. A. Kaushik, M. Singh, and G. Verma, *Carbohydrate Polymers*, 2010, doi:10.1016/j.carbpol.2010.04.063.
73. B. Wang, and M. Sain, *Bioresources*, Vol. 2, p. 371, 2007.
74. D.G. Hepworth, and D.M. Bruce, *J. Mater. Sci.*, Vol. 35, p. 5861, 2000.
75. M.A. Hubbe, O.J. Rojas, L.A. Lucia, and M. Sain, *Bioresources*, Vol. 3, p. 929, 2008.

17

Nanocomposites Based on Starch and Fibers of Natural Origin

Kestur Gundappa Satyanarayana[1], Fernando Wypych[2], Marco Aurelio Woehl[2], Luiz Pereira Ramos[2] and Rafael Marangoni[2]

[1]Acharya R & D Center, Acharya Institute of Technology, Bangalore, India, BMS College of Engineering and Poornapragna Institute of Scientific Research, Bangalore; India Formerly Visiting Researcher, Federal University of Paraná, Curitiba, Brazil
[2]Research Center for Applied Chemistry (CEPESQ), Department of Chemistry, Federal University of Parana (UFPR), –Curitiba, Brazil

Abstract

Persistent search for new materials to meet growing needs of human life has led to increasing interest for the development of high performance materials such as nanocomposites, which exhibit appropriate properties to meet the requirements of emerging applications. Together with such developments, environmental safety by the use of renewable materials such as biopolymers and natural materials will have great impact on the world economy and on the improvement of our quality of life. These become important driving forces of technological advancement world over. This Chapter discusses nanocomposites based on starches and biofillers highlighting some details about biopolymers and nanosized natural fibers, as well as processing methods of bionanocomposites using these. Some results on structure, properties applications and potentials of bionanocomposites are also presented, which lead to the opportunities they provide with offering new technology and business opportunities for all sectors of industry in addition to being environmentally friendly.

Keywords: Nanocomposites, natural fibers, biopolymers, cellulose whiskers, morphology, properties, applications

17.1 Introduction

17.1.1 Historical Developments

There has been worldwide growing search for new materials from time immemorial starting from Stone Age days. This has led to the development of some advanced materials and their processing methods. These have great impact on the world economy and on the improvement of our quality of life. It is therefore not surprising that certain eras are named after such important and dominant materials as shown in Scheme 17.1.

Srikanth Pilla (ed.) Handbook of Bioplastics and Biocomposites Engineering Applications, (471–510)

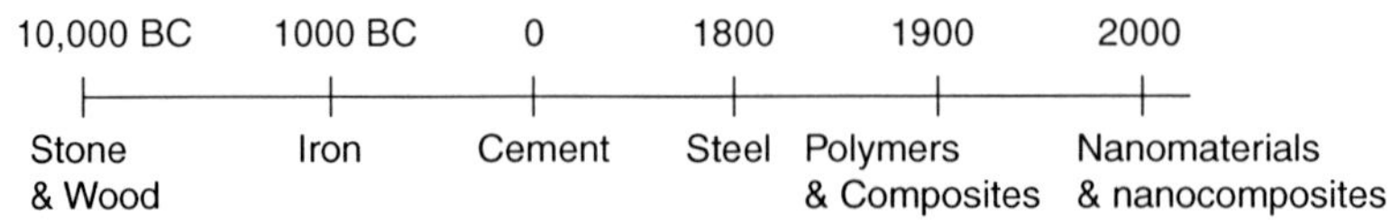

Scheme 17.1 Time scale of materials development.

Accordingly, search for new sources of raw materials has been on the rise particularly due to the low premium on availability as many of them being non-renewable (such as petroleum based feedstocks), growing cost of some of these conventional materials used, and increasing environmental concerns leading to governmental laws and societal push. One such man made material is "composite," which pervades almost all application fields required by the society starting from human habitat, furniture, food packaging, transportation, entertainment, etc. Composite materials are macroscopic combination of two or more distinct materials, having discrete and recognizable interface separating them and exhibiting synergetic effects in their properties. However, IUPAC has defined composites as "Multicomponent material comprising of multiple, different (non-gaseous) phase domains in which at least one type of phase is continuous." These materials have been extensively developed since the 20th century due to the limitations of performance of monolithic materials. Starting from almost 200 B.C. and later till.1200 A.D. onwards, composite materials have gone through significant progress in terms of developments and use of different raw materials, processes and even applications. Some important milestones in this field are listed in Table 17.1.

Composites have gone through at least "five generations." In the early days of their development (first generation), composites were produced using either petroleum feedstock based materials for both matrix (polymers) and reinforcements (carbon fibers, etc) or other nonrenewable materials such as silica (glass fibers). Most of these were produced by conventional plastic processing methods. Many applications demanded combination of properties, which could be obtained by a combination of either two polymers or a number of reinforcements. Such composites were called "hybrid composites," which may be termed as "second generation" composites.

Since all these were based on petroleum products, therefore nonrenewable and expensive particularly for civilian applications, use of natural fibers and other renewable resources attracted the attention of scientists and technologists. This development may be termed "third generation." However, the concomitant increase in the use of plastic-based materials resulted in the generation of wastes also leading to environmental pollution. This attracted the attention of several governments resulting in bringing suitable laws for the use of renewable resources particularly in large scale applications such as automotives. This has led to the search/development of new polymers from natural sources in addition to increased use of natural fibers, which led to the development of biodegadable composites, which may be termed as "fourth generation."

Simultaneously, during 90s, other development was taking place demanding multifunctional materials with paradigm shift taking place from structural materials to functional materials. This was due to limitations on properties and performance of all the existing conventional and processed materials, including composite systems despite the tremendous strides made in the development of new materials for

Table 17.1 Some landmarks in manmade composites.

Year	Land Mark
200 B.C.	Reinforced clay like deposits of Nile valley with grass plant fibers for use in temple walls, tombs and houses—Egyptians Watchtowers of the far Western Great wall of China-supposed to have been built with straw reinforced bricks—Han Dynasty
1500 B.C.	Use of laminated wood furniture-high quality wood veneers bonded to the surfaces of cheaper woods—Egyptians
1200 A.D.	Bows made with adhesively bonded laminates of horns of buffaloes or antilope, wood or silk and ox-neck tendons—Mangolians Cloissone ware—wire reinforced ceramics—Ancient China
Beginning of 20th Century	Phenolic resin-asbestos
1942	First Fiber-Glass Boat/Fiber Reinforced Plastic (FRP) in Aircraft and Electrical Components
1946/50	Invention of Filament Winding and Its Use for Missile Applications
1960s	First Introduction of B_f and High Strength C_f
1968	Use of advanced Composites in Aircrafts
1970	Development of Kevlar fibers and introduction of Metal Matrix Composites (MMCs)
1980	Significant Increase in Utilization of High Strength and High Modulus Fibers
1990 onwards	Newer MMC/Ceramic Matrix Composite (CMC)/Bio and Nanocomposites/New Areas of Applications/Discovery of Carbon Nanotubes (CNT)
2003–04	Textile containing two Carbon nanotube fiber super capacitors woven in orthogonal direction/Composites with CNT

all types of applications. In this context, nanocomposites have emerged as suitable alternatives to overcome these limitations just like the emergence of micro composites to overcome the limitations of monolithics. These materials can be termed "fifth generation" composites, whose arrival on the horizon of materials may be one of the latest landmarks in such developments. In fact, from the first reports on nanocomposites by Japanese workers in 1990, these materials are emerging as materials of 21st Century (1, 2). This is because they exhibit certain unusual property combinations and design uniqueness that is not found in conventional composites. Besides, they pose challenges in their preparation through control of elemental composition and stoichimetry of the nanocluster phase. In view of these attractive attributes, increasing research is directed towards the synthesis of nanomaterials in recent years (3). For example, the number of publications related with nano-prefixed word such as nanoscience, nanotechnology, nanomaterials, etc. have increased over the years with about 18,800 papers published so far (1986–2010) as per the literature found through the ISI Web of Knowledge. Of these, about 2300 are during the current year-to-date (8th September 2010) including special issues of some journals that are devoted exclusively to this emerging science and technology of nanomaterials. Similarly, the number of patents filed on nanocomposites during the same period is about 1200. Further, based on the ISI Web of Knowledge, the number of publications has also shown a steady increase over the years with almost none from 1986–1993 and then to about 100 from 1993-2010 in the case of biopolymers-based nanocomposites, compared to about 5820 in the same period (4).

It is interesting to note that starch, one of the most abundantly and easily available natural polymers at low cost, possibility of its plastisizing and its processing by conventional methods used for polymeric materials, has attracted much attention for the development of biodegradable composites. But it has some negative attributes such as high water sensitivity, limited mechanical properties (5) and high brittleness. Application of nanotechnology to this material may open up new avenues with possibly improved properties in addition to being low cost efficiency (6). Similarly, natural fibers, which are the sources of cellulose exhibiting unusual strength properties, have received increased attention due to their renewability, low cost and environmental concerns that are associated to the use of synthetic fibers to develop biodegradable composites. In fact, these cellulose fibers are known to improve the properties as in the case of synthetic fibers (7), but their impact on barrier properties (lowering of oxygen permeability) was not reported till recently for a biocomposite of PLA and cellulose microcrystals (8, 9).

Considering the above discussions, this Chapter presents the fundamentals about nanocomposites, biomaterials, bionanocomposites based on starches and biofillers. Some details about biopolymers and nano-sized natural fibers will be highlighted. Besides, processing methods for preparing nano-biocomposites and some results on structure and properties will also be presented together with potential applications of bionanocomposites. This will underline possible opportunities for the developing new technology and business opportunities for all sectors of industry such as construction, food packaging, transportation (particularly automotives), etc., in addition to being environment friendly.

17.1.2 Nanocomposites

Though "nano" level is the first level of all biological and man-made systems, Richard Feyman was the first to introduce the concept of "nanotechnology" in his celebrated lecture of 1959 at the American Physical Society (10). From then, on nanotechnology, which can be defined as creation, characterization and use of materials structures, devices and systems possessing superior properties resulting from the nanostructures, has developed into a multidisciplinary area of applied science and technology (11). Nanocomposites are two phase materials (both being in solid phases) in which at least one of the phases has dimensions in the nanometer range (1-100nm; 1 nm = 10^{-9} m). The nanometer-size features of one of the components are reflected in superior electronic (affecting optical, catalytic, etc), physical, mechanical, thermal, gas barrier and other properties of these new materials leading to their some potential applications. This is expected since it is well known that, when the feature size is less than a particular level, called "the critical size," changes would result in many properties (12) and unusual property combinations are observed without sacrificing other properties, even at very low concentrations (~5wt. %) of reinforcements. Compared to conventional composites, they even provide better recyclability and low density (8). Besides, a large improvement in the interaction at the interface between the different phases is expected as dimensions reach the nanometer level. Hence, understanding of the structure–property relationship of these materials depends on the surface area/volume ratio of the reinforcement materials. Another new and interesting dimension emerging in the area of nanocomposites, making further inroads for their processing and applications, is the discovery of carbon nanotubes (CNTs) in 1991 (13), the possibility to spin these into composite products and textiles (14) and the use of CNTs to fabricate composites exhibiting at least one of the unique mechanical, thermal and electrical properties of CNTs (15–17). In addition to being environment friendly, all the above mentioned attributes are leading science and technology of nanocomposites offering new technology and business opportunities for aerospace, automotive, electronics, packaging and biotechnology applications (18).

17.1.3 Biopolymers

The worldwide consumption of plastics and polymers may reach about 230 million tons by 2015 according to the plastics industry, the majority of which will be derived from limited reserves of fossil fuels [Plastics industry: http://www.Cipet.gov.in / plastics_statics. html-Accessed on 9th Aug.2010]. If they were to be disposed to environment after use (for example, 70% of the short term products such as packaging materials are thrown in landfills), it would lead to pollution. This has resulted in several governments resorting to regulations for the safety of environment. Recognizing the pressing need to reduce the human impact on environment and the constant increase in fossil fuel costs, alternate chemistry has been developed during the last few years to develop environment friendly and biodegradable systems with better properties. This includes polymers, which are the backbone of many applications, leading to the search of biopolymers, which are naturally occurring polymers such as starches, polylactides, polyhydroxyalkanoates, etc. These biopolymers may be renewable and/or fossil resources based or only renewable resource based. The latter is produced with low energy consumption and thus become appropriate materials to replace conventional petroleum-based polymers particularly for short-life range of applications such as food packaging, agriculture, etc. These materials have advantages over the conventional polymers such as renewability, biodegradability and biocompatibility. Many of these, though expensive, are produced commercially and many products have been developed, manufactured on industrial scale and are being used (19).

There are a number of definitions of biopolymers, some of which are ambiguous. According to ASTM standard D-54894d, these are biodegradable materials, which can undergo decomposition either by enzymatic degradation or by microorganisms (20). Accordingly, biopolymers have been classified into four main categories depending on their synthesizing methods (21), which is shown in Figure 17.1 (21).

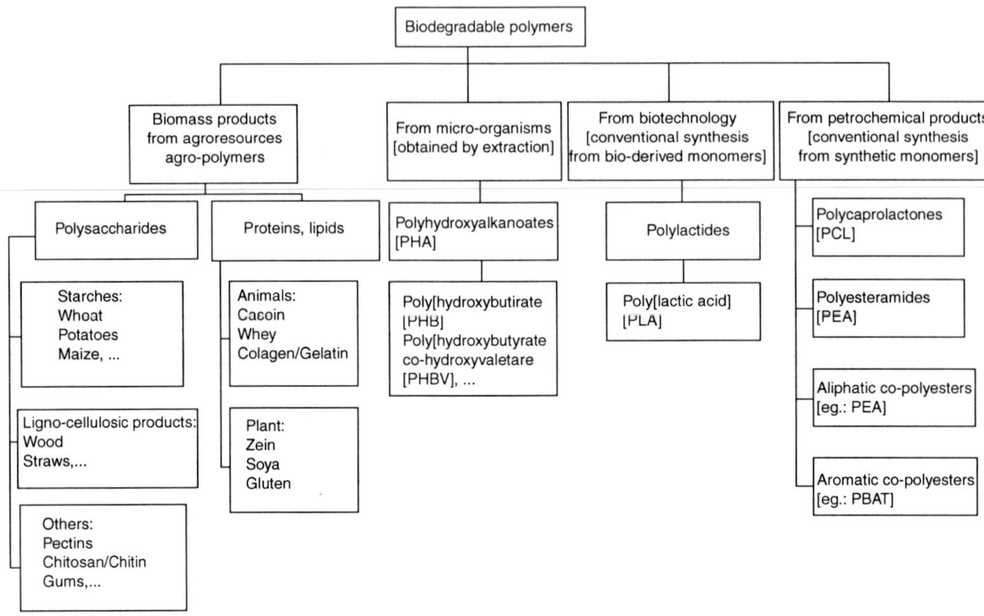

Figure 17.1 Classification of the biodegradable polymers and their nomenclatures. Adapted from Ref.21.

17.1.4 Market, Perspectives, Potentials of and Opportunities in Bionanocomposites

Let us first look at the possible market, potentials and the general opportunities bionanocomposites with polymer matrices provide before going into details regarding their processing, structure, properties and applications.

It is well known that biopolymer materials are getting wider acceptability for use in industry due to their unique characteristics such as renewability and lightweight, although presently they are expensive. However, they have some disadvantages, such as brittleness and water absorption. A very effective approach for their wider use will be to improve mechanical properties through blending or by the addition of nano sized fibers, whiskers, platelets or particles as reinforcements as is being done with conventional polymeric materials. This is due to the fact that nanoscale reinforcements have an exceptional potential to generate new phenomena, which leads to special properties in these materials as will be seen later. These will exhibit unusual increase in various properties such as heat resistance, mechanical strength and impact resistance, electrical conductivity and gas-permeability concerning oxygen and water vapor, etc. In fact, it is surprising to note that even the lignocellulosic fibers at nanosizes would provide gas barrier properties (8, 22). Besides, the reinforcing efficiency of these composites, even at low volume fractions, is comparable to 40–50% for fibers in microcomposites (23). Such nanosized additions of reinforcements to a wide variety of biopolymers produces a dramatic improvement in their properties including biodegradability. This underlines a good example of polymer matrix bionanocomposites as promising systems for ecofriendly applications. It is also pertinent at this juncture to point out that information on nanomaterials, nanoindustries and a host of possible A to Z applications of polymer nanocomposites are available (24).

Bionanocomposites is an emerging class of extraordinarily versatile materials that can be prepared by a combination of a variety of largely available natural polymeric matrices and reinforcements/fillers, both from renewable resources. Recent published reports (4) have shown that composites even stronger than the existing materials derived from fossil fuels can be obtained only by the right combinations of matrix and reinforcement. Accordingly, there have been many attempts in recent times (4, 24) to develop many products for different applications of nanobiopolymer composites compared to well developed and marketed biopolymer products. Some of these include automotive parts, electronic devices smart papers communication devices, packaging materials for food and pharmaceuticals, ID tags and flexible substrates for portable and foldable display systems. These indicate that the market and perspectives are very promising for the future not only from the point of view of the abundant resources and exciting properties that can be tailored, but also due to the biocompatibility and biodegradability. Taking into consideration an estimated 100,000 metric tons per annum of lignocellulosic fibers for composites in the European automotive sector alone (25, 26), compared to the estimated global market of about 900,000 metric tons of wood plastics and natural fiber composites (27), an exciting and bright market with high potentials can be expected. Besides, bionanocomposites will be highly acceptable by the people at large, since they can be easily disposed after use, thus stimulating the development of beneficial microorganism for fertilizing the soils.

17.2 Biomaterials

17.2.1 Cellulose

Cellulose is the most widespread and one of the largest occurring natural biopolymer. It is available in most of the living species including plants, animals and, to a lesser degree in algae, fungi, bacteria, invertebrates and even amoeba (protozoa), mainly acting as reinforcing agent (28–30). After Anselme Payen discovered and isolated cellulose in 1838, a number of studies have been continuously made with respect to its biosynthesis, assembly, and structural features that have inspired a number of research efforts among a broad number of disciplines. Several physical and chemical aspects of cellulose have been studied (31).The size of the cellulose molecule is normally given in terms of its degree of polymerization (DP), i.e., the number of anhydroglucose units present in a single chain and its value is usually estimated in higher plants 8,000 to 10,000 (32). This determines significantly the mechanical properties of plant fibers (33). However, the conformational analysis of cellulose (Figure 17.2) indicated that cellobiose (4-O-β-D-glucopyranosyl-β-D-glucopyranose) rather than glucose as its basic structural unit (34, 35).

Cellulose, being a renewable organic material, is produced in the biosphere at an estimated annual production over 7.5×10^{10} tons (30).

Cellulose has 6 different polymorphs of which the first one termed "Cellulose I" is called "native cellulose." Due to its important characteristics, this cellulose has

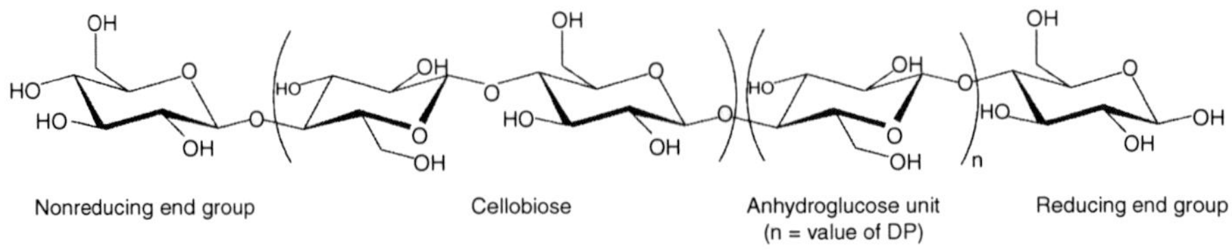

Figure 17.2 Cellulose structure. (Modified from references 34, 35).

been the subject of intense research starting from elucidation of its crystal structure to its final use after suitable preparation by scientist all over the world (28).

It is interesting to note that various crystal structure have been proposed for the cellulose by different researchers, mostly due to its origin and the experimental conditions in which they are produced from the source. Further, the cellulose is found to consist of both crystalline (about 80%) and amorphous parts. The latter is due to the disordered domains, while the native cellulose is semi-crystalline. In fact, the crystalline part of cellulose is tightly packed slender-rod like structure (sometimes termed "microfibrils"), while the amorphous cellulose is loosely packed and has a lower density (36). According to Samir *et al.*, 2005 (28), the first report on the existence of crystalline zones inter-spread in an amorphous structure was made as early as in 1877. Attempts have been made to separate these two parts, which give two types of nano-sized materials for further utilization of cellulose as reinforcement in polymeric and other matrices. The cellulose available from plants and animals resources are in the form chains, which can be biosynthesized in to continuous form as well as into aggregates to form what are known as "microfibrils." These microfibrils will be long thread-like molecular bundles with lateral stabilization by hydrogen bonding. Figure 17.3 shows the internal structure of a cellulose microfibril.

Dimensions (length, width/diameter/thickness) of these cellulosic microfibrils vary with the origin of cellulose within 2–20nm diameter and length in micron levels. Accordingly, their aspect ratio (l/d) varies from 1–100 with colloidal behavior. During the biosynthesis process, all the chains in microfibrils will have to be elongated at the same rate thus suggesting the range of load carrying capacity of these microfibrils. Accordingly, reported values of Young's modulus (YM) and tensile strength (TS) of cellulose indicate that they are dependent on the source of cellulose although earliest values reported (38) are of the order of 130GPa for modulus (YM), while the recent ones mentioned in the literature are of the order of 150GPa for YM and 10 GPa for TS (28). More details on two types of cellulose will be discussed in a later section (2.3).

17.2.2 Bio Matrix Materials

17.2.2.1 Starch

Starch is the main reserve carbohydrate in plants, being responsible for 70–80% of the calories in human alimentation. This native starch normally occurs in the form

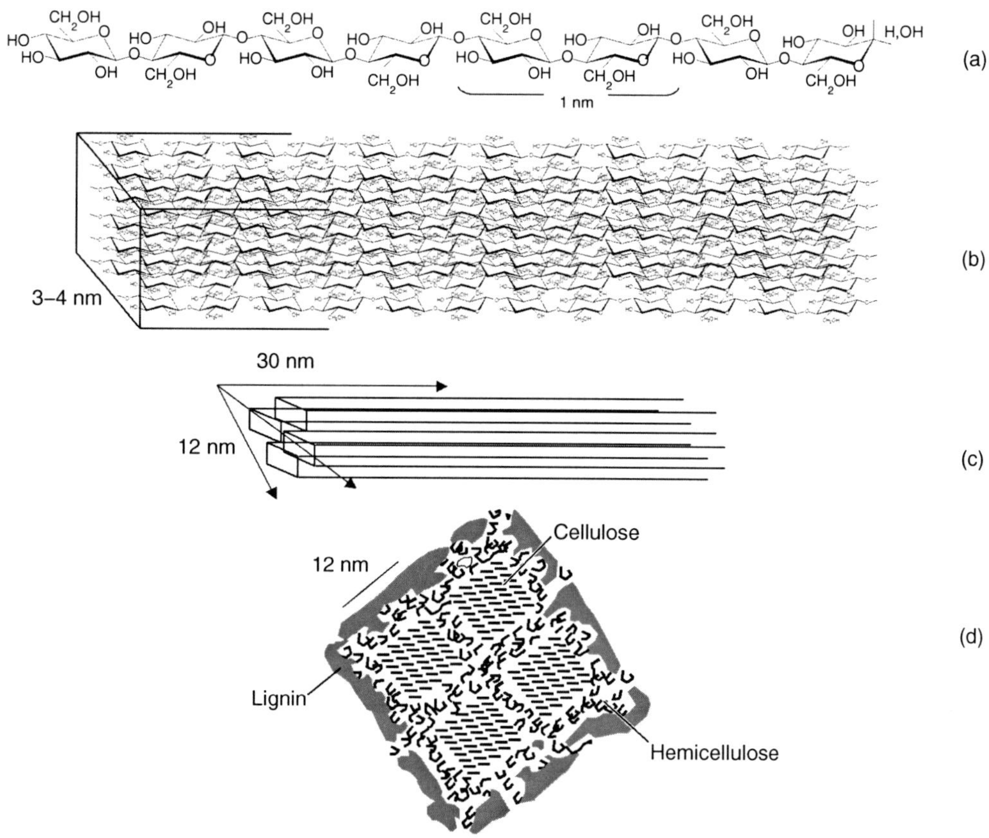

Figure 17.3 Chemical association in the plant cell wall: (a) the cellulose backbone, with an indication the length of its basic structural unit, cellobiose; (b) framework of cellulose chains in the elementary fibril; (c) cellulose crystallite; (d) microfibril cross section. [Reproduced with the kind permission of the Publishers of Ref. 37].

of discrete and partially crystalline microscopic granules, which are held together by an extended micellar network of associated molecules. Starch is composed of two polysaccharides: *amylose* and *amylopectin*. The former shown in Figure 17.4a is an essentially linear polymer composed of α-D-glucopiranose units linked mainly by α-(1→4) linkages, with a very small amount (0.3–0.5%) of α-(1→6) linkages.

Amylose chains tend to form a single helix on which the hydroxyl groups are turned outward, which is responsible for its hydrophilic character.

The interior of the helix is lipophilic and form a dark blue complex with poly-iodine ions in aqueous solution. This interaction allows a distinction to be made between amylose and amylopectin and the determination of the amylose content on native starch (39–43).

On the other hand, the latter (*amylopectin*) shown in Figure 17.4b is a highly ramified polymer, containing 5-6% of α-(1→4) linkages. The amylopectin content in native starch varies with the plant source. In cereal endosperms, amylopectin accounts for 72–82% of the starch mass. However, starch from some

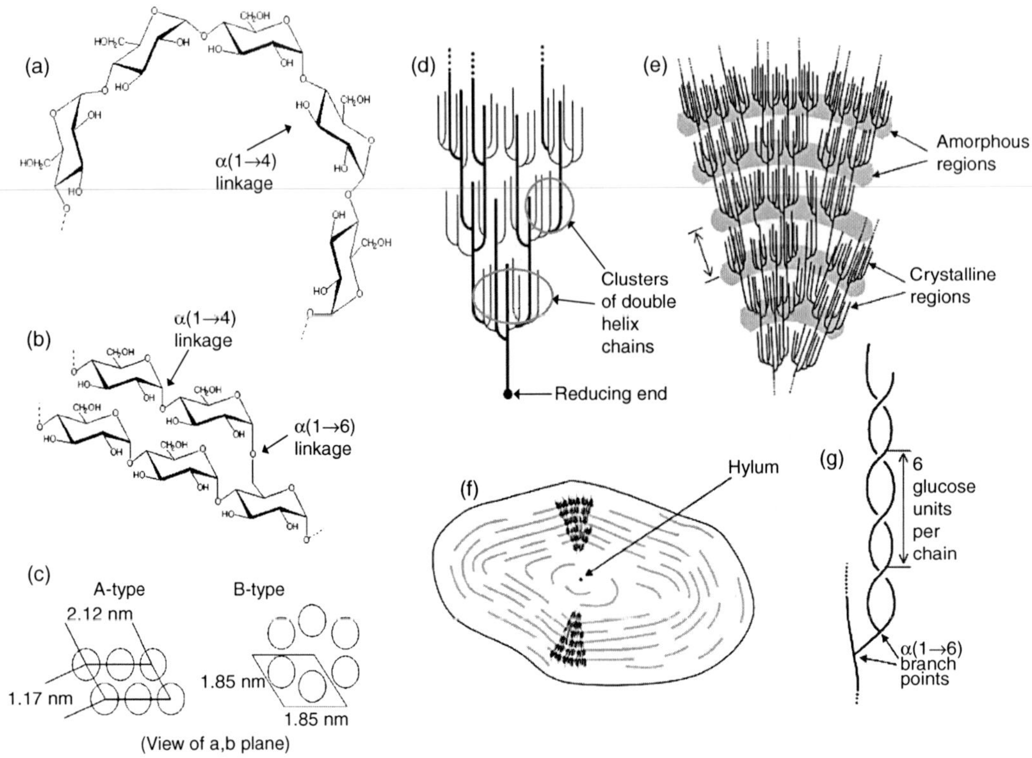

Figure 17.4 Structural organization of starch. (a) Amylose; (b) amilopectin; (c) A- and B-type amylopectin crystalline packing (with unit cells a and b parameters); (d) organization of the adjacent ramifications in a amylopectin chain; (e) packing of the chains in a starch granule; (f) Starch granule; (g) Amylopectin double helix. (Adapted from 46 & 47 and Modified).

wheat genotypes contains more than 70% amylose and the starches termed *waxy* are constituted almost totally of amylopectin, containing less than 1% amylose (42–44).

In the plant cells, the packing of the chains in the crystalline regions occurs either in the *A-type* crystalline pattern, predominant in cereal starches, or in the *B-type* crystalline pattern, common in root starches. The A-type crystals are more compact, with 8 molecules of water per unit cell. The B-type packing is more open, with a hexagonal arrangement containing 36 molecules of water per unit cell (Figure 17.4c). The C-type starch, found in cassava roots and in leguminous seeds, is a combination of the A- and B-type allomorphs. Each of these crystalline patterns is easily characterized by X-ray diffraction (39, 43, 45, 46).

Furthermore, starch is organized in *granules* (Figure 17.4f) ranging from 1 to 100 μm in diameter. The inner structure of these granules with alternate crystalline and amorphous regions constituted mainly of amylose (Figure 17.4e). In the latter, amylopectin organizes in double helixes (Figure 17.4g), packed antiparallel (Figure 17.4) (39, 42, 44, 46).

Some of the properties of starches mostly depend on amylopectin. For example, although linear, the amylopectin shows a conformation that hinders its regular association with other chains. Thus the crystallinity of the starch granules is attributed to this characteristic. However, in the presence of hot water, the starch grains are hydrated in a process called *gelatinization*. Initially, the grains are swollen and subsequently disrupted and the subsequent release of the starch chains in the aqueous medium results in the formation of a gel. On cooling, the gel undergoes *retrogradation*, which results in the formation of a compact three dimensional network whose structure is based on the interaction between adjacent starch chains (45, 46, 48).

17.2.2.2 Thermoplastic Starch (TPS)

Starch, in granular form, is not a thermoplastic polymer and hence exhibits poor melting processability. Also, it is highly water soluble and difficult to process. Films formed exclusively with this starch will be brittle due to anarchical growth of one of their constituents (amylopectin and amylose). The granular structure of starch can be disrupted by heating or by adding plasticizers. (49, 50). The plasticizers (e.g., water or glycols) interact with the polymer chains, reducing the interaction between adjacent chains and enhancing film flexibility (44), thus making starch suitable for engineering applications such as extrusion, injection or molding (49, 50). However, the addition of plasticizers lowers the degree of crystallinity of the films [48] and increases the permeability to gases and liquids (51).

The above material thus obtained is termed *"thermoplastic starch"* (TPS). The TPS can be produced by dry mixing and heating/shearing or by a casting process, in which an aqueous dispersion of starch and plasticizer is gelatinized in the dry form to get a thin film (52, 53).

In this process of making starch flexible, the melting point and the glass transition temperature (Tg) of starch decrease. Accordingly, various factors influence the morphology and mechanical behavior of TPS. These include processing conditions, amylose-amylopectin ratio, type and amount of plasticizer and moisture content. For example, the production of TPS films under high temperature and low relative humidity result in a low degree of crystallinity of amylopectin and amylose. In general, the degree of crystallinity of the films increases with amylose content (48). Amylose films have better ductility and barrier properties than amylopectin films (51). Also, depending on the source and processing conditions of starch, it is possible to tailor the mechanical properties of these thermoplastics to suit structural applications by using blends or fillers (e.g., Cellulose micro/nanofibrils and natural "macro" fibers) (54–56).

When water is used as a plasticizer of starch, it should be in very small amounts since its high mobility in the matrix (starch) impedes its practical utilization (48, 51). The mechanism of plasticization of biopolymers by water is based reducing the hydrogen bonds between macromolecules and also in the reduction of dipole-dipole interactions due to alterations in the dielectric constant (57). On the other hand, glycerol is the mostly used plasticizer in the TPS production due to its availability, biodegradability and affinity with starch (48, 51, 58–61). The glycerol content affects the TPS properties. Figure 17.5 shows schematic representation of the interactions between the components of TPS and TPS-cellulose nanocomposites, where Figure 17.5a shows the plasticizing effect of glycerol. Low contents of glycerol would lead to an antiplasticization effect (Figure 17.5b). If the amount of plasticizer is insufficient to provide extra free volume, the interactions between plasticizer and polymer hinder the mobility of chain segments and the resulting film will be brittle (48, 62). A high content of glycerol (Figure 17.5c) leads to the segregation of a glycerol-rich phase (63).

The reorganization of the starch network upon ageing in moist conditions leads to a new crystalline pattern, the V_H-*type*, resulting from alignment of single helix chains (64, 65). The presence of cellulose nanostructures in the interior of the matrix also alters the molecular organization of TPS. An amylopectin crystalline region may be formed in the starch-cellulose interface (Figure 17.5e). The glycerol may also accumulate in that interface (Figure 17.5f). This competitive interaction between the components reduces the structural reinforcement expected for the addition of cellulose (60).

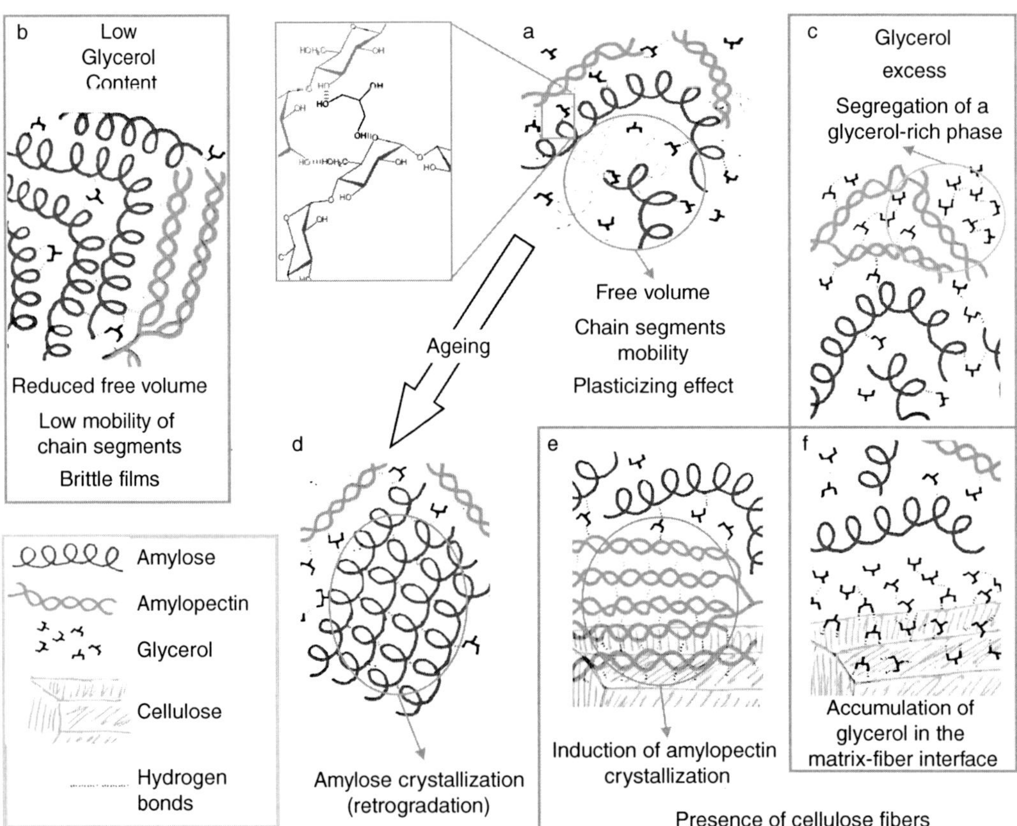

Figure 17.5 Scheme of the interactions between the components of the TPS and TPS-cellulose nanocomposites. (a) plasticizing effect of glycerol; (b) antiplasticizing effect of low glycerol concentrations; (c) segregation of a glycerol-rich phase in high glycerol concentrations; (d) retrogradation of amylose upon ageing; (e) formation of transcrystalline amylopectin regions induced by the presence of cellulose nanostructures; (f) accumulation of excess glycerol in the cellulose-matrix interface (66).

The plasticizing effect has been explained by the free volume theory (67). According to this theory, the plasticizer generates free volume regions that allow higher mobility of polymeric chain segments (Figure 17.5a). The chemical affinity of a substance with the polymer is a necessary characteristic to be a plasticizer. In the case of starch, this condition is satisfied by polyols like glycerol or sorbitol, whose hydroxyls establish hydrogen bonds with starch hydroxyls (44).

The other method of modification to make useful materials for structural applications is by blending with other polymers. This has been studied by many researchers over the years. These include twin-screw extruded blends of wheat starch with PLA or low density polyethylene (LDPE), poly(lactic acid) (PLA) or LDPE with starch granules (63, 68). In both cases, both glycerol and water were used as plasticizers, which resulted in retention of biodegradability in the former and increased modulus with decreased ductility in the latter. Thus, plasticization can lead to a wide range of properties, including variable biodegradation, depending on the plasticizer content (glycerol and water) and its dispersion in blending polymer such as PLA. More details on this can be seen in the references mentioned. Similarly, potato starch with or without plasticizer but blended with PHB produce films with good thermal stability, while Tg showed slight increase. This showed marked increase along with higher thermal stability when the thermoplastic was used (69).

Due to the inherent lower strength and stiffness, the TPS shows relatively lower mechanical properties compared to conventional polymers (70). For example, YM, UTS and strain at break for TPS range from 75–160 MPa, 5.5–7.5 MPa and 75–180% respectively, compared to 210–2700 MPa, 55–11,000 MPa and 120–380 for some of the petroleum based polymers such as LDPE, HDPE, PP and PET.

17.2.3 Cellulose Based Nano-bioreinforcements/Fillers

In principle, almost any cellulosic material can be a potential source of producing nanofibers. Thus, two types of nano-reinforcements have been obtained using the cellulose from different sources, namely, microfibrils and whiskers. The former are obtained from the amorphous regions of cellulose, while the latter are from the crystalline part of the microfibril. The whiskers are pure single crystals obtained under controlled conditions. These are highly ordered structures and hence exhibit unusually high strength and significant changes in other physical properties such as electrical, optical, conductivity, etc. Various authors have called whiskers as "nanorods," "rod-like cellulose microcrystal," "cellulose nanocrystals" and "microfibrillated cellulose (MFC)."

Agriculture residues from which the microfibrils and whiskers can be obtained include softwood (e.g. pine, *Picea abies* of Norway), sugarcane bagasse, cassava bagasse, rice hull, maize straw, sugar peet, potato tuber, etc. (71). On the other hand, animal based microfibrils are obtained from an edible tunican (a sea animal) and chitin, while bacterial cellulose (BC) is produced when *Acetobacter* species are cultivated in the presence of nutrients and sugar. This is a popular material for medical applications (72).

17.2.3.1 *Plant-based Cellulose*

Plant-based cellulose is a fibrous, tough, water-insoluble substance that plays an essential role in maintaining the structure of plant cell wall (30). It is the major component in the rigid cell walls in plants, where it occurs as a linear homopolysaccharide consisting of glucose (D-glucopiranose) units linked together by β-(1-4) glycosidic bonds (β-D-glucan), consisting of carbohydrate polymers of tens to hundreds to several thousand monosaccharide units. All of the common polysaccharides synthesized by plants and animals as well as by humans to be stored for food, structural support, or metabolized for energy, contain glucose as the main monosaccharide unit (73).

Several models have been proposed to explain the internal structure of cellulose within the plant cell wall. In general, the cell wall is subdivided into middle lamella, primary wall and secondary wall. The distribution of cellulose, hemicelluloses and lignin varies considerably among these layers (74). The middle lamella is the outermost layer and is almost entirely composed of lignin in the mature wood. This layer fixes the cells together and is responsible for the structural integrity of the plant tissue. The primary wall in a mature cell is also highly lignified and consists of a thin layer that is initially deposited around the plant cell. Two adjacent primary walls and the intervening middle lamella are often called the compound middle lamella. Inside the primary wall, the secondary wall is formed by a sequence of three lamellae, S_1, S_2, and S3 (35, 73, 74) where the central layer is usually thicker than the others. As a result, most of the fiber properties, particularly those of interest for the pulp and paper industry, are derived from the characteristics of this layer. Each layer of the secondary wall contains cellulose microfibrils that lie more or less parallel to one another. This common orientation results in a helical disposition that can be characterized according to the angle displayed by the microfibrils in relation to the longitudinal axis of the cell. Since the microfibril angle varies between two adjacent lamellae, a crossed microfibrilar structure is observed. The S_2 layer is generally characterized by small microfibril angles, resulting in a steep helix, whereas flat helices are usually found in the S_1 and S_3 layers (72, 73, 74). It is to be noted that adjacent cellulose chains form a framework of water-insoluble aggregates of varying length and width (Figure 17.3) and these elementary fibrils contain both ordered (crystalline) and less ordered (amorphous) regions, because of the linearity of the cellulose backbone (35, 75). The lattice forces that are responsible for maintaining the crystalline regions are basically the result of extensive inter- and intramolecular hydrogen bonding. According to Fengel and Wegener (35), the cellulosic fibers are constituted of fibrils with an average thickness of 3.5 nm, which can be associated with one another form of cellulose crystallites whose dimensions depend on the origin and treatment of the sample. Four of these basic crystalline aggregates are then held together by a monolayer of hemicelluloses, generating 25 nm wide thread-like structures which are enclosed in a matrix of hemicellulose and lignin (Figure 17.3). The natural composite that results from this close association is referred to as cellulose "microfibril"(73).

Considering the positive aspects including the sustainable and environmental friendly aspects of using the renewable resource-based reinforcements, plant-based fibers have attracted greater attention due to their easy availability being pervasive in all parts of the world. In addition, these fibers have been found to be viable alternates to inorganic/mineral reinforcements to develop composites, provided that processing conditions are proper and higher water absorption is not so critical in their application. Besides they have environmental benefits from the point of view of disposability. One of the specific characteristics of the lignocellulosic compounds is their high density of hydroxyl groups, which provides them a hydrophilic nature. These composites have been successfully used in many applications, particularly in the automotive industry (75, 76–79). Also, their potential in railways, irrigation, furniture and entertainment items are being explored [80]. These plant fibers are referred to by other names such as "cellulosic fibers" as cellulose is one of the main chemical components, or as "lignocellulosic fibers" as they often contain natural polyphenolic polymer called lignin in their structure. These fibers are considered themselves as natural composites with cellulose fibrils being held together by lignin and hemicelluloses (a group of polysaccharides having several sugar units with non-crystalline nature). The cellulose microfibrils have a complex layered structure with two walls, primary and secondary, the latter having three layers of which the middle layer determines the mechanical strength of the fiber (81). This middle layer is consisting of a series of helically wound microfibrils of size 2–30nm (depending on their origin) in cylindrical manner forming about long cellulose molecule (30–100) chains of several tens of microns (28). Schematic representation of simple model of cellulose microfibril has been reported (81).

Although cellulose crystallites are known since 1960s, structured cellulose elements have gained importance only in the last few years in view of the interest in the development of new materials and nanomaterials based on renewable resources (34). These microfibril chains are found to contain only small number of defects and cannot be folded. Therefore, they can be considered highly organized cellulose bundles of high strength linked to the another amorphous domains. Their modulus (~150GPa) as mentioned earlier is close to that of pure crystals of natural cellulose (28).

In general, biomass is subjected to acid hydrolysis to obtain stable aqueous suspensions of cellulose monocrystals. These cellulose colloidal materials can be of different shapes (spheres, rods and even as platelets) (81, 82, 83) and therefore are referred to by different nomenclatures (28) such as "whiskers," "monocrystals," "nanocrystals," "nanocrystallites." These nanocrystals offer a wide variety (from particles to 100) of aspect ratios depending on their source. In view of their colloidal suspension characteristic, they can be used to prepare nanocomposites.

Microcystalline Cellulose (MCC) are commercially available as stable thixotropic-gel systems of hydrolyzed aqueous colloidal dispersions. In fact, these stable gel systems can be obtained when the mechanical energy is introduced into the aqueous suspension of level-off DP (degree of polymerization) cellulose having ~5% or more of total solid content, provided the mechanical energy is so strong that it could liberate minimum number of monocrystals.

The sizes of these cellulose nanomaterials include: spherical shapes with 200nm–2μm in length, 4–15nm in diameter; rod-like highly crystalline nanocrystals obtained from maize starch are in the form of platelets with 20-40nm long, 15–30nm wide and 5–7 nm thick (83, 84). On the other hand, MFC had thickness of

10–100nm with low crystallinity index due to the amorphous domains of cellulose remaining intact. Nanocrystals (whiskers) of cassava bagasse were rod like crystalline nanocrystals with 1150 nm long, 15 nm in diameter giving an aspect ratio of 76. This ratio is high compared to those reported for crystalline whiskers obtained from other agricultural residues such as cotton (10) and wheat straw (~45) and even animal (tunicin) cellulose (~67) (83). Crystallinity index of these crystals is reported to be 54.1%.

Since particulate fillers can improve some of the mechanical properties such as stiffness, creep resistance and impact resistance, attempts have also been made to produce carbon nanospheres of 85% pure graphite from plant cellulose through pyrolysis of cellulose char (85).

These nanospheres are found to be hollow with density value of 1950 kgm^{-3} compared to heavy mineral spherical particulates normally used as fillers in many polymeric matrices to achieve the improvement in the above mentioned properties. Besides, they also affect properties such as polymorphism and rate of crystallization in polymeric materials, which in turn result in the improvement of their mechanical properties. Another advantage of using such carbon nanospheres from renewable materials is the cost of feedstock.

17.2.3.2 *Bacterial Cellulose*

Another variety of cellulose, which has caught the attention of biocomposite researchers, is the "bacterial cellulose" (BC). Known since long by the vinegar producers as vinegar's "mother" or "plant," the BC consists of a gelatinous and translucent pellicle formed in the surface of *Acetobacter* cultures, mainly *Acetobacter xylinus*. A scanning electron micrograph of the pellicle (Figure 17.6) reveals a random network of cellulose microfibrils with a width of less than 100 nm (86).

The cellulose chains are aligned parallel to each other in the so-called "crystalline" regions of the microfibrils. As mentioned in the case of plant cellulose, "amorphous" regions, which occupy nearly 90% of the volume of the microfibrils,

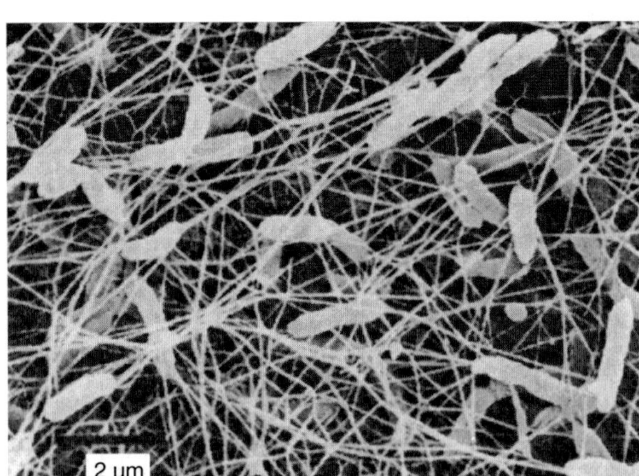

Figure 17.6 Scanning electron micrograph of bacterial cellulose nanofibrils (Reproduced from Ref. 86, with kind permission from Springer Science+Business Media B.V.).

are much less organized and contains cellulose chains that are more dispersed and do not possess any definite orientation. Since these amorphous regions are highly hydrated, they contribute very little to the overall mass of cellulose (87).

The BC consists of a combination of the allomorphic forms, namely, Iα and Iβ (88). It is reported that this Iα cellulose fraction is about 44%, much higher than the fraction present in superior plants, while being close to algae like *Cladophora*. On the other hand, upon enzymatic hydrolysis, the amount of Iβ fraction increases, indicating a higher susceptibility of the Iα allomorph to cellulase attack (89, 90).

Figure 17.7 shows the photograph of the bacterial cellulose mat, before and after drying and a schematic representation of the BC structure.

17.2.3.3 Preparation of Cellulose Microfibrils/Whiskers

Several methods are available for preparing the cellulose microfibrills and whiskers. General principle followed for this is as follows.

Pre-treatment of cellulose fiber bundles by swelling followed by chemical treatments, viz., acid hydrolysis, which removes the pectin and hemicelluloses. Then an alkali treatment is given to remove the lignin. This is followed by mechanical treatments such as cryocrushing, where high impact is used to liberate the microfibrils from the cell walls. Finally, one can resort to defibrillation by high impact and high shear to obtain individual fibers with improved dispersion of fibers.

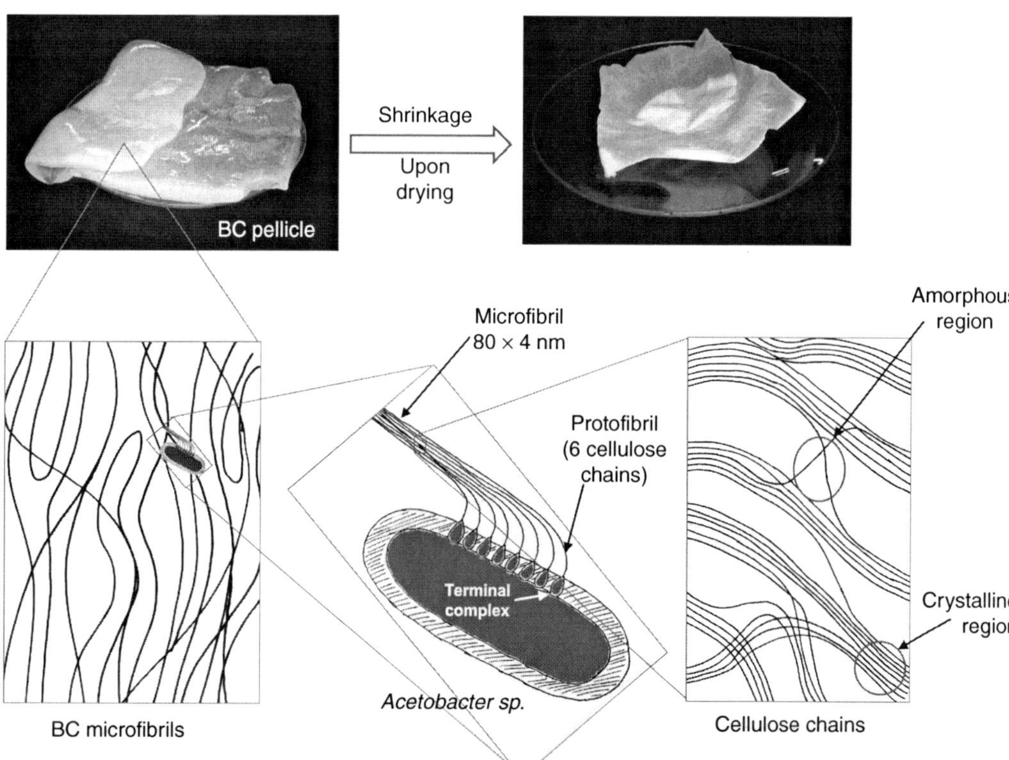

Figure 17.7 Photograph of the bacterial cellulose mat, before and after drying (top). Schematic representation of the BC structure (below) (Reproduced from Ref. 66)

Alternately, combination of chemical, biological and mechanical treatments can also be used with any cellulosic source. Studies made by some researchers are given below:

Saxena and Ragauskas (91) have obtained cellulose whiskers by first grinding the source raw material (cellulose fiber) in a mechanical mill to a size that can pass through 0.5 mm mesh screen. This milled pulp was hydrolyzed with sulfuric acid of certain weight for about 45 min at about 45°C. This resulted in a milky colloidal suspension. After terminating the hydrolysis by the dilution of with deionized water, the suspension was kept overnight at room temperature. Then, the settled cellulosic material was collected with a minimum amount of water. Excess acid was removed by water washing. Then, the material was centrifuged initially at lower speed (3000rpm) and then at higher speed (11000 rpm) for about 20 min. After discarding the aqueous phase, the washing cycle was repeated for another two times. This was followed by dialyzing the sample against water till the pH of the whiskers containing slurry became neutral. The suspension was finally sonicated before keeping it over a mixed bed resin (Sigma TMD 8) for 2 days, followed by filtration and freeze-drying. This process yielded about 35% whiskers. Finally, measured bulk charge on both the starting material and the obtained whiskers were determined.

Samir *et al* (28) have reported another method of processing plant based micro fibrils/whiskers by controlled chemical treatment to destroy the molecular bonds from the network structure in which microcrystals are embedded. Mechanical means are subsequently used to disperse sufficient amount of unattached microcrystals in the aqueous solution.

Vazquez and coworkers (92) have used two different methods to extract cellulose from sisal fibers (93). First, the fibers are washed, chopped and dewaxed before subjecting them to extraction. In the first method, sisal fibers were pretreated with alkali (NaOH) at certain concentration at a warm temperature for a fixed time. This was then treated with hydrogen peroxide, followed by another alkali treatment. Then the material was treated with $Na_2B_4O_7$ and a final treatment with a nitric acid + HAc mixture at higher temperature for a short duration. In the second method, the fibers were first treated with $NaClO_2$ at certain pH and boiled for a fixed time, followed by an acid treatment with $NaHSO_3$ and again alkali treated. In both cases the final product was purified cellulose. This had average diameter of ~30 and about 9 μm for the first and second method, respectively, compared to the initial fiber diameter of ~220 μm.

Using the cellulose thus obtained, cellulose nanofibers were prepared by resorting to acid hydrolysis using H_2SO_4 at high temperature for neutralization followed by washing and sonification. The resulting product was freeze dried. This nanofiber had dimensions of 10–25nm (diameter), and ~300nm (length).

Teixeira *et al* (83) and Pasquini *et al* (84) have extracted the cassava bagasse cellulose nanofibrils as well as whiskers (CBW) from cassava bagasse (CBN) by acid hydrolysis method. Both cellulose nanofibers and whiskers were prepared from cassava bagasse fibers by first treating the fibers with sulfuric acid and mechanically stirred vigorously at 60°C. By centrifugation for about 10 min, the acid excess was removed and the resulting suspension was subjected to dialysis against distilled water using a cellulose membrane in the case of nanofibers and an organic solvent in the case of whiskers. The resulting suspension was sonicated for about 5 min to eliminate the aggregates and stored in freeze drier.

Some of the other methods used for the preparation of cellulose nanofibers (94) involve the combination of chemical and mechanical treatments. Chemical treatments include acid hydrolysis using strong acids to break down the amorphous cellulose into nanosized crystals or resorting to enzymatic treatment, which also breaks down the amorphous regions of the cellulose making them easy to separate into smaller sizes. The latter method yields high aspect ratio nanofibers. Other methods used are solvent swelling of microcrystalline cellulose, which separates the fibers easily or resorting to newly developed electro-spinning, which yields extremely fine fibers and threads. Even a homogenizer (Microfluidizer) for the production of microfibrillated cellulose has also been used (95).

On the other hand, bacterial cellulose is mostly preferred starting material by many researchers as it exhibits higher widths even before processing (11). Dimensions of chitin whiskers reported are 500nm (length), 50nm (diameter) (96) and 417nm (length), 33nm (diameter) (97).

In the case of synthesis of the cellulose microfibrils, a proteic complex termed *terminal complex* (TC), situated in the external surface of the bacterial cell membrane is used. Rows of up to 60 TCs have been used to synthesize the cellulose chains in the form of nanofibrils of 2-4 nm diameters, which in turn aggregate in to microfibrils of width of 70–150 nm and a height of 4–7 nm (29, 85, 98).

17.2.3.4 *Properties of Microfibrils/Whiskers*

In general, cellulose microfibrils/whiskers exhibit high crystallinity, high tensile strength, high melting temperature, about 200 times more surface area than the starting material and finer web-like network (30). They impart significant strength and directional rigidity to a composite. Etched molecular pattern at the surface (C6 -OH groups) of these provides modification opportunities.

Though the characteristics of both microfibrils and whiskers depend on their sources and producing method, several researchers working in the area of cellulose nanomaterials have reported the following values for some of their characteristics:

For microfibrils, the dimensions reported are: diameter of 2–20nm and length in micrometer range (28, 99); 5–10nm (diameter) and 100nm-micron range (length) (99). Microfibrils from Cotton: 15nm (diameter), 250nm (length) (71), microfibrils from tunicates: 10–15nm (diameter, 1200nm-to several μm with aspect ratio [l/d] of 67, YM; 143 GPa and interfacial /surface area of $150m^2/g$ to several hundred m^2/g (100).

Teixeira *et al* (83) extracted the cassava bagasse cellulose nanofibrils from cassava bagasse (CBN) by acid hydrolysis. These were characterized for their dimensions, crystallinity, thermal stability and components present (various types of sugars) in the suspensions of CBN. XRD patterns obtained for as received cassava bagasse (CB) and that of solid residues from the suspensions after 40 minutes hydrolysis (CW_{40}) indicated higher crystallinity for CW_{40} than that of CB.

This suggests that the hydrolysis not only narrows both peaks, but also increases their magnitude, which may be due to the higher crystallinity of the hydrolyzed cellulosic residue compared to the original fibers present in CB. Similar results have been observed (83) in the case of cassava bagasse nanofibers (CBN), indicating higher crystallinity for CBN than that of CB.

For whiskers, the dimensions reported are: 8–20nm or less (diameter), 500 to 1–2μm length [9, 101]; Young's modulus~150GPa and TS:10GPa, which are about 7 times lower than those of single wall carbon nanotubes (102); 5–20nm (diameter), 300–2000nm (length) and values of TS and YM being the same (9, 103–107).

While the whiskers are not available commercially, microcrystalline cellulose (MCC) is available. This is produced by the removal of amorphous region of cellulose by the acid hydrolysis. This results in less accessible crystalline regions in the form of fine crystals stiff rod like. MCCs are stable thixotropic gels with aqueous colloidal dispersions of appropriate degree of polymerization (DP) at high solid concentrations. This was patented first by Battista & Smith in 1961 (108). Some of the resources used to produce these are valonia, cotton, wood pulp, sugar beet pulp and tunicans (28). Depending on the source, their typical sizes reported are 20–40nm (diameter), 200–400nm (length) to give an aspect ratio of 10 and DP of 140–400 (109). For example, cotton-based MMC is 200nm long and 50Å in lateral dimension and for tunican-based MMCs, they are 1μ m and 150Å, respectively (28).

Reported elastic modulus (Young modulus) of dried bacterial cellulose pellicles is in the range of 15–35 GPa, which is extraordinarily high for bi-dimensional biologic materials (29, 86). This makes BC a good choice as reinforcement material in paper pulp or in the manufacture of diaphragms in high-fidelity speakers (86). Hsieh *et al.* (110) estimated its value as 114 GPa by monitoring the shift upon straining in the 1095 cm^{-1} Raman band of cellulose, corresponding to the stretching of C-O bonds in the cellulose chain. This shift was linearly proportional to the fibril deformation up to 0.8% and the rate with it varies with deformation can be related to the Young modulus of the fibrils (111).

Exceptional mechanical properties of BC are frequently attributed to its high degree of crystallinity. However, estimated crystallinity index of BC obtained from X-ray diffraction are 63% for BC produced in agitated medium and 71% in static medium. These numbers are not particularly high (110).

X-ray powder diffraction (XRPD) has been used to verify possible changes in crystallinity, crystalline organization and relative dimensions of the cellulose crystallites in bacterial cellulose. For this purpose, the moisture present in both untreated and partially hydrolysed bacterial cellulose was exchanged using tert-butanol by a stepwise gradient method with the last washing steps being carried out with pure tert-butanol (38, 53). With a view to minimize recrystallization of the cellulose chains during drying, the fiber suspension was centrifuged for 20 min at 10000 rpm and the recovered fibers were lyophilized in a bench-top freeze-drier. This led to the consequent alteration in its crystallinity index. While the estimation of the cellulose crystallinity followed the method of Chen *et al.*, 2007 (112), crystallite dimensions were determined using the Scherrer equation (37).

The molecular weight distribution of both the untreated and the partially hydrolysed bacterial cellulose has been determined by gel permeation chromatography (GPC) after per-carbanylation (53) using tetrahydrofuran (THF) as the eluting solvent. The elution profile was monitored by UV absorbance using a UV detector at an appropriate wavelength (say, 254 nm). The GPC calibration curve was generated from polystyrene standards with narrow Mw distributions with the Mark-Houwink coefficients for calibration could be Kp = 1.18×10^{-4} and αp = 0.74 for polystyrene and Kc = 2.01×10^{-5} and αc = 0.74 for cellulose tricarbanylate (113). The degree of polymerization (DP) of cellulose has been obtained by dividing the molecular weight (Mw) of the tricarbanylated polymer by the Mw of the tricarbanyl derivate of anhydroglucose (DP = Mw / 519).

17.2.3.5 Morphology Studies of Microfibrils/Whiskers

To understand the morphology of the cellulose prepared, scanning electron microscopy (SEM) and transmission microscopy have been used with appropriate preparation and operating conditions (53, 71, 83, 84, 99, 114). For example, both SEM of plant based cellulose and of bacterial cellulose as well as TEM of nanowhiskers have been obtained. SEM of BC with and without hydrolysis obtained by Woehl *et al* (53) is shown in Figure 17.8.

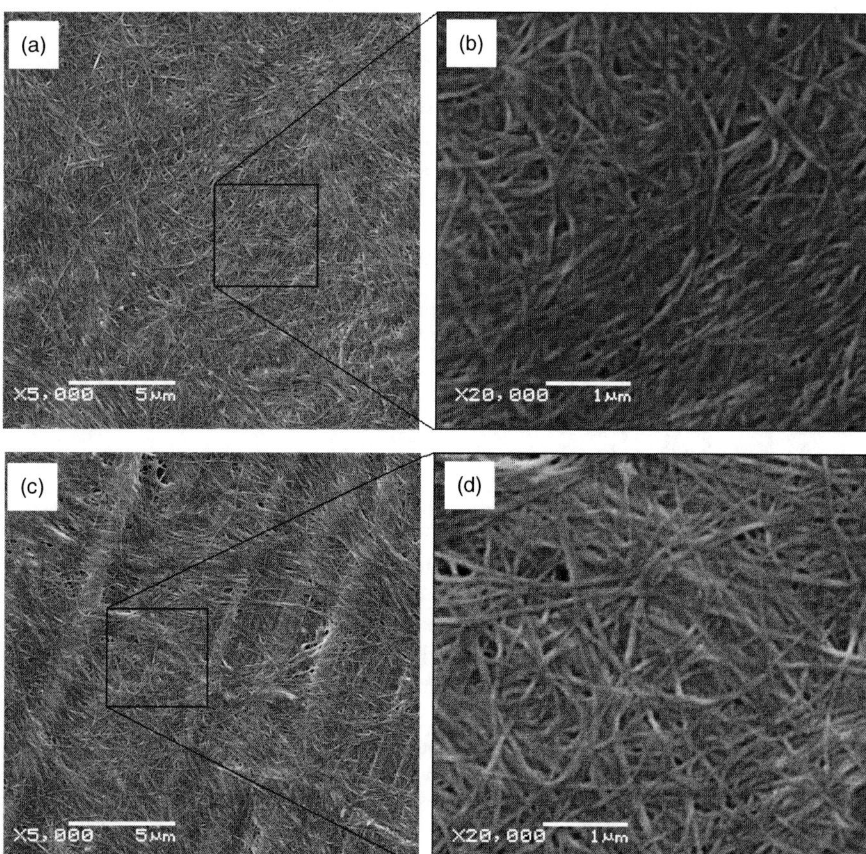

Figure 17.8 Scanning electron micrograph of the bacterial cellulose fibers: (a) untreated, (b) higher magnification of marked portion in "a" (c) hydrolysated for 60 min, (d) higher magnification of marked part in "c" (Reproduced with the kind permission of the Publishers of Ref. 53).

Atomic force microscopy (AFM) has been used (53, 115) to study the phase contrast and also to obtain the height profile and width of the fiber images. Appropriate software available with the device could be used for this purpose. For example, AFM analysis of bacterial cellulose before and after enzymatic hydrolysis was very elucidative. The untreated material (Figure 17.9a) shows more organized fiber bundles of 2 to 10 μm in width. Even the individual fibers show a tendency to aggregate into small bundles. After 60 min of enzymatic hydrolysis (Figure 17.9b), the orientation of the fibers was lost and their distribution seemed almost completely random. Also, the fibers seemed to be much shorter, although the authors reported inability for the precise length measurements due to the diffused image and partial fiber overlapping. Higher magnification of the untreated bacterial cellulose (Figure 17.9c) showed fibers with irregular surface (blue arrows) and "elbows" (red arrows) with the fiber direction changed abruptly.

These features disappeared almost completely in the partially hydrolyzed fibers. Analysis of the height profiles (Figure 17.9d) reveals that hydrolysis brings in

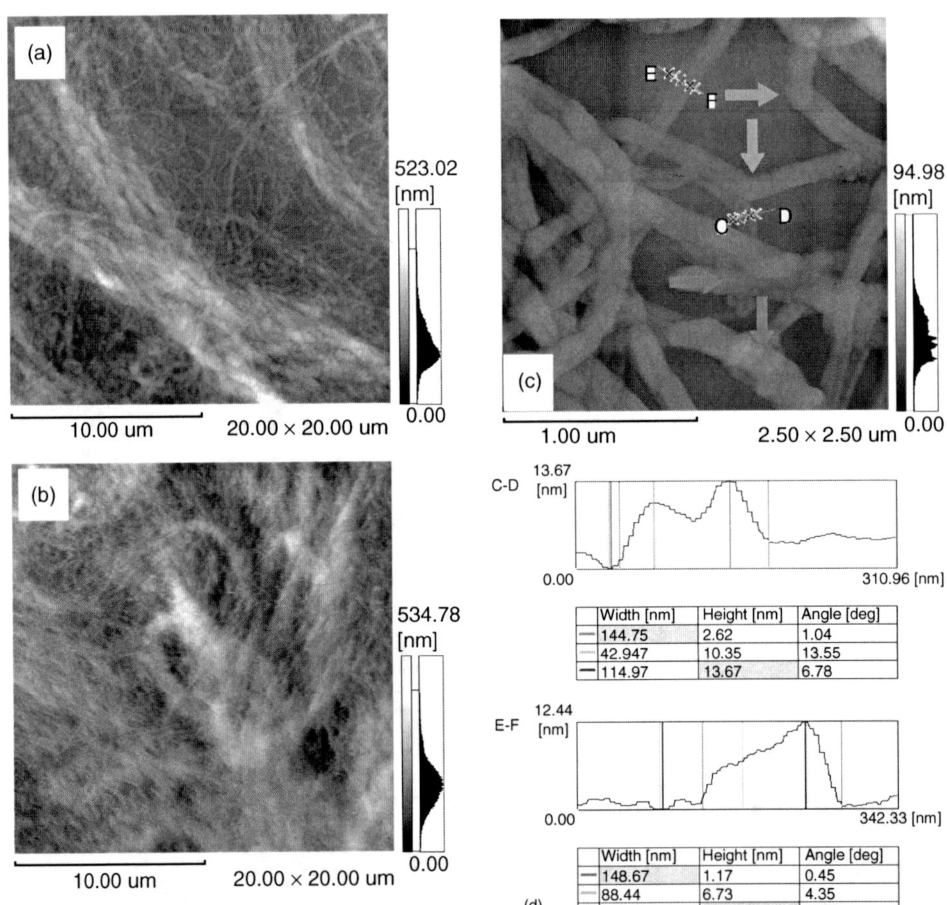

Figure 17.9 AFM micrographs of bacterial cellulose (a) before, (b) after enzymatic hydrolysis for 60 min., (c) higher magnification of "b," (d) height profiles of the original bacterial cellulose. (Reproduced with the kind permission of Ref. 53).

smoothness in the initially sharp profile of the fibers, as if the fiber edges were cut by the concerted action of endoglucanases.

Some authors have shown that changes in fiber morphology are related to subtle increases in fiber mechanical properties (116). The superior mechanical properties of microfibrillated cellulose fibers have been attributed to the elimination of fiber defects that might act as crack initiators. If the irregularities observed in the original bacterial cellulose correspond to those crack initiators, an additional reinforcing mechanism may occur when the hydrolysis eliminates those regions, which are probably less organized than the rest of the fiber.

Woehl *et al* (53) suggested that the smoothening of the fiber surface could also be a confirmation of the proposed disaggregation mechanism. They did not observe clearly the amorphous regions in the AFM images due to the image broadening caused by the microscope tip width. However, they observed their disappearance by the ease with which the tip could approach the fiber surface.

17.3 Bionanocomposites Based on Plasticized Starch Reinforced with Plant Based Cellulose/Bacterial Cellulose Nanofibers

Bionanocomposites have been increasingly accepted as a hybrid nanostructured material consisting of naturally occurring materials for both matrix and the reinforcing phases, with one of these having nanosize and the product showing nanosize at least in one dimension (4). It is also interesting to note that bionanocomposite is not a new concept as there are a number of materials occurring in nature such as enamel and dentine in teeth, bone, ivory etc., which are examples bionanocomposites. However, since 1994, when preparation of these materials started at CERMAV-CNRS (Grenoble, France) by casting in aqueous solutions (99, 103, 105, 106), interest in these materials have been on the rise. This is due to the reasons mentioned in earlier Sections. In the following Sections, some details about the processing, structure and properties of bionanocomposites will be discussed.

17.3.1 Processing Aspects

17.3.1.1 Preparation of the Bionanocomposite Using Plant Based Cellulose

Incorporation of cellulose nanofibers/whiskers is limited to hydrophilic matrices due to similar surfaces. For hydrophobic matrices, surface modification is needed as in the case of microcomposites, which will add to the cost of final product (117). Hence, the cost of the whiskers limits the preparation of bionanocomposites to laboratory scale only. As in the case of microcomposites, some of the advantages of surface modification are (i) improving wettability and adhesion between fibers and matrix, (ii) improvement in fiber distribution and fiber-matrix interfacial adhesion, (iii) reduction of the hydrophobic tendency of fiber and, of course (iv) possibility of imparting new properties to the composite. Several methods for this have been reported (94), wherein surface energy measurements are used as tools to evaluate the compatibility of fiber-matrix system. Some of the methods include: ionic group formation, hydrophobic surface creation and adsorption. The first method involves sulfonation (mild acid treatment forms sulfonic acid groups), carboxylation (forming carboxylic acid groups) and electrical discharge (Corona, which activates surface and plasma, which increases the surface roughness) and, grafting. It is reported that the chemical grafting of nanocellulose surfaces is a promising method to control the dispersion during processing the nanocellulose-polymer interaction and also the formation of a network in nanocellulose composite systems (118). The second method involves acetylation/alkylation to impart hydrophobic nature to the fiber surface (isocyanate, MAPP) and silane treatment by which desired functional groups can be attached. Finally, the third method involves adsorption by the addition of surfactants or through polyelectrolytes treatment

It should be noted that processing techniques of a composite are dictated by the intrinsic properties (solubility, dispersibility, degradation, etc.) of both constituents. This rule applies to bionanocomposites also. Accordingly, some of the methods used for the preparation of nanocomposites are solution intercalation, in situ intercalative polymerization, melt intercalation, solution casting, mixing of cellulose whiskers with surfactants followed by freeze-drying and then disperse in some organic solvent (99, 119) and preparation of gels followed by extrusion.

A simple method for the preparation of nanocomposites of cellulose microfibrils or whiskers is by casting. This involves incorporation of these reinforcements into selected polymeric matrices (hydrophilic/aqueous systems) using a processing medium such as water or solvents. Then, a freeze dryer can be used to speed up the evaporation process and a magnetic field can be applied to achieve high fiber alignment. Liu *et al* (120) have prepared pea starch-bamboo cellulose crystals by this method. They observed good and comparable reinforcing effect even at as low a fiber content as 8 wt. % to that of composites containing 30–50 wt. % cellulosic fibers (104, 121).

Another method is by melt-compounding, which involves both the compounding and extrusion processes. In this process, the cellulose nanofibers or whiskers are dispersed in an organic medium by first coating the whiskers using a surfactant or by surface modification whereby the surface of the reinforcement is chemically modified (119, 122–124). Both processes improve wetting and also the fiber distribution. It should be noted that high temperatures and shear forces should be avoided during compounding and extrusion in order to prevent fiber degradation. This has been used by Bondeson and Oksman, (117) to prepare bionanocomposites using commercially available microcrystalline cellulose with PLA. They first compounded the mixture using a co-rotating and intermeshing twin extruder with low shear forces to minimize degradation of materials. After this process, the mixture was extruded in the form of rods at a suitable temperature (170–200°C) with certain speed (150 rpm) and pressure (5–8 bars). The extruded rods were then compression molded into plates.

One of the problems encountered in the preparation of bionanocomposites by casting method, compared to conventional melt processing method using cellulose nanocrystals, is the complexity involved (83). These include the four constituents, viz., starch, cellulose, main plasticizer and water, as well as the competitive interactions involved between these constituents. For example, the accumulations of plasticizers (glycerol and water) near the interface zone of amylopectin/cellulose interferes with hydrogen bonding of inter whiskers, which results in lowering of the mechanical properties of nanocomposites.

Teixeira *et al.* (83) prepared cassava starch-cellulose whiskers with sorbitol and glycerol mixture as plasticizer. First, a homogeneous mixture of whiskers and the plasticizer was obtained in a Rheomixer by processing for fixed time and speed. This mixture was then subjected to the ultrasonic treatment for a given time (~ 5 min) followed by the addition of required amount of aqueous suspension of whiskers. The mixture was then hot pressed at a temperature, which should not volatilize glycerol and also not thermally degrade the whiskers, into suitable plates. Samples were then dried at 60–70°C to constant weight before any testing was done.

17.3.1.2 Preparation of the Bionanocomposite Films Using Bacterial Cellulose

The use of BC as a structural reinforcement in composites has been recently investigated. The fibrils can be incorporated in a polymer matrix (53, 125) or the polymer can be added to the culture medium (126). Another alternative is adding mineral nanoparticles (e.g. silica) to the culture medium (127).

Bionanocomposites of BC nanocrystals and cellulose acetate butyrate were prepared by solution casting (128). Tunican whiskers have been dispersed in dimethylformamide (DMF) without the use of additives or any surface modifications (129). This has demonstrated the possibility for using hydrophobic matrices for such a purpose, as well as the establishment of surface modifications of whiskers, which are incompatible with the presence of water.

Woehl *et al.,* (53) used mats of *Acetobacter xylinum* bacterial cellulose (Membracel Produtos Tecnológicos Ltda, Almirante Tamandaré, PR, Brazil) and cassava starch having 17 wt. % amylose and 83 wt. % amylopectin (Corn Products Brasil, Balsa Nova, PR, Brazil), with comercially pure glycerol obtained (Sigma-Aldrich, St. Louis, Missouri, USA) used as a plasticizer. The BC having 99.5 % moisture was first dispersed in $0.1 mg.mL^{-1}$ sodium azide and homogenized in a blender kept under refrigeration. Then, it was subjected to enzymatic hydrolysis followed by immersion into boiling water to inactivate the enzymes and then cooled in ice bath. Using these aqueous suspensions of BC (both with and without surface treated with *Trichoderma reesei* endoglucanases) with cassava starch plasticized with 30 % glycerol (in relation to starch), films of bionanocomposites were prepared by casting method. Before casting, the mixture was heated under reflux and stirred at 90–95 °C for 30 min to ensure complete gelatinization of starch. It was observed that high temperatures and/or long drying periods without vacuum produced inhomogeneous materials with air bubbles visibly trapped in the films. Even under optimized conditions, the films showed some opacity, demonstrating that the dispersion of the cellulose nanofibers was not ideal (130). The resulting films were maintained for 10 days at 43% relative humidity over a saturated K_2CO_3 solution in a desiccator at room temperature, according to ASTM E 104 before subjecting them for various tests.

17.3.2 Properties of Bionanocomposites

As in the characterization of microcomposites, various techniques have been used to characterize of bionanocomposites. Morphological characterization has been done using field emission SEM whereby homogeneity of dispersion, presence of voids presence of aggregates, orientation of whiskers, among others can be determined. Sedimentation can also be determined by wide angle XRD where one can compare the patterns obtained by the two faces. Polarized microscopy can be used to compare the homogeneity obtained by different processing techniques through the birefringence of the resulting composites. For example, it was observed that casting/evaporation method gives better homogeneity than other techniques (28). Small angle neutron scattering (SANS) is used to determine the nature of dispersion, whereas AFM is a powerful technique used to understand the morphology and topography of films containing nanofibers and also to understand fracture surfaces of nanocomposites. TEM is used to understand the morphology and also the nanoscale surface chemistry.

17.3.2.1 Properties of the Bionanocomposite Films Using Plant Based Cellulose

Very limited studies have reported on the properties of biodegradable nanocomposites containing lignocellulosic and other natural materials (82, 131–138).

Teixeira *et al* (83) have determined the tensile properties as function of CBN content in both type of matrices (TPSG and TPSGS). They observed an increase in TS from 1.8 to 4.8 MPa for 5% incorporation of CBN in cassava starch (CS), where after it decreased at higher CBN loading of 10-20%. Similarly, a greater increase in elongation was observed in cassava bagasse cellulose nanofibrils (CBN) containing cassava starch (CS).

This is attributed to the presence of carbohydrates in CBN and to a weak reinforcing effect. They also observed significant decrease in YM of thermoplastic starch (cassava starch + glycerol-TPSG) compared to that of thermoplastic cassava starch with glycerol+ sorbitol mixture (TPSGS). YM increased from 16.8 MPa for the pure matrix to 84.3MPa for composite containing 5 % CBN, followed by a drastic decrease. This is attributed to the possible transcrystallization observed in XRD studies, which may induce lower interactions between nanofibrils and starch, hindering the reinforcing effect.

The dynamic mechanical analysis (DMA) for the matrices as function of temperature showed partial miscibility with two main relaxation phenomena in tan δ curves. However, these values were not affected by the introduction of CBN in TPSGS while it showed greater decrease with its incorporation. Thermal stability as determined by TGA indicated that the initial decomposition temperatures of 280 and 220°C for CB and CBN, respectively, are attributed to starch and cellulose depolymerization, while the decomposition of CBN is found to be lower than that of CB, since acid hydrolysis lowers thermal stability as observed in bacterial cellulose (139). They also observed a decreased crystallinity (determined by XRD) with the incorporation of CB and CBN. On the other hand, water absorption (32–37% for TPGS and 15–18% for TPSGS) was found to decrease with the incorporation of these nanoreinforcements, but it remained constant up to 20% of their

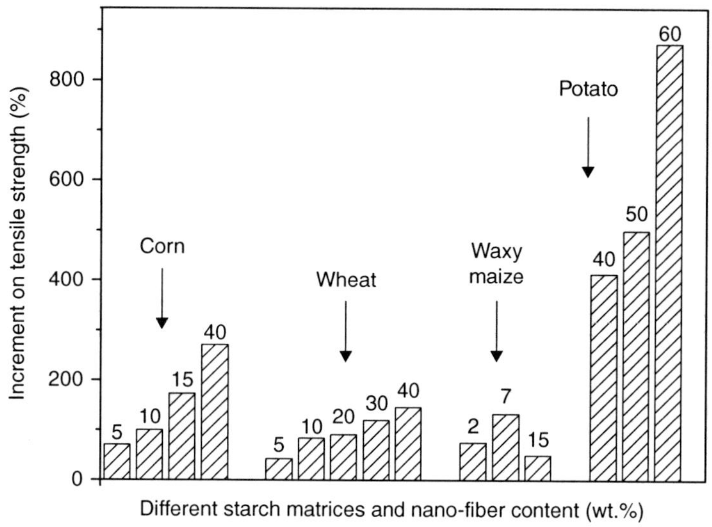

Figure 17.10 Variation of increase in tensile strength vs nanofiber content in different starch matrices (140).

incorporation. This is attributed to network formation of cellulose nanoparticles, preventing swelling of starch and its water absorption. Also, a decrease in hydrophilicity was observed due to incorporation of CBN in thermoplastic starch.

Vázquez and Alvarez (92) have reported the tensile properties of biodegradable composites containing different amounts (6–50 wt. %) of cellulose nanofibers in various starches such as corn, wheat, potato and waxy maize. Figure 17.10 shows plots of different amounts of nanofibers in different starches versus % increase in tensile strength of composites over the respective matrices. It can be seen that potato starch based composites showed the highest increase in tensile strength (~ 400%) followed by that of corn starch composites (~300%) for the same amount of fiber content (40 wt. %). (92). There are also reports of up to 250% increase in tensile strength with 5–10% loading of cellulose nanofibers in different matrices (72).

17.3.2.2 *Properties of the Bionanocomposite Films Using Bacterial Cellulose*

Woehl *et al.*, (unpublished results) have studied the effect of enzymatic hydrolysis of BC and duration of this on the tensile properties using Aminex HPX-87H column (Bio-Rad) at 65 °C with 8 mmol/L of H_2SO_4 as mobile phases. Using differential refractometry, they detected soluble sugars (glucose and cellobiose). Then, one can calculate the total hydrolysis mass loss in relation to the dry weight of the starting material by converting obtained results to anhydroglucose equivalents. Powder X-ray diffraction method was also used to verify the crystallinity developed in the nanocomposites using both the control and the composite samples maintained at controlled humidity conditions (53).

The authors have also observed that freshly prepared TPS films show only an amorphous halo in the X-Ray diffraction pattern (Figure 17.11), indicating

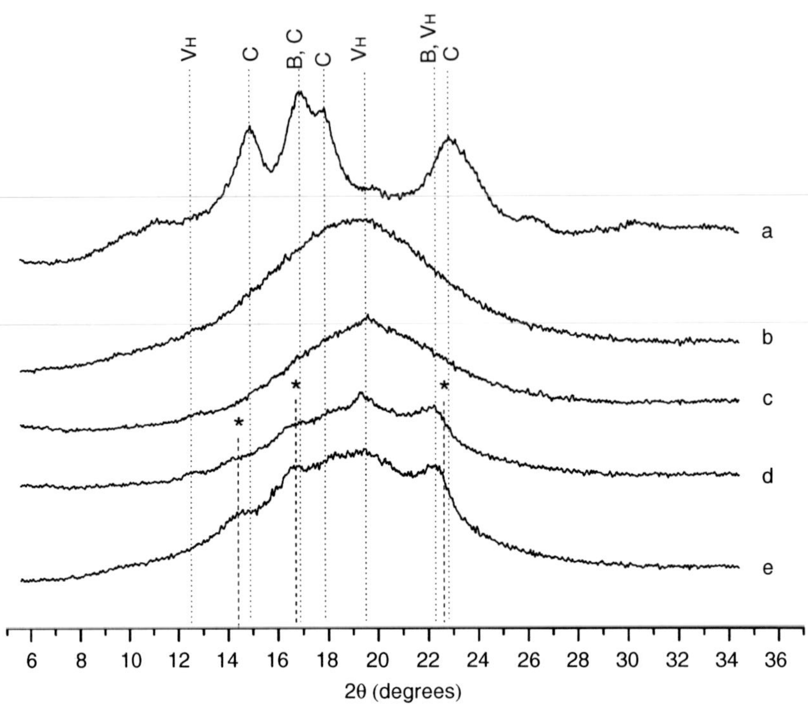

Figure 17.11 X-ray diffraction patterns of cassava starch (a), cassava starch film plasticized with 30% glycerol (b), the same film after conditioning at 43% RH (c), film reinforced with 2.5% untreated bacterial cellulose (d) and film with 2.5% bacterial cellulose hydrolyzed with endoglucanases for 60 min (e). Traced lines (*) indicate characteristic cellulose I peaks. Dotted lines indicate peaks corresponding to the B, C and V_H starch crystalline patterns. [Reproduced from Ref. 66].

complete gelatinization of the starch. However, after conditioning for 10 days at 43% RH, a crystalline pattern developed. They also found that peaks at $2\theta = 19.5°$ and $12.5°$ are consistent with the expected V_H crystallinity pattern resulting from starch retrogradation (64, 65). Both peaks appeared with higher intensity on the bacterial cellulose incorporated films (Figure 17.11d), suggesting that the presence of the fibers induced starch crystallization. They also observed peaks that were attributed to recrystallization of amylopectin around the fibers in a B-type pattern (59). The same peaks were more intense in the film containing enzyme-treated cellulose (Figure 17.11e), which suggested that the development of B-type crystallinity was favored by the treated fibers. They have explained the broadening of these peaks as due to smaller size of the formed crystals, which would have contributed to improved mechanical properties of the treated cellulose composite in comparison with the untreated one (see Figure 17.12). They have also suggested another possibility of the apparent broadening as due to the superposition of peaks from another crystalline pattern, possibly of C-type.

Woehl *et al.* (53) have also determined the mechanical properties of starch-BC incorporated bionanocomposites, following the ASTM D 882-95a method by testing a minimum of five samples for each condition with 10 mm wide samples at a

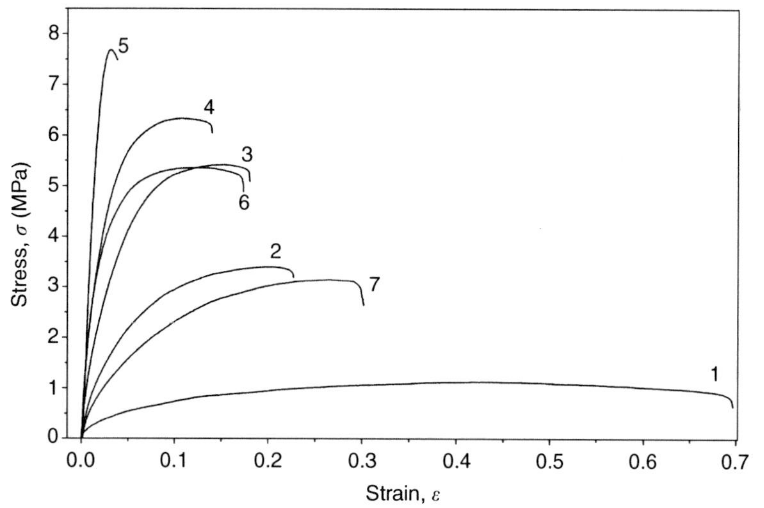

Figure 17.12 Stress/strain curves of the matrix and its nanocomposites with BC: (1) TPS, (2) TPS- 2.5 wt. % untreated BC; TPS- 2.5 wt. % treated BC for different duration (minutes.): (3) 20; (4) 40, (5) 60, (6) 80 and (7) 120. (Reproduced with the kind permission of the Publishers of Ref. 53).

cross-head speed of 4 mm min^{-1}. Typical stress-strain curves obtained are shown in Figure 17.12 for the matrix (cassava starch + glycerol), and its composites containing 2.5 wt% of untreated and enzyme-treated BC nanofibers for different duration ranging from 20 to 120 minutes.

The curves show an increasing slope with increasing time of treatment of BC fibers compared to an almost flat nature of the matrix control. Young's modulus, ultimate tensile strength and % strain are evaluated from these curves, which are shown in Figure 17.13a–c.

Hence, incorporation of partially hydrolysed BC into the cassava TPS matrix showed an initial slow increase in both tensile strength and Young's modulus with increasing time of hydrolysis. The data also suggested a substantial effect of incorporation on both YM and UTS, but the extent of this effect depended upon the duration of hydrolysis with *Trichoderma reesei* endoglucanases, reaching a maximum for about 60 minutes in both cases. After this time, both values decreased with further increasing time of hydrolysis reaching almost the same values as those of TPS matrix itself after 120 minutes.

On the other hand, for short enzymatic treatment times, the values of strain at break were found to be higher, but more scattered than those at the optimum duration of hydrolysis (60 min). The authors have concluded that this was due to a poor dispersion of the nanofibers into the TPS matrix.

It is also interesting to note that the observed Young's modulus of bionanocomposites containing hydrolysed BC fibers is four times (575.7±166.7 MPa) higher than that containing untreated BC fibers (140.6±40.3 MPa) and about seventeen times higher than that of the plasticized starch matrix (33.4±4.3 MPa). Likewise, the tensile strength of bionanocomposites containing hydrolysed BC fibers is almost double (8.45±2.35 MPa) compared to that of nanocomposites containing

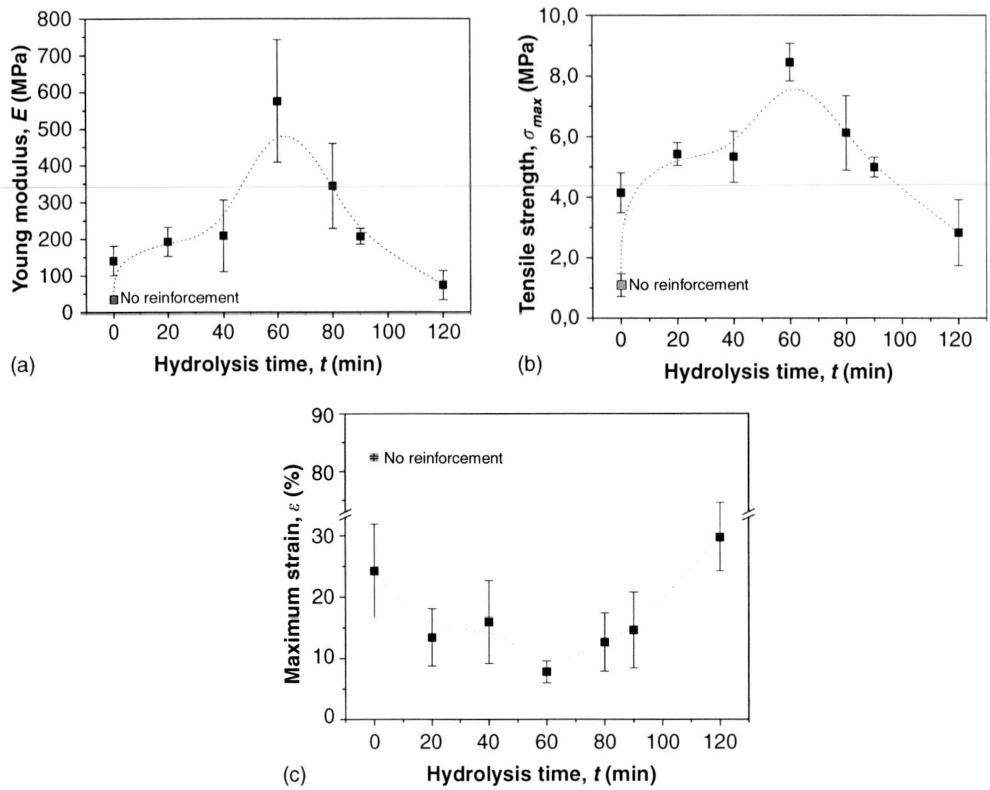

Figure 17.13 Tensile Properties of bionanocomposite films of cassava starch + glycerol + BC as a function hydrolysis of BC (2.5 wt. %) with *Trichoderma reesei* endoglucanases. (a) Young's modulus and (b) Ultimate tensile strength (c) % Strain at break. Values of bothYM, UTS and % strain at break of the matrix TPS film is shown in gray color. (Reproduced with the kind permission of the Publishers of Ref. 53).

untreated fibers (4.15±0.66 MPa) and eight times higher than that of the TPS matrix (1.09±0.39 MPa).

These results underlined that the enzymatic treatment of the BC nanofibers significantly enhances their reinforcement capacity. They have explained the above results as due to the disruption of the amorphous regions within the BC nanofiber bundles up to the hydrolysis of 60 minutes, which led to the better dispersion of these fibers in the TPS matrix. However, longer times of hydrolysis would lead to the generation of defects on the surface of the fibers due to extensive breaking of chains, reversing the preliminary gains in mechanical properties. Values of strain at break also reinforces the above proposed mechanism indicating that longer hydrolysis times of BC nanofibers leads to damages in the supramolecular structure of the fibers, with shorter fibers generated by this process loosing part of their entanglement capacity, leading to higher strains.

The effect of BC nanofiber content on the tensile properties of nanopolymer composites showed (Figure 17.14a–d) that the addition of even a very small amount (0.25 wt. %) of nanofibers decreases the strain at break of the TPS (~ 80 %) to less

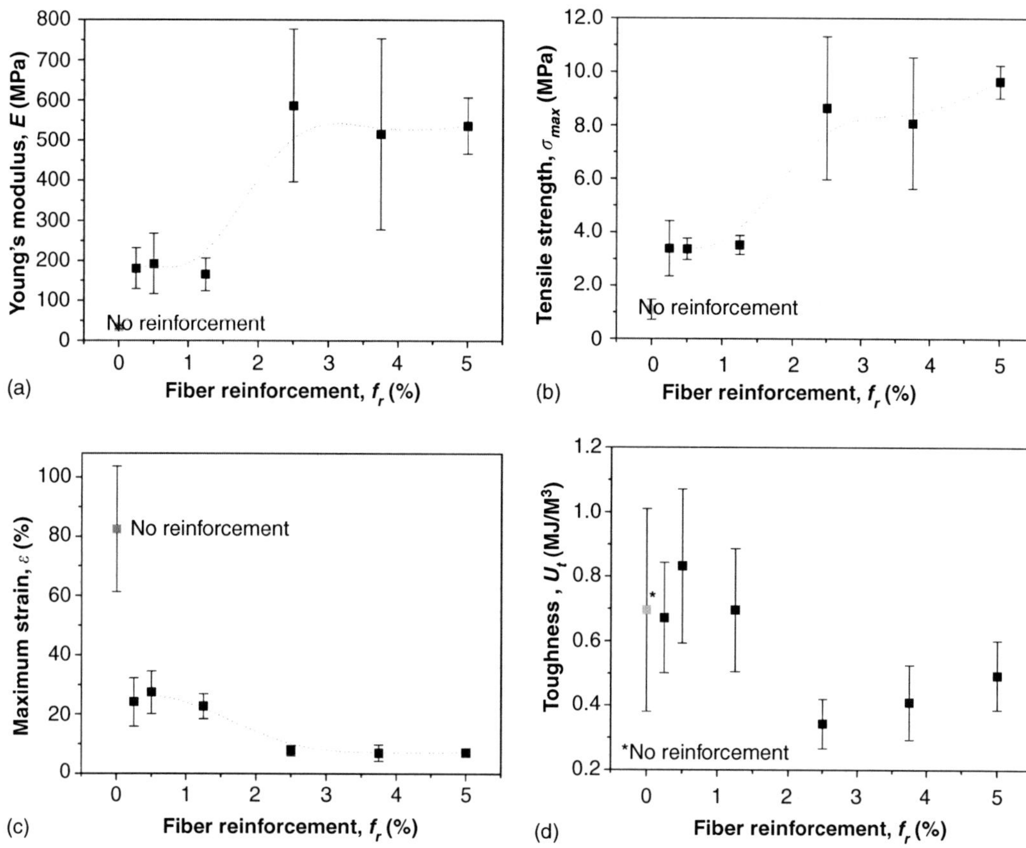

Figure 17.14 Tensile properties as functions of fiber content. (a)-Young's Modulus; (b)-Ultimate Tensile strength, (c)- % strain at break and (d)-Toughness (Woehl *et al.*-Unpublished).

than 30%. Higher fiber contents of 0.5 and 1.25 % did not show any significant change but, with 2.5 % fibers, the % strain at break decreased to about 7 %. This clearly brings out the importance of the dispersion and entanglement to optimize nanofibers on their reinforcement capacity. Both YM and UTS do not show any change (~175 MPa and ~.3.5 MPa respectively) till 2 wt.% addition of BC fibers, while these increased dramatically (~ 550-570MPa and ~9.6MPa, respectively) at ~ 2.5% BC fiber incorporation (Figure 17.14a-b). Both YM and UTS almost remained constant at those values or showed a slightly decreasing trend up to 5 wt. % fiber content. These have been attributed to the effects of formation of a percolation network (30, 141) or to the fiber entanglement (142). Also, large deviations of both YM and UTS values at the higher cellulose contents were attributed to a poor dispersion of the fibers in the TPS matrix.

It was also shown that the entanglement of the fibers reaches a critical threshold at about 2.5 % fiber concentration from which difficulty in dispersion of fibers is expected due to a greater fiber-fiber interaction. Therefore, the values of YM, UTS and % strain at break remained practically constant for higher amounts of nanofiller, a behavior frequently observed in nanocomposites (143). Besides, the heterogeneity in the fiber dispersion accounts for the lowering of strength properties since the failure of the composite, being a stochastic process is governed mainly by the crack initiation and crack propagation (144). The observations have been supported (66) by SEM studies of fractured samples (Figure 17.15). Figure 17.15a, b shows the plasticized starch matrix, while Figure 17.15c, d is referred to composites with enzymatically treated BC fibers for 60 min.

The projections of films created by the tensile testing are shown by arrows in Figure 17.15c. It can be seen that (i) there is no significant features in the fracture surface (Figure 17.15a, b) of the TPS matrix alone due to his amorphous nature and (ii) the presence of nanofibers induces partial crystallization of the starch, resulting in fractures with a granular aspect (Figure 17.15c, d) that are not due to the crystalline structure of the cellulose itself (arrows are indicating holes and protuberances due to the pull-out of fiber bundles). This is in similar to the SEM of TPS/cellulose whisker composites (59, 105), which showed bright spots being attributed to the charge build-up at the extremity of the cellulose whiskers. Similar features are observed in composites with partially hydrolyzed (60 min) cellulose fibers with *Trichoderma reesei* endoglucanases.

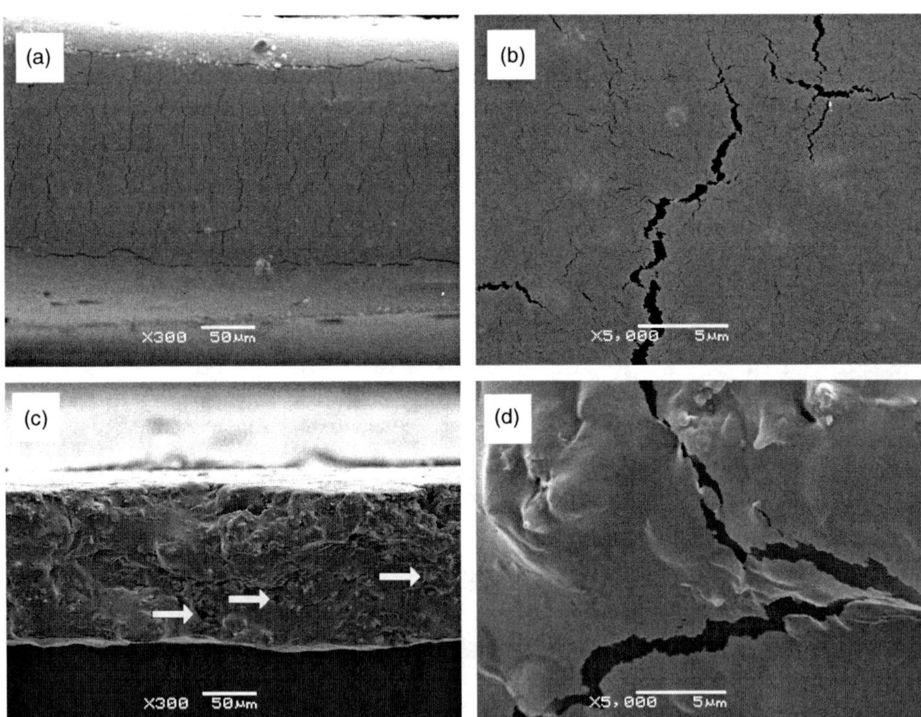

Figure 17.15 Scanning electron micrographs after tensile testing. Non-reinforced TPS (a,b); TPS reinforced with 2,5 wt% of bacterial cellulose enzymatically treated for 60 min (c,d). The crack patterns are artifacts generated by the cracking of the gold sputtered layer during the application of the microscope vacuum. (Reproduced with the kind permission of the Publishers of Ref. 53).

17.4 Applications and Products of Bionanocomposites

There are a number of potential applications for polymer nanocomposites. These include automotive parts and building blocks, electronic devices such as capacitors, inductors and transistors, smart papers such as sensors, communication devices, electromagnetic shields, paper-based displays, packaging materials for food and pharmaceuticals, radio frequency ID tags (printed on cellulosic paper) and flexible substrates for portable and foldable display systems (145, 146) In the last case, a piece of fully doped filter paper was used as the carrier of a chemical necessary to activate the generation of electricity, which had a maximum voltage of 1.56V after 10s activation. Some of the composite systems for the above applications have cellulose nanofibers as a reinforcing filler in the desired polymer matrices (automotive parts and building blocks), wood microfibers incorporated in anionic poly (3,4-ethylenedioxythiophene)–poly(styrenesulfonate) and cellulosic nanofibers and optical polymer matrices (flexible substrates). In the latter case, it is reported that the films would offer excellent mechanical properties, low thermal expansion and high light transmittance (72).

17.5 Concluding Remarks

Bionanocomposites, the latest generation of composites, may be termed as relatively new composites even though they have been known almost for two decades. Also, bionanocomposite is not a new concept as there are a number of materials occurring in nature such as enamel and dentine in teeth, that are readily classified as such. Interest in these materials is due to their unique characteristics such as biocompatibility, biodegradability and even functional properties such as the ability to act as a gas barrier and a high thermal stability. Nanobiocomposites consist of different types of cellulose nanofibers, or microfibrills, or nanorods, or crystals, or whiskers incorporated in biopolymers such as starches, polyhydroxynoates, poly(lactic acid), etc. A number of processing methods for the production of cellulose nanofibers/whiskers have been developed. Depending on the source and processing method used, dimensions of cellulose nanofibers or whiskers vary considerably. Cellulose whiskers are highly ordered structures and hence exhibit unusually high strength and physical properties such as electrical, optical, conductivity, among others. Reported values of Young's modulus and tensile strength indicate that they are dependent on the source of cellulose and are in the range of 110–150 GPa and 10–12 GPa, respectively. In fact, the YM values are higher than those of aluminum (70GPa) and glass fibers (76GPa), while the estimated tensile strength value of cellulose nanofiber is about 7 times that of steel (72). One of these, named Microcrystalline cellulose (MCC), has been used in pharmaceutical, food and paper industries in addition to being used in the preparation of composites. Various processing techniques have been used for the preparation of nanocomposites such as solution intercalation, in situ intercalative polymerization, melt intercalation and solution casting, but the most commonly used methods are film casting extrusion and hot pressing after obtaining homogenous mixture of the constituents with appropriate treatments of the nanoreinforcements, irrespective of the type of cellulose fibers/whiskers used. Physical, mechanical, thermal and other properties have been determined for these resulting bionanocomposites. Possible applications, particularly for cellulose whiskers, include their use in security paper and polymer electrolytes in lithium batteries. Considering the increasing research going on in the area of bionanocomposites, it appears that future development will focus on improved properties and multi-functionality with greater possibilities in view of the abundant availability of renewable lignocellulosic materials for both matrix and reinforcements. These materials, when properly disposed, will safely decompose into CO_2, humus, etc, which may again produce lignocellulosic materials by photosynthesis. Hence more studies on this topic should be pursued considering the advantages nanocomposites would offer in general. Furthermore, the concept of nanocomposites for load bearing applications being new, commercialization of nanostructured reinforcements such as cellulose microfibrils will be challenging in view of their disintegration while extracting them from plant cell walls and also in polymer matrices, despite high cost involved in the former (82).

Accordingly, some of the areas for future research should include (i) the use of nanofibers with or without spinning in various synthetic but biodegradable polymers to produce composites as superior structural components (lighter than their micro counterparts); (ii) use of nanofibers in different areas such as biomedical, electrical and optical as a component for various functional devices; (iii) basic research on structure-property correlations in nanocomposites, which may pose new challenges in the development of suitable fabrication techniques to reduce the production costs and improve understanding about chemical interactions at such sizes, (iv) understanding and control of thermal degradation, and (iv) modelling and simulation of mechanical properties of nanofiber-containing composites. Last but not the least, a serious attention must be focussed on the social implications of nascent and potential nanotechnology towards the safety aspects due to nanosized particles and their composites. With all these, it is hoped that these new materials will not only pave the way for wide range of applications and open new dimensions for biopolymers and their composites, but also receive larger acceptance by the society and its political leadership to make human life more enjoyable.

Acknowledgements

The authors are grateful to all the authors of the papers and publishers of the journals and other web sites from where Figures have been reproduced, for their courtesy and kind permission. Particular mention should be made for the following: M/s. Elsevier publishers, and M/s. Springer Science+Business Media B.V.for giving permission free of cost to reproduce the figures from their esteemed Journals. The financial support of the Brazilian agencies (FINEP, CNPq and CAPES) during the preparation of this work is also acknowledged. One of the authors (Dr. KGS) also sincerely thank the three institutions with which he is currently associated with in Bangalore (India) for their encouragement and interest in this work.

References

1. A. Okada, M. Kawasumi, A. Usuki, Y. Kojima, T. Kurauchi, and O. Kamigaito, *Materials Research Society Symposium*, p. 171, p. 45, 1990.
2. A. Usuki, Y. Kojima, M. Kawasumi, A. Okada, T. Kurauchi and O. Kamigaito, *Abstracts of papers of the American Chemical Society*, Vol. 200, p. 218. 1990.
3. C. Suryanarayana, *Bulletin of Materials Science*, Vol. 17, p. 307, 1994.
4. M. Darder, P. Aranda, and E. Ruiz-Hitzky, *Advanced Materials*, Vol. 19, p. 1309, 2007.
5. U.R. Vaidya, and M. Bhattacharya, *Journal of Applied Polymer Science*, Vol. 52, p. 617, 1994.
6. A. Sorrentino, G. Gorrasi, and V. Vittoria, *Trends in Food Science and Technology*, Vol. 18, p. 84, 2007.
7. D.N. Sahed, and J.P. Jog, *Advances in Polymer and Technology*, Vol. 18, p. 351, 1999.
8. L. Petersson, and K. Oksman, *American Chemical Society Symposium Series*, Vol. 938, p. 132, 2006.
9. M.A.S. Azizi Samir, F. Alloin, J.Y. Sanchez, and A. Dufresne., *Polymer*, Vol. 45, p. 4149, 2004.
10. A. Khademhosseini, and R. Langer, *Chemical Engineering Progress*, Vol. 102, p. 38, 2006.
11. H.M.C. de Azeredo. *Food Research International*, Vol. 42, p. 1240, 2009.
12. O.J. Kamigaito, *Journal of the Japan Society of Powder and Powder Metallurgy*, Vol. 38, p. 315, 1991.
13. S. Iijima, *Nature*, Vol. 354, p. 56, 1991.
14. A.B. Dalton, S. Collins, E. Munoz, J.M. Razal, V.H. Ebron, J.P. Ferraris, J.N. Coleman, B.G. Kim, and R.H. Baughman, *Nature*, Vol. 423, p. 703, 2003.
15. M.J. Biercuk, M.C. Llaguno, M. Radosvljevic, J.K. Hyun, A.T. Johnson, and J.E. Fischer, *Appled Physics Letters*, Vol. 80, p. 2767, 2002.
16. Z. Ounaies, C. Park, K.E. Wise, E.J. Siochi, and J.S. Harrison, *Composites Science and Technology*, Vol. 63, p. 1637, 2003.
17. M.C. Weisenberger, E.A. Grulke, D. Jacques, T. Rantell, and R. Andrews, *Journal of Nanoscience and Nanotechnology*, Vol. 3, p. 535, 2003.
18. Y.H. Choa, J.K Yang, B.H. Kim, Y.K Jeong, J.S Lee, T. Nakayama, T. Sekino, and K. Niihara, *Journal of Magnetism and Magnetic Materials*, Vol. 266, p. 12, 2003.
19. K.G. Satyanarayana, G.G.C. Arizaga, F. Wypych, *Progress in Polymer Science*, Vol. 34, p. 982, 2009.
20. F. Chivrac, E. Pollet, and Luc Averous, *Materials Science & Engineering Reports*, Vol. 67, p. 1, 2008.
21. P. Bordes, E. Pollet, and L. Averous, *Progress in Polymer Science*, Vol. 34, p. 125, 2009.
22. M.D. Sanchez-Garcia, E. Gimenez, and J.M. Lagaron, *Carbohydrate Polym*ers, Vol. 71, p. 235, 2008.
23. S.S. Ray, and M. Bousmina, *Progress in Materials Science*, Vol. 50, Issue 8, p. 962, 2005.
24. P.H.C. Camargo, K.G. Satyanarayana and F. Wypych, *Materials Research*, Vol. 12, Issue 1, 1, 2009.
25. A.K. Bledzki, V.E. Sperber and O. Faruk, "Natural and Wood Fiber Reinforcement in Polymers," *Rapra Review Reports*, Vol.13, Issue 8, Report No.152 (ISBN: 13-978185973594; ISBN: 1859573592) United Kingdom: Rapra Technology Ltd.; 2002.
26. B.C. Suddell and W.J. Evans, "The increasing use and application of natural fiber composite materials within the automotive industry." In: Proceedings of 7th International Conference on Wood Fiber-Plastic Composites. 2003.
27. J. Marikarian, *Plast Addit. Compound*, Vol. 10, Issue 4, p.20, 2008.
28. M.A.S. Azizi Samir, F. Allion, and A. Dufresne., *Biomacromolecules*, Vol. 6, p. 612, 2005.
29. D. Klemm, B. Heublein, H.P. Fink, and A. Bohn, *Angewandte Chemie International Edition*, Vol. 44, p. 3358, 2005.
30. Y. Habibi, L.A. Lucia, and O.J. Rojas, *Chemical Reviews*, Vol. 110, Issue 6, p. 3479, 2010.
31. Y. Habibi and A. Dufresne, *Biomamocromolecules*, vol. 9, p. 3479, 2008.
32. G.T. Toumis, *Science and Technology of Wood: Structure, Properties, Utilization*, New York, Van Nostrand Reinfold, 1991.
33. A.K. Bledzki, and J. Gassan, *Progress in Polymer Science*, Vol. 24, p. 221, 1999.
34. D. Klemm, B. Philipp, T. Heinze, U. Heinze, and W. Wagenknecht, *Comprehensive Cellulose Chemistry; Volume 1: Fundamentals and Analytical Methods*, Wiley-VCH Verlag, 1998.
35. D. Fengel and G. Wegener, *Wood: Chemistry, Ultrastructure, Reactions*, Berlin, Walter de Gruyter, 1989.

36. J.S. Han and J.S. Rowell, "Chemical Composition of Fibers", R.M. Rowell, R.A. Young and J.M. Rowell, eds. in *Paper and Composites from Agro-based Resources*, Boca Raton, CRC/Lewis Publishers, p. 83-132, 1996.
37. L.P. Ramos, *Química Nova*, Vol.26, p. 863,2003.
38. I. Sakurada, Y. Nukushina, and T. Ito, *Journal of Polymer Science*, Vol. 57, p. 651, 1962.
39. A. Buleon, P. Colonna, V. Planchot, and S. Ball, *International Journal of Biological Macromolecules*, Vol. 23, p. 85, 1998.
40. R.A. Freitas, R.C. Paula, J.P.A. Feitosa, S. Rocha, and M.R. Sierakowski, *Carbohydrate Polymers*, Vol. 55, p. 3, 2004.
41. D. Le Corre, J. Bras, and A. Dufresne, *Biomacromolecules*, Vol. 11, p. 1139, 2010.
42. S. Perez, P. Baldwin, and D.J. Gallant, "Structure features of starch granules". in J.N. BeMiller and R.L. Whistler, eds, *Starch: Chemistry and Technology*, 3rd edition, Academic Press, 2009.
43. R.F. Tester, J. Karkalas, and X. Qi, *Journal of Cereal Science*, Vol. 39, p.151, 2004.
44. C.A. Romero-Bastida, L.A. Bello-Perez, M.A. García, M.N. Martino, J. Solorza-Feria, and N.E. Zaritzky, *Carbohydrate Polymers*, Vol. 60, p. 235, 2005.
45. N.W.H. Cheetham, and L.P. Tao, *Carbohydrate Polymers*, Vol. 36, p. 277, 1998.
46. R. Parker, and S.G. Ring, *Journal of Cereal Science*, Vol. 34, p. 1, 2001.
47. T. P. Coultate, *Food: The Chemistry of its Components*, Paperback Edition (2009), Publishers: Royal Society of Chemistry.
48. R.A. Talja, H. Helen, Y.H. Roos, and K. Jouppila, *Carbohydrate polymers*, Vol. 67, p. 288, 2007.
49. L. Yu, K. Dean, and L. Li, *Progress in Polymer Science*, Vol. 31, p. 576, 2006.
50. R. He, X.L. Wang, Y.Z. Wang, K.K. Yang, H.B. Zeng, and S.D. Ding, *Carbohydrate Polymers*, Vol. 65, p. 28, 2006.
51. A. Lopez-Rubio, J.M. Lagaron, M. Ankerfors, T. Lindstrom, D. Nordqvist, A. Mattozzi, and M.S. Hedenqvist, *Carbohydrate polymers*, Vol. 68, p. 718, 2007.
52. S. Fakirov, and D. Bhattacharayya, *Handbook of Engineering Biopolymers*, Cincinatti, Hanser Publishers, 2007.
53. M.A. Woehl, C.D. Canestraro, A. Mikowski, M.R. Sierakowski, L.P., and F. Wypych, *Carbohydrate Polymers*, Vol. 80, p. 866, 2010.
54. C. Bourban, E. Karamuk, M.J. deFondaumiere, K. Ruffieux, J. Mayer, and E. Wintermantel, *Journal of Environmental Polymer Degradation*, Vol. 5, p; 159, 1997.
55. A. Dufresne, and M.R. Vignon. *Macromolecules*, Vol. 31, p. 2693, 1998.
56. A.A.S. Curvelo, A.J.F. de Carvalho, and J.A.M. Agnelli. *Carbohydrate Polymers*, Vol. 45, p. 183, 2001.
57. Y.I. Matveev, V.Y. Grinberg, and V.B. Tolstoguzov, *Food Hydrocolloids*, Vol. 14, p. 425, 2000.
58. H.M. Wilhelm, M.R. Sierakowski, G.P. Souza and F. Wypych, *Carbohydrate Polymers*, Vol. 52, p. 101, 2003.
59. M.N. Angles, and A. Dufresne, *Macromolecules*, Vol. 33, p. 8344, 2000.
60. M.N. Angles, and A. Dufresne, *Macromolecules*, Vol. 34, p. 2921, 2001.
61. L. Averous, A. Fringant, and L. Moro, L., *Polymer*, Vol. 42, p. 6565, 2001.
62. D. Lourdin, H. Bizot, and P. Colonna, *Journal of Applied Polymer Science*, Vol. 63, p. 1047, 1997.
63. F.J. Rodriguez-Gonzalez, B.A. Ramsay, and B.D. Favis, *Polymer*, Vol. 44, p. 1517, 2003.
64. R.M.S.M. Thire, C.T. Andrade, and R.A. Simão, *Polímeros: Ciência e Tecnologia*, Vol. 15, p. 130, 2005.
65. J.J.G. Van Soest, and J.F.G. Vliegenthart, *Trends in Biotechnology*, Vol. 15, p. 208, 1997.
66. M.A.Woehl, MSc Thesis, Universidade Federal do Paraná, 2009; Available on http://hdl.handle.net/1884/19558.
67. C.H. Carraher Jr., *Polymer chemistry: an introduction*. 3rd ed. New York, Marcel Dekker, 1996.
68. M. A. Hunealt, H. L, N. Chapleau, and B.D. Favis. "Processing and properties of poly (lactic acid)/plasticized starch blends". In: World Polymer Congress and 41st International Symposium on Macromolecules. paper 841, 2006.
69. S. Godbole, S. Gote, M. Latkar, and T. Chakrabarti, *Bioresource Technology*, Vol. 86, p. 33, 2003.
70. R. Zhao, P. Torley, and P.J. Halley, *Journal of Materials Science*, Vol. 43, p. 3058-3071, 2008.
71. J.P. de Mesquita, C.L. Donnici, and F.V. Pereira, *Biomacromolecules*, Vol. 11, p. 473, 2010.

72. Prof. M. Saini, University of Toronto, Canada (Private Communication).
73. E.N.J. Ford, S.K. Mendon, S.F. Thames, and J.W. Rawlins, *Journal of Engineered Fibers and Fabrics*, Vol. 5, p. 10, 2010.
74. S.A. Rydholm, *Pulping Processes*, Interscience: New York (1965) 3.
75. H. Abe, R. Funada, "The Orientation of Cellulose Microfibrils in the Cell Walls of Tracheids" in *Conifers, IAWA Journal*, Vol. 26, issue No. 2, 161–174, 2005.
76. R.H. Atalla, "Structural transformations in celluloses" in "Steam explosion techniques: fundamentals and industrials applications" B. Focher, A. Marzetti and V. Crescenzi, eds., Philadelphia, Gordon and Breach, p. 97, 1988.
77. S. Hill, *New Scientist*, Vol. 153, p. 36, 1997.
78. R. Kozlowski and B. Mieleniak, "New trends in the utilization of byproducts of fiber crops residue in pulp and paper industry, building engineering, automotive industry and interior furnishing," In *Proceedings of the 3rd International Symposium on Natural Fibers and Composites* (ISNAPOL 2000), p. 504, 2000.
79. A.L. Leao, R Rowell, N.Tavares, "Application of Natural Fibers in Automotive Industry in Brazil—Thermoforming Process". Eds. Prasad, P. N., Mark, J. E., Kandil, S., Kafafi, Z. H., In *Science and technology of polymers and advanced materials*. Plenum Press: New York, 1998; pp. 755–760.
80. B. Dahlke, H. Larbig, H.D. Scherzer, and R. Poltrock, *Journal of Cellular Plastics*, Vol. 34, p. 361, 1998.
81. A.S. Herrmann, J. Nickel, and U. Reidel. *Polymer Degradation and Stability*, Vol. 59, p. 251, 1998.
82. M.J. John, and S. Thomas, *Carbohydrate Polymers*, Vol. 71, p. 343, 2008.
83. E.D. Teixeira, D. Pasquini, A.A.S. Curvelo, E. Corradini, M.N. Belgacem, and A. Dufresne, *Carbohydrate Polymers*, vo. 78, p. 422, 2009.
84. D. Pasquini, E.M. Teixeira, A.A.S. Curvelo, M.N. Belgacem, and A. Dufresne, *Industrial Crops and Products*, 2010, In Press, doI: 10.1016/j.indicrop.2010.06.022.
85. M.J. Sobkowicz, J.R. Dorgan, K.W. Gneshin, A.M. Herring, and J.T. McKinnon, *Journal of Polymers and the Environment*, Vol. 16, p 131, 2008.
86. M. Iguchi, S. Yamanaka, and A. Budhiono, *Journal of Materials Science*, Vol. 35, p. 261, 2000.
87. S. Koizumi, Z. Yue, Z., Y. Tomita, T. Kondo, H., Iwase, D. Yamaguchi, and T. Hashimoto, *The European Physical Journal E*, Vol. 26, p. 137, 2008.
88. J. Sugiyama, R. Vuong, and H. Chanzy, *Macromolecules*, Vol. 24, p. 4168, 1991.
89. N. Hayashi, T. Kondo, and M. Ishihara, *Carbohydrate Polymers*, Vol. 61, p. 191, 2005.
90. N. Hayashi, J. Sugiyama, T. Okano, and M. Ishihara, *Carbohydrate Research*, Vol. 305, p. 261, 1997.
91. A. Saxena, and A.J. Ragauskas, *Carbohydrate Polymers*, Vol. 78, p. 357, 2009.
92. A. Vazquez and V. A. Alvarez, "Starch–Cellulose Fiber Composites" Chapter 11, in, Long Yu, ed., *Biodegradable Polymer Blends and Composites from Renewable Resources*, John Wiley & sons Inc. Publications, 2008.
93. J.I. Morán, V.A. Alvarez, V.P. Cyras, and A. Vázquez, *Cellulose, Extraction of cellulose and preparation of nanocellulose from sisal fibers*, Springer Verlag, 2007.
94. M.A. Hubbe, O.J. Rojas, L.A. Lucia, and M. Saini,. *Bioresources*, Vol. 3, Issue 3, p. 929, 2008.
95. S. Kamel, *Express Polymer Letters*, Vol. 1, p. 546, 2007.
96. S. K. Lee, M. Sheridan, and A. Mills, *Chemistry of Materials*, Vol. 17, p. 2744, 2005.
97. J. Sriupayo, P. Supaphol, J. Blackwell, and R. Rujiravanit, *Carbohydrate Polymers*, Vol. 62, p. 130, 2005.
98. R.M. Brown, J.H.M. Willison, and C.L. Richarson, *Proceedings of the National Academy of Sciences of the USA*, Vol. 73, p. 4565, 1976.
99. K. Oksman, A.P. Mathew, D. Bondeson and I. Kvien, *Composites Science and Technology*, Vol. 66, p. 2776, 2006.
100. N. Ljungberg, J.Y. Cavaillé and L. Heux, "Nanocomposites of isotatic polypropylene reinforced with rod-like cellulose whiskers," *Polymer*, Vol. 47, Issue 18, 6285–6292, 2006.
101. M.M.D. Lima, and R. Borsali, *Macromolecular Rapid Communications*, Vol. 25, p. 771, 2004.
102. P. Podsiadlo, S.Y. Choi, B. Shim, J. Lee, M. Cuddihy, and N.A. Kotov, *Biomacromolecules*, Vol. 6, p. 2914, 2005.

103. W. Helbert, C.Y. Cavaillé, and A. Dufresne, *Polymer Composites*, Vol. 17, Issue 4, p. 604, 1996.
104. S.J. Eichhorn, C.A. Baillie, N.Zafeiropoulos, L.Y. Mwaikambo, M.P. Ansell, A.Dufresne, K.M. Entwistle, P.J. Herrera-Franco, G.C.Escamilla, L.Groom, M. Hughes, C. Hill T.G. Rials and P.M. Wild, *Journal of Materials Science*, Vol.36, Issue 9, p. 2107, 2001.
105. A. P. Mathew, and A. Dufresne, *Biomacromolecules*, Vol.3, p. 609, 2002.
106. A. Morin and A. Dufresne, *Macromolecules*, Vol. 35, Issue 6, p. 2190, 2002.
107. W. Hamad, *Cellulsoic Materials, Fibers, Networks and Composites*, Boston / Dordrecht / London: Kluwer Academic Publishers, 2002.
108. O.A. Battista and P.A. Smith. "Level-off d.p. cellulose products" U. S. Patent 2978446 (April 4, 1961).
109. Q.J. Wu, M. Henriksson, X. Liu, and L.A. Berglund, *Biomacromolecules*, Vol. 8, p. 3687, 2007.
110. Y.C. Hsieh, H. Yano, M. Nogi, and S.J. Eichhorn, *Cellulose*, Vol. 15, p. 507, 2008.
111. A. Sturcova, G.R. Davies, and S.J. Eichhorn, *Biomacromolecules*, Vol. 6, p. 1055, 2005.
112. Y. Chen, A. J. Stipanovic, W.T. Winter, D.B.Wilson, and Y.-J. Kim, *Cellulose*, Vol. 14, p. 283, 2007
113. L. Valtasaari, L., and K. Saarela,. *Paperi ja Puu*, Vol. 57, p. 5–10, 1975.
114. L. Jiang, E. Morelius, J. Zhang, and M. Wolcott, *Journal of Composite Materials*, Vol. 42, No. 24, P. 2629–2645, 2008.
115. I. Kvien, B.S. Tanem, and K. Oksman, *Biomacromolecules*, Vol. 6, p. 3160, 2005.
116. A.N. Nakagaito, and H. Yano, *Applied Physics A*, Vol. 78, p. 547, 2004.
117. D. Bondeson, and K. Oksman, *Composites*, Vol. 38, p. 2486, 2007.
118. H. Lönnberg, L. Fogelström, M. A. S. A. Samir, L.Berglund, E. Malmström, A. Hult, *European Polymer Journal*, Vol. 44, p. 2991, 2008.
119. L. Heux, G. Chauve, and C. Bonini, *Langmuir*, Vol. 16, p. 8210, 2000.
120. D. Liu, T. Zhong, P.R. Chang, K. Li and Q. Wu., *Bioresource Technol.* Vol.101, p. 2529, 2010.
121. G. Bogoeva-Gaceva, M. Avella, M. Malinconico, A. Buzarovska, A. Grozdnov, G. Gentile, M.Errico, *Polymer Composites*, Vol. 28, Issue 1, p. 98, 2007.
122. J. Araki, M. Wada, S. Kuga and T. Okano, *Colloids and Surfaces A: Physicochemical and Engineering Aspects*, Vol. 142, p. 75, 1998.
123. C. Bonini and L. Heux, French Patent FR 99.07493, 1999/2000.
124. C. Gousse, H. Chanzy, G. Excoffier, L. Soubeyrand and E. Fleury, *Polymer*, Vol. 43, p. 2645, 2002.
125. W. Gindl, and J. Keckes, *Composites Science and Technology*, Vol. 64, p. 2407, 2004.
126. E.E. Brown, and M.P.G. Laborje, *Biomacromolecules*, Vol. 8, p. 3074, 2007.
127. S. Yano, H. Maeda, M. Nakajima, T. Hagiwara, and T. Sawaguchi, *Cellulose*, Vol. 15, p. 111, 2008.
128. M. Grunert, W.T. Winter, *Journal of Polym Environment*, Vol. 10, Issue 1–2, p. 207, 2002.
129. T. Ebeling, M. Paillet, R. Borsali, O. Diat, A. Dufresne, J.-Y. Cavaille´, H. Chanzy. *Langmuir*, Vol. 15, Issue 19, p. 6123, 1999.
130. N. Ljungberg, Bonini, C. Bortolusi, F. and Boisson, C and Heux, L. *Biomacromolecules*, Vol. 6, Issue 5, p. 2732, 2005.
131. H. Takagi and A. Asano,.*Composites Part A*, Vol. 39, p. 685, 2008.
132. K.G. Satyanarayana, L.P. Ramos, F. Wypych, "Comparative study of Brazilian natural fibers and their composites with others." In: Thomas S, Pothan LA, editors, *Natural Fiber Reinforced Polymer Composites*, Philadelphia: Maney Publishers, p. 473, 2008.
133. I. Kvien, and K Oksman, *Applied Physics A: Materials Science and Process*, Vol. 87, p. 641, 2007.
134. B. Chen, J.R.G. Evans, H.C. Greenwell, P. Boulet, P.V. Conveney, A.A. Bowden and Whiting, A. *Chemical Society Review*, Vol. 37, p. 568, 2008.
135. N.L.G. Rodriguez, W. Thielemans and A. Dufresne, *Cellulose*, Vol. 13, p. 261, 2006
136. D.M. Bruce, R.N. Hobson, J.W. Farrent and D.G. Hepworth,. *Composites Part A*, Vol. 36, p. 486, 2005.
137. H. Yano, A. Hirose, P.J. Collins and Y.J. Yazaki, *Journal of Materials Science Letters*, Vol. 20, p. 1125, 2001.
138. C. Xiaodong, D. Hua and L. Chang-Ming, *Biomacromolecules* Vol. 8, p. 899, 2007.
139. M.Roman and W.T. Winter, *Biomacromolecules*, Vol. 5, p. 1671, 2004.
140. Vazquez and Alvarez, Private Communication.

141. V. Favier, R. Dendievel, G. Canova, J.-Y. Cavaille and P. Gilormini, *Acta Materialia*, Vol. 45, p. 1557, 1997.
142. F. Dalmas, L. Chazeau, C. Gauthier, J.-Y.Cavaille and R. Dendivel, *Polymer*, Vol. 47, p. 2802, 2006.
143. S.C. Tjong, *Materials Science and Engineering Reports, Vol. 53*, p. 73, 2006.
144. H. Varela-Rizo, M. Weisenberger, D.R. Bortz and I. Martin-Gullon, *Composite Science and Technology*, Vol. 70, 1189, 2010.
145. S. E Zweig, *Biopharm International*, Vol. 18, p. 36, 2005.
146. K.B. Lee, *Journal of Micromechanics and Microengineering*, Vol. 16, p. 2312, 2006.

18

Biogenic Precursors for Polyphenol, Polyester and Polyurethane Resins

Ali Harlin

VTT Technical Research Centre of Finland, Espoo, Finland

Abstract
Attention is widely paid to the use of biomass as a source of energy and transportation fuels [1–4] due both to the Rio declaration [5], and the EU target of 30% of chemicals should be prepared from renewable resources by 2025 [6, 8]. Especially with thermoplastics the target is understandable, as both the preparation of the required components and waste management of polymer materials are demanding. In this review more details are discussed: the value added class of resins, also called reaction polymers, which include thermoset epoxies, unsaturated polyesters, and phenolics as well as thermoplastic polyurethanes [9].

Keywords: Reaction polymers, biogenic, monomers, catalytic synthesis, glyserols, diols, hydroxy acids, caorboxylic acids, plastizers, polyols, furans, terpenes, phenols

18.1 Composite Materials

18.1.1 Reaction Polymers

The actual resin formations are complex requiring many additives and supplemental treatments. Complex systems required long development before they become useful materials in applications like composites, glues, coatings, elastomers and textiles. However, they provide marked potential for partial bio-replacement and bio-based additives especially through monomers and plastizers.

Urethanes or carbamates are formed in reaction between a diol and a diisocyanate, like that of 1,4-butanediol and hexamethylene diisocyanate, is a good example of a system that is partially bio-replaceable. Typically the diols are relatively easy to find, extract or convert from natural sources, while diisocyanates are the totally opposite.

Unsaturated polyesters are typically pre-crosslinked polyesters of an anhydride or diacid with diols. Diols are considered as potential bio-based components, where the issue is to either find synthetic routes to bio-based ethylene and propylene glycol or find other bio-based replacements. Unsaturated anhydrides

Srikanth Pilla (ed.) Handbook of Bioplastics and Biocomposites Engineering Applications, (511–554)

are still produced from basic petrochemicals, like maleic anhydride produced from benzene or butane in an oxidation process [10].

Epoxyresins are typically derived from epichlorhydrin with aromatic amines and phenols. The intermediate epichlorhydrin can be produced of bio-based raw materials and especially natural carboxylic acids, such as succinic acid, and phenols are also possible to find, while amines and anhydrides are more challenging to replace.

Phenol-formaldehyde resins are based on phenol pre-polymers called novolacs, which are possble to be replaced with several natural polymers, like tannins, lignans and lignins. Formaldehyde has been considered a problem for polymers, but especially natural phenolic polymers enable higher reactivity reducing residual formaldehyde, which could also be replaced with other aldehydes.

Amino-formaldehyde resins may be prepared using formulations including formaldehyde, glycerin and at least one amino compound selected from the group consisting of urea, melamine, and mixtures thereof under reaction conditions sufficient to prepare a resin. Bio-oil based glycerin is available, but even if urea is a metabolic molecule it is typically produced industrially through dehydration of ammonia and carbon dioxide as well as melamine from urea through von Liebig synthesis.

Polyimines are formed of polyamic acids which are reaction products of diamines and dianhydrides. Even if certain diamides, like cadaverine is the decarboxylation product of the biochemically available amino acid lysine, the useful compounds like hydrazine and ethyldiamine are not, and that is why polyimines are not discussed. Table 18.1 show certain possible monomers.

18.1.2 Hybrid Materials and Composites

Hybrid materials and composites can be made of the reactive polymers described above. A hybrid material is a material that includes two moieties blended on the molecular scale. Many natural materials follow this pattern when they consist of inorganic and organic building blocks distributed on the (macro) molecular or nano-scale. They are both materials that show weak interactions between the two phases, such as van der Waals, hydrogen bonding or weak electrostatic interactions, as well as those that show strong chemical interactions between the components [11–13].

As an example, in the formation of biopolymer-clay nanocomposites intercalation with monomers can be applied followed by *in situ* polymerization [14–17]. The hybrid composites containing more than one type of fiber reinforcement are motivated by the ability to combine advantageous features of various fiber systems—improved performance as well as reduced weight and cost [18]. This may lead to unexpected results like the incorporation of a moderate amount of carbon nano fiber into an ultra high modulus polyethylene that significantly improves the compressive strength, flexural modulus, and flexural strength [19]. Further examples of this system can be seen in nano-cellulose composites such as reinforced films with biopolymers [20], and polyurethane based shape memory polymers (SMPs) [21].

Table 18.1 Monomers for reactive polymers.

Nr	Substnce Name	Net Formula	GAS Number	Mw g mol^{-1}	Tm °C	Tb °C	Application
1	Glycerol	$C_3H_8O_3$	56-81-5	92.0944	17.8	290	Seeding monomer for branched polyesterification like rubbery polyhydroxy acids
6	3-Hydroxypropionic acid	$C_3H_6O_3$	503-66-2	90.0786	–25	143	Polymerization to polyester oligomers
9	Acrylic acid	$C_3H_4O_2$	79-10-7	72.06	14	141	Acrylic monomer
10	Acryl amide	C_3H_5NO	79-06-1	71.0786	84.5	125 at 25 mm Hg	Modifer and acceleration of urea formaldehyde resins
11	Succinic acid	$C_4H_6O_4$	110-15-6	118.089	183 – 185	235	Precursor for polyesters, polyurethanes, and polyethers
18	Ethylene glycol						Production of polyesters
19	1,3-Propylene glycol	$C_3H_8O_2$	504-63-2	76.095	–27	214	Production of polyesters
20	1,2-Propylene glycol	$C_3H_8O_2$	57-55-6	76.095	–59	–59	Optional production of polyesters
21	1,4-Butanediol	$C_4H_{10}O_2$	110-63-4	90.1218	16	230	Production of polyesters
22	*1,5-pentanediol*	$C_5H_{12}O_2$	111-29-5	104.1486	–16	242	Optional production of polyesters
23	Methyl-1,4-butanediol	$C_5H_{12}O_2$	626-95-9	104.1486		127 at 14 mm Hg	Optional production of polyesters
24	1,6-hexanediol	$C_6H_{14}O_2$	629-11-8	118.1754	45	250	Production of polyesters
25	Erythritol	$C_4H_{10}O_4$	10030-58-7	122.12	121	329–331	**Initiation monomer for urethanes**

Table 18.1 (cont.) Monomers for reactive polymers.

Nr	Substnce Name	Net Formula	GAS Number	Mw g mol^{-1}	Tm °C	Tb °C	Application
28	Epichlorohydrin	C_3H_5ClO	106-89-8	92.5249	–57	117.9	Converted to Bisphenol A diglycidyl ether for epoxy resins
29	Furfural	$C_5H_4O_2$	98-01-1	96.08	–36.5	161.7	Furfural resins
31	2,5-Furandicarboxylic acid	$C_6H_4O_5$	3238-40-2	156.0946	310–305	419.2	Replacement of tereftalic acid
34	Camphene	$C_{10}H_{16}$	79-92-5	136.24	45–46	159	Intermediate of fragrance materials, acrylates, terpene-phenol resins, and other derivatives
35	Limonene	$C_{10}H_{16}$	5989-27-5	136.24	–74.35	176	Polymerizable intermediate
42	Cardanol	$C_{21}H_{27}O$	37330-39-5		–20	225	Phenolic novolac, coating, lamination and as friction materials
45	Acrolein	C_3H_4O	107-08-2	56.06	–88	53	Preparation of polyester resin, polyurethane, propylene glycol, acrylic acid, acrylonitrile, and glycerol
45	3-Methyltetrahydrofuran	$C_5H_{10}O$	13423-15-9	86.13		86–87	Furan comonomer
47	Isosorbide	$C_6H_{10}O_4$	652-67-5	146.14	60–63		Diol monomer
48	Benzoazines				80–85		Phenol azine monomer

18.2 Biogenic Raw Materials

There are three strategies to convert the biogenic precursors, namely gasification to synthesis gas, thermal conversions to simple precursors, and utilization of highly selective catalysis [22]. In a direct analogy to a petroleum refinery, an economically attractive biorefinery should produce multiple products, including fuel, power, and bulk or fine chemicals, from biomass [23]. In addition, valorization of all components of biomass is essential for a viable biorefinery [24].

18.2.1 Sugar Platform

Biomass carbohydrates are the main renewable resources available even beyond lignin, and they are currently viewed as a feedstock for the green chemistry of the future [25–27]. D-glucose is the most abundant aldose occurring in nature. In many parts of the world D-glucose is an abundant carbon source, produced enzymatically from starch, sucrose or even cellulose, which can be metabolized by aerobic and anaerobic organisms. The use of microbial cellulose to generate D-glucose from cellulosic wastes is of considerable commercial interest and, as a consequence, much research is being done on the enzymes required. If it is possible to produce these aldopentoses economically from D-glucose through whole microbial cell or enzyme biocatalysts, it will also be possible to use them as valuable starting materials for high-value products [28–30]. Biotechnology is providing new, low-cost and highly efficient fermentation processes for the production of chemicals from biomass resources [31–34].

Applying bioconversion processes predicates the total use of lignocellulosic sugars obtained by acid or enzymatic hydrolysis cellulose hexoses and a complex mixture of hemicellulose pentoses (xylose, arabinose), hexoses (glucose, mannose, galactose) as well additional uronic acids, acetate, furfural and other aromatic compounds [35] metabolically convertible to chemicals [36–42]. Further rare sugars are widely studied and may be considered optional precursors as well [43–47].

Depolymerization of wood results in the formation of low-molecular mass components (sugars, phenols, furfural, various aromatic and aliphatic hydrocarbons, etc.) which are unique building blocks for further chemical synthesis. Such depolymerization can be done by hydrolysis in the presence of homogeneous acid catalysts (sulfuric acid). Wood biomass also contains many valuable raw materials for producing fine and specialty chemicals. These raw materials are carbohydrates, fatty acids, terpenoids, and polyphenols, such as stilbenes, lignans, and tannins [48].

18.2.2 Lipid Platform

Fats and oils obtained from vegetable and animal sources could become one of the major players in the chemical industry in the near future [49, 50]. The raw-materials are formed by mixed triglycerides having fatty acid moieties and are available in a large proportion of vegetable oils, such as coconut, palm, and palm kernel oils, soybean, rapeseed, and sunflower oils as well as from animal fat obtained from the meat industry, with beef tallow being the most abundant fat, and fish oil coming from the fishing industry.

There are various methods of oil splitting to produce fatty acids and glycerol: low-temperature Twitchell process with catalyst [51], medium-pressure autoclave splitting with catalyst [52, 53], continuous counter-current uncatalyzed high pressure method, and enzymatic fat-splitting [54]. Recently, use of solid acids such as zeolite, exchange resin, or silica-alumina for the hydrolysis of methyl esters for production has become more pronounced with fatty acids [55].

The oxidation of glycerol is a common process used for the formation of oxygenated compounds, and it generally occurs using stoichiometric mineral acid or fermentation routes. Glyserol is also proposed for a seeding monomer for branched polyesterification [56]. In general the derivatives of glyserol are numerous, but the market for these products has not developed yet because of the low selectivity and yield of the current oxidation processes.

18.2.3 Bio-based Aromates

When considered, the energy content of the different biomass products, terpenes top the list, followed by vegetable oils, and lignin. Since the production of terpenes is too low to meet the requirements for biofuels, it is not surprising that the most attention has been focused on vegetable oils. This leads to availability of vegetable oil based feed for the exiting petrochemical production, but terpenes require novel routes to be developed for chemicals.

Metabolic and enzymatic conversions have not yet developed viable routes to produce aromates. Even if shikimate is known, the toxicity of the compounds on the production organism is not solved [57–59]. Lignin is the only renewable source of an important and high-volume class of the aromatics. When considering the different kinds of structural motifs present in lignin from various biomass sources. The resources can be used also for modifiers and intermediates shown in Table 18.2.

18.2.4 Biogenic Olefin Platform

Ethylene is a main platform chemical in petrochemistry leading to various arenas of chemistry. Ethanol has been industrially produced from ethylene [60, 61] for a long time but the biogenetic route to produce ethylene through catalytic conversion of ethanol was applied by Sovay in 1960 in Brazil [62]. Ethanol is catalytically dehydrated to produce ethylene in endothermic reaction with conversion of ethanol with several proposed typically zeolite based catalytic systems [63–67]. With recycling ethanol in process [68] and developing process [69] it has been possible to reach production of 99.95% ethylene from a 95 wt% ethanol feed [70, 71]. Methanol is another alternative to catalytic conversion to ethylene and propylene [72–74]. With modified silicoaluminophosphate (SAPO)-34 molecular sieves have been achieved for feasible methanol to olefins (MTO) reaction [75, 76]. Technology is under commercialisation [77, 78]. Further methanol and ethylene metathesis results propylene.

Table 18.2 Promoters for reactions.

Nr	Substnce Name	Net Formula	GAS Number	Mw g mol^{-1}	Tm °C	Tb °C	Application
2	Glyceryl carbonate	C4H604 C4H803: Glycerol Formal is the mixture of 5-hydroxy-1,3-d ioxane and 4-hydroxymethyl-1,3- dioxolane (60:40)	931-40-8	118.088	152	360,4	Reacted with anhydrides to form ester linkages or with isocyanates to formurethane linkages typically in coatings or epoxide resins, dissolves polyemrs
3	Glycerol formal		225-248-9 and 226-758-4	104.11		193 –195	Seeding compound in phenol resins, adhesives, metal cleaning, textiles, leather processing
4	Glycolic acid	$C_2H_4O_3$	79-14-1	76.0518	70–74	112	biodegradable polymers and personal care, adhesives, metal cleaning, textiles, leather processing
5	Lactic acid	$C_3H_6O_3$	50-21-5	90.0786			biodegradable polymers and personal care
7	γ-valero lactone	$C_5H_8O_2$	108-29-2	100.117	–31	207–208	Intermediate, γ-Hydroxybutyric acid
8	α-methyl-γ-valero lactone	$C_6H_{10}O_2$		114.1424			

Table 18.2 (cont.) Promoters for reactions.

Nr	Substnce Name	Net Formula	GAS Number	Mw g mol^{-1}	Tm °C	Tb °C	Application
12	Levulinic acid	$C_5H_8O_3$	123-76-2	116.1164	33–35	245–246 115 at 1 mm Hg	Precursor for valerolactones 2,5-Furandicarboxylic acid precursor, optional bio fuel additive
30	5-hydroxymethyl furfural	$C_6H_6O_3$	67-47-0	126.1116	32–35		Used to achieve or improve specific properties in phenols,
32	2,5-Bis (hydro xymethyl) furan	$C_6H_8O_3$	1883-75-6	128.13	74–77	275	formaldehyde, urea, furfural res ins.Intermediate Metal-casting cores and moulds, corrosion-resistant coatings, polymer concretes, wood adhesives and binders, sand consolidation, low flammability and smoke materials, graphitic electrodes.
33	Furfyryl alcohol	$C_5H_6O_2$	98-00-0	98.10	–29	170	Used to isomerize trans-limonene into isoterpinolene, and
37	Terpinolene	$C_{10}H_{16}$	586-62-9	136.234	115	186	oxidation to terpin-olene erythritol
38	Cymene	$C_{10}H_{14}$ Mixture containg	99-87-6	134.21	–68	177	Solvent for dyes and varnishes
41	Cashew nut shell liquid	Cardenol	na	318,5	55–65	250 at 100 mmHg	Particleboard adhesive, cardanol precursor

18.3 Glyserols

18.3.1 Glyserol

Glyserol is a product of lipid platform. There are various methods of oil splitting to produce fatty acids and glycerol 1: low-temperature Twitchell process with catalyst [81], medium-pressure autoclave splitting with catalyst [82, 83], continuous counter-current uncatalyzed high pressure method, and enzymatic fat-splitting [84]. Recently, use of solid acids such as zeolite, exchange resin, or silica-alumina for the hydrolysis of methyl esters for the production has become more pronounced with fatty acids [85].

The oxidation of glycerol is a common process used for the formation of oxygenated compounds, and it generally occurs using stoichiometric mineral acid or fermentation routes. Glyserol is also proposed for a seeding monomer for branched polyesterification [86]. But in general the derivatives of glyserol are numerous although the market for these products has not been developed yet because of the low selectivity and yield of the current oxidation processes.

18.3.2 Epichlorohydrin

Ephichlorhydrin 28 is the main component in the production of epoxy resins with phenols like bisphenol-A. Crude glycerol could be converted economically into chlorinated compounds to epichlorohydrin [87]. Glycerol of first generation biodiesel production is considered as a future raw material for biogenetic epichlorhydrin, but current cost and availability of glycerol are demotivating for large scale investments. The same technology is possible to be used in the production of dichloropropanol as well.

18.3.3 Glyceryl Carbonate

Glyceryl carbonate 2 is a key bifunctional compound employed as a solvent, additive, monomer, and chemical intermediate. Glyceryl carbonate possesses a cyclic carbonate group [88] and a primary nucleophilic hydroxymethyl group that may be reacted with anhydrides [89] to form ester linkages or with isocyanates to form urethane linkages [90, 91]. The alkylene carbonate materials produced may be reacted with diamines to form polyurethane, which is used as a protective coat for wood and metal substrates [92].

Glyceryl carbonate can be produced by transesterification of ethylene carbonate with glycerol, using an alkaline base (Na_2CO_3) as a catalyst, at 298–308 K. The process needs neutralization steps and further distillation in order to recover GC [93]. It can also be produced by transesterification of dimethyl carbonate with glycerol in the presence of tetra-*n*-butylammonium bromide at 393 K, after 6 h, with 92% a yield [94]. Another method for the preparation of glyceryl carbonate consists of reacting glycerol with phosgene or diethyl carbonate in pyridine [95].

18.3.4 Glycerol Formal

Glycerol formal 3 is a mixture of 5-hydroxy-1,3-dioxane and 4-hydroxymethyl-1,3-dioxolane (60:40), which are cyclic ether compounds having two oxygen atoms in the ring structure and substituted by an alcohol group. The mixture is a viscous, colorless liquid having very little odor. The acetalization of glycerol with formaldehyde has been reported in the presence of different homogeneous acids, such as H2SO4, with or without of benzene, [96–98] or PTSA at a reflux of benzene with a yield of 90% [99, 100].

18.4 Acid Platform

Organic acids are reasonably easy to produce in biotechnical ways which also enables emerging production routes of valuable intermediates and vinyl monomers, like acrolein (45), acrylic acid (9) and acryl amide (10). See Figure 18.1.

18.4.1 Acrolein

Production of acrolein (45) has been investigated for the potential of glycerol dehydration as a versatile intermediate for the production of acrylic acid esters, superabsorber polymers or detergents [101]. Glyserol can be converted catalytically to acrolein, with an hydration and hydrogenation process the Degussa Company of Germany has developed as an industrialized technology for the production of 1, 3-propylene glycol by using acrolein as raw material [102]. Acrolein process is characterized by relatively mild reaction conditions, simple technology, mature hydrogen addition process, a simple catalyst system and low requirements for facilities; however acrolein itself is also an important organic intermediate which is high in cost and falls under severe toxic, inflammable and high explosive substances, hard to store and transport.

18.4.2 Hydroxy Acids

Lactic acid is the most known hydroxy acid. The D-lactic acid monomer is produced by fermentation and applied in production of thermoplastic biopolyesters [445].

Biodegradable lactic acid (5) based poly(ester-urethanes), PEU are thermoplastic elastomers like poly(L-lactic acid-co-DL-mandelic acid-urethanes) and poly(L-lactic acid-co-ε-caprolactone-urethane) having reasonable strength combined with significant elongation [103–109]. They are basically produced through an initiator (e.g. stannous octoate) catalyzed ring-opening polymerization (ROP) of L- or D,L-lactide in the presence of co-monomer, like linear poly(lactic acid) (BHMBA-PLA) using bis(hydroxymethyl) butyric acid (BHMBA) to possessing a pendent carboxylic acid group [110].

18.4.2.1 Glycolic Acid

Glycolic acid (4) is perhaps the best-known member of a group of chemicals called fruit acids or alpha-hydroxy acids (AHA) derived from sugar cane. Glycolic acid

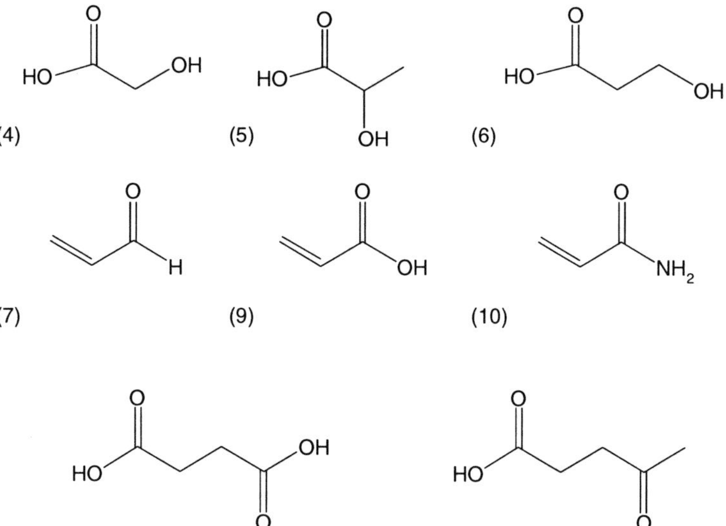

(4) (5) (6) (7) (9) (10) (11) (12)

Figure 18.1 Acids and acrylates: 4) glycolic acid, 5) lactic acid, 6) 3-hydroxypropionic acid, 7) acrolein, 9) acrylic acid, 10) acryl amide, 11) succinic acid and 12) levulinic acid.

is one of the most important fine chemicals, extensively used in adhesives, metal cleaning, textiles, leather processing [111], biodegradable polymers [112], and as a component in personal care product [113]. Ethylene glycol (18) is one of the cheap starting materials for the production of glycolic acid through an oxidation reaction. Microbial conversion of ethylene glycol to glycolic acid was expected to be an attractive alternative method for the value-added production of glycolic acid with no by-production [114]. Although the production of glycolic acid by microbial means was very attractive, the inhibition of glycolic acid was a key limitation for industrial application. The end-product inhibition by glycolic acid resulted in several problems, where the use of adsorbent resin system offered a simple way to remove products from an aqueous phase into a second solid phase [115, 116]. Invention of Gluconobacter oxydans DSM 2003 marked capacity to incompletely oxidize polyol substrates has led to numerous production processes for the synthesis of compounds [117–120], applied also for high productivity conversion of ethylene glycol to glycolic acid using anion exchange resin D315 as the adsorbent for selective removal of glycolic acid from the reaction mixture [121].

18.4.2.2 3-Hydroxypropionic Acid

3-Hydroxypropionic acid (6) is a structural isomer of lactic acid also produced from glucose fermentation. At the moment there is not a commercially viable production route from fossil fuel feedstocks. Like lactic acid, 3HPA has a bifunctionality that allows for multiple chemical transformations. The alcohol function of 3HPA can be dehydrated, leading to unsaturated compounds. Moreover, the bifunctional nature of 3HPA also allows polymerization to polyesters, oligomers, and cyclization to propiolactone and lactides.

18.4.3 Valerolactones

Levulinic acid (LA) can be easily and cheaply produced from lignocellulosic materials by using a simple and robust hydrolysis process [122–124]. Levulic acid can be converted to a γ-valero lactone (7) monomer by cyclization and hydrogenation at 215 °C and 55 bar in dioxane resulting in 96.1% selectivity and 100% conversion with a Rhenium catalyst and 98.5% selectivity and 98.7% conversion with an Iridium catalyst [125]. Further, the γ-valero lactone can be alkylated to provide an acrylic α-methyl- γ-valero lactone (**8**) monomer at 340 °C with formaldehyde in gas phase applying Ba-acetate/SiO2 catalyst. Also, this process has excellent yield over 95% [126].

18.4.4 Acrylic Acid

Acrylic acid (**9**) and its esters, salts, or amides are important compounds used as monomers in the manufacture of polymers and copolymers with numerous applications such as surface coatings, absorbents, textiles, papermaking, sealants, adhesives, etc. Acrylic acid has been obtained by thermal dehydration in the liquid phase of 3HPA at reduced pressure (4–5 kPa) using sulfuric or phosphoric acid catalysts in the presence of copper powder as a polymerization inhibitor at temperatures between 413 and 433 K. Yields of acrylic acid around 80% were obtained [127], but the process can also be run in similar conditions in the presence of alcohol [128]. Various heterogeneous catalysis has been applied, like NaH_2PO_4 supported on silica gel [129], in solution, and high-surface-area γ-Al2O3, Nafion NR50, montmorillonites, and EM-1500 zeolite in gas phase [130]. Conversion was high, namely 98% for the solution and 88% for gas phase processes.

There is also a method for producing acrylamides (10) and N-substituted acrylamides by heating mixtures of 3HPA and an amine in liquid or in vapor phase with or without the use of a catalyst which enhances the rate of the dehydration reaction. The preferred catalysts are solid acid catalysts such as high-surface-area SiO_2. However, low yields of acrylamides (between 20 and 50%) are generally obtained [131].

18.4.5 Succinic Acid

Succinic acid (11) is a versatile compound able to undergo a variety of reactions to useful products, and its production uses and reactions have been extensively reviewed in the literature. The development of improved fermentation microorganisms and separation technology reduces the overall cost of bio-based succinic acid [132–134]. Direct hydrogenation of succinic acid, succinic anhydride, and succinates leads to the formation of the product family consisting of BDO of great interest as a starting material for the production of important polymers such as polyesters, polyurethanes, and polyethers [135]. In order for fermentation-derived succinates to compete with butane-derived maleic anhydride, the production cost for succinic acid must approach the production cost for maleic anhydride [136].

18.5 Diols

Naturally occurring polyols like castor oil and sucrose can also be used to make synthetic polymeric polyols. Typical polyols used are glycols such as ethylene glycol (18), glyserol (1) or erythrol (25). See also Figure 18.2.

18.5.1 Ethylene Glycol

Ethylene glycol (18) is also a starting material in the production of polyesters and it is produced from ethylene via the intermediate ethylene oxide. Naturally, the used ethylene can also be of biogenic sources. Ethylene oxide reacts with water to produce ethylene glycol. This reaction can be catalyzed by either acids or bases, or can occur at neutral pH under elevated temperatures. The highest yields of ethylene glycol occur at acidic or neutral pH with a large excess of water. Under these conditions, ethylene glycol yields of 90% can be achieved. The major byproducts are the ethylene glycol oligomers diethylene glycol, triethylene glycol, and tetraethylene glycol [137].

18.5.2 Propylene Glycol

1,3-Propanediol (19) is also a starting material in the production of polyesters. It is used together with terephthalic acid to produce polytrimethylene terephthalate (PTT), which is in turn used for the manufacture of fibers and resins. This polymer is currently manufactured by Shell Chemical (Corterra polymers) and DuPont (Sorona 3GT). Natural glycerol could be converted to propylene glycol through an

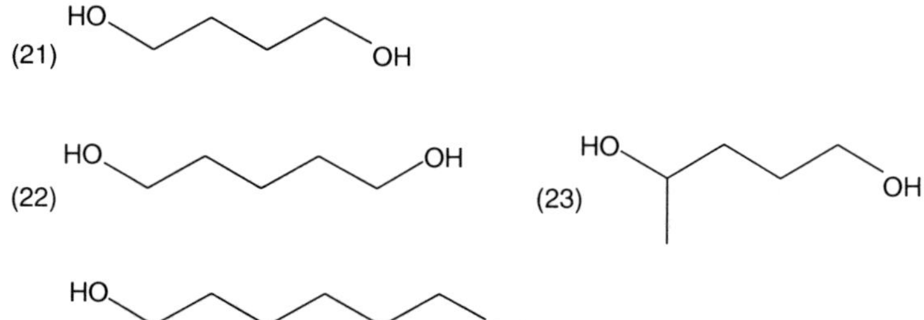

Figure 18.2 Diol compounds for polyurethanes and polyesters 18) ethylene glycol, 19) 1,3-propylene glycol, 20) 1,2-propylene glycol, 21) 1,4-butanediol, 22) 1,5-pentanediol, 23) methyl-1,4-butanediol, and 24) 1,6-hexanediol.

acetol intermediate at temperatures and pressures in the range of 423–523K and 1–25 bar, respectively [138].

Industrially, propylene glycol is a product of propylene oxide [139]. Used propylene can also be from biogenic sources. Manufacturers use either non-catalytic high-temperature processes at 473 K to 493 K, or a catalytic method, which proceeds at 423 K to 453 K in the presence of an ion exchange resin or a small amount of sulfuric acid or alkali. Propylene glycol can also be converted from glycerol, a biodiesel byproduct. The method is great in equipment investment, difficult in technology, complex in catalyst systems, rigorous in production technology and its ligand is severe toxic.

1,3-Propanediol (1,3-PD) is one of the oldest known fermentation products. It was reliably identified as early as 1881 by August Freund, in a glycerol-fermenting mixed culture obviously containing Clostridium pasteurianum as the active organism [140]. Research remained around wine-spoiling bacillus and enterobacteria nearly one-hundred years [141, 142] until the 1960s, when interest shifted to the glycerol-attacking enzymes and coenzyme B12 leading description [143] of 1,3-PD-forming clostridia in 1983 as part of a process to obtain a specialty product from glycerol-excreting algae[144].

Today there is considerable industrial interest in microbial 1,3-PD as it could compete with 1,3-PD made by petrochemistry [145–150], which was claimed in 1993 by Henkel [151]. However, the yields and productivities for such metabolically tailored pathways are still insufficient. It is therefore a challenge for both biochemical and metabolic engineering to develop improved biotechnological processes. These processes could be based on either two genetically and physiologically optimized organisms in one or two-stage fermentation, or a single-stage fermentation with one organism having the combined pathways together with improved gene regulation and the desired cellular functions.

Several strategies are therefore being pursued to reduce the costs of the biotechnological process. In one extended improvement of already existing 1,3-PD fermentations by increasing the gene dosage for limiting steps and/or by knocking-out genes responsible for undesired results: Glycerol dehydratase is limiting by-a product formation enzyme for 1,3-PD production in C. butyricum and K. pneumoniae respectively. [152, 153] On the other hand use of glucose should be considered, which is considerably cheaper than glycerol. The processes are mild in condition, simple in operation, less in accessory substance and environmentally friendly. Further options for the preparation method of 1,3-propylene glycol are as follows.

There are few catalytic processes useful for preparing 1,3-propanediol from 3-hydroxypropionic esters [154–157]. Recently, conversions and selectivity to 1,3-butanediol (100%) has been reached by applying a nano CuO/SiO2-based catalyst at temperatures between 393 and 473 K and at hydrogen pressures between 10 and 136 atm by using a liquid-phase slurry process for the hydrogenation of 3-hydroxy esters, using as a solvent a mixture of an alcohol and a high-boiling-point solvent such as tetraethylene glycol dimethyl ether or sulfolane. Under these reaction conditions, only a small amount of lactone is formed [158].

DuPont and Tate & Lyle PLC have undertaken to produce 1,3-propanediol using a proprietary fermentation and purification process based on glucose. In 2006 DuPont opened a factory in London, Tennessee, to produce commercial-scale quantities of bio-based 1,3-propanediol from corn sugar.

18.5.3 1,2-Propylene Glycol

Industrial production of 1,3-propylene glycol products includes 20% 1,2-propanediol, 1.5% of dipropylene glycol and small amounts of other polypropylene glycols. Pure 1,2-propylene glycol (20) can be prepared from bio-based glycerol. In this method, a CuO—CeO_2—SiO_2 catalyst is filled into a fixed bed reactor, a glycerol solution is flowed into the reactor together with hydrogen gas in a manner of top feeding, and controlling the reaction temperature to be 443–473° C., the reaction pressure to be 1.0–5.0 MPa, so as to realize the production of 1,2-propylene glycol from the hydrogenation of glycerol. The catalyst used in this invention can sustain a high selectivity for the target product and a high conversion for glycerol for 500 hours [159].

18.5.4 1,4-Butanediol (BDO)

The majority of 1,4-butanediol (21) BDO is currently produced commercially by the Reppe process in which acetylene is reacted with formaldehyde [160]. However, the process suffers several disadvantages, such as severe reaction conditions and the use of acetylene (with explosion hazard) and formaldehyde (with carcinogenic effects). A promising alternative to this process is the hydrogenation of maleic anhydride to BDO via a multistep reaction. Maleic anhydride is hydrogenated to succinic anhydride, which is then converted to GBL. The step enables utilization of biogenic succinic acid.

The hydrogenation of GBL leads to 1,4-butanediol in a reversible reaction, and depending on the reaction conditions, the dehydration of BDO to THF is observed. Byproducts are propionic acid and butyric acid, the corresponding aldehydes and alcohols, ethanol, and acetone [161]. The maleic anhydride process performed on a Pd/Ag/Re catalyst on carbon provides a 93% yield of BDO [162]. 149 However, a route that is widely used commercially starts with the fast formation of diethyl maleate from maleic anhydride and ethanol, catalyzed by an ion-exchange resin. Diethyl maleate is hydrogenated to GBL and then to BDO in two reaction steps in the vapor phase over bulk Cu-Cr or Cu-Zn reduced mixed-oxide catalysts. The reaction is carried out at temperatures around 473 K, mild pressures (30–40 bar), and high molar hydrogen/ester ratios [163].

18.6 Higher Diols

18.6.1 1,5-Pentadiol

Bio based 1,5-pentadiol (22) or higher are not yet commercially available. As an example glutamic acid may enable production of 1,5-pentadiol, but hydrogenations of carboxylic acids and esters require, in general, very high pressures, temperatures, and hydrogenations [164]. Glutamic acid is a non-essential amino acid which is found in abundance in plant and animal proteins. Currently, most of the production of glutamic acid is based on bacterial fermentation. In this method, bacteria are grown aerobically in a liquid nutrient medium containing sugar as a carbon source, a nitrogen source such as ammonium ions or urea, and mineral ions and growth factors. The bacteria excrete glutamic acid into the medium, which accumulates there. The glutamic acid is separated from the fermentation broth by filtration, concentration, acidification, and crystallization [165–167].

18.6.2 Methyl-1,4-butanediol

Also, 2-methyl-1,4-butanediol (23) and 3-methyltetrahydrofuran (46) are available through hydrogenation of itaconate esters which has been described using Pt-Re250 and Cu-based catalysts. [168, 169] Itaconic acid is produced by the filamentous fungi *Aspergillus terreus* and *Aspergillus itaconicus* from carbohydrates like sucrose, glucose, and xylose [170–172].

18.6.3 1,6-Hexanediol

Glucose hydrogenation over Raney nickel [173] or recently, Ruthenium catalyst [174, 175], is an important reaction for production of sorbitol [176], whose annual production was more than one million tons in 2000 [177]. Sorbitol has been reported to be converted to 1,6-hexanediol (24) by means of catalytic hydrogenolysis at 513 K and 130 bar in water and presence of a CuO/ZnO catalyst with only 35% selectivity with 98.4% conversion [178].

18.6.4 Isosorbide

Isosorbide (47) is a heterocyclic compound derived from glucose by acid catalyzed protonation on primary hydoxyls in four steps trough sorbitol to isosorbide [179]. The monomer is applied for various polymers[180], such as polyamides [181, 182] and polyesters [183, 184].

18.7 Polyols

18.7.1 Erythritol

Erythritol (25) is a biological sweetener which is a possible initiation monomer for urethanes. Large-scale production of erythritol uses fermentative processes with pure glucose, sucrose and dextrose from chemically and enzymatically hydrolyzed wheat and corn starches used as major carbon sources [185, 186]. Erythritol is produced by fermentation involving yeast-like fungi such as Trigonopsis variabilis [187], Trichosporon sp [188]. Torula sp [189]., Moniliella sp [190]., and Candida magnoliae [191]. Further Leuconostocoenos can also produce erythritol but only under anaerobic conditions [192]. A high initial concentration of glucose favors erythritol production by osmophilic microorganisms. Generally, an increase in the initial glucose concentration increases the production rate and yield in a batch process if the microorganisms can tolerate a higher concentration of sugar and a higher osmotic pressure. Erythritol has been produced commercially using a mutant of Aureobasidium [193]. Recent investigations have shown that an acetatenegative mutant of Y. lipolytica Wratislavia K1, can produce simultaneously high amounts of erythritol and citric acid under nitrogen-limited conditions [194]. This development has led to high-yield production of erythritol from raw glycerol [195].

18.7.2 Polyols

Polyols (17) produced from epoxidized fats (16) and alcohols are of interest for the production of foams, dispersants, and fluid polyurethane resins [196]. Lipids are a topic for further development of surfactants and bolamampiphiles useful in polysaccharide compatibilization to hydrophobic polymers [197].

18.7.3 Polyglyserols

Catalytic polymerization of glyserols (26) lead to several diglycerols as well as to tri-, tetra-, and higher glycerols. The reactions can be catalyzed by hydroxides, carbonates, and oxides of several metals [198], but the alkaline polymerization of glycidol offers more selective processes [199, 200]. Further alkoxylated polyglycerols can be formed through condensation of glycerol in the presence of NaOH, followed by alkoxylation with ethylene oxide and/or propylene oxide [201]. Also, use of solid catalysts such as high-alumina zeolites has also been described [202, 203].

18.7.4 Polyol Modification

Polyesters with carbohydrate or polyol repeat units in the chain can be produced by chemical methods [204]. However, elaborate protection-deprotection steps [205–208] are needed to avoid cross-linking between polyol units. Multistep routes to non-cross-linked polyol polyesters limit the potential of their practical use. Lipases and proteases are well known to provide regioselectivity during esterification reactions at mild temperatures [209, 210].

These characteristics motivated their study as catalysts for selective polyol polymerizations. The activation of carboxylic acids with electron-withdrawing groups was thought to be necessary for enzymecatalyzed copolymerizations with polyols [211–215]. An obstacle to lipase- or protease-catalyzed polymerizations of polyols is their insolubility in nonpolar organic media. Polyols are soluble in polar solvents [216–219] such as pyridine, dimethyl sulfoxide, 2-pyrrolidone, and acetone. However, these solvents cause large reductions in enzyme activity [220]. High molecular weight soluble polymers can be produced with highly active and selective lipase as the catalyst as well as adjusting the reaction mixture without the need to activate monomer acid groups or add an organic solvent [221].

18.8 Plastizers

18.8.1 Terpene Phenolic Resin

Terpene phenolic resins (10) are a low softening point resin with broad polymer compatibility. These thermoplastic resins are tackifiers, hot melt and pressure sensitive adhesives providing improved tack, peel and flexibility, for example, EVA, SIS and acrylic based systems. When formulated as a co-resin they significantly improve the performance properties of rubber-based adhesives with limited impact on heat resistance. Because of these properties they are also useful as aliphatic polyester softeners [222].

18.8.2 Sterols

Sterols are unsaponifiable neutral components in tall oil, also called pitch (15) which can be utilized in further transformations for production of fine and specialty chemicals. The unsaponifiable mixture of sterols in crude soap from the sulphate cellulose process contains 5% α-sitosterol and 95% β-sitosterol [223], the latter can be isolated from this mixture. The sterols can be used optionally as plastizers but typically they are hydrogenated and esterified to sitosterol, which find major applications in health promoting products [224].

18.8.3 Rosin Acids

The higher plants contain terpenoid-based substances commonly known as resins, like abietic acid. These resins may polymerize on exposure to air, and the polymerization *in situ* in dead plants gives rise to the resinite material found in almost all coals [225, 226]. The most important application areas for resin acids are in paper sizing to control water absorptivity, production of synthetic adhesives and surface coatings, as well as the production of synthetic rubbers, paints, and pharmaceuticals [227].

Rosin acids are easily oxidized and in order to avoid oxidation they can be hydrogenated over a Raney nickel catalyst [228]. The products in the hydrogenation of abietic acid over a Raney Ni catalyst at 443K and 60 bar hydrogen were dihydro- and tetrahydro-resin acids as well as dehydroabietic acid. Most of the literature on catalytic resin hydrogenations considers hydrogenation of the ethylenic double bonds in abietic acid, since it can be easily oxidized causing the color degradation of resin acids [229]. Hydrogenation of resin acids (14) over heterogeneous catalysts has been investigated extensively [230–238].

The chemistry of the hydrocracking of the rosin acids has been addressed markedly little in the literature. Diterpenoid compounds can be successfully hydrogenated and cracked to cycloakanes and hydroaromatics using supported NiMo and Ni-Y catalysts. With careful tailoring of the process temperature and time of reaction in the region of 20 minutes at 723K, high quality distillates could be obtained containing toluene and cymene, which are possible precursors for typical petrochemical aromate synthesis [239].

18.8.4 Epoxidized Plant Oils

Besides the traditional uses of oleochemicals [240] as surface active compounds (alkanoamides, alkyloamines, sulfonic derivatives), soaps, detergents (mono- and di-aliphatic acid glycerides, sulfonated monoglycerides), coatings and in the cosmetics, textile, and pharmaceutical industries, derivatives such as epoxidized plant oils (16) find uses as plasticizers and stabilizers in the production of PVC [241].

The epoxides change the solubility and flexibility of the PVC resins and react with hydrochloric acid liberated from the PVC resins under the prolonged action of light and heat. The applicability of an epoxidized oil depends on its purity, oxirane number, and iodine number; the commercial specifications are an oxirane number of 6.5% and an iodine number below 2.5 (iodine number is the mass of iodine in grams that is consumed by 100 grams of a chemical substance) [242]. The most important and most applicable epoxidation processes follow the Prileschajew reaction [243], which uses organic peracids, such as peracetic or performic acid, obtained through the catalyzed oxidation of corresponding carboxylic acids [238]. A concurrent way to prepare peracids [245, 246] is that immobilized lipases from *Candida antarctica*, such as Novozym 435, become active for the conversion of fatty acids with hydrogen peroxide to peroxy fatty acids yielding 72–91% after 16 h [247].

18.9 Furans

Thermal dehydration of pentoses and hexoses in acid media leads to the formation of three important nonpetroleum basic chemicals: furfural, or 2-furancarboxaldehyde (29) arising from dehydration of pentoses, 5-hydroxymethylfurfural HMF (30) arising from dehydration of hexoses, and levulinic acid arising from hydration of HMF [248]. See Figure 18.3.

Furfural (29) has been an industrial commodity for many decades because it can be prepared quite readily and economically from a vast array of agricultural or forestry wastes, all containing *pentoses* in sufficient amounts to justify a commercial exploitation: corn cobs, oat and rice hulls, sugar-cane bagasse, cotton seeds, olive husks, wood chips, etc [249].

It appears to be the only unsaturated large-volume organic chemical prepared from carbohydrate resources and is a key derivative for the production of important nonpetroleum-derived chemicals competing with crude oil [250, 251]. The reaction involves hydrolysis of pentosan into pentoses, mainly xylose, which under high temperatures (473–523 K) and in the presence of acid catalysts, mainly sulfuric acid [252]. Under these conditions, the selectivity to furfural is not higher than 70%, and only when continuous extraction with supercritical CO_2 is performed is 80% selectivity reached [253]. Most of the furfural produced worldwide is converted into furfuryl alcohol (33) by simple reduction processes [254].

This compound finds a variety of applications in the chemical industry [255]. It is mainly used in the manufacture of resins as a starting material for the

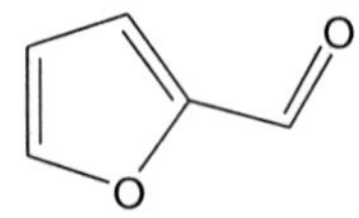

(29)

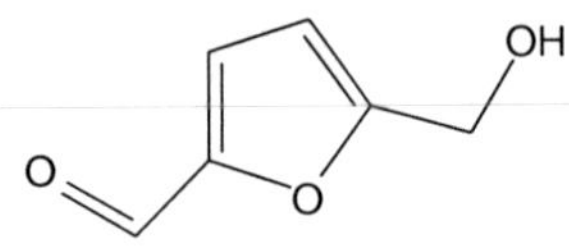

(30)

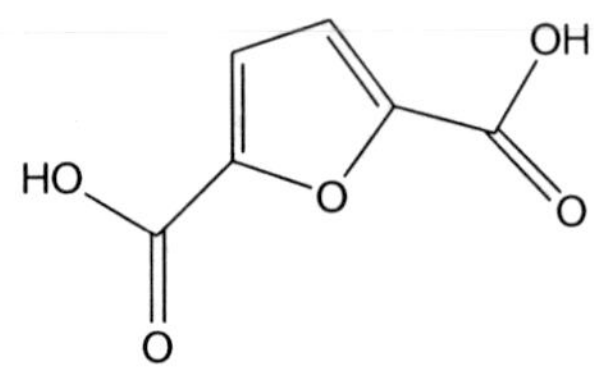

(31)

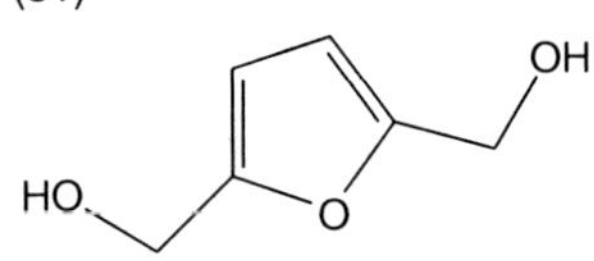

(32)

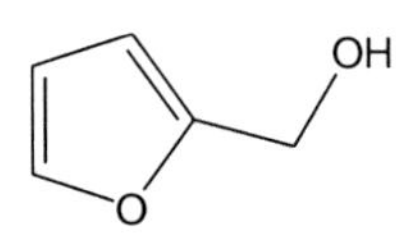

(33)

Figure 18.3 Furfural an furan resin components proposed several times also as a replacement of aromatic compounds 29) furfural, 30) 5-hydroxymethylfurfural, 31) 2,5-furandicarboxylic acid, 32) 2,5-bis(hydroxymethyl) furan and 33) furfyryl alcohol.

manufacture of furfuryl alcohol (33), and it is also an important chemical intermediate for the manufacture of fragrances, vitamin C, and lysine. In gas-phase hydrogenation, selectivities to furfuryl alcohol between 35% and 98% have been reported [256, 257], while in liquid-phase hydrogenation, selectivities in the order of 98% were found [258–260].

In recent years the chemistry of hydroxymethyl furfural HMF (30) and its derivatives has been more pronounced in research [261–265]. The synthesis of HMF is based on the acid-catalyzed triple dehydration of hexoses (mainly glucose and fructose), but oligo- and polysaccharide wastes can be used as well [266]. The formation of HMF is very complex with about 37 products; besides dehydration, it includes a series of side reactions such as isomerization, fragmentation, and condensation which strongly influence the yield of the process [267, 269].

HMF (30) especially possesses a high potential industrial demand, and it has been called a "sleeping giant" [269] and one of the new "petrochemicals readily accessible from re-growing resources" [270] with most versatile intermediate chemicals of high industrial potential and simple large-scale transformations like 2,5-furandicarboxylic acid (31) replacing tereftalic acid.

18.9.1 2,5-Furandicarboxylic Acid

2,5-Furandicarboxylic acid FDCA (31) is a compound with high potential applications in the polymers field because it can replace terephthalic, isophthalic, and adipic acids in the manufacture of polyamides, polyesters, and polyurethanes [271, 272]. HMF has been oxidized to FDCA using conventional oxidants such as nitric acid [273, 274]. The diacid is found to be the exclusive product [275], but found that the oxidation was not selective and significant amounts of byproducts (mainly 5-formyl-2-furancarboxylic acid) were also formed. Better results were found in the electrochemical oxidation of HMF using a nickel oxide-hydroxide electrode in an alkaline aqueous solution, and a 71% yield of FDCA was reported [276, 277].

18.9.2 2,5-Bis(hydroxymethyl)furan

2,5-Bis(hydroxymethyl) furan BHMF (32) is a valuable product in the furan family, useful as an intermediate in the synthesis of drugs [278], crown ethers [279], and polymers [280, 281]. BHMF is generally produced by two catalytic routes: the hexose route through reduction of HMF, and the pentose route through the hydroxymethylation of furfuryl alcohol with formaldehyde. Although there are various reports on the reduction of HMF with sodium borohydride [282], BHMF has been mainly obtained by catalytic hydrogenation of HMF. Thus, copper chromite [283], nickel, platinum oxide, cobalt oxide, molybdenum oxide, sodium amalgam [284], and Cu/Cr catalysts [285] have been used to perform the hydrogenation of HMF to BHMF.

18.9.3 Furfyryl Alcohol

The literature on the resinification of furfyryl alcohol (33) and on the properties and applications of the materials is wide, and is a topic of several articles and patents [286–291]. Areas in which these polymers find a successful and sometimes irreplaceable usage include metal-casting cores and moulds, corrosion-resistant coatings, polymer concretes, wood adhesives and binders, sand consolidation and well plugging, materials possessing low flammability and low smoke release, and carbonaceous products comprising industrial graphitic electrodes. Although the major component of all these resins is furfyryl alcohol, many "comonomers" like phenols, formaldehyde, urea, furfural, and 2,5-bis(hydroxymethyl)furan have been used to achieve or improve specific properties. The "simplest" system of furfyryl alcohol is that of an acidic medium [292]. The actual mixture of products has a brown color and already contains more complex structures, including unexplained aliphatic moieties [293, 294].

18.9.4 Furfural Resins

Polycondensation reactions represent clear-cut step polymerizations with well-understood mechanisms and macromolecular architectures, whereas others, often termed 'resinifications", are characterized by ill-defined reaction pathways and consequently by polymer structures which are far from being straightforward. Furfural (29) gives self-condensation products when initially anhydrous conditions submitted to a prolonged thermal treatment at 373–523 K in the dark [295]. The net result of this process was a slow, but progressive accumulation of the black crosslinked product. Dynamic equilibration is reached between the formation of soluble colored condensation products and their subsequent precipitation induced by further growth, and no evidence for ring-opening reactions is detected.

Furfural (29) is particularly sensitive to resinification, a feature which has been known for decades [296], and the classical product of this generic process, induced by acids and bases including zeolites [297], but also, to a lesser extent, by high temperatures in neutral conditions, is a black insoluble solid. When aqueous acidic media are used, the condensation reactions are accompanied by an hydrolytic ring opening, which is a general feature of the furan heterocycle. The latter reaction is best described as the conversion of the unsaturated cyclic structure into aliphatic open-chain products bearing functions which depend on the specific furan derivative used. Thus, for example, with 2,5-disubstituted furans, but also with the corresponding monosubstituted homologues and with furan itself.

Much work has been devoted to the study of resins in which furfural is coupled with other reagents, like phenols, bisphenols and acetone. For commercial use prepolymers and its application involves *in situ* crosslinking during processing, just like with more widespread resins such as formaldehyde-based compositions [298]. The major mechanism in these cationic polymerizations involves the vinyls of 2-furfurylidene methyl ketone and its homologues, but condensation reactions involving carbonyl groups and electrophilic substitution at a fifth carbon also occur [299, 300]. These resins have found several applications as adhesives, and corrosion-resistant coatings and floors. Furfural-based resins also include compositions with comonomers like quinacetophenone coupled with various substituted benzoic acids [301], hydroxyquinoline [302], and cardanol [303], which were examined in terms of their chelating properties towards metal ions. However their industrial success is lagging.

18.10 Terpenes

Numerous catalytic chemical processes have been developed for the production of valuable products from terpenes through hydrogenation, oxidation, isomerization/rearrangement, hydration, hydroformylation, condensation, cyclization, ring contraction, etc. see Figure 18.4. The main terpenes and terpenoids that we have considered as building blocks are pinene, limonene (35), carene, geraniol/nerol, citronellol, citral, and citronellal. The most important sources of terpenes are the turpentine oleoresins extracted from coniferous trees and terebinth and the essential oils obtained from citrics.

The isomerization of α-pinene in the presence of acid catalysts has been widely studied, and it produces a complex mixture of mono-, bi-, and tricyclic terpenes. The main products obtained are camphene and limonene, with selectivity and

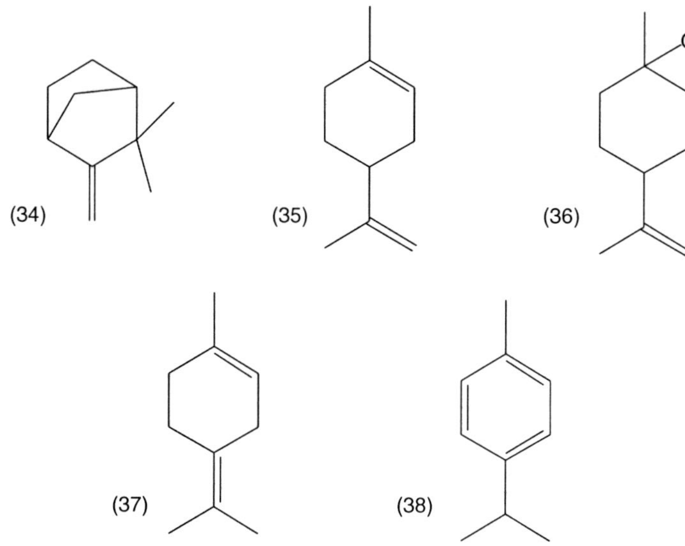

Figure 18.4 Examples of pinene based composites: 34) camphene, 35) limonene, 36) limonene oxide, 37) terpinolene and 38) cymene.

conversion depending on the nature, strength, and number of acid sites of the catalyst [314].

18.10.1 Camphene

Camphene (34) is used as an intermediate in the chemical industry for production of fragrance materials, acrylates, terpene-phenol resins, and other derivatives. Industrial methods for the isomerization of pinene over TiO2 catalysts under normal pressure at temperatures above 373 K yields camphene, limonene, tricyclene, and small amounts of flenchenes and bornylene around 75–80% [315].

Solid acids such as zeolites and modified clays as well as ZMS-5 have been largely used and studied as catalysts for the isomerization of pinene [316–319]. Typically selectivity to camphene was constantly around 30% independently of the conversion level up to 90% [320].

Kaolin[321] based catalysts are more promising , like kaolinitic acid-treated clay [322] for the pinene isomerization at 373 K and obtained a pinene conversion of 67–94% and selectivities to camphene and limonene of 65 and 23%, respectively.

18.10.2 Limonene

Limonene (35) is achieved from essential oils like citrus oil, but from pinene more preferably in metal catalytic liquid- and gas-phase isomerization [323] of α- and β-pinene over metal(IV) (Sn, Ti, Zr) phosphate polymer at 438 and 573 K. The maximum conversion of pinene obtained in the gas phase was 88–94%, and the main products formed were limonene and pironene, and camphere only in low amounts. The isomerization reactions gave generally complex mixtures because of the occurrence of secondary reactions of disproportionation, aromatization, and polymerization, together with aromatization of cyclic olefins [324, 325].

18.10.3 Limonene Oxide

Limonene oxide (36) is an active cycloaliphatic epoxide with low viscosity, and it may also be used with other epoxides in applications including metal coatings, varnishes, and printing inks. Marked amount of work has been done on the heterogeneous catalytic epoxidation of limonene. The catalysts used include heterogenized Co- and Mn-salen, porphyrins, ruthenium and cobalt complexes, as well as polyoxometalates, Titanium substituted zeolites and mesoporous materials, and hydrotalcites [326–351].

18.10.4 Terpinolene

The preparation of terpinolene (37) at 448 K gives selectivities of 70.5% at 64.2% conversion [352], while somewhat higher selectivity 75% at 41% conversion to terpinolene was obtained by isomerization of limonene in the presence of orthotitanic acid [353]. A basic catalyst was, however, used to isomerize trans-limonene into isoterpinolene. Thus, using a high-surface-area sodium/alumina, which was partially deactivated, a 28–31% yield of isoterpinolene was achieved [354].

18.10.5 p-Cymene

Bicyclic 3-carene occurs naturally in turpentine in contents around 60%, together with α- an β-pinene. The main drawback of turpentine stability is the easy oxidation of 3-carene on exposure to air. Thus a preferable alternative is its conversion into a mixture of cymenes (38), which finds a large number of applications in chemicals for example, as a solvent for dyes and varnishes. Pure acid catalysts such as partially exchanged Y and ZSM-5 zeolites, 3-carene could be converted into cymenes but with a low selectivity [355]. Better results were obtained with a Cr2O3-on-Al2O3 catalyst that gave 37% *m*-cymene and 49% *p*-cymene when starting from 2-carene, and 43% of *m*-cymene and 53% of *p*-cymene from 3-carene [356, 357].

However, *p*-cymene's actual main use involves its conversion to cresol. In the literature, there are examples of aromatization of pinene using solid catalysts. Pinene conversion over bifunctional calcined, impregnated H3PW12O40-*x*H2O (33%)-mesoporous provided at 313–433 K conversion of pinene close to 100%, with a yield of cymenes of 70% [358]. Transition metal-based materials with Pt or Pd constitute excellent catalysts for hydrogenation and dehydrogenation processes, while zeolites gave marked amounts of menthenes and carvomenthenes [359, 360]. When the acidity of ZSM-5 was eliminated by the presence of Na+ and a Pd-Ce/Na-ZSM-5 catalyst was used to catalyze the transformation of limonene into *p*-cymene, a selectivity up to 80% was obtained without *m*- or *o*-cymenes, the remaining 20% being *p*-menthanes and *p*-menthenes [361–363]. The yield was increased to 92% in the presence of olefins (1-decene and 1-undecene) as hydrogen acceptors, working at 453 K [364].

18.10.6 Benzoazines

Condensation reaction of primary amines with formaldehyde and substituted phenols for the synthesis of benzoxazine monomers (48) with Mannich bridge is welldefined [304–306]. Various types of benzoxazine monomer can be synthesized using various phenols and amines, such as p-cresol based benzoxazine by using aniline, formaldehyde and p-cresol as starting materials in dioxane [307–309]. Polybenzoxazine have a wide range of interesting features compared to conventional novolac and resole type phenolic resins [310–312], such as (i) near-zero volumetric change upon curing, (ii) low water absorption, (iii) for some polybenzoxazines Tg much higher than cure temperature, (iv) high char yield, (v) no strong acid catalysts required for curing, and (vi) release of no toxic by-product during curing [313].

18.11 Phenols

The oil crisis of the 1970s led to increased interest in plant based polymeric resins, and significant research developments on tannin-based resins were achieved in South America, Australia and South Africa [365]. Tannins and cashew nut shell liquid (CNSL) are groups of natural resins that are receiving wide attention as substitutes to synthetic binders in the production of biocomposites [366]. See Figure 18.5 and Table 18.3.

18.11.1 Novolac-type Phenolic Resins

Different types of cardanol-based novolac-type phenolic resins are produced under a wide range of operating conditions for application in resin producing industries, like special phenolic resins for coating, lamination and as friction materials [367]. Cashew nut shell liquid CNSL (41) is an agricultural byproduct from the cashew tree (Anacardium Occidental), which is a source of unsaturated hydrocarbon phenol and behaves as an excellent monomer for thermosetting polymer production [368–369]. CNSL polymerizes either by polycondensation with electrophilic compounds, such as formaldehyde, furfuraldehyde or by chain polymerization through the unsaturation, presents in the side chain using acid catalysts or by the functionalization at the hydroxyl group and subsequent oligomerization to obtain a functionalized pre-polymer [370–374].

During the extraction process of CNSL, cardanol results in different mixtures of saturated and unsaturated phenol compounds [375–378]. Substantially cardanol free cashew nut shell liquid (CNSL) and/or bhilavan nut shell liquid (BNSL) is available in process for producing phosphated polyols from CNSL/BNSL, where cardanol is removed from CNSL/BNSL by heat treating the same either in the presence of a catalyst or directly under vacuum [379]. On the other hand cardenols are used to modify phenol formaldehyde resins like laminated papers [380].

Phenalkamines epoxy curing agents are obtained through amination of cardanol [381]. Epoxide-containing polycardanol was also synthesized enzymically

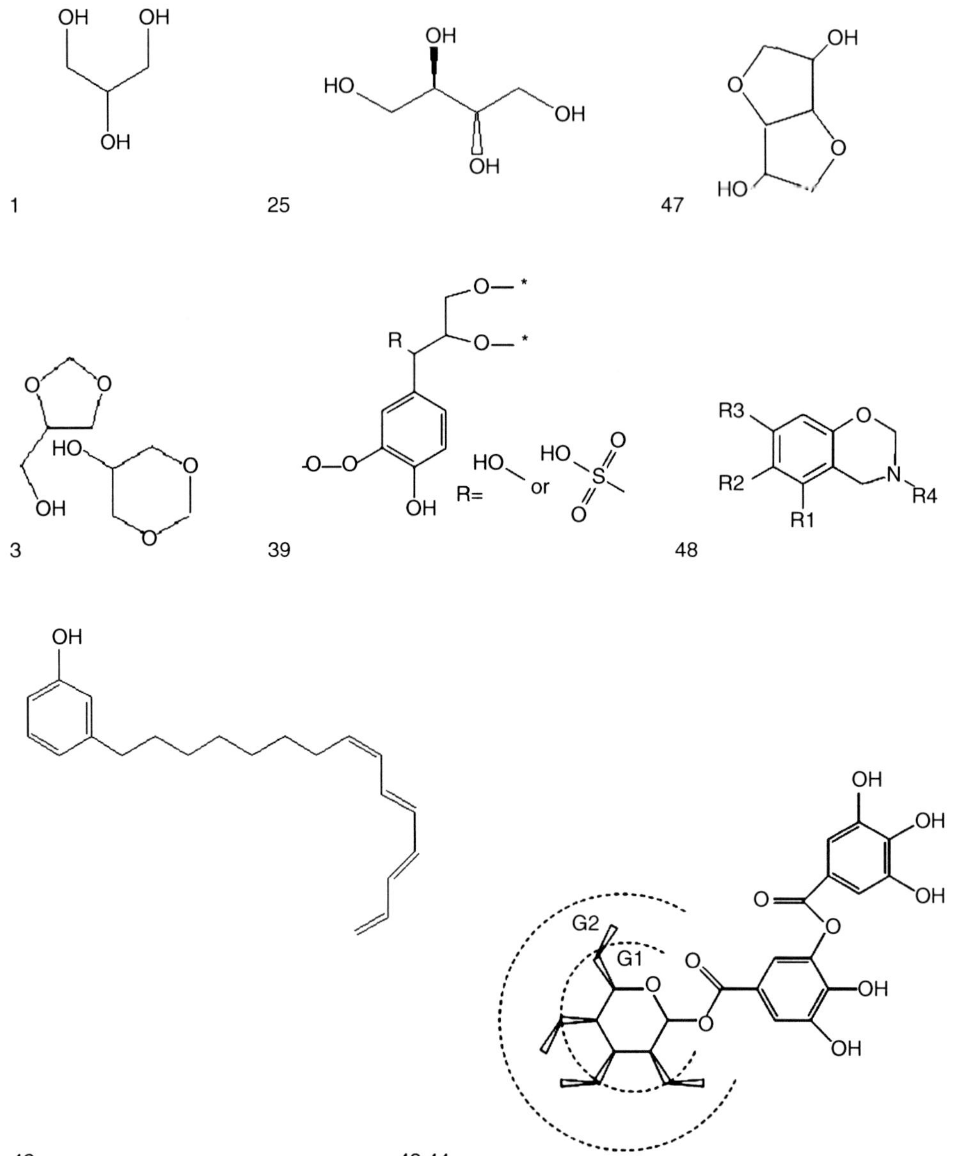

Figure 18.5 Examples of phenolic compounds useful in urethane resins seeding and phenolic resing growth 1) glycerol, 3) glycerol formal mixture, 25) erythritol, 39) lignins/lignoosulfonates, 42) cardanol and 43, 44) tannin dendrimeric structures, 47) isosorbide and 48) benzazine.

via two routes using two different enzymes, namely lipase and peroxidase. Lipase catalysis was used for the epoxidation of the unsaturated alkyl chains of both cardanol and polycardanol. Peroxidase catalysis was used for the polymerizationn of both cardanol and epoxide-containing cardanol [382, 383]. Traditional Japanese natural coating of Urushi, an enzymically crosslinked material of phenol derivitates, from urushi trees, urushiols, are targeted to be replaced accordingly [384, 385].

Table 18.3 Polymers and additives.

Nr	Substnce Name	Net Formula	GAS Number	Mw g mol^{-1}	Tm °C	Tb °C	Application
13	Terpene phenolic resins Hydrogenated rosinacids and polymers	Polymer	68648-57-7	10000	110–120	na	Tackifiers, hot melt, pressure sensitive adhesives and aliphatic polyester softeners
14	Unsaponifiable neutrals, tall oil	C20H34	na	350–2000	75	>250	Paper sizing, adhesives, coatings, syntheticrubbers, paints, and pharmaceuticals Rubber softening agents, and as an adhesion promoter of rubber to metal cord. Optional plastizers
15	pitch	mixture of sterols	na	250–1000	19	na	Plasticizers and stabilizers in the production of vinyl polymer
16	Epoxidized plant oils	Product of epoxided plant oil	e.g 68082-35-9	24–280	–12		Foam stabilizer, dispersants, and fluid polyurethane resins

Table 18.3 (cont.) Polymers and additives.

Nr	Substnce Name	Net Formula	GAS Number	Mw g mol^{-1}	Tm °C	Tb °C	Application
17	Polyol fats	Polymer					
26	Polyglyserols						
27	Dichloropropanol	$C_2H_6Cl_2O$	96-23-1	128.98	–4	172–176	
36	Limonene oxide	$C_{10}H_{16}O$	1195-92-2	152.23344		197–198	Metal coatings, varnishes, and printing inks. Adhesives, resins and binders, phenol replacement in phenol formaldehyde resins,
39	Ligno sulfonate	(C9H907S)n	8062-15-5	< 140000		>200	Polyelectrolyte
40	Muconic acid	$C_6H_6O_4$	3588-17-8	142.11	194–195	320	Acrylic monomer precursors
43	Condensed tannins						Adhesive
44	Tannic acid	$C_{76}H_{52}O_{46}$	1401-55-4	1701.2			Wet adhesives

Cardanol (42) was also reacted with formaldehyde in a particular mole ratio in the presence of glutaric acid catalyst to give a high-ortho novolac resin. The polyol was condensed with diphenylmethane diisocyanate to produce rigid polyurethanes [386].

18.11.2 Tannins

Tannins are polyphenol substances found in plants, characterized by the presence of more than one phenol unit or building block per molecule. Polyphenols are generally divided into hydrolyzable tannins (gallic acid esters of glucose and other sugars) and phenylpropanoids, such as lignins, flavonoids, and condensed tannins. Tannin is a component in a type of industrial particle board adhesive [387].

The simple polyphenolic units derived from secondary plant metabolism of the shikimate pathway [388]. Star anise has been reported to yield 3 to 7% shikimic acid. Typically tannins are found as plant protective minor compounds in bark and leaves. Recently biosynthetic pathways in *E. coli* have been enhanced to allow the organism to accumulate enough material to be used commercially [389, 390]. Tannins are mainly physically located in the vacuoles or surface wax of plants found in leaf, bud, seed, root, and stem tissues. The main current source is acacia, but industrially tannins could be separated e.g. from debarking waters or extracted from bark as a forest industry side stream product.

Chemically, tannins (43) are made up of complex phenolic compounds of high molecular weight, ranging from 500 to 20,000. There are two main categories of tannins: (a) hydrolysable tannins HT and (b) condensed tannins (CT). Generally, tannins are soluble in water, with exception of some very high molecular weight compounds. HT are readily soluble in water, making it possible for them to react with other substances to yield a wide range of water-soluble chemicals such as gallic acid (-gallotannins) or ellagic acid (-ellagitannins) [391].

Gallo- and ellagitannins can be considered dendrimers [392]. Also, tannic acids are mimicking dendrimers as small intestine submucosa stabilizing nanomordants [393]. Dendrimers are repeatedly branched, roughly spherical large molecules [394, 395]. Their structure is spheroid or globular nanostructures from 1 to over 10nm that are typically precisely engineered to carry molecules encapsulated in their interior void spaces or attached to the surface [396–398]. Size, shape, and reactivity are determined by generation (shells) and chemical composition of the core, interior branching, and surface functionalities [399], which provide polyvalent interactions between surfaces and bulk materials for applications such as adhesives, surface coatings, or polymer cross-linking [400–403].

18.11.3 Tannic Acid

Commercial tannic acid (44) (pK_a around 6, average formula of $C_{76}H_{52}O_{46}$) is a mixture of related compounds glucose esters of gallic acid having high reactivity due to the numerous phenol groups. It is a yellow to light brown amorphous powder which is highly soluble in water; one gram dissolves in 0.35 mL of water. Gallic acid (3,4,5-trihydroxybenzoic acid) is found in gallnuts, sumac, witch hazel, tea leaves, oak bark, and other plants [404].

18.12 Lignin

Almost all lignin in current commercial use is as lignosulfonates (39) produced from spent sulfite pulping liquors. Westvaco and LignoTech Sweden also produce lignosulfonates by sulfonation of kraft lignin, but this process produces a product with a much lower molecular weight.

18.12.1 Lignin as Chemical Source

Lignin constitutes one of the major components of lignocellulosic biomass [405], but its three dimensional amorphous polymer structure consisting of methoxylated phenylpropane makes it complicated to apply [406]. The dominant linkage in both softwood and hardwood lignin is the β-O-4 linkage, consisting of approximately 50% of spruce linkages and 60% of birch and eucalyptus linkages [407]. The β-O-4 ether bond is readily cleaved where the fragmentation tends to lead to the generation of water-soluble compounds containing phenolic hydroxyl groups. Common subsequent reactions involve oxidation of coniferyl alcohol to form vanillin or oxidation of the aromatic ring to form quinones. Coniferyl alcohols constitute approximately 90% of softwood lignin, whereas roughly equal proportions of coniferyl alcohol and sinapyl alcohol appear in hardwood lignin, although many exceptions are known [408].

The strategy to apply lignin fragments can be based on functionality on surfaces and ignoring the inner structures as a black box. The control of the fragments is not as requested in the chemical industry due to formulation for high and controlled performance. However, lesser remanding applications are possible. Adhesives, resins and binders are the main polymeric uses for lignosulfonates. Foundry casting and moulding resins take advantage of the water adsorption, dispersing, adhesion and lubricating properties of lignosulfonates. The Karatex process, developed by the Finnish Pulp and Paper Institute, has shown that a high-molecular weight fraction of kraft lignin can be used to displace up to 70% of the phenol required for phenol-formaldehyde (PF) resins [409]. At this high level of displacement, curing times were somewhat lengthened. However, the bonding strength was comparable to pure PF resin. PF resins used in electrical, household, and automotive moulding compound applications could provide another opportunity. Although PF moulding compounds are losing market share in automotive applications due to increased heat resistance requirements [410].

High performance products require substances, such as *p*-substituted benzyl alcohols or aldehydes, which are useful for the production of plastics and other polymers, pigments, dyes, resins, and many other products [411]. Generally, lignin reduction catalytic systems produce bulk chemicals with reduced functionality, whereas lignin oxidation catalytic systems produce fine chemicals with increased functionality [412]. Lignin is converted directly to fine chemicals, such as vanillin [413], that resembles the lignin structure, but more complicated target molecules may also be produced with additional improvements in catalytic technology [414]. Lignin represents the only viable source to produce the renewable aromatic compounds on which society currently depends, but intensified efforts should involve the development of new catalyst materials specifically designed to meet these challenges rather than simply applying the "old" catalyst technology, developed for petroleum refining, to new substrates.

18.12.2 Lignin Pyrolysis

Lignin may be pyrolyzed to yield a liquid product and an ash. The current state of the art in pyrolysis processing of biomass is called fast pyrolysis or flash pyrolysis. It is accomplished in a number of reactor types, most commonly in fluidized beds or circulating fluidized beds. Typical processing conditions are 0.5 seconds at 773 K at 1 atm with an inert atmosphere and inert solid particle heat carrier. Dry feedstock is a prerequisite for this process; therefore wet lignin from most pulping processes or as a byproduct from an aqueous-based biorefinery is incompatible with pyrolysis and must be dried before entering a pyrolysis process. Through fast pyrolysis, wood can be liquefied to form pyrolysis oils containing biopolymer components in water and light oxygenates such as acetic acid, hydroxyacetaldehyde and hydroxyacetone among many others including guaiacol and alkylated guaiacols.

18.12.3 Lignin Cracking

Technology developments of lignin may also arise from depolymerization in the form of C-C and C-O bond rupture to aromatics in order to form benzene, toluene and xylene BTX fraction added with phenol and includes aliphatics in the form of C1 to C3 fractions. There is the possibility of forming some C6-C7 cycloaliphatics as well. These products could be easily and directly used by conventional petrochemical processes. Development of the required non-selective chemistries is part of the long-term opportunity but is likely to be achievable sooner than highly selective depolymerizations like vanilin. In fact, some of the past hydroliquefaction work with lignin suggests that, with further development, this concept is a good possibility [415].

NREL has developed a process where in the first reaction step of hydroprocessing, the depolymerized lignin is contacted with a hydroeoxygeneation catalyst to produce a hydrodeoxygenated (HDO) intermediate product. The second reaction step utilizes a hydrocracking catalyst (HCR) at 653 K to convert the hydrodeoxygenerated intermediate into the reformulated gasoline additive. The final product is a mixture of naphthenic and paraffinic organics in the gasoline range. In principle the system is operative but recovery of lignin dropper efficiency is below 50% [416].

Various methods for conversion of lignin and lignocellulose materials to produce fuel gas and chemicals have been patented, claiming cracking in different reactor systems targeting the petrochemical feedstocks [417, 418]. Stemcracking lignin in supercritical conditions in steam produced markedly different cresols. In the catalytic cracking with HZSM-5 at temperatures of 500–650°C was achieved with total aromates of 89,4–84,4% where BTX was 73,2–79,1% respectively. In this case it was markedly notable that high temperatures prefer toluene [419].

18.12.4 Lignin Oxidation

The natural breakdown of lignin is initiated by fungi such as *Phanerocheate chrysosporium* using extracellular lignin peroxidise enzymes [420], but the smaller aromatic fragments generated by this breakdown are then degraded by both fungi and bacteria. Many simple aromatic compounds such as benzene, toluene, xylenes, benzoic acid, phenylacetic acid and phenylpropionicacid are degraded by aerobic soil bacteria such as *Pseudomonas, Acinetobacter* (Gram negative) and *Rhodococcus* (Gram positive) [421]. A number of man-made aromatic compounds have found their way into the environment, for example, as pesticides, detergents, oils, solvents, paints or explosives, which can be fairly readily degraded by microorganisms using the same enzymes used for the degradation of naturally occurring compounds [422]. For example, the carbamate insecticide carbaryl can be degraded via the bacterial naphthalene degradative pathway [423].

Unsubstituted aromatics such as benzene, naphthalene, and biphenyl are converted into the corresponding 1,2-dihydroxycatechols *via* conversion to the *cis*-1,2-dihydrodiol, followed by oxidation. The dihydroxylation reaction is carried out by a family of three-component dioxygenase enzymes comprising: (1) an NADH-dependent flavin reductase; (2) a ferredoxin electron transfer protein containing two Rieske [2Fe/2S] clusters; and (3) a terminal dioxygenase subunit, containing a mononuclear iron(II) cofactor and two further [2Fe/2S] clusters [424]. Catechols are also commonly produced by *ortho*hydroxylation of phenols, carried out by FAD-dependent monooxygenases, whose mechanism of action has been reviewed [425]. However, the enzymatic routes have not yet been shown to be attractive for industrial manufacturing of aromatic chemicals.

Lignin produced in the pulping process contains chemical contaminants like sulphonates in kraft process and sodium salt in soda process. Unlike these lignins, the modern biomass conversion processes are relatively clean [426], opening possible routes to phenolic chemicals [427]. Processing and conditions has a marked influence on the multitude of possible products [428]. Possible processes are thermal liquefaction in solvents [429, 430], thermal hydrolysis in aqueous phase [431–433], pyrolysis [434, 435], hydrocracking [436–438], and alkaline oxidation [439].

In sulfite pulping removed sulfonates have several uses. Lignosulfonates, or sulfonated lignin, (CAS number 8062-15-5) are water-soluble anionic polyelectrolyte polymers: they are byproducts from the production of wood pulp using sulfite pulping [440]. Most delignification in sulfite pulping involves acidic cleavage of ether bonds, which connect many of the constituents of lignin [441].

Various approaches to develop new and more specific chemical lignin depolymerizing processes for industrial purposes are reported in the literature, like using free copper ions as an additive in peroxide bleaching [442]. In some cases the idea was to mimic the redox cycle of the lignin polymerizing/depolymerizing enzymes. Laccase is a copper containing enzyme which is oxidized by molecular oxygen while lignin peroxidase and manganese peroxidase are iron containing enzymes, oxidized by hydrogen peroxide [443].

The so-called GIF system, a reaction of copper with lignin including peroxide, has been reported to yield toxic pyridine derivatives such as 2,2'-bipyridyl, 2,3'-bipyridyl, methyl pyridine and pyridinone. The high amount of pyridine per se as well as the creation of large amounts of toxic derivatives makes it impossible to utilize this process for industrial purposes [444].

Fenton reaction based on Fe2+ with hydrogen peroxide is another reaction mimicking biological processes of brown rot fungi decomposes carbohydrates in lignocellulosic material. Likewise, manganese is known as a degrader of lignin when oxidized to Mn (III) by manganese peroxidase.

Lignin can be oxidized effectively with a coordinated copper ion or fenton reagent to unsaturated acids like *cis,cis*-muconic acid, produced by some bacteria by the enzymatic degradation of various aromatic chemical compounds. The product could be further converted to acrylic monomers.

18.13 Conclusions

Wood biomass also contains many valuable raw materials for producing fine and specialty chemicals. These raw materials are carbohydrates, fatty acids, terpenoids, and polyphenols, such as stilbenes, lignans, and tannins. Further biobased lipids and oils are markedly valuable raw materials. There are three strategies to convert the biogenic substances to precursor and further to value added chemicals, namely gasification to synthesis gas, thermal conversions to simple precursors, and utilization of highly selective catalysis and bioconversion, but diversity of the biomaterials and structural complexity leads to demanding fractionation and chemical conversion.

Lignin represents the only viable source to produce the renewable aromatic compounds on which society currently depends, but intensified efforts should involve the development of new catalyst materials for refining new substrates. Strategies to apply lignin fragments can be based on functionality on surfaces and ignoring the inner structures as a black box. As an example, a high-molecular weight fraction of kraft lignin can be used to displace up to 70% of the phenol required for phenol-formaldehyde (PF) resins. Further different types of cashew nut based cardanol novolac phenolic resins find a wide range of industrial uses, like resins for coating, lamination and friction materials.

Gasification of biomass as well as alcohols provide direct routes to petrochemical processes, which may reduce required investment costs but are not yet considered cost efficient. Meanwhile there are several examples of possible side streams, enabling routes to polyurethanes, epoxies, unsaturated polyesters, and phenolics. One example is glycerol of first generation biodiesel production, considered as a raw material for biogenetic epichlorhydrin, but numerous emerging applications for glycerol are demotivating investments. However, several diols and polyols are readily available in natural sources encouraging development of aliphatic polymers instead.

Alternatively available is the well known chemistry of furans. The limitations involved are in the complex reaction routes with multiple side reactions. Meanwhile developments based on glucose are emerging, like the 1,3-propanediol production using a proprietary fermentation and purification process. The next breakthrough could be 3-hydroxypropionic acid (3HPA, isomer of lactic acid), enabling production of acrylates. Bioconversions of sugars may solve the problems of complexity but also lead to issues related to the source yet competing with food, where cellulose based sugars should become more available.

References

1. Huber, G. W.; Chheda, J. N.; Barrett, C. J.; Dumesic, *J. A. Science*, 2005, 308, 1446.
2. Huber, G. W.; Iborra, S.; Corma, *A. Chem. Rev.*, 2006, 106, 4044.
3. Petrus, L.; Noordermeer, M. A. *Green Chem.* 2006, 8, 861.
4. Ragauskas, A. J.; Williams, C. K.; Davison, B. H.; Britovsek, G.; Cairney, J.; Eckert, C. A.; Frederick, W. J.; Hallett, J. P.; Leak, D. J.; Liotta, C. L.; Mielenz, J. R.; Murphy, R.; Templer, R.; Tschaplinski, T. *Science*, 2006, 311, 484.
5. Report of the United Nations Conference on Environment and Development, Rio de Janeiro, June 3–14, 1992. http://www.un.org/esa/sustdev, date 2.5.2007.
6. Technology Vision 2020, The US Chemical Industry, 1996. http://www.ccrhq.org/vision/index.html, date 2.5.2007.
7. Workshop Report on New Process Chemistry, Technology Roadmap, July 2001, Council for Chemical Research, p. 9. http://www.ccrhq.org, USA, date 3.5.2007.
8. CEFIC, www.cefic.be. Date 2.5.2007.
9. Gum, Wilson; Riese, Wolfram; Ulrich, Henri (1992). *Reaction Polymers*, New York: Oxford University Press.
10. *Journal of Chemical Education*. Vol. 84, No. 12, December 2007.
11. Guido Kickelbick, *Hybrid Materials. Synthesis, Characterization, and Applications*. Wiley-VCH Verlag GmbH & Co., 2007.
12. Special issue of *J. Mater. Chem.* 2005, 15, 3543–3986.
13. K. G. Sharp, "Inorganic/organic Hybrid Materials," *Adv. Mater.* 1998, 10, 1243–1248.
14. Special issue of *J. Nanosci. and Nanotechn.* 2006, 6, 265–572.
15. A. Usuki, N. Hasegawa, M. Kato, "Polymer-Clay Nanocomposites," *Adv. Polym. Sci.*, 2005, 179, 135–195.
16. R. M. Laine, "Nanobuilding blocks based on the [OSiO1.5]x (x = 6, 8, 10) octasilsesquioxanes," *J. Mater. Chem.* 2005, 15, 3725–3744.
17. M. Smolander, International Forum on Emerging Technologies in Food Processing, Illinois, 2009
18. Y.M Chiang Martin, X Wang, C R Schultheisz, J He; *Composites Science and Technology*, 60, 5231–5239 (2005).
19. Amit Desai, et al. "Composites & Polycon," *American Composites Manufacturers Association*. October 17–19, 2007.
20. Ruhul A. Khan et al., *J. Agric. Food Chem.*, 2010, 58 (13), pp 7878–7885.
21. Maria L. Auad, et al., 9th Int. Conf. Wood & Biofiber Plastic Composites, Madison WI, 2007.
22. Gallezot, P. *Catal. Today*. 2007, 121, 76.
23. Zhang, Y.-H. P., *J. Ind. Microbiol. Biotechnol.* 2008, 35, 367.
24. Bozell, J. J.; Holladay, et al., "Top Value Added Candidates from Biomass," Volume II: *Results of Screening for Potential Candidates from Biorefinery Lignin;* Pacific Northwest National Laboratory: Richland, WA, 2007.

25. Lichtenthaler, F. W. *Acc. Chem. Res.* 2002, 35, 728.
26. Lichtenthaler, F. W. Peters, S. C. R. *Chim.* 2004, 7, 65.
27. Lichtenthaler, F. W. *Carbohydr. Res.* 1998, 313, 69.
28. Eriksson, K.E. and Wood, T.M. (1985). In *Biosynthesis and Biodegradation of Wood Components*, Higuchi, T. ed. Academic Press, London, UK. pp. 469–503.
29. Wood, T.M. (1994). In *Enzymatic Conversion of Cellulose into D-glucose*, James, N.B. ed., *Methods in Carbohydrate Chemistry*. John Wiley and Sons, New York, USA. pp. 219–230.
30. Wood, T.M. (1991). In *Biosynthesis and Biodegradation of Cellulose and Cellulosic Materials*, Weimer, P.J. and Hagler, C. eds. Dekker, New York, USA. pp. 491–534.
31. Lipinsky, E. S.; Sinclair, R. G. *Chem. Eng. Prog.* 1986, 82, 26.
32. Litchfield, J. H., *AdV. Appl. Microbiol.* 1996, 42, 45.
33. Louwrier, A., *Biotechnol. Appl. Biochem*, 1998, 27, 1.
34. Varadarajan, S.; Miller, D. *J. Biotechnol. Prog.* 1999, 15, 845.
35. Ladisch, M.R., Svarczkopf, J.A., 1991. *Biores. Technol.* 36, 83- 95.
36. Nigam, P., Singh, D., 1995. *Process Biochem.* 30, 117–124.
37. Roberto, I.C., Sato, S., Mancilla, I.M., 1996. *J. Ind. Microbiol.* 16, 348±350.
38. Furlan, S.A., Bouilland, P., Strehaiano, P., Riba, J.P., 1991., *Biotechnol. Lett.* 13, 203–206.
39. Horitsu, H., Yahashi, Y., Takamizawa, K., Kawai, K., Suzuki, T., Watanabe, N., 1992, *Biotechnol. Bioeng.* 40, 1085–1091.
40. Vandeska, E., Amartey, S., Kuzmanova, S., Jeffries, T., 1995. *World J. Microbiol. Biotechnol.* 11, 213±218.
41. Roseiro, J.C., Peito, M.A., Gfrio, F.M., Amaral-Collaco, M.T., 1991. *Microbiol.* 156, 484–490.
42. Parajo, J.C., Dominguez, H., Dominguez, J.M., 1996., *Biotechnol. Lett.* 18, 593–598.
43. Moses, V. and Ferrier, R.J. 1962, *Biochemical Journal* 83:8–14.
44. Ahmed, Z.; Sasahara, H.; Bhuiyan, S.H.; Saiki, T.; Shimonishi, T.; Takada, G. and Izumori, K. *Journal of Bioscience and Bioengineering* (1999) 88:676–678.
45. Ahmed, Z. and Bhowmik, P.K. 2000. *Pakistan Journal of Biological Sciences* 3:1492–1495.
46. Ahmed, Z. and Izumori, K., *Bangladesh Journal of Scientific and Industrial Research* (2001).
47. Zakaria Ahmed, *Process Biotechnology*, Vol. 4 No. 2, Issue of August 15, 2001
48. Päivi Mäki-Arvela, et al., *Catalysis Reviews*, 49:197–340, 2007
49. Biermann, U., et. al. *Chem., Int. Ed. Engl.* 2000, 39, 2206.
50. Jenck, J. F.; Agterberg, F.; Droescher, M. *J. Green Chem.* 2004, 6, 544.
51. Sonntag, N. O. V. *J. Am. Oil Chem. Soc.* 1979, 56, A729.
52. Muckerheide, V. J. *Industrial Production of Fatty Acids: Fatty Acids; Their Chemistry, Properties, Production and Uses*, 2nd ed.; Interscience: New York, 1967.
53. Sonntag, N. O. V. *J. Am. Oil Chem. Soc.* 1979, 56, A861.
54. Linfield, W. M. et al., *J. Am. Oil Chem. Soc.* 1984, 61, 191.
55. Packet, D. EP Patent 1352891, 2003.
56. EP Patent 1444285B1, 2009.
57. Marco Krämer, et al., *Metabolic Engineering* 5 (2003) 277–283.
58. Luise Johansson, et al., *Biotecnology and Bioengineering*, Vol. 92, no 5, Dec 5, 2005.
59. Andrew Knaggs, "The Biosynthesis of Shikimate Metabolites," *Nat. Prod. Rep.*, 2001, 18, 334–355.
60. Direct hydration of ethylene to ethanol, United States Patent 3686334, 1972.
61. Hennell, H. 1828., *Philosophical Transactions* 118: 365.
62. Antonio Morschbacker, "Bio-Ethanol Based Ethylene" *Polymer Reviews*, Vol. 49, Issue 2 April 2009 , p.79–84.
63. Ravindra Datta, *Ind. Eng. Chem. Res.*, 36 (11), 4466 -4475, 1997.
64. US Patent 4873392.
65. WO 200610091073.
66. Guangwen Chen, et al., *Catalysis Today*, Vol. 125, Issues 1–2, 15 July 2007, Pages 111–119.
67. Takahara, Isao; Saito, Masahiro; Inaba, Megumu; Murata, Kazuhisa, *Catalysis Letters*, Vol. 105, Nr 3–4, Dec. 2005 , pp. 249–252(4).
68. United States Patent 4134926.
69. SRI consulting: PEP Review 79-3-4.

70. SRI consulting: PEP Review 79-3-4.
71. United States Patent 4134926.
72. Juguin, B. Hughes, F., and Hamon, C. 1992. in: "Novel production methods for ethylene light hydrocarbonsand aromatics," *Chemistry Industry*. 46, Dekker, New York, 381–390.
73. Tabak, S. and Yurchak, S.1990, *Catalysis Today* 6, 307–327.
74. Sawa, M., Niwa, M., and Murakami, Y. 1989, *Applied Catalysis A*. 53, 169–181.
75. UOP technical information. MTO reaction and compared with that of unmodified SAPO-34.
76. U.S. Patent 6,534,692 assigned to UOP
77. LLC UOP, Honeywell business information.
78. SRI 2003 PEP Review 2001-11.
79. Biermann, U. et al., *Chem., Int. Ed. Engl.* 2000, 39, 2206.
80. Jenck, J. F.; Agterberg, F.; Droescher, M. *J. Green Chem.* 2004, 6, 544.
81. Sonntag, N. O. V. *J. Am. Oil Chem. Soc.* 1979, 56, A729.
82. Muckerheide, V. J. *Industrial Production of Fatty Acids: Fatty Acids; Their Chemistry, Properties, Production and Uses,* 2nd ed.; Interscience: New York, 1967.
83. Sonntag, N. O. V. *J. Am. Oil Chem. Soc.* 1979, 56, A861.
84. Linfield, W. M. et al., *J. Am. Oil Chem. Soc.* 1984, 61, 191.
85. Packet, D. EP Patent 1352891, 2003.
86. EP Patent 1444285B1, 2009.
87. EP1770081 (2007).
88. Shaikh, A. A. G.; Sivaram, S. *Chem. Rev.* 1996, 96, 951.
89. D'Alelio, G.; Huemmer, T. EP Patent 0328150, 1967.
90. Gillis, H. R.; Stanssens, D. US Patent 5703136, 1997.
91. Whelan, J. M., Jr.; Cotter, R. R. US Patent 3072613, 1963.
92. Whelan, J. M., Jr.; Cotter, R. J. US Patent 3072613, 1963.
93. Bell, J. B., Jr.; Arthur, C. V. US Patent 2915529, 1959.
94. Grey, R. A. US Patent 5091543, 1992.
95. Kawabata, O. JP Patent 6009610, 1994.
96. Burghelea, T. RO Patent 78145B, 1982.
97. Spassova, M. K. *Coll. Czech. Chem. Commun.* 1986, 51, 1512.
98. Teregulova, G. T. *Zh. Prikl. Khim.* 1991, 64, 1591.
99. Kopp, E.; Smidt, J. Ann. *Chem. Justus Liebig* 1966, 693, 117.
100. Radi, S.; Lazrek, H. B. Bull. *Korean Chem. Soc.* 2002, 23, 437.
101. Otto L, Bicker M and Vogel H, *Green Chem* 8:214–220 (2006).
102. U. S. patent 5093537.
103. Kylmä, J., Härkönen, M., and Seppälä, J. V., *J. Appl. Polym. Sci.* 63 (1997) 1865–1872.
104. Kylmä, J. and Seppälä, J. V., "Synthesis and Characterization of a Biodegradable Thermoplastic Poly(ester-urethane) Elastomer," *Macromolecules* 30 (1997) 2876–2882.
105. Hiljanen-Vainio, M., Kylmä, J., Hiltunen, K., and Seppälä, J. V., *J. Appl. Polym. Sci.* 63 (1997) 1335–1343.
106. Kylmä, J., Hiljanen-Vainio, M., and Seppälä, J. V., *J. Appl. Polym. Sci.* 76 (2000) 1074–1084.
107. Kylmä, J. and Seppälä, J. V., *J. Appl. Polym. Sci.* 79 (2001) 1531–1539.
108. Helminen, A., Kylmä, J., Tuominen, J., and Seppälä, J. V., *Polym. Eng. Sci.* 40 (2000) 1655–1662.
109. Kylmä, J., Tuominen, J., Helminen, A., and Seppälä, J., *Polymer* 42 (2001) 3333–3343.
110. Tim R. Cooper and Robson F. Storey, *Macromolecules* 2008, 41, 655–662.
111. J.M. Krochta, S.J. Tillin, J.S. Hudson, *Appl. Biochem. Biotechnol.* 17 (1988) 23–32.
112. M.J. Hayes, M.D. Lauren, *J. Appl. Biomater.* 5 (1994) 215–220.
113. Y. Yamamoto, et al., *J. Dermatol.* 33 (2006) 16–22.
114. M. Kataoka, et al., *Biosci. Biotechnol. Biochem.* 65 (2001) 2265–2270.
115. E. Bonnin, L. Lesage-Meessen, M. Asther, J.F. Thibault, *J. Sci. Food. Agric.* 79 (1999) 484–486.
116. D.L. Hua, et al., *Appl. Microbiol. Biotechnol.* 74 (2007) 783–790.
117. J. De Ley, M. Gillis, J. Swings, "The Genus Gluconobacter," in: N.R. Krieg, J.G. Holt (Eds.), *Bergey's Manual of Systematic Bacteriology*, vol. 1, Williams and Wilkins, Baltimore, 1984, pp. 267–278.

118. U. Deppenmeier, M. Hoffmeister, C., *Appl. Microbiol. Biotechnol.* 60 (2002) 233–242.
119. C. Prust, et al., *Nat. Biotechnol.* 23 (2005) 195–200.
120. K. Matsushita, H. Toyama, O. Adachi, *Adv. Microb. Physiol.* 36 (1994) 247–301.
121. Guodong Weia, et al., *Biochemical Engineering Journal* 47 (2009) 127–131.
122. D. L. Klass, *Biomass for Renewable Energy, Fuels and Chemicals*, Academic Press, London, 1998.
123. G.W. Huber, S. Iborra, A. Corma, *Chem. Rev.* 2006, 106, 4044 – 4048.
124. J.-P.Lange in *Catalysis for Renewables: From Feedstock to Energy Production* (Eds.: G. Centi, R. A. van Santen), Wiley-VCH, Weinheim, 2007, pp. 21–51.
125. US 6617464, 9.9.2003.
126. Manzer, L.E., 2004, *Appl. Catal. A: Gen.*, 272: 249–256.
127. Amherst, B. C. R. US Patent 2469701, 1946.
128. Amherst, B. C. R.; Griffin, G. R. US Patent 246768, 1946.
129. Tsobanakis, P.; Meng, X.; Abraham, T. W. WO Patent 2003082795, 2003.
130. U. US Patent 2005222458, 2005.
131. US Patent 2005222458, 2005.
132. Gokarn, R.; Eitemann, M. A.; Sridhar, J. *ACS Symp. Ser.* 1997, 666,224.
133. Nghiem, N. P.; Davison, B. H.; Suttle, B. E.; Richardson, G. R. *Appl.Biochem. Biotechnol.* 1997, 63–5, 565.
134. Fumagalli, C. Succinic acid and succinic anhydride. In *Kirk-Othmer Encyclopedia of Chemical Technology*, 4th ed.; Kroschwitz, J., Home-Grant, M., Eds.; John Wiley and Sons: New York, 1997; Vol. 22, p1074.
135. Weissermel, K.; Arpe, H. J. *Industrial Organic Chemistry;* Wiley-VCH: Weinheim, 2003.
136. Chem. Mark. Rep.1999, 255, 34.
137. Siegfried Rebsdat1 and Dieter Mayer "Ethylene Glycol" in *Ullmann's Encyclopedia of Industrial Chemistry*, 2002, Wiley-VCH, Weinheim.
138. EP1727875 (2007).
139. Chauvel, Alain; Lefebvre, Gilles. "Petrochemical Processes," Volume 1: *Synthesis-Gas Derivatives and Major Hydrocarbons*. Editions Technip. pp. 26.
140. Freund A, *Monatsh Chem 2:* (1881), 636–641.
141. Braak HR, Onderzoeking over vergisting von glycerine. Dissertation, Delft (1928).
142. Mickelson MN, Werkman CH, *Enzymologia* 8: (1940) 252–256.
143. Lin ECC "Glycerol dissimilation and its regulation in bacteria," *Annu Rev Microbiol* 30: (1976) 535–578.
144. Nakas JP, et al., *Appl Environ Microbiol* 46: (1983) 1017–1023.
145. DE 37 34764 A1, 1987.
146. H. Biebl, K. Menzel A.-P. Zeng, W.-D. Deckwer, *Appl Microbiol Biotechnol* (1999) 52: 289–297.
147. EP 0373230A1. 1990.
148. Deckwer W-D "Microbial Conversion of Glycerol to 1,3-propanediol," *FEMS Microbial Rev* 16: (1995) 143–149.
149. WO96/53796.
150. McCoy M "Chemical Makers Try Biotech Paths" *Chem Eng News* 22 (June): (1998) 13–19.
151. US 5254467, 1993.
152. Abbad-Andaloussi S, Guedon E, Spiesser E, Petitdemange H, *Lett Appl Microbiol* 22: (1996) 311–314.
153. Ahrens K, Menzel K, Zeng A-P, Deckwer W-D, *Biotechnol Bioeng* 59: (1998) 544–552.
154. Crabtree, S. P.; Henderson, R. K. WO Patent 2001070659, 2001.
155. Crabtree, S. P.; Henderson, R. K. GB Patent 2362157, 2000.
156. Forschner, T. C.; Powell, J. B.; Slaugh, L. H.; Weider, P. R. WO Patent 2000018712, 1999.
157. Lee, B. N.; Jung, I. S.; Jang, E. J.; Lee, J. H.; Kim, H. R.; Han, Y. H. US Patent 2003069456, 2002.
158. Lee, B. R.; Lee, J. H.; Kim, H. R.; Han, K. S.; Jang, E. S. JP Patent 2004034021, 2003.
159. US Patent Application 20090156866, Method for producing 1,2-propylene glycol using bio-based glycerol.
160. Szmant, H. H. *Organic Building Blocks of the Chemical Industry;* John Wiley and Sons: New York, 1989.

161. Herrmann, U.; Emig, G. *Ind. Eng. Chem. Res.* 1997, 36, 2885.
162. Budge, J. R.; Attig, T. G.; Pedersen, S. E. US Patent 5473086, 1995.
163. US Patent 4751334, 1988.
164. Broadbent, H. S.; Campbell, G. C.; Bartley, W. J.; Johnson, J. H. *J.Org. Chem.* 1959, 24, 1847.
165. Hara, Y.; Izui, H.; Ito, H. US Patent 2005196846, 2005.
166. Sato, M.; Akiyoshi, N. EP Patent 1233070, 2002.
167. Ueda, H.; Koda, T.; Sato, M. EP Patent 1233069, 2002.
168. DE Patent 4231782, 1992.
169. US Patent 5536854, 1995.
170. Kautola, H. *Appl. Microbiol. Biotechnol.* 1990, 33, 7.
171. Reddy, C. S. K.; Singh, R. P. *Bioresour. Technol.* 2002, 85, 69.
172. Willke, T.; Vorlop, K. D. *Appl. Microbiol. Biotechnol.* 2001, 56, 289.
173. Hoffer, B.W., et al., *Appl. Catal. A: Gen.*, 253: 437–452.
174. Crezee, E., et al., *Appl. Catal. A: Gen.*, 251: 1–17.
175. US 0171889, 2.9.2004.
176. Arun, N. *Synth. Fibres* 2003, 32, 18.
177. Kusserow, B., Schimpf, S., and Claus, P. 2003, *Adv. Synth. Catal.*, 345 (1–2): 289–299.
178. Blanc, B., Bourrel, A., Gallezot, P., Haas, T., and Taylor, P., 2000., *Green Chem.*, 89–91.
179. Stoss P, Hemmer R, *Adv. Carbonhydr. Chem Biochem*, 1991; 49, p 119.
180. Rupp-Dahlem C., *Greentech* 2002, Netherlands, 2002.
181. A. Caoutur, et al., *Polym. J.*, Vol. 43, 2007, p 220–230.
182. Karla A. et al., *Macromol. Symp.* 2009, 283–284, 144–151.
183. Catalani, L. H., et al., *Macromolecules* 2010, 43, 10315–10319.
184. BioPerspectives 2005 –BREW Symposium –Wiesbaden -May 11th 2005–1.
185. Aoki MAY, Pastore GM, Park YK, *Biotechnol Lett* (1993) 15:338–383.
186. Marina AYA, Glaucia MP, Park YK, *Biotechnol Lett* (1993) 15:383–388.
187. Kim SY, Lee KH, Kim JH, Oh DK, *Biotechnol Lett* (1997) 19(8):727–729.
188. Park J, Seo B, Kim J, Park J., *J Ferment Bioeng* (1998) 86(6):577–580.
189. Kim KA, Noh BS, Lee JK, Kim SY, Park YC, Oh DK, *J Microbiol Biotechnol* (2000) 10:69–74.
190. Lin SJ, Wen CY, Liau JC, Chu WS, *Process Biochem* (2001) 36:1249–1258.
191. Ryu YW, Park CY, Park JB, Kim SY, Seo JH. *J Ind Microbiol Biotechnol* (2000) 25:100–103.
192. Veigada-Cunha M, Firme P, San-Romao MV, Santos H. *Appl Environ Microbiol* (1992) 58: 2271–2279.
193. Ishizuka H, Wako K, Kasumi T, Sasaki T, *J Ferment Bioeng* (1989) 68:310–314.
194. Rymowicz W, Rywinska A, Gładkowski W, *Chem Pap* (2008) 62(3):239–246.
195. Waldemar Rymowicz, et al., *Biotechnol Lett* (2009) 31:377–380.
196. Gruber, B.; Hofer, R.; Kluth, H.; Meffert, A. Fett Wissenschaft. *Technol.-Fat Sci. Technol.* 1987, 89, 147.
197. Jae-Woo Kim, et al., Proceeding 64th International Convention, Forest Products Society, 2010 Madison WI.
198. Garti, N.; Aserin, A.; Zaidman, B. *J. Am. Oil Chem. Soc.* 1981, 58, 878.
199. Dobson, K. S.; Williams, K. D.; Boriack, C. J. *J. Am. Oil Chem. Soc.* 1993, 70, 1089.
200. DE Patent 3118417, 1982.
201. Patent 2002066136, 2002.
202. WO Patent 9325511, 1993.
203. US Patent 5349094, 1993.
204. Kumar, R.; Gao, W.; Gross, R. A. *Macromolecules* 2002, 35, 6835.
205. Shen, Y.; Chen, X.; Gross, R. A. *Macromolecules* 1999, 32, 2799.
206. Tian, D.; Dubois, P.; Grandfils, C.; Jerome, R. *Macromolecules* 1997, 30, 406.
207. Haines, A. H. *Adv. Carbohydr. Chem. Biochem.* 1981, 39, 13.
208. Haines,A. H. *Adv. Carbohydr. Chem. Biochem.* 1976, 33, 11.
209. Therisod, M.; Klibanov, A. M. *J. Am. Chem. Soc.* 1986, 108, 5638.
210. Patil. D. R.; Dordick, J. S.; Rethwisch, D. G. *Macromolecules* 1991, 24, 3462.
211. Kline, B. J.; Beckman, E. J.; Russell, A. *J. Am. Chem. Soc.* 1998, 120, 9475.

212. Tsujimoto, T.; Uyama, H.; Kobayashi, S. *Biomacromolecules* 2001, 2, 29.
213. Uyama, H.; Inada, K.; Kobayashi, S. *Macromol. Biosci.* 2001, 1, 40.
214. Uyama, H.;Inada, K.; Kobayashi, S. *Macromol. Rapid Commun.* 1999,20, 171.
215. Chaudhary, A. K.; Lopez, J.; Beckmann, E. J.; Russell, A. *J. Biotechnol. Prog.* 1997, 13, 318.
216. Kim, D.-Y.; Dordick, J. S. *Biotechnol. Bioeng.* 2001, 76, 200.
217. Park, O.; Kim, D.-Y.; Dordick, J. S. *Biotechnol. Bioeng.* 2000, 70, 208.
218. Uyama, H.; Klegraf, E.; Wada, S.; Kobayashi, S. *Chem. Lett.* 2000, 800.
219. Morimoto, T.; Murakami, N.; Nagatsu, A.; Sakakibara, *J. Chem. Pharm. Bull.* 1994, 42, 751.
220. Patil, D. R.; Rethwisch, D. G.; Dordick, J. S. *Biotechnol. Bioeng.* 1991, 37, 639.
221. Ajay Kumar, et al., *Macromolecules*, Vol. 36, No. 22, 2003, Communications to the Editor 8221.
222. Arizona Chemicals, Technical Service: Savannah, GA, 2010.
223. Hamunen, A. (1983) Process for the Purification of Beta-sitosterol Isolated from the Unsaponifiables in Crude Soap from the Sulphate Cellulose Process, FR2510118, 28.1.1983.
224. Pollak, O.J. (1953) Reduction of blood cholesterol in man. Circulation, 7 (5): 702–706.
225. Simoneit, B.R.T., et al., *OrgOnic Geochemisny*, 1988,13, p.677.
226. Wilson, MA., Collin, P.J., Vassallo, A.M., Russel, N.J. *J. Org. Geochem.* 1986,7, 161.
227. *Ullmann's Encyclopedia of Industrial Chemistry*, 6th ed; Wiley-VCH: Weinheim; Vol. 35, 2003, p. 451–464.
228. *Ullmann's Encyclopedia of Industrial Chemistry*, 6th ed.; Wiley-VCH: Weinheim; Vol. 31, 2003, p. 347–355.
229. Duan, W., Chen, X., and An, X. (2003), *Sepu*, 21 (2): 174–177.
230. Tahara, K., et al., 1997, *Appl. Catal. A: Gen.*, 154: 75–86.
231. Duan, W., Chen, X., and An, X., 2003, *Sepu*, 21 (2): 174–177.
232. Chen, X.-P., Wang, L.-L., Ma, J., and Ynag, C.-L., 2002, *Linchan Huaxue Yu Gongye*, 22: 12–16.
233. JP 2001181400, 3.7.2001.
234. Yang, C.-L., et al., *Gaoxiao Huaxue Gongcheng Xuebao*, 17 (1): 65–70.
235. Chen, X.-P., et al., 2003, *Yingyong Huaxue*, 20: 520–523.
236. Yang, C.-L., et al., *Gaoxiao Huaxue Gongcheng Xuebao*, 17 (1): 65–70.
237. Chen, X.-P., et al., 2004., *Guangxi Daxue Xuebao, Ziran Kexueban*, 29: 161–164.
238. Chen, X., et al., 2002 *Huagong Xuebao*, (Chinese Edition), 53: 522–527.
239. Hydrogenation/dehydrogenation reactions of rosin, Richard P. Dutta and Harold H. Schobert Fuel Science Program, Pennsylvania State. University, the Air Force W A Propulsion and Power Doktorate. Wright-Patterson AFB.1992.
240. Biermann, U.; et al., *Chem., Int. Ed. Engl.* 2000, 39, 2206.
241. Kuo, M. C.; Chou, T. C. *Ind. Eng. Chem. Res.* 1987, 26, 277.
242. Snell, F. D.; Ettre, L. S. *Encyclopedia of Industrial ChemicalAnalysis;* Intersciences: New York, 1971; Vol. 12, p 230.
243. Biermann, U. et al., *Chem., Int. Ed. Engl.* 2000, 39, 2206.
244. FMC Corp. GB Patent 765903, 1957.
245. Bjorkling, F.; Frykman, H.; Godtfredsen, S. E.; Kirk, O. *Tetrahedron* 1992, 48, 4587.
246. Bjorkling, F.; Godtfredsen, S. E.; Kirk, O. *J. Chem. Soc., Chem.Commun.* 1990, 1301.
247. Warwel, S.; Klaas, M. R. G. *J. Mol. Catal. B: Enzymatic* 1995, 1, 29.
248. Corma, A.; Iborra, S; Velty, A. *Chemical Reviews*, 2007, Vol. 107, No. 6 2427.
249. Dunlop, A. P. and Peters, F. N., *The Furans*. Reinhold Publishing Co., New York, 1953.
250. Lichtenthaler, F. W. *Carbohydr. Res.* 1998, 313, 69.
251. Zeitsch, K. J. *The Chemistry and Technology of Furfural and its Many By-Products*, 1st ed.; Elsevier: Amsterdam, 2000; Vol. 13.
252. Antal, M. J.; Leesoboom, T.; Mok, W. S. Richards, G. N. *Carbohydr. Res.* 1991, 217, 71.
253. Sako, T.; et al., *J. Chem. Eng. Jpn.* 1992, 25, 372.
254. *Kirk-Othmer Encyclopedia of Chemical Technology*,Vol. 11, ed. M. Grayson. John Wiley & Sons, Inc., New York, 1981, p. 499.
255. *Kirk-Othmer Encyclopedia ofChemical Technology*, 4th ed.; Kroschwitz, J., Home-Grant M., Eds.; John Wiley and Sons: New York, 1997; volume supplement, p 155.
256. Rao, R.; Dandekar, R.; Baker, R. T. K.; Vannice, M. A. *J. Catal.* 1997, 171, 406.

257. Seo, G.; Chon, H. *J. Catal.* 1981, 67, 424.
258. Frainier, L. J.; Fineberg, H. DE Patent 3007139, 1980.
259. Haidegger, E. DE Patent 2740057, 1977.
260. Hinnekens, H. DE Patent 3425758, 1984.
261. Cottier, L.; Descotes, G. *Trends Heterocycl. Chem* 1991, 2, 233.
262. Faury, A.; Gaset, A.; Gorrichon, J. P *Inf. Chim.* 1981, 214, 203.
263. Gaset, A.; Gorrichon, J. P.; Truchot, E. *Inf. Chim.* 1981, 212, 179.
264. Kuster, B. F. M. *Starch-Starke* 1990, 42, 314.
265. Lewkowski, J. *ArkiVoc* 2001, 2, 17.
266. Rapp, M. K. DE Patent 3601281, 1987.
267. Van Dam, H. E.; Kieboom, A. P. G.; Van Bekkum, H. *Starch-Starke* 1986, 38, 95.
268. Antal, M. J.; Mok, W. S. L.; Richards, G. N. *Carbohydr. Res.* 1990,199, 91.
269. Bicker, M.; Kaiser, D.; Ott, L.; Vogel, H. *J. Supercrit. Fluids* 2005, 36, 118.
270. Lichtenthaler, F. W.; Cuny, E.; Martin, D.; Ronninger, S. In *Carbohydrates as Organic Raw Materials;* Lichtenthaler, F. W., Ed.; VCH: Weinheim/New York, 1991.
271. Gandini, A.; Belgacem, M. N. *Prog. Polym. Sci.* 1997, 22, 1203.
272. Moreau, C.; Belgacem, M. N.; Gandini, A. *Top. Catal.* 2004, 27, 11.
273. El Hajj, T.; Masroua, A.; Martin, J. C.; Descotes, G. *Bull. Soc. Chim.Fr.* 1987, 855.
274. US Patent 4320043, 1982.
275. El Hajj, T.; Masroua, A.; Martin, J. C.; Descotes, G. *Bull. Soc. Chim.Fr.* 1987, 855.
276. Grabowski, G.; Lewkowski, J.; Skowronski, R. PL Patent 161831,1990.
277. Grabowski, G.; Lewkowski, J.; Skowronski, R. Electrochim. Acta 1991, 36, 1995.
278. *Jpn. Kokai* 1973, 73, 763.
279. Timko, J. M.; Cram, D. J. *J. Am. Chem. Soc.* 1974, 96, 7159.
280. FR Patent 2556344, 1985.
281. Gandini A *ACS Symp. Ser.* 1990, 433, 195.
282. Moye, C. *J. Rev. Pure Appl. Chem.* 1964, 14, 161.
283. US Patent 3083236, 1958.
284. Faury, A.; Gaset, A.; Gorrichon, J. P *Inf. Chim.* 1981, 214, 203.
285. US Patent 3847952, 1974.
286. Gandini, A., *Adv. Polym. Sci.*, 1977, 25,47.
287. Gandini, A., in *Comprehensive Polymer Science, First Supplement,* eds. S. L. Aggarwal and S. Russo. Pegamon Press, Oxford, 1992, p. 527.
288. McKillip, W. J., *ACS Symp. Ser.*, 1989, 385,408.
289. *Kirk-Othmer Encyclopedia of Chemical Technology,* Vol. 11, ed. M. Grayson. John Wiley & Sons, Inc., New York, 1981, p. 499.
290. Gandini, A., in *Encyclopedia of Polymer Science and Engineering,* Vol. 7, eds. H. F. Mark, N. M. Bikales, C. G. Overberger and G. Menges. Wiley, New York, 1988, p. 454.
291. Schmitt, C. R., *Polym. Plast. Technol. Eng.*, 1974, 3, 121.
292. Choura, M., Belgacem, M. N. and Gandini, A., *Macromolecules*, 1996, 29, 3839.
293. Gonzalez, R., Martinez, R. and Ortiz, P., *Makromol. Chem.*, 1992,193,l.
294. Gonzalez, R., Martinez, R. and Ortiz, P., *Makromol. Chem.*, Rapid Commun., 1992, 13, 5 17.
295. Galego, N. and Gandini, A., *Revista Cenic,* 1975, 6, 163; Galego, N. and Gandini, A., *Revista Cenic,* 1984, 15, 143.
296. Nguyen, C. and Zavarin, E., *J. Wood Sci. Technol.*, 1986, 6, 15.
297. Roque-Malherbe, R. and De Onate-Martinez, J., *J. Muter. Sci. L&t.*, 1993, 12, 1037.
298. McKillip, W. J., *ACS Symp. Ser.*, 1989, 385,408.
299. Rodriguez, V. J. and Gandini, A., *Revista Cenic,* 1974, 5, 29.
300. Rodriguez, V. J. and Gandini, A., *Revistu Cenic,* 1975, 6, 155.
301. Burma, S. P. N. S., Lenka, X and Nayak, P. L., *J. Appl. Polym. Sci.*, 1992, 44, 1311.
302. Prusty, J., Lenka, S. and Nayak, P. L., *Mucromol. Rep.*, 1993, A30, 27.
303. Swain, S. K., Shadoo, S., Mohapatra, D. K., Mishra, B. K., Lenka, S. and Nayak, P. L., *J. Appl. Polym. Sci.*, 1994, 54, 14 13.
304. Holly FW, Cope AC. *J Am Chem Soc* 1944;66:1875–9.

305. Burke WJ., *J Am Chem Soc* 1949;71:609–12.
306. Burke WJ, Bishop JL, Glennie ELM, Bauer WN., *J Org Chem* 1965;30:3423–7.
307. Kimura H, Matsumoto A, Hasegawa K, Ohtsuka K, Fukuda A., *J Appl Polym Sci* 1998;68:1903–10.
308. Ning X, Ishida HS., *J Polym Sci B: Polym Phys* 1994;32:921–7.
309. Ning X, Ishida H., *J Polym Sci A: Polym Chem* 1994;32:1121–9.
310. Liu J, Ishida H. In: Salamone JC, editor. "A New Class of Phenolic Resins with Ring-Opening polymerisation," *The Polymeric Materials Encyclopedia.* Florida: CRC Press; 1996. p. 484–94.
311. Takeichi T. "Polybenzoxazine as a Novel Phenolic Resin," *Kobunshi* 2001;50:786.
312. Takeichi T, Agag T, Guo Y, *Research Signpost* 2000;4:85–105.
313. N.N. Ghoshl, B. Kiskan, Y. Yagci, *Prog. Polym. Sci.* 32 (2007) 1344–1391.
314. Rudakov, G. A.; Ivanova, L. S.; Pisareva, T. N.; Borovskaya, A. G. Gidroliz. *Lesokhim. Prom-st* 1975, 4, 7.
315. Severino, A.; Vital, J.; Lobo, L. S. *Stud. Surf. Sci. Catal.* 1993, 78, 685.
316. Yilmaz, S.; Ucar, S.; Artok, L.; Gulec, H. *Appl. Catal. A-General* 2005, 287, 261.
317. Gunduz, G.; Dimitrova, R.; Yilmaz, S.; Dimitrov, L. *Appl. Catal. A-General* 2005, 282, 61.
318. Gunduz, G.; Dimitrova, R.; Yilmaz, S.; Dimitrov, L.; Spassova, M. *J. Mol. Catal. A* 2005, 225, 253.
319. Dimitrova, R.; Gunduz, G.; Spassova, M. *J. Mol. Catal. A* 2006, 243, 17.
320. Akpolat, O.; Gunduz, G.; Ozkan, F.; Besun, N. *Appl. Catal. A-General* 2004, 265, 11.
321. Volzone, C. et al., *Mater. Chem. Phys.* 2005, 93, 296.
322. Besun, N.; Ozkan, F.; Gunduz, G. *Appl. Catal. A-General* 2002,224, 285.
323. Costa, M. C.; Johnstone, R. A.; Whittaker, D. *J. Mol. Catal. A* 1996, 104, 251.
324. Tanabe, K.; Hattori, S. JP Patent 2292230, 1991.
325. Accrombessy, G. et al., *Bull. Soc. Chim. Fr. II-Chim. Mol. Org. Biol.* 1974, 705.
326. Venkatathri, N.; Srivastava, R. *Catal. Commun.* 2005, 6, 177.
327. Bhattacharjee, S.; Dines, T. J.; Anderson, J. A. *J. Catal.* 2004, 225, 398.
328. Maschmeyer, T.; Rey, F.; Sankar, G.; Thomas, J. M. *Nature* 1995, 378, 159.
329. Skrobot, F. C. et al., *J. Mol. Catal. A* 2003, 201, 211.
330. Joseph, T.; Sawant, D. P.; Gopinath, C. S.; Halligudi, S. B. *J. Mol.Catal. A* 2002, 184, 289.
331. Buffon, R.; Schuchardt, U. *J. Braz. Chem. Soc.* 2003, 14, 347.
332. van Vliet, M. C. A. et al., *Green Chem.* 2001, 3, 243.
333. Mandelli, D. et al., *J. Mol. Catal. A* 2001, 168, 165.
334. Saladino, R.; Andreoni, A. A.; Neri, V.; Crestini, C. *Tetrahedron* 2005, 61, 1069.
335. Joseph, T.; Halligudi, S. B. *J. Mol. Catal. A* 2005, 229, 241.
336. Tangestaninejad, S.; Mirkhani, V. *Chem. Lett.* 1998, 1265.
337. Cheung, W. H.; Yu, W. Y.; Yip, W. P.; Zhu, N. Y.; Che, C. M. *J.Org. Chem.* 2002, 67, 7716.
338. Cagnoli, M. V. et al., *Appl. Catal.A-General* 2005, 287, 227.
339. Cagnoli, M. V. et al., *Catal. Today* 2005, 107–108, 397.
340. van der Waal, J. C.; Rigutto, M. S.; Van Bekkum, H. *Appl. Catal.A-General* 1998, 167, 331.
341. Hutter, R.; Mallat, T.; Baiker, A. *J. Catal.* 1995, 153, 177.
342. Herrero, E.; Casuscelli, S.; Fernandez, J.; Poncio, C.; Rueda, M.;Oyola, O. *Molecules* 2000, 5, 336.
343. Villa de P., A. L.; Sels, B. F.; De Vos, D. E.; Jacobs, P. A. *J. Org.Chem.* 1999, 64, 7267.
344. Villa de P., A. L.; Taborda, F.; de Correa, C. M. *J. Mol. Catal. A* 2002, 185, 269.
345. Casuscelli, S. G. et al., *Appl. Catal. A-General* 2004, 274, 115.
346. Oliveira, P.; Ramos, A. M.; Fonseca, I.; do Rego, A. B.; Vital, *J.Catal. Today* 2005, 102, 67.
347. Prada, N. Q.; Stashenko, E.; Paez, E. A.; Martinez, J. R. *Rev. Colombiana Quim.* 1999, 28, 45.
348. Pillai, U. R.; Sahle-Demessie, E.; Varma, R. S. *Synth. Commun.* 2003, 33, 2017.
349. Bussi, J. et al., *Appl. Catal. A-General* 2003, 253, 177.
350. Aramendia, M. A. et al., *Stud. Surf. Sci. Catal.* 2000, 130B, 1667.
351. Aramendia, M. A. et al., *Appl. Catal. A-General* 2001, 216, 257.
352. Morikawa, T.; Takahashi, T. JP Patent 1319431, 1989.
353. Johnson, J. R.; Walter, E. US Patent 4551570, 1985.
354. Misra, A. N.; Sarma, M. R.; Soman, R.; Dev, S. IN Patent 146086, 1979.
355. Eswaramoorthy, M. et al., *Indian J. Chem. Sect. A-Inorg. Bio-Inorg. Phys. Theor. Anal. Chem.* 2001, 40, 264.

356. Krishnasamy, V.; Yeddanapalli, L. *Can. J. Chem.* 1976, 54, 3458.
357. Verghese, J. J. *Indian Chem. Soc.* 1959, 36, 151.
358. Jaramillo, H.; Palacio, L. A.; Sierra, L. *Stud. Surf. Sci. Catal.* 2002, 142B, 1291.
359. Roberge, D. M.; Buhl, D.; Niederer, J. P. M.; Holderich, W. F. *Appl. Catal. A-General* 2001, 215, 111.
360. Weyrich, P. A.; Trevino, H.; Holderich, W. F.; Sachtler, W. M. H. *Appl. Catal. A-General* 1997, 163, 31.
361. Hoelderich, W. F. *Catal. Today* 2000, 62, 115.
362. Buhl, D.; Roberge, D. M.; Holderich, W. F. *Appl. Catal. A-General* 1999, 188, 287.
363. Hoelderich, W.; Sachtler, W. WO Patent 9825873, 1998.
364. Lesage, P.; Candy, J. P.; Hirigoyen, C.; Humblot, F.; Basset, J. M. *J. Mol. Catal. A* 1996, 112, 431.
365. R.W. Hemingway, *Wood and Pulping Chemistry*, TAPPI (1989), pp. 377–386.
366. E.T.N. Bisanda and M.P. Ansell, *J. Mater. Sci.* 27 (1992), pp. 1690–1700.
367. Ranjana Yadav, Deepak Srivastava, *European Polymer Journal* 45 (2009) 946–952.
368. Menon ARR, et al., *J Sci Ind Res* 1985;44:324–38.
369. Tyman JHP. *Synthetic and Natural Phenols Studies in Organic Chemistry*, vol. 52. Amsterdam: Elsiever; 1975. p. 518.
370. Attanasi OA, Bunatti SB. *La Chimica e I'Industria* 1996;78:693–6.
371. Prabhakaran K, Narayan A, Pvithram C. *J Eur Cer Soc* 2001;21:2873–8.
372. Pillai CKS, Prasad VS, Sudha JD, Bera SC, Menon ARR. *J Appl Polym Sci* 1990;41: 2487–501.
373. Nimuru N, Miyakoshi T., *Int J Polym Anal Char* 2003;8:47–66.
374. Hulkko, Janne; Koskimies, Salme; Kylaekoski, Reetta; Laine, Aki; Luttikhedde, Hendrik; Willberg-Keyrilaeinen, Pia. (Ciba Holding Inc., Switz.). *PCT Int. Appl.* (2008).
375. Lee YK, et al., *J Appl Polym Sci* 2003;89:2589–96.
376. U.S. Pat. 2464207; 1949.
377. Jayabalan M, Rajadurai S. *Popular Plast* 1985;30:30–5.
378. Panasare V, Kulkarni A., Indian Pat. Appl. (2008).
379. Venkateswarlu, Konduri. (Solvosol Paints Pvt. Ltd., India), Indian Pat. Appl. (2008).
380. Yang, Weiguo; et al., Faming Zhuanli Shenqing Gongkai Shuomingshu (2008).
381. Dallons, Jean-Luc. *Asia Pacific Coatings Journal* (2004), 17(5), 38–39.
382. Kim, Yong Hwan et al., *Journal of Molecular Catalysis B: Enzymatic* (2007), 45(1–2), 39–44.
383. Won, Keehoon et al., *Biomacromolecules* (2004), 5(1), 1–4.
384. Kobayashi, Shiro, "Polymer Preprints," *Am. Chem. Soc. Div. of Polym. Chem.*, (2006), 47(2), 290–291.
385. Kobayashi, Shiro et al., 225th ACS National Meeting, New Orleans, LA, March 23–27, 2003 (2003).
386. C. V. Retna, A. Malar; Gopalakrishnan, S. *India. Bulletin of Materials Science* (2004), 27(3), 235–241.
387. E. T. N. Bisanda, W. O. Ogola and J. V. Tesha, *Cement and Concrete Composites*, Vol. 25, Issue 6, August 2003, Pages 593–598.
388. Knaggs, AR (2001). "The Biosynthesis of Shikimate Metabolites," *Natural Product Reports* 18 (3): 334–55.
389. Marco Krämer et al., *Metabolic Engineering* 5 (4): 277–283.
390. Johansson Louise, et al., *Biotechnology and Bioengineering* 92 (5): 541–552.
391. A. Pizzi, Editor, *Wood Adhesives—Chemistry and Technology*, Marcel Dekker Inc. (1983).
392. Niemetz & Gross (2005) *Phytochemistry* 66: 2001–2011.
393. Kasanov, V., et al., *Biomaterials.* 2006 Feb;27(5):745–51. Epub 2005 Aug 18.
394. Nanjwade, Basavaraj K.; Hiren M. Bechraa, Ganesh K. Derkara, F.V. Manvia, Veerendra K. Nanjwade (2009). "Dendrimers: Emerging polymers for drug-delivery systems," *European Journal of Pharmaceutical Sciences* (Elsevier) 38 (3): 185–196.
395. A. W. Bosman, H. M. Janssen, E. W. Meijer, *Chem. Rev.* 1999, 99, 1665–168.
396. http://dnanotech.com/dendrimerOverview.pdf.
397. J. M. J. Fréchet, D. A. Tomalia, *Dendrimers And Other Dendritic Polymers*, John Wiley & Sons, New York, U.S.A., 2001.
398. G. R. Newkome, C. N. Moorefield, F. Vögtle, *Dendrimers and Dendrons: Concepts, Synthesis, Applications*, Wiley-VCH, Weinheim, Germany, 2001.
399. Egon Buhleier, Winfried Wehner, Fritz Vögtle. *Synthesis* 1978: 155–158.

400. U.S. Patent 4,289,872.
401. U.S. Patent 4,410,688.
402. U.S. Patent 4,507,466.
403. D. A. Tomalia, et al., *Polymer Journal* 17: 117.
404. LD Reynolds and NG Wilson, *Scribes and Scholars* 3rd Ed. Oxford: 1991. pp193–4.
405. Hill, C. *Wood Modification. Chemical, Thermal, and Other Processes;* John Wiley & Sons: Chichester, U.K., 2006.
406. Chakar, F. S.; Ragauskas, A. J. *Ind. Crops Prod.* 2004, 20, 131.
407. Balakshin, M. Y.; Capanema, E. A.; Chang, H.. In *Characterization of Lignocellulosic Materials;* Hu, T. Q., Ed.; Blackwell: Oxford, U.K., 2008; p 148.
408. Brunow, G. In *Biorefineries—Industrial Processes and Products;* Kamm, B., Gruber, P. R., Kamm, M., Eds.; Wiley-VCH Verlag:Weinheim, Germany, 2006; Vol. 2, p151.
409. Kilpelainen, H., Forss, K., Fuhrmann, A. Proc. IUFRO Conf. Wood Gluing (1976), U. S. Forest Prod. Lab., Madison, Wis, US.
410. Northey, R.A. in *Chemical Modification and Usage of Lignin*, Hu, T. H. ed., Kluwer Academic Press, New York (2002), Chapter 8.
411. Kim, Y. S.; Chang, H.-m.; Kadla, J. F. *J. Wood Chem. Technol.* 2008, 28, 1.
412. Weckhuysen B.M., et al., *Chemical Reviews*, 2010, Vol. 110, No. 6.
413. Rao, S. R.; Ravishandkar, G. A. *J. Sci. Food Agric.* 2000, 80, 289.
414. Petrocelli, F. P.; Klein, M. T. *Ind. Eng. Chem. Prod. Res. DeV.* 1985, 24, 635.
415. *Kirk-Othmer Encyclopedia of Chemical Technology*, 4th Edition.
416. Lynn Montague. Harris Group, Inc., Seattle, Washington, Report 42002/02: Review of Design, October 2003, NREL/SR-510-31579.
417. US Pat 4164397 Hunt et al.
418. US Pat 4166830 Guth et al.
419. US Pat. 4409416 bu Snell et al. 1983.
420. T. K. Kirk and R. L. Farrell, *Annu. Rev. Microbiol.*, 1987, 41, 465.
421. Bugg and Winfield: "Enzymatic Cleavage of Aromatic Rings," *Natural Product Reports*, 1998.
422. S. Dagley, *Essays Biochem.*, 1975, 11, 81.
423. J. I. Davies and W. C. Evans, *Biochem. J.*, 1964, 91, 251.
424. J. R. Mason and R. Cammack, *Annu. Rev. Microbiol.*, 1992, 46, 277.
425. V. Massey, *J. Biol. Chem.*, 1994, 269, 22 459.
426. H.H. Yoon, Z. Wo and Y.Y.Lee, *Appl. Biochem. Biotech*, 51/52, 5 (1995).
427. I.S.Goldstein, *Appl. Polym. Symp.*, 28, 259, (1975).
428. H.L. Chum, D.K. Johnson, S. Black, M. Ratcliff and D.W.Goheen, *Advantages in Solar Energy, Annual Review of Research and Development*, K.W. Boer, ed. pp.91–171, Plenum Press (1988).
429. M.Heitz, A.Brown, and E.Chornet, *Can J.Chem Eng*, 72, (1994), p1021.
430. U.Schuchardt, A.R.Rodrigues, A.R.Cotrim, and L.M.Costa, *Bioresource Technol.*, 44, (1993),p.123 .
431. M.Heitz, G.Wu, and J.Lapointe, *J. Wood. Chem. Technol*, 15 (1995) p.515.
432. A.Oasmaa and A.Raimo., *Bioresource Technol*, 40 (1992) p.189.
433. D.Meier, R.Ante, and O.Faix, *Bioresource Technol*, 40, (1992), p.171.
434. W.J.Frederick, M.Hupa, and T.Uusikartano, *Bioresource Technol*, 48, (1994), p.59.
435. M.Kudsy, H.Kumazawa, and E.Sada, *Can. J. Chem. Eng*, 73 (1995), p.411.
436. R.P.Overend, *Can. J. Chem. Eng.*,71 (1993) 107.
437. M.Koyama, *Bioresource Technol.* 44, (1993), p.209.
438. R.W.Thring and E.Chronet, *J.Chromatography*, 467 (1994) p.441.
439. G.Wu, M.Heitz, and E.Chronet, *Ind. Eng. Chem. Res*, 33, (1994) p718.
440. *Kirk-Othmer Encyclopedia of Chemical Technology.* John Wiley & Sons, Inc.
441. E. Sjöström, *Wood Chemistry: Fundamentals and Applications.* Academic Press, 1993.
442. G. C. Hobbs and J. Abbot *J. Wood Chem. Technol.*, 14, 195–125, 1994.
443. S. Kawai and H. Ohashi *Holzforschung*, 47, 97, 1993.
444. M. M. Santos et al. *J. Braz. Chem. Soc.*, 6, 257, 1995.
445. S. Zhou, T.B. Causey, A. Hasona, K.T. Shanmugam and L.O. Lugram, *Appl. Environ Microbial*, 69, p. 399–407, 2003.

19

Long Biofibers and Engineered Pulps for High Performance Bioplastics and Biocomposites

Alan Fernyhough and Martin Markotsis

Biopolymer & Green Chemical Technologies Team, Scion, Rotorua, New Zealand

Abstract

Various biocomposite products based on natural fiber mat reinforcements, textile technologies and compounding (for injection or compression moulding) have been developed to extend the performance of plastics and bioplastics. The utilization of long fiber reinforcements, either as mats or mouldable long fiber pellets, can advance performance even further, particularly with respect to improving strength and impact properties. Developments in long natural fiber reinforcements for biocomposites, including long wood fibers, are described. Wood fibers are generally not described as "long" but when genuine wood fibers are utilized, and appropriately processed, significant improvements in properties can be achieved compared to "conventional" wood flour (particle) composites and indeed compared to other biofiber composites. Examples of Scion technologies for long biofiber and long wood/pulp fiber reinforcements, or engineered pulps, are described for the improvement of thermoplastics, including bioplastics. The wood (pulp) fibers provide a highly cost effective route to property enhancements.

Keywords: Long fiber, wood fiber, biofiber, bioplastic, biocomposite, pulp

19.1 Introduction to Long Fiber Reinforced Plastics and Processes

Fiber reinforced thermoplastics are widely used in many industrial applications. They offer a combination of high mechanical properties, light-weight, mouldability and design flexibility. The relentless demand for higher performance and light-weighting from plastics (indeed materials in general) has led to a demand of increasingly higher mechanical performance from moulded reinforced plastic parts. The performance benefits of using fiber reinforcements in plastics are well known and, if suitably developed and applied, can impart higher stiffness, higher strength and/or higher impact resistance, among other attributes, to many plastics [1–4]. Fundamental composite theory, and indeed practical applications indicate that, other things being equal, a higher fiber (reinforcement) aspect ratio (that is length: diameter ratio) can lead to higher mechanical properties [2].

Srikanth Pilla (ed.) Handbook of Bioplastics and Biocomposites Engineering Applications, (555–580)

Numerous processes and fiber forms have been developed for long fiber reinforced composites. In thermoset resin composites processes such as hand lay-up, prepreg moulding, spray-up moulding, vacuum-bag/autoclave moulding, resin transfer moulding, filament winding and pultrusion are common. Continuous fiber rovings, yarns, textile fabrics, various fiber mats are common in such processes [3, 4]. However, the focus of this article is thermoplastic matrix processes, particularly as applied to natural and wood fiber reinforced thermoplastics.

Long fiber thermoplastic moulding processes and materials have been developed to further advance the performance of reinforced plastics[1–4, 5]. Examples of this approach include polyolefin (e.g. polypropylene) and polyamide plastics reinforced with long glass (primarily) or carbon (or other) fibers. As referred to above, fiber length in a reinforced plastic matrix influences the final performance of a part in terms of stiffness, strength and impact strength. Residual fiber length, however, is only one aspect. The improvement of mechanical properties also depends on the fiber distribution, fiber orientations, other fiber breakage or damage (not just by length attrition), and the interfacial bonding or interactions of the fibers and the polymer matrix. The latter is most often tailored through reasonably well known compatibilization strategies such as, for example, the use of functional interfacial polymers and/or fiber surface modifications [6, 7]. The inherent fiber and matrix properties are, of course, also important.

The most common processes relevant to fiber reinforced thermoplastics are glass mat thermoplastic moulding (GMT) using mat reinforcements, and compounding long fiber reinforced thermoplastics (LFRT) processes. One example is a process that provides long fiber pellet feedstock for injection moulding [8]. For GMT composites typically two main routes are used to produce glass mat thermoplastic intermediates (GMT):

- a wet slurry process to produce a non-woven web of intimately mixed fibers and polymer which is then dried passed and consolidated;
- a random fiber mat process wherein fiber mats are impregnated with molten polymer in a layered arrangement.

GMT materials can be processed into final parts by stamping or high speed compression moulding. Products based on polypropylene (PP) and moderate loadings of relatively long glass fibers in a random array have proven to be successful in the automotive industry. GMTs are typically produced on large laminating machines wherein a partially consolidated composite sheet is typically produced. GMT is well suited to large parts with low tooling costs and fast cycle times using stamping or compression moulding. In some cases, for the most demanding applications, textile-reinforced GMTs can be used, often as a sandwich or within a layered construction. GMT can thus be a combination of various fiber mat forms (usually with a consistent resin matrix) and can be used for the manufacture of finished parts with great flexibility in terms of design and fiber deployment. Examples of global GMT suppliers include Quadrant Plastic Composites AG and AZDEL Inc. and a number of other producers or regional suppliers [9].

LFRTs have been developed from short-glass injection moulding concepts. In injection moulding of conventional (short) fiber reinforced thermoplastics a relatively low residual fiber length after processing limits the mechanical properties [10, 11]. Long fiber thermoplastic pellets have been developed by providing a higher fiber aspect ratio (length/diameter ratio). Typically, pellets of LFRTs are formed into more complicated part designs and the pellets in the glass or carbon fiber examples are usually 10 mm or 12 mm (or longer) in length. Initial (feed) fiber length is often determined (or the same as) pellet length. Such pellets are typically made by either pultrusion of continuous fiber through molten resin and cutting the small diameter rod profiles into pellets, or via cross-head extrusion coating of continuous fiber feedstock (akin to cable coating) and pelletizing [12].

The first LFRTs were launched in the 1990s. Today, major suppliers of pre-compounded LFRTs include SABIC (GE Plastics)/LNP Specialty Compounds, Ticona, Dow, Chisso, RTP, Borealis, Technyl and others. Pellets can be made highly loaded (up to 70 % fiber) and can be let down as a masterbatch at moulding to adjust fiber levels. The most common forms of compounded LFRTs are glass-reinforced polypropylenes (PPs) though a broad range of semi-crystalline and amorphous resins have been used. After PP, nylon (polyamide or PA) 6 and 6,6 are the next most common matrices. A number of factors influence preservation of reinforcement integrity and composite performance. These factors include the pellet feedstock material, its composition, delivery, processing/moulding variations and manufacturing equipment and designs [11, 12]. The initial fiber fraction and aspect ratio clearly have a major effect on the residual fiber length after moulding.

Direct long-fiber thermoplastic (D-LFT) processes essentially combine compounding and moulding into an integrated process. In D-LFT systems the moulder can introduce resin, reinforcement and additives at moulding prior to feeding, directly, to injection or compression moulding for part processing. D-LFT is relatively capital intensive and less common than pre-compounded LFRT pellets. Some D-LFT systems, for injection or compression, feed glass roving (or other fiber) into a twin-screw extruder where it is chopped by the screws as it is mixed with molten polymer and additives that have been metered in separately. Examples of machinery developers include Krauss-Maffei, Plasticomp, Dieffenbacher, Coperion Werner & Pfleiderer and Composite Products Inc. Other variations exist such as Pushtrusion (Plasticomp) and special approaches for liquid (reactive) polyurethane long fiber moulding, particularly pioneered by Krauss Maffei [13].

19.2 Introduction to Biofibers, Bioplastics and Biocomposites

Industry is seeking more environmentally friendly materials such as natural fibers, biobased polymers and biodegradable materials as a result of growing environmental awareness and changing regulations. Conventional petrochemical plastics and non–renewable fibers or fillers are coming under increased criticism because of their less favourable impacts on the global environment, for example due to their lack of renewability and/or "biodegradability" features. Renewably resourced polymers (biopolymers, or bioplastics in the current context) and natural fiber (biofiber) composites (especially in combination with bioplastics) offer a possible, and increasingly attractive, alternative to traditional non-renewable options. Some also offer added benefits such as desirable end-of-life options such as compostability, ability to be incinerated, as well as recycling [14].

19.2.1 Biofibers

The use of natural fibers dates back almost to the start of civilization, with people such as the ancient Egyptians using grass and straw to reinforce mud-bricks [15]. Plants, such as flax, cotton, hemp, jute, sisal, kenaf, coir, pineapple, ramie, bamboo, banana, as well as wood, and indeed other plants, are sources of biofibers. Their availability, renewability, low density and price, in combination with appropriate mechanical properties, make them attractive reinforcing fibers in the manufacture of composites [16, 17].

In more recent history the development of novel synthetic materials rapidly displaced the use of natural materials, slowing the development of biobased materials almost to a complete stand-still. Recent drivers for more environmentally friendly, sustainable products and materials are leading to greater interest in replacing inorganic fillers (e.g. calcium carbonate, talc) and reinforcement materials (e.g. glass fiber) with renewable organic (biobased) materials such as biofibers [18].

Natural (bio-) fibers can be classified in various ways. One classification approach is depicted Figure 19.1, wherein some property data ranges are included as an indication of possible performance comparisons (fiber properties) [19].

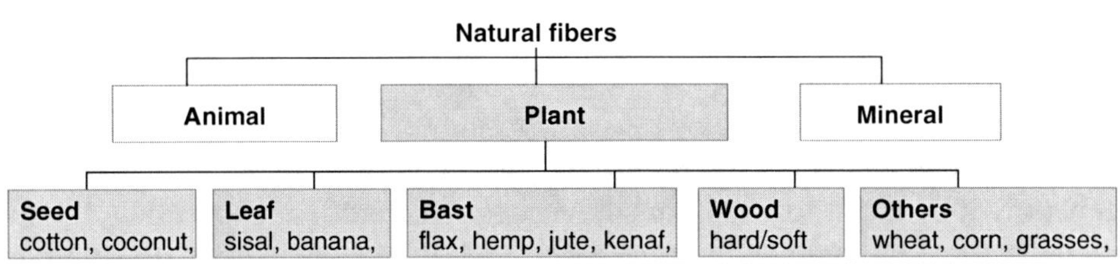

	Density g/cm^3	Tensile strength MPa	Specific tensile strength MPa/gcm^{-3}	Tensile modulus GPa	Specific modulus GPa/gcm^{-3}
E glass	*2.52*	*2400*	*950*	*72*	*29*
Linen flax	1.4-1.5	500-1500	350-1030	50-75	30-50
Hemp	1.4-1.5	500-1000	350-700	40-70	27-47
Jute	1.3-1.5	200-800	150-600	20-55	13-37
Sisal	1.4-1.5	300-800	200-600	10-40	7-27
Kenaf	1.4-1.5	300-500	200-350	25-50	17-33
Wood	0.6-1.4	100-300	100-300	5-30	5-30
Cotton	1.5	400	270	12	8

Figure 19.1 Classification of reinforcing natural fibers (biofibers).

It can be seen that natural fibers have desirable properties such as good stiffness, while offering lower density, together with potential economic and ecological advantages compared to glass fibers. Most natural fibers are available at relatively low cost. Since they have lower densities than glass, they can be used at higher loadings to reduce the overall part cost. Although natural fibers have been grown and used for centuries, a revision of the traditional harvesting and preparation techniques is generally required to produce a consistent fiber source suitable for the current composites industry [20, 21]. Natural fibers, including flax, hemp, jute, sisal, and kenaf, have been used commercially at least since the 1990s in interior automotive applications [22].

The properties of plant derived fibers are harvest-dependent and influenced by climate, location, soil characteristics etc [23]. In addition, of course, the fibers need to be extracted from the plant resources, and processed and presented in forms suited to composite fabrication. The end properties are thus also affected by fiber extraction and processing methods (for example, refining, retting, scutching, bleaching, spinning, etc.) and by the methods applied for their incorporation into composites (fiber handling, mat/fabric formation, impregnation and consolidation).

Wood as a source of fiber reinforcement has recently attracted much interest. The development of wood–plastic composites (WPC) in the United States has been significant and WPCs have become increasingly popular materials due to the abundance, low cost, and processability of wood as a filler [24, 25]. Several authors have reviewed the history and growth of WPCs, particularly in the US where they are widely used in building applications for example [26, 27]. Issues of high moisture content, a tendency to absorb moisture, together with a low bulk density and low thermal stability, have been overcome through various innovations. These developments have led to US demand for wood-plastic composites and plastic lumber projected to grow at 9.2% pa to $5.3 billion (1.5 million tons) by 2013 [28]. Furthermore, wood is typically not grown as a food crop, and is harvested and processed efficiently, in high volumes and consistencies, into many product forms, including structural solid wood products, pulp and paper, composite fiberboards, and packaging products [29, 30]. With plastics, wood has primarily been used as wood flour for filling plastics (conventional wood plastic composites, WPCs) and is often derived from residue streams. Although many use the description of "wood fiber", in such circumstances actually the "fibers" are very short in most cases and are "flours" or particles with typical dimensions less than 1mm in any direction. Thus, commercial WPCs are typically made from wood flour, with a low aspect (l/d) ratio, typically much less than 10.

Low l/d ratios often limit or reduce strength compared to neat polymer. Both strength and stiffness increases can be realized with increasing fiber length, if: (1) adhesion between wood fibers and the plastic matrix is good; (2) fibers are uniformly dispersed in the matrix; and (3) fibers are adequately oriented. As described later, technologies for deploying genuine wood fibers (with longer aspect ratios) have been more recently developed and represent a major advancement over wood flour filled composites. Further, as with natural fiber and indeed glass or carbon fiber composites, wood fiber reinforcements can, in principle, come in a variety of forms and can be processed via a variety of processes.

Long wood fibers have slightly different connotations compared to long glass, carbon or indeed hemp, flax or many other non-wood natural fibers. Typically "long wood fibers" or genuine wood fibers may be 2–3 mm long. While this is not "long" compared to the other fiber types the difference compared to a wood flour is still significant [31]. An overall aspect ratio of softwood refiner fibers has been indicated as approximately 25 while aspect ratios of commercial pine wood flour (mesh size between 20 and 140, i.e. between 0.85 mm and 0.106 mm) are only between 3.4 and 4.5 [32]. Hence, wood fibers from commercial refining have the potential for greater reinforcement of WPC due to their high aspect ratio. In addition, modification of fiber surfaces during the refining process may lead to improved fiber-matrix adhesion.

Common limitations for the use of biofibers in plastics and composites include lower processing temperatures, due to the possibility of biomass degradation. This constraint limits their applications to some matrices (thermoplastic polymers) and restricts use with those polymers processed at ~ 200°C or more. In addition, the hydrophilic and polar nature of biofibers can limit their use since the relatively higher moisture absorption of natural fibers often requires extensive pre-drying regimes, can lead to in-service water uptake and durability concerns, especially if not well processed, and can also affect compatibility or binding of the fibers to the polymer matrix. Furthermore, biofibers can impart a relatively lower impact strength compared to glass fiber options [33]. However, the likely increasing availability and favourable economics for using biofibers will almost inevitably drive application developments and innovations to overcome limitations, and indeed already is. It is interesting to note, also, the parallel future trends in biofuels or biorefinery operations - almost all using biomass composed primarily of cellulose, hemicelluloses and lignins to produce fuels and chemicals. It will increasingly be important to find more uses for, and extract value from, the co-products whether they are sources of fibers, biopolymers or biochemicals. The future seems to be bright for biofibers and biobased materials.

19.2.2 Bioplastics

Bioplastics can be defined as either:

- plastics derived from renewable biobased materials and/or
- plastics which are able to be described as biodegradable (as certified according to certain standards) [34].

Numerous reports identify the massive potential for bioplastics and many suggest that by 2020 bioplastics utilization could be at least 10% (one of the more conservative estimates) of the global plastics industry. Some reports suggest up to 30% or more penetration rates [35, 36] (Similar, if somewhat less aggressive, forecasts are predicted for natural fiber reinforcements as alternatives to, for example, glass fiber reinforcements [37]).

Renewably resourced bioplastics can be produced via a number of routes (Figure 19.2) such as:

a. production (growing) and direct extraction of polymers as they occur naturally in biomass/plants, then modified or formulated as necessary (e.g. plasticized) for use in plastics processes (e.g. plasticized starch based plastics);
b. production of polymers by polymerization of biobased intermediates (monomers/precursors which are bio-derived via direct extraction, thermo-chemical modification, or fermentation of biomass materials - often of starches). Polylactic acid (a type of polyester made from the chemical polymerization of lactic acid intermediates derived from fermented starch), or polymers derived from bioethanol platform chemicals are leading examples;
c. production (and extraction) of biobased polymers made using micro-organisms and biomass/waste feedstocks (e.g. polyhydroxy-alkanoate – a type of polyester made in bacteria).

The leading types of bioplastics are either polysaccharides or biopolyesters [38]. **Starch** based bioplastics, typically blends or plasticized natural starches (e.g. derived from corn, wheat, potato, cassava etc), and **cellulosic esters** (derived from wood pulp) are the leading polysaccharide bioplastics. **Polylactic acid** (also polylactide-PLA – derived from fermented corn starch primarily), **polyhydroxy-alkanoates** (PHA, derived from bacteria) and other polyesters (including polysuccinate esters) are the other leading commercial class of bioplastics. Other types are emerging. In the future, lignocellulosic materials and more non-food biomasses

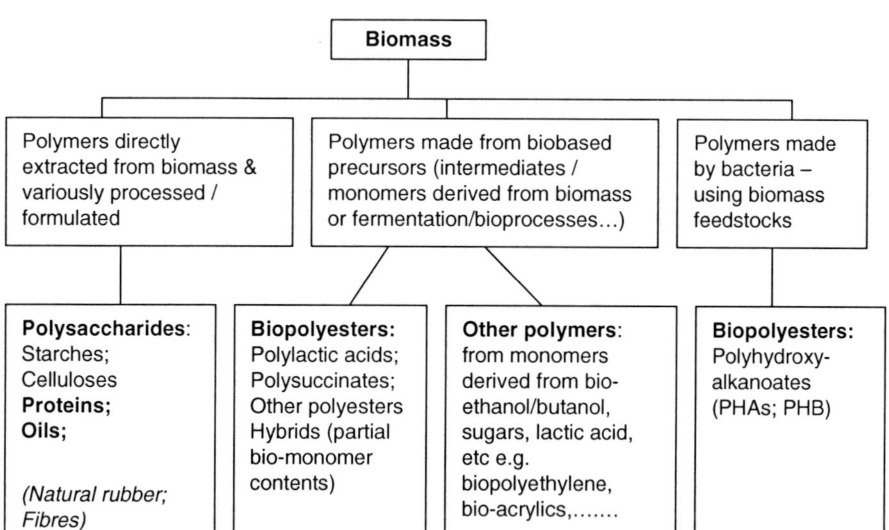

Figure 19.2 Major types of commercial bioplastics (2010) – derived from renewable biomass resources.

are forecast to be increasingly used as sources of monomers or precursors to bioplastics. New routes to traditional petroleum based monomers will likely become available from developments in bioethanol and other biofuels technologies, sugar conversion technologies and the application to lignocellulosics or other bioresources. The recent developments of bio-succinic acid (a precursor to polysuccinate esters), and bio-polyethylene (using ethylene derived from bioethanol) is evidence of such trends [39].

Polylactic acid (PLA, also polylactide) is the single most commercially available bioplastic. NatureWorks, currently the largest PLA producer, manufacture the biopolymer from corn, via a well known process. PLA can also be derived from fermentation (or other) process of any starch or other precursors, which generate lactic acid. PLA is well known for being biocompatible and compostable and has received considerable industrial attention as a replacement for existing petrochemical-based polymers due to its processability and properties which in some circumstances can match or exceed the performance of conventional plastics such as polypropylene [40, 41]. PLA grades are available for injection moulding, extrusion, film or sheet, and emulsion coating and it is used in a wide variety of applications including packaging (e.g. water bottles, clam shells), textiles (clothing, furnishings, non-wovens), paper coating, and electronics casings. Widespread application of PLA is still limited however, primarily because of its relatively high cost and property limitations such as brittleness, low heat deflection temperature and gas barrier performance [42]. Although the resin price for PLA has become much more competitive it is still priced relatively highly for many commodity products. Modification of PLA is gaining popularity and the addition of groups such as maleic anhydride to the PLA chain enables better mixing with other polymers or fillers [43, 44]. Work at Scion has observed similar improvements in the tensile modulus and strength with the use of modified PLAs.

Polyhydroxyalkanoates (PHAs) are synthesized biochemically by microbial fermentation. The most common PHA is poly(3-hydroxybutyrate) PHB which is produced as a carbon/energy store in a wide variety of bacteria [45]. PHAs are generally stable under normal usage conditions, but in soil, aqueous or composting conditions they are readily broken down into carbon dioxide and water by a variety of bacteria. The most common route to improving the processability range and/or properties is via the synthesis of copolymers by adjusting the bacteria's "feed". Poly(3-hydroxybutyrate-co-3-hydroxyvalerate) (PHBV), first manufactured by ICI in the 1980s, is made this way and can have melting points and glass transition temperatures which vary according to composition and which can be similar to polypropylene. However, the commercial uptake of PHAs has been limited by the prohibitive cost, in addition to the limited processing temperature range between melting and thermal degradation, and some property limitations [46].

Other polyesters include polysuccinate esters, which, while often compostable are largely petroleum derived currently. However, such polymers are likely to be increasingly renewably resourced due to future bio-succinic acid availability. A variety of hybrid or partial bioderived polyesters are also available (for example due to the use of bioderived diol feedstocks).

Polyamides (PA) are an emerging further class of bioplastics, more commonly used in engineering applications. Nylon developments date back to the 1930s and commercial processes were developed using petrochemical sources. The first bio-based polyamides were actually developed in the 1950s, however the cost-performance ratio was regarded as unacceptable [47]. With growing consumer pressure for more environmentally friendly solutions some of the larger polymer companies are now re-investigating the development of engineering plastics from renewable resources and including some polymers partially derived from renewable sources (hybrids). BASF has developed Ultramid Balance nylons which consist of up to 65% bio-based content. DSM have developed high temperature processing EcoPaXX material, Evonik offers a range of biobased nylons under the trade name Vestamid Terra, and Arkema offer Polyamide 11 with the trade name Rilsan®. Some, such as polyamide 11, are wholly bio-derived using castor oil as a source of monomer.

19.2.3 Biocomposites

Biocomposites is a very broad field with developments ranging from wood flour filled petrochemical polymers, natural fiber reinforced polymers, through to fully renewable bioplastic resins reinforced with natural biofibers, and more recently, wood fibers. By definition, a composite comprises two or more different, distinct, material components and is typically a matrix or binder with a reinforcement or filler. Thus, biocomposites comprise one or more such components derived from biobased resources (either a matrix and/or reinforcement and/or filler) [48].

As indicated above, natural fiber reinforcement of petroleum based polymers, especially as a glass fiber replacement, is gaining popularity and commercial materials based on natural fiber reinforced composites are finding wider applications [49]. Wood is increasingly used as a very effective biofiller for plastics and wood plastic composites are now well established in many building products and indeed wider applications.

The development of fully biodegradable "green" composites by combining biofibers with biodegradable resins is also growing, particularly for some packaging applications. Similarly, renewably resourced biocomposites comprising biobased matrix and a biofiber – though not necessarily "biodegradable" - are increasingly attracting interest as durable, wholly biobased, biocomposites.

In order to increase the range of applications for bioplastics such as PLA (and others), reducing material cost while also maintaining or advancing material performance is key. Biocomposites based on biobased fillers or fibers are a route to achieving this goal. Although moisture sensitivity and biological attack (durability) remains a concern for natural fiber or bio-filled composites, the literature contains references to various technologies designed to improve natural fiber biocomposite water resistance [50, 51]. Another known problem in natural fiber reinforced composites is the poor interface quality between the fibers and the polymer matrix. To enhance the adhesion between both components, chemical pre-treatments are often applied to the fibers during fiber processes – and/or additives are included in the resins and/or onto fibers to enhance compatibility and interfacial adhesion.

Various studies have indicated that simple methods such as alkali treatments can be effective fiber treatments for improved composite properties [52].

In recent years, the Nova-Institut has conducted numerous studies in the field of biofibers and biocomposites [53]. The automotive industry (mainly compression moulding with natural fiber-reinforced thermosets), packaging (biodegradable packaging) and the construction industry (extruded WPC products) are dominant users. As for production methods, injection moulding is continuing to gain in importance and is making it possible to penetrate new markets, mainly in the furniture, consumer goods and electrical/electronics products industries. In the future, further use of natural fibers in reinforced bioplastics is expected. During the last two years in particular, positive trends have become evident. For example, German WPC production volume has risen from 5 000 tons in 2005 to about 20 000 tons in 2007. Nova-Institut estimates that the present production volume in Europe is about 100 000 tons. Germany/Austria, Benelux and Scandinavia are the leading regions. These figures refer to the building, construction and furniture industries. In the European automotive industry, another 50 000 tons of WPCs are used per annum.

19.3 Natural Fiber Mat & Wood Fiber Sheet Moulding for Composites

The GMT (glass mat thermoplastic) type processes as described above have been adapted to NMT ("Natural Mat Thermoplastic") composites [54, 55]. Commercial natural mat forming technologies have been available for a number of years. Indeed NMTs are well reported and are offered as a commercial alternative to GMT. The lower density, improved environmental image and impacts, among other benefits, have driven some of these developments.

Much of the early work in NMT was with flax or hemp fibers – and polyolefins, particularly polypropylene (PP), due to the predominant use or preference for PP in automotive composite parts. The mechanical properties of glass and flax NMT (sheet-moulded or film stacked for example) were shown to be comparable in stiffness, though the flax materials were typically poorer in impact strength and slightly lower in tensile or flexural strength [56, 57, 58]. However, overall the properties compare well with GMT composites and particularly on a specific basis since the NMT composites typically have densities lower than 1 – about 25% lower than GMT composites. Flax and hemp continue to be important natural fibers for mat reinforcements though others have been developed. The processing of such fibers has been primarily developed for random non-woven mats for use in composite products with moderate mechanical properties.

Oskman and others have extended the NMT studies to bioplastic matrices such as polylactic acid (PLA). These thermoplastic mat composites combine renewable biofibers with matrices which also originate from renewable raw materials [59]. Oskman showed that PLA works very well as matrix material for natural fiber mat composites and obtained promising mechanical properties with PLA and flax fiber mat composites. Although results show that composite densities are somewhat greater than PP – natural fiber equivalents. The tensile strength (53MPa) of their composite was about 50% greater than equivalent PP/flax fiber composites and modulus was 8.4 GPa, with 30 wt% flax fibers. Cyras *et al* studied starch/PCL (polycaprolactone)/sisal fiber composites and reported a tensile modulus of 0.7 GPa and a strength of 14.4 MPa with a 30 wt.% sisal fiber content [60]. These values are very low compared to reported PLA/flax data. Riedel *et al* studied different biocomposites, many incorporating unidirectional reinforcements, and reported very high mechanical properties of composites [61]. As yet unpublished work at Scion has indicated flexural strengths and moduli of 95–135MPa and 5–6.4 GPa respectively for PLA-ramie mat/sheet composites at 33-39 wt% fiber contents.

For higher performance in natural fibers some have looked to textile fabric fiber forms. Highly structured textile reinforcements, such as interwoven fabrics and unidirectional fabrics made from natural fiber yarns perform considerably better than random non-woven mats in natural fiber composites [62, 63]. Although the production of natural fiber yarns and fabrics from flax, hemp and other natural fibers such as ramie is a somewhat expensive process requiring careful controls, such fiber forms have found their way into commercial reinforcements. Also, as in other fiber reinforced composites interfacial bonding between the fiber and the polymer is required for achieving maximal strength in the final composites. A number of other routes to improve the performance of natural fiber mat thermoplastics have been described and these include optimizing fiber fineness and mat construction for higher impact performance [64].

Examples of commercial products include those from Biotex [65] who have created a family of natural fiber reinforcements and biocomposite materials, based on textile technologies including commingled flax/PP, flax/PLA and flax yarns, fabrics and pre-consolidated sheets. The materials use a twistless technology to provide high levels of performance and processability normally associated with glass reinforced materials. Using twistless technology, the flax fibers in the yarns are highly aligned and claimed to give up to 50% better fiber efficiency over conventional twisted yarns. Thereby easier to impregnate, they give improved fiber/matrix interaction and better performance. Typical properties for flax/PLA composites (at a density of 1.33g/cm^3) include a tensile modulus of 13.2 GPa, a tensile strength of 102 MPa and a Charpy impact strength of 32.8kJ/m^2 (unnotched, flatwise). Related developments have been undertaken by others, such as the Biopolymer Network Ltd, who also developed unique harakeke fiber (*Phormium tenax*, also known as "New Zealand flax", although a leaf fiber) yarns [66]. They have also developed novel pulp based reinforcement mats for composites, again based on harakeke [67].

However, the handling and processing of natural fibers has limitations which are well documented and which still pose challenges. A recent review summarizes the fabrication options for kenaf–polypropylene sheets as an example of natural fiber reinforced composites [68]. They claim the optimal fabrication method for these material types is a compression moulding process utilizing a layered sifting of a microfine polypropylene powder and chopped kenaf fibers. A fiber content of both 30% and 40% by weight has been shown to provide adequate reinforcement to increase the strength of the polypropylene powder. The use of a coupling agent, 3% Epolene (maleic anhydride- polypropylene copolymer) enabled good fiber–matrix adhesion. The optimal compression moulded kenaf–PP composites showed benefits over other compression moulded natural fiber composites such as other kenaf, sisal, and coir reinforced thermoplastics. With the elastic modulus data from testing, it was also possible to compare the economic benefits of using this kenaf composite over other natural fibers and E-glass. The kenaf–maleated polypropylene composites manufactured in this study have a higher modulus/cost and a higher specific modulus than sisal, coir, and even E-glass thereby providing an opportunity for replacing existing materials with a higher strength, lower cost alternative that is environmentally friendly.

Mitchell [69] has described common approaches to **wood fiber thermoplastic sheet** manufacture. Examples include dry formed wood fiber sheets with plastics, wood fiber sheets prepared by organic solvent based impregnation, wood fiber sheets impregnated by dipping with emulsions in water, and molten polymer impregnated sheets. Other approaches include formable pulp compositions from aqueous wood and natural fiber slurries [70]. Extrusion of fibers and plastics into a sheet is another common approach. Subsequent compression moulding or thermoforming can then take place to produce consolidated sheets or moulded shapes using matched die moulds. Various other patents describe further variations [71]. Early work by Bhattacharyya *et al* indicated the formability of genuine wood fiber reinforced thermoplastics [72].

Research at Scion has developed several approaches to the production of thermoformable fiber-plastics composite sheet or film products based on wood fibers or indeed other lignocellulosic or natural fibers [73]. One favoured process conveys a stream of wood fibers produced via thermomechanical refining, and coats the fibers in a blowline mixing process with a thermoplastic binding agent. This process is common in a medium density fiberboard manufacturing plant, The impregnated fibers are then formed into a solid or semi-solid product such as a panel or sheet, which is thermoformable, or stampable into shapes. Additional thermoplastic polymer(s), as powder, pellets, film, or fibers can be added at various points of the whole process. The medium density fiberboard (MDF) process uses a high temperature thermomechanical pulp process to produce MDF fibers which are wood cells (tracheids, vessels, fibers and fiber-tracheids), rather than particles and is a low cost form of wood fiber which has an aspect ratio to allow reinforcing of composites (for example, with *radiata* pine, approx 2.5 mm×30 μm). The process provides a very cost-effective method for producing wood fiber plastic sheets which are thermoplastically processable.

Sheets made from MDF fiber (as a representative fiber) and PP (polypropylene – as a representative thermoplastic polymer), together with added compatibilizers exhibited properties such as those in Table 19.2. For example, a modulus (flexural) of 3–6 GP, and a strength (flexural) of up to 82 MPa were achievable with wood fiber polypropylene sheets.

The data in Table 19.1 compare well with published example data for NMT (natural fiber mat thermoplastics) [57–61,74] where, for example ~ 60 MPa is typical for a flexural strength for hemp or flax fiber composites with fiber contents ranging from 60 to 70 wt.-% and density values between 0.90 and 0.96 gcm^{-3}.

Table 19.1 Properties of wood fiber sheet mouldings.

Thermoformable Sheet Sample Description (all using radiate pine TMP fibers – except PP reference)	Density g/cm³	Flexural Modulus GPa	Flexural Strength MPa	Impact Strength (Charpy) J/m
MDF-PP sheet (80 wt% MDF fiber) 20M1	0.96	4.42	40.8	155.0
MDF-PP sheet (70 wt% MDF fiber) 37FH	0.96	3.69	45.5	118.5
MDF-PP sheet (40 wt% MDF fiber) 4MF1	0.95	2.99	58.2	100.4
MDF-PP sheet (40 wt% MDF fiber) 4M37F2	1.03	5.50	81.8	89.4
MDF-PLA sheet (40 wt % MDF fiber) M-PLA1	1.15	4.57	65.7	65.5
PP (polypropylene reference; unreinforced polymer only): PPF	0.86	1.30	38.6	40.0

PP= polypropylene; MDF= thermomechanical pulp (MDF fibers); PLA= polylactic acid. More details and data in WO/2007/073218 [73].

This means that the specific strength and stiffness properties of these wood fiber thermoplastics can compete with NMTs and GMTs used in the automotive industry with added advantages of light weight, and for wood fiber composites in particular, potential lower cost.

The hot-pressing of wood fiber panels has also been described by others, though via different routes [75]. For example, Gehrmann *et al* [76] performed extrusion trials with wood fiber composites using longer refiner fibers and panels were hot-pressed immediately following extrusion of round WPC strands. Another process for engineered wood composites was developed by van Dyk *et al* [77] who combined two non-woven textile technologies, bicomponent fiber and needle punching. Here, hardwood fiber was blended with urea formaldehyde resin and formed into mats. The mats were then sandwiched with polypropylene/polyester bicomponent fibers and then needle punched. Needle punching was achieved via the use of barbed needles that oscillated in a vertical direction relative to the fiber mat. The barbed needles mechanically interlaced the bicomponent web to the wood-fiber mat and pulled some of the polymer fibers through the thickness direction of the mat. During hot pressing, the polypropylene sheath of the bicomponent fiber flows, bonded with adjacent wood fibers, and coalesced with the sheath of the adjacent bicomponent fibers. The mats were pressed until the urea formaldehyde was fully cured. Bending and tensile properties of the needle-punched wood composite were assessed and compared with medium-density fiberboard (MDF). A mean longitudinal tensile modulus of 923 MPa was measured for the laminate panels.

Faurecia have developed wood fiber composites for interior-trim components based on "the Lignotock family". For example, wood fiber is processed with phenol-formaldehyde binder resin, reducing weight and cost compared to glass-reinforced thermoplastics while boosting sound-deadening performance [78]. While this is not thermoplastic product, a newer more ductile grade called Lignoflex uses 70 wt% wood fiber, plus a mixture of polyethylene terephthalate (PET) fibers and binder. Faurecia has developed additional materials ("Lignotock Plus" and "Lignoprop") that feature BASF's Acrodur polymer as low VOC options. Such products also make significant use of reclaimed wood fibers.

19.4 Natural Fiber & Wood Fiber Injection Moulding Compounds

In recent years there has been a growth in the development of natural fiber reinforced injection moulding compounds. The Nova-Institut has undertaken several studies of natural fiber use in German and European markets such as construction, furniture, automotive, etc.) and have presented data showing high growth in such markets [79]. A new generation of natural and wood fiber composites, compounded or processed as long fibers in conjunction with bioplastics for use in injection moulding is now emerging [80]. For example hemp [81], flax, [82] ramie [83], bamboo [84] and many other biofibers [85] have been applied to reinforce plastics and bioplastics in injection moulded applications.

The handling and feeding of natural fibers in extruders and injection moulding machines has been a challenge for many years. Compounding natural fibers can lead to excessive friability, which causes feeding difficulties. Furthermore, high moisture contents need venting and have the potential to slow extrusion rates. Potential end-use problems include low thermal stability of many of the fibers, which limits use to polymers that process at lower temperatures (typically < 200°C), and in-service water absorption, which may cause properties to vary depending on humidity. Pre-treating the fibers and increasing their bulk density allows natural fibers to be fed into the extruder at industrial throughputs. Although feeders or crammers have been proposed for such fibers they are often not effective in commercial production. A number of other approaches which essentially provide pellet or granulate fiber feedstocks have been developed for extrusion and injection moulding in particular. The granules or pellets often seek to retain a reasonable fiber length within them. As discussed above, greater fiber length has beneficial effects on mechanical properties and this is equally true of natural fiber mouldings as for glass or carbon fiber. Indeed, as will be seen below it is also very significant in wood fiber (pulp) mouldings wherein even moving from sub- 1mm fiber lengths to lengths of ~ 2–3mm has very significant property improvements if appropriately formulated and processed.

In terms of injection moulding there have been several commercial developments of natural fiber reinforced polyolefins, other plastics and, more recently, for bioplastics such as polylactic acid. A number of major companies such as Toyota, NEC, and Samsung have used and/or have developed kenaf fiber (a bast fiber) reinforced bioplastic mouldings in recent product launches.

Toyota has developed the use of kenaf in automotive applications such as a door trim base material first produced in 2000, and in 2008 kenaf was being used for five components in a total of 27 car models, mainly high-end cars. Toyota models now use about 3 000 tpa of kenaf fiber in auto interior parts made typically from polypropylene-kenaf, or polylactic acid-kenaf [86].

Ford have indicated future use of polypropylene reinforced with 30% sisal fibers for injection moulding in automotive parts [87, 88]. The 30% sisal fiber reinforced parts have already passed crash and head impact test requirements. A centre console made using the material weighs 20% less than talc filled PP. Other advantages include a 20% lower melt temperature and a 10% faster cycle time. Ford have also developed 50% kenaf fiber reinforced PP for injection moulding, used in Ford Mondeo, Focus and Fiesta door panels. Ford is also looking at using 30% hemp fiber reinforced PP. Ecological comparisons of glass fiber and hemp reinforced battery trays showed that the global warming potential over 100 years is 45% lower in terms of the equivalent weight of carbon dioxide for the natural fiber option – and part weight is also lower. Examples of other commercial suppliers of long fiber products include:

- TITK institute have developed NAFARU natural fiber composites, which are injection mouldable pellets based on long flax or hemp fibers.
- GreenGran and AFT Plasturgie have developed commercial pellets or granules [89].
- GreenCore NCell™ Natural Fiber reinforced Thermoplastics are a family of high-performance natural fiber - thermoplastic compounds for injection moulding and extrusion also supplied in pellet forms. 40 wt% fiber in PP is a standard product used in automotive components (replacing glass fiber PP, and in furniture or consumer goods [90].
- Creafill market a series of high alpha cellulose fibers (and powders) for thermoplastics [91].
- Rettenmaier produce a range of biofillers or biofibers some as granulates, for plastics and other uses [92].

More recently other leading plastic compound suppliers have also added long natural fiber pellets to their existing product range including RTP and LNP/ Sabic [93].

A recent study has compared injection moulded PP natural fiber composites with bioplastic-natural fiber composites, in particular with poly(hydroxybutyrate-co-valerate) (PHBV) and PLA biopolymers [94]. One of the main drawbacks concerning the engineering applications for bioplastics is their low impact strength and improvements have been demonstrated by using long natural fibers as reinforcements. Wong *et al* studied PHB composites with flax fibers and showed that poly(ethylene glycol) and tributyl citrate are effective plasticizers for PHB, although some reductions in glass transition temperatures were observed [95]. Mohanty *et al* has described the importance of compatibilization for PHB composites with natural fibers [96]. Other cellulose based fibers ("man-made cellulose", "viscose" or "rayon") are also being considered as options for reinforcement [97].

This is due to the sustainability of the fiber source (native cellulose), and the high reproducibility of the "man-made" cellulose fiber characteristics (e.g. diameter, stiffness, tenacity, etc.) and improvements in impact strength.

Work at Scion, New Zealand, has shown the advantages of long biofibers in plastics and bioplastics. In related approaches, for example for non-wood long fibers such as flax and ramie etc, a manufacturing concept of continuous fiber impregnation-chopping and extrusion/moulding was developed as a route to making high fiber content mouldable pellets. From such compounds high performance mouldings were produced based on long natural reinforced polylactic acid, polyolefins, and engineering plastics (acetal; nylons). Proprietary treatments or process concepts were developed for retained fiber length, compatibilization and high temperature co-processing with engineering plastics such as nylons. Example data is reported below (Table 19.2).

Wood is a low cost source of fibers and relatively widely used in extruded and injection moulded plastics (wood plastic composites,WPCs) [98]. Much has already been developed for injection moulded wood flour (particle) composites. Indeed many studies of wood fiber-thermoplastic composites have reported on the use of wood fibers as low cost reinforcing fillers in several thermoplastics. As discussed previously, an important parameter is the fiber aspect ratio, which has an influence on the mechanical properties of the composite. A study by Bledzki *et al* [99] showed differences between wood fiber type and processing method in wood fiber reinforced polypropylene composites. Composites containing different types of wood fiber (hard and softwood fiber) and prepared by an injection moulding and a compression moulding process were compared. Injection moulding showed better tensile and flexural properties compared to compression moulding, though it was evident aspect ratios had an effect. As in other fiber composites compatibilization is a key aspect to achieving a good balance of properties. This is particularly important for wood fibers, which are not necessarily well matched to conventional, less polar, plastics. There is much reported on compatibilizers for wood fibers or flour and plastics. Typically acid functional or other reactive polymers (epoxy, isocyanate etc) are used to enhance the interfacial bonding between wood and polymer [100, 101].

Recently, long wood fiber reinforced compounds for injection moulding or extrusion have been developed by a small number of research groups. In extrusion, compounding or injection moulding approaches early work exploring wood fibers was undertaken by Sears *et al* who have described the use of wood fibers with an alpha cellulose content purity >80%, typical of some pulps which are kraft pulps or chemically pulped [102, 103].

More recently Schirp reported on WPCs based on extruded 70% (wt) refiner (TMP – thermomechanical pulp) wood fibers and mechanically processed hemp fibers, in a two-step process [104]. However, during extrusion, both natural fiber types were severely shortened due to strong shear forces, and homogeneous dispersion of fibers in the matrix was not achieved. Composites based on hemp fibers displayed the best strength properties of the formulations tested in this work which also suggested that for the wood fibers the current extruder screw and die configurations need to be modified to achieve improved fiber reinforcement and to create new, structurally demanding applications for natural fiber and wood fiber plastic composites.

WPCs based on refiner (long) wood fibers have been processed by Scion and others. Migneault *et al* extruded WPC based on CTMP (chemithermomechanical pulp) fibers in a two-step process, though the fibers used were short (between 0.196mm and 0.481 mm) [105]. The processing of loose wood fibers in extrusion is known though fiber crammers and feeders have proven unsuccessful and this has driven development of the novel wood fiber pellet technologies developed at Scion. Research at Scion has focused in areas which deliberately target the use of long wood (and other bio-) fibers as genuine reinforcements, rather than conventional wood flour fillers [106]. Thus, typically fiber lengths will be longer than 1mm and often 2–3 mm. These developments are summarized below:

- Wood fiber sheets or mats for plastic and bioplastic mouldings (see above 19.3) [73].
- Discontinuous long wood fiber plastic/bioplastic pellets and moulding via MDF and related manufacturing technologies [106, 107].
- Discontinuous long wood and other biofiber moulding compounds via other pulp-(bio)plastic manufacturing routes.

Scion has developed technologies for longer wood fibers and their convenient incorporation into plastics and bioplastics often via novel pellet intermediates and extrusion compounding (formulation [additives] and process conditions) to maximize fiber length retention.

Table 19.2 shows data on some modified long wood (and non-wood) fiber reinforced acetal and nylon, as examples of higher temperature or engineering plastics, compared to glass fiber references. Example PLA and PP data is also presented for comparison. Proprietary processes were developed to achieve long fiber lengths and/or thermal stability. The data indicates again the promising potential for long fibers, including long wood fibers, as low cost effective reinforcements when appropriately used. The data compare favourably to others who have evaluated cellulosic fibers/fillers with nylon matrices such as via plasticized or modified nylon polymers, or using lower temperature processable nylons [108, 109].

In another approach, and using a proprietary technology, the commercial MDF fiber process, arguably the lowest cost route to convert wood chips to impregnated fibers in significant volume and reproducibly, has been adapted to produce wood fiber polymer pellets suitable for plastic (and bioplastic) processes such as extrusion and injection moulding [106].

Data in Figure 19.3 shows the benefits of long MDF ("M") fibers, and indeed long kraft ("K") fibers, appropriately delivered as long fiber wood-plastic pellets, to an extruder and then injection moulded within a polypropylene matrix. The data is compared to polypropylene ("PP") and to hemp ("H") fiber reinforced PP and wood flour ("SD") filled PP, all equivalently formulated and processed at 40wt% loadings. Clear benefits in properties arise from the long wood fibers – with MDF fiber pellets being a particularly effective, a low cost, fiber reinforcement - if appropriately processed.

Table 19.2 Properties of wood (WF) and natural fiber (NF) injection mouldings.

Biocomposite Material Composition	Flexural Modulus (GPa)	Flexural Strength (MPa)	Tensile Modulus (GPa)	Tensile Strength (MPa)
Nylon	1.38	65.17	2.64	61.29
Nylon + 30% NF1	4.10	119.91	6.72	84.77
Nylon + 30% NF2	4.10	111.62	5.74	69.07
Nylon + 30% WF	3.44	105.61		
Nylon + 15% GF	2.97	129.46	4.97	94.14
Nylon + 30% GF	5.31	189.87	8.30	137.49
Acetal	1.96	78.33	2.52	56.48
Acetal + 30% NF1	3.39	61.42	6.00	65.35
Acetal + 30% NF2	5.31	102.26		
Acetal + 30% WF	3.68	84.97	4.60	53.39
PLA	3.42	103.62	3.92	63.04
PLA + 30% NF1	6.82	109.66	8.03	65.14
PLA + 30% NF2	6.15	103.97	7.12	62.71
PLA + 30% GF	9.47	98.05	10.34	63.10
PLA + 30% WF	6.59	86.97		
PP	869.78	25.66	1160.09	19.78
PP + 30% GF			2862.43	17.61
PP + 30% NF1	2173.06	32.06	2974.50	18.61
PP + 30% WF	1861.13	33.58	2691.14	21.59
PP + 30% NF2	2298.81	31.90		

PP = polypropylene; PLA = polylactic acid; Nylon = nylon 6; GF = glass fiber; WF = treated wood fiber; NF- agrifiber such as hemp, flax, ramie etc, variously treated. More details and data from authors (Scion).

The same benefits are shown in other Scion research with commercial bioplastics reinforced with long fibers. Example data are shown in Figures 19.4 and 19.5 with polylactic acid (PLA) and variously prepared long wood fibers using extrusion compounding and injection moulding under conditions which minimize fiber damage.

Addition of long wood fibers with different treatments, and/or modified PLAs, enabled fiber reinforcement such that both strength and stiffness were increased

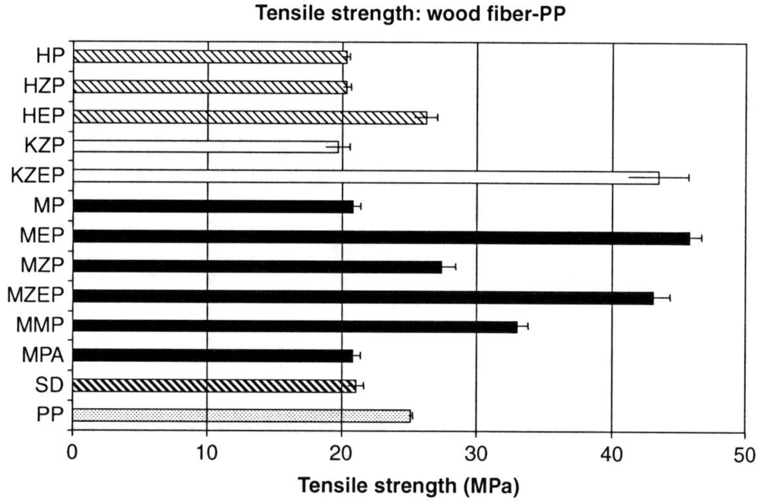

Figure 19.3 Properties of wood-fiber reinforced polypropylene (injection moulded; Scion).
Notes: PP = polypropylene; SD = wood flour; M** = wood fibers (MDF type) variously modified/processed; K**= wood fibers (kraft type) variously modified/processed; H** = hemp fibers variously modified/processed.

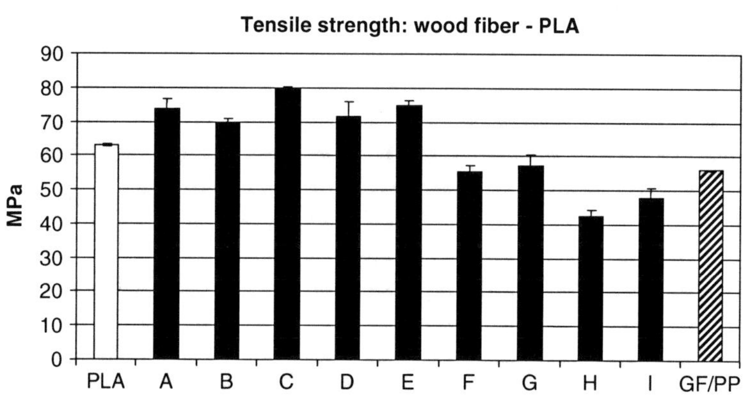

Figure 19.4 Tensile strength of wood fiber polylactic acid (Scion).
Notes: PLA= polylactic acid; GF/PP = glass fiber reinforced polypropylene (20wt% fiber); A–I = variously modified wood/pulp fiber (MDF, kraft,..)- PLA systems, at ~40wt% loadings.

to levels higher than un-filled PLA—and significantly better than glass filled polypropylene. Routes to improving the impact resistance were also identified within the Scion studies and are the subject of ongoing development and possible patent protection. Other bioplastics have been similarly studied with various long wood fibers showing, in some cases, somewhat similar effects, the details of which will form the subject of future publications.

Scion's work has also extended to the manufacture of prototype and commercial mouldings. Example products are shown in Figures 19.6 and 19.7 and have included furniture parts such as seat rest supports or chair bases typically made

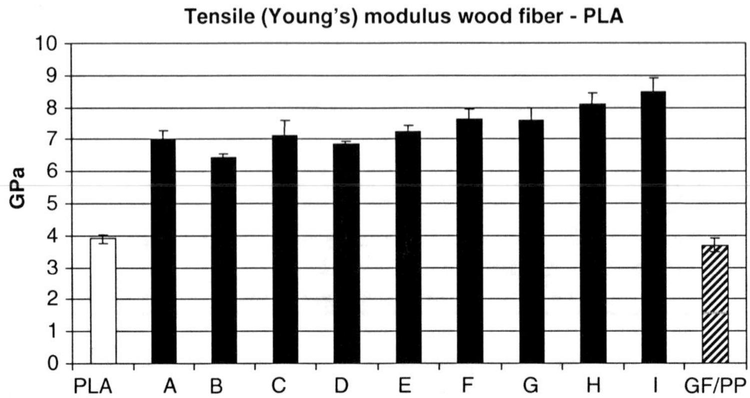

Figure 19.5 Tensile modulus of wood fiber polylactic acid (Scion).
Notes: PLA= polylactic acid ; GF/PP = glass fiber reinforced polypropylene (20wt% fiber); A–I = variously modified wood/pulp fiber (MDF, kraft,..) - PLA systems, at ~40wt% loadings.

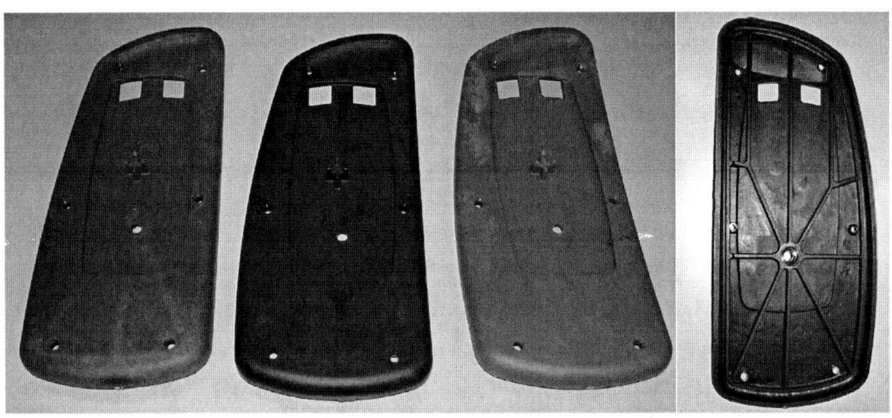

Figure 19.6 Chair part (support) from biofiber PLA (Scion/Axiam Plastics).

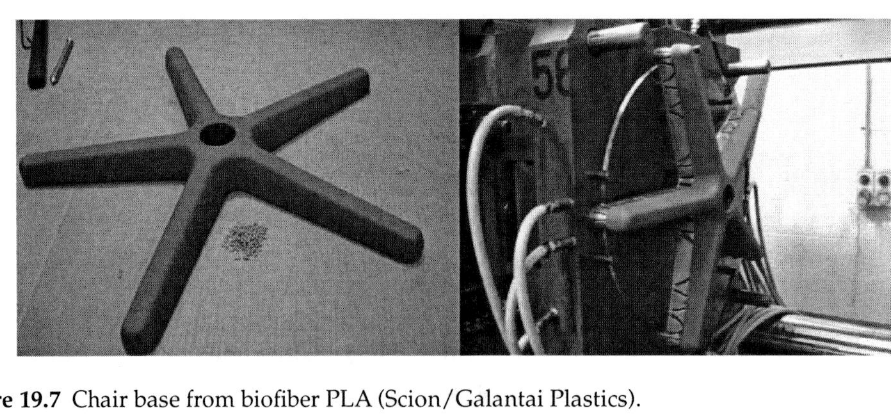

Figure 19.7 Chair base from biofiber PLA (Scion/Galantai Plastics).

from glass fiber polypropylene or glass fiber polyamide, though shown here in biofiber reinforced bioplastics (modified PLAs). The parts perform well in other tests and together with data presented here, and in future publications, show the great potential for appropriately processed genuine wood fibers, as engineered pulps, for enhancing the performance of commercial plastics and bioplastics while, also, potentially lowering overall costs.

Acknowledgements

The authors would like to acknowledge contributions from Jeremy Warnes, Damien Even, Ross Anderson, Daniel Parker, Stephanie Weal, Fabien Venon, Karl Murton, Nancy Hati, Brendan Lee, Armin Thumm, and Michael Witt for their contributions to aspects of the data presented. In addition the authors acknowledge the New Zealand Foundation for Research Science & Technology for funding of some aspects.

References

1. R. S. Bailey, M. Davies and D. R. Moore, "Processing Property Characteristics for Long Glass Fiber Reinforced Polyamide," *Composites*, Vol. 20, 5, 453–460, 1989.
2. J. L. Thomason, M. A. Vlug, *Comp. Part A: App. Sci. Manuf.*, Vol. 27, 6, 477–484, 1996.
3. I.Y. Chang, J.K. Lees, *Therm. Comp. Mat.* Vol. 1, 3, 277–296, 1988.
4. G. Akovali, ed., *Handbook of Composite Fabrication*, Rapra Technology Ltd., 2001.
5 M. Metten, M. Cremer, "Injection Moulding of Long Fiber- Reinforced Thermoplastics." *Kunststoffe*, Vol. 90, 80– 83, 2000.
6. D. Maldas, and B. V. Kokta, *Compos. Interf.*, Vol. 1, 87, 1993.
7. R. Rowell, Proc. 8th Pacific Rim Bio-Based Composites Symposium, 2–11. Advances and Challenges of Wood Polymer Composites, Kuala Lumpur, 23–26 November 2006.
8. P. Malnati, *Comp, Tech.*, August 2007.
9. Quadrant Plastic Composites: Glass Mat Reinforced Thermoplastics (Processing Guidelines) http://www.quadrantcomposites.com/English/pdf_e/Processing%20Guidelines.pdf, 2002.
10. U. Yilmazer, M. Cansever, "Effects of Processing Conditions on the Fiber Length Distribution and Mechanical Properties of Glass Fiber Reinforced Nylon-6," *Polymer. Composites*, Vol. 23, 61–71, 2002.
11. E. Lafranche, P. Krawczak, J.P. Ciolczyk, P. Maugey, "Injection Moulding of Long Glass Fiber Reinforced Polyamide 66: Processing Condition/Microstructure/Flexural Properties Relationship," *Advances in Polymer Technology*, Vol. 24, 114–131 (2005).
12. K. S. Kumar, A. K. Ghosh, N. Bhatnagar, *Polymer Composites*, Vol. 28, 2, 2007.
13. Krauss Maffei Showcasing KraussMaffei's strength in polyurethane processing, http://www.kraussmaffei-technologies.info/en/News/Press-releases/2008__919/NewsDetail__4219/, 2008.
14. P.A. Sreekumar, S. Thomas, "Matrices for natural-fiber reinforced composites" in: *Properties and Performance of Natural-Fiber Composites*, K. L. Pickering, ed., Cambridge, Woodhead Publishing Limited, (2008).
15. R. M. Rowell, "Natural fibers: types and properties" in *Properties and Performance of Natural-Fiber Composites*, K. L. Pickering, ed., Cambridge, Woodhead Publishing Limited, (2008).
16. R. Kozlowski, M. Wladyka-Przbylak, "Uses of Natural Fiber Reinforced Plastics," In *Natural Fibers, Plastics and Composites*, F. T. Wallenberger and N. Weston eds., Boston, Kluwer Academic Publishers, 2004.
17. J. Summerscales, N. Dissanayake, A. Virk , Wayne Hall, "A Review of Bast Fibers and Their Composites, Part 2 – Composites," *Comp.A: App. Sci. Manuf.*, Vol. 41, 10, 1336–1344, 2010.
18. G.T. Pott, "Natural Fibers with Low Moisture Sensitivity," in *Natural Fibers, Plastics and Composites*, F. T., Wallenberger and N. Weston, eds. Boston, Kluwer Academic Publishers, 2004.

19. A. Fernyhough, Developments in Polymer & Composite Materials from Renewable Resources, Presented at Composites Association of New Zealand Annual Meeting/Conference, September 2004.
20. K. Oksman , J.F. Selin, "Plastics and Composites from Polylactic Acid" in *Natural Fibers, Plastics and Composites*, F. T. Wallenberger and N. E. Weston, Editors, Kluwer Academic Publishers, 149–165, Dordrecht/ Boston/London, 2004.
21. N. Reddy , Y .Yang, "Biofibers from Agricultural By-products for Industrial Applications," *Trends in Biotechnology*, Vol. 23, 1, 22–27, 2005.
22. A.K. Bledzki, J. Gassan, *Prog. Polym. Sci.*, Vol. 24, 2 221–274, 1999.
23. I. Van de Weyenberg, T. Chi Truong, B. Vangrimde, I. Verpoest, *Comp. A: App. Sci. Manuf.*, Vol. 37, 9, 1368–1376, 2006.
24. A K. Bledzki, O. Faruk, *App. Comp. Mat.*, Vol. 10, 6, 365–379, 2003.
25. C. Clemons, "Wood-plastic Composites in the United States: The Interfacing of Two Industries," *Forest Prod. J.*, Vol. 52, 6, 10–18, June 1 2002.
26. M.P.Wolcott, and K. Englund, A technology review of wood-plastic composites, Proceedings 33rd International Particleboard/Composite Materials Symposium, Washington State University, Pullman, pp. 103–111, 1999.
27. D. F. Caulfield, C. Clemons, R. E. Jacobson, R. M. Rowell, "Chapter 13, Wood Thermoplastic Composites" in *Handbook of Wood Chemistry and Wood Composites*, Roger Rowell, ed., CRC Press, Boca Raton, FL, 2005, 365–378 http://www.fpl.fs.fed.us/documnts/pdf2005/fpl_2005_caulfield001.pdf
28. Wood-Plastic Composite & Plastic Lumber to 2013 (US industry forecasts for 2013 & 2018) http://www.the-infoshop.com/press/fd92369_en.shtml The Freedonia Group, 2009.
29. UNECE/FAO: Wood resources availability and demands -implications of renewable energy policies A first glance at 2005, 2010 and 2020 in European countries, http://www.unece.org/timber/docs/tc-sessions/tc-65/policyforum/Wood_availability_and_demand.pdf, 2007.
30. MAF (Ministry of Agriculture & Forestry, New Zealand, The Forestry Sector in New Zealand, http://www.maf.govt.nz/forestry/publications/forind/
31. U. Lohmann, *Holz-Lexikon*, 4th Edition, DRW-Verlag, Leinfelden-Echterdingen, 2003.
32. N.M. Stark, R.E. Rowlands, "Effects of Wood Fiber Characteristics on Mechanical Properties of Wood/Polypropylene Composites," *Wood Fiber Sci*, Vol. 35(2):167–174, 2003.
33. D. Brosius, "Natural Fiber Composites Slowly Take Root," *Comp. Tech.*, February 2006.
34. www.europeanbioplastics.org
35. R. Harrington, Food & Production Daily. Strong growth predicted for US bioplastics. http://www.foodproductiondaily.com/Product-Categories/Packaging-Materials/Strong-growth-predicted-for-US-bioplastics 01Sep2010.
36. A Yezza, The Global Market For Bioplastics, http://www.scribd.com/doc/4899724/The-global-market-for-bioplastic-
37. Accelerating the Development of the Market for Bio-based Products in Europe Report Of The Taskforce On Bio-Based Products, {COM(2007) 860 final}, http://ec.europa.eu/enterprise/policies/innovation/files/lead-market-initiative/bio_based_products_taksforce_report_en.pdf
38. U.B. Queiroz and F.P. Collares-Queiroz, "Innovation and Industrial Trends in Bioplastics," *J. Macromolecular Sci., Part C: Polymer Reviews*, Vol. 49, 65–78, 2009.
39. L. Sibley, A growing future for bioproducts from sugar, December 7, 2009, Cleantech Group. http://cleantech.com/news/5391/growing-future-bioproducts-sugar
40. R.l Auras, B. Harte, S. Selke, "An Overview of Polylactides as Packaging Materials," *Macromol Biosci*," Vol. 4, 9, 835–864, 2004.
41. M.S. Huda, L. T. Drzal, *J. App.Pol. Sci.*, Vol. 102, 5: 4856–4869, 2006.
42. S. Pilla, S. Gong, *J. App. Pol. Sci.*, Vol. 111, 1, 37–47, 2009.
43. M.A. Huneault, M. A. H. Li, "Morphology and Properties of Compatibilized Polylactide/Thermoplastic Starch Blends," *Polymer*, Vol. 48: 270–280, 2007.
44. D.Carlson, L. Nie, et al, "Maleation of Polylactide (PLA) by Reactive Extrusion," *J. App. Pol. Sci.*, Vol. 72: 477–485, 1999.

45. J. M. Luengo, B. García, A. Sandoval, G. Naharro, E. R. Olivera, "Bioplastics from Microorganisms," *Current Opinion in Microbiology*, Vol. 6, 3. 251–260, 2003.
46. P A Sreekumar, S. Thomas, "Matrices for natural-fiber reinforced composites" in: *Properties and Performance of Natural-Fiber Composites*, K. L. Pickering, ed., Cambridge, Woodhead Publishing Limited, 2008.
47. A. Jaszkiewicz, A. K. Bledzki, High Strength Bio-Based Composites - Mechanical Performance and Capability Study - PP vs. PLA vs. PA6.10. 8th Global WPC and Natural Fiber Composites Congress and Exhibition, Stuttgart / Germany, 2010.
48. A.K. Mohanty, M. Misra, G. Hinrichsen, "Biofibers, Biodegradable Polymers and Biocomposites: an Overview," *Macromol. Mat. Eng.*, Vol. 276–277, 1–24, 2000.
49. A.K. Bledzki, A. Jaszkiewicz and D. Scherzer, "Natural Fiber and Biocomposites for Technical Applications," *Bioplastics. Mag.*, Vol. 3,2 12–15, 2008.
50. H. Deng, C.T. Reynolds, N.O. Cabrer, N. M. Barkoul, , B. Alcockaand T. Peijs, *Comp.B: Engi.*, Vol. 41, 4, 268–275, 2010.
51. R. Newman, "Development of non-wood natural-fiber composites," in *Properties and Performance of Natural-Fiber Composites*, K. L. Pickering, ed., Cambridge, Woodhead Publishing Limited, 2008.
52. I. Van de Weyenberg, et al, *Comp. Sci. Technol.*, Vol. 63, 9, 1241–1246, 2003.
53. www.nova-institut.de
54. J. C. M. De Bruijn, "Natural Fiber Mat Thermoplastic Products from a Processor's Point of View," *App. Comp. Mat.*, Vol. 7: 415–420, 2000.
55. M. Pervaiz, M. M. Sain, "Sheet-Molded Polyolefin Natural Fiber Composites for Automotive Applications," *Macromol. Mat. Eng.*, Vol. 288, 7, 2003.
56. B. van Voorn, H. H. G. Smit, R. J. Sinke, B. de Klerk, *J. Compos., Part A*, Vol. 32, 1271, 2001.
57. S. K. Garkhail, R. W. H. Heijenrath, T. Peijs, *J. Appl. Compos. Mater.*, Vol. 7, 351, 2000.
58. K. Oksman, *J. Appl. Compos. Mater.*, Vol. 7, 403, 2000.
59. K. Oksman, M. Skrifvars, J.F. Selin, *Comp. Sci. Tech.*, Vol. 63, 9, 2003, 1317–1324.
60. V.P. Cyras, S. Innace, J.M. Kenny and A. Vazques, "Relationship Between Processing and Properties of Biodegradable Composites Based on PCL/Starch Matrix and Sisal Fibers," *Polymer Composites*, Vol. 22, 1, 104–110, 2001.
61. A.S. Herrmann, J. Nickel and U. Riedel, "Construction Materials Based upon Biologically Renewable Resources—from Components to Finished Parts," *Polymer Degradation and Stability*, Vol. 59, 251–261, 1998.
62. S. Goutianos, T. Peijs, B. Nystrom and M. Skrifvars, "Development of Flax Fiber Based Textile Reinforcements for Composite Applications," *Appl. Comp. Mat.*, Vol. 13,4, 199–215, 2006.
63. L. Zhang, M. Miao, "Commingled Natural Fiber/Polypropylene Wrap Spun Yarns for Structured Thermoplastic Composites," *Comp. Sci. Tech.*, Vol. 70, 1, 2010, 130–135, 2010.
64. D. H. Mueller, A. Krobjilowski, Improving the Impact Strength Of Natural Fiber Reinforced Composites By Specifically Designed Material and Process Parameters, INJ Winter, 31–34, 2004.
65. www.compoistesevolution.com
66. J.E.P. Carpenter, M. He Miao, P. Brorens, *J. Adv. Mat. Res.*, Vol. 29 – 30, 263–266. (Advanced Materials and Processing IV), November, 2007.
67. M. Le Guen , R. Newman, "Pulped *Phormium Tenax* Leaf Fibers as Reinforcement for Epoxy Composites," *Comp. Part A: App. Sci. Manuf.*, Vol. 38, 2109–2115, 2007.
68. M. Zampaloni, F. Pourboghrat , J. Moore, S.A. Yankovich, B.N. Rodgers, L.T. Drzal, A.K. Mohanty, M. Misra, "Kenaf Natural Fiber Reinforced Polypropylene Composites: A Discussion on Manufacturing Problems and Solutions," *Comp. A, App. Sci. Manuf.*, Vol. 38A, 6, 1569–1580, 2007.
69. J Mitchell, *Forestry & Forest Products, Newsletter*, Vol. 4, 2, Autumn CSIRO, Australia, 1–4, 1990.
70. US patent 4,562,218.
71. GB-A-1 010 043, U.S. Pat. No. 3,396,142, EP0235794B1, EP0312090B1, U.S. Pat. No. 4,612,238.
72. D. Bhattacharyya, M. Bowis, K Jayaraman, "Thermoforming of Wood Fiber–Polypropylene Composite Sheets," *Comp. Sci. Tech.*, Vol. 63, 353–365, 2003.
73. A. Fernyhough, J.M. Warnes, C.R. Anderson, B. J. Lee, M. R. J. Witt, PCT Patent Application WO/2007/073218. Method For Producing Wood Fiber Composite Products, 2007.

74. M. Pervaiz, M.M. Sain, "Sheet-Molded Polyolefin Natural Fiber Composites for Automotive Applications," *Macromol. Mater. Eng.*, Vol. 288, 7, 553–557, 2003.
75. R. Marutzky, P. Buchholzer, V. Thole, Fiber-based WPC with enhanced mechanical properties: New manufacturing processes and first research results, Proceedings, 2nd Wood Fiber Polymer Composites Symposium, Bordeaux, 24–25 March 2005.
76. K. Gehrmann, R. Starke, G. Wilczek, F. Nagel, K. Morawietz, S. Meinicke, M.Busch, Perspektiven mit Profilextrusion – Erste Ergebnisse über den Einfluss unterschiedlicher Holzfaserstoffe und Polypropylen-Typen auf mechanische Kennwerte von Composites, *Holz- Kunststoffverarb (HK)*, Vol. 4:50–52, 2003.
77. H. van Dyk, P. Peralta, I. Peszlen, P. Banks-Lee, *Forest Products Journal*, Vol. 59, 11–12, 11(7) 2009.
78. P. Malnati, "Interior innovation: The value proposition: OEMs and Tier suppliers ask materials and molding processes to go 'green' and do more for less," *Comp. Tech*, August 2010 http://www.compositesworld.com/articles/interior-innovation-the-value-proposition
79. Nova Institut: http://www.nova-institut.de/nr/index.php?tpl=startlist&id=&aid=&lng=en
80. J. Markarian, "Strengthening Compounds Through Fiber Reinforcement," *Reinforced Plastics*, Vol. 14 February 2007.
81. M.A. Sawpan, K.L. Pickering, A. Fernyhough, *Advanced Materials Research*, Vol. 29 – 30, IV, 337–340, 2007.
82. B. Bax, J. Müssig, *Comp. Sci. Tech.*, Vol. 68, 7–8, 1601–1607, 2008.
83. T. Yu, J. Ren, S.Li, H. Yuan, Y. Li, *Comp. A: App. Sci Manuf.*, Vol. 41, 4, 495–505, 2009.
84. R. Tokoro, D. Minh Vu, K. Okubo, T. Tanaka, T. Fujii, T. Fujiura, *J. Mat. Sci.*, Vol. 43, 2, 775–787, 2008.
85. N. Graupner, A. S. Herrmann, J.Müssig, *Comp. A: App. Sci. Manuf.*, Vol. 40, 6–7,810–821, 2009.
86. Toyota Bushoku Ltd., http://wardsauto.com/ar/supplier_green-up_hybrids_100901
87. D. Vink, "Ford Focuses on Sustainable Mouldings," *Plastics & Rubber Weekly*, 21st July 2010 http://www.prw.com/subscriber/headlines2.html?cat=1&id=1279701293&q=Ford
88. G. Mougin, M. Magnani, N. Eikelenberg, "Natural-fibers Composites for the Automotive Industry: Challenges, Solutions and Applications," *International Journal of Materials and Product Technology*, Vol. 36, 1–2, 176–188 (13), 2009.
89. http://www.agrofibercomposites.com/GreenGran%20NF305070%20fact%20sheet%20and%20comparison%20chart.pdf
90. http://greencorenfc.com/products.htm
91. http://www.creafill.com/index.htm
92. http://www.jrs.de/cgi-bin/wPermission.cgi?file=/wEnglisch/unternehmen/kernkompetenz.shtml&navid=10
93. www.reinforcedplastics.com
94. A.K. Bledzki, A. Jaszkiewicz, "Mechanical Performance of Biocomposites Based on PLA and PHBV Reinforced with Natural Fibers – a Comparative Study to PP," *Comp. Sci. Tech.*, Vol. 70, 12, 1687–1696, 2010.
95. S. Wong, R. Shanks and A. Hodzic, "Properties of Poly(3-hydroxybutyric acid) Composites with Flax Fibers Modified by Plasticiser Absorption," *Macromol Mater. Eng.*, Vol. 287 (10) 647–655, 2002.
96. A.H Mohanty, L..T. Drzal, P.Mulukut , S.M. Desai M . Misra M., "Green" composites from biofibers and poly(3-hydroxybutyrate), Proceedings of Annual Technical Conference (ANTEC), Navy Pier and Sheraton Chicago Hotel & Towers, Chicago, Illinois, USA; p1483–7, 2004.
97. J. Ganster H. P. Fink, "Novel Cellulose Fiber Reinforced Thermoplastic Materials," *Cellulose*, Vol. 13, 271–280, 2006.
98. A.K. Bledzki S. Reihmane, J. Gassan, Thermoplastics Reinforced with Wood Fillers: a Literature Review, *Polym Plast Technol Eng.*, Vol. 37 (4): 451–468, 1998.
99. A.K. Bledzki, O. Faruk, "Wood Fiber Reinforced Polypropylene Composites: Compression and Injection Molding Process," *Pol. Plast. Tech. Eng.*, Vol. 43, 3, 871– 888, 2004.
100. Gaylord, U.S. Pat. Nos. 3,765,934, 3,869,432, 3,894,975, 3,900,685, 3,958,069: Casper et al, U.S. Pat. No. 4,051,214; Raj et al, U.S. Pat. No. 5,120,776.
101. T. Yu, J. Ren, S. Li, Hu Yuan, Yan Li, *Comp A App. Sci Manuf.*, Vol. 41, 4, 499–505, 2010.

102. K.D. Sears et al, Proc. 6th International Conference on Woodfiber Plastics Composites, Forest Products Society, 27–34, 2001 and K Sears U.S. Pat. No. 6,270,883, 1999.
103. R. Jacobson, D, Caulfield, K. Sears, and J. Underwood, Low temperature processing (LTP) of ultra-pure cellulose fibers into nylon 6 and other thermoplastics, Proceedings, Sixth International Conference on Woodfiber/Plastic Composites, Madison WI, 2002.
104. A. Schirp, J. Stenter, *Eur. J. Wood & Wood Prod.*, Vol. 68,2 ,219–231, 2010.
105. S. Migneault, A. Kouba, F. Erchiqui, A. Chaala, K. Englund, C. Krause, M. Wolcott, "Effect of Fiber Length on Processing and Properties of Extruded Wood-Fiber/HDPE Composites," *J App. Pol. Sci.*, Vol. 110: 1085–1092, 2008.
106. A. Fernyhough, M. Witt, J .Warnes, C.R. Anderson, B.J. Lee, PCT Patent Application: WO/2006/001717: Method For Producing Wood Fiber Pellets, Pub: 05.01.2006.
107. J. Warnes, "High Aspect Ratio Fibers," 9th Pacific Rim Bio-Based Composites Symposium, Rotorua, New Zealand, 5–8 November, 2008.
108. Y. Amintowlieh, "Nylon-6/Agricultural Filler Composites," Masters Thesis: Waterloo University, Ontaraio, September 2010. Abstract available at; http://uwspace.uwaterloo.ca/bitstream/10012/5476/1/Amintowlieh_Yasaman.pdf
109. X. Xu, "Cellulose fiber reinforced nylon 6 or nylon 66 composites" PhD thesis: Georgia Institute of Technology, December 2008.

Index

Also of Interest

Check out these published and forthcoming related titles from Scrivener Publishing

Biopolymers: Biomedical and Environmental Applications

Edited by **Susheel Kalia and Luc Avérous**

Forthcoming August 2011. ISBN 978-0-470-63923-8

Plastics Sustainability

Michael Tolinski

Forthcoming December 2011. ISBN 978-0-470-93878-2

Polymers from Renewable Resources

Ram Nagarajan

Forthcoming Spring 2012. ISBN 9780470626092

Green Chemistry for Environmental Remediation

Edited by **Rashmi Sanghi and Vandana Singh**

Forthcoming September 2011 ISBN 978-0-470-94308-3

Polymer Nanotube Nanocomposites: Synthesis, Properties, and Applications

Edited by **Vikas Mittal**

Published 2010. ISBN 978-0-470-62592-7

Handbook of Engineering and Specialty Thermoplastics

Part 1: Polyolefins and Styrenics
By Johannes Karl Fink
Published 2010. ISBN 978-0-470-62483-5

Part 2: Water Soluble Polymers
By Johannes Karl Fink
Published 2011. ISBN 978-1-118-06275-3

Part 3: Polyethers and Polyesters
Edited by Sabu Thomas and Visakh P.M.
Published 2011. ISBN 978-0-470-63926-9

Part 4: Nylons
Edited by Sabu Thomas and Visakh P.M.
Forthcoming August 2011. ISBN 978-0-470-63925-2

A Concise Introduction to Additives for Thermoplastic Polymers

By **Johannes Karl Fink**

Published 2010. ISBN 978-0-470-60955-2

Introduction to Industrial Polyethylene: Properties, Catalysts, Processes

By **Dennis P. Malpass**

Published 2010. ISBN 978-0-470-62598-9

The Basics of Troubleshooting in Plastics Processing

By **Muralisrinivasan Subramanian**

Published 2011. ISBN 978-0-470-62606-1

Miniemulsion Polymerization Technology

Edited by **Vikas Mittal**

Published 2010. ISBN 978-0-470-62596-5